MODERN NONLINEAR OPTICS
Part 2

ADVANCES IN CHEMICAL PHYSICS

VOLUME LXXXV

EDITORIAL BOARD

MODERN NONLINEAR OPTICS
Part 2

Edited by

MYRON EVANS
Department of Physics
The University of North Carolina
Charlotte, North Carolina

STANISŁAW KIELICH
Nonlinear Optics Division
Institute of Physics
Adam Mickiewicz University
Poznań, Poland

ADVANCES IN CHEMICAL PHYSICS
VOLUME LXXXV

Series Editors

ILYA PRIGOGINE
University of Brussels
Brussels, Belgium
and
University of Texas
Austin, Texas

STUART A. RICE
Department of Chemistry
and
The James Frank Institute
University of Chicago
Chicago, Illinois

AN INTERSCIENCE® PUBLICATION
JOHN WILEY & SONS, INC.
NEW YORK • CHICHESTER • BRISBANE • TORONTO • SINGAPORE

CONTRIBUTORS TO VOLUME LXXXV
Part 2

DAVID L. ANDREWS, School of Chemical Sciences, University of East Anglia, Norwich, England

H. J. CAULFIELD, Department of Physics, Alabama A&M University, Normal, Alabama

W. T. COFFEY, School of Engineering, Department of Microelectronics & Electrical Engineering, Trinity College, Ireland

MYRON W. EVANS, Department of Physics, University of North Carolina, Charlotte, North Carolina

AHMED A. HASANEIN, Department of Chemistry, Faculty of Science, Alexandria University, Alexandria, Egypt

H. JAGANNATH, Department of Physics, Alabama A&M University, Normal, Alabama

YU. P. KALMYKOV, Institute of Radio Engineering and Electronics, Russian Academy of Sciences, Fryazino, Russia

AKHLESH LAKHTAKIA, Department of Engineering Science and Mechanics, The Pennsylvania State University, University Park, Pennsylvania

E. S. MASSAWE, School of Engineering, Department of Microelectronics & Electrical Engineering, Trinity College, Dublin, Ireland

A. S. PARKINS, Department of Physics, University of Waikato, Hamilton, New Zealand

G. E. STEDMAN, Department of Physics, University of Canterbury, New Zealand

JEFFREY HUW WILLIAMS, Institut Max von Laue-Paul Langevin, Grenoble, France

INTRODUCTION

Few of us can any longer keep up with the flood of scientific literature, even in specialized subfields. Any attempt to do more and be broadly educated with respect to a large domain of science has the appearance of tilting at windmills. Yet the synthesis of ideas drawn from different subjects into new, powerful, general concepts is as valuable as ever, and the desire to remain educated persists in all scientists. This series, *Advances in Chemical Physics*, is devoted to helping the reader obtain general information about a wide variety of topics in chemical physics, a field that we interpret very broadly. Our intent is to have experts present comprehensive analyses of subjects of interest and to encourage the expression of individual points of view. We hope that this approach to the presentation of an overview of a subject will both stimulate new research and serve as a personalized learning text for beginners in a field.

ILYA PRIGOGINE
STUART A. RICE

PREFACE

Statistical molecular theories of electric, magnetic, and optical saturation phenomena developed by S. Kielich and Piekara in several papers in the late 1950s and 1960s clearly foreshadowed the developments of the next thirty years. In these volumes, we as guest editors have been honored by a positive response to our invitations from many of the most eminent contemporaries in the field of nonlinear optics. We have tried to give a comprehensive cross section of the state of the art of this subject. Volume 85 (Part 1) contains fourteen review articles by the Poznań School and associated laboratories, and volume 85 (Part 2 and Part 3) contain a selection of reviews contributed from many of the leading laboratories around the world. We thank the editors, Ilya Prigogine and Stuart A. Rice, for the opportunity to produce this topical issue.

The frequency with which the work of the Poznań School has been cited in these volumes is significant, especially considering the overwhelming societal difficulties that have faced Prof. Dr. Kielich and his School over the last forty years. Their work is notable for its unfailing rigor and accuracy of development and presentation, its accessibility to experimental testing, the systemic thoroughness of the subject matter, and the fact that it never seems to lag behind developments in the field. This achievement is all the more remarkable in the face of journal shortages and the lack of facilities that would be taken for granted in more fortunate centers of learning.

We hope that readers will agree that the contributors to these volumes have responded with readable and useful review material with which the state of nonlinear optics can be measured in the early 1990s. We believe that many of these articles have been prepared to an excellent standard. Nonlinear optics today is unrecognizably different from the same subject in the 1950s, when lasers were unheard of and linear physics ruled. In these two volumes we have been able to cover only a fraction of the enormous contemporary output in this field, and many of the best laboratories are not represented.

We hope that this topical issue will be seen as a sign of the ability of scientists all over the world to work together, despite the frailties of human society as a whole. In this respect special mention is due to Professor Mansel Davies of Criccieth in Wales, who was among the first in the West to recognize the significance of the output of the Poznań School.

MYRON W. EVANS

Charlotte, North Carolina
October 1993

CONTENTS

xi

HOLOGRAPHY AND DOUBLE PHASE CONJUGATION

H. JAGANNATH AND H. J. CAULFIELD

Department of Physics, Alabama A & M University, Normal, AL

CONTENTS

Modern Nonlinear Optics, Part 2, Edited by Myron Evans and Stanisław Kielich. Advances in Chemical Physics Series, Vol. LXXXV.
ISBN 0-471-57546-1 © 1993 John Wiley & Sons, Inc.

I. INTRODUCTION

Two- and four-wave mixing processes in nonlinear optics have proved to be extremely useful. Several applications in optical image processing, optical communications, real-time holography, and opto-electronic neural networks have been developed in recent years. Both processes are accurately and appropriately described in terms of ordinary holography occurring in real time. The spatial or temporal interference between coherent beams spatially modulates the characteristics of the recording medium, storing information. This information can be retrieved simultaneously or later with another coherent beam. Wave mixing between mutually incoherent beams is a new phenomenon which occurs in the same materials. Double phase conjugation is one such mutually incoherent beam-coupling process, which also produces accurate three-dimensional images. Though there are some similarities between real-time holography and double phase conjugation, we will show that there are distinct differences between the two processes, requiring a new holographic interpretation of double phase conjugation. The advantage of this interpretation is that double phase conjugation can be viewed as a more general form of holography—holography with mutually incoherent sources.

    This chapter is devoted to double phase-conjugate mirror (DPCM). Real-time holography and all multiwave mixing processes including mutually incoherent beam coupling (MIBC) are dynamic self-diffraction processes. We present a comprehensive review of holography (Section II), volume holography and dynamic self-diffraction (Section III), two- and four-wave mixing (Sections IV–VI), and mutually incoherent beam coupling (Section VII) to establish a link between holography and double phase conjugation. Emphasis is placed mostly on the studies with photorefractive materials, because they are the materials of choice for MIBC/DPCM and real-time holography. The double phase-conjugate mirror is reviewed in Section VIII. The holographic interpretation of the two- and four-wave mixing processes and the physical mechanism responsible for MIBC lead us to believe that DPCM is a new type of holography in which the primary holograms interact to create another hologram. This "second-order" holography is explored in Section IX. Finally, the relaxed stability and coherence requirements of the incident beams lead to new applications in optical imaging and optical communications, some of which are presented in Section X. Due to the limited scope of this chapter, the interested reader is referred to other chapters in the book and to the following for additional information: photorefractive materials and their applications,[1-3, 7] volume holography,[4-6, 23] two-wave mixing,[17] degenerate four-wave mixing and phase conjugation,[3, 53] and photorefractive oscillators.[17, 40, 68]

II. HOLOGRAPHY

Holography is an optical process by which storage and retrieval of optical wavefronts is possible. Spatial information about the object is contained in the time evolution of the two-dimensional wavefronts emanating from the object. Complete information about the phase and amplitude of the wavefronts is needed to reconstruct the original object. Since recording materials do not respond to the phase directly, interferometric techniques are used to record the phase information. This places severe restrictions on the illuminating beams. For good contrast fringes and hence good efficiency, the illuminating source must have good coherence properties. High mechanical stability of the source and the object is required because of the time requirements of recording. The availability of monochromatic coherent laser sources has made possible the three-dimensional imaging of objects and has led to a large number of industrial applications. Holographic processing involves two steps: (1) recording and (2) reconstructing of the object wavefront. We review each briefly.

A. Holographic Recording

In the recording process, mutually coherent object and reference wavefronts are brought together to interfere (Fig. 1). A recording medium, such as a photographic plate, placed in the overlap region of the two wavefronts records the interference pattern.

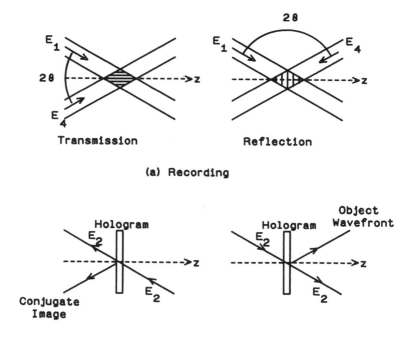

(a) Recording

(b) Reconstruction

Figure 1.   Holographic recording and wavefront reconstruction. E_1 and E_4 are the object and reference beams, respectively; E_2 is the reconstructing beam.

Assuming that the two beams are plane waves with parallel polarizations, the electric field of the object wavefront can be written as

$$E_1(\mathbf{r}) = A_1 \exp\left[-i(\omega t - \mathbf{k}_1 \cdot \mathbf{r})\right] \qquad (1)$$

and that of the reference wavefront as

$$E_4(\mathbf{r}) = A_4 \exp\left[-i(\omega t - \mathbf{k}_4 \cdot \mathbf{r})\right] \qquad (2)$$

The intensity I_T of the light in the overlapping region is given by

$$I_T = \left| E_1(\mathbf{r}) + E_4(\mathbf{r}) \right|^2 \qquad (3)$$

For a recording medium in the xy plane, if the direction of propagation is along the z axis, and 2ϑ is the angle between the wave vectors $\mathbf{k}_1$ and $\mathbf{k}_4$,

$$I_T = A_0^2 \left[1 + m \cos\left(\frac{2\pi y}{d} \right) \right] \tag{4}$$

where m is the modulation depth given by

$$m = \frac{2A_0 A_R}{A_0^2 + A_R^2} \tag{5}$$

and $d = \lambda/2 \sin \vartheta$ is the spacing between the inteference maxima.

The spatial variance in the light intensity corresponding to the total field is recorded in the medium in the form of either an optical density pattern leading to an absorption grating or a phase shift pattern leading to a phase hologram. The transmission $T(y)$ of the electric vector through the absorption grating generated by I_T can be expressed as[5]

$$T(y) = T_0 + \beta E_0 m \cos Ky \tag{6}$$

where $K = 2\pi/d$ is the grating vector, T_0 is constant field transmission, E_0 is the exposure, and β is the transfer characteristic of the recording material.

In a phase grating the field transmission is given by[5]

$$T(y) \propto \left\{ J_0(\phi_1) + 2\sum \left[J_n(\phi_1) \frac{\cos(nKy)}{i^n} \right] \right\} \tag{7}$$

where $J_n(\phi_1)$ are the Bessel functions of the phase modulation ϕ_1. $J_0(\phi_1)$ contributes to the constant background, while $J_1(\phi_1)$ contributes to the first-order diffracted intensity in reconstruction.

B. Wavefront Reconstruction

To reconstruct the object wavefront, we illuminate the holographic recording or hologram by a reconstruction beam E_2 (Fig. 1b). The transmitted wave-front field for the case of absorption grating is given by

$$T(y) = E_2 e^{-i\omega t} \left[T_0 e^{iky} + \tfrac{1}{2}\beta E_0 m [e^{-i(K-k)y} + e^{i(K+k)y}] \right] \tag{8}$$

The first term is a plane wave propagating in the direction of the reconstruction beam. The second term corresponds to the primary image beam. The third term corresponds to a beam traveling in the opposite direction to the object beam, and forms the conjugate image. In the case of phase gratings, reconstruction by the illuminating beam leads to the undeviated, object, and conjugate beams.

The diffraction efficiency of the hologram is defined as the ratio of the intensity of the reconstructed wavefront to the intensity of the reconstruction beam. The maximum efficiency of transmission hologram with a single-layer absorption grating is 0.0625, and that of a phase grating is 0.339. A single-layer reflective hologram is a reflection grating that can be blazed to obtain high diffraction efficiency. Reflection holograms with reconstruction efficiencies of 0.85 have been reported.

III. VOLUME HOLOGRAPHY

The efficiency of a hologram can be improved by increasing the thickness of the recording medium. This allows for more diffraction gratings to be written in the recording medium. The hologram with multiple layers of recording is called a "thick" hologram. If d is the thickness and Λ is the grating spacing, then for a thick hologram $d > n\Lambda^2/2\pi\lambda$. For sufficiently thick recording materials, the total beam overlap volume can be used to write a volume hologram. Because of the multiple planes at which light diffraction occurs in a thick hologram during reconstruction process, diffraction at or near the Bragg angle leads to efficient wave-front reconstruction. The higher diffraction orders are quenched by interference. This is true for both transmission and reflection holograms.

A.   Thick Gratings and Beam Coupling

In thick gratings, the illuminating beam is strongly depleted as it propagates through the material due to large diffraction efficiency. Within the hologram, two mutually coherent beams are traveling: the incoming reference beam E_1 and the outgoing signal beam E_4 (Fig. 2). As the beams propagate, each beam is diffracted in the direction of the other beam at the grating surfaces. Energy exchange takes place between the beams and the beams are coupled. Kogelnik[6] analyzed these spatially dynamic interaction processes in terms of the coupled wave theory and obtained expressions for diffraction efficiencies for reflection and transmission holograms with absorption and phase gratings.

Wave propagation in the holographic grating is described by the scalar wave equation

$$\nabla^2 E + k^2 E = 0 \qquad (9)$$

For small amplitudes of spatial modulations in the refractive index n_1 and

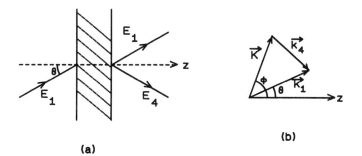

Figure 2. Beam coupling in a volume hologram: (a) geometry of the interacting beams; (b) wave-vector diagram. **K** is the grating vector.

the absorption coefficient α_1, the propagation constant k in the material can be expressed in terms of the dielectric constant ε_0, the average conductivity σ_0, the grating vector **K**, and, the coupling constant κ as

$$k^2 = \beta^2 - 2i\alpha\beta + 2\kappa\beta(e^{i\mathbf{K}\cdot\mathbf{r}} + e^{-i\mathbf{K}\cdot\mathbf{r}}) \qquad (10)$$

where

$$\beta = \frac{2\pi\varepsilon_0^{1/2}}{\lambda} \qquad \alpha = \frac{\mu c \sigma_0}{2\varepsilon_0^{1/2}} \qquad \kappa = \frac{\pi n_1}{\lambda} - \frac{i\alpha_1}{2} \qquad (11)$$

κ describes the coupling between the reference beam E_1 and signal beam E_4. Amplitudes $A_1(z)$ and $A_4(z)$ vary along the propagation direction as a result of the coupling. The wave vector $\mathbf{k}_4$ is forced by the grating to satisfy the relation $\mathbf{k}_4 = \mathbf{k}_1 - \mathbf{K}$.

If the reference beam is incident at an angle ϑ with respect to the direction of propagation z, and the grating wave vector **K** is slanted at an angle ϕ, the Bragg condition is given by

$$\cos(\phi - \vartheta) = \frac{K}{2\beta} \qquad (12)$$

From the scalar wave equation (9), assuming that the $A_1(z)$ and $A_4(z)$ vary slowly, the following set of coupled equations are obtained:

$$c_R \frac{dA_1}{dz} + \alpha A_1 = -i\kappa A_4 \qquad (13)$$

$$c_S \frac{dA_4}{dz} + (\alpha + i\theta)A_4 = -i\kappa A_1 \qquad (14)$$

where $c_R = \cos \vartheta$, $c_S = \cos \vartheta - K/2\beta$, and θ is the dephasing measure given by

$$\theta = K \cos(\phi - \vartheta) - \frac{K^2 \lambda}{4\pi n} \tag{15}$$

The physical picture of the diffraction process is reflected in the coupled equations (13)–(15). As the waves propagate in the material, the amplitudes of the waves change due to coupling to the other waves ($\kappa A_1, \kappa A_4$), to absorption ($\alpha A_1, \alpha A_4$), or to both. For deviation from the Bragg condition, the two beams are forced out of synchronization and the interaction decreases.

A complete analysis of the wave propagation in lossless, lossy, and slanted, transmission, and reflection holograms was given by Kogelnik.[6] The results show that the diffraction efficiencies are less for absorption gratings than for phase gratings. The maximum diffraction efficiencies are 0.037 and 0.072 for lossless, unslanted absorption gratings in transmission and reflection holograms, respectively. The grating slant improves the efficiencies of the lossy absorption gratings, though the maximum efficiencies are still below those of the lossless gratings. The maximum diffraction efficiency for lossless, unslanted phase grating is ~ 1.00 both for transmission and reflection holograms. As in the case of absorption gratings, the efficiency of the lossy phase gratings can be improved by grating slant. All these results have been verified experimentally.

One of the important applications of volume holography is in information storage in holographic memory systems. The diffraction limited storage densities attainable in volume holograms are very high. Several materials, such as photographic emulsions, photochromic glasses, dichromate sensitized gelatin, alkali halide, and electro-optical crystals, have been used in these applications. The condition for the thick holograms, $Q = 2\pi\lambda d/n\Lambda^2 \gg 1$, can be easily met for thickness d of a few micrometers. Of all these recording materials, the electro-optic crystals are of particular interest since the phase grating in these materials is phase shifted from the sinusoidal light interference pattern. The result is a strong coupling of the writing beams during the recording process. This leads to nonreciprocal energy exchange and self-diffraction of the writing beams.

B. Photorefractive Materials

The electro-optic materials in which the refractive indices are changed by light-induced electric fields are called photorefractive materials. A large number of materials have been observed to show photorefraction at moderately low illuminations; $BaTiO_3$, $Bi_{12}SiO_{20}$(BSO), $Bi_{12}GeO_{20}$(BGO), $Bi_{12}TiO_{20}$(BTO), $KNbO_3$, $LiNbO_3$, and $Sr_xBa_{1-x}Nb_2O_6$(SBN). The materials are intrinsically transparent in the visible region. But the presence of impurities permits the generation of photoexcited charges which lead to light-induced fields under nonuniform illumination. The materials are erasable and can be used for read–write applications.

The generation and transportation of charge carriers in electro-optic materials have been investigated by several workers.[7] Two charge transportation mechanisms have been proposed. In the charge hopping model,[8] the charge carriers that are excited from the donor sites in the presence of light, hop to adjacent sites with a probability proportional to the intensity of the light. The drift of the carriers by hopping continues in all directions until the carriers are out of the illuminated areas, resulting in a net electric field. In the band-conduction model[9] (Fig. 3a), the charge carriers are excited from the donor levels to the conduction band. The charges

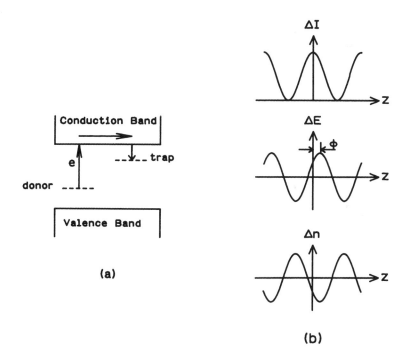

Figure 3. (a) Energy-level diagram of a photorefractive material in the band conduction model; (b) the spatial distribution of the intensity I, the space-charge field E, and the refractive index n under sinusoidal illumination. Φ is the phase shift between the light-intensity pattern and the space-charge field.

then diffuse or drift in the presence of the external fields and are trapped at the acceptor sites. This charge redistribution leads to the space-charge field. The band-conduction model has been widely accepted because it explains space-charge field formation in a variety of materials in which the transportation lengths vary widely. Both positive and negative charge carriers have been observed to take part in the charge transportation processes. The one-carrier, one-donor model discussed below has been extended to simultaneous electron–hole transportation models.[10]

The equation for generation and transportation of the charge carriers can be written as[11-14]

$$e\frac{\partial N_D^+}{\partial t} = \frac{\partial J}{\partial z} \qquad \text{Continuity equation} \qquad (16)$$

$$\frac{\partial N_D^+}{\partial t} = sIN_D - \gamma_R nN_D^+ \qquad \text{Rate equation} \qquad (17)$$

$$J = e\mu nE_s - k_BT\mu\frac{\partial n}{\partial z} \qquad \text{Equation of motion} \qquad (18)$$

$$\frac{\partial E_S}{\partial z} = \frac{e}{\varepsilon_0\varepsilon_s}(N_A - N_D^+) \qquad \text{Poisson's equation} \qquad (19)$$

where n is the electron density, N_D is density of donors, N_A is the density of acceptors, N_D^+ is the density of ionized donors, J is the current density, E_S is the space-charge field, I is the light intensity, β_e is the rate of thermal generation, s is the rate of photoionization, γ_R is the recombination rate, μ is the mobility, and, ε_s is the relative static dielectric constant. In the formulation of the above set of equations, the following assumptions have been made: $n \ll N_D^+$; $\beta_e \ll sI$; $N_D^+ \ll N_D$; spatial variations are only along the direction of propagation z; there is no photovoltaic effect; and ε_s is independent of z.

Equations (16)–(19) are nonlinear and have been solved under various approximations for spatially periodic light illumination. In the two cases of short time limit and saturation time limit, the equations have been solved in terms of a Fourier series.[15] Another approach, developed by Kukhtarev et al.,[12, 13] assumes solutions of the form

$$a(z, t) = a_0 + \left[a_1(z, t)e^{iKz} + \text{c.c.}\right]/2 \qquad (20)$$

where $a = N_D^+$, n, J, and E_s, and all higher order terms are dropped. If we assume that the modulations in the parameters are small, the material equations can be reduced to[14]

$$\frac{\partial E_s}{\partial t} + gE_s = hm \qquad (21)$$

where

$$g = (E_q + E_D + iE_0)/E_q D\tau_d, \quad h = (E_0 + iE_D)/D\tau_d,$$

$$D = (E_M + E_D + iE_0)/E_M$$

$$E_D = KTk_B/e, \quad E_q = eN_A/\varepsilon_0\varepsilon_s K, \quad E_M = \gamma_R N_A/\mu K, \qquad (22)$$

$$\tau_d = \varepsilon_0\varepsilon_s/e\mu n_0, \quad n_0 = sI_0 N_D/\gamma_R N_A$$

and m is the fringe modulation.

The steady-state space-charge field is given by $E_s = hm/g$. The phase shift Φ_g between the light intensity pattern and the space-charge field is given by

$$\tan \Phi_g = \frac{E_0^2 + E_D(E_D + E_q)}{E_0 E_q} \qquad (23)$$

In general, if both diffusion and drift fields are present, there are two components of the E_s: one that is in phase with the intensity modulation and another whose phase is $\pi/2$ shifted with respect to the intensity modulation. In barium titanate, the charge migration is due to diffusion only. The space-charge field is $\pi/2$ phase shifted from E_s. BSO and other materials have shifted and unshifted space-charge field components. In the absence of external field, and for a modulation $m = 1$, the steady-state space-charge field can be written as

$$E_s = -i\frac{k_B T}{e} \frac{K}{1 + (K/k_0)^2} \qquad (24)$$

where $k_0^2 = (e^2 N_A/k_B T\varepsilon_0\varepsilon_s)$.

The measured parameters of relevance for the photosensitivity of some of the photorefractive materials are given in Table I. GaAs, BSO, and BGO have better response times for the grating formation than other materials. The efficiency of beam coupling is larger for barium titanate and SBN.

The mechanism of hologram formation in photorefractive materials is the following.[16] When the material is illuminated by two beams that

TABLE I
Relevant Optical Parameters of Some Photorefractive Materials

Material	λ (μm)	r (pm/V)	n	$\varepsilon/\varepsilon_0$	α (cm^{-1})	τ (s)	γ (cm^{-1})	Q^a (pm/Vε_0)
BaTiO$_3^{20}$	0.5	$r_{42} = 1640$	$n_e = 2.4$	$\varepsilon_1 = 3600$	1.0	1.3^{18}	20^{18}	6.3
SBN:75[19]	0.5	$r_{33} = 1400$	—	$\varepsilon_{33} = 3000$	—	0.6^{19}	—	5.6
SBN:60[19]	0.5	$r_{33} = 40$	—	$\varepsilon_{33} = 900$	—	0.05^{19}	14^{19}	6.3
SBN:Ce[20]	0.5	$r_{33} = 235$	$n_e = 2.33$	$\varepsilon_c = 880$	0.7	0.8^{20}	14^{20}	4.8
BSKNN-2[19]	0.5	$r_{33} = 170$	—	$\varepsilon_{11} = 360$	—	0.6^{19}	—	6.0
BSO[20]	0.6	$r_{41} = 5$	$n = 2.54$	$\varepsilon = 56$	0.13	0.015^{14}	10^{14}	1.5
BGO[7]	0.5	$r_{41} = 3.4$	$n = 2.55$	$\varepsilon = 47$	2.1	0.015^7	$3\text{-}5^7$	1.2
BTO[21]	0.6	$r_{41} = 5.2$	$n = 2.25$	$\varepsilon = 47$	—	—	2^{21}	1.3
KNbO$_3^{20}$	0.6	$r_{42} = 380$	$n = 2.3$	$\varepsilon_3 = 240$	3.8	0.1^7	$1\text{-}5^7$	19.3
LiNbO$_3^{20}$	0.6	$r_{33} = 31$	$n_e = 2.2$	$\varepsilon_3 = 32$	0.1	$\sim 10^2$	5^7	10.3
GaAs[20]	1.1	$r_{12} = 1.4$	$n_e = 3.4$	$\varepsilon = 123$	1.2	8×10^{-5}	0.4^{22}	4.7

$^a Q = n^3 r/\varepsilon$ is defined as the figure-of-merit of a photorefractive material.

produce sinusoidally modulated light intensity pattern, the donors in the regions of bright illumination are excited (Fig. 3b). The excited charge carriers migrate to regions of dark illumination by diffusion or by drift and are retrapped. There is a charge redistribution and a space-charge field is set up. This field modulates the refractive index of the material by Δn through Pockels effect:

$$\Delta n = r_{eff} n_0^3 |E_s| \tag{25}$$

and a phase grating is formed. The index modulation can be erased by illuminating the material uniformly. Methods have been developed to fix the holograms permanently.

C. Dynamic Self-diffraction

The volume nature of the hologram that is written in photorefractive materials leads to interaction between the writing beams. Writing and reading of the hologram occur simultaneously. The refractive index modulation induced by the incident beams affect the intensity and phase of the incident beams, which, in turn, modify the index grating. Self-diffraction of the writing beams causes a continuous recording of a new grating which is nonuniform through the thickness of the material. In addition, the gratings may be phase shifted with respect to the light-intensity pattern and slanted or even bent. The dynamic gratings are considerably different from the static gratings.

Dynamic self-diffraction in highly nonlinear media has been of interest for many years.[23] The phenomena of self-focusing, self-defocusing, and other self-action effects have been investigated in many transparent media with cubic and higher order nonlinearities. Strong optical fields incident on the medium bring about local or nonlocal, instantaneous or inertial responses in the medium. The light-induced nonlinearity $\Delta\varepsilon$, in general, can be written as[23]

$$\frac{\partial}{\partial t}(\Delta\varepsilon) = D\frac{\partial^2}{\partial t^2}(\Delta\varepsilon) - v\frac{\partial}{\partial z}(\Delta\varepsilon) - \frac{\Delta\varepsilon}{\tau_\varepsilon} + \beta I + \gamma F\left(\frac{\partial}{\partial z}, \int I\,dz\right) \quad (26)$$

where the first two terms on the right side describe diffusion and drift components of pump flux along the z axis. These components are responsible for the variation in the number of current carriers or excitons in semiconductors or ferroelectrics, or the fluxes of heat and liquids, if $\Delta\varepsilon$ is proportional to the temperature. The third term describes the relaxation of the excitations. The fourth term corresponds to the local response of the medium determined by the polarizability, and the last term corresponds to the nonlocal response of the medium to the incident radiation. The change in the permittivity $\Delta\varepsilon$ gives rise to complex grating formation. The interaction between the medium and the incident beams are described by time-dependent Maxwell's equations. The theoretical and experimental work on these dynamic interactions was reviewed by Vinetskii et al.[23] We summarize some of the results below.

In static unshifted gratings, the diffraction of two coherent beams is always accompanied by a change in their intensities, except for beams of equal intensity. However, in dynamic gratings, two coherent beams writing unshifted gratings do not take part in energy transfer for any ratio of the intensities. The absence of energy transfer is due to the fact that in the interaction between the gratings and the beams, the gratings adjust in such a way that the energy transfer in the two beams is the same and is mutually compensated for any intensity. The dynamic self-diffraction is always accompanied by energy exchange between the beams if the material response is nonlocal and the dynamic gratings are phase shifted from the light-intensity pattern.

Self-diffraction and transient energy exchange have been observed in materials having noninstantaneous (inertial) local response. The energy redistribution between the beams occurs in time intervals compared with the relaxation time of the light-induced refractive index change. Such energy transfer has been observed in electro-optical crystals in which drift mechanism is operative, and in absorbing liquids in which thermal gratings are formed by the writing beams. An artificial phase mismatch may be created between the index grating and the interference field in local response media if the grating relaxation time is large. The detection medium may be spatially displaced during the writing process to create mismatch. The moving medium method has been used to write dynamic holograms in absorbing media such as liquids and in transparent Kerr media.[17] Beam coupling and energy transfer have also been observed in nondegenerate two-wave mixing processes and in nondegenerate nonlinear processes of SRS and SBS. The universal method of achieving dynamic holography in nonlinear media with instantaneous local response is multiwave mixing. These processes are discussed in later sections.

IV. TWO-BEAM COUPLING IN PHOTOREFRACTIVE MATERIALS

As we saw in Section III, beam coupling in photorefractive materials is a dynamic process in which the incident beam and the material interact with each other. The beam coupling processes are described by Maxwell's equations. The material equations are obtained from the charge carrier generation and transportation relations modeled by Kukhtarev for a single-carrier and single-donor/trap system. The temporal and spatial response of the material is described by the solution to the set of equations. Since the equations are nonlinear, analytical solutions have been obtained for steady-state conditions and under undepleted pump approximation for transient conditions. Numerical methods have been used for obtaining exact solutions to the set of equations.

A. Steady-State Analysis[17]

For two waves E_1 and E_2 propagating in the crystal (Fig. 4), the spatial intensity distribution is given by

$$I = |A_1|^2 + |A_2|^2 + A_1^* A_2 \, e^{-\mathbf{K} \cdot \mathbf{r}} + A_1 A_2^* \, e^{i\mathbf{K} \cdot \mathbf{r}} \qquad (27)$$

The space-charge field-induced index modulation can be written as

$$n = n_0 + \frac{n_1}{2I_0} \, e^{i\Phi} A_1^* A_2 \, e^{-i\mathbf{K} \cdot \mathbf{r}} + \text{c.c.} \qquad (28)$$

where $I_0 = |A_1|^2 + |A_2|^2$ is the total intensity, and Φ is the phase shift of the index grating with respect to the interference pattern. The refractive index change n_1 is given by Eq. (25).

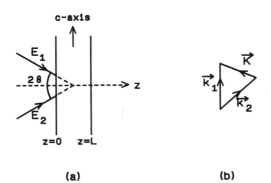

Figure 4. Two-beam coupling in photorefractive materials: (a) beam-interaction geometry; (b) wave-vector diagram.

Wave propagation is described by the scalar wave equation (9). Under SVEA, and from Eq. (28), we obtain the following coupled wave equations for $I_1 = |A_1 \exp(-i\psi_1)|^2$ and $I_2 = |A_2 \exp(-i\psi_2)|^2$:

$$dI_1/dz = -\gamma I - \alpha I_1 \tag{29}$$

$$dI_2/dz = \gamma I - \alpha I_2 \tag{30}$$

$$d\psi_1/dz = \beta I/I_1 \tag{31}$$

$$d\psi_2/dz = \beta I/I_2 \tag{32}$$

where

$$I = \frac{I_1 I_2}{I_1 + I_2} \qquad \gamma = \frac{2\pi n_1}{\lambda \cos(\vartheta/2)} \sin \Phi \qquad \beta = \frac{\pi n_1}{\lambda \cos(\vartheta/2)} \cos \Phi \tag{33}$$

The solutions for Eqs. (29) and (30) are

$$I_1(z) = I_1(0) \frac{1 + m^{-1}}{1 + m^{-1} e^{\gamma z}} e^{-\alpha z} \tag{34}$$

$$I_2(z) = I_2(0) \frac{1 + m}{1 + m e^{-\gamma z}} e^{-\alpha z} \tag{35}$$

where $m = I_1(0)/I_2(0)$ is the beam intensity ratio.

The energy exchange between the coupling beams depends on γ. The sign of γ is determined by the orientation of the c axis. If γ is positive and large enough to overcome the absorption, beam 2 is amplified. For sufficiently strong coupling, beam 1 is completely depleted. If γ is negative, the direction of energy transfer process is reversed. As a result, beam 1 is attenuated. The phase shift between the beams $\psi(z)$ is

$$\psi(z) = \psi_2(z) - \psi_1(z) = \psi(0) + \frac{\beta}{\gamma} \ln\left[\frac{e^{\gamma z}(1 + m)^2}{(m + e^{\gamma z})^2}\right] \qquad (36)$$

The beam coupling gain Γ is

$$\Gamma = \frac{I_2(L)}{I_2(0)} = \frac{1 + m}{1 + m\,e^{-\gamma L}}\,e^{-\alpha L} \qquad (37)$$

For contradirectional beam coupling, the signs of the right-hand-side expressions for I_2 and ψ_2 (Eqs. (30) and (32)) are reversed. The transmittance of the two beams are similar to Eqs. (34) and (35).

B.   Transient Analysis

The temporal behavior of the two interacting beams in the transient two-wave mixing (2WM) process depends on the dynamics of the optical fields as well as the dynamics of the nonlinear material in a self-consistent manner. The time-dependent material equation is given by Eq. (21). For small beam interaction lengths, such as occur in many practical situations, we can treat the optical fields as adiabatically following the space-charge field. Under these assumptions, the equations for the optical fields are Eqs. (29) and (30).

The coupled equations have been solved analytically for the case of undepleted pump approximation by several methods in which the boundary conditions are applied in the frequency domain[24-26] as well as in spatial coordinates,[27, 28] and by numerical methods for exact solutions.[29-31] The analytical results predict oscillations in the signal beam intensities in the transient regime, which has been experimentally observed.[27] The buildup and decay of the signal beam are complex functions of time even in the case of undepleted pump beams. For positive γL, the decay rate decreases with increasing γL. At high amplification, the signal energy is drawn mostly from the pump beam which is scattered by the grating. The grating strength is not altered at the moment the input signal is cut off, and the output drops off slowly. For negative γL, the steady signal is attenuated by the destructive interference between the signal and the pump beam that is scattered in the direction of the pump beam. After the input signal is turned off, the phase of the signal in the interaction zone reverses. The new signal generates a new grating of opposite phase compared with the existing grating, leading to a rapid erasure of the grating. The decay rate in this case increases as γL is decreased. Similar qualitative behavior is observed in the case of the grating buildup: The fastest buildup rate corresponds to the lowest γL.

A different approach for numerical solution was devised by Heaton and Solymar[29] based on the following physical picture. At time $t = 0$, the optical fields are determined by the boundary conditions and the space-charge field E_S, which is zero. For a short interval Δt both A_1 and A_2 remain unchanged. However, as a result of the interference of A_1 and A_2, a finite E_S is established which can be evaluated from Eq. (21). This space-charge field couples the beams in accordance with Eqs. (29) and (30). The spatial dependence of the optical fields for the next time interval Δt may then be obtained from Eqs. (29) and (30) using numerical integration techniques. The new optical field intensity modulation changes E_S in the next time interval Δt. This process continues for further time steps. The accuracy of the whole solution depends on the magnitude of the steps Δt and Δz in the numerical integration. The solutions are assumed to be correct when the results become independent of the step sizes. This method has been successfully used in many of the numerical studies in transient behavior of the 2WM and 4WM processes in photorefractive materials.

C. Nondegenerate Two-Beam Coupling

As in Kerr media, nondegenerate 2WM in photorefractive materials can be achieved by moving grating technique. The time evolution of the space-charge field E_S is given by the linear equation (21). The solution of the equation is partly decaying and partly oscillating. It is possible to improve the response of the system by introducing an external oscillatory term with a frequency that is chosen to optimize the imaginary part of g, i.e., the $\pi/2$ phase-shifted component of E_S.[32] A number of experimental and theoretical investigations have been reported in beam coupling in BSO using moving grating methods.[14, 33, 34] Both frequency[14] and phase shifting[35] techniques have been used. These methods are also applicable to other materials.

D. Two-Beam Coupling in Other Crystals

The anisotropic barium titanate and SBN have large coupling constants and have been extensively investigated. These crystals are used predominantly with extraordinary polarized light beams, because the electro-optic coefficients are favorable for those orientations. Other materials, such as cubic GaAs, have smaller coupling constants, but have much faster responses compared to barium titanate. The optical isotropy and tensor nature of cubic crystals allows beam coupling in several orientations, and cross-polarization coupling has been observed. A general theory of 2WM in cubic crystals was reported by Yeh.[36]

E. Applications

Two-beam coupling in photorefractive materials has a wide range of applications: in real-time holography,[37-39] in phase-conjugate mirrors,[40] in laser gyros,[41] and in image amplification.[42, 43] We will touch upon a few applications.

The beam coupling constants of photorefractive materials are large. In barium titanate, amplification of ~ 4000 has been experimentally attained in two-beam coupling.[42] Such large amplification factors allow detection of very low-level (picowatt) optical signals.[46] Optical amplification of signal beams in image processing and in opto-electronic neural networks is implemented by two-beam coupling. One of the problems encountered in image amplification is optical noise. The photoinduced scattering in the crystal reduces the contrast in the amplified image.[44] In barium titanate, this scattered beam emerges from the crystal as a large fan of light. Several techniques have been developed to reduce beam fanning.[44-46]

The origin of beam fanning is light scattering by stimulated two-wave mixing process.[47] Random inhomogeneities in the crystal give rise to Raleigh scattering. These scattered photons form noise gratings with the pump beam and undergo parametric processes, resulting in a unidirectional gain. In back scattering, the two-wave mixing process is similar to stimulated Raman scattering and stimulated Brilloiun scattering, and the scattered beam is the phase conjugate of the incident beam. The scattering also induces fluctuations in the index of refraction.[48] Though stimulated photorefractive scattering is a nuisance in image amplification, the scattered beams initiate grating formation in phase-conjugate oscillators and in mutually incoherent beam coupling.

The other important application of two-beam coupling is in real-time holography. Photoconductive electro-optic crystals such as BSO, BGO, BTO, and reduced $KNbO_3$ are ideally suited for real-time applications. The holographic recording times in these materials are $\sim 10^{-3}-10^{-2}$s for beam powers of $15\,mW\,cm^{-2}$. Vibrational analysis of structures and vibrating objects[49] have been investigated using double-exposure[37] and time-averaged holographic techniques.[38]

V. MULTIWAVE MIXING AND PHASE CONJUGATION

One of the problems in optical image transmission is distortion of the signal during propagation. Phase distortion can be corrected if the conjugated signal is available. Multiwave mixing is a way of generating phase-conjugate waves[50, 51] whereby the signal beam is mixed with the pump beams in a nonlinear medium. In this dynamic interaction process, self-diffraction and energy transfer also occur with the generation of the conjugate beam.

A. Degenerate Three-Wave Mixing

Optical phase conjugation and distortion correction were first demonstrated by Zeldovich et al.[52] by stimulated Brilluoin scattering. This led to the investigation of phase conjugation in other nonlinear mixing processes. If two optical fields E_1 and E_2 are incident on a medium with a quadratic nonlinearity d (Fig. 5), the polarization P^{NL} is[51]

$$P^{NL} = dA_1 A_2^* \exp\{i[(\omega_1 - \omega_2)t - (\mathbf{k}_1 - \mathbf{k}_2) \cdot \mathbf{r}]\} \qquad (38)$$

This polarization generates a field E_3 at $\omega_3 = \omega_1 - \omega_2$, and $\mathbf{k}_3 = \mathbf{k}_1 - \mathbf{k}_2$,

$$E_3 \propto dA_1 A_2^* \exp\{-i[\omega_3 t - \mathbf{k}_3 \cdot \mathbf{r}]\} \qquad (39)$$

If $\omega_3 = \omega_2 = \omega$ and $\mathbf{k}_3 = \mathbf{k}_2 = \mathbf{k}$, then $E_3 \propto E_2^*$ and is the phase conjugate of the original field E_2. The spatial and temporal evolution of E_3 is obtained by solving the Maxwell's equation,

$$\nabla \times \nabla \times \mathbf{E} - \mu\varepsilon \frac{\partial^2 \mathbf{E}}{\partial t^2} = \frac{\partial^2}{\partial t^2} \mathbf{P}^{NL} \qquad (40)$$

For a pump beam E_1 at 2ω and the signal beam E_2 at ω, under SVEA and undepleted pump approximation, Eq. (40) is reduced to a set of

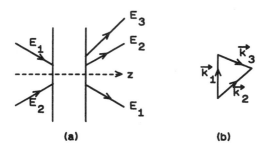

Figure 5. Three-wave mixing. E_1 and E_2 are the incident fields, and E_3 is the out-going field generated in the interaction process. (a) Beam-interaction geometry; (b) wave-vector diagram.

coupled equations:

$$\frac{dA_2^*}{dz} = igA_3\,e^{i\Delta kz} \qquad \frac{dA_3}{dz} = -ig^*A_2^*\,e^{-i\Delta kz} \tag{41}$$

where $g = \omega(\mu/\varepsilon)^{1/2}\,dA_1$ and $\Delta\mathbf{k} = \mathbf{k}_1 - \mathbf{k}_2 - \mathbf{k}_3$.

Solving Eq. (41) with the boundary conditions $A_2^*(z=0) = A_{20}^*$, $A_3(z = 0) = 0$, and $A_1(z = 0) = A_{10}$, gives

$$A_3(L) = -2i(g^*/b)\sinh(bL/2)\,A_{20}^*\exp(-i\Delta kL/2) \tag{42}$$

where $b = [4|g|^2 - (\Delta k)^2]^{1/2}$. Parametric amplification is possible if $g \gg \Delta k/2$. The output wave E_3 is the conjugate of the input field E_2 if $\Delta kL \ll \pi$. The basic disadvantage of three-wave mixing is the phase-matching requirement for efficient generation of conjugate waves. In crystals, this condition can be satisfied along one direction only, which limits the amount of distortion that can be corrected.

B. Degenerate Four-Wave Mixing

The most universal method for investigating dynamic holography is four-wave mixing. The basic arrangement for 4WM is shown in Fig. 6. Consider a nonlinear medium with local instantaneous response characterized by the third-order nonlinearity $\chi^{(3)}$. If E_1, E_2, and E_4 are three optical fields incident on the medium satisfying the relation $\omega_1 = \omega_2 = \omega_4$, and $\mathbf{k}_1 + \mathbf{k}_2 = 0$, then the nonlinear polarization is given by[50]

$$P^{NL}(\omega_3 = \omega_1 + \omega_2 - \omega_4) = \chi^{(3)}A_1A_2A_4^*\,e^{[-i(\omega t + k_4 z)]} + c.c. \tag{43}$$

This polarization generates a wave at ω traveling in the $-z$ direction,

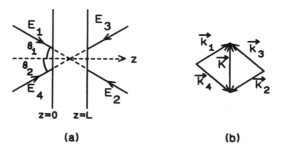

Figure 6. Four-wave mixing: (a) beam-interaction geometry; (b) wave-vector diagram.

whose amplitude is proportional to A_4^*; thus, it is the complex conjugate of E_4.

The spatial and temporal evolution of E_3 is obtained by solving Maxwell's equation (40). Under the slowly varying envelope approximation (SVEA), and if $\mathbf{k}_3 + \mathbf{k}_4 = 0$, Eq. (40) is reduced to a set of coupled scalar equations:

$$\frac{dA_3}{dz} = i\kappa^* A_4^* \quad \text{and} \quad \frac{dA_4^*}{dz} = i\kappa A_3 \qquad (44)$$

where

$$\kappa^* = \omega \left(\frac{\mu}{\varepsilon}\right)^{1/2} \chi^{(3)} A_1 A_2$$

When Eq. (44) is solved with the boundary conditions $A_4(z = 0) = A_{40}$ and $A_3(z = L) = 0$, the outgoing $A_3(0)$ and $A_4(L)$ are given by

$$A_3(0) = -i\left(\frac{\kappa^*}{|\kappa|} \tan|\kappa|L\right) A_4^*(0) \qquad (45)$$

$$A_4(L) = \frac{A_4(0)}{\cos|\kappa|L} \qquad (46)$$

The results show that (1) the reflected wave $A_3(0)$ is proportional to the complex conjugate of the incident field $A_4(0)$, and (2) the reflected wave intensity exceeds that of the input wave for $\pi/4 < |\kappa|L < 3\pi/4$. This shows that in 4WM, dynamic self-diffraction and energy exchange are possible in media with local response. Unlike in the case of 3WM, phase-matching conditions are always satisfied automatically for arbitrary angles between pairs of input beams. Similar sets of coupled equations can be obtained for media with local inertial response. The coupling κ depends on the mechanism of index grating formation (Eq. (26)).

Four-wave mixing is analogous to real-time holography. If we assume that E_1 is the reference beam and E_4 is the signal beam, the resulting transmission T of the hologram that is formed is

$$T \propto |E_4 + E_1|^2 = |E_4^2| + |E_1^2| + E_4 E_1^* + E_4^* E_1 \qquad (47)$$

In the reconstruction step, the hologram is illuminated by the reference beam E_2 propagating in the direction opposite to that of E_1. The diffracted field E_3 is

$$E_3 = TE_2 \propto \left(|E_4^2| + |E_1^2|\right)E_1^* + (E_1^*)^2 E_4 + |E_1 E_1^*| E_4^* \qquad (48)$$

The first term on the right side is proportional to the incident field $E_2(= E_1^*)$. The second term is not phase matched and will not radiate. The last term is

$$E_3 \propto |E_1|^2 E_4^* = |E_1 E_2| E_4^* \qquad (49)$$

which for $z < 0$ corresponds to a "time-reversed" phase-conjugate replica of the original object wave E_3. This is similar to what was obtained in Eq. (8).

The analogy between 4WM and real-time holography is helpful in extending all the conventional holographic techniques in optical information processing to 4WM. This opens up an immense application potential of 4WM in many areas of image processing,[1-3] such as real-time holography, image amplification, optical information processing, holographic storage, and nonlinear holographic associative memories. In media with local response, the diffraction efficiency depends on $\chi^{(3)}$, which, in general, is small and requires very high optical fields. Photorefractive materials have better responses to low-intensity optical fields. However, there is a trade-off in terms of the response times. The photorefractive materials are slow in responding to optical fields ($> 10^{-3}$s) compared with the materials with local response ($> 10^{-10}$s).

VI.   FOUR-WAVE MIXING IN PHOTOREFRACTIVE MATERIALS

High photosensitivities, low power requirements, and large beam coupling constants of the photorefractive materials make them attractive for optical image processing applications. The analogy between 4WM and conventional holography is more complete in photorefractives, since the responses of the materials are nonlocal and noninstantaneous. The processes of writing and reading the holographic gratings can be identified and studied separately.

As in two-beam coupling, the dynamic interactions between the beams are described by Maxwell's equations, while coupling between optical fields and the photorefractive material is described by Eq. (21). Because of their complexity, the equations can be solved only under various approximations.

A. Steady-State DFWM[53]

The basic interaction geometry for 4WM is shown in Fig. 6. Beams 1, 2, 3, and 4 are of the same frequency ω and polarization state and are propagating through the medium. The electric field amplitudes are assumed to be

$$E_j(r, t) = A_j(r)\exp\left[i(\mathbf{k}_j \cdot \mathbf{r} - \omega t)\right] + \text{c.c.} \tag{50}$$

The propagation directions satisfy the relations $\mathbf{k}_1 = -\mathbf{k}_2$ and $\mathbf{k}_3 = -\mathbf{k}_4$. The interference between the beams gives rise to six sets of spatial intensity modulations in the material, each with a different fringe spacing. Two of the sets, between A_1 and A_4 and between A_2 and A_3, form transmission gratings, while the other four form reflection gratings. The resulting space-charge field induced index is

$$
\begin{aligned}
n = n_0 &+ \frac{n_\mathrm{I} \exp(i\Phi_\mathrm{I})}{2} \frac{(A_1^* A_4 + A_2 A_3^*)}{I_0} e^{i\mathbf{k}_\mathrm{I} \cdot \mathbf{r}} \\
&+ \frac{n_\mathrm{II} \exp(i\Phi_\mathrm{II})}{2} \frac{(A_1 A_3^* + A_2^* A_4)}{I_0} e^{i\mathbf{k}_\mathrm{II} \cdot \mathbf{r}} \\
&+ \frac{n_\mathrm{III} \exp(i\Phi_\mathrm{III})}{2} \frac{A_1 A_2^*}{I_0} e^{i\mathbf{k}_\mathrm{III} \cdot \mathbf{r}} \\
&+ \frac{n_\mathrm{IV} \exp(i\Phi_\mathrm{IV})}{2} \frac{A_3^* A_4}{I_0} e^{i\mathbf{k}_\mathrm{IV} \cdot \mathbf{r}} + \text{c.c.}
\end{aligned} \tag{51}
$$

where $I_0 = \Sigma I_j$, Φ_j is the phase shift between the index modulation with respect to the intensity modulation, $\mathbf{K}_\mathrm{I} = \mathbf{k}_4 - \mathbf{k}_1 = \mathbf{k}_2 - \mathbf{k}_3$, $\mathbf{K}_\mathrm{II} = \mathbf{k}_1 - \mathbf{k}_3 = \mathbf{k}_4 - \mathbf{k}_2$, $\mathbf{K}_\mathrm{III} = 2\mathbf{k}_1$, and $\mathbf{K}_\mathrm{IV} = 2\mathbf{k}_4$. The expressions for Φ and n are given by Eqs. (23) and (25).

From Eqs. (9) and (51), under SVEA approximation, we obtain the coupled wave equations:

$$\frac{2c}{\omega} \cos \vartheta_1 \frac{dA_1}{dz} = t_1 A_4 + r_1 A_3 + r_2 A_2 - \frac{2c}{\omega} \cos \vartheta_1 \alpha_1 A_1 \tag{52}$$

$$\frac{2c}{\omega} \cos \vartheta_1 \frac{dA_2}{dz} = t_1^* A_3 + r_1^* A_4 + r_2^* A_1 + \frac{2c}{\omega} \cos \vartheta_1 \alpha_1 A_2 \tag{53}$$

$$\frac{2c}{\omega} \cos \vartheta_2 \frac{dA_3}{dz} = -t_1 A_2 + r_1^* A_1 - r_3 A_4 + \frac{2c}{\omega} \cos \vartheta_2 \alpha_2 A_3 \tag{54}$$

$$\frac{2c}{\omega} \cos \vartheta_2 \frac{dA_4}{dz} = -t_1^* A_1 + r_1 A_2 - r_3^* A_3 - \frac{2c}{\omega} \cos \vartheta_2 \alpha_2 A_4 \tag{55}$$

where

$$t_1 = -\frac{in_1 \exp(-i\Phi_1)}{I_0}(A_1 A_4^* + A_2^* A_3)$$

$$r_1 = -\frac{in_{II} \exp(i\Phi_{II})}{I_0}(A_1 A_3^* + A_2^* A_4) \tag{56}$$

$$r_2 = -\frac{in_{III} \exp(i\Phi_{III})}{I_0}A_1 A_2^* \quad \text{and} \quad r_3 = -\frac{in_{IV} \exp(-i\Phi_{IV})}{I_0}A_3 A_4^*$$

Consider the case of transmission grating under undepleted pump approximation. Equations (52)–(55) reduce to

$$\frac{dA_1}{dz} = -QA_4 - \alpha A_1 \tag{57}$$

$$\frac{dA_2^*}{dz} = -QA_3^* + \alpha A_2^* \tag{58}$$

$$\frac{dA_3}{dz} = QA_2 + \alpha A_3 \tag{59}$$

$$\frac{dA_4^*}{dz} = QA_1^* - \alpha A_4^* \tag{60}$$

where $Q = \gamma(A_1 A_4^* + A_2^* A_3)/I_0$, $\gamma = i\omega\Delta n \exp(-i\Phi)/2c \cos \vartheta$, and Δn is given by Eq. (25).

The equations can be solved in terms of Bessel functions. For $\alpha = 0$, the phase conjugate reflectivity can be written as

$$\rho = \frac{A_3(0)}{A_4^*(0)} = -\left(\frac{A_1 A_2}{A_1^* A_2^*}\right)^{1/2} \frac{\sinh(\gamma L/2)}{\cosh[(\gamma L/2) + (\ln r)/2)]} \tag{61}$$

where r is the pump beam ratio, $r = I_2(L)/I_1(0)$. From Eq. (61), we see that for $\gamma L = \pi$ and $r = 1$, self-oscillation occurs when phase shift is zero. The optimum phase-conjugate reflectivity is obtained for unequal beam intensities. This is in contrast with the situation in media with local response, where the interaction of the pumping beams should be equal. Also, in materials in which charge migration is predominantly due to diffusion, the phase shift is $\pi/2$, and self-oscillation is not possible. However, by detuning the probe beam from the pump beam or by applying electric fields, the holographic phase shift can be modified to obtain self-oscillation.

The results of the reflection gratings and the transmission gratings are the same under undepleted pump approximation. With pump depletion, the results for the transmission gratings show that phase-conjugate reflectivity exhibits multistability for a certain range of parameters. The phase-conjugate reflectivity remains finite even when the pump beam 2 is absent. In the case of reflection gratings, the reflectivity goes to zero in the limit of the infinite pump beam ratio I_1/I_2. This has important consequences for passive phase-conjugate mirrors.

Phase conjugation in the simultaneous presence of transmission and reflection gratings is a complex process. There is competition between the two coupling mechanisms in the generation of the phase conjugate waves. The spatial and temporal evolution of the optical fields have been investigated by numerical methods.[54–57] The results show that the presence of a second mechanism improves the efficiency of the phase-conjugation process. Improvement depends not only on the relation between the coupling strengths, but also on their absolute values. The absorption of the material modifies the multigrating interaction, and phase conjugation improves.[54] The multistability of the solution to the coupled equations leads to chaos,[55] which has been experimentally observed.[58]

One of the peculiarities of the standard 4WM geometry with parallelly polarized pump beams is the limitation on the maximum attainable phase-conjugate reflectivity in the transmission grating. The two transmission gratings written by A_1 and A_4 and by A_2 and A_3 are out of phase with each other, reducing the overall efficiency of the grating. Equation (61) shows that unsymmetric pumping improves the grating efficiency. This also results in the directionality of the power flow. The efficiency may also be improved by shifting the phase of one of the transmission gratings with respect to the other. This may be accomplished by utilizing different frequencies for the writing beams, thereby creating running gratings (nearly degenerate 4WM).[35, 62, 63] The other method involves tilting the gratings with respect to each other, which introduces a spatial phase shift between the gratings. This can be done by tilting one of the pump beams with respect to the other pump beam.[62, 64] The phase shift and the negative interaction between the transmission gratings are also avoided by using orthogonally polarized pump beams.[65–67] Improvement in the case of barium titanate and SBN is not substantial because the coupling constants are not favorable.[66] However, in selenite-type crystals, such as BSO and BGO, it is possible to obtain symmetric gain in the forward as well as backward directions.[67]

B. Photorefractive Oscillators

We saw in Section VI.A that in 4WM, the phase-conjugate reflectivity remains finite for infinite pump beam ratio. This has been effectively used

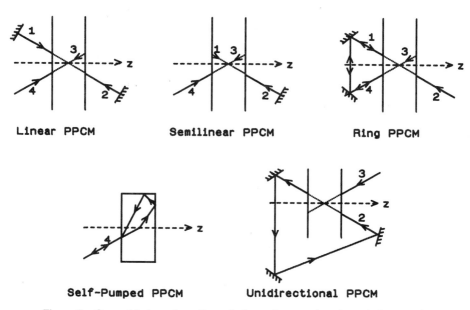

Linear PPCM   **Semilinear PPCM**   **Ring PPCM**

Self-Pumped PPCM   **Unidirectional PPCM**

Figure 7. Geometrical configurations of the various passive phase-conjugate mirrors (PPCM). The self-pumped PPCM is also known as "Cat" mirror.

in the development of passive phase-conjugate mirrors (PPCM) with large reflectivities and in phase-conjugate resonators.[17, 40, 53, 68] The need for high-quality external pump beams is obviated in PPCMs. The input signal beams generate the remaining beams required for phase conjugation in the four-wave mixing process. The optical configuration of some of the devices is shown in Fig. 7.

The physical basis for the buildup of oscillations in PPCMs is as follows. Light scattered by the impurities or defects in the crystal are amplified by two-beam coupling process. The scattered light is fed back to the crystal by external mirrors or by total internal reflection at the corners of the crystal to enhance the interaction, as in a unidirectional ring resonator, or to initiate a 4WM process, as in ring resonators and other PPCMs. The oscillating beams continue to build up until a steady state is reached, and then start pumping the crystal as a phase conjugator for the input beams. The threshold coupling strengths and the beam amplitudes are obtained by solving the beam-coupling equations (21) and (52)–(55) with appropriate boundary conditions. The ring resonator and the linear PPCM have self-starting thresholds. The self-pumped phase conjugator (SPPCM)[69] is a special kind of ring resonator in which the crystal corners act as feedback surfaces. Due to the ease of alignment and relaxed coherence requirements, the SPPCM has numerous applications.

The optical feedback present in the PPCMs and PC resonators results in self-frequency detuning.[17, 68] As the gratings and the oscillating beams build up in time, the amplitudes and the phases of the beams must satisfy the oscillating conditions for each round-trip. The phase transferred to the coupling beams in the interaction depends on the complex beam-coupling constant. An additional phase shift occurs due to the optical feedback path. The oscillation beam is therefore detuned from the pump beam to satisfy the oscillation condition on phase. Self-detuning has been observed in several PC resonators. In ring PPCM, in which the optical paths for the two oscillating beams are equal, a nonreciprocal phase shift due to residual electric fields has been observed to lead to detuning of the beams. Self-detuning effects in PPCMs have been used in scanning dye lasers[70] and in optical gyros.[41]

C. Transient Analysis

As in the case of transient two-beam coupling, the coupled wave equations for transient 4WM, in general, can be written as

$$dA_1/dz = Q_1 A_4 + Q_2 A_3 + Q_3 A_2 - \alpha A_1 \tag{62}$$

$$dA_2^*/dz = Q_1 A_3^* + Q_2 A_4^* + Q_3 A_1^* + \alpha A_2^* \tag{63}$$

$$dA_3/dz = -Q_1 A_2 + Q_2^* A_1 - Q_4 A_4 + \alpha A_3 \tag{64}$$

$$dA_4^*/dz = -Q_1 A_1^* + Q_2^* A_2^* - Q_4 A_3^* - \alpha A_4^* \tag{65}$$

under SVEA approximation and assuming that the optical fields follow the space-charge field adiabatically. The amplitudes Q_1, Q_2, Q_3, and Q_4 are given by

$$\frac{\partial Q_1}{\partial t} + g_1 Q_1 = \frac{\gamma_1}{I_0} g_1 (A_1 A_4^* + A_2^* A_3) \tag{66}$$

$$\frac{\partial Q_2}{\partial t} + g_2 Q_2 = \frac{\gamma_2}{I_0} g_2 (A_1 A_3^* + A_4 A_2^*) \tag{67}$$

$$\frac{\partial Q_3}{\partial t} + g_3 Q_3 = \frac{\gamma_3}{I_0} g_3 A_1 A_2^* \tag{68}$$

$$\frac{\partial Q_4}{\partial t} + g_4 Q_4 = \frac{\gamma_4}{I_0} g_4 A_3 A_4^* \tag{69}$$

The expressions for g and γ are given by Eqs. (21) and (56).

Equations (66)–(69) have been solved analytically for transmission gratings under undepleted pump approximation using Fourier transform techniques with the boundary conditions specified in frequency domain,[26] and uses Bessel functions with the boundary conditions specified in spatial coordinates.[27] The results show that after the signal A_4 is turned on, the phase conjugate wave A_3 buildup rate increases with increasing $|\gamma|L$ for a specific beam ratio r, and increases with increasing r for a specific γL. The temporal decay of the phase-conjugate beam from steady state after the signal beam is turned off depends on r and $|\gamma|L$. The decay rate decreases as $|\gamma|L$ is increased for negative γL, and vice versa for positive γL.

There are regions of instability in the transient regime as γL is varied. For diffusion-dominated photorefractives, the stability condition is $|\gamma|L < 2\pi$. In the other limiting case of drift-dominated photorefractives, for a pump beam ratio of 1, the stability condition is $|\gamma|L < \pi$. Horowitz et al.[27] showed that 4WM is analogous to 2WM in a unidirectional ring cavity and that the self-pulsation frequency in the unstable regime of the 4WM corresponds to the frequency detuning of the oscillation in the unidirectional ring resonator. These instabilities have also been investigated under the pump depletion condition using numerical techniques.[58-61] The results show that in addition to periodic oscillations, chaos and multistability are possible. Such large instabilities have been observed in several of the phase-conjugate resonators.[58-61, 71] The double phase-conjugate mirror (DPCM) is an interesting case of this instability in which self-oscillation occurs even in the absence of the signal beam.[72] The photoinduced noise in the crystal sets up stable oscillations, and two phase-conjugate beams are generated with a total depletion of the pump beams.

VII.   MUTUALLY INCOHERENT BEAM COUPLING

We have seen in Section IV that 4 WM is the most widely used technique for generating phase-conjugate waves. The configuration requires three coherent beams, two of which are the pump beams and the other a signal beam whose phase conjugate is sought. However, this is not the only coupling process that generates phase-conjugate beams in photorefractives. Two mutually incoherent beams have also been observed to interact strongly in the presence of photorefractive materials, generating phase conjugates of each other. The beam threshold and the coupling strengths required for this interaction are high. Mutually incoherent beam coupling (MIBC) was observed initially in barium titanate, whose beam-coupling strength is high.

There are several geometries in which MIBC has been investigated[72-77] (Fig. 8). The differences are in the directions of the propagation of the input beams in relation to the c axis of the crystal. The underlying physical mechanism, however, is the same for all geometries.[74] Each of the two input beams incident on the crystal produces its own beam fanning. The fanned light has a large number of fanned gratings formed between the incident light and the scattered light amplified in the two-beam coupling process. The mutually incoherent beams Bragg diffract off the fanning gratings due to the other beam, enhancing in the process only the phase-matched gratings and erasing the rest. In steady state, a set of mutually shared gratings remain after a repeated erase–enhance process. Since the beams are mutually incoherent, all the gratings are transmissive.

Mamaev and Zozulya[77] reported MIBC in the geometry of two interconnected ring mirrors in which any combination of reflection and transmission gratings may be utilized. This geometry is also applicable to SBS-active media. MIBC has been observed in other local transient response media[78] and under stimulated diffusive backscattering.[79] In Kerr media also[80] numerical investigations show that Stokes scattering of the incident photons results in the formation of cross-readout gratings which can couple the beams.

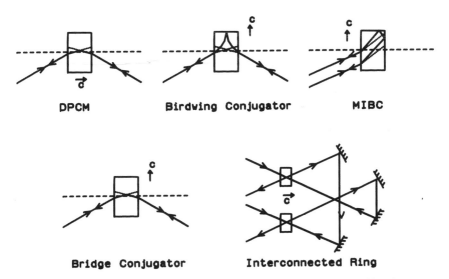

Figure 8. Mutually incoherent beam coupling geometries. DPCM, double phase-conjugate mirror; MIBC, mutually incoherent beam conjugator.

Of the optical geometries shown in Fig. 8, the double phase-conjugate mirror is the simplest configuration in which there are no internal reflections at the surfaces. The beam overlapping volume is a single interaction region and the 4WM model can be applied with the appropriate boundary conditions. DPCM is discussed in the next section. The ability to couple two coherent sources of different frequencies through MIBC has led to the development of new applications in adaptive optics and optical communications. These applications are discussed in a later section.

VIII.   DOUBLE PHASE-CONJUGATE MIRROR

Simultaneous generation of two phase-conjugate waves in a transmission geometry was first proposed by Cronin-Golomb et al.[53] However, Weiss et al.[72] were the first to demonstrate the device and develop the theory and applications. The basic optical configuration of DPCM is shown in Fig. 9. Input beams 2 and 4 are incident on the opposite faces of the photorefractive crystal. The beams fan inside the crystal, generating a set of mutually shared gratings. In the steady state, beam 4 diffracts off the shared gratings along the direction of beam 2, generating beam 1, which is the phase conjugate of beam 2; beam 2 diffracts off the shared gratings along the direction of beam 4, generating beam 3, which is the phase conjugate of beam 4. Beam 3 (beam 1) carries the spatial information of beam 4 (beam 2) and the temporal information of beam 2 (beam 4). In general, if the beams are mutually incoherent, only transmission gratings are involved in the beam-coupling process. However, both reflection and transmission gratings are operative in mutually coherent beam coupling. The competi-

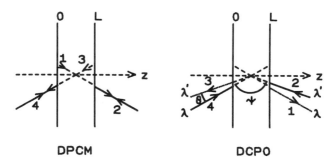

Figure 9.   The four-wave mixing geometry in a double phase-conjugate mirror (DPCM) and a double color pump oscillator (DCPO). ψ is the angle between pump beams 2 and 4; ϑ is the angular detuning.

tion between the two types of gratings results in large fluctuations in the output beams.[81, 82]

A.   Steady-State Analysis of DPCM

In the steady state,[72] DPCM can be modeled as a 4WM process in which transmission gratings dominate. Under undepleted pump approximations and negligible absorption, the coupled equations can be written as

$$\frac{dA_1}{dz} = -\frac{\gamma}{I_0}\left[|A_4|^2 A_1 + (A_2^* A_4) A_3\right] \tag{70}$$

$$\frac{dA_3}{dz} = \frac{\gamma}{I_0}\left[|A_2|^2 A_3 + (A_2 A_4^*) A_1\right] \tag{71}$$

where γ is the coupling constant, and $I_0 \simeq I_2 + I_4$. Solving Eqs. (70) and (71) with the boundary conditions $A_1(z = 0) = A_{10}$ and $A_3(z = L) = A_{3L}$, we obtain for the output beams,

$$\rho = \frac{A_3(0)}{A_{3L}} = \frac{(1 - q) - \alpha\beta(e^{p\gamma L} - 1)}{e^{p\gamma L} - q}$$

$$t = \frac{A_1(L)}{A_{10}} = \frac{(1 - q)e^{p\gamma L} - q(\alpha\beta)^{-1}(e^{p\gamma L} - 1)}{e^{p\gamma L} - q} \tag{72}$$

where $q = I_4/I_2$, $p = (1 - q)/(1 + q)$, $\alpha = A_4^*/A_2^*$, and $\beta = A_{10}/A_{3L}$.

We see from Eq. (72) that there exists γL at which zero $A_1(0)$ and $A_3(L)$ build up into nonzero outputs. The oscillation ($\rho = t = \infty$) occurs at

$$\gamma L = (\ln q)(1 + q)/(1 - q) \tag{73}$$

The threshold value of the coupling constant $|\gamma L|_{th}$ is 2. The solution of the coupled equations for depleted pump beams gives the same result for the threshold γL. The results also show that the complex amplitude transmittivities for each of the counterpropagating beams are the same in both directions. The symmetric transmission is given by

$$T = \frac{I_3(0)}{I_2(L)} = \frac{I_1(L)}{I_4(0)} = \frac{a^2\left[q^{1/2} + q^{-1/2}\right]^2 - \left[q^{-1/2} - q^{1/2}\right]^2}{4} \tag{74}$$

where a is defined by the relation $a = \tanh(-\gamma La/2)$.

The reflectivities of the device are

$$R_0 = \frac{I_3(0)}{I_4(0)} = \frac{T}{q} \quad \text{and} \quad R_L = \frac{I_1(L)}{I_2(L)} = Tq \qquad (75)$$

At $q = 1$, the transmission is maximum and equal to a^2. The range q for oscillation is

$$\frac{1-a}{1+a} < q < \frac{1+a}{1-a} \qquad (76)$$

The maximum reflectivity $R_{max} = a^2/(1 - a^2)$ at $q = (a^2 + 1)/(1 - a^2)$, and is greater than one for $|\gamma L| > 2.49$. Thus, DPCM can act as an image amplifier in the dynamic range $q_{max} - q_{min}$. For the same γL, an externally pumped phase conjugate mirror and a 2WM amplifier have much larger dynamic ranges.[83] MIBC with two interaction regions has a higher threshold coupling constant and a lower dynamic range.[84]

The wave vectors of the four beams in DPCM satisfy the relation $\mathbf{k}_1 - \mathbf{k}_4 = \mathbf{k}_3 - \mathbf{k}_2$. This relation requires only that A_1 and A_2 be coherent with A_4 and A_3, respectively. Because the beams A_3 and A_1 are generated internally in the beam-coupling process, there are infinite sets of solutions satisfying the wave-vector relation. Thus, for plane wave inputs, output beams A_1 and A_3 emerge as cones of light.[53, 68, 85] In barium titanate, the oscillation process usually selects the spatial mode in which the oscillating beams have maximum spatial overlap with respect to the input beams and thus the highest gain. The conical scattering collapses as the buildup continues to give rise to two well defined beams. However, if beams 2 and 4 are spatially modulated, $\mathbf{k}_1$ and $\mathbf{k}_3$ are constrained to the plane containing $\mathbf{k}_2$ and $\mathbf{k}_4$, and the conical scattering is absent.

Experimentally, it was found that the fidelity of mutual phase conjugation is high for spatially modulated beams.[86, 87] In BTO, it was observed that spatial modulation is a necessary condition for observing DPCM.[88] In other mutually pumped phase conjugators (MPPC), such as the bird-wing conjugator, due to the presence of more interaction regions, only one solution exists for the wave vectors of the output beams, and the conical scattering is absent.[74]

B. Double Colored Pump Oscillator

From $\mathbf{k}_1 - \mathbf{k}_4 = \mathbf{k}_3 - \mathbf{k}_2$, it can be seen that since $\mathbf{k}_1$ and $\mathbf{k}_3$ are free, the wavelengths of pump beams 2 and 4 need not be the same. The gratings written by beams 4 and 1 with wavelength λ_1 can be mutually shared by beams 2 and 3 with wavelength λ_2 as long as $\mathbf{k}_1 - \mathbf{k}_4 = \mathbf{k}_3 - \mathbf{k}_2$ is satisfied. Such a device, the double colored pump oscillator (DCPO), was demonstrated and analyzed by Sternklar and Fischer.[87] Output beams 3 and 1 in DCPO are angularly detuned with respect to input beams 2 and 4, respectively, and, unlike in DPCM, these beams are not phase conjugates of the input beams (Fig. 9). However, there is no cross-talk. If ψ is the angle between k_4 and k_2, the beam deflection angle ϑ inside the crystal is given by

$$\vartheta = \tan^{-1}\left[\frac{\sin\psi}{(\lambda'/\lambda) - \cos\psi}\right] - \tan^{-1}\left[\frac{\sin\psi}{(\lambda/\lambda') - \cos\psi}\right] \quad (77)$$

DCPO has been used for image color conversion and for beam steering with automatic Bragg matching.[89]

C. 4WM and DPCM Geometries

We saw in Section VI.A that in 4WM transmission geometry with parallelly polarized pump beams, the gratings written by 1 and 4 are antiphase with those written by 2 and 3. The sign of the secondary grating can be inverted by changing the polarization of one of the pump beams. The coupled wave equations for the transmission gratings with orthogonal polarization are[66]

$$\frac{dA_1}{dz} = QA_4 \qquad \frac{dA_2^*}{dz} = -QA_3^* \qquad \frac{dA_3}{dz} = QA_2 \qquad \frac{dA_4^*}{dz} = -QA_1^* \tag{78}$$

where $Q = \gamma(A_1 A_4^* + A_2^* A_3)/I_0$. If A_1 and A_3 are the pump beam amplitudes, then under undepleted pump approximation

$$\frac{dA_3}{dz} = \gamma\left[(A_1 A_2)A_4^* + |A_2|^2 A_3\right]/I_0 \tag{79}$$

$$\frac{dA_4^*}{dz} = -\gamma\left[(A_1 A_2)^* A_3 + |A_1|^2 A_4^*\right]/I_0 \tag{80}$$

These equations are similar to Eqs. (70) and (71), if we make the correspondence $A_4^* \leftrightarrow A_1$. Solving the equations, we get for the oscillation condition $\gamma L = (\ln q)(1 + q)/(1 - q)$, which is also the threshold condition (73) for DPCM. The two geometries, DPCM with parallelly pumped polarized pump beams and 4WM with orthogonally polarized pump beams, are interchangeable.

Self-oscillation and double phase conjugation have been reported in cubic BTO in the 4WM geometry with orthogonally polarized pump beams.[90] For the 4WM geometry with parallelly polarized pump beams, self-oscillation is not expected for materials that operate by pure diffusion.[53, 91] In BTO, however, in the equivalent configuration of DPCM with orthogonally polarized pump beams, two oscillating beams have been observed, which are tilted ($\sim$ mrad) with respect to the pump beams.[90]

D. DPCM as a Building Block

Beam coupling and phase conjugation with MIBC can be achieved in many configurations involving more than one interaction region. However, all the configurations can be unfolded and analyzed in terms of beam coupling in a single interaction region.[92, 93] Weiss et al.[93] have shown that MIBC configurations with one/two internal reflections are equivalent to two cascaded DPCMs. The effect of the reflections in MIBC can be taken into account by including a transfer function in the cascaded DPCMs. As may be expected, MIBC has a higher threshold compared to DPCM. The response time of a MIBC conjugator, however, is less than that of DPCM for comparable incident beam powers.[74] The model gives a reasonable agreement with the experimental results.

The DPCM may be considered as a basic building block in the implementation of optical architectures for image-processing applications. Two or more DPCMs may be cascaded to achieve multiple phase-conjugate beams, each carrying its own spatial information (Fig. 10). In these configurations, the fanned light from one DPCM is channeled to a second facing DPCM (2F-DPCM). The speed of formation of the conjugate beams may be improved by channeling the transmitted light also (2Z-DPCM).[94] Multiple phase conjugation can also be achieved by using a single pump beam in a cascaded PPCM oscillator (2C-PPCM).[95]

Double phase conjugation has been observed in many other photorefractive materials, for example, GaAs,[96] BGO:Cu,[97] BSO,[98] SBN,[76] and BSKNN.[76] Unlike in barium titanate where the pump beams propagate along the c axis, in SBN:60, the pump beams propagate at right angles to the c axis, giving rise to a "bridge" geometry, taking advantage of the large r_{33}. BSKNN has both large r_{33} and r_{42}, and conjugates efficiently in many geometries.

Double phase conjugation and mutually incoherent beam coupling have been theoretically investigated in the media with local response.[80] As we saw earlier, DPCM/MIBC is a threshold process, and local scattering in the crystal lowers the power requirements for achieving beam coupling. This, however, is not an optimum geometry for media with local response, which have high thresholds. Mamaev and Zozulya[77] proposed a geometry

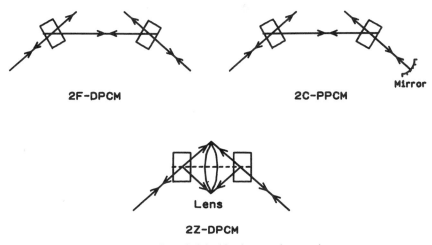

2F-DPCM   **2C-PPCM**

2Z-DPCM

Figure 10.   Cascaded double phase-conjugate mirrors.

of interconnected ring mirrors to achieve lower thresholds. In this geometry, the scattered radiation is obtained from the cross-readout of the transmission/reflection gratings formed in the ring oscillations. This improves the efficiency of the generation of the shared gratings for MIBC. Mamaev and Zozulya[77] used these geometries to obtain DPCM in SBN and $KNbO_3$. Vallet et al.[99] used a similar approach of interconnected ring oscillations to achieve DPCM in sodium, a local response medium. The first model proposed for DPCM,[100] in fact, was based on the cross-readout of holograms formed by the two pump beams with the two counterpropagating oscillations in an interconnected ring resonator.

E.   Transit Analysis

The temporal response of a DPCM, as well as other PPCMs, is poor compared with the material response for 4WM and phase conjugation. Typical experimental response in the case of barium titanate is shown in Fig. 11. The phase-conjugate wave generation involves dynamic interaction between a large number of gratings in the transient regime until the mutually shared gratings are established. There is little phase-conjugate output during this stage. The interaction is later dominated by 4WM in a small set of transmission gratings. This leads to a thresholding behavior, as is evident from Fig. 11.

A complete analysis of the temporal response of DPCM requires the inclusion of all the transmission gratings involved in the buildup process. This is not an easy task, given the nonavailability of suitable models for

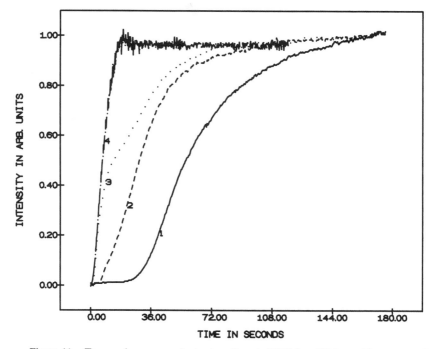

Figure 11. Temporal response of a barium titanate DPCM at 632.8 nm. The powers of beams 2 and 4 are 10.0 and 8.0 mW, respectively. The other experimental parameters are crystal dimensions, $10 \times 10 \times 10$ mm^3; beam diam, 1.5 mm, angle between the beams, 171°; and orientation of the crystal with respect to the c axis, 8.5°. The four curves shown are DPCM responses: 1, without external seeding; 2, with an external seed beam of power 0.001 mW and 0.5-s duration; 3, with an external seed beam of power 0.001 mW and 1.0-s duration; 4, beam fanning.

beam fanning. However, the time-dependent coupled equations can be solved for DPCM with one transmission grating for understanding the basic features of the transient behavior. The coupled wave equations are given by Eqs. (62)–(69). The boundary conditions for the device are $A_2(z = L) = A_{2L}$ and $A_4(z = 0) = A_{40}$. In addition, we assume that beam fanning provides seed beams A_{3L} and A_{10}. In barium titanate in which the beam coupling is highly asymmetric, A_3 is amplified by a much larger factor than A_1. Therefore, A_{3L} is more effective than A_{10} in initiating the phase-conjugation process.

The analytical solution for the coupled wave equation was obtained by Horowitz et al.[27] under undepleted pump approximation in terms of Bessel functions. The results show that as the amplification approaches the oscillation threshold, the phase-conjugate signal increases and the normalized buildup rate decreases. Exact solutions for Eqs. (62)–(69) have been obtained by numerical methods.[101] The results show that the buildup time for the phase-conjugate signal is inversely proportional to the total intensity of the beam. The buildup time also (1) decreases as γL increases and tends to saturate at high coupling gain, (2) increases with the pump beam ratio, and (3) decreases as the amplitude of the seed beam increases. The numerical calculations also show that the nonconjugated beams present in beam fanning build up rapidly first, forming the conical scattering observed experimentally. This scattering collapses gradually when the phase-conjugate waves dominate the 4WM process.[102]

The response time of DPCM is one of the concerns in the applications. The material response is less of a problem, since beam fanning in the material is 1–2 orders of magnitude faster than the buildup of phase-conjugate beams. The pump beams are, in general, loosely focused into the interacting volume to decrease the response time. The other factor that influences the phase-conjugate buildup time is the amount of scattering. Sharp et al.[76] reported that in SBN, it is easier to form MIBC in bird-wing geometry at high dopant concentrations than at low dopant concentrations. Beam fanning is greater at high dopant concentrations, allowing beam coupling in shorter interaction lengths.

External seeding is another way of improving the response of DPCM.[103, 104] Unlike the internal seeding present in beam fanning, the seed direction can be controlled to form seed gratings that are parallel to the mutually shared gratings. This speeds up the shared grating formation by reducing the competition for charge carriers by other nonconjugated beams. Once the shared gratings are established, the diffracted pump beams and the beam fanning act as the seed beams, and the external seeding becomes redundant. External seeding for short durations has been found to improve DPCM response by an order of magnitude.[104] The seed beam to pump beam ratio is $\sim 10^{-3}$–10^{-4}. The seed duration times are a fraction (0.01–0.02) of the response times of DPCM without the seed (Fig. 11). Since external seeding is present only for a short duration, the phase-conjugate nature of the two conjugate beams is not affected.

IX. DOUBLE PHASE CONJUGATION AND
SECOND-ORDER HOLOGRAPHY

We have seen so far that the dynamic interaction between the coherent beams propagating through a medium with local or nonlocal, instantaneous or noninstantaneous response in which the permittivity ε is modified leads to self-diffraction. Dynamic self-diffraction becomes more interesting when it is accompanied by energy-transfer processes. In such a case, opto-electronic control of optical beams is possible, and several such applications have been developed. The primary requirement of energy transfer in dynamic self-diffraction is the nonzero phase shift between the optical field and the material response. This is possible in media with nonlocal and noninstantaneous response. In other media, methods have been devised to obtain the dynamic energy transfer. These, in general, are degenerate or nearly degenerate multiwave mixing processes requiring mutually coherent beams. Conventional holography is a two-wave mixing process in which the spatial interference pattern of the beams is recorded as an absorptive or phase grating. Since holographic material responds photochemically, the dynamical effects are not seen in real time. The dynamical self-diffraction in multiwave mixing processes can be interpreted in terms of the basic holographic grating formation and grating readout.

Of all the dynamical interactions of coherent beams, MIBC is a special class. In DPCM, which is the lowest order process in MIBC, involving only one interaction region, the holographic requirements of mutual coherence and simultaneous presence of the beams are completely relaxed.[105] Source and object stability and path matching are not a concern. Even the wavelengths of the two beams can be different. Degenerate or nearly degenerate two-beam coupling, however, follows all the requirements of holography. If we neglect for a moment the physical process leading to the beam coupling, this suggests that MIBC cannot be interpreted in terms of holography as can all other mutually coherent processes. However, the very signature of holography, the transduction of one wavefront into another, is also present in DPCM, albeit in a different form. In dynamic self-diffraction in a volume hologram, all of the interacting beams are diffracted by their mutual gratings reproducing the other wavefront, while in DPCM the reproduced wavefronts are the phase conjugates of each other. Three-dimensional imaging by MIBC is another example where the holographic signature is evident.

We now reinterpret MIBC in terms of holography. The physical process that leads to the beam coupling in DPCM is the formation of the elementary transmissive holograms by each of the input beams with the fanned (scattered and diffracted) beams. These are the first-order holograms. The two sets of transmissive holograms interact differently with the input beams, depending on their relative orientations. This leads to a cooperative behavior (second-order constructive interference) in some orientations, and opposing behavior (second-order destructive interference) in others. The primary holograms are enhanced whenever the two sets of transmissive holograms behave cooperatively. The net effect is the formation of a second-order transmissive hologram that bears the common information between the first-order holograms from both the beams.[105] Given two beams O and R, the first-order holograms can be represented by $T_R = |R + S_R|^2$ and $T_O = |O + S_O|^2$, where S_R and S_O correspond to the scattered beams of R and O, respectively. The effective transmission due to the presence of T_O and T_R is

$$T_H \propto T_R T_O = \left(|R|^2 + |S_R|^2 + RS_R^* + R^*S_R \right)$$
$$\times \left(|O|^2 + |S_O|^2 + OS_O^* + O^*S_O \right) \qquad (81)$$

T_H contains the interesting term $R^*O^*S_R S_O$. For maximum coupling, we require that $S_R \sim S_O^*$, and, $S_O \sim S_R^*$. For point scatterers, this implies that the back scattering due to beam O should be the phase conjugate of the forward scattering due to beam R, and vice versa.

In the dynamic self-diffraction process, input beams R and O generate

$$R[R^*O^*S_R S_R^*] \propto O^* \qquad (82)$$

and

$$O[R^*O^*S_R S_R^*] \propto R^* \qquad (83)$$

The input beams R and O generate the phase conjugates of each other in the second-order holographic readout process. Thus, MIBC can be considered a second-order holographic process.

The elementary holograms that are formed between the scatterers and the input beams are within the beam-crossing volume of the crystal. These are similar to local reference beam holograms in which the object beam generates its own reference beam. Local reference beam holograms are less restrictive due to reduced temporal and spatial coherence and stability requirements.[105] To generate a second-order hologram, the sole requirement is that the first-order holograms from both beams coexist long enough for the material to adjust internally. In the currently available materials, for achievable irradiance levels, this process may take long time —up to seconds.

X. APPLICATIONS

As we saw in Section VIII, the interacting beams in DPCM need not be coherent, of the same wavelength, or present at the same time. The beams can carry separate spatial, temporal, and phase information. Through the interaction, each beam picks up the spatial and phase information of the other input beam in a "hand-shaking" mode. By cascading the DPCMs, it

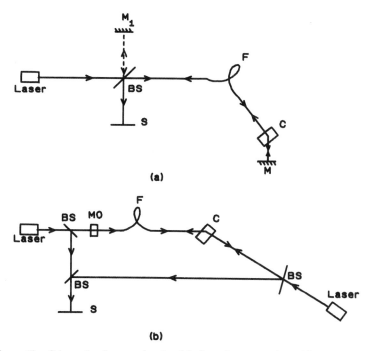

Figure 13. Schematic diagram showing (a) distortion correction of images transmitted through multimode fibers using a semilinear PPCM, and (b) one-way transmission of a uniform phase using the Mach-Zehnder arrangement. The arrangement (a) can be converted to a Michelson interferometer by retroreflecting the split input beam by mirror M_1. C, crystal; F, fiber bundle; BS, beamsplitter; S, screen; MO, phase modulator.

real-time image-processing applications. One such implementation using 2C-PPCM was demonstrated by Fischer et al.[95] Optical thresholding,[83] optical switching,[83, 110] optical edge enhancement,[83] and beam coupling and locking[111] are other applications in which DPCMs have been used.

B. Optical Communication

1. Two-Way Communication

The bidirectional phase-conjugate capability of DPCM is an attractive feature for long-distance optical communications.[112–115] A typical bidirectional transmitter–receiver scheme is shown in Fig. 14a. Beam 1 at station 1 may be used as a beacon to call information from station 2. Any temporal modulation of beam 2 is carried by the phase-conjugate beam 4

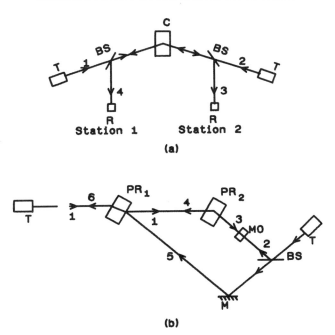

Figure 14. Schematic diagram showing two-way optical communication between two stations (a) without amplifier stage and (b) with optical amplifier stage included at one of the stations. C, PR_1, PR_2, crystals; BS, beamsplitter; MO, modulator; M, mirror; T, transmitter; R, receiver.

back to the source at station 1. Similarly, if beam 1 is temporally modulated, the information will be transferred to station 2 via beam 3. There is no cross-talk between the two information channels as long as the modulated frequency exceeds the photorefractive time response. For very low frequencies, the link can be kept open by applying a bias.

A major limitation of this configuration is the power loss at the various optical elements. Also, for efficient coupling in DPCM, the intensities of the pump beams should be identical. This can be achieved by including a beam amplifier in the communication link. One configuration for implementing amplified phase conjugate signal transmission is shown in Fig. 14b.[113] The beam 1 from station 1 is conjugated in the crystal PR2 with a small fraction of the laser beam power at station 2. The conjugate beam 4 is amplified by the beam 5 in a two-beam coupling process in PR1. This amplified beam is returned to station 1 as beam 6. A similar arrangement may be used at station 1 to obtain efficient two-way transmission. The configuration shown in Fig. 14b can be modified to obtain spatial separation between the amplified output and the incoming beams.

2. Broadcasting

The station-to-station communication link can be extended to multiple links for broadcasting or for combining signals[115] (Fig. 15). The signal beam from different stations can be imaged into different interaction regions of the crystal, establishing communication with the receiving stations. To establish a communication link, the station about to receive the broadcast signal has to transmit a signal to keep the link open in a kind of "hand-shaking" communication protocol. By switching the beams on and off, communication can be established with different receiving stations. The device thus acts as a dynamic splitter, and the interconnection can be reconfigured in real time.

3. Optical Interconnects

The broadcasting scheme discussed above can be used as point-to-point interconnects, where the link is established by switching the beams on/off at the points of interest.[115] This time division multiplexing (TDM) has the drawback of imposing constraints on the communication protocol. This can be avoided by using a tunable wavelength source for the sender and assigning different wavelengths to the receiving stations. The messages are now self-routed. The efficiency of this system is lower than in TDM. The TDM can be extended for interconnecting an array of points in two dimensions (space division multiplexing). The interconnections in 2-D can also be achieved by wavelength division multiplexing, in which each source has a different wavelength. All sources are mixed and coupled to a single channel for transmission, and are separated at the end of the transmission

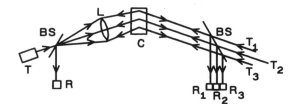

Figure 15. Schematic diagram showing optical implementation of broadcasting and point-to-point interconnects. C, crystal; BS, beamsplitter; L, lens; T, R, transmitter and receiver at the broadcasting station; T_1, T_2, T_3, transmitters at the receiving stations; R_1, R_2, R_3, receivers at the receiving stations. For point-to-point communication, the receiver beams are turned on and off. Full point-to-point interconnections in two dimensions can be established by using an array of transmitters instead of one (space division multiplexing).

channel, where they are picked up by different receivers. Multiplexing and demultiplexing can be accomplished with DPCMs.

4. Adaptive Interconnects

The relaxed stability requirements and the self-aligning capability of DPCM allow bidirectional interconnections under thermal, vibrational, and other environmental fluctuations. Such adaptive optical interconnections were proposed by Schamshula et al.[114]

5. Coherent Communication

An advantage of the DPCM's phase conjugation is its ability to be used for coherent detection. Coherent heterodyne detection requires that the incoming signal beam and the local oscillator beam have a common transversal phase. In atmospheric propagation, the wavefronts are distorted. This imposes severe restrictions on the etendu of the collecting optics, limiting it to $\sim \lambda^2$. We saw in Section VIII that in DPCM, nonuniform spatial modulations on the pump beam are not transferred to the phase-conjugate beam of the second pump beam. Phase distortions can be completely avoided in the transmission of temporally modulated signals. Hence, with DPCM, we can use a much larger throughput, and heterodyne detection of the incoming signals is possible.[113, 116, 117]

An optical configuration for coherent optical communication through distorting media is shown in Fig. 16. The incoming temporally modulated signal beam 1 is phase conjugated with the local unmodulated oscillation beam in PR1. The conjugated signal beam 4 preserves the uniform phase

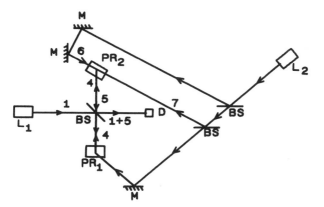

Figure 16.   Schematic diagram showing coherent communication with DPCM. L_1, signal source; L_2, local oscillator; M, mirror; PR_1, PR_2, crystals; BS, beamsplitter; D, detector.

of the input beam 1, but without any temporal modulation. This beam is phase conjugated again by the 4WM process in PR2. Beams 1 and 5 are now identical except for the temporal modulations in beam 1. The hetero-dyne signal is detected by D.

C. 3-D Imaging

The holographic interpretation of DPCM (Section IX) opens up new areas where MIBC/DPCM can be used instead of first-order holography. The advantages of MIBC over conventional holography are (1) minimum stability and coherence requirements, (2) no restriction on object depth, and (3) real-time processing. The materials are reusable. This allows us, for example, to use a cw laser of limited coherence to record a 2-D image of an object at unknown range and size illuminated by a pulsed laser of limited coherence. Industrial testing with DPCM interferometry was demonstrated by Petrov et al.[118]

XI. CONCLUSIONS

Self-diffraction of light beams has found extensive applications in optical information processing. The multiwave mixing of mutually coherent beams which leads to self-diffraction can be interpreted in terms of dynamic holography. Mutually incoherent beam coupling is different from two-beam coupling and four-wave mixing processes. Many of the limitations on the optical characteristics of the interacting beams are relaxed. This makes double phase conjugation, the lowest order MIBC process, more useful than ordinary holography. The unique properties of DPCM have opened up new possibilities in optical communications. Though at present, the photorefractives are the only materials in which MIBC has been observed, recent experiments in sodium and new configurations proposed by Mamaev and Zozluya suggest other possibilities. If larger, faster, less expensive, and more sensitive materials can be found, we may find second-order holography reaching and even exceeding conventional first-order holography in importance.

Acknowledgments

The work of H. Jagannath was supported by NSF Grant 8802971.

References

1. J. P. Huignard and P. Gunter (Eds.), *Photorefractive Materials and Their Applications*, Vol. 1, Springer, New York, 1983.
2. J. P. Huignard and P. Gunter (Eds.), *Photorefractive Materials and Their Applications*, Vol. 2, Springer, New York, 1983.

3. R. A. Fisher (Ed.), *Optical Phase Conjugation*, Academic, New York, 1983.

4. L. Solymar and D. J. Cooke *Volume Holography and Volume Gratings*, Academic, New York, 1981.

5. H. J. Caulfield and Sun Lu, *Holography*, Wiley Interscience, New York, 1970.

6. H. Kogelnik, *Bell Syst. Tech. J.* **48**, 2909 (1969).

7. P. Gunter, *Phys. Rep.* **93**, 201 (1982).

8. J. Feinberg, D. Heiman, A. R. Tanguay, and R. W. Hellwarth, *J. Appl. Phys.* **51**, 1297 (1980).

9. F. S. Chen, J. T. La Macchia, and D. B. Fraser, *Appl. Phys. Lett.* **13**, 223 (1968).

10. G. C. Valley, *J. Appl. Phys.* **59**, 3363 (1986).

11. L. Young, W. K. Y. Wong, M. L. W. Thewalt, and W. D. Cornish, *Appl. Phys. Lett.* **24**, 264 (1974).

12. N. V. Kukhtarev, *Sov. Tech. Phys. Lett.* **2**, 438 (1976).

13. N. V. Kukhtarev, V. B. Markov, S. G. Odulov, M. S. Soskin, and V. L. Vinetskii, *Ferroelectrics* **22**, 949 (1979).

14. Ph. Refrigier, L. Solymar, H. Rajbenbach, and J. P. Huignard, *J. Appl. Phys.* **58**, 45 (1985).

15. M. G. Moharram, T. K. Gaylord, R. Magnusson, and L. Young, *J. Appl. Phys.* **50**, 5642 (1979).

16. D. L. Staebler and J. J. Amodei, *J. Appl. Phys.* **43**, 1042 (1972).

17. P. Yeh, *IEEE J. Quantum Electron.* **25**, 484 (1989).

18. D. Rak, I. Ledoux, and J. P. Huignard, *Opt. Commun.* **49**, 302 (1984).

19. R. R. Neurgaonkar, W. K. Cory, J. R. Oliver, M. D. Ewbank, and W. F. Hall, *Opt. Engg.* **26**, 392 (1987).

20. P. Yeh, *Appl. Opt.* **26**, 602 (1987).

21. S. I. Stepanov and S. L. Sochava, *Sov. Phys. Tech. Phys.* **32**, 1054 (1987).

22. M. B. Klein, *Opt. Lett.* **9**, 350 (1984).

23. V. L. Vinetskii, N. V. Kukhtarev, S. G. Odulov, and M. S. Soskin, *Sov. Phys. Usp.* **22**, 742 (1979).

24. L. Solymar and J. Heaton, *Opt. Commun.* **51**, 76 (1984).

25. M. Cronin-Golomb, A. M. Biernacki, C. Lin, and H. Kong, in *Tech. Digest*, Topical Meeting on Photorefractive Materials, 1987, p. 142.

26. G. C. Papen, B. E. A. Saleh, and J. A. Tataronis, *J. Opt. Soc. Am.* **B5**, 1763 (1988).

27. M. Horowitz, D. Kligler, and B. Fischer, *J. Opt. Soc. Am.* **B8**, 2204 (1991).

28. D. R. Erbschwe and T. Wilson, *Opt. Commun.* **72**, 135 (1989).

29. J. M. Heaton and L. Solymar, *Opt. Acta* **32**, 397 (1985).

30. D. C. Jones and L. Solymar, *Opt. Commun.* **85**, 372 (1991).

31. J. M. Heaton and L. Solymar, *IEEE J. Quantum Electron.* **34**, 558 (1988).

32. S. I. Stepanov, K. Kolikov, and M. Petrov, *Opt. Commun.* **44**, 19 (1982).

33. G. C. Valley, *J. Opt. Soc. Am.* **B1**, 868 (1984).

34. J. Goltz and T. Tschdi, *Opt. Commun.* **67**, 414 (1988).

35. P. J. Johansen, *J. Phys. D, Appl. Phys.* **22**, 247 (1989).

36. P. Yeh, *J. Opt. Soc. Am.* **B4**, 1382 (1987).

109. S. Sternklar, S. Weiss, M. Segev, and B. Fischer, *Appl. Opt.* **25**, 4518 (1986).
110. Q. C. He, J. G. Duthie, and D. A. Gregory, *Opt. Lett.* **14**, 575 (1989).
111. S. Sternklar, S. Weiss, M. Segev, and B. Fischer, *Opt. Lett.* **11**, 528 (1986).
112. H. J. Caulfield, J. Shamir, and Q. He, *Appl. Opt.* **26**, 2291 (1987).
113. J. Shamir, H. J. Caulfield, and B. M. Hendrickson, *Appl. Opt.* **27**, 2912 (1988).
114. M. P. Schamshula, H. J. Caulfield, and C. M. Verber, *Opt. Lett.* **16**, 1421 (1991).
115. S. Weiss, M. Segev, S. Strenklar, and B. Fischer, *Appl. Opt.* **27**, 3422 (1988).
116. Q. He, J. Shamir, and J. G. Duthie, *Appl. Opt.* **28**, 306 (1989).
117. L. E. Adam and R. S. Bondurant, *Opt. Lett.* **16**, 832 (1991).
118. M. Petrov, E. U. Mokrushina, and H. J. Caulfield, unpublished data.

LASER AND PULSED-LASER NMR SPECTROSCOPY

M. W. EVANS*

Cornell Theory Center, Cornell University, Ithaca, NY
and
Materials Research Laboratory, The Pennsylvania State University,
University Park, PA

CONTENTS

*Present address: Department of Physics, University of North Carolina, Charlotte, NC 28223.

Modern Nonlinear Optics, Part 2, Edited by Myron Evans and Stanisław Kielich. Advances in Chemical Physics Series, Vol. LXXXV.
ISBN 0-471-57546-1 © 1993 John Wiley & Sons, Inc.

I. INTRODUCTION

It was realized soon after the first demonstration of lasing action that circularly polarized electromagnetic radiation can produce bulk magnetization. Pershan and coworkers developed the effect theoretically[1] and experimentally,[2-4] and named it the inverse Faraday effect. Thereafter, Kielich[5, 6] produced a general theory of the relevant magneto-optics in terms of molecular property tensors, and Atkins and Miller[7] extended the theoretical considerations with quantum field theory. This important work has in common the induction by a circularly polarized laser of an electronic magnetic dipole moment ($m_i^{(\mathrm{ind})}$) in the atom or molecule under consideration

$$m_i^{(\mathrm{ind})} = \tfrac{1}{4}{}^m\chi_{ijk}^{\mathrm{ee}}(0; \omega, -\omega)E_j(\omega)E_k^*(\omega) \tag{1}$$

Here ${}^m\chi_{ijk}^{\mathrm{ee}}(0; \omega, -\omega)$ is the relevant susceptibility tensor,[8-10] and $E_jE_k^*$ is the tensor product of the electric field vector of the laser (in volts per meter) with its complex conjugate $\mathbf{E}^*$. Later, Manakov et al.[8-10] developed the general quantum structure of the susceptibility in Eq. (1) between eigenstates of different magnetic quantum number M. In this quantum theory, worked out analytically for atoms, the mean value of the permanent electronic magnetic dipole moment operator is

$$\mathbf{m}_0 = \langle JM|\hat{m}_0|JM'\rangle \equiv \langle \psi_m(t)|\hat{m}_0|\psi_{m'}(t)\rangle \tag{2}$$

where ψ_m and $\psi_{m'}$ are solutions of the Schrödinger equation corresponding to the initial and final atomic eigenstates $|JM\rangle$ and $|JM'\rangle$, both of which are M-fold degenerate. The initial and final eigenstates correspond in general therefore to different projections of M, i.e., different spatial quantization. The total electronic magnetic dipole moment in the presence of the laser can therefore be expressed as

$$\mathbf{m} = \mathbf{m}_0 + \mathbf{m}^{(\mathrm{ind})} = \langle JM|\hat{m}_0|JM'\rangle + \tfrac{1}{4}{}^m\chi_{ijk}^{\mathrm{ee}}E_jE_k^* \tag{3}$$

to order two in $\mathbf{E}$. The susceptibility also depends on M and M', and must therefore be spatially quantized. The induced magnetic dipole moment $\mathbf{m}^{(\mathrm{ind})}$ of Eq. (1) is therefore also spatially quantized when there is M degeneracy of this type. This is an important point in second-order laser NMR effects, because it means that the laser-induced electronic magnetic dipole moment $\mathbf{m}^{(\mathrm{ind})}$ can couple within the framework of quantum angular momentum theory[11-14] with the permanent nuclear magnetic dipole moment to shift and split existing NMR resonance lines, leading to a new spectral fingerprint. The same is true of second-order effects in laser ESR. The J and M quantization structure of the susceptibility ${}^m\chi_{ijk}^{ee}$ of the inverse Faraday effect is given by Manakov et al.[8-10] in atoms when net electronic angular momentum. The interested reader is referred to Eq. (5.7), p. 607, of Ref. 9 for details. The quantization structure of the induced magnetic dipole moment of Eq. (1) for these atoms is clearly the same, and quantum transitions are allowed between M states governed by selection rules determined by the symmetry of the susceptibility tensor.

When there is M degeneracy, therefore, the induced electronic magnetic dipole moment of the inverse Faraday effect is spatially quantized, its expectation value depends explicitly on M and J through the Wigner-Eckart theorem,[8-10] and the induced magnetic dipole operator is proportional to a quantum mechanical electronic angular momentum operator, which can couple with another quantized nuclear or electronic angular momentum to produce spectral shifting and splitting in laser NMR and ESR.

II. FIRST-ORDER EFFECT IN LASER NMR

The theories mentioned in the introduction do not deal with an important first-order interaction energy that is dominant[14-18] in laser NMR. This first-order interaction will be developed before we return to second-order effects. The origin of this interaction energy can be found in the antisymmetric part of the tensor product $E_i E_j^*$:

$$\text{antisym}(E_i E_j^*) = \tfrac{1}{2}(E_i E_j^* - E_j E_i^*) \tag{4}$$

which can be written as the vector cross-product

$$\Pi^{(A)} = \mathbf{E} \times \mathbf{E}^* = \pm 2E_0^2 i\mathbf{k} \tag{5}$$

the antisymmetric conjugate product of a circularly polarized laser. It has been shown elsewhere[19] that $\Pi^{(A)}$ is negative to motion reversal symmetry (T), is positive to parity inversion (P), is an axial vector, and changes sign when the circular polarity (handedness) of the laser is switched from left to right. The IUPAC standard expressions for $\mathbf{E}$ and $\mathbf{E}^*$ are (Appendix A)

$$\begin{aligned}
\mathbf{E}_L &= E_0(\mathbf{i} + \mathbf{ij})\exp(-i\phi_L) & \mathbf{E}_L^* &= E_0(\mathbf{i} - \mathbf{ij})\exp(i\phi_L) \\
\mathbf{E}_R &= E_0(\mathbf{i} - \mathbf{ij})\exp(-i\phi_R) & \mathbf{E}_R^* &= E_0(\mathbf{i} + \mathbf{ij})\exp(i\phi_R)
\end{aligned} \tag{6}$$

and from these definitions:

$$\mathbf{E}_L \times \mathbf{E}_L^* = -\mathbf{E}_R \times \mathbf{E}_R^* = -2E_0^2 i\mathbf{k} \qquad (7)$$

a purely imaginary quantity. It is important to note that $\mathbf{k}$ in this equation is a P-positive, T-negative, unit axial vector in the propagation axis of the circularly polarized laser, denoted Z, and should not be confused with the propagation vector of the laser (denoted $\boldsymbol{\kappa}$), which is negative to P and T.[20] In Eq. (7), E_0 is the scalar electric field strength amplitude of the laser in volts per meter. Clearly, if the laser is linearly polarized, it is 50% right and 50% left circularly polarized, and $\mathbf{\Pi}^{(A)}$ vanishes.

Using the fundamental electrodynamical relations

$$I_0 = \tfrac{1}{2}\varepsilon_0 c E_0^2 \qquad (8)$$

and

$$E_0 = cB_0 \qquad (9)$$

where B_0 is the scalar magnetic flux density amplitude of the laser in tesla, c is the speed of light (a scalar), I_0 is the laser's scalar intensity in watts per square meter, and

$$\varepsilon_0 = 8.854 \times 10^{-12} J^{-1} C^2 m^{-1} \qquad (10)$$

is the scalar permittivity in vacuo in S.I. units, the vector $\mathbf{\Pi}^{(A)}$ can be rewritten as

$$\mathbf{\Pi}^{(A)} = \left(\frac{8I_0 c}{\varepsilon_0}\right)^{1/2} (\pm B_0 \mathbf{k}) i \qquad (11)$$

which is proportional to the product of B_0 and $\mathbf{k}$ through a T- and P-positive scalar quantity. The product

$$\mathbf{B}_\Pi = B_0 \mathbf{k} \qquad (12)$$

is the laser's equivalent static magnetic flux density vector, because (1) it has the units of magnetic flux density (tesla), and (2) it has the correct fundamental symmetries of magnetic flux density, being a T-negative, P-positive axial vector. We therefore arrive at the important relations

$$\mathbf{\Pi}^{(A)} = i\left(\frac{8I_0 c}{\varepsilon_0}\right)^{1/2} (\pm \mathbf{B}_\Pi) \qquad (13)$$

and

$$|\mathbf{B}_\Pi| = B_0 = \left(\frac{2I_0}{\varepsilon_0 c^3}\right)^{1/2} \tag{14}$$

between $\Pi^{(A)}$ and $\mathbf{B}_\Pi$ of a circularly polarized laser, and find that such a laser generates a static magnetic field $\mathbf{B}_\Pi$ in its propagation axis. Since $\mathbf{B}_\Pi$ is negative to motion reversal T, it must change sign if the motion of the laser is reversed, i.e., if the laser is made to propagate in the opposite direction in Z. However, $\mathbf{B}_\Pi$ is positive to parity inversion P because the conjugate product $\Pi^{(A)}$ is positive to P:

$$P(\mathbf{E}) = -\mathbf{E} \qquad P(\mathbf{E}^*) = -\mathbf{E}^* \qquad P(\mathbf{E} \times \mathbf{E}^*) = \mathbf{E} \times \mathbf{E}^* \tag{15}$$

confirming that $\Pi^{(A)}$ and $\mathbf{B}_\Pi$ have the same fundamental symmetries and are both axial vectors, proportional to each other through a T- and P-positive scalar quantity. It is important to note that the conjugate product $\Pi^{(A)}$ changes sign with circular polarity of the laser and so does $\mathbf{B}_\Pi$. The origin of the sign change in $\Pi^{(A)}$ in this context is neither the operation P nor T, but the different operation $(\hat{H})$ on the components $\mathbf{E}$ and $\mathbf{E}^*$ (Appendix A):

$$\left[E_L = E_0(\mathbf{i} + \mathbf{ij})e^{-\mathbf{i}\phi_L}\right] \xrightarrow{\hat{H}} \left[E_R = E_0(\mathbf{i} - \mathbf{ij})e^{-\mathbf{i}\phi_R}\right]$$

an operation that switches the laser from left to right circular polarization (c.p.). In a left c.p. laser, $\Pi^{(A)}$ is formed by multiplying $-\mathbf{B}_\Pi$, and in a right c.p. laser, it is formed by multiplying $+\mathbf{B}_\Pi$.

We conclude that a circularly polarized laser can act as a light magnet, generating a static magnetic flux density $\pm\mathbf{B}_\Pi$ in the axis of propagation. This is of key importance to the first-order effect of such a laser on NMR spectra, and probably also to many other effects, because a circularly polarized laser can always be used as a magnet, and a magnet is used in many interesting investigations.

For I_0 of $10\,000$ W/m^2 (1.0 W/cm^2)

$$\mathbf{B}_\Pi \doteq 10^{-5} \, \mathbf{k} \, \mathrm{T} \tag{16}$$

and the equivalent magnetic flux density of the laser is 10^{-5} T. Pulsed lasers can deliver up to 10^{16} W/m^2 (Ref. 20) over subpicosecond time intervals, delivering 10.0 T. This is about the same order of magnitude as the most powerful contemporary superconducting magnets.

Since $\mathbf{B}_{\Pi}$ is a real static magnetic flux density, it forms a T- and P-positive scalar interaction energy with a nuclear magnetic dipole moment $\mathbf{m}^{(N)}$. This is the key to laser and pulsed laser NMR spectroscopy, because this is the big, first-order mechanism. Note that in this context we have defined $\mathbf{m}^{(N)}$ to be a real axial vector, T negative and P positive. The quantum-mechanical nuclear magnetic dipole moment operator $\hat{\mathbf{m}}^{(N)}$ has the same symmetries, but is a purely imaginary quantity.[21] Therefore, $\mathbf{m}^{(N)}$ is to be regarded as the real observable generated by the operator $\hat{\mathbf{m}}^{(N)}$. The interaction energy

$$E_1 = -\mathbf{m}^{(N)} \cdot \mathbf{B}_{\Pi} \tag{17}$$

is quantized, in the same way as the usual interaction energy generated by $\mathbf{m}^{(N)}$ and the static magnetic flux density $\mathbf{B}_0$ of the big permanent magnet of an NMR spectrometer.

If we now direct a circularly polarized laser parallel to $\mathbf{B}_0$, the combined interaction energy becomes

$$E_2 = -\mathbf{m}^{(N)} \cdot (\mathbf{B}_0 \pm \mathbf{B}_{\Pi}) \tag{18}$$

i.e., the original NMR resonance frequency is shifted by the laser by an amount

$$\Delta f = \pm(\mathbf{m}^{(N)} \cdot \mathbf{B}_{\Pi})/h \tag{19}$$

in hertz, where h is Planck's constant. It turns out (vide infra) that experimentally such a shift is site specific, i.e., the shift is different for each resonating nucleus, essentially because the $\mathbf{B}_{\Pi}$ at the nucleus is different from the applied $\mathbf{B}_{\Pi}$ due to the well-known chemical shift effect, and this leads to a valuable new fingerprint of a complex system such as a folded protein in solution.[22-30]

A. Bulk Magnetization Due to $\mathbf{B}_{\Pi}$: Comparison with Data

It is useful as a check to calculate the order of magnitude of magnetization $\mathbf{M}_{\Pi}$ (in amps per meter) expected from a laser delivering $\mathbf{B}_{\Pi}$ in tesla, and to compare this with experimental data. The basic relation between $\mathbf{M}_{\Pi}$ and $\mathbf{B}_{\Pi}$ is given by Atkins[31]:

$$\mathbf{M}_{\Pi} = \frac{1}{\mu_0}\left(\frac{\kappa_m}{1+\kappa_m}\right)\mathbf{B}_{\Pi} \tag{20}$$

where κ_m is the dimensionless mass susceptibility of the sample, and μ_0 is the permeability in vacuo,

$$\mu_0 = 4\pi \times 10^{-7} NA^{-2} \qquad (21)$$

Atkins gives a useful model calculation for an ensemble of atoms with and without net electronic spin. In the latter case (Ref. 31, Eq. (14.2.27a)),

$$\kappa_m = -\frac{e^2\mu_0}{6m_e} N\langle r^2 \rangle \qquad (22)$$

where e is the electronic charge, N the number density (atoms per cubic meter), m_e the mass of the electron, and $\langle r^2 \rangle$ is about 10^{-20} m^2. For a model atomic weight of 20, this gives a mass susceptibility of the order 2×10^{-9} and a magnetization from Eq. (20) of the order 10^{-5} A/m for B_Π of 0.03 T. The latter is derived using Eq. (14) from the peak intensity of 10^{11} W/m^2 recorded by Pershan et al.,[2-4] who observed a magnetization of the order 10^{-5} G (10^{-2} A/m) in 3.1% Eu^{2+}-doped calcium fluoride glass due to a circularly polarized giant ruby laser pulse of 30 ns. The calculated figure is therefore about 300 times smaller than the experimental one, but the experimental sample was paramagnetic, whereas Eq. (22) is for a diamagnetic atomic sample. Furthermore, the effective molecular weight of the doped glass sample is greater than 20, and Pershan et al.[2-4] observed a dominant paramagnetic contribution that was temperature dependent. Following the argument on p. 386 of Atkins,[31] the magnetization for one gram of sample with unpaired spin (such as Na) will be of the order 1000 times bigger. This argument leads to a calculated magnetization M_Π fortuitously of the same order of magnitude as that observed by Pershan et al.[2-4]

The concept of equivalent static magnetic flux density leads to a reasonable description of the only available data in the literature on magnetization by a circularly polarized laser pulse. This estimate is dominated, however, by electronic mass susceptibility in a sample with net electronic angular momentum, and for optical NMR it is necessary to consider the interaction of B_Π with both nuclear and electronic spins in the sample. The existence of B_Π is the key to the fundamental interaction Hamiltonian at first order in optical NMR, and this is developed in the following section.

B. The Fundamental Equations of the First-Order Effect

Because B_Π is a static magnetic flux density, it is governed by fundamental equations from which the first-order optical NMR Hamiltonian can be constructed. The static magnetic flux density B_Π is related formally through

1. The relation

$$\nabla \times \mathbf{A}_\Pi = \pm \frac{i}{2} \left(\frac{\varepsilon_0}{2I_0 c} \right)^{1/2} (\mathbf{E} \times \mathbf{E}^*) \qquad (33)$$

shows that the vector potential $\mathbf{A}_\Pi$ curls around the propagation axis of a circularly polarized laser.

2. On the Argand diagram $\mathbf{E}$ is perpendicular to its conjugate $\mathbf{E}^*$, and both vectors are rotating in the same direction at a fixed relative orientation. Their vector product is therefore mutually perpendicular (in the propagation axis) and is time independent, because the relative orientation of $\mathbf{E}$ and $\mathbf{E}^*$ does not change.

3. The P and T symmetries of the well-known Poynting vector[31]

$$\mathbf{N} = \mathbf{E} \times \mathbf{H}^* \doteq \frac{1}{\mu_0} \mathbf{E} \times \mathbf{B}^* \qquad (34)$$

are not the same as that of the conjugate product $\mathbf{E} \times \mathbf{E}^*$. The Poynting vector measures the instantaneous energy flow in the direction of propagation of an electromagnetic wave, and is P- and T-negative. It has the same P and T symmetries as the propagation vector $\boldsymbol{\kappa}$ of a laser, and is a polar vector (one that changes sign with P). In contrast, the conjugate product $\Pi^{(A)}$ and the $\mathbf{B}_\Pi$ vector are axial vectors, which do not change sign with P. It is important to make this distinction. Clearly, the Poynting vector cannot be a static magnetic flux density vector because of its negative P symmetry.

The reason for this is not straightforward. Consider the complete set of electromagnetic plane wave equations:

$$
\begin{array}{lll}
\mathbf{E}_L = E_0(\mathbf{i} + i\mathbf{j})e^{-i\phi_L} & P(\mathbf{E}_L) = -\mathbf{E}_L & T(\mathbf{E}_L) = \mathbf{E}_R^* \\[4pt]
\mathbf{E}_R = E_0(\mathbf{i} - i\mathbf{j})e^{-i\phi_R} & P(\mathbf{E}_R) = -\mathbf{E}_R & T(\mathbf{E}_R) = \mathbf{E}_L^* \\[4pt]
\mathbf{E}_L^* = E_0(\mathbf{i} - i\mathbf{j})e^{i\phi_L} & P(\mathbf{E}_L^*) = -\mathbf{E}_L^* & T(\mathbf{E}_L^*) = \mathbf{E}_R \\[4pt]
\mathbf{E}_R^* = E_0(\mathbf{i} + i\mathbf{j})e^{i\phi_R} & P(\mathbf{E}_R^*) = -\mathbf{E}_R^* & T(\mathbf{E}_R^*) = \mathbf{E}_L \\[4pt]
\mathbf{B}_L = B_0(\mathbf{j} - i\mathbf{i})e^{-i\phi_L} & P(\mathbf{B}_L) = \mathbf{B}_L & T(\mathbf{B}_L) = -\mathbf{B}_R^* \\[4pt]
\mathbf{B}_R = B_0(\mathbf{j} + i\mathbf{i})e^{-i\phi_R} & P(\mathbf{B}_R) = \mathbf{B}_R & T(\mathbf{B}_R) = -\mathbf{B}_L^* \\[4pt]
\mathbf{B}_L^* = B_0(\mathbf{j} + i\mathbf{i})e^{i\phi_L} & P(\mathbf{B}_L^*) = \mathbf{B}_L^* & T(\mathbf{B}_L^*) = -\mathbf{B}_R \\[4pt]
\mathbf{B}_R^* = B_0(\mathbf{j} - i\mathbf{i})e^{i\phi_R} & P(\mathbf{B}_R^*) = \mathbf{B}_R^* & T(\mathbf{B}_R^*) = -\mathbf{B}_L
\end{array}
\qquad (35)
$$

The effect of P and T on each of these has been given explicitly, and to understand why the operators have the effect they do requires the following realization. The unit vectors $\mathbf{i}$ and $\mathbf{j}$ for the electric components are polar unit vectors that are T positive, P negative. Almost universally (and unfortunately), the same $\mathbf{i}$ and $\mathbf{j}$ notation is also used for the T-negative, P-positive, axial unit vectors of the magnetic components. However, the two sets of unit vectors denote quite different symmetries, which is revealed when we take the limit $\omega \to 0$, i.e., when the angular frequency of the wave is vanishingly small, so that the $\mathbf{E}$ and $\mathbf{B}$ vectors of the plane wave must become static electric field strength and magnetic flux density vectors. In this limit, the plane wave equations reduce to

$$\mathbf{E}_L \xrightarrow{\omega \to 0} E_0(\mathbf{i} + \mathrm{i}\mathbf{j}) \text{ etc.} \qquad \mathbf{B}_L \xrightarrow{\omega \to 0} B_0(\mathbf{j} - \mathrm{i}\mathbf{i}) \text{ etc.} \qquad (36)$$

and the real parts must have the correct P and T symmetries of static $\mathbf{E}$ and $\mathbf{B}$. It follows that the unit vectors in the usual plane wave descriptions also have different fundamental symmetries, as described. From this the fundamental symmetries of the Poynting vector emerge as follows:

$$P(\mathbf{E}_L \times \mathbf{B}_L^*) = -(\mathbf{E}_L \times \mathbf{B}_L^*) \text{ etc.}$$
$$T(\mathbf{E}_L \times \mathbf{B}_L^*) = -(\mathbf{E}_R^* \times \mathbf{B}_R) \qquad (37)$$
$$= -(\mathbf{E}_L \times \mathbf{B}_L^*) \text{ etc.}$$

It is important to realize that the equivalent magnetic flux density vector $\mathbf{B}_\Pi$, introduced in this chapter, is fundamentally different from the Poynting vector and propagation vector. For example, the latter two exist in linearly polarized radiation, because

$$\mathbf{E}_L \times \mathbf{B}_L^* = \mathbf{E}_R \times \mathbf{B}_R^* \qquad (38)$$

while $\mathbf{B}_\Pi$ vanishes.

Having been careful to make these key distinctions, it is now possible to develop the first-order interaction Hamiltonian of optical NMR from an argument parallel with that given by Atkins,[31] for example:

$$H = H^{(0)} + H^{(1)} + H^{(2)} \qquad (39)$$

where the first-order term is

$$H^{(1)} = -\gamma_e \mathbf{B}_\Pi \cdot \mathbf{L} \qquad (40)$$

with $\mathbf{L}$ denoting the orbital angular momentum, and γ_e the electronic gyromagnetic ratio, and the second-order term is (see also Appendix B)

$$H^{(2)} = \frac{e^2}{8m_e} B_\Pi^2 (X^2 + Y^2)$$

$$= \frac{e^2}{2m_e} A_\Pi^2$$

(41)

This is a nonrelativistic treatment, so that the electronic spin component in Eq. (40) is missing. Reinstating the spin term leads to the familiar-looking first-order Hamiltonian describing the interaction between $\mathbf{B}_\Pi$ and an electronic magnetic dipole moment.

An entirely analogous procedure leads to the first-order Hamiltonian describing the interaction between $\mathbf{B}_\Pi$ and a nuclear magnetic dipole moment $\mathbf{m}^{(N)}$ and this has the same form as the right side of Eq. (17). It becomes clear, therefore, that the equations of conventional NMR, which use, for example, a homogeneous superconducting magnet to deliver a static $\mathbf{B}_0$ of up to 20.0 T, can be adapted easily for use with laser and pulsed-laser NMR using the new concept of $\mathbf{B}_\Pi$ (Appendix C).

III. CONTINUOUS-WAVE LASER NMR, OR LASER-ENHANCED NMR SPECTROSCOPY (LENS)

The LENS technique relies in the simplest case on the use of a continuous-wave, circularly polarized laser directed into the sample tube of a conventional NMR spectrometer. Following the initial (second-order) theoretical predictions of the present author,[22-30] the first series of LENS experiments[32, 33] has successfully demonstrated the ability of a circularly polarized laser to cause site-specific shifts in NMR resonances. It was found that the shifts were different for each resonating nucleus, for example, ^{1}H and [2]D. This means that the technique is capable of providing an entirely new fingerprint of a sample, something that can be interpreted empirically in the analytical, industrial, and biochemical laboratory, in the usual way, without the immediate need for detailed theoretical work of a fundamental nature. For example, the laser-induced shift pattern, as a function of laser intensity and resonance site could lead to an easier interpretation of a very complicated pattern of resonances from identical amino acids in slightly different environments in a protein in solution, a pattern that in the absence of the laser might be indistinguishable, but that in the presence of a laser might become clearly different. This would identify different amino acids residues on the same protein.

The first series of LENS experiments, carried out by Warren and cowork-ers,[32, 33] set out to detect small shifts caused by a cw argon ion laser of up to 3.0-W/cm^2 intensity, and great care was taken to remove artifacts due to heating. In this section this experiment is described in some detail.

Using data from a series of careful measurements, a small selection of results was presented[32, 33] with 514-nm laser light from a Coherent Innova 200 argon ion laser propagated about 20 m across the laboratory and aimed directly down a 5-mm capillary tube, spinning, as usual in NMR, in a commercial (JEOL GX-270) spectrometer. The laser beam was designed to provide a uniform 10.0-mm diameter at the sample tube, and the wavelength of 514 nm was chosen to be in a transparent part of the sample's visible absorption spectrum (Fig. 3) as a precaution against heating by absorption, and to demonstrate that LENS patterns can be produced with a laser frequency tuned to a transparent[32, 33] as well as an absorbing part of the sample's spectrum.

The results reproduced in Fig. 4 show both bulk and local frequency shifts, and it was found that shifts for a given solute were different in different solvents. Each resonance site has its own shift pattern as a function of laser intensity, which is one example of the new LENS fingerprint. This pattern must, theoretically, be made up of both first-order shift and several types[22-30] of second-order shift, all present simultane-ously. In other words, cw laser NMR, even with very low laser intensity, is potentially rich in analytical information. For example, the H resonances of "identical" amino acid residues in different environments of a protein in solution would probably give quite different LENS spectra of the type. Without the need for any further interpretation, it would already have become clear that the two different amino acid residues occupied different sites in the protein, thus the name "laser enhancement."

In future, LENS might well be extended to multidimensional NMR, in which resolution is optimal prior to laser enhancement.[34, 35] In this context, a circularly polarized cw laser provides, in principle, that extra power of analysis that makes the difference between an interpretable LENS fingerprint and an uninterpretable conventional NMR spectrum of heavily overlapping resonance lines in one more dimensions.

Care was taken by Warren and coworkers[32, 33] to minimize heating artifact. This type of artifact is eliminated completely in pulsed-laser NMR (vide infra), being in that case even less of a problem than possible heating caused by the universally employed[34, 35] pulsed radio-frequency field of Fourier transform NMR spectroscopy. In cw laser NMR, however, a laser beam can heat the interior of an NMR tube, a possibility that was countered[32, 33] with specially designed capillary sample tubes, and by gating the laser with a shutter that was controlled by the spectrometer's

ated chloroform is also shifted by the circularly polarized laser. All shifts appeared to go in the direction of increased shielding, independently of the sense of circular polarization. This result probably points toward a dominant first-order mechanism, but much further work is needed to confirm this in other samples and to clarify the role of second-order mechanisms. The observed laser-induced shifts were solvent dependent [33] and were different for the same solute in chloroform and benzene, for example. Local shifts (i.e., shifts in resonances from local protons) were observed which depended on the position of the proton in the molecule and on the handedness of the enantiomer. These important experimental results provide clear first evidence therefore for the theoretical predictions[22-30] that an off-resonance circularly polarized laser can shift (and under certain circumstances split) NMR resonances, and enhance NMR resolution without heating.

The concept of $\mathbf{B}_\Pi$ appears to provide a framework for a simple analysis of these results. The permanent magnetic flux density reported[32, 33] (270 MHz JOEL NMR spectrometer's permanent magnet) is 6.4 T. The extra static magnetic flux density imparted by a circularly polarized laser of 1.0-W/cm^2 intensity is therefore about 2 ppm in absolute terms, about an order of magnitude greater than the relative change, in ppm, in the chemical shift recorded in Fig. 4 in parts per million.[33] In Fig. 4 this is 0.008 ppm for a particular H resonance (methoxy resonance, laevorotatory enantiomer), which converts into 2.18 Hz for the 272-MHz spectrometer used. The experimental data are consistent in the first approximation with an interaction energy of the form

$$E_3 = -\mathbf{m}^{(N)} \cdot \left[\mathbf{B}_0(1 - \sigma_1) \pm \mathbf{B}_\Pi(1 - \sigma_2) \right] \qquad (42)$$

where σ_2 is the chemical shift constant in the presence of the laser for a particular resonating nucleus. The shielding constant σ_2 appears to be different in general from its equivalent without the laser because the latter can also generate an extra magnetic field at the nucleus due to Fermi contact interaction[23] and other second-order effects.[24-30] Without these second-order mechanisms there is no reason why the constant σ_1 should change, but the absolute value of the chemical shift in hertz is increased because the effective external magnetic field is increased from $\mathbf{B}_0$ to $\mathbf{B}_0 + \mathbf{B}_\Pi$ by the laser. Therefore, the laser increases the spectrometer's effective resolution and gives a new pattern of chemical shifts.

On the basis of Eq. (42) the simplest type of interaction energy is an absolute proton resonance shift of 425.78 Hz, equivalent to a $\mathbf{B}_\Pi$ of 10^{-5} T. (Note that the shift in Fig. 4 is expressed in parts per million, and is a unitless ratio[34] of resonance frequencies, not an absolute laser-induced shift expressed in hertz.) On this same scale, a permanent NMR magnet of flux density of 6.4 T is equivalent to 272.5 MHz. Therefore, it is expected that the absolute value in hertz of the chemical shift, originally $\delta_1 B_0$, is changed by the laser to $\delta_2 B_0$. On the basis of the first-order interaction (42), parameter δ_1 itself, for a given resonance site, should not change to first order, but to second order such a change is expected due to Fermi contact[23] and other mechanisms. The result reproduced in Fig. 4 indicates that the expected absolute, laser-induced frequency shift for the unshielded proton resonance (425.78 Hz for a B_{Π} of 10^{-5} T), is different for different proton sites. In the case of the methoxy proton in Fig. 4, the laser-induced shift relative to the standard resonance frequency is changed by 2.18 Hz from the equivalent in the absence of the laser. This is the type of change that forms the useful site-selective, laser-induced, resonance fingerprint.

Theoretically, a very intense laser pulse is capable of increasing dramatically the absolute frequency separation of the original NMR resonances, and therefore capable of increasing the effective resolution of the conventional NMR spectrum of any type (one or n dimensional). If this were realized in practice it might be a useful development in several types of NMR spectroscopy.

In summary, a laser of $B_{\Pi} = 10^{-5}$ T is expected to produce an unscreened proton shift of 425.78 Hz in the first-order mechanism.[22] In addition, there are other, second-order, mechanisms that are expected to produce smaller shifts, as described in the literature. Among these are shifts due to the vectorial polarizability,[24] hyperpolarizability shifts,[25, 26] and shifts due to Fermi contact interactions.[23] The various types of shift are not mutually independent, but result in angular momentum coupling, illustrated briefly in the following example for atoms with net electronic spin such as H and various metal atoms in the ground state.[27]

A. Angular Momentum Coupling: Atomic Model

We have seen that a circularly polarized laser generates the magnetic flux density B_{Π} in its axis of propagation, and that the pure imaginary conjugate product $\Pi^{(A)}$ is built up from the pure real B_{Π} by multiplication by i and a T- and P-positive scalar constant. The real, static, flux density appears to be a novel fundamental property of circularly polarized electromagnetic radiation (of any frequency), and forms an interaction energy with any type of magnetic dipole moment, and in LENS the latter is a nuclear magnetic dipole moment $m^{(N)}$. However, the conjugate product $\Pi^{(A)}$ is also capable of forming a scalar interaction energy (Appendix B):

$$E_4 = i\alpha'' \cdot \Pi^{(A)} \tag{43}$$

where $i\alpha''$ is the imaginary part of the atom's or molecule's electronic electric polarizability. Clearly, α'' must be T negative and P positive, and also an axial vector if this energy is to be a T- and P-positive scalar. The required vectorial form is obtained[36, 37] from the usual tensor α''_{ij} by tensor multiplication with the rank three, totally antisymmetric unit tensor, known as the Levi-Civita symbol[36, 37]:

$$\alpha''_k = \varepsilon_{ijk}\alpha''_{ij} \tag{44}$$

Neglecting the chemical shift screening constants in the first approximation, the complete (first- plus second-order) interaction Hamiltonian is therefore

$$E_5 = -\mathbf{m}^{(N)} \cdot (\mathbf{B}_\Pi + \mathbf{B}_0) + i\alpha'' \cdot \mathbf{\Pi}^{(A)} \tag{45}$$

and in general there is angular momentum coupling (for example, Landé coupling[31]) between the two terms. This is a quantum-mechanical phenomenon, which further enriches the LENS spectrum in theory. It is akin to the family of angular momentum coupling phenomena such as spin–orbit and coupling and spin–rotation coupling, measurable in such phenomena as the anomalous Zeeman effect.[31] The fundamental reason for the phenomenon in the context of the energy (45) is that the vectors $\mathbf{m}^{(N)}$ and α'' are both proportional to a quantized angular momentum, and these two angular momenta do not commute in general, in the same way that orbital and spin electronic angular momentum vectors do not commute in the anomalous Zeeman effect Hamiltonian.[31] It is revealing to write the vectorial polarizability in terms of an induced electronic magnetic dipole moment:

$$\mathbf{m}^{(ind)} = 2c^2 B_0 \alpha'' \tag{46}$$

showing clearly that $\mathbf{m}^{(ind)}$ and α'' have the same fundamental symmetries, i.e., T-negative, P-positive axial vector symmetry. Being an electronic magnetic dipole moment, $\mathbf{m}^{(ind)}$ can be written as

$$\mathbf{m}^{(ind)} = 2c^2 B_0 \gamma_\Pi \mathbf{J} \tag{47}$$

where $\mathbf{J}$ is the electronic angular momentum and γ_Π the gyroptic ratio.[24] The interaction energy[45] therefore becomes

$$E_5 = -\gamma_N \mathbf{I} \cdot (\mathbf{B}_\Pi + \mathbf{B}_0) - 2c^2 B_0 \gamma_\Pi i(\mathbf{L} + 2.002\mathbf{S}) \cdot \mathbf{B}_\Pi \tag{48}$$

implying that the laser-induced electronic magnetic dipole moment is proportional to the sum $L + 2.002S$ of orbital and spin electronic angular momenta components in the laser's propagation axis. The interaction energy (48) therefore consists in general of three noncommuting angular momenta.[29, 30] For atoms, these are I, the nuclear spin angular momentum, L, the electronic orbital angular momentum, and $2.002S$, the electronic spin angular momentum. The electronic spin angular momentum is the only contribution to the vectorial polarizability[29, 30] in the $J = \frac{1}{2}$ ground state of an atom, for example, and for this reason the electron has spin polarizability,[38] a property that has been computed ab initio in some cases by Manakov et al.[8-10] To understand this property, recall that electronic spin is a relativistic quantity in theory, and the term does not imply a physically spinning object. A fuller understanding of spin polarizability must therefore be sought in the Dirac equation, taking into account such effects as Thomas precession.[31]

Using the standard methods of quantum mechanics, the interaction energy can be developed in terms of coupled eigenstates, as usual:

$$
\begin{aligned}
E_5 = & -\langle LSJIFM_F | \gamma_N I_Z (B_{\Pi Z} + B_{0Z}) \\
& + 2B_0 c^2 \gamma_\Pi (L_Z + 2.002 S_Z) B_{\Pi Z} | L'S'J'I'F'M_F' \rangle
\end{aligned}
\tag{49}
$$

in the coupling scheme

$$
J = L + S \qquad F = J + I \tag{50}
$$

from which the LENS resonance frequency can be calculated[30] to be

$$
\hbar \omega_R = \frac{|2B_0 c^2 \gamma_\Pi B_{\Pi Z}(g_1 + 2.002 g_2) + g_N \gamma_N (B_{0Z} + B_{\Pi Z}) g_3|}{[F(F+1)(2F+1)]^{1/2}} \tag{51}
$$

where the g factors are evaluated with the usual 9-j symbols. For reasonable values of the quantum numbers L, S, and I in atoms with net electronic spin, we obtain results that show that the c.p. laser causes a shift to higher resonance frequencies, a shift that is linear in the square root of the intensity to first order, and linear in the intensity to second order. The combined effect of the laser's properties B_Π and $\Pi^{(A)}$ is to enrich greatly the original resonance spectrum.

B. Interrelation of Some Molecular Property Tensors

In Section III.A, an example was given of a novel interrelation between molecular property tensors, Eq. (46), which shows that the antisymmetric (or vectorial) polarizability is directly proportional to an induced electronic magnetic dipole moment $\mathbf{m}^{(\text{ind})}$. It follows that the irreducible representations of these vectors are the same in all molecular and atomic point groups, and in the group of all rotations and reflections[39] of an ensemble in the laboratory frame. Furthermore, it is reasonable to conclude that a permanent magnetic dipole moment (nuclear or electronic), must also transform as a higher rank molecular property tensor of some kind, because any axial vector (i.e., rank one tensor) is simultaneously definable as a second rank, antisymmetric, polar tensor.[36] In the case of a nuclear magnetic dipole, this conforms with the fact that a nucleus has an electric polarizability, whose antisymmetric component is proportional to its magnetic dipole moment, which can occur in both protons and neutrons.[31]

Whenever a magnetic dipole moment occurs in a given spectroscopic situation it must have the same fundamental symmetries as vectorial electric polarizability, and the same is true of the corresponding quantum-mechanical operators. On the grounds of symmetry and tensor algebra, such interrelations can also be constructed between other ranks of tensors, such as rank three hyperpolarizabilities of various kinds, by replacing the magnetic dipole moment wherever it occurs by the vectorial electric polarizability. These relations provide insight to the fundamental nature of both the magnetic dipole moment and the antisymmetric electric polarizability.

A relation can also be forged between the vectorial polarizability and the magnetizability by considering the sample in the simultaneous presence of both a circularly polarized laser and permanent magnetic flux density. In this case the interaction energy between the laser-induced magnetic dipole moment and the magnetic flux density of the permanent magnet can be written in the simplest case as (see also Appendix B)

$$E_6 = -\mathbf{m}^{(\text{ind})} \cdot \mathbf{B}_0 = -2c^2 B_0 \alpha_Z'' B_{0Z} \tag{52}$$

Comparing this expression with Eq. (14.2.9) of Atkins[31] yields a relation between the vectorial polarizability and the magnetizability

$$\xi_Z^{(\text{II})} = 2c^2 \mu_0 \alpha_Z'' \tag{53}$$

which can be written equivalently as

$$\xi_{XY}^{(\text{II})} = -\xi_{YX}^{(\text{II})} = 2c^2 \mu_0 \alpha_{XY}'' \tag{54}$$

where $\xi_{XY}^{(\text{II})}$ is the laser-induced magnetizability of the sample in the presence of a static magnetic field. We have therefore forged relations between the magnetic electronic dipole moment, the vectorial electric electronic polarizability, and the magnetizability.

IV. PULSED-LASER NMR AND ESR SPECTROSCOPY

The LENS method as developed to date uses a low-intensity cw laser in the sample tube of an NMR spectrometer. By looking at Eq. (23) it can be seen that a pulsed laser delivering a nonzero $\partial \mathbf{B}_{\Pi}/\partial t$ over an interval t also generates an electric field $\mathbf{E}_{\Pi}$ whose curl is no longer zero. This electric field can be picked up as a signal and stored in the computer of a contemporary Fourier transform NMR spectrometer. It is well known that laser pulses can be very short and intense. For example,[40] a passively mode-locked dye laser give 0.5 GW of power in about 0.4 ps at a frequency of about 600 THz in the visible. For a laser aperture of 1.0 cm^2 this gives an intensity I_0 of 5.0 TW/m^2. If focused to 1.0 mm^2 the intensity is increased 100 times. In contrast, a continuous-wave argon ion laser such as the one used in the first LENS demonstration[33] provides an intensity twelve orders of magnitude lower, up to about 3.0 W/cm^2, but still results in easily measurable and very useful NMR shifts. We have seen that first-order shifts due to $\mathbf{B}_{\Pi}$ are proportional to the square root of the laser's intensity, and second-order shifts are proportional to the intensity itself. In principle, therefore, very large shifts to higher frequency (enormous increases in effective resolution) are possible using a circularly polarized laser. There appear to be several other ways, however, of incorporating a laser pulse in NMR spectroscopy, relying specifically on the use of the Fourier transform,[34, 35, 41] which appear to be technically less of a challenge, and which are introduced as follows.

Contemporary Fourier transform NMR and ESR technology is mature and highly diversified after 50 years of development. Consequently, it is of advantage to incorporate a laser pulse as directly as possible into the contemporary designs,[35, 41] and it is shown in this section that the 180° radio-frequency pulse generated in a conventional spectrometer can be augmented by a pulse of magnetization from a circularly polarized laser radiation.

It is well known[35, 41] that the essence of Fourier transform NMR spectroscopy is the measurement of free induction decay following radio-frequency pulses and pulse sequences. In the simplest case a sample is excited with a 90° pulse, and the free induction decay is Fourier transformed, as first demonstrated by Ernst and Anderson.[42] The magnetization pulse is achieved with a rotating and linearly polarized radio-frequency field at right angles to the static magnetic field of the bit NMR magnet. One of the circularly polarized components of the rf field rotates in the

A pulse if circularly polarized laser radiation traveling in Z through a sample tube of a contemporary Fourier transform spectrometer, parallel or antiparallel with $\mathbf{B}_0$ of the big magnet, will cause $\mathbf{B}_0$ to change to $\mathbf{B}_0 \pm \mathbf{B}_\Pi$, depending on the direction of travel of the laser and its circular polarization (right or left). The more intense the laser, the greater is $\mathbf{B}_\Pi$ (Eq. (14)). In this situation the Larmor precession frequency changes to

$$\omega_1 = \gamma_1 B_{0Z} \tag{60}$$

where

$$\gamma_1 = \gamma_N \left(1 \pm \frac{B_{\Pi Z}}{B_{0Z}} \right) \tag{61}$$

is the new effective gyromagnetic ratio (nuclear (NMR), or electronic (ESR)). If a pulse of circularly polarized laser radiation passes through the sample, its pulse frequency is

$$f = \frac{\gamma_N B_{\Pi Z}}{2\pi} \text{ hertz} \tag{62}$$

and the angle θ is defined by

$$\theta = \gamma_N B_{\Pi Z} t \tag{63}$$

in radians, where t is the pulse duration in seconds.

If $\mathbf{B}_\Pi$ suddenly increases $\mathbf{B}_0$ of the permanent magnet of the instrument, the extra $\mathbf{B}_\Pi$ will change the minimum energy alignment in the field $\mathbf{B}_0$, but will not create mean transverse magnetization, as we have argued. This is precisely what happens with the conventional 180° rf pulse. We conclude that a laser can act in contemporary NMR spectrometers in exact analogy with the widely used 180° rf pulse, leading in principle to numerous technological possibilities provided that there is simultaneous synchronization of spins. If $I = \frac{1}{2}$, for example, there are two nuclear spin states, as usual. If $\mathbf{B}_0$ is suddenly increased to $\mathbf{B}_0 \pm \mathbf{B}_\Pi$ by the laser pulse, the energy separation between the two spin states is increased, and therefore the number of spins is expected to remain in the lower energy state,[34] i.e., the new equilibrium value of longitudinal magnetization $\mathbf{M}$ is increased. The system will respond to the laser pulse by means of spins undergoing transitions from the upper to lower energy levels, defined by the laser-on Boltzmann distribution

$$\frac{N_{\text{higher}}}{N_{\text{lower}}} = \exp\left[- \frac{2\mathbf{m}^{(N)} \cdot (\mathbf{B}_0 + \mathbf{B}_{\Pi})}{kT} \right] \tag{64}$$

where N_{higher} and N_{lower} are the numbers of higher and lower energy state spins, respectively, and $\mathbf{m}^{(N)}$ is the permanent nuclear magnetic dipole moment. This process involves a loss of energy by the system over time T_1, the longitudinal spin–lattice relaxation time. It is well known[34] that T_1 can be as long as 1000 s, many orders of magnitude longer than a typical high energy laser pulse (vide infra). Conversely, if $\mathbf{B}_{\Pi}$ opposes $\mathbf{B}_0$, spins will go from lower to upper energy levels, a process involving a gain of energy over a time T_1.

In a laser-generated 180° pulse, therefore, there is no mean transverse magnetization, and the lower energy spins are converted to higher energy ones by a laser producing $-\mathbf{B}_{\Pi}$. After the laser pulse passes through the spinning sample tube of the NMR spectrometer, the spins relax to their normal state in $\mathbf{B}_0$ of the permanent magnet with a characteristic longitudinal relaxation time T_1. This process in itself produces no resonance effects and no signal, as with a conventional 180° pulse.[34] However, the resonance effect of the laser pulse can then be detected a time $t = \tau$ after the pulse by the application of a conventional 90° radio-frequency pulse. A simple pulsed NMR sequence is therefore

$$180° \text{ laser pulse} - \tau - 90° \text{ radio-frequency pulse} \tag{65}$$

A. Numerical Estimates

Consider a pulse of circularly polarized giant ruby laser radiation, with an intensity of 10^{11} W/m² and a pulse duration of 30 ns (Refs. 1–4). These were the characteristics used[1-4] to demonstrate the inverse Faraday effect. The laser generates a $\mathbf{B}_{\Pi}$ of about 0.03 T from Eq. (14) over 30 ns, which rotates the spin by an angle of about 0.1 radians (about 6°) from Eq. (63). This is increased to about 60° by focusing the laser to an intensity 100 times greater. Following the passage of the pulse through the sample, there is a laser-induced longitudinal relaxation, followed τ seconds later by the 90° measuring radio-frequency pulse tuned as usual to resonance.

Clearly, any flip angle θ can be generated by adjusting the duration and/or intensity of the circularly polarized laser pulse. The terminology "180° laser pulse" should therefore not be taken to imply that the angle θ must be exactly 180°, because free induction decay will always occur if the spins are tilted out of their initial equilibrium in the big magnet's static imparts magnetization to the ensemble of nuclear spins, information that can be used to analyze in a new way the nature of the sample under investigation.

C. Application to Molecular Dynamics and Chemical Reactions, or Similar Dynamical Processes

It it well known that a laser pulse on the femtosecond or picosecond scale acts as camera shutter on molecular dynamical processes[44] taking place on a similar time scale, and effectively freezes them for measurement. In the particular context of pulsed-laser NMR, the duration and intensity of the circularly polarized laser pulse determines the flip angle and the extent of magnetization produced prior to free induction decay. The nature of the free induction decay following the pulse is therefore determined by the dynamics of the molecules in which the nuclear spins are situated. The laser pulse does not cause the nuclear magnetization to rise immediately (in an infinitely short time) from zero to a final level, but initiates a rise transient of nuclear spins, a process that occupies a finite interval of time, presumably on the femtosecond scale. Nothing appears to be known about the nuclear spin rise transient, which can be investigated in principle by varying the length and intensity of the laser pulse and measuring the resulting free induction decay with a 90° rf pulse to produce the resonance spectrum.

Molecular dynamical rise transients are well known in several different contexts and can be simulated by computer,[46] for example. Similarly, the degree of magnetization for a given pulse depends on the B_Π field of the laser to first order, an don its $\Pi^{(A)}$ vector to second order, and the final level attained by the transient is governed by a Langevin or Langevin–Kielich function.[47] In much the same way as the T_1 decay time is affected by Brownian motion (more accurately, molecular dynamics[43]) so is the T_1 rise time in response to a laser pulse. This aspect appears not to be considered in conventional Fourier transform NMR spectroscopy, so that it is apparently assumed that the rise transient of nuclear magnetization must be effectively very short on the microsecond scale of the conventional radio-frequency pulse. This is no longer tenable if the nuclear spin–rise transient and the laser pulse both occur on the same time scale, as seems inevitable when the laser pulse is only femtoseconds in duration.

In this eventuality the degree of longitudinal magnetization is presumably proportional to the saturation level of the nuclear spin–rise transient, a level that in a molecular context is well known[46, 47] to depend on the nature of the dynamics (saturation values of rise transients form Langevin or Langevin–Kielich functions). Conversely therefore, the degree of magnetization following a laser pulse provides valuable information in principle on the molecular dynamical processes affecting nuclear magnetization, processes that occur on the picosecond or femtosecond scale in gases and liquids, for example. Again, a chemical reaction taking place on this time scale can be monitored by using a series of LASEI measurements, varying the laser pulse duration. The result is a series of longitudinal free induction decay curves that can be picked up for analysis by the 90° rf pulse and interpreted to give information on the progress of the reaction on the femtosecond time scale.

D. Laser Spin-Echo NMR Spectroscopy

In this type of pulsed laser NMR spectroscopy the 180° laser pulse would be used to refocus phase coherence.[35] To modify the conventional and well-known Carr–Purcell sequence for use with a pulsed circularly polarized laser, transverse magnetization would first be created with a conventional rf 90° pulse magnetization, which is allowed to decay over an interval of time τ, during which the phase coherence of individual spin vectors is being lost.[35] A 180° laser pulse is then applied in the $+Y$ axis (perpendicular to the Z axis of the big magnet's field) to recreate the phase coherence, which is detected as a laser-induced spin echo. This method would implicitly have all the advantages of the laser 180° pulse discussed already, for example, an effectively infinite bandwidth in hertz.

E. Selective Population Transfer with a Laser Pulse

When a selective 180° laser pulse is applied to an I spin transition,[35] the spin populations are changed, resulting in a change in the intensities of S spin transition resonance lines of a weakly coupled two-spin $\frac{1}{2}$ system IS.[35] This is a laser-induced selective population transfer, a phenomenon that can be used to demonstrate the magnetization effect of a laser pulse in, for example, a system with a proton attached directly to a resonating ^{13}C nucleus. The selective population transfer experiment has been replaced in conventional technology by two-dimensional NMR methods, but appears to be useful in the above context.

F. Solid-State Laser NMR Spectroscopy

In conventional solid-state NMR spectroscopy, resonances are very broad and conventional methods such as magic angle spinning[35] are used in an attempt to narrow them and to resolve the underlying resonance spectrum. Efficient narrowing of solid-state NMR spectra would open up many new areas of investigation in the condensed state of matter. A conventional method that could be adopted for use with lasers in principle is pulse spinning,[35] where magnetic spin–spin interactions (the source of broadening) are averaged out by forcing the spins in the solid lattice to change quickly and selectively about a magnetic field direction by applying 90° rf pulses along both the X and Y axes and alternating their phases.[35] Conventionally, this requires "short" (microsecond) and intense rf pulses, and the technology needed for this is difficult to implement, requiring rf pulse trains of up to about 52 individual pulses. The data are sampled during intervals in the pulse train[35] and stored in a computer. The essence of the method is that individual chemical-shift directional anisotropies in, for example, a crystal are determined by both the parallel and perpendicular magnetic field components during a pulse train.

In principle, it appears much easier to implement a train of suitably aimed 180° laser pulses to "scramble" the nuclear spins and average out the spin–spin coupling and magnetic anisotropies that cause solid-state NMR broadening. Laser pulses are much more intense and of much shorter duration than rf pulses, and contemporary laser pulse train technology is highly developed. The effect of the laser pulses could be picked up using a suitable LASEI sequence, for example, and the technique, if developed, appears to have considerable merits.

V. LASER-IMAGING NMR SPECTROSCOPY

The key to laser-imaging NMR spectroscopy is to introduce deliberate inhomogeneities into the $\mathbf{B}_0$ of the big permanent magnet by scanning the sample (for example, a tissue mounted on a slide) with a circularly polarized laser, so that at each spot the effective field is increased to $\mathbf{B}_0 \pm \mathbf{B}_\Pi$ from $\mathbf{B}_0$, introducing an effective change

$$\delta f = \pm \frac{\gamma_N \mathbf{B}_{\Pi Z}}{2\pi} \text{ hertz} \qquad (66)$$

in the resonance frequency at that spot. In principle, spatial information about the sample (an image) becomes available through the laser-induced shift in the effective applied magnetic field at that spot. This is similar to conventional NMR imaging, where the frequency shift is introduced with deliberately imposed magnetic field gradients, but the image in laser NMR is built up with a different scanning principle. As the laser passes across a given spot, the field $\mathbf{B}_0 \pm \mathbf{B}_\Pi$ induces a resonance whose intensity at that spot is proportional to the number of equivalent spins per unit volume contained there. The resonance signal recorded by the instrument is therefore proportional to the density of spins at that spot, which is recorded in a computer equipped with state of the art imaging software.

Using well-developed laser scanning technology, a sample can be scanned many thousands of times if needs be, and the average image built up in two dimensions. A three-dimensional image is constructed as in conventional[4] projection–reconstruction imaging NMR by rotating the sample.

It appears at this stage that the image might well be optimized in quality by making use of Fourier transform imaging techniques[41] with the incorporation of laser pulses. Each spot in the sample could be examined in this way with optimization of image quality using a LASEI sequence for each spot. The free induction decay following the 180° laser pulse at that spot would be unique, and an image could be computed from a knowledge of the details of the LASEI spectrum at each locality in the specimen. It appears possible to use trains of scanning laser pulses, whose average effect would be recorded at any given spot in the sample, data reduction being a matter of software ingenuity. Laser echo line imaging is another possibility,[4] the key idea of which would be to use a 180° selective laser pulse in the presence of a conventional 90° rf detection/resonance pulse in the presence of a conventional g gradient.[4] This pulse rotates the spins of a selected plane perpendicular to the Z axis, and the difference between two induction signals, one with and one without the 180° laser pulse, is used to give the required signal. In this method, the 180° laser pulse has all the advantages of being an effectively perfect impulse function.

VI. MULTIDIMENSIONAL LASER NMR SPECTROSCOPY

The incorporation of 180° laser pulses into the highly developed contemporary technology in this field relies again on the replacement of the 180° rf pulse. In one-dimensional Fourier transform NMR[35] the free induction decay is a function of a single time variable,[35] which is supplemented in two-dimensional NMR by the introduction of a second time variable τ_2. A series of free induction decay curves is collected as a function of τ_2 and double Fourier transformation[35] gives a spectrum that is a function of two frequency axes, with many advantages. An example of how a laser pulse might be incorporated with minimum design change is as follows. The first pulse is a conventional 90° rf pulse, whose free induction decay is not recorded. After a variable interval τ_2, this pulse is followed by a 180° laser refocusing pulse, so that the nuclear magnetization is coherently reestablished to reach a maximum (the echo) at double the chosen interval.[35] At this point the data collection is initiated. The experiment is then repeated, as in the conventional method, with τ_2 incremented. This method is essentially the reverse of LASEI, because its key feature is a conventional 90° rf pulse followed by a 180° laser pulse. Of basic scientific interest in this type of experiment would be the first experimental demonstration that a 180° laser pulse can refocus nuclear spins to produce an echo whose extent would depend on the laser-induced flip angle θ, so that the echo is maximized for a flip angle of exactly 180°. However, there should be an echo present with any flip angle, because the laser is always being used to rebuild magnetization to some measurable extent. In this type of experiment the refocusing effectively introduces high resolution[41] and this would be the role played by the laser.

Whenever a 180° radio-frequency pulse is used in contemporary NMR technology (which is mature and highly developed), it can be replaced in principle by a circularly polarized laser pulse. Among many applications would be the use of a laser pulse in the conventional technology[35] used in creating multiple-quantum coherence, for example, double and triple quantum correlated spectroscopy. Another example in which a 180° rf pulse can be replaced by a 180° laser pulse is INEPT,[35] the technique in which insensitive nuclei are enhanced by polarization transfer. Here the 180° laser pulse would be sandwiched between two conventional 90° rf pulses. Yet another possibility is the use of a 180° laser pulse in the Müller sequence[35, 41] in proton-detected COSY experiments. These are more advanced pulse designs, however; the simplest experiment is the LASEI sequence, which at this early stage would be important in demonstrating for the first time the ability of a laser pulse to generate free induction decay, i.e., to produce nuclear magnetization through the $\mathbf{B}_\Pi$ vector, a magnetization that decays characteristically with time after the laser pulse has passed through the sample. The rate of decay can be orders of magnitude slower, and the characteristic longitudinal decay time T_1 orders of magnitude longer, than the pulse duration.

APPENDIX A: SENSE OF ROTATION AND HANDEDNESS OF AN ELECTROMAGNETIC PLANE WAVE

In this appendix we stress the difference between the handedness and the sense of rotation of an electromagnetic plane wave. The subscripts R and L in Eqs. (6), (7), and (38), for example, refer to sense of rotation, i.e., leftwise or rightwise of the rotating vectors $\mathbf{E}$ and $\mathbf{B}$. Thus, κ_L should be read as "the linear momentum of a photon whose sense of rotation is leftwise," i.e., has a left state of spin, corresponding to the well-known photon quantum number $M_S = +1$. The spin of the photon is intrinsic and unremovable from fundamental postulate, and the parity operator P cannot change the sense of this or any other spin, i.e., cannot change the spin of the photon's angular velocity vector. Therefore, in this notation

$$P(\kappa_L) = -\kappa_L \qquad (A.1)$$

because the direction of linear momentum of the photon is changed by P, but the leftwise sense of spin is not.

Clearly, the subscript L must be understood to signify the sense of spin of the photon. The handedness of the photon can be thought of as the product of the photon's linear and angular velocities, and the handedness operator $\hat{H}$ of the text is the product

$$\hat{H}(\mathbf{v}\omega) = P(\mathbf{v})P(\omega) = -(\mathbf{v}\omega) \qquad (A.2)$$

Therefore, the handedness operator can be thought of as equivalent to two parity operations, the product (A.2).

The existence of a left- and right-spinning photon with respect to the direction of its propagation implies, inter alia, the existence[31] of left and right circularly polarized electromagnetic plane waves, this being a type of particle/wave duality. It follows that the P operator cannot change the sense of rotation of the tip of a vector such as $\mathbf{E}$ which traces out a circle[36] in a fixed plane perpendicular to the direction of propagation of the plane wave. Thus,

$$\mathbf{E}_L = E_0(\mathbf{i} + \mathbf{ij})\exp\left[-\mathrm{i}(\omega t - \kappa_L \cdot \mathbf{r})\right] \qquad (A.3)$$

means a left rotating plane wave in our notation, propagating in $+\kappa_L$. The P operator changes κ_L to $-\kappa_L$ and produces a left rotating plane wave propagating in $-\kappa_L$. The handedness operator $\hat{H}$ produces a right rotating $\mathbf{E}$ vector propagating in $+\kappa_R$.

Therefore, we prefer to make a distinction between the parity operator P and the handedness operator $\hat{H}$ when dealing with electromagnetic plane waves or, equivalently, photons. We emphasize that the subscripts L and R must be understood to mean "rotating leftwise or rightwise" throughout this chapter and in related articles in the literature,[14-20] and that they do *not* mean "left-handed" or "right-handed." This is a subtle but fundamentally important distinction.

Note that the motion reversal operator T must reverse the sense of any motion, and so must reverse the sense of spin of a photon or the sense of rotation of a vector such as $\mathbf{E}$ in its equivalent electromagnetic plane wave. Also, T must reverse the linear momentum of the photon, and so must also reverse the sign of the propagation vector κ in the plane wave equivalent to the photon. So

$$T(\kappa_L) = -\kappa_R \qquad (A.4)$$

and it follows that[19]

$$T(\mathbf{E}_L) = \mathbf{E}_R^* \text{ etc.} \qquad (A.5)$$

as in the text. The handedness of the photon or wave is unaffected by the T operator, because the latter reverse both the sense of rotation and direction of propagation simultaneously:

$$T(\mathbf{v})T(\boldsymbol{\omega}) = +(\mathbf{v}\boldsymbol{\omega}) \qquad (A.6)$$

APPENDIX B: DEVELOPMENT OF THE SECOND-ORDER INTERACTION ENERGY

In this appendix we show that the second-order electronic interaction energy $i\boldsymbol{\alpha}'' \cdot \boldsymbol{\Pi}^{(A)}$ can be rewritten as a term quadratic in $\mathbf{B}_\Pi$ and proportional to the paramagnetic electronic susceptibility χ'', a T-negative, P-positive axial vector. This demonstrates clearly that $i\boldsymbol{\alpha}'' \cdot \boldsymbol{\Pi}^{(A)}$ is a term describing an induced electronic magnetic dipole moment multiplied by the square of $\mathbf{B}_\Pi$.

The first steps are

$$E_4 = i\boldsymbol{\alpha}'' \cdot \boldsymbol{\Pi}^{(A)}$$
$$= -(2|\boldsymbol{\alpha}''|cB_0\mathbf{k}) \cdot (cB_0\mathbf{k}) \qquad (B.1)$$
$$= -2|\boldsymbol{\alpha}''|c^2\mathbf{B}_\Pi \cdot \mathbf{B}_\Pi$$

where $|\boldsymbol{\alpha}''|$ is the magnitude of the axial vector $\boldsymbol{\alpha}''$, and $\mathbf{k}$ is a unit axial vector as in the text.

The sequence (B.1) shows immediately that E_4 is quadratic in $\mathbf{B}_\Pi$, so that the quantity $2|\boldsymbol{\alpha}''|c^2$ must have the units of electronic magnetic susceptibility,[36] denoted

$$\chi_{ij} = \chi'_{ij} + i\chi''_{ij} \qquad (B.2)$$

It is well established[36] that the imaginary part χ''_{ij} is a T-negative, P-positive, rank two polar tensor, which by fundamental tensor algebra is also an axial vector:

$$\chi'' \equiv \chi''_i = \varepsilon_{ijk}\chi''_{jk} \qquad (B.3)$$

It follows by symmetry that

$$|\chi''| = 2\zeta|\boldsymbol{\alpha}''|c^2 \qquad (B.4)$$

where ζ is a T- and P-positive unitless scalar. Thus,

$$E_4 = -\frac{|\chi''|B_{\Pi}^2}{\zeta} = -\frac{|\chi''|B_0^2}{\zeta} \tag{B.5}$$

where $|\chi''|$ is the magnitude of the vector χ''. Comparing (B1) and (B4) yields

$$i\boldsymbol{\alpha}'' \cdot \boldsymbol{\Pi}^{(A)} = -\frac{|\chi''|B_0^2}{\zeta} \tag{B.6}$$

This relation provides several insights to the meaning of the term $i\boldsymbol{\alpha}'' \cdot \boldsymbol{\Pi}^{(A)}$, which on first sight may appear to some readers to be abstract.

1. E_4 is simply an energy of interaction of $\mathbf{B}_{\Pi}$ with a laser-induced electronic magnetic dipole moment of magnitude

$$|\mathbf{m}^{(\text{ind})}| = |\chi''|B_0/\zeta \tag{B.7}$$

where $|\chi''|$ is the magnitude of the sample's electronic paramagnetic susceptibility, a molecular property tensor defined from time-dependent second-order perturbation theory[31, 36]:

$$\chi''_{\alpha\beta} = -\chi''_{\beta\alpha} = -\frac{2}{\hbar} \sum_{j \neq n} \frac{\omega}{\omega_{jn}^2 - \omega^2}$$
$$\times \mathrm{I}_m\!\left(\langle n|\hat{m}_\alpha|j\rangle\langle j|\hat{m}_\beta|n\rangle\right) \tag{B.8}$$

where ω_{jn} is a transition frequency between states n and j, and ω is the frequency of the electromagnetic radiation, i.e., the frequency that appears in $\mathbf{E}$ and $\mathbf{E}^*$. In Eq. (B.8) $\hat{m}_\alpha$ and $\hat{m}_\beta$ are magnetic dipole moment operators.

From Eq. (B.1),

$$i\boldsymbol{\Pi}^{(A)} \cdot \mathbf{k} = 2c^2\mathbf{B}_{\Pi} \cdot \mathbf{B}_{\Pi} \tag{B.9}$$

where $\mathbf{k}$ is a unit axial vector, meaning that

$$i|\mathbf{E} \times \mathbf{E}^*| = 2c^2\mathbf{B}_{\Pi} \cdot \mathbf{B}_{\Pi} \tag{B.10}$$

This shows that the vector cross product of $\mathbf{E}$ with $\mathbf{E}^*$ is proportional to the vector dot product of $\mathbf{B}_{\Pi}$ and $\mathbf{B}_{\Pi}$.

2. Using the relations

$$\mathbf{B}_{\Pi} = \nabla \times \mathbf{A}_{\Pi} \tag{B.11}$$

and the vector identity

$$(\mathbf{F} \times \mathbf{G}) \cdot (\mathbf{H} \times \mathbf{I}) = (\mathbf{F} \cdot \mathbf{H})(\mathbf{G} \cdot \mathbf{I}) - (\mathbf{F} \cdot \mathbf{I})(\mathbf{G} \cdot \mathbf{H}) \tag{B.12}$$

gives

$$\mathbf{E}_4 = -2|\alpha''|_c^2 \nabla^2 A_{\Pi}^2 = -\frac{|\chi''|}{\zeta} \nabla^2 A_{\Pi}^2 \tag{B.13}$$

which brings out a formal similarity with Eq. (41) of the text. However, Eq. (B.13) refers to a T-negative, P-positive, paramagnetic susceptibility vector, while Eq. (41) relates[31] to a T-positive, P-positive, symmetric, diamagnetic, polar susceptibility tensor of rank two, the real part of χ_{ij} of Eq. (B.2), a tensor that does not have a rank one vector equivalent. Note that in Eq. (B.13), ∇^2 operates on A_{Π}^2, while in Eq. (41), A_{Π}^2 is not operated upon. We have

$$\nabla^2 = \nabla \cdot \nabla = \frac{\partial^2}{\partial X^2} + \frac{\partial^2}{\partial Y^2} + \frac{\partial^2}{\partial Z^2} \tag{B.14}$$

and, self-consistently,

$$\nabla^2 A_{\Pi}^2 = B_{\Pi}^2 = B_0^2 \tag{B.15}$$

3. Finally, from semiclassical, time-dependent, second-order perturbation theory,[36]

$$\alpha''_{\alpha\beta} = -\alpha''_{\alpha\beta} = -\frac{2}{\hbar} \sum_{j \neq n} \frac{\omega}{\omega_{jn}^2 - \omega^2} \mathrm{Im}\left(\langle n|\hat{\mu}_\alpha|j\rangle\langle j|\hat{\mu}_\beta|n\rangle\right) \tag{B.16}$$

where $\hat{\mu}$ are electric dipole moment operators. This shows that

$$\chi'' \propto (\text{Velocity})^2 \alpha'' \tag{B.17}$$

because an electric dipole moment has the units of charge times distance, and magnetic dipole moment has the units of charge times distance times linear velocity.[31] This is consistent with Eq. (B.4) because c has the units of velocity, and ζ is unitless.

Therefore, the $\mathbf{B}_{\text{II}}$ concept leads to a self-consistent development of the energy E_4, and to a simple physical interpretation, Eq. (B.4).

APPENDIX C: RELATING THE $\mathbf{B}_{\text{II}}$ VECTOR TO FUNDAMENTAL ELECTROMAGNETIC QUANTITIES

It is well known[36] that the polarization state of a monochromatic electromagnetic plane wave can be described in terms of the four, real, scalar Stokes parameters. Of particular interest is the third Stokes parameter S_3, defined by

$$S_3 = -i(E_X E_Y^* - E_Y E_X^*) \qquad (C.1)$$

which is none other than the magnitude of the conjugate product vector $\mathbf{\Pi}^{(A)}$ of the text. It follows immediately that

$$\mathbf{\Pi}^{(A)} = iS_3\mathbf{k} = i\left(\frac{8I_0 c}{\varepsilon_0}\right)^{1/2} (\pm\mathbf{B}_{\text{II}}) \qquad (C.2)$$

an equation that defines $\mathbf{B}_{\text{II}}$ in terms of the well-known S_3 of fundamental electromagnetic theory.

This is a useful link because S_3 is well defined in terms of polarization properties. For example,

$$S_3 = 2E_0^2 \sin 2\eta \qquad (C.3)$$

where η is the ellipticity of the plane wave. For circular polarization, $\eta = \pm\pi/4$. The ellipticity can be conversely related to S_3 by[36]

$$\eta = \frac{1}{2}\tan^{-1}\left[\frac{S_3}{(S_1^2 + S_2^2)^{1/2}}\right] \qquad (C.4)$$

Furthermore, if $I(\sigma, \tau)$ denotes the intensity of light transmitted through a retarder which subjects the Y component to a retardation τ with respect to the X component, followed by an analyzer with its transmission axis oriented at an angle σ to the X axis, then

$$S_3 \propto I\left(\frac{3\pi}{4}, \frac{\pi}{2}\right) - I\left(\frac{\pi}{4}, \frac{\pi}{2}\right) \qquad (C.5)$$

so that S_3 is the excess in intensity transmitted by a device that accepts right circularly polarized light over one that accepts left circularly polarized light.[36] This shows that an excess of circular polarization is needed to produce $\mathbf{B}_\Pi$ in a laser beam, as in the text. $\mathbf{B}_\Pi$ is maximized if the beam is fully left or right circularly polarized, because S_3 is maximized. S_3 vanishes in linear polarization and so does $\mathbf{B}_\Pi$.

The polarization state of a light beam can also be defined by a matrix[36]:

$$\rho_{\alpha\beta} = \frac{E_\alpha E_\beta^*}{E_0^2} \tag{C.6}$$

with scalar elements. This is a Hermitian polarization density matrix, or a coherency matrix, the off-diagonals of which contain S_3:

$$\rho_{\alpha\beta} = \frac{1}{2S_0} \begin{bmatrix} S_0 + S_1 & -S_2 + iS_3 \\ -S_2 - iS_3 & S_0 - S_1 \end{bmatrix} \tag{C.7}$$

Clearly, this leads to Eq. (C.1) because the off-diagonals contain iS_3 and $-iS_3$. It follows that the off-diagonal elements of the coherency matrix of a plane wave also contain the magnitude of the vector $\mathbf{B}_\Pi$, because this magnitude is simply proportional to S_3. The coherency matrix is a second-rank spinor,[36] a real four-dimensional vector in a complex two-dimensional space. The components of the vector are related to the four Stokes parameters, and the coherency matrix is analogous with a density matrix in quantum mechanics. An incoherent superposition of pure quantum-mechanical states is a mixed quantum-mechanical state[36] must be represented by a density matrix. These analogies are useful for the demonstration of the existence of $\mathbf{B}_\Pi$ in rigorous quantum field theory, relativistic electrodynamics, and gauge theory. It is also clear that $\mathbf{B}_\Pi$ has a rigorous definition in stochastic electrodynamics, as discussed in several articles of this book.

In quantum optics, the third Stokes parameter S_3 (Ref. 48) becomes a Hermitian operator, $\hat{S}_3$, and the electric field vector becomes an operator ($\hat{E}$) of the electromagnetic field, being proportional to an annihilation operator, labeled $\hat{a}_+$ or $\hat{a}_-$ according to the sense of polarization.[48] In this notation, the third Stokes operator can be shown to be the difference

$$\hat{S}_3 = \hat{a}_+^+ \hat{a}_+ - \hat{a}_-^+ \hat{a}_- \tag{C.8}$$

where the annihilation operators satisfy the commutation relation

$$[\hat{a}_+, \hat{a}_+^+] = \delta_{++} \tag{C.9}$$

The quantum-field Stokes operators themselves obey a commutation law:

$$\left[\hat{S}_1, \hat{S}_2\right] = 2i\hat{S}_3 \qquad (C.10)$$

and the classical third Stokes parameter S_3 is the expectation value of the quantum-field operator $\hat{S}_3$. In Cartesian coordinates we have

$$\hat{a}_\pm = \frac{1}{\sqrt{2}}\left(\hat{a}_X \mp i\hat{a}_Y\right) \qquad (C.11)$$

In the quantum electromagnetic field of a laser, therefore, the $\mathbf{B}_\Pi$ vector becomes an operator proportional to the third Stokes $\hat{S}_3$, i.e., $\mathbf{B}_\Pi$ can be quantized, unlike a conventional flux density vector generated by a magnet.

A coherent eigenstate of the quantum field can be defined with respect to the annihilation operator $\hat{a}$ by[48]

$$\hat{a}|\alpha\rangle = \alpha|\alpha\rangle \qquad (C.12)$$

and it can be shown[48] that $|\alpha|^2$ is a mean number of photons in the quantum field. The classical third Stokes parameter S_3 is the expectation value

$$S_3 = \langle\alpha|\hat{S}_3(Z)|\alpha\rangle \qquad (C.13)$$

which can be written[48] as

$$S_3 = |\alpha_+|^2 - |\alpha_-|^2 \qquad (C.14)$$

which is a difference in the mean number of photons in the $+$ circular mode of the operator $\hat{E}_+$ and the $-$ circular mode.

It follows that the classical $\mathbf{B}_\Pi$ vector is an expectation value of the quantized $\hat{B}_\Pi$ operator:

$$\mathbf{B}_\Pi = \langle\alpha|\hat{B}_\Pi(Z)|\alpha\rangle \qquad (C.15)$$

The number of photons in these circular modes are constants[48] of the motion, which implies inter alia that $\hat{B}_\Pi$ is frequency independent. Therefore, $\mathbf{B}_\Pi$ has a clearly defined meaning in quantum-field theory, and is generated in general by an electromagnetic field that is made up of photons, i.e., is quantized. This is yet another form of particle wave duality.

Finally, it is interesting to note that the Heisenberg uncertainty principle can be expressed in terms of[48]

$$\left(\left\langle \left(\Delta \hat{S}_1 \right)^2 \right\rangle \left\langle \left(\Delta \hat{S}_2 \right)^2 \right\rangle \right)^{1/2} \geq |\langle \hat{S}_3 \rangle| \qquad \text{(C.16)}$$

and is fundamentally dependent, therefore, on the existence of the quantum-field operator $\hat{B}_\Pi$.

APPENDIX D: SIMPLE PHYSICAL PROPERTIES OF THE B_Π VECTOR

In this appendix we discuss a few simple physical properties expected of the B_Π vector. Because B_Π is a static, uniform, magnetic field an electronic charge in B_Π will move (classically) along a helix under the influence of B_Π, a helix having its axis along the direction of the magnetic field with a radius r. The linear velocity of the electron would be constant. There is a striking similarity between the motion of the electron in B_Π and the motion of the tip of the E vector of a circularly polarized laser. The tips of the E vectors at a given instant distributed along the direction of propagation of a circularly polarized light beam form[36] a helix that moves along the direction of propagation but does not rotate. An electronic charge will follow the same pattern of motion (because the force on the electron is eE), and as the helix moves through space, the electronic charge e rotates in a plane according to the sense of rotation (Appendix A). This examination of the motion of an electron in a circular polarized laser shows that the B_Π vector is closely connected with the rotational motion of the E vector of the laser (Appendix B).

If the B_Π vector is spatially inhomogeneous, specifically, if its magnitude varies in the Z axis of the laser's propagation, the inhomogeneity is expected to lead to a slow transverse drift of the guiding center of the helical trajectory of an electron placed in the field. Since the magnitude of B_Π is directly proportional to the square root of the laser's intensity by Eq. (14), a gradient $\partial B_\Pi / \partial Z$ can be established by varying the laser's intensity along the direction of propagation Z. This is accomplished experimentally by expanding or focusing the beam. If a beam of electrons (or atoms such as silver) is directed at a target in the same axis Z, colinear with the expanding or focused laser beam, we expect a deflection of the electron or atom beam by the gradient $\pm \partial B_\Pi / \partial Z$ of the laser, in direct analogy with the well-known Stern-Gerlach experiment. In order to detect the deflection, it is an advantage to make $\partial B_\Pi / \partial Z$ as large as possible, and this can be accomplished by pulsing the laser and using a fast detector system, readily available from contemporary technology. As in the original Stern-Gerlach experiment, such a deflection shows the existence inter alia of electronic spin and the gradient $\partial \mathbf{B}_\Pi / \partial Z$ of a focused or expanded laser beam. Unlike the original Stern-Gerlach experiment, however, the gradient $\partial \mathbf{B}_\Pi / \partial Z$ is quantized, as discussed in Appendix C.

In classical electrodynamics, $\mathbf{B}_\Pi$ is a uniform magnetostatic field and, formally, there must be a source of this field located at infinity. This source may be an electric current density or magnetic charge density. However, since $\mathbf{B}_\Pi$ is created by a circularly polarized electromagnetic field, the source of $\mathbf{B}_\Pi$ and the source of the electromagnetic field must be the same. This implies that the classical $\mathbf{B}_\Pi$ vector exists in the absence of charges, because a classical electromagnetic field exists in the absence of charges, and thus propagates in vacuo. In Appendix C we saw that $\mathbf{B}_\Pi$ is the expectation value of an operator $\hat{B}_\Pi$ proportional to the third Stokes operator of the quantized electromagnetic field: thus, $\mathbf{B}_\Pi$ is clearly defined in terms of photons. It follows that the source of $\mathbf{B}_\Pi$ is the same as the source of these pohtons, and is therefore well defined. The classical magnetostatic $\mathbf{B}_\Pi$ is an expectation value of a quantized electromagnetic field operator $\hat{B}_\Pi$ directly proportional to the third Stokes operator.

APPENDIX E: THE QUANTIZED $\hat{B}_\Pi$

In this appendix quantum-field theory is used to show that a beam of N photons generates quantized static magnetic flux density $(\hat{B}_\Pi)$, defined by the operator equation

$$\hat{B}_\Pi = B_0 \frac{\hat{J}}{\hbar} \qquad \text{(E.1)}$$

where B_0 is the scalar magnetic flux density amplitude (tesla) of the beam and $\hat{J}$ is the intrinsic angular momentum operator of one photon. The operator $\hat{B}_\Pi$ is related directly to the angular momentum operator $\hat{J}$ of the photon.

Here $\hat{B}_\Pi$ is the quantum-field description of the classical $\mathbf{B}_\Pi$ vector. Equation (E.1) appears to be a new fundamental property of photons, indicating immediately the existence of $\hat{B}_\Pi$ given that of $\hat{J}$. Clearly, $\hat{J}$ and $\hat{B}_\Pi$ are zero only if the (laser) beam of N photons has no element of coherent circular polarity (e.g., if it is linearly polarized).

The transition from classical to quantized field theory is made through the third Stokes operator $\hat{S}_3$, recently introduced by Tanaś and Kielich[48]:

$$\hat{S}_3 = -\left(\frac{2\pi\hbar\omega}{n^2(\omega)V}\right)i\left(\hat{a}_X^+\hat{a}_Y - \hat{a}_Y^+\hat{a}_X\right) \qquad (E.2)$$

Here $n(\omega)$ is the refractive index, V is the quantization volume, and $\hat{a}^+$ and $\hat{a}$ denote respectively the creation and annihilation operators. Defining a coherent state of a beam of N photons by the Schrödinger equation[48]

$$\hat{a}|\alpha\rangle = \alpha|\alpha\rangle \qquad (E.3)$$

provides the expectation value

$$\langle\alpha|\hat{S}_3|\alpha\rangle = |\alpha_+|^2 - |\alpha_-|^2 \qquad (E.4)$$

with

$$\alpha_+ = \frac{1}{\sqrt{2}}(\alpha_X \mp i\alpha_Y) \qquad (E.5)$$

We define the operator

$$\hat{B}_\Pi \equiv \left(\frac{\varepsilon_0}{8I_0 c}\right)^{1/2}\hat{S}_3 \qquad (E.6)$$

and arrive at the equation

$$\hat{B}_\Pi = \left(\frac{\varepsilon_0}{8I_0 c}\right)^{1/2}\left(\frac{2\pi\omega}{n^2(\omega)V}\right)\hbar\left(\hat{a}_+^+\hat{a}_+ - \hat{a} \pm \hat{a}_-\right) \qquad (E.7)$$

with

$$\hat{a}_\pm = \frac{1}{\sqrt{2}}(\hat{a}_X \mp i\hat{a}_Y) \qquad (E.8)$$

The quantity

$$\hbar\left(\hat{a}_+^+\hat{a}_+ - \hat{a}_-\hat{a}_-\right) \equiv \hbar\left(\hat{n}_+ - \hat{n}_-\right) \qquad (E.9)$$

has the units of quantized angular momentum because $(\hat{n}_+ - \hat{n}_-)$ is dimensionless. Here

$$\hat{n}^+ = \hat{a}_+^+ \hat{a}_+$$
$$\hat{n}^- = \hat{a}_-^+ \hat{a}_- \qquad \text{(E.10)}$$

are the number of photons operators.[48]

We know that the total angular momentum of a beam of N photons propagating in Z is $NM_J\hbar$, where M_J is the azimuthal quantum number associated with $\hat{J}$. Defining the angular momentum eigenfunction of a single photon by $|JM_J\rangle$ we arrive at the Schrödinger equation:

$$\hbar(\hat{n}_+ - \hat{n}_-)|JM_J\rangle = \hbar N M_J |JM_J\rangle \qquad \text{(E.11)}$$

From (E.7) in (E.11)

$$\frac{n^2(\omega)V}{2\pi\omega}\hat{S}_3|JM_J\rangle = \hbar N M_J |JM_J\rangle \qquad \text{(E.12)}$$

and with the identity

$$\hat{J} = \left[\frac{n^2(\omega)V}{2\pi\omega N}\right]\hat{S}_3 \qquad \text{(E.13)}$$

Eq. (E.12) becomes the standard Schrödinger equation describing the angular momentum of one photon:

$$\hat{J}|JM_J\rangle = \hbar M_J |JM_J\rangle \qquad \text{(E.14)}$$

From (E.6) and (E.13)

$$\hat{B}_\Pi = \left(\frac{\varepsilon_0}{8I_0 c}\right)^{1/2}\left[\frac{2\pi\omega N}{n^2(\omega)V}\right]J \equiv \zeta\hat{J} \qquad \text{(E.15)}$$

showing that $\hat{B}_\Pi$ is directly proportional to $\hat{J}$. Considerable insight to the nature of the constant ζ is obtained with the results of Tanaś and Kielich[48]:

$$S_0 = \frac{2\pi\omega\hbar}{n^2(\omega)V}\langle\alpha|\hat{S}_0|\alpha\rangle$$
$$\langle\alpha|\hat{S}_0|\alpha\rangle = N \qquad \text{(E.16)}$$

where $\hat{S}_0$ is the zeroth Stokes operator, and S_0 is the zeroth Stokes parameter. Furthermore, we make use of the classical result[36]

$$S_0 = 2E_0^2 \tag{E.17}$$

From (E.16) and (E.17)

$$E_0^2 = \frac{\pi \omega \hbar N}{n^2(\omega) V} \tag{E.18}$$

Using (E.18) in (E.15) gives, finally, the fundamental and simple operator equation

$$\hat{B}_\Pi = B_0 \left(\frac{\hat{J}}{\hbar} \right) \tag{E.19}$$

It is clear that Eq. (E.19) defines $\hat{B}_\Pi$ simply and directly in terms of $\hat{J}$, the angular momentum operator of one photon. It follows that

$$\hat{B}_{\Pi Z} |JM_J\rangle = B_0 M_J |JM_J\rangle \tag{E.20}$$

so that the expectation value of the Z component of $\hat{B}_\Pi$ is $B_0 M_J$. It appears that there are many implications of this result, in principle, a few of which are as follows.

1. A beam of N photons propagating in Z generates the magnetic field $B_0 M_J$, whose classical value is $\pm B_0 \mathbf{k}$.
2. This magnetic field forms a scalar interaction energy with any quantized magnetic dipole moment, nuclear or electronic, and this is the basis of optical NMR and ESR.[33]
3. Conventionally,[31] a right-handed photon carries the minimum intrinsic angular momentum $M_J = -1$, and the left photon $M_J = +1$. However, a photon's M_J can exceed unity.[31]
4. Since $\hat{J}$ is an intrinsic, unremovable property of the photon, then $\hat{B}_\Pi$ is an intrinsic, unremovable property of the beam of N photons.
5. It follows that a circularly polarized beam of N photons must always generate $\hat{B}_\Pi$, which invariably forms a scalar interaction energy with a quantized magnetic dipole moment operator following the theory of angular momentum coupling.[21]

6. Finally, whenever a static magnetic field is used experimentally, it is also possible, in principle, to use a circularly polarized laser beam, at any frequency, from radio frequencies to those of gamma rays. The magnitude of the classical B_Π can be estimated from the simple equation (14).

Acknowledgments

The Leverhulme Trust is thanked for a Fellowship for the development of optical resonance techniques, and Rex Richards, F.R.S., for correspondence. The Cornell Theory Center is ackowledged for additional support, and the Materials Research Laboratory of Penn State University for the award of a Visiting Senior Research Associateship. Correspondence and many interesting discussions are acknowledged with Warren S. Warren and colleagues at Princeton, Jack L. Freed and Keith A. Earle at Cornell, Stanisław Kielich at Poznań, Ray Freeman, F.R.S., at Cambridge, and Laurence Barron at Glasgow. Richard Ernst is thanked for a seminar invitation to ETH, Zürich, and for interesting and positive discussions on the optical resonance ideas presented in this article.

References

1. P. S. Pershan, *Phys. Rev. A* **130**, 919 (1963).
2. Y. R. Shen, *Phys. Rev. A* **134**, 661 (1964).
3. J. P. van der Ziel, P. S. Pershan, and L. D. Malmstrom, *Phys. Rev. Lett.* **15**, 190 (1965).
4. P. S. Pershan, J. P. van der Ziel, and L. D. Malmstrom, *Phys. Rev.* **143**, 574 (1966).
5. S. Kielich, *Proc. Phys. Soc.* **86**, 709 (1965).
6. S. Kielich, *Acta. Phys. Pol.* **29**, 875 (1966); **30**, 851 (1966).
7. P. W. Atkins and M. H. Miller, *Mol. Phys.* **15**, 503 (1968).
8. N. L. Manakov, V. D. Ovsiannikov, and S. Kielich, *Acta Phys. Pol.* **A53**, 581 (1978).
9. N. L. Manakov, V. D. Ovsiannikov, and S. Kielich, *Acta Phys. Polon.* **A53**, 595 (1978).
10. N. L. Manakov, V. D. Ovsiannikov, and S. Kielich, *Acta Phys. Polon.* **A53**, 737 (1978).
11. B. L. Silver, *Irreducible Tensor Methods*, Academic, New York, 1976.
12. R. Zare, *Angular Momentum*, Wiley Interscience, New York, 1984.
13. C. P. Slichter, *Principles of Magnetic Resonance*, 2nd ed., Springer, Berlin, 1978.
14. M. W. Evans, *Phys. Rev. Lett.* **64**, 2909 (1990).
15. M. W. Evans, *Opt. Lett.* **14**, 863 (1990).
16. M. W. Evans, *Phys. Lett. A* **146**, 475 (1990); **147**, 364 (1990).
17. M. W. Evans, *J. Mol. Spectrosc.* **143**, 327 (1990); **146**, 351 (1991).
18. M. W. Evans, *Physica B*, **168**, 9 (1991).
19. M. W. Evans, in I. Prigogine and S. A. Rice (Eds.), *Advances in Chemical Physics*, Vol. 81, Wiley Interscience, New York, 1992, pp. 361–702.
20. M. W. Evans, *Phys. Lett. A* **146**, 185 (1990).
21. A. R. Edmonds, *Angular Momentum in Quantum Mechanics*, 2nd ed., Princeton University Press, Princeton, NJ, 1960.
22. M. W. Evans, *Physica B*, **179**, 237 (1992).
23. M. W. Evans, *Physica B*, **182**, 118 (1992).

and is related to $\mathbf{A}_\Pi$ through

$$\mathbf{B}_\Pi = \nabla \times \mathbf{A}_\Pi \qquad (4)$$

Section IV interprets the cross product $\mathbf{E} \times \mathbf{E}^*$ with the antisymmetric part of Maxwell's electromagnetic stress tensor,[14] which is part of the electromagnetic energy/momentum four tensor. Thus, $\mathbf{B}_\Pi$ is proportional to a vorticity in the classical electromagnetic field. This is illustrated by determining the equations of motion of an electron in $\mathbf{B}_\Pi$ by solving the novel Lorentz equation

$$\mathbf{p} + 2e\mathbf{A}_\Pi = \text{constant} \qquad (5)$$

where $\mathbf{p}$ is the electron's momentum and e is its charge. The field $\mathbf{B}_\Pi$ drives the electron forward in a helical trajectory, with constant linear velocity in z. It is shown finally that this is the same trajectory as that of the electron in a circularly polarized plane wave, obtained by solving the Lorentz equation for this case. In other words, the solutions of the Lorentz equation (5) are also solutions of the Lorentz equation of motion of the electron in a circularly polarized plane wave, showing that a plane wave can generate the characteristics of a magnetostatic flux density $\mathbf{B}_\Pi$, which takes meaning as it interacts with the electron, driving the latter in a helical trajectory.

The paper ends with a short discussion of the interpretation of $\mathbf{B}_\Pi$ in the required relativistic context.

II. THE CONTINUITY EQUATION FOR $\mathbf{B}_\Pi$

Maxwell's phenomenological equations lead to the following well-known[15] continuity equation when electromagnetic radiation interacts with matter:

$$\nabla \cdot \mathbf{N} = -\frac{\partial U}{\partial t} - \mathbf{E} \cdot \mathbf{J}^* = 0 \qquad (6)$$

where $\mathbf{J}^*$, the current density, is zero in free space. The energy density U is defined in free space as the time-averaged quantity:

$$U = \tfrac{1}{2}(\mathbf{H}^* \cdot \mathbf{B} + \mathbf{E}^* \cdot \mathbf{D}) \qquad (7)$$

with $\mathbf{B} = \mu_0 \mathbf{H}$ and $\mathbf{D} = \varepsilon_0 \mathbf{E}$. Here μ_0 is the permeability of free space.

Clearly, in free space, the wave does no work on matter, such as electronic charge, because the density $\mathbf{E} \cdot \mathbf{J}^*$ of power lost from the fields

$\mathbf{E}$ and $\mathbf{B}^*$ is zero. U is therefore a field energy density, which takes meaning only if there is field–matter interaction.

By considering the product $\mathbf{E} \times \mathbf{E}^*$, which is proportional, as we have seen, to the antisymmetric vector part of light intensity, it can be demonstrated as follows that there exists a novel continuity equation linking $\mathbf{B}_\Pi$ and U_Π. Using the vector relation,

$$\nabla \cdot (\mathbf{E} \times \mathbf{E}^*) = \mathbf{E}^* \cdot (\nabla \times \mathbf{E}) - \mathbf{E} \cdot (\nabla \times \mathbf{E}^*) \tag{8}$$

and the Maxwell equations in free space,

$$\nabla \times \mathbf{E} = -\frac{\partial \mathbf{B}}{\partial t} \qquad \nabla \times \mathbf{E}^* = -\frac{\partial \mathbf{B}^*}{\partial t} \tag{9}$$

implies

$$\nabla \cdot (\mathbf{E} \times \mathbf{E}^*) = \nabla \cdot (\mathbf{B} \times \mathbf{B}^*) = 0 \tag{10}$$

that is,

$$\nabla \cdot \mathbf{B}_\Pi = 0 \tag{11}$$

which shows that $\mathbf{B}_\Pi$ is uniform and divergentless. From Eqs. (9) and (10), we can write

$$\nabla \cdot (\mathbf{E} \times \mathbf{E}^*) = 2E_0 c i \nabla \cdot \mathbf{B}_\Pi = -\mathbf{E}^* \cdot \frac{\partial \mathbf{B}}{\partial t} + \mathbf{E} \cdot \frac{\partial \mathbf{B}^*}{\partial t} \tag{12}$$

Integration by parts of the right side of Eq. (12) gives the result

$$\int \mathbf{E}^* \cdot \frac{\partial \mathbf{B}}{\partial t} \, dt - \int \mathbf{E} \cdot \frac{\partial \mathbf{B}^*}{\partial t} \, dt = \frac{1}{2} (\mathbf{E}^* \cdot \mathbf{B} - \mathbf{E} \cdot \mathbf{B}^*) = -2i E_0 B_0 \tag{13}$$

Defining the quantity

$$U_\Pi = \frac{1}{2E_0 c i} \left(\int \mathbf{E}^* \cdot \frac{\partial \mathbf{B}}{\partial t} \, dt - \int \mathbf{E} \cdot \frac{\partial \mathbf{B}^*}{\partial t} \, dt \right)$$
$$= -\frac{B_0}{c} \tag{14}$$

M. W. EVANS

implies

$$\nabla \cdot \mathbf{B}_\Pi = -\frac{\partial U_\Pi}{\partial t} = 0 \tag{15}$$

which is a free space continuity equation for $\mathbf{B}_\Pi$. Equation (15) is the precise counterpart of the free space continuity equation (6) for $\mathbf{N}$:

$$\nabla \cdot \mathbf{N} = -\frac{\partial U}{\partial t} = 0 \tag{16}$$

Equation (15) is novel to this work, whereas Eq. (16) is standard in classical electrodynamics.[15]

The continuity equations (15) and (16) have the same structure and must be interpreted in the same way. Thus, in classical electrodynamics, the Poynting vector $\mathbf{N}$ is the flux of the density (U) of the electromagnetic energy. Similarly, the novel $\mathbf{B}_\Pi$ is the flux of the magnetostatic density of the electromagnetic plane wave, a flux that is uniform and does not vary with time. U is the electromagnetic field energy density in free space (i.e., electromagnetic power per unit volume[15]), and therefore U_Π is the magnetostatic density in free space generated by an electromagnetic plane wave, or the magnetostatic power of the wave per unit volume occupied by that wave in free space.

Both Eqs. (15) and (16) are continuity equations relating[15] the time rate of change of a density (the scalar fields U or U_Π) to the divergence of a flux (the vector fields $\mathbf{N}$ or $\mathbf{B}_\Pi$). It is well established[15] that the notion of electromagnetic field energy density U takes meaning only when there is interaction between the field and matter. Similarly, U_Π takes meaning only when there is wave–matter interaction. Equations (15) and (16) are both statements based on the existence in free space of the light intensity tensor I_{ij}, which is Hermitian.[16] Equation (16) is concerned with the scalar part of I_{ij}, and Eq. (15) with its vector part, which, as we have seen, is the quantity $\varepsilon_0 c \mathbf{E} \times \mathbf{E}^*$ proportional to $\mathbf{B}_\Pi$. The existence of $\mathbf{B}_\Pi$, and of Eq. (15), is an inevitable consequence of the fact that I_{ij} is a tensor,[16] with a vector (i.e., antisymmetric) component that is purely imaginary as a consequence of the Hermitian nature of I_{ij} (Ref. 16). It follows that the classical free space electromagnetic plane wave generates $\mathbf{B}_\Pi$ and its associated scalar field U_Π.

III. THE VECTOR POTENTIAL A_Π IN FREE SPACE

Since B_Π is a magnetostatic flux density, it is assumed that there exists a vector potential A_Π such that

$$B_\Pi = \nabla \times A_\Pi \qquad (17)$$

Since B_Π and U_Π are well defined in free space, it follows from the assumption (17) that A_Π would also be well defined in free space. If Eq. (17) were true, it follows that A_Π would be a function[17] of the type

$$A_\Pi = -\tfrac{1}{2} r \times B_\Pi \qquad (18)$$

where r is a positive coordinate vector in frame (X, Y, Z). From Eq. (1),

$$A_\Pi = -\tfrac{1}{2} r \times (E \times E^*)/(2E_0 ci) \qquad (19)$$

Defining

$$r \equiv Xi + Yj + Zk \qquad (20)$$

and using the vector relation

$$r \times (E \times E^*) = E(r \cdot E^*) - E^*(r \cdot E) \qquad (21)$$

with[17-19]

$$\begin{aligned} E &= E_0(i - ij)e^{i\phi} \\ E^* &= E_0(i + ij)e^{-i\phi} \end{aligned} \qquad (22)$$

where ϕ is the phase[17] of the plane wave, we have

$$A_\Pi = \frac{B_0}{2}(Xj - Yi) \qquad (23)$$

Since

$$|B_\Pi| = B_0 \qquad (24)$$

it is clear that B_Π and A_Π of Eq. (24) are related by Eq. (17).

In other words there exists a vector potential A_Π in free space whose curl is B_Π that is defined by the vector field $(B_0/2)(Xj - Yi)$, a field that is generated in free space by the classical electromagnetic plane waves

(22), both solutions of Maxwell's equations. Note that $\mathbf{B}_\Pi$ is also a solution of Maxwell's equations in vacuo. In the same way, $\mathbf{N}$ is not in itself a solution of Maxwell's equations, but is generated therefrom through a cross product $\mathbf{E} \times \mathbf{B}^*/\mu_0$.

It has been demonstrated that $\mathbf{B}_\Pi$ is related to a well-defined $\mathbf{A}_\Pi$ in free space, and to a well-defined scalar field U_Π. The next section considers the interaction of $\mathbf{B}_\Pi$ with an electron, i.e., wave–matter interaction.

IV. INTERACTION OF $\mathbf{B}_\Pi$ WITH AN ELECTRON

From Eq. (18) it is clear that

$$\dot{\mathbf{A}}_\Pi = -\tfrac{1}{2}\mathbf{v} \times \mathbf{B}_\Pi \tag{25}$$

where $\mathbf{v} = \partial\mathbf{r}/\partial t$ is a velocity in frame (X, Y, Z). Consider the interaction of $\mathbf{B}_\Pi$ with an electron. Since action and reaction are equal and opposite, the action of $\mathbf{B}_\Pi$ on the electronic charge e is balanced by a reaction of e upon $\mathbf{B}_\Pi$. However, as usual in classical electrodynamics, we assume that $\mathbf{B}_\Pi$ is not changed greatly by the action of e upon it,[13-15] so that the Lorentz equation of motion applies. Since there is no $\mathbf{E}_\Pi$ present,

$$\dot{\mathbf{p}} = e\mathbf{v} \times \mathbf{B}_\Pi \tag{26}$$

where $\mathbf{p}$ is the momentum of the electron, and $\mathbf{v}$ is its velocity $(\mathbf{p}/m)$ in the field $\mathbf{B}_\Pi$. With Eq. (25) the Lorentz equation can be rewritten directly in terms of the novel vector potential $\mathbf{A}_\Pi$:

$$\dot{\mathbf{p}} + 2\dot{\mathbf{A}}_\Pi e = 0 \tag{27}$$

The equations of motion of e in $\mathbf{B}_\Pi$ thus become[15] those of Coriolis acceleration:

$$\left.\begin{matrix} \dot{v}_X = \Omega v_Y \\ \dot{v}_Y = -\Omega v_X \end{matrix}\right\} \qquad \dot{v}_Z = 0 \tag{28}$$

where the parameter Ω is defined as

$$\Omega = \left(1 - \frac{v^2}{c^2}\right)^{1/2}\left(\frac{e}{m}\right)B_{\Pi Z} \tag{29}$$

Equations (28) can be rewritten[15] as

$$\frac{d}{dt}(v_X + iv_Y) = -i\Omega(v_X + iv_Y)$$

$$v_X + iv_Y = a \exp(-i\Omega t) \tag{30}$$

where a is complex and defined by

$$a = v_{0t} \exp(-i\alpha) \tag{31}$$

where v_{0t} and α are real. Here

$$v_{0t} = \left(v_X^2 + v_Y^2\right)^{1/2} \tag{32}$$

is the velocity of the electron in the XY plane. The trajectory of the electron in the field $\mathbf{B}_\Pi$ is therefore that of a helix

$$X = X_0 + r\sin(\Omega t + \alpha)$$
$$Y = Y_0 + r\cos(\Omega t + \alpha) \tag{33}$$
$$Z = Z_0 + v_{0Z}t$$

where the radius of the helix $\mathbf{r}$ is defined by

$$r = \frac{p_t}{eB_{\Pi Z}} = \frac{mv_{0t}}{eB_{\Pi Z}}\left(1 - \frac{v^2}{c^2}\right)^{-1/2} \tag{34}$$

where p_t is the projection of the momentum on to the XY plane. Note that the same result can be obtained directly from the equation

$$\dot{\mathbf{p}} = -2\mathbf{A}_\Pi e \tag{35}$$

to give

$$\dot{v}_X = -\Omega v_Y \text{ etc.} \tag{36}$$

We now demonstrate that the motion of an electron in a circularly polarized plane wave is precisely the same as the motion of an electron in the novel magnetic field $\mathbf{B}_\Pi$. Assume that the reaction of the electron upon the circularly polarized electromagnetic field is negligible, so that the

Lorentz equation again applies:

$$\dot{\mathbf{p}} = e(\mathbf{E} + v \times \mathbf{B}) \tag{37}$$

Here $\mathbf{E}$ and $\mathbf{B}$ are plane wave solutions of Maxwell's equations in free space:

$$\mathbf{E} = E_0(\mathbf{i} + \mathrm{i}j)\mathrm{e}^{\mathrm{i}\phi} \quad \mathbf{B} = B_0(\mathbf{j} - \mathrm{i}\mathbf{i})\mathrm{e}^{-\mathrm{i}\phi} \tag{38}$$

i.e., they are the usual oscillating, phase-dependent, electric and magnetic field vectors of the wave. Note carefully that $\mathbf{B}$ is quite different in nature from $\mathbf{B}_\Pi$. The velocity of the electron in frame (X, Y, Z) is

$$\mathbf{v} = v_X\mathbf{i} + v_Y\mathbf{j} + v_Z\mathbf{k} \tag{39}$$

Substituting Eqs. (38) and (39) into the Lorentz equation (37) and using $|\mathbf{B}_\Pi| = B_0 = E_0/c$ gives the equation of motion:

$$\dot{v}_X = \Omega(c - v_Z)\mathrm{e}^{-\mathrm{i}\phi} \tag{40}$$

$$\dot{v}_Y = \mathrm{i}\Omega(c + v_Z)\mathrm{e}^{-\mathrm{i}\phi} \tag{41}$$

$$\dot{v}_Z = \Omega(v_X + \mathrm{i}v_Y)\mathrm{e}^{-\mathrm{i}\phi} \tag{42}$$

Multiplying Eq. (41) by i and adding to Eq. (40) gives

$$\frac{\mathrm{d}}{\mathrm{d}t}(v_X + \mathrm{i}v_Y) = -2v_Z\Omega\mathrm{e}^{-\mathrm{i}\phi} \tag{43}$$

If we assume a solution of the type

$$v_Z = \frac{\mathrm{i}}{2}(v_X + \mathrm{i}v_Y)\mathrm{e}^{\mathrm{i}\phi} \tag{44}$$

then separation of variables occurs

$$\frac{\mathrm{d}}{\mathrm{d}t}(v_X + \mathrm{i}v_Y) = -\mathrm{i}\Omega(v_X + \mathrm{i}v_Y) \tag{45}$$

$$\frac{\mathrm{d}}{\mathrm{d}t}v_Z = -2\mathrm{i}\Omega v_Z\mathrm{e}^{-2\mathrm{i}\phi} \tag{46}$$

Equation (44) is a consistent solution of Eqs. (40)–(42), a solution that implies that the time average of v_Z must be zero

$$\langle v_Z \rangle_t = 0 \tag{47}$$

because the time average of the oscillating function

$$\langle e^{i\phi} \rangle_t = \langle \cos \phi \rangle_t + i \langle \sin \phi \rangle_t \tag{48}$$

is zero.[15] This is consistent with the fact that the unaveraged v_Z itself is in general nonzero and complex. Equations (42) and (46) both imply that $\langle \dot{v}_Z \rangle_t$ is also zero. Equation (44) is therefore a solution of Eqs. (40)–(42) when both $\langle v_Z \rangle_t$ and $\langle \dot{v}_Z \rangle_t$ vanish.

Note that Eq. (45) is the same as Eq. (30), derived when considering the motion of one electron in $\mathbf{B}_\Pi$, and can be solved to give the trajectory of the electron in a circularly polarized plane wave. This is, from Eq. (45), a circle:

$$X = X_0 + r \sin(\Omega t + \alpha)$$
$$Y = Y_0 + r \cos(\Omega t + \alpha) \tag{49}$$
$$Z = Z_0$$

a conclusion that is consistent with that of Landau and Lifshitz[15] from the relativistic Hamilton-Jacobi equations of the electromagnetic field, but derived in a different way by solving the Lorentz equation assuming Eqs. (38).

For an initially stationary electron, i.e., for $v_{0Z} = 0$, Eqs. (49) are identical with Eqs. (33), describing the trajectory of the electron in the novel field $\mathbf{B}_\Pi$. The trajectories (49) are consistent with the assumption (44), which implies that the time averages of v_Z and $\dot{v}_Z$ both vanish.

In summary, the trajectory of an initially stationary electron in the field $\mathbf{B}_\Pi$ is the same as that in the fields $\mathbf{E}$ and $\mathbf{B}$: An initially stationary electron moves in a circle under the influence either of $\mathbf{B}_\Pi$ or of $\mathbf{E}$ and $\mathbf{B}$. In the former, the velocity in Z vanishes identically; in the latter this component is zero on the average.

An observer noting this trajectory would not be able to define unambiguously the influence that causes the electron to move as it does in a circle, be this the wave or the field $\mathbf{B}_\Pi$. The influence upon an initially stationary electron of a circularly polarized electromagnetic plane wave is identical in all respects in the plane perpendicular to the propagation axis with the influence of $\mathbf{B}_\Pi$ upon that electron. Therefore, the motion in this plane of an electron in a circularly polarized electromagnetic plane wave can be represented exactly by a magnetostatic field B_Π, which influences an electron to move in the same trajectory. B_Π is therefore a real, physically meaningful, influence.

Furthermore, we have shown that B_Π is in units of tesla, is $\hat{T}$-negative and $\hat{P}$-positive, and is accompanied by a well-defined scalar field U_Π and a well-defined vector potential A_Π. It follows that B_Π has several characteristics of a magnetostatic field. However, B_Π is clearly not identical with an ordinary magnetostatic field, because it is a property of light. Apparently there is no E_Π, and E_Π cannot be generated from B_Π by Faraday induction. If there were an E_Π an electron's trajectory in that E_Π would be a catenary[15]; clearly, from Eqs. (33) and (49), this is not the case. An electric field E_Π cannot be generated from products such as $E \times E^*$, $B \times B^*$, $E \times B^*$, or $B \times E^*$ on the grounds of fundamental $\hat{P}$ and $\hat{T}$ symmetry.

Finally, further physical interpretation can be placed upon $E \times E^*$ by considering the Maxwell stress tensor[15]:

$$\sigma_{\alpha\beta} = -\varepsilon_0 E_\alpha E_\beta^* - \mu_0 H_\alpha H_\beta^* + \tfrac{1}{2}\delta_{\alpha\beta}\big(\varepsilon_0 E_\alpha E_\beta^* + \mu_0 H_\alpha H_\beta^*\big) \quad (50)$$

which is the momentum flux density of electromagnetic radiation, and part of its energy momentum four tensor.[15] It is clear that the antisymmetric component of $\sigma_{\alpha\beta}$ is proportional to $E \times E^*$ in vector notation, so that the field B_Π is also proportional to the antisymmetric part of $\sigma_{\alpha\beta}$. In this context the antisymmetric part of stress in mechanics is a vorticity, with the same symmetry as angular momentum, so we deduce that B_Π is proportional to the angular momentum of classical radiation, as in our operator equation $\hat{B}_\Pi = B_0 \hat{J}/\hbar$ (Ref. 2) of quantum field theory. Clearly, angular momentum takes meaning in the energy momentum four tensor of the electromagnetic field through the antisymmetric part of the Maxwell stress tensor, a vorticity, i.e., the antisymmetric part of the momentum tensor per unit volume.

V. DISCUSSION

It has been argued that the notion of B_Π is consistent with several properties of a magnetostatic field, but it must be borne in mind that B_Π is generated by a photon that travels at the speed of light. The classical theory with which we have been concerned must come to terms with the fact that B_Π is relativistic in nature. In so doing[8, 15] it becomes clear that the only relativistically invariant component of B_Π is that in the propagation axis, which is, of course, consistent with Eq. (1). Furthermore, it can be shown[8] that there is no Faraday induction by a time-modulated B_Π;

i.e., the hypothetical $\mathbf{E}_\Pi$ cannot be generated from $\partial \mathbf{B}_\Pi / \partial t$ through Faraday's law. This is a consequence of the Lorentz transformations. Furthermore, the existence of a $\mathbf{B}_\Pi$ is consistent with the fact that a valid solution of the Maxwell equations in free space is

$$\mathbf{B} = B_0(\mathbf{j} - i\mathbf{i})e^{-i\phi} + B_Z\mathbf{k} \tag{51}$$

where $B_Z k$ is in general a magnetostatic field such as $\mathbf{B}_\Pi$. It has been shown that there exists a vector potential $\mathbf{A}_\Pi$ such that $\mathbf{B}_\Pi = \nabla \times \mathbf{A}_\Pi$, and there exists a scalar field U_Π linked to $\mathbf{B}_\Pi$ by a continuity equation. Since $\mathbf{A}_\Pi$ is defined in terms of X and Y coordinates it is relativistically invariant, i.e., does not change under Lorentz transformation. This is consistent with the fact that $\mathbf{B}_\Pi$ is defined in Z, and is a magnetic flux density, and so is also invariant to Lorentz transformation.

Finally, it is straightforward to deduce that

$$\mathbf{N} = \pm c U_n \qquad \mathbf{B}_\Pi = \pm c U_\Pi \mathbf{k} \tag{52}$$

where $\mathbf{n}$ is a propagation vector whose magnitude is refractive index, and $\mathbf{k}$ is a unit axial vector. The first of these equations shows that the relation between energy and momentum in an electromagnetic field is the same as that in a particle moving at the speed of light,[15] i.e., the photon of quantum field theory. The second equation shows that there exists a similar proportionality between the magnetic flux density $\mathbf{B}_\Pi$ and U_Π.

VI. CONCLUSION

The classical theory of fields has been used to develop and interpret the concepts $\mathbf{E} \times \mathbf{E}^*$ and $\mathbf{B}_\Pi$, and it has been shown that these concepts are self-consistent and physically meaningful. For example, the motion of an electron in $\mathbf{B}_\Pi$ is the same as that in $\mathbf{E}$ and $\mathbf{B}$ of a plane wave. The latter can therefore be thought of as generating a magnetostatic field.

Acknowledgments

The Materials Research Laboratory of Penn State University and the Cornell Theory Center are thanked for research support. Faramarz Farahi and Yasin Raja are thanked for many interesting communications.

References

1. M. W. Evans, *Physica B*, **182**, 118 (1992).
2. M. W. Evans, ibid., **182**, 227 (1992).
3. M. W. Evans, ibid., **183**, 103 (1993).

4. M. W. Evans, ibid., **179**, 342 (1992).

5. M. W. Evans, ibid., **182**, 237 (1992).

6. M. W. Evans, ibid., **179**, 6 (1992).

7. M. W. Evans, in press, (1993).

8. M. W. Evans, **179**, 157 (1992).

9. P. W. Atkins, *Molecular Quantum Mechanics*, 2d ed., Oxford University Press, 1983.

10. R. Tanaś and S. Kielich, *J. Mod. Opt.* **37**, 1935 (1990).

11. W. S. Warren, S. Mayr, D. Goswami, and A. P. West, Jr., *Science* **255** 1683 (1992).

12. L. D. Barron, *Molecular Light Scattering and Optical Activity*, Cambridge University Press, Cambridge, UK, 1982.

13. S. Kielich, in M. Davies (Senior Reporter), *Dielectric and Related Molecular Processes*, Vol. 1, Chem. Soc., London, 1972.

14. S. Kielich, *Nonlinear Molecular Optics*, Nauka, Moscow, 1981.

15. L. D. Landau and E. M. Lifshitz, *The Classical Theory of Fields*, 4th ed., Pergamon, Oxford, UK, 1975.

THE ELEMENTARY STATIC MAGNETIC FIELD*
OF THE PHOTON

I. INTRODUCTION

It is well known that the photon has an intrinsic, unremovable spin, which can be expressed as its quantized angular momentum operator $\hat{J}$ (Refs. 1–4). This is the essential explanation in quantum-field theory for the existence of classical left and right circular polarization in electromagnetic plane waves. In this paper it is argued that the photon also generates an intrinsic and unremovable static magnetic field (flux density in tesla) which can be described through the operator equation

$$\hat{B}_\Pi = B_0 \frac{\hat{J}}{\hbar} \tag{1}$$

where B_0 is the scalar magnetic flux density amplitude of a beam of N photons (for example, a circularly polarized laser beam). The expectation value of the component of the operator $\hat{B}_\Pi$ in the propagation axis of the laser is $B_0 M_J$, where M_J is the azimuthal equantum number associated with the operator $\hat{J}$. Classically, this expectation value is $\pm \mathbf{B}_\Pi \cdot \mathbf{k}$, where $\mathbf{k}$

---

*Printed by permission, based on Physica B, **182**, 227 (1992).

is a unit axial vector in the propagation axis, axis Z of the laboratory frame of reference (X, Y, Z). The derivation of the fundamental operator equation (1) is given in Section II, followed in Section III by a suggestion for a key experiment to test the theory, which consists of reflecting at right angles a circularly polarized laser beam from a beam of electrons, and of measuring the frequency shift in the reflected laser due to the interaction

$$\Delta H_\Pi = -\hat{m} \cdot \hat{B}_\Pi \tag{2}$$

between $\hat{B}_\Pi$ and the electron's magnetic dipole moment operator $\hat{m}$. Section IV develops some consequences in spectroscopy of the existence of $\hat{B}_\Pi$, specifically a quantum field theory of the optical Zeeman effect, splitting due to $\hat{B}_\Pi$ in spectra at visible frequencies, and optical NMR and ESR,[5] in which $\hat{B}_\Pi$ shifts and splits conventional resonance features in liquids and condensed matter. Finally, a discussion is given of some other immediately interesting consequencies of the existence of $\hat{B}_\Pi$, for example, an optical Stern-Gerlach effect.

II. DERIVATION OF THE OPERATOR EQUATION FOR $\hat{B}_\Pi$

It is seen immediately that the operator equation (1) can be derived on the basis of symmetry and dimensions alone, and in this section the rigorous quantum field theoretical derivation is given using the recent results of Tanaś and Kielich[6] and of the present author.[7-10] Before embarking on this it is instructive to note the role of fundamental symmetries, namely the motion reversal operator $\hat{T}$, and the parity inversion operator $\hat{P}$. The operators $\hat{B}_\Pi$ and $\hat{J}$ have the same $\hat{P}$ and $\hat{T}$ symmetries, respectively, positive and negative, so one is proportional to the other through a T- and P-positive scalar quantity. Furthermore, the unit of the operator $\hat{J}$ in quantum mechanics is the reduced Planck constant $\hbar$, and therefore the scalar proportionality constant must be a scalar magnetic flux density amplitude, B_0, in tesla. In a laser beam of N photons, the constant is the laser's scalar flux density amplitude in tesla. When $N = 1$ (one photon), B_0 remains finite, and it follows that the single photon generates a quantum of magnetostatic flux density, described in quantum field theory by $\hat{B}_\Pi$ of Eq. (1).

To derive this result rigorously it is convenient to consider first the classical equivalent of $\hat{B}_\Pi$, which is a vector quantity in the propagation axis of the laser:

$$\mathbf{B}_\Pi = B_0 \mathbf{k} \tag{3}$$

114 M. W. EVANS

where **k** is a unit axial vector. The classical $\mathbf{B}_\Pi$ is proportional[7-10] to the conjugate product[11-15]

$$\mathbf{\Pi}^{(A)} = \mathbf{E} \times \mathbf{E}^* \tag{4}$$

a vector cross product of the electric field strength **E** of a circularly polarized laser with its complex conjugate **E***:

$$\mathbf{E} = E_0(\mathbf{i} + i\mathbf{j})\exp(i\phi)$$
$$\mathbf{E}^* = E_0(\mathbf{i} - i\mathbf{j})\exp(-i\phi) \tag{5}$$

Here, as usual, E_0 is the scalar electric field strength amplitude in volts per meter of the laser, **i** and **j** are unit polar vectors in X and Y of the laboratory frame (X, Y, Z), mutually orthogonal to the propagation axis Z, and ϕ is the phase. From Eq. (5)

$$\mathbf{\Pi}^{(A)} = -2E_0^2\mathbf{ki} \tag{6}$$

where **k** is a unit axial vector in Z. The product $\mathbf{\Pi}^{(A)}$ is an axial vector which is also negative to T and positive to P. Equation (6) can be rewritten using the fundamental in vacuo relation

$$E_0 = cB_0 \tag{7}$$

as

$$\mathbf{\Pi}^{(A)} = -2E_0c(B_0\mathbf{k})\mathbf{i} \equiv -2E_0c\mathbf{B}_\Pi\mathbf{i} \tag{8}$$

with the definition

$$\mathbf{B}_\Pi \equiv B_0\mathbf{k} \tag{9}$$

where c is the (scalar) speed of light.

Clearly, $\mathbf{\Pi}^{(A)}$ and $\mathbf{B}_\Pi$ must have the same T and P symmetries, and so does the unit axial vector **k**. The classical quantity $\mathbf{B}_\Pi$ has been defined[7-10] as equivalent magnetostatic flux density in vacuo of a circularly polarized plane wave. It has no dependence on the phase ϕ of the wave, and therefore none on the angular frequency ω and propagation vector **k**. Using the relation between intensity (watts per square meter) and electric field strength

$$I_0 = \tfrac{1}{2}\varepsilon_0 cE_0^2 \tag{10}$$

where ε_0 is the permittivity in vacuo,[1]

$$\varepsilon_0 = 8.854 \times 10^{-12} C^2 J^{-1} m^{-1} \tag{11}$$

we arrive at

$$\mathbf{B}_{\Pi} = i\left(\frac{\varepsilon_0}{8I_0 c}\right)^{1/2} (\mathbf{E} \times \mathbf{E}^*) \tag{12}$$

which shows clearly that $\mathbf{B}_{\Pi}$ is directly proportional to the conjugate product $\mathbf{E} \times \mathbf{E}^*$, and that $\mathbf{B}_{\Pi}$ is a real quantity. The classical third Stokes parameter is the real scalar

$$S_3 = -i(E_X E_Y^* - E_Y E_X^*) \tag{13}$$

so that

$$\mathbf{B}_{\Pi} = -\left(\frac{\varepsilon_0}{8I_0 c}\right)^{1/2} S_3 \mathbf{k} \tag{14}$$

Equation (14) defines $\mathbf{B}_{\Pi}$ in (8). Equation (10) defines $\mathbf{B}_{\Pi}$ in terms of S_3, and shows that the former is a real axial vector that changes sign with the laser circular polarization (left to right). The transition from classical to quantized field theory is made through the third Stokes operator $\hat{S}_3$, recently introduced by Tanaś and Kielich[6]:

$$\hat{S}_3 \equiv -\left[\frac{2_\pi \hbar \omega}{n^2(\omega)V}\right] i\left(\hat{a}_X^+ \hat{a}_Y - \hat{a}_Y^+ \hat{a}_X\right) \tag{15}$$

Here $n(\omega)$ is the refractive index, V is the quantization volume, and $\hat{a}^+$ and $\hat{a}$ denote, respectively, the creation and annihilation operator. Defining a coherent state of a laser beam of N photons by the Schrödinger equation[6]

$$\hat{a}|\alpha\rangle = \alpha|\alpha\rangle \tag{16}$$

provides the expectation value

$$\langle \alpha|\hat{S}_3|\alpha\rangle = |\alpha_+|^2 - |\alpha_-|^2 \tag{17}$$

operator equation we seek to prove:

$$\hat{B}_\Pi = B_0\left(\frac{\hat{J}}{\hbar}\right) \tag{33}$$

III. A KEY EXPERIMENT FOR $\hat{B}_\Pi$

The theoretical existence of $\hat{B}_\Pi$ implies many different things experimentally, because a circularly polarized laser acts as a simple magnet, and delivers equivalent (or "latent" or "potential") static magnetic flux density through a vacuum, flux density which is able to form a scalar interaction Hamiltonian with a dipole moment operator $\hat{m}$:

$$\Delta H = -\hat{m} \cdot \hat{B}_\Pi$$

The electron carries an elementary $\hat{m}$:

$$\hat{m} = g_e \gamma_e \hat{I} \tag{34}$$

where g_e is the electron's g factor[1] (2.002), and $\hat{I}$ is the electron's angular momentum operator. The key experiment devised in this section isolates the effect of the Hamiltonian (2) on a circularly polarized visible frequency laser reflected at right angles from an electron beam. There is a frequency shift

$$\Delta f = \frac{\langle \hat{m} \cdot \hat{B}_\Pi \rangle}{h} \tag{35}$$

in the reflected beam, which provides a method of measuring $\hat{B}_\Pi$ spectrally with a high-resolution spectrometer. The derivation of Eq. (35) is based on conservation of momentum and kinetic energy when a circularly polarized laser beam of N photons is reflected at right angles from the electron beam. Consider the collision of one photon of the beam with one electron, whereby the former is reflected by an angle θ and the latter by an angle θ'. Initially, the electron is at rest with relativistic energy $m_e c^2$, where m_e is its mass.[1] After the collision, the electron's linear momentum magnitude is p and it translational kinetic energy is $(p^2 c^2 + m_e^2 c^4)^{1/2}$. The initial linear momentum of the photon is h/λ_i where λ_i is its wavelength, and its initial energy is hc/λ_i. The photon strikes the electron, considered stationary,[1] is deflected through an angle θ, and emerges with linear momentum h/λ_f and translational kinetic energy hc/λ_f. The electron after collision moves off at an angle θ' to the incident photon's trajectory. Conserving linear momentum and translational kinetic energy gives three equations

$$p \cos \theta' + \frac{h}{\lambda_f} \cos \theta = \frac{h}{\lambda_i} \tag{36}$$

$$p \sin \theta' + \frac{h}{\lambda_f} \sin \theta = 0 \tag{37}$$

$$m_e c^2 + \frac{hc}{\lambda_i} = \left(p^2 c^2 + m_e^2 c^4 \right)^{1/2} + \frac{hc}{\lambda_f} \tag{38}$$

which can be solved simultaneously to give the standard Compton equation for the wavelength shift

$$(\lambda_f - \lambda_i) = \frac{2h}{m_e c} \sin^2 \frac{\theta}{2} + \frac{h}{2m_e c} \frac{\lambda_f}{\lambda_i} \cos^2 \theta \tag{39}$$

At $\theta = 90°$

$$\lambda_f - \lambda_i = \frac{h}{m_e c}, \tag{40}$$

a result that has no classical counterpart.[1] The wavelength shift is in the X-ray region of the spectrum.

The theory of Compton's effect, embodied in Eqs. (36)–(38), takes no account of the interaction energy

$$\Delta E_\Pi = -\langle \hat{m} \cdot \hat{B}_\Pi \rangle \tag{41}$$

In consequence, Eq. (38) for the conservation of kinetic energy must be modified to

$$m_e c^2 + \frac{hc}{\lambda_i} = \left(p^2 c^2 + m_e^2 c^4 \right)^{1/2} + \frac{hc}{\lambda_f} + \Delta E_\Pi \tag{42}$$

i.e., ΔE_Π contributes to the total energy after collision. Also the theory

(36) to (38) takes no account of the intrinsic angular momenta of either photon or electron, and there must also be conservation of rotational kinetic energy and angular momentum. However, equations (36), (37), and (42) suffice to solve for $\lambda_f - \lambda_i$ in the presence of ΔE_Π. For an electromagnetic beam of N photons (a circularly polarized laser) reflected at 90° off the electron beam, solving (36), (37), and (42) gives

$$\lambda_f - \lambda_i = \frac{h/m_e c + \lambda_i \lambda_f (\Delta E_\Pi/hc)(1 - \Delta E_\Pi/2m_e c^2)}{1 - \Delta E_\Pi/m_e c^2} \tag{43}$$

We consider the order of magnitude of ΔE_Π compared with $m_e c^2$. The magnitude of the observable associated with the operator $\hat{m}$ is the Bohr magneton multiplied by the electron's g factor (2.002), and is about 10^{-23} J T^{-1}. The magnitude of the observable with the elementary photon operator $\hat{B}_\Pi$ is[7]

$$|\mathbf{B}_\Pi| \sim 10^{-7} I_0^{1/2} \tag{44}$$

and for I_0 of 1.0 W cm^{-2} (10 000 W m^{-2}) is of the order 10^{-5} T. Therefore ΔE_Π is of the order 10^{-28} J for this intensity. However, $m_e c^2$ is of the order 10^{-13} J per electron, so that to an excellent approximation

$$\frac{\Delta E_\Pi}{m_e c^2} \ll 1 \tag{45}$$

and Eq. (43) reduces to

$$\lambda_f - \lambda_i = \frac{h}{m_e c} + \lambda_i \lambda_f \frac{\Delta E_\Pi}{hc} \tag{46}$$

which is consistent with Eq. (40) for $\Delta E_\Pi = 0$. It is useful to express Eq. (46) in terms of wave numbers:

$$\bar{\nu}_f = \frac{m_e c \bar{\nu}_i}{m_e c + h\bar{\nu}_i} - \frac{m_e \Delta E_\Pi}{m_e hc + h^2 \bar{\nu}_i} \tag{47}$$

We now compare the order of magnitude of $m_e hc$ (about 10^{-55} kg J m) with $h^2 \bar{\nu}_i$, which is about $4.4 \times 10^{-67} \bar{\nu}_i$ kg J m; and find that for $\bar{\nu}_i \leq 10^9$

m^{-1} (10^7 cm^{-1})

$$\bar{\nu}_f - \bar{\nu}_i = -\frac{\Delta E_\Pi}{hc} \tag{48}$$

to a very good approximation. In terms of frequency in hertz

$$\Delta f = -\frac{\Delta E_\Pi}{h} = \frac{\langle \hat{m} \cdot \hat{B}_\Pi \rangle}{h} \tag{49}$$

which is Eq. (35).

This equation shows that at electromagnetic frequencies well below 10^7 cm^{-1} (wave numbers), for example, at visible frequencies, the change in frequency in hertz in a circularly polarized laser reflected at right angles off an electron beam is given by Eq. (49). This is based on the interaction energy $\langle \hat{m} \cdot \hat{B}_\Pi \rangle$ between $\hat{m}$ of the electron and $\hat{B}_\Pi$ of the photon, two elementary properties of quantized matter. In general, the frequency shift is proportional to the square root of the incident laser intensity I_0. The interaction energy is quantized according to the quantum theory[19-21] of operator products, and can be expressed as the expectation value

$$\Delta E_\Pi = -\big\langle IJFM_F | \hat{m} \cdot \hat{B}_\Pi | I'J'F'M_F' \big\rangle$$
$$\hat{m} = -2.002\gamma_e \hat{I} \equiv -g_e\gamma_e \hat{I} \tag{50}$$

where I is the angular momentum quantum number of the electron ($I = \frac{1}{2}$), and J is the angular momentum quantum number of the photon (a positive integral quantity greater than zero[1]). Here γ_e is the gyromagnetic ratio and g_e the electronic g factor.[1] The quantum number F is given by the Clebsch-Gordan series

$$F = J + I, \ldots, |J - I| \tag{51}$$

and the expectation value of the Z component of the resultant angular momentum operator $\hat{F}$ is given by M_{F^h}, with the selection rule

$$\Delta M_F = \pm 1 \tag{52}$$

and M_F having $(2F + 1)$ values from $-F$ to F as usual. Therefore, depending on the value of J, there are several different values possible of the frequency shift Δf; i.e., an analysis of the reflected laser beam will reveal their presence as a spectrum. The diagonal matrix elements of (50) can be worked out analytically,[19-21] giving the frequency shift

$$\Delta f = -\frac{10^{-7}I_0^{1/2}g_e g \gamma_e}{(2\pi)}$$

$$g = [3(2F + 1)I(I + 1)(2I + 1)J(J + 1)(2J + 1)]^{1/2} \quad (53)$$

$$\times \begin{bmatrix} I & I & 1 \\ J & J & 1 \\ F & F & 0 \end{bmatrix}$$

Here the quantity in braces is the well-known 9-j symbol. For $I = \frac{1}{2}$, $J = 1$, and $F = \frac{3}{2}$, this is -0.05, and for an incident circularly polarized laser intensity of 1.0 W cm^{-2} the frequency shift in hertz from Eq. (19) is $\sim 20\,000 M_F$ Hz. For $I = \frac{1}{2}$, $J = 1$, $F = \frac{1}{2}$, the shift is $\sim -10\,000 M_F$ Hz. For modest incident laser intensity the shift is already in the kilohertz range, easily measurable, and has the following characteristics:

1. It should change sign with respect to the incident laser frequency if the incident laser's circular polarity is switched from left to right (i.e., if the azimuthal expectation value of the photon's static magnetic flux density operator is changed from $B_0|M_J|$ to $-B_0|M_J|$).

2. There should be no frequency shift or spectral detail if the incident laser is linearly polarized, because the net static flux density delivered by the photon beam is zero, being 50% $B_0|M_J|$ and 50% $-B_0|M_J|$.

3. The shifts with respect to the incident laser frequency should be proportional to the square root of the laser's incident intensity I_0.

These features should provide adequately for the measurement of the novel elementary property $\hat{B}_\Pi$, and the method can be extended to other elementary particles by using, for example, a neutron beam in place of the laser beam. This would allow an experimental determination of whether neutrons also have elementary magnetic flux density similar to $\hat{B}_\Pi$ of the photon (the elementary "magneton" of electromagnetic radiation).

IV. DISCUSSION

In addition to the frequency shift phenomenon introduced in this paper, it is possible to predict novel phenomena due to $\hat{B}_\Pi$ wherever the photon interacts with matter, one of these being the optical Zeeman effect and another the optical Faraday effect. Both effects have been suggested in a semiclassical context recently[22-26] and their existence is reinforced by that of $\hat{B}_\Pi$ on a fundamental level. In the optical Zeeman effect the magneton $\hat{B}_\Pi$ plays the role taken by a static magnetic flux density in the conventional Zeeman effect and its relatives, the anomalous Zeeman effect, and Paschen-Back effect. In the optical Faraday effect the magneton rotates the plane of polarization of a linearly polarized problem. Clearly, the interaction energy in these magneton-based effects must be constructed from the quantum theory of operator products, as in Eq. (53), and off-diagonal components of the matrix elements so obtained are important in general, as well as diagonal elements. In molecules, these off-diagonal elements must probably be worked out numerically, but for this purpose there are many standard ab initio packages available.

Clearly, the magneton $\hat{B}_\Pi$ is capable also of generating optically induced resonance spectra (optical NMR and ESR), and evidence for this has been obtained recently,[27] although a full theoretical description is not yet available, and must probably be generated ab initio, using software packages such as HONDO or GAUSSIAN 90 by taking into account the interaction of the magneton $\hat{B}_\Pi$ with the large and complicated chiral test molecule used by Warren et al.[27] to show interesting, site-specific effects of a low-power, circularly polarized laser on a conventional one- and two-dimensional NMR spectrum. This technique appears to have considerable promise, especially if the laser intensity could be increased by pulsing. In principle, considerable increase in resolution of conventional resonance spectra is obtainable.[7-10, 27]

Finally, the optical equivalent of the Stern-Gerlach experiment is possible in principle by using an expanding or focused laser beam to generate the optical equivalent of a magnetic field gradient in the axis of propagation of a circularly polarized laser beam coaxial with a beam of atoms, such as silver atoms. The magneton theory of this will be the subject of future work.

APPENDIX

The neutron has a magnetic moment and quantum number $I = \frac{1}{2}$, but is approximately 10 000 times heavier than the electron. The theory of spectral detail in a circularly polarized laser beam reflected off a beam of neutrons can be set up in the same way as for an electron beam, but the expected splitting in the reflected laser beam is much smaller and much more difficult to detect with a spectrometer. However, the presence of such detail would be further evidence for the existence of the magneton $\hat{B}_\Pi$. It is also possible to replace the electron beam by a beam of atoms with net electronic dipole moment, for example, and for any material with net electronic or nuclear dipole moment the interaction with the magneton produces spectral detail in the reflected laser beam, which can be analyzed spectroscopically. The experiment is not confined, furthermore, to beams, but can also proceed, in principle, by reflecting the circularly polarized laser beam from a material of interest with a net magnetic dipole moment. In this context, the behavior of superconducting surfaces is particularly interesting[28] and the magneton $\hat{B}_\Pi$ could well provide an entirely novel way of analyzing type I and II superconductors by reflecting a laser beam from the surface of the material at right angles, and looking for the specific magnetic effects due to $\hat{B}_\Pi$. Type II superconductors are of particular interest[28] because they remain superconducting in the presence of magnetic flux density, which is known to propagate in such material in the form of quantized flux lines, each carrying one quantum[28] of magnetic flux. In this case it might be expected that the magneton $\hat{B}_\Pi$ would be converted in the type II superconductor to the quantum of magnetic flux hc/m_e. Furthermore, Bitter imaging techniques[28] can be utilized in superconductors to detect the presence of magnetization due to the magneton $\hat{B}_\Pi$ of the circularly polarized laser, which can also be used to scan the surface of the sample and induce individual vortices of magnetization in the superconducting sample.

More generally and fundamentally it is interesting to speculate on the possibility that elementary particles with spin are also capable of generating magnetons of flux akin to $\hat{B}_\Pi$ of the photon. The latter is massless and travels at c in vacuo, whereas the neutron, for example, has mass and does not travel at c. The electron also has mass and does not travel at c. Nevertheless, the electron and neutron both have intrinsic, irremovable spin, essentially in the same way as the photon, and in terms of symmetry and dimensionality, both electron and neutron can generate magnetons of flux through equations identical in structure to Eq. (1) of the text. However, neither electron nor neutron are electromagnetic plane waves, but different types of wave, and the question comes down to whether a beam of electrons or neutrons carries a finite, scalar flux density amplitude akin to $\hat{B}_\Pi$ of the photon. It is known that the neutron, for example, obeys the Planck relation between energy and frequency, but there appears to be no evidence that the Maxwell equations can be written for neutrons or electrons, and solved to generate plane waves akin to electromagnetic waves. It appears at present that the electron and neutron generate elementary magnetic dipole moments and that the photon generates the elementary magnetic field $\hat{B}_\Pi$.

These speculations can be extended to other elementary particles with intrinsic spin (i.e., angular momentum operators) and experiments can be devised to test the speculations. For example, if the electron does indeed generate its own magneton, a quantized magnetic flux density operator,

$\hat{B}_e$, a beam of electrons reflected off a beam of neutrons will generate the interaction Hamiltonian

$$\Delta H_1 = -\hat{m}_n \cdot \hat{B}_e \qquad (A.1)$$

where $\hat{m}_n$ is the magnetic dipole moment of the neutron. This is quantized as in Eq. (53), and consequently the energy of the emerging electron beam must record in some way the presence of ΔH_1. If the electron beam has wave properties, it should be analyzable spectrally, and the spectral pattern due to the interaction ΔH_1 should be measurable experimentally. Electron diffraction is evidence that electrons can behave as waves as well as particles, which is a result of the de Broglie principle. Proceeding with the speculative logic in this way, it becomes clear that reflecting a beam of any particle with intrinsic elementary spin from any other particle beam with intrinsic elementary magnetic dipole moment could, in principle, result in an interaction energy of type (A.1). In other words, we speculate on the possibility that elementary particles in general can each generate its own magneton.

References

1. P. W. Atkins, *Molecular Quantum Mechanics*, 2d ed., Oxford University Press, Oxford, UK, 1983.

2. P. A. M. Dirac, *The Principles of Quantum Mechanics*, 4th ed., Oxford University Press, Oxford, UK, 1982.

3. J. M. Jauch and F. Rohrlich, *The Theory of Photons and Electrons*, Addison Wesley, London, 1959.

4. M. D. Levenson, R. M. Shelby, and D. F. Walls, *Phys. Rev. Lett.* **57** 2473 (1986).

5. M. W. Evans, *J. Phys. Chem.* **95** 2256 (1991).

6. R. Tanaś and S. Kielich, *J. Mod. Opt.* **37**, 1935 (1990).

7. M. W. Evans, in M. W. Evans and S. Kielich (Eds.), *Modern Non-linear Optics*, a special issue in three volumes of *Advances in Chemical Physics*, I. Prigogine and S. A. Rice (Series Eds.), Wiley Interscience, New York, 1993, in preparation, collection of forty-five review articles.

8. M. W. Evans, *J. Mol. Spectrosc.* **146**, 351 (1991).

9. M. W. Evans, in I. Prigogine and S. A. Rice (Eds.), *Advances in Chemical Physics*, Vol. 81, Wiley Interscience, New York, 1992.

10. M. W. Evans, *Physica B* **168**, 9 (1991).

11. N. L. Manakov, V. D. Ovsiannikov, and S. Kielich, *Acta Phys. Pol.* **A53**, 581, 595 (1978).

12. S. Kielich, in M. Davies (Senior Rep.), *Dielectric and Related Molecular Processes*, Vol. 1, Chem. Soc., London, 1972.

13. K. Knast and S. Kielich, *Acta Phys. Pol.* **A55**, 319 (1979).

14. S. Kielich, *Proc. Phys. Soc.* **86**, 709 (1965).

15. R. Tanaś and S. Kielich, *Quantum Opt.* **2**, 23 (1990).

16. M. W. Evans, *Opt. Lett.* **15**, 863 (1990).

17. W. Heitler, *The Quantum Theory of Radiation*, 3d ed., Oxford University Press, Oxford, UK, 1953.

18. M. Born and E. Wolf, *The Principles of Optics*, 6th ed., Pergamon, Oxford, UK, 1975.

19. B. L. Silver, *Irreducible Tensor Methods*, Academic, New York, 1976.

20. R. Zare, *Angular Momentum*, Wiley, New York, 1988.

21. L. D. Barron, *Molecular Light Scattering and Optical Activity*, Cambridge University Press, Cambridge, UK, 1982.

22. S. Woźniak, M. W. Evans, and G. Wagnière, *Mol. Phys.* **75**, 81 (1992).

23. S. Woźniak, M. W. Evans, and G. Wagnière, *Mol. Phys.* **75**, 99 (1992).

24. M. W. Evans, *Int. J. Mod. Phys. B* **5**, 1963 (1991) (review).

25. M. W. Evans, *Chem. Phys.*, **157**, 1 (1991).

26. M. W. Evans, S. Woźniak, and G. Wagnière, *Physica B* **173**, 357 (1991); **175**, 416 (1991).

27. W. S. Warren, S. Mayr, D. Goswami, and A. P. West, Jr., *Science*, **255**, 1683 (1992).

28. D. J. Bishop, P. L. Gammel, D. A. Huse, and C. A. Murray, *Science*, 10th January, 1992, pp. 165 ff.

THE PHOTON'S MAGNETOSTATIC FLUX QUANTUM: SYMMETRY AND WAVE PARTICLE DUALITY—FUNDAMENTAL CONSEQUENCES IN PHYSICAL OPTICS

I. INTRODUCTION

It has recently been demonstrated theoretically that there exists an operator $\hat{B}_\Pi$ of the quantized electromagnetic field that describes the photon's magnetostatic flux density:

$$\hat{B}_\Pi = B_0 \frac{\hat{J}}{\hbar} \tag{1}$$

Here B_0 has been interpreted[2-5] as a scalar magnetic flux density amplitude of a beam of circularly polarized light consisting of one photon, and $\hat{J}$ is the boson operator[6,7] describing that photon's quantized angular momentum. The classical equivalent of $\hat{B}_\Pi$ is a novel axial vector $\mathbf{B}_\Pi$, which is directed in the propagation axis of the beam. In this paper it is demonstrated using elementary tensor algebra, and from inspection of the Maxwell equations of the classical field, that there is another possible interpretation of the scalar amplitude B_0, designated henceforth by $(B_0)_+$, where the $+$ subscript is to be interpreted as "positive to parity inversion." It turns out that B_0 can be interpreted both as a scalar and as a pseudoscalar quantity, designated $(B_0)_-$, where the minus subscript means "negative to parity inversion." This is designated "symmetry duality," and is shown in this work to imply that $\hat{B}_\Pi$ can be defined simultaneously in terms of the photon's angular momentum operator $\hat{J}$ and linear momentum operator $\hat{p}$, a result that is a generalization of a keystone of wave mechanics, the de Broglie wave particle duality.[6] The latter is linked through $\hat{B}_\Pi$ to a symmetry duality in Maxwell's classical equations.

It has already been shown theoretically[2-5] and experimentally[8,9] that circularly polarized light can magnetize, leading, for example, to the inverse Faraday effect[10-13] and novel, potentially very useful, light-induced shifts in NMR spectroscopy[8,9] in one and more dimensions. The existence of the operator $\hat{B}_\Pi$ and its classical equivalent $\mathbf{B}_\Pi$ makes it much easier to interpret these magnetization effects by treating circularly polarized light as a "magnet" generating this novel flux quantum per photon. The $\hat{B}_\Pi$ concept also makes it relatively straightforward to forecast the existence of novel spectral phenomena, such as optical Zeeman, anomalous Zeeman, and Paschen-Back effects,[3] an optical Faraday effect and optically induced magnetic circular dichroism,[4] and optical Stern-Gerlach effect, using a focused laser beam to produce a light-induced magnetic field gradient, optical ESR effects, optically induced effects in interacting beams, such as a beam of circularly polarized photons reflected[5] from a beam of polarized electrons, and so on. All these effects can be thought of as arising from the replacement (or augmentation) of an ordinary magnet by or with a circularly polarized laser. These theories allow scope for the development of several novel analytically useful methods.

In this paper it is shown that $\mathbf{B}_\Pi$ is related directly to the ubiquitous,[14] pseudoscalar, third Stokes parameter S_3 of the classical electromagnetic plane wave, which becomes in quantum-field theory the third Stokes operator of Tanaś and Kielich.[1] Therefore, it follows immediately that several well-known phenomena of physical optics can be reinterpreted fundamentally in terms of the operator $\hat{B}_\Pi$, or its classical equivalent $\mathbf{B}_\Pi$. Examples include ellipticity in the plane wave, ellipticity developed in the measuring beam of the electrical Kerr effect, and circular dichroism, which are shown in this work to be magneto-optic phenomena. Therefore, not only does $\hat{B}_\Pi$ allow this reinterpretation, in both classical and quantum field theory, it also allows a link to be made between de Broglie wave particle duality and symmetry duality in the classical Maxwell equations. It appears, therefore, to go to the root of physical optics and field theory.

In Section II we develop the mathematical basis of symmetry duality with elementary vector and tensor algebra, before embarking in Section

III on a discussion of symmetry duality in the link between $\mathbf{B}_\Pi$ and S_3. In Section IV we develop the link between wave particle duality and the symmetry duality in Maxwell's equations demonstrated in Section III, and discuss qualitatively the implications for elementary particle theory. In Section V we develop the link between $\mathbf{B}_\Pi$ and S_3 into a novel explanation for ellipticity and circular dichroism in physical optics.

II. SYMMETRY DUALITY IN THE VECTOR PRODUCT OF TWO POLAR VECTORS

It is well known that the components of a vector that can be written as the cross product of two polar vectors do not change sign under parity inversion ($\hat{P}$) and that the vector so formed is an axial vector,[15] or pseudovector. The conjugate product of the classical electromagnetic field[2-5]

$$\mathbf{\Pi}^{(A)} = \mathbf{E} \times \mathbf{E}^* = 2\left(E_0^2\right)_+ i\mathbf{e}_+ \tag{2}$$

where $\mathbf{E}^*$ is the polar complex conjugate of the polar electric field strength vector $\mathbf{E}$, is an axial vector, therefore. Here, $\mathbf{e}_+$ is an axial unit vector, positive to $\hat{P}$, and the quantity $(E_0^2)_+$ is a scalar, also positive to $\hat{P}$. The overall motion reversal ($\hat{T}$) symmetry of $\mathbf{\Pi}^{(A)}$ is negative, and it is natural to define $\mathbf{e}_+$ as a $\hat{T}$-negative unit vector, so that $(E_0^2)_+$ is a $\hat{T}$ positive scalar.

It appears at first sight that these definitions are both necessary and sufficient for the complete definition of the axial vector $\mathbf{\Pi}^{(A)}$; but mathematically, there is an alternative, which is revealed through writing any arbitrary axial vector as

$$\mathbf{C} = C_+\mathbf{e}_+ = C_-\mathbf{e}_- \tag{3}$$

where C_+ and $\mathbf{e}_+$ are respectively $\hat{P}$-positive scalar and unit axial vector quantities, and where C_- and $\mathbf{e}_-$ are respectively $\hat{P}$-negative pseudoscalar and $\hat{P}$-negative polar unit vector quantities. The overall $\hat{P}$ symmetry of the complete axial vector $\mathbf{C}$ is positive in both cases.

This seemingly mundane observation in elementary vector analysis has far-reaching consequences in the theory of the classical and quantized electromagnetic fields. In tensor algebra, the general vector cross product $\mathbf{C} = \mathbf{A} \times \mathbf{B}$ is written with the third rank antisymmetric (or alternating) unit tensor, $\varepsilon_{\alpha\beta\gamma}$, known as the Levi-Civita symbol[7, 15]:

$$C_\alpha = \tfrac{1}{2}\varepsilon_{\alpha\beta\gamma}\left(A_\beta B_\gamma - A_\gamma B_\beta\right) \equiv \tfrac{1}{2}\varepsilon_{\alpha\beta\gamma}C_{\beta\gamma} \qquad (4)$$

where the $\hat{P}$ symmetry of $\varepsilon_{\alpha\beta\gamma}$ is negative, so that $C_{\beta\gamma}$ is a $\hat{P}$-negative antisymmetric polar tensor of rank two. Evidently, C_α must be $\hat{P}$-positive, and is the rank one axial tensor (i.e., an axial vector). However, $C_{\beta\gamma}$ can also be written[15] as

$$C_{\beta\gamma} = -i\varepsilon_{\beta\gamma}C_- \qquad (5)$$

where $\varepsilon_{\beta\gamma}$ is the $\hat{P}$-positive, axial, unit antisymmetric tensor of rank two, and C_- is the pseudoscalar of Eq. (3). Equation (5) shows that the polar antisymmetric tensor of rank two can be reduced, quite generally, to a pseudoscalar, a particular result of a generalization in the relativistic theory of the classical electromagnetic field.[15] Note that $C_{\beta\gamma}$ is purely imaginary from the Hermitian properties of the general second-rank tensor, which can always be written as a sum of real symmetric and imaginary antisymmetric parts.[15]

Therefore,

$$C_\alpha = -\frac{i}{2}\varepsilon_{\alpha\beta\gamma}\varepsilon_{\beta\gamma}C_- \qquad (6)$$

or

$$C_\alpha \equiv -iC_+\varepsilon_{\alpha+} = -\frac{i}{2}\varepsilon_{\alpha\beta\gamma}\varepsilon_{\beta\gamma}C_- \qquad (7)$$

where $\varepsilon_{\alpha+}$ is the rank one axial unit tensor, positive to $\hat{P}$, and C_+ is a $\hat{P}$-positive scalar. Recall that C_- is a $\hat{P}$-negative pseudoscalar. Equations (3) and (7), using vector and tensor notation, respectively, are expressions of symmetry duality, a purely mathematical result that shows that a scalar and pseudoscalar may both be used to define an axial vector. Clearly, if we take the magnitude ($|\mathbf{C}|$) of the axial vector $\mathbf{C}$ in Eq. (3), we obtain

$$C^2 \equiv \mathbf{C} \cdot \mathbf{C} = C_+^2 = C_-^2$$
$$|\mathbf{C}| = \left|(C^2)^{1/2}\right| = |C_+| = |C_-| \qquad (8)$$

so that the positive parts of the scalar C_+ and pseudoscalar C_- are equal in absolute magnitude. This same result can be obtained from the tensor

Equation (7) by taking a particular Z component:

$$C_Z = -iC_+\varepsilon_{Z+} = -\frac{i}{2}C_-(\varepsilon_{ZXY}\varepsilon_{XY} + \varepsilon_{ZYX}\varepsilon_{YX}) \qquad (9)$$

where the Einstein convention of summation over repeated indices has been used on the right side. With the component definitions $\varepsilon_{ZXY} = 1$, $\varepsilon_{XY} = 1$, $\varepsilon_{ZYX} = -1$, and $\varepsilon_{YX} = -1$, we obtain

$$C_+\varepsilon_{Z+} = C_-\varepsilon_{Z-} \qquad (10)$$

where

$$\varepsilon_{Z-} \equiv \varepsilon_{ZXY}\varepsilon_{XY} + \varepsilon_{ZYX}\varepsilon_{YX} \qquad (11)$$

is the Z component of the $\hat{P}$-negative polar unit tensor of rank one, $\varepsilon_{\alpha-}$. Note that Eqs. (3) and (10) are identical in symmetry character for the considered Z components of C. Equation (10), which is a direct and fundamental consequence of elementary tensor algebra, again shows the symmetry duality between scalar and pseudoscalar in the definition of the axial, or pseudo, vector. It is now possible to apply the purely mathematical principle of symmetry duality to the classical, nonrelativistic (or relativistic) field to obtain novel information of fundamental importance in physical optics, particularly in respect of a $\hat{P}$-positive, $\hat{T}$-negative axial vector, a novel magnetostatic field, $\mathbf{B}_{\Pi}$,[2-5] associated with the electromagnetic plane wave or in the quantized field, the magnetostatic flux density operator $\hat{B}_{\Pi}$ of the photon.

III. AN EXAMPLE OF SYMMETRY DUALITY: THE RELATION BETWEEN $\mathbf{B}_{\Pi}$ AND THE STOKES PARAMETER S_3

Consider the classical electromagnetic wave in free space, so that the real scalar refractive index is unity. It follows from Maxwell's equations for a plane wave that

$$E_0 = cB_0 \qquad (12)$$

where E_0 and B_0 are $\hat{P}$- and $\hat{T}$-positive scalars, amplitudes, respectively, of the electric field strength and magnetic flux density. The intensity of the wave is defined in free space by

$$I_0 = \varepsilon_0 cE_0^2 \qquad (13)$$

where ε_0 is the free space permittivity[6] in S.I. units, and c is the speed of light in vacuo. With Eqs. (12) and (13), Eq. (2) can be rewritten as

$$\Pi^{(A)} = 2(E_0)_+ c i \mathbf{B}_\Pi \tag{14}$$

where we have defined the magnetostatic flux density vector $\mathbf{B}_\Pi^{2-5}$ of the classical electromagnetic plane wave in free space:

$$\mathbf{B}_\Pi \equiv (B_0)_+ \mathbf{e}_+ \tag{15}$$

where $\mathbf{e}_+$ is a $\hat{P}$-positive unit axial vector. The overall $\hat{T}$ symmetry of $\mathbf{B}_\Pi$ is negative, and the overall $\hat{P}$ symmetry is positive. In the introduction we have given an account of the role of $\mathbf{B}_\Pi$ in the reinterpretation of well-known effects, such as circular dichroism and ellipticity, and its mediating role in new effects such as optical NMR and ESR,[9] optical Faraday[4] and Zeeman[3] effects, optical Stern-Gerlach effects, optical Compton scattering,[5] and so on.[2-5] Its quantized equivalent is the magnetostatic flux density operator of Eq. (1), in which $(B_0)_+$ is defined as the $\hat{P}$- and $\hat{T}$-positive scalar magnetic flux density amplitude of one photon.

Again, as in Section II, it would appear at first sight as if the definition of the seemingly mundane quantity B_0 as a $\hat{P}$- and $\hat{T}$-positive scalar is sufficient. Remarkably, however, this is not the case, there is an alternative definition possible of the novel classical vector $\mathbf{B}_\Pi$ which uses B_0 as a pseudoscalar. Not only does this emerge naturally from the Maxwell equations for the plane wave, it also provides a natural link between $\mathbf{B}_\Pi$ and the third Stokes parameter S_3.[1, 7, 14, 15]

These conclusions emerge straightforwardly from the equations linking the E and B vectors of the classical electromagnetic plane wave in a medium of refractive index n, defined through the classical wave vector κ, a $\hat{T}$- and $\hat{P}$-negative polar vector directed in the propagation axis Z of the plane wave[7]:

$$\kappa = \frac{\omega}{c}\mathbf{n} \qquad n = \frac{c}{v} \tag{16}$$

Here ω is the angular frequency in radians per second of the plane wave, as usual. Maxwell's equations give[7]

$$\mathbf{B} = \frac{1}{c}\mathbf{n} \times \mathbf{E} \qquad \mathbf{E} = -\frac{c}{n^2}\mathbf{n} \times \mathbf{B} \tag{17}$$

In free space, the positive absolute magnitude of the $\hat{P}$- and $\hat{T}$-positive scalar n is unity. Using Equation (17) yields the conjugate product

$$\mathbf{E} \times \mathbf{E}^* = -\frac{c}{n^2}\mathbf{E} \times (\mathbf{n} \times \mathbf{B}^*) = -\frac{\mathbf{n}}{n}\frac{c(\mathbf{E} \cdot \mathbf{B}^*)}{n} \tag{18}$$

We note that the vector $\mathbf{n}$ is a $\hat{P}$-negative, $\hat{T}$-negative polar vector, defined as usual[7] as a propagation vector whose scalar magnitude is equal to the $\hat{T}$-positive scalar refractive index n; the dot product $\mathbf{E} \cdot \mathbf{B}^*$ is a $\hat{T}$-positive pseudoscalar. Equation (18) reduces to

$$\underset{\text{Scalar}}{\frac{E_0 B_0 c}{n}} + \underset{\substack{\text{Axial} \\ \text{unit} \\ \text{vector}}}{\mathbf{e}_+} = \underset{\text{Pseudoscalar}}{\frac{E_0 B_0 c}{n}} - \underset{\substack{\text{Polar} \\ \text{unit} \\ \text{vector}}}{\frac{\mathbf{n}}{|\mathbf{n}|}} \tag{19}$$

in which we have designated the various symmetries. It follows algebraically that

$$(B_0)_+\mathbf{e}_+ = (B_0)_-\frac{\mathbf{n}}{n} \tag{20}$$

which can be rewritten in the notation of Section II as an example of symmetry duality in the Maxwell equations:

$$(B_0)_+\mathbf{e}_+ = (B_0)_-\mathbf{e}_- \qquad \mathbf{e}_- \equiv \frac{\mathbf{n}}{n} \tag{21}$$

This shows that classical vector $\mathbf{B}_\Pi$ can be defined simultaneously in terms of the unit axial vector $\mathbf{e}_+$ and the unit polar vector $\mathbf{e}_-$, which is related to the propagation vector $\boldsymbol{\kappa}$, the photon linear momentum. In free space, with $n = 1$,

$$\mathbf{B}_\Pi = (B_0)_+\mathbf{e}_+ = (B_0)_-\mathbf{e}_- = (B_0)_-\frac{c}{\omega}\boldsymbol{\kappa} \tag{22}$$

demonstrating a duality between the classical angular and linear momentum of the plane wave. We shall see that this is none other than the classical equivalent of the de Broglie wave particle duality for the photon in the quantized field. Before making the transition to the quantized field, however, another fundamentally new result emerges when we consider the definition[15] of the Stokes parameter S_3:

$$E_\alpha E_\beta^* - E_\beta E_\alpha^* = -i\varepsilon_{\alpha\beta}(S_3)_- \tag{23}$$

so that

$$(S_3)_- \equiv (E_0^2)_- \tag{24}$$

is a pseudoscalar quantity, implying inter alia the symmetry duality

$$\Pi^{(A)} = 2(E_0^2)_+ \, ie_+ = 2(S_3)_- \, ie_- \tag{25}$$

It follows directly that the magnetostatic vector $\mathbf{B}_\Pi$ can be defined in free space ($n = 1$) in terms of $(S_3)_-$ as follows:

$$\mathbf{B}_\Pi = \frac{(S_3)_-}{2E_0 c}\mathbf{e}_- \equiv (B_0)_- \, \mathbf{e}_- \tag{26}$$

and we find that the role of B_0 as pseudoscalar is none other than the Stokes parameter S_3 scaled by an appropriate $\hat{P}$- and $\hat{T}$-positive scalar quantity. Thus, $\mathbf{B}_\Pi$ can be defined in free space through the symmetry duality

$$\mathbf{B}_\Pi = (B_0)_+ \, \mathbf{e}_+ = \frac{(S_3)_-}{2E_0 c}\mathbf{n} = \frac{(S_3)_-}{2E_0 c}\frac{c}{\omega}\boldsymbol{\kappa} \tag{27}$$

where the unit polar vector $\mathbf{n}$ can be identified with the unit vector $\mathbf{e}_-$ of this section. We thus forge a novel and fundamental link between the pseudoscalar magnitude of $\mathbf{B}_\Pi$ and the pseudoscalar S_3.

VI. SYMMETRY DUALITY AND WAVE PARTICLE DUALITY FOR THE PHOTON

Equation (1) shows that the photon's novel magnetic field operator $\hat{B}_\Pi$ is directly proportional to its well-defined[6] angular momentum boson operator $\hat{J}$ through B_0 in its scalar representation $(B_0)_+$, interpreted as the magnetic flux density amplitude of a single photon. The latter is a massless lepton that propagates at the speed of light and is not localized in space[16] unlike a massive lepton such as the electron or proton. These well-known properties are contained in Eq. (1), in that B_0 varies with intensity I_0 for a beam of circularly polarized light containing one photon, and therefore B_0 for one photon depends on the beam cross section, a finite area. The eigenvalues of the operator $\hat{J}$ are known to be $M_J\hbar$, $M_J = \pm 1$; there is no $M_J = 0$ component from relativistic considerations.[6,7] Therefore, the eigenvalues of $\hat{B}_\Pi$ are $\pm(B_0)_+$, where $(B_0)_+$ is a scalar, the positive eigenvalue corresponds to one particular circular polarization, and vice versa,[7] as in the convention for the operator $\hat{J}$.

We now use the result[6,7] that the eigenvalue of the linear momentum operator $\hat{p}$ of the photon is

$$\mathbf{p} \equiv \langle \hat{p} \rangle = \hbar\boldsymbol{\kappa} \tag{28}$$

where $\boldsymbol{\kappa}$ is the wave vector as defined classically in the preceding section. It follows straightforwardly from Eqs. (22) and (28) that in free space $(n = 1)$

$$\hat{B}_\Pi = (B_0)_+ \frac{\hat{J}}{\hbar} = \frac{(B_0)_-}{n} \frac{c}{\omega} \frac{\hat{p}}{\hbar} \tag{29}$$

which expresses the duality of Eq. (22) in terms of quantum field theory, and shows that the $\hat{B}_\Pi$ operator of the photon is simultaneously proportional to both its angular and linear momentum operators. Equation (29) summarizes a duality in symmetry, linear/angular momentum, and wave–particle character with the results

$$\hat{T}[(B_0)_+] = +\hat{T}[(B_0)_-]$$
$$\hat{P}[(B_0)_+] = -\hat{P}[(B_0)_-] \tag{30}$$

Equation (29) implies the free space relation

$$\hat{p} = n\frac{\omega}{c}\hat{J} \qquad n = 1 \tag{31}$$

The expectation value of $\hat{p}$ is therefore given by the expectation value of $\hat{J}$, which is $\pm\hbar$. Taking without loss of generality the positive eigenvalue $\hbar$, we have, with $n = 1$,

$$p = \frac{\omega}{c}\hbar \tag{32}$$

which is the de Broglie wave particle duality for the photon.

We have therefore succeeded in relating directly the de Broglie wave-particle duality of quantum mechanics to the novel symmetry duality (22) of classical electromagnetic field theory. It has also been shown that the novel flux quantum $\hat{B}_{\Pi}$ is definable simultaneously in terms of $\hat{J}$ and $\hat{p}$, one operator being directly proportional to the other, implying that both must be quantized in the same way. In a sense, therefore, $\hat{B}_{\Pi}$ is the keystone of de Broglie's concept of duality for the photon.

Furthermore, contemporary elementary particle theory argues that the photon is a chiral entity, a massless lepton that travels in any frame of reference at c, and whose chirality, in consequence,[17] is well defined in terms of the eigenvalues of Dirac's $\hat{\gamma}_5$ operator. The chirality of a lepton with mass (i.e., a massive lepton) such as the electron is not well defined, leading to the idea[17] that mass itself is ill-defined chirality. Well-defined chirality in the photon can be thought of as a consequence of superimposed linear and angular momentum, and Eq. (29) shows that there is a duality between these two fundamental quantities. It appears therefore that the novel $\hat{B}_{\Pi}$ operator of the photon is a true chiral influence as defined by Barron,[17] and is therefore fundamentally different in nature from a magnetostatic flux density, such as a magnetic field generated in an electromagnet. The latter is now known to be an example of a false chiral influence,[17] and cannot, for example, be a cause of enantioselective synthesis. This is in contrast to the circularly polarized electromagnetic field, which le Bel[18] in 1874 conjectured to be a truly chiral influence, and which is now known to influence enantioselectivity in chemical reactions. The definition of the $\hat{B}_{\Pi}$ operator in Eq. (29) also allows insight to the symmetry of natural optical activity, i.e., circular dichroism and optical rotatory dispersion, as developed in the next section.

It may be conjectured that a magnetostatic flux quantum $\hat{B}_{\Pi}$ is always carried by a massless lepton whose chirality can be precisely defined as the eigenvalues of the Dirac $\hat{\gamma}_5$ operator; and, conversely, that the massive lepton does not support $\hat{B}_{\Pi}$ and does not have precisely defined eigenvalues of $\hat{\gamma}_5$. This conjecture would imply that fundamentally, $\hat{B}_{\Pi}$ is always a consequence of the absence of mass. It would therefore follow that the neutrino (and antineutrino) carries a $\hat{B}_{\Pi}$ field, but that the electron, neutron, and proton do not. However, it is not clear whether the neutrino has a classical counterpart such as the classical electromagnetic plane wave, the counterpart of the photon. If the parallel between photon and neutrino can be carried further, it would appear that the neutrino must also be thought of as unlocalized in space. This would imply inter alia that localization in space implies the presence of mass and the absence of well-defined chirality (or well-defined eigenvalues of $\hat{\gamma}_5$), and that the absence of mass implies the absence of space localization. Carrying the argument further, wave particle duality in a massive lepton such as the electron has been observed, because an electron beam can be diffracted, for example, but since the electron is localized and does not have well-defined chirality, its wave nature must be fundamentally different from that of the photon, and in consequence, no $\hat{B}_\Pi$ can be constructed or defined for the electron. Wave particle duality in the electron is therefore fundamentally different in nature from duality in the photon. The electron has a magnetic dipole moment that is proportional to the electron's spin angular momentum operator through the gyromagnetic ratio. We therefore conjecture that a massless lepton cannot support a magnetic dipole moment, because its effective gyromagnetic ratio would be infinite, but can support a magnetostatic flux quantum. The opposite is true for a massive lepton. With these assumptions, the $\hat{B}_\Pi$ operator of a massless lepton would always be able to form an interaction Hamiltonian operator to first order with the magnetic dipole moment operator of a massive lepton, an example being a photon beam interacting with an electron beam,[5] or a neutrino beam with a neutron beam and so on, giving rise to measurable effects in principle. The inference overall, therefore, is that a beam of massless leptons, for example, photons or neutrinos, can magnetize but cannot be magnetized, whereas a beam of massive leptons cannot magnetize but can be magnetized.

The charge conjugation symmetry operator can be defined as $\hat{C}$ (which operates to reverse the sign of charge), and with this definition we recall the fundamental Luders-Pauli-Villiers theorem[17]:

$$\hat{C}\hat{P}\hat{T} = \hat{1} \tag{33}$$

The violation of $\hat{P}$ has been observed[17] in a number of different ways, whereas the violation of $\hat{T}$ has been observed in only one critical experiment.[17] The violation of $\hat{P}$ leads to the result that the space-inverted enantiomers of a truly chiral entity such as the photon or neutrino are not degenerate, or exactly the same in energy, because of the existence of the $\hat{P}$ violating electroweak force.[17] In contrast, the space-inverted "enantiomers" of a falsely or pseudo chiral entity, such as an ordinary magnetic field, are precisely the same in energy.[17] Thus, it is important to note that the true enantiomer of the photon, or neutrino, is not generated by space inversion or by application of the $\hat{P}$ operator, i.e., by reversing the linear momentum and keeping the angular momentum the same. Assuming that the photon is uncharged, so that $\hat{C}$ has no effect, its true or exact enantiomer must be generated by simultaneous $\hat{P}$ and $\hat{T}$ violation in order to conserve the validity of the Luders-Pauli-Villiers theorem. (33). The true enantiomer of the left-handed photon is presumably, therefore, an object that must be designated the right-handed "antiphoton," and there is a very small, but nonzero, energy difference between the left-handed photon and the right-handed photon. If the right-handed photon is to be regarded as having a different energy from the left-handed photon, then either $\hat{P}$ has been violated and $\hat{T}$ and $\hat{C}$ have been conserved, or $\hat{T}$ has been violated and $\hat{P}$ and $\hat{C}$ have been conserved. Assuming that $\hat{C}$ has no effect on the photon, because it is uncharged, the combined operation $\hat{P}\hat{T}$ must be used to generate the right antiphoton from its true enantiomer, the left photon, and vice versa. The photon is an object whose chirality is generated only as a result of its simultaneous translational and rotational motion, and the novel $\hat{B}_{\Pi}$ operator is a fundamental manifestation of this chirality. The latter is conserved, furthermore, in the photon and antiphoton, because both travel at the speed of light and both are massless. It is also known[17] that neutrinos conserve chirality, in that only left-handed neutrinos and right-handed antineutrinos exist. The $\hat{P}$ violating weak force is known to play a critical part in the interaction of left-handed neutrinos with left, but not with right, spin-polarized relativistic electrons, and of right-handed antineutrinos with right, but not left, polarized relativistic electrons.[17]

These arguments lead to the interesting possibility that a beam of, say, left photons, each carrying a flux quantum $\hat{B}_{\Pi}$, may interact differently with a beam of left and right polarized relativistic electrons, each carrying the magnetic dipole moment $\hat{m}$, through the interaction Hamiltonian operator

$$\Delta \hat{H} = -\hat{m} \cdot \hat{B}_{\Pi}$$

This difference may be picked up by observation of the Zeeman splitting caused by $\Delta \hat{H}$ in, for example, a circularly polarized visible laser beam reflected from a polarized, relativistic, electron beam. Such an experiment has been proposed recently to evaluate the effect of $\hat{B}_{\Pi}$ (Ref. 5).

V. THE ROLE OF B_{Π} IN ELLIPTICITY AND ASSOCIATED EFFECTS IN PHYSICAL OPTICS, FOR EXAMPLE, CIRCULAR DICHROISM

The link between $|\mathbf{B}_{\Pi}|$ and the third Stokes parameter $(S_3)_-$ can be expressed through the intensity I_0 as

$$|\mathbf{B}_{\Pi}| = (B_0)_- = \left(\frac{\varepsilon_0}{4I_0 c} \right)^{1/2} (S_3)_- \tag{34}$$

so that it follows that whenever $(S_3)_-$ occurs in physical optics, it can be replaced by the pseudoscalar quantity B_0, multiplied by the scalar $(4I_0 c/\varepsilon_0)^{1/2}$. This is a key link between the photon's magnetostatic field operator $\hat{B}_\Pi$, in its classical limit, and the ubiquitous $(S_3)_-$, revealing immediately the root cause of several well-known phenomena in physical optics.

As an example, the ellipticity η of an elliptically polarized beam of light is related to $(S_3)_-$ by

$$(S_3)_- = 2E_0^2 \sin 2\eta \tag{35}$$

with

$$\eta = \tan^{-1}\frac{b}{a} \tag{36}$$

where a and b are respectively the major and minor axes of the polarization ellipse.[7] This shows that there is a direct link between ellipticity and the vector $\mathbf{B}_\Pi$, which is the classical equivalent of the operator $\hat{B}_\Pi$. In the theory of the electrically induced Kerr effect,[7] for example, ellipticity is developed in an initially circularly polarized measuring beam after it has passed through a material to which a static, uniform, electric field has been applied perpendicular to the propagation direction of the beam and at 45° to the aximuth of an incident linearly polarized beam. For the emerging beam in the electric Kerr effect it can be shown that

$$|\mathbf{B}_\Pi| = (B_0)_- = \left(\frac{I_0}{\varepsilon_0 c^3}\right)^{1/2} \sin(2\eta) \tag{37}$$

showing that the root cause of ellipticity in the Kerr effect is the pseudoscalar magnitude $(B_0)_-$ of $\mathbf{B}_\Pi$. Note that for the incident, linearly polarized beam, $\mathbf{B}_\Pi$ is zero, but that in the transmitted, elliptically polarized beam it is nonzero.

Another example of the fundamental role of the pseudoscalar $(B_0)_-$ is the phenomenon of circular dichroism, which is a manifestation of optical activity, whereby the intensity of initially linearly polarized electromagnetic radiation transmitted by a structurally chiral material contains an excess of left over right circularly polarized components, or vice versa. In this context[7]

$$\frac{(S_3)_-}{(S_0)_+} = \frac{I_L - I_R}{I_L + I_R} \tag{38}$$

where $(S_0)_+$ is the zeroth Stokes parameter, a scalar quantity defined by

$$(S_0)_+ = 2E_0^2 \qquad (39)$$

Therefore, the root cause of circular dichroism is the pseudoscalar magnitude $(B_0)_-$ of the vector $\mathbf{B}_{\Pi}$:

$$|\mathbf{B}_{\Pi}| = (B_0)_- = \left(\frac{1}{\varepsilon_0 c^3 I_0}\right)^{1/2} (I_R - I_L) \qquad (40)$$

an equation that is valid at all electromagnetic frequencies.

The origin of circular dichroism therefore resides in the photon's magnetostatic flux quantum $\hat{B}_{\Pi}$. In other words, circular dichroism is magneto-optic in origin, and the observable $(I_R - I_L)$ is a spectral consequence of the interaction of $\hat{B}_{\Pi}$ with structurally chiral material. From Eq. (40), $I_R - I_L$ is proportional to the real pseudoscalar quantity $(B_0)_-$ after each photon of the beam emerges from the chiral material through which the beam has passed, i.e., after interaction has occurred between the incident flux quantum $\hat{B}_{\Pi}$ per photon and the appropriate molecular property tensor of the material.[7] This leads to a new way of describing the fundamental mechanism of natural optical activity by considering the mechanism of interaction of $\hat{B}_{\Pi}$ with a structurally chiral molecule, or center of optical activity. A quantum $\hat{B}_{\Pi}$ per photon is evidently absorbed and reemitted with different characteristics imparted by the chiral structure.

For a beam consisting of one photon, the observable $I_R - I_L$ provides an experimental measure of the transmitted elementary $\hat{B}_{\Pi}$ at each frequency of that beam. Although $\mathbf{B}_{\Pi}$ is itself independent of the phase of the beam, the interacting molecular property tensor depends on the beam frequency through semiclassical perturbation theory,[7] which gives

$$|\mathbf{B}_{\Pi}| = (B_0)_- = \left(\frac{I_0}{\varepsilon_0 c^3}\right)^{1/2} \tanh\left[\omega\mu_0 c l N \zeta''_{XYZ}(g)\right] \qquad (41)$$

where μ_0 is the permeability in vacuo, ω the angular frequency of the beam, l the length of sample through which the beam has passed, and ζ''_{XYZ} an appropriately averaged molecular property tensor component, a pseudoscalar.[7] Equation (41) shows that all circularly dichroic spectra are signatures of the reemitted $\hat{B}_{\Pi}$ property of the photon.

More generally, any property in physical optics that involves $(S_3)_-$, in classical or quantized[1] form, necessarily involves $\mathbf{B}_{\Pi}$ or the quantized $\hat{B}_{\Pi}$

per photon. There are several of these phenomena, each of whose origin can be traced to the novel elementary flux quantum $\hat{B}_\Pi$ of the photon. Rayleigh refringent scattering theory, for example,[7] shows that $(S_3)_-$ is associated with a change $d\eta/dz$ in ellipticity in a beam passing through a sample of thickness z. It is immediately possible to say, therefore, that $d\eta/dz$ measures changes in the flux quantum $\hat{B}_\Pi$ per photon as it passes through the sample, i.e., as $\hat{B}_\Pi$ is absorbed and reemitted, a process from Eq. (29) that must involve changes in the incident photon's angular and linear momentum. Ellipticity is therefore magneto-optic in fundamental origin.

VI. DISCUSSION

One of the interesting consequences of the development in the preceding sections is that the speed of light c must be regarded as a $\hat{T}$-positive scalar quantity. This is because c is a universal constant that is relativistically the same in any frame of reference, and cannot be reversed by the motion reversal operator $\hat{T}$, because c is independent of motion. However, a velocity v that is less than c is $\hat{T}$-negative, because it is reversed by motion reversal in a given reference frame. In consequence, the scalar refractive index n, defined by c/v in a material, must be a $\hat{T}$-positive quantity. The value of n in vacuo is numerically unity and is the mathematical limit of c/v as $v \to c$. It is proper to regard n as being $\hat{T}$-positive in this limit, and this is the point of view utilized in this paper.

It follows that the unit polar vector $\mathbf{n}/n$ must be $\hat{T}$-negative, because it is the quotient of $\hat{T}$-negative/$\hat{T}$-positive quantities. In Eq. (20), for example, $\hat{T}[(B_0)_-] = +$, $\hat{T}(\mathbf{e}_+) = -$, $\hat{T}[(B_0)_+] = +$, $\hat{T}(\mathbf{n}/n) = -$, so that there is a balance of net $\hat{T}$ symmetries on either side of the equation.

To interpret rigorously the generalization, Eq. (29), of the de Broglie equation (32) in its "textbook" form, it must be borne in mind that the de Broglie duality rigorously implies the symmetry duality summarized in Eq. (30) for $\hat{P}$. Equation (31), therefore, is more rigorously expressed as

$$\hat{P} = n\frac{(B_0)_+}{(B_0)_-}\frac{\omega}{c}\hat{j} \qquad n = 1 \qquad (42)$$

and Eq. (32) as

$$p = n\frac{(B_0)_+}{(B_0)_-}\frac{\omega}{c}\hbar \qquad n = 1 \qquad (43)$$

The quotient $(B_0)_+/(B_0)_-$, and the $\hat{T}$-positive free space refractive index ($n = 1$) are missing or implied in the usual textbook definition of the de Broglie wave particle duality.

In conclusion, it has been demonstrated that there is an inherent symmetry duality in the definition of the magnetostatic flux quantum $\hat{B}_\Pi$, which is the root of the de Broglie wave particle duality for the photon. The operator $\hat{B}_\Pi$ can be defined simultaneously in terms of the angular and linear momentum operators of the photon. This type of symmetry duality occurs throughout physical optics, and is inherent in the fact that $\hat{B}_\Pi$, or its classical equivalent B_Π, is at the root of several well-known effects, such as circular dichroism and ellipsometry of various kinds. The operator $\hat{B}_\Pi$ can also be used straightforwardly to predict and describe novel and useful spectroscopic effects that depend on magnetization by circularly polarized light.

Acknowledgments

The Leverhulme Trust is thanked for a Fellowship, and the Cornell Theory Center and the Materials Research Laboratory of Penn State is thanked for additional support.

References

1. R. Tanaś and S. Kielich, *J. Mod. Opt.*, **37**, 1935 (1990).

2. M. W. Evans, *Physica B*, **182**, 227 (1992).

3. M. W. Evans, *Physica B*, **182**, 237 (1992).

4. M. W. Evans, *Physica B*, **183**, 103 (1993).

5. M. W. Evans, *Physica B*, in press.

6. P. W. Atkins, *Molecular Quantum Mechanics*, 2d ed., Oxford University Press, Oxford, UK, 1983.

7. L. D. Barron, *Light Scattering and Optical Activity*, Cambridge University Press, Cambridge, UK, 1982.

8. M. W. Evans, *J. Phys. Chem.* **95**, 2256 (1991).

9. W. S. Warren, S. Mayr, D. Goswami, and A. P. West, Jr., *Science* **255**, 1683 (1992).

10. P. S. Pershan, *Phys. Rev. A* **130**, 919 (1963).

11. J. P. van der Ziel, P. S. Pershan, and L. D. Malmstrom, *Phys. Rev. A* **15**, 190 (1965).

12. P. S. Pershan, J. P. van der Ziel, and L. D. Malmstrom, *Phys. Rev. A* **143**, 574 (1966).

13. Y. R. Shen, *Phys. Rev. A* **134**, A661 (1964).

14. M. Born and E. Wolf, *Principles of Optics*, 6th ed., Pergamon, Oxford, UK, 1975.

15. L. D. Landau and E. M. Lifshitz, *The Classical Theory of Fields*, 4th ed., Pergamon, Oxford, UK, 1975.

16. B. W. Shore, *The Theory of Coherent Atomic Excitation*, Vols. 1 and 2, Wiley, New York, 1990.

17. L. D. Barron, *Chem. Soc. Rev.* **15**, 189, (1986).

18. S. F. Mason, *Bio Systems* **20**, 27 (1987).

EXPERIMENTAL DETECTION OF THE PHOTON'S FUNDAMENTAL STATIC MAGNETIC FIELD OPERATOR: THE ANOMALOUS OPTICAL ZEEMAN AND OPTICAL PASCHEN-BACK EFFECTS

I. INTRODUCTION

In this paper we continue a systematic theoretical search for a method of detecting and measuring unequivocally the photon's fundamental static magnetic field operator[1-5]

$$\hat{B}_{\Pi} = B_0 \frac{\hat{J}_{\Pi}}{\hbar} \tag{1}$$

where B_0 is the scalar magnetic flux density amplitude in tesla of a circularly polarized laser beam, made up of N photons, $\hat{J}_{\Pi}$ is the photon's angular momentum operator, and $\hbar$ is the reduced Planck constant $h/2\pi$.

Fragmentary experimental evidence for the existence of $\hat{B}_{\Pi}$ is available through the inverse Faraday effect (IFE),[6-9] and the recent emergence of optical NMR (ONMR), or laser-enhanced nuclear magnetic resonance spectroscopy (LENS).[10, 11] Both techniques measure the ability of circularly polarized laser radiation to magnetize. The IFE measures bulk magnetization and ONMR measures light-induced resonance shifts, which are different for each resonating site and are therefore useful for sample identification and spectral analysis. However, the theoretical existence of $\hat{B}_{\Pi}$ also implies other effects, such as an optical Zeeman effect, in which the magnetic effect of a circularly polarized laser splits electric dipole transitions in atoms occurring in the visible frequency range. This paper provides a fairly rigorous quantum theory of the anomalous optical Zeeman effect and the optical Paschen-Back effect, in which both spin and orbital electronic angular momenta are considered in various coupling schemes.

The existence of an optical Zeeman effect in atoms appears to have been implicit in the theory of the inverse Faraday effect proposed by Pershan et al.,[6-9] Kielich et al.,[12] and Atkins and Miller.[13] The present author independently arrived at the existence of an optical Zeeman effect in a series of papers[14-17] based on symmetry considerations and semiclassical theory, considerations that also led to ONMR.[18-21] Recently, he has proposed theoretically the existence of the photon's $\hat{B}_{\Pi}$ operator, a fundamental property of the photon itself, whose classical equivalent is a static magnetic field, $\hat{B}_{\Pi}$, produced by circularly polarized electromagnetic radiation at all frequencies.[1-4] It is important to realize that the operator $\hat{B}_\Pi$ (or the classical $\hat{B}_\Pi$) is different fundamentally from the usual **B** vector of electromagnetic plane waves.[22] The **B** vector is frequency dependent, whereas the **B** vector is not. The **B** vector has components in X and Y directions mutually perpendicular to the propagation axis Z of the laser, and no component in the Z axis, whereas $\mathbf{B}_\Pi$ is directed in the Z axis only. The **B** vector depends on the photon's linear momentum vector κ (i.e., the propagation vector), whereas $\mathbf{B}_\Pi$ does not. Again, $\mathbf{B}_\Pi$ is positive to the parity inversion operator $\hat{P}$ and negative to the motion reversal operator $\hat{T}$, and is therefore fundamentally different in symmetry from the Poynting vector[23] and the propagation vector.[24] It appears that $\mathbf{B}_\Pi$ (and its quantum field equivalent $\hat{B}_\Pi$) is a fundamentally new concept in electromagnetic field theory.

The optical Zeeman effect appears to be a promising method of detecting the effects of $\hat{B}_\Pi$ experimentally. Electric dipole transitions in atoms are readily measured and identified spectrally.[24] The key to the optical Zeeman effect is to replace the magnet of the conventional Zeeman effect[24] by a circularly polarized laser. In the ordinary Zeeman effect, where there is no consideration given to the role of net electronic spin angular momentum,[22] a singlet $^1S \rightarrow {}^1P$ transition is split by a magnet into three lines. Its optical equivalent has recently been proposed theoretically[5] using the concept summarized in Eq. (1), and produces a splitting of the original $^1S \rightarrow {}^1P$ electric dipole transition, a splitting pattern whose details are different in quantum-field theory (in which the operator $\hat{B}_\Pi$ forms a Hamiltonian with the electronic magnetic dipole moment operator $\hat{m}$ of the atom) and in semiclassical theory (in which the Hamiltonian is formed from a product of $\hat{m}$ and the classical vector $\mathbf{B}_\Pi$). The pattern also depends on the type of angular momentum interaction used in the quantum-field theory, i.e., whether a coupled or uncoupled representation of $\hat{B}_\Pi$ and $\hat{m}$ is used.[22] The semiclassical result is recovered[5] only in the uncoupled representation. In the coupled representation, the quantum field theory produces three lines, but the central line is displaced in frequency. In the uncoupled representation of the quantum field theory of the optical Zeeman effect, and in the semiclassical representation of the same problem, the splitting pattern obtained[5] is the same as that in the conventional Zeeman effect, i.e., two lines each side of a central line situated at the original frequency of the electric dipole transition $^1S \rightarrow {}^1P$.

In Section II, these findings are augmented by the consideration of electronic spin angular momentum in quantum field and semiclassical approaches using in the former different coupling models for the angular momenta involved in the interaction Hamiltonians. In Section III, the details are given of the spectral splitting due to the circularly polarized laser for each theoretical approach of Section II. Finally, some experimental details are discussed and estimates of the splittings in hertz are given for each theoretical approach, namely the quantum field theory in the fully coupled, semicoupled, and uncoupled representations, and the semiclassical approach where $\mathbf{B}_\Pi$ is considered as a classical field vector. Conditions are discussed under which the anomalous optical Zeeman effect gives way to the optical Paschen-Back effect in the quantum field and semiclassical representations of the same phenomena.

II. QUANTUM FIELD AND SEMICLASSICAL INTERACTION HAMILTONIANS

The core of the description of the anomalous optical Zeeman effect and the optical Paschen-Back effect in atoms is the construction of the interaction Hamiltonians between the novel photon property $\hat{B}_\Pi$ (quantum field theory) or $\mathbf{B}_\Pi$ (semiclassical theory) and the atom's net electronic magnetic dipole moment operator $\hat{m}$. In this section, first-order interaction Hamiltonians are constructed in the framework of quantum field and semiclassical descriptions of the same problem.

A. Quantum Field Theory

The interaction Hamiltonian is the operator product

$$\Delta \hat{H}_1 = -\hat{m} \cdot \hat{B}_\Pi \tag{2}$$

from which the energy of interaction is calculated from an expectation value such as

$$\Delta E_1 = -\frac{\gamma_e B_0}{\hbar} \langle SLJJ_\Pi FM_F | \hat{L} + 2.002\hat{S} | S'L'J'J'_\Pi F'M'_F \rangle \tag{3}$$

In this expression, the magnetic dipole moment of the atom is developed as

$$\hat{m} = \gamma_e (\hat{L} + 2.002\hat{S}) \tag{4}$$

where γ_e is the electronic gyromagnetic ratio, $\hat{L}$ is the operator describing the net electronic orbital angular momentum, and $2.002\hat{S}$ is the operator description of the net electronic spin angular momentum.[22] The $\hat{B}_\Pi$ operator is developed in terms of the photon's angular momentum operator in Eq. (1). The angular momentum quantum numbers S, L, and J_Π are associated with the operators $\hat{S}$, $\hat{L}$, and $\hat{J}_\Pi$, respectively. The interaction energy (3) is one in which a coupled representation[22] is considered for the three angular momenta just introduced. In this case the coupling scheme is

$$\mathbf{J} = \mathbf{L} + \mathbf{S} \qquad \mathbf{F} = \mathbf{J} + \mathbf{J}_\Pi \qquad (5)$$

so that the values of the quantum number J, associated with the operator $\hat{J}$ are given as usual by the Clebsch-Gordan series

$$J = L + S, \ldots, |L - S| \qquad (6)$$

Similarly, the overall quantum number F is defined by

$$F = J + J_\Pi, \ldots, |J - J_\Pi| \qquad (7)$$

i.e., from a coupled representation of the $\hat{J}$ operator of the atom and the novel[1-5] $\hat{J}_\Pi$ operator of the photon whose effects we are attempting to describe.

There are other ways of writing the interaction energy for the given Hamiltonian (2), these being the semicoupled and uncoupled representations. In the former, the operators $\hat{L}$ and $2.002\hat{S}$ of the atom are combined in a coupled angular momentum representation[22] to give the $\hat{J}$ operator, but the interaction energy is formed as follows:

$$\Delta E_2 = -\frac{\gamma_e B_0}{\hbar} \langle SLJM_J; J_\Pi M_{J_\Pi} | \hat{L} + 2.002\hat{S} | S'L'J'M'_J; J'_\Pi M'_{J_\Pi} \rangle \quad (8)$$

i.e., with $\hat{J}$ and $\hat{J}_\Pi$ considered in an uncoupled representation, in which the projections $\hat{M}_J$ and $\hat{M}_{J_\Pi}$ onto the azimuthal axis (Z, the laser's propagation axis) are well defined, but in which the net angular momentum operator $\hat{F}$ is not. In the latter, the interaction energy is written as

$$\Delta E_3 = -\frac{\gamma_e B_0}{\hbar} \langle SM_S; LM_L; J_\Pi M_{J_\Pi} | \hat{L} + 2.002\hat{S} | S'M'_S; L'M'_L; J'_\Pi M'_{J_\Pi} \rangle$$
$$(9)$$

in which all three angular momentum operators—$\hat{L}$ and $2.002\hat{S}$ of the atom, and $\hat{J}_\Pi$ of the photon—are considered in a fully uncoupled representation. All three representations are possible theoretically, and which is the most appropriate can be determined only by independent consideration of the physics of the problem.

1. The Coupled Representation

The first stage is the usual one. The Wigner-Eckart theorem is used to separate out the M_F dependence[22, 25]:

$$\Delta E_1 = -(-1)^{F-M_F}\begin{pmatrix} F & 0 & F \\ -M_F & 0 & M_F \end{pmatrix}\langle SLJJ_\Pi F \| \hat{m} \cdot \hat{B}_\Pi \| SLJJ_\Pi F \rangle \quad (10)$$

We have restricted our consideration to diagonal elements of the interaction energy, and in this case, the 3-j symbol is[25]

$$(-1)^{F-M_F}\begin{pmatrix} F & 0 & F \\ -M_F & 0 & M_F \end{pmatrix} = (2F + 1)^{-1/2} \quad (11)$$

The quantum nature of the interaction energy is therefore contained within the reduced matrix element in Eq. (10). This is a problem of the type first considered by Curl and Kinsey[26] and which is summarized in Eq. (13.8) of Silver,[25] one in which there are three types of commuting (independent) angular momenta, described by operators $2.002\hat{S}$, $\hat{L}$, and $\hat{J}_\Pi$ in spaces 1, 2, and 3, in the fully coupled representation of angular momentum quantum theory.[22] It is helpful to write the interaction energy (10) as

$$\Delta E_1 = -\frac{\gamma_e B_0}{\hbar}(2F + 1)^{-1/2}\left(\langle SLJJ_\Pi F\|\left[\left[\hat{1}^0 \otimes \hat{L}^1\right]_0^1 \otimes \hat{J}_{\Pi 0}^1\right]_0^0\|SLJJ_\Pi F\rangle\right.$$

$$\left. + 2.002\langle SLJJ_\Pi F\|\left[\left[\hat{S}^1 \otimes \hat{1}^0\right]_0^1 \otimes \hat{J}_{\Pi 0}^1\right]_0^0\|SLJJ_\Pi F\rangle\right) \quad (12)$$

in terms of tensor products of the type illustrated in Eq. (13.7) of Silver,[25] to which we refer the reader for background and details of irreducible tensorial methods. These methods allow the reduced matrix element to be written in terms of the 9-j symbols of atomic quantum mechanics,[25] allowing the interaction energy to be expressed simply as

$$\Delta E_1 = -\gamma_e B_0 \hbar g_1 \quad (13)$$

where the g_1 factor is a complicated combination of terms defined through the individual angular momentum quantum numbers:

$$g_1 = (2J + 1)\left[3(2F + 1)J_{\Pi}(J_{\Pi} + 1)(2J_{\Pi} + 1)\right]^{1/2}\begin{bmatrix} J & J & 1 \\ J_{\Pi} & J_{\Pi} & 1 \\ F & F & 0 \end{bmatrix}$$

$$\times \left[(2S + 1)L(L + 1)(2L + 1)\right]^{1/2}\begin{bmatrix} S & S & 0 \\ L & L & 1 \\ J & J & 1 \end{bmatrix} \qquad (14)$$

$$+ 2.002\left[(2L + 1)S(S + 1)(2S + 1)\right]^{1/2}\begin{bmatrix} S & S & 1 \\ L & L & 0 \\ J & J & 1 \end{bmatrix}$$

Note at this stage that there are several energy levels, because there are several allowed combinations of quantum numbers through the appropriate Clebsch-Gordan series. For each energy level there will be an individual g_1 factor. The physical meaning of this coupled representation is discussed later, and in Section III the result (13) is used in the context of electric dipole transitions in atomic states split by the photon property $\hat{B}_{\Pi}$ generated by a circularly polarized laser.

2. The Semicoupled Representation

This is, perhaps, the most realistic representation of the problem in quantum field theory, because of the nature of the photon operator $\hat{B}_{\Pi}$. The photon propagates at the speed of light and is massless, so that the azimuthal components of the angular momentum operator $\hat{J}_{\Pi}$ are always well defined (i.e., specified)[22] in terms of the azimuthal quantum numbers

$$M_{J_{\Pi}} = \pm 1$$

(A sign change in this context denotes switching from left to right circular polarization.) Relativity theory forbids any component of the photon angular momentum perpendicular to the azimuthal (propagation) axis Z. It appears natural, therefore, to combine the angular momentum operators $\hat{J}$ and $\hat{J}_{\Pi}$ in the uncoupled representation of the quantum theory of angular momentum coupling,[22, 25] a representation in which the azimuthal components of the angular momenta are specified, but in which the resultant angular momentum is not.[22]

In the semicoupled representation, the interaction energy (8) can therefore be written

$$
\Delta E_2 = -\frac{\gamma_e B_0}{\hbar}\left(\langle SLJM_J|\left[\hat{1}^0 \otimes \hat{L}^1\right]_0^1|S'L'J'M_J'\rangle\right) + 2.002
$$

$$
\times \langle SLJM_J|\left[\hat{S}^1 \otimes \hat{1}^0\right]_0^1|S'L'J'M_J'\rangle\Big)\langle J_\Pi M_{J_\Pi}|\hat{J}_{\Pi 0}^1|J_\Pi' M_{J_\Pi}'\rangle \tag{15}
$$

an expression that can be reduced using tensorial methods (using Eqs. (14.10) ff. of Silver[25]) to the form

$$
\Delta E_2 = -\left(\frac{\gamma_e B_0}{\hbar}\right)M_J\left[\frac{3J(J+1)-L(L+1)+S(S+1)}{2J(J+1)}\right](-1)^{J_\Pi - M_{J_\Pi}}
$$

$$
\times \begin{pmatrix} J_\Pi & 1 & J_\Pi' \\ -M_{J_\Pi} & 0 & M_{J_\Pi}' \end{pmatrix}\langle J_\Pi\|\hat{J}_\Pi\|J_\Pi'\rangle \tag{16}
$$

This can be reduced further to the simple result

$$
\Delta E_2 = -\gamma_e B_0 g_L M_J M_{J_\Pi}\hbar \tag{17}
$$

using the following results[25] and notation:

$$
g_L = \frac{3J(J+1)-L(L+1)+S(S+1)}{2J(J+1)} \tag{18}
$$

$$
(-1)^{J_\Pi - M_{J_\Pi}}\begin{pmatrix} J_\Pi & 1 & J_\Pi \\ -M_{J_\Pi} & 0 & M_{J_\Pi} \end{pmatrix} = M_{J_\Pi}\left[J_\Pi(J_\Pi+1)(2J_\Pi+1)\right]^{-1/2} \tag{19}
$$

$$
\langle J_\Pi\|\hat{J}_\Pi\|J_\Pi\rangle = \hbar\left[J_\Pi(J_\Pi+1)(2J_\Pi+1)\right]^{1/2} \tag{20}
$$

In this semicoupled representation, therefore, the interaction energy in quantum field theory becomes the product of a g_L factor which depends only on the quantum numbers J, L, and S, with the azimuthal quantum number product $M_J M_{J_\Pi}$. The g_L factor in this case is recognizable as the Landé factor of atomic theory.[22, 25]

3. The Uncoupled Representation

This is a possible representation of the same problem, in which the three operators $2.002\hat{S}$, $\hat{L}$, and $\hat{J}_\Pi$ are decoupled operators acting independently on decoupled states, each operator acting independently on states built from independent sets of coordinates in spaces 1, 2, and 3 (Ref. 22). The interaction energy (9) is therefore written as

$$\Delta E_3 = -\frac{\gamma_e B_0}{\hbar}\Big(\langle SM_S; LM_L; J_\Pi M_{J_\Pi}|\hat{1}_0^0\hat{L}_0^1\hat{J}_{\Pi0}^1|S'M'_S; L'M'_L; J'_\Pi M'_{J_\Pi}\rangle$$

$$+2.002\langle SM_S; LM_L; J_\Pi M_{J_\Pi}|\hat{S}_0^1\hat{1}_0^0\hat{J}_{\Pi0}^1|S'M'_S; L'M'_L;\quad (21)$$

$$J'_\Pi M'_{J_\Pi}\rangle\Big)$$

and the Wigner-Eckart theorem applied three times to give a superficially complicated result:

$$\Delta E_3 = -\gamma_e B_0\hbar(-1)^{S-M_S+L-M_L+J_\Pi-M_{J_\Pi}}$$

$$\times\Bigg[\begin{pmatrix} S & 0 & S' \\ -M_S & 0 & M'_S \end{pmatrix}\begin{pmatrix} L & 1 & L' \\ -M_L & 0 & M'_L \end{pmatrix}\begin{pmatrix} J_\Pi & 1 & J'_\Pi \\ -M_{J_\Pi} & 0 & M'_{J_\Pi} \end{pmatrix}$$

$$\times \langle S\|\hat{1}\|S'\rangle\langle L\|\hat{L}\|L'\rangle\langle J_\Pi\|\hat{J}_\Pi\|J'_\Pi\rangle \qquad (22)$$

$$+2.002\begin{pmatrix} S & 1 & S' \\ -M_S & 0 & M'_S \end{pmatrix}\begin{pmatrix} L & 0 & L' \\ -M_L & 0 & M'_L \end{pmatrix}\begin{pmatrix} J_\Pi & 1 & J'_\Pi \\ -M_{J_\Pi} & 0 & M_{J_\Pi} \end{pmatrix}$$

$$\times\langle S\|\hat{S}\|S'\rangle\langle L\|1'\|L'\rangle\langle J_\Pi\|J_\Pi\|J'_\Pi\rangle\Bigg]$$

However, with standard results, [22, 25] such as

$$\langle S\|\hat{S}\|S\rangle = [S(S+1)(2S+1)]^{1/2}\hbar \qquad (23)$$

$$\langle S\|\hat{1}\|S\rangle = (2S+1)^{1/2} \qquad (24)$$

$$\begin{pmatrix} S & 1 & S \\ -M_S & 0 & M_S \end{pmatrix} = (-1)^{S-M_S}M_S[S(S+1)(2S+1)]^{-1/2} \qquad (25)$$

$$\begin{pmatrix} S & 0 & S \\ -M_S & 0 & M_S \end{pmatrix} = (-1)^{S-M_S}(2S+1)^{-1/2} \qquad (26)$$

the interaction energy in the decoupled representation of quantum field theory collapses to

$$\Delta E_3 = -\gamma_e B_0 \hbar M_{J_{\Pi}} (M_L + 2.002 M_S) \qquad (27)$$

in which there is no g factor at all, and which is a simple product of azimuthal quantum numbers of the atom and the photon's novel $\hat{B}_{\Pi}$ operator in which we are interested.

B. Semiclassical Theory

The semiclassical representation of the anomalous optical Zeeman and Paschen-Back effects depends on the interaction Hamiltonian

$$\Delta \hat{H}_2 = -\hat{m} \cdot \mathbf{B}_{\Pi} \qquad (28)$$

where $\mathbf{B}_{\Pi}$ is now a classical field vector,[1-5] not a quantum-mechanical operator. The interaction energy in this case is

$$\Delta E_4 = -\gamma_e |\mathbf{B}_{\Pi}| \langle SLJM_J|\hat{L} + 2.002\hat{S}|S'L'J'M'_J \rangle \qquad (29)$$

which can be reduced to

$$\Delta E_4 = -\gamma_e |\mathbf{B}_{\Pi}| g_L \hbar M_J \qquad (30)$$

where g_L is the same Landé factor as in Eq. (17).

III. APPLICATION TO ELECTRIC DIPOLE TRANSITIONS IN ATOMS

In this section the results of Section II are applied to predict the splitting of a visible frequency electric dipole transition in an atom by a circularly polarized laser generating the flux quantum $\hat{B}_{\Pi}$ of Eq. (1). The selection rules governing such a transition in the conventional theory of the anomalous Zeeman effect in atoms are well known.[22, 25, 26] They are determined by rules on the existence of the 3-j symbol in the Wigner-Eckart expansion of the matrix elements of the transition electric dipole moment operator $\hat{\mu}$. For the Z component

$$\langle SLJM_J|\hat{\mu}_0^1|S'L'J'M'_J \rangle = (-1)^{J-M_J} \begin{pmatrix} J & 1 & J' \\ -M_J & 0 & M'_J \end{pmatrix} \langle SLJ\|\hat{\mu}_0^1\|S'L'J' \rangle$$

$$(31)$$

LONGITUDINAL FIELDS AND PHOTONS

and the selection rules are

$$\Delta J = 0, \pm 1 \qquad \Delta M_J = 0 \qquad (32)$$

Similarly, for the X and Y components of $\hat{\mu}$,

$$\Delta J = 0, \pm 1 \qquad \Delta M_J = \pm 1 \qquad (33)$$

However, in the anomalous optical Zeeman effect, the atomic terms between which the electric dipole transition takes place are each being considered in the presence of the operator $\hat{B}_{\Pi}$, in the various coupling schemes of Section II. Therefore, the electric dipole transition selection rules must also be derived in the appropriate coupling scheme.

We shall consider an atomic transition[22] between the atomic terms $^2P_{1/2}$ and $^2D_{3/2}$. In the former, $L = 1$, $S = \frac{1}{2}$, and $J = \frac{1}{2}$; and in the latter, $L = 2$, $S = \frac{1}{2}$, and $J = \frac{3}{2}$. The Laporte (or parity) selection rule is also obeyed in such a transition, i.e., $\Delta L = 1$ in this case. The transition occurs at a frequency that is determined from the appropriate electric dipole selection rules,[22] and the spectrum in the absence of $\hat{B}_{\Pi}$ consists of a single line which can be measured at visible frequencies in a spectrometer.

We are specifically interested in how this line is affected by the presence of an additional, circularly polarized laser, generating the flux quantum $\hat{B}_{\Pi}$ of Eq. (1), and substituting for the usual magnet of the Zeeman effect.[22] In examining the effect of $\hat{B}_{\Pi}$ we use the four results, Eqs. (13), (17), (27), and (30), in turn. In each case the effect of $\hat{B}_{\Pi}$ is first determined on the $^2P_{1/2}$ atomic term, and then on the $^2D_{3/2}$ term. Each of these two terms is split into nondegenerate energy levels by the addition of quantized energy such as ΔE_1, described by Eq. (13), for example. Various electric dipole transitions can then occur between the split $^2P_{1/2}$ term and the split $^2D_{3/2}$ term according to the transition electric dipole selection rules appropriate for the coupling scheme. Overall, therefore, we expect that the novel flux quantum $\hat{B}_{\Pi}$ splits the original line corresponding to the transition $^2P_{1/2} \to {}^2D_{3/2}$ in an atom. The details of the splitting pattern depend on which of the various schemes of Section II are chosen. This procedure is similar to the standard theory of the conventional Zeeman effect,[22-24] but in the optical Zeeman effect, $\hat{B}_{\Pi}$ is a quantum-mechanical operator. In the conventional Zeeman effect, the applied magnetic field $\mathbf{B}_0$ is always a classical, magnetostatic field vector, whose origin is not electromagnetic.

A. Quantum Field Theory, Coupled Representation, Eq. (13)

The atomic $^2P_{1/2}$ term is split from Eq. (13) into two levels by the novel operator $\hat{B}_{\Pi}$ of the photon:

$$g_1\left(F = \tfrac{1}{2}, J_{\Pi} = 1, J = \tfrac{1}{2}, L = 1, S = \tfrac{1}{2}\right)$$

$$g_2\left(F = \tfrac{3}{2}, J_{\Pi} = 1, J = \tfrac{1}{2}, L = 1, S = \tfrac{1}{2}\right)$$

and there is a different g factor for each level. The atomic $^2D_{3/2}$ term is split into the three levels:

$$g_3\left(F = \tfrac{1}{2}, J_{\Pi} = 1, J = \tfrac{3}{2}, L = 2, S = \tfrac{1}{2}\right)$$

$$g_4\left(F = \tfrac{3}{2}, J_{\Pi} = 1, J = \tfrac{3}{2}, L = 2, S = \tfrac{1}{2}\right)$$

$$g_5\left(F = \tfrac{5}{2}, J_{\Pi} = 1, J = \tfrac{3}{2}, L = 2, S = \tfrac{1}{2}\right)$$

each with a different g factor. In general the five g factors (two in the lower term and three in the upper) are all different. Electric dipole transitions within the atom can now occur between the two lower levels and three upper levels with selection rules determined as follows.

The transition electric dipole moment operator is developed using the Wigner-Eckart theorem between coupled states to give

$$\langle SLJJ_{\Pi}FM_F|\hat{\mu}_0^1|S'L'J'J'_{\Pi}F'M'_F\rangle$$

$$= (-1)^{F-M_F}\begin{pmatrix} F & 1 & F' \\ -M_F & 0 & M'_F \end{pmatrix}\langle SLJJ_{\Pi}F\|\hat{\mu}_0^1\|S'L'J'J'_{\Pi}F'\rangle \qquad (34)$$

This procedure yields immediately the selection rules on the Z component of $\hat{\mu}$:

$$\Delta F = 0, \pm 1 \qquad \Delta M_F = 0 \qquad (35)$$

Similarly, for X and Y components of $\hat{\mu}$,

$$\Delta F = 0, \pm 1 \qquad \Delta M_F = \pm 1 \qquad (36)$$

All selection rules now refer to the net quantum number F.

There are six possible spectral lines generated by electric dipole transitions between the two $^2P_{1/2}$ levels and the three $^2D_{3/2}$ levels, but one of these, from the $F = \tfrac{1}{2}$ level of the split $^2P_{1/2}$ term to the $F = \tfrac{5}{2}$ level of the split $^2D_{3/2}$ term, is forbidden by the selection rule (35, 36) just derived, i.e., by the fact that the maximum change in F must be $+1$.

Discussion of the physical meaning of this result is given later. The quantum field theory of the anomalous optical Zeeman effect in the coupled representation splits the original visible frequency spectral line into five, each displaced from the original frequency.

B. Quantum Field Theory, Semicoupled Representation, Eq. (17)

In this case the selection rules on the electric dipole transitions are obtained by the development (for the Z component):

$$\langle SLJM_J; J_\Pi M_{J_\Pi} | \hat{\mu}_0^1 \hat{1}_0^0 | S'L'J'M'_J; J'_\Pi M'_{J_\Pi} \rangle$$

$$= \langle SLJM_J | \hat{\mu}_0^1 | S'L'J'M'_J \rangle \langle J_\Pi M_{J_\Pi} | \hat{1}_0^0 | J'_\Pi M'_{J_\Pi} \rangle \tag{37}$$

$$= (-1)^{J-M_J} \begin{pmatrix} J & 1 & J' \\ -M_J & 0 & M'_J \end{pmatrix} \langle SLJ \| \hat{\mu}_0^1 \| S'L'J' \rangle$$

so that the 3-j symbol is nonzero if and only if

$$\Delta J = 0, \pm 1 \qquad \Delta M_J = 0 \tag{38}$$

Similarly, for X and Y components of $\hat{\mu}$,

$$\Delta J = 0, \pm 1 \qquad \Delta M_J = \pm 1 \tag{39}$$

The Landé g_L factor of Eq. (17) is the same for each level of the $^2P_{1/2}$ term. For each level of the split $^2D_{3/2}$ term the Landé factor is again the same, g_{L1}. Transitions between the levels are controlled by the selection rule $\Delta M_J = 0, \pm 1$. The resulting spectral pattern is three groups of doublets, i.e., six lines. This is recognizable as the same pattern observed in the conventional semiclassical theory of the anomalous Zeeman effect, as illustrated, for example, in Fig. 9.27 of Ref. 22.

C. Quantum Field Theory, Uncoupled Representation, Eq. (27)

In this case the electric dipole (Z component) transition selection rules are determined from the development

$$\langle SM_S; LM_L; J_\Pi M_{J_\Pi} | \hat{1}_0^0 \hat{\mu}_0^1 \hat{1}_0^0 | S'M'_S; L'M'_L; J'_\Pi M'_{J_\Pi} \rangle$$

$$= (-1)^{L-M_L} \begin{pmatrix} L & 1 & L' \\ -M_L & 0 & M'_L \end{pmatrix} \langle L \| \hat{\mu}_0^1 \| L' \rangle \tag{40}$$

Assuming, as usual,[22-24] that the spin selection rule

$$\Delta S = 0 \qquad (41)$$

is obeyed, we obtain

$$\Delta L = 0, \pm 1 \qquad \Delta M_L = 0 \qquad (42)$$

Similarly, for the X and Y components of $\hat{\mu}$,

$$\Delta L = 0, \pm 1 \qquad \Delta M_L = \pm 1 \qquad (43)$$

In this case there are no g factors in either term, and both $^2P_{1/2}$ and $^2D_{3/2}$ terms are split to the same extent.[22] The result is a spectral pattern of three lines, which can be thought of as three coincidental doublets. This is recognizable as the same pattern obtained in the conventional theory of the Paschen-Back effect.[22] In the uncoupled representation of quantum field theory, therefore, the novel $\hat{B}_\Pi$ operator is expected to produce the optical Paschen-Back effect.

D. Semiclassical Theory, Eq. (30)

It is straightforward to see that in this case the transition electric dipole moment selection rules are those given by Eqs. (38) and (39), and that the splitting pattern is the same as that in the conventional theory of the anomalous Zeeman effect, consisting of three doublets.

IV. DISCUSSION

In four different schemes we have deduced that the novel property $\hat{B}_\Pi$ of the photon[1-5] splits electric dipole transitions occurring at visible frequencies in atoms. It is appropriate to ask which scheme is likely to be the most realistic. It is well known[22] that in the quantum theory of angular momentum coupling, the uncoupled representation leaves the magnitude of the total angular momentum undefined and says nothing about the relative orientation of the contributing individual angular momenta, but defines individual components. The coupled representation defines the total angular momentum but leaves individual components undefined. Either scheme is equally valid and acceptable mathematically. In Section II we found that there is also a third scheme, which we have called the semicoupled representation. All three are valid in the quantum-field theoretical description of the effect of $\hat{B}_\Pi$ on atomic transitions.

However, it is independently known that the photon propagates at the speed of light, which implies that the component of $\hat{B}_\Pi$ in the azimuthal axis be well defined, because there cannot be any perpendicular components from the theory of relativity. Therefore, it appears that in our coupled representation of Section II, there is a conflict of reasoning, in that the total angular momentum is defined as well as the azimuthal component of the novel field operator of the photon $\hat{B}_\Pi$. Therefore, the commutator $[F^2, J_{\Pi_Z}]$ is not zero. However, it is well known[22] that this type of "paradox" can be resolved by remembering that the commutator is an operator, which acts on a wave function, ψ, and if the result

$$\left[F^2, J_{\pi_Z}\right]\psi = 0 \tag{44}$$

is true, then F^2 and M_{J_Π} can be simultaneously well defined in the quantum theory of angular momentum coupling.[22]

This is the mathematical basis for our coupled representation of the problem in Section II. In physical terms, the coupled representation leads to five lines, instead of six as in the semicoupled representation, and this can be tested experimentally to reveal which is the truer representation.

The semicoupled representation treats $\hat{J}$ and $\hat{J}_\Pi$ in an uncoupled scheme, so that azimuthal components of both are well defined, but their resultant $\hat{F}$ is not. Therefore, F does not appear in Eq. (17) and there is clear definition of directionality, in that the azimuthal quantum number M_J does appear in Eq. (17) and controls the selection rules as described in Section III. The directionality comes from the presence of the circularly polarized laser, generating the quantity $\hat{B}_\Pi$, a laser that propagates in the azimuthal axis Z. In the coupled representation that gives Eq. (13) no azimuthal quantum number appears, but F is well defined and selection rules on F now govern the effect of $\hat{B}_\Pi$ on atomic transitions.

In the uncoupled representation of Section II, the only difference from the semicoupled representation is that the operator $2.002\hat{S}$ has been decoupled from $\hat{L}$, leading to the optical Paschen-Back effect. This is therefore a type of strong field limit, in which $2.002\hat{S}$, $\hat{L}$, and $\hat{J}_\Pi$ precess independently[22] about the propagation or azimuthal axis Z.

In the semiclassical representation, $\mathbf{B}_\Pi$ is a classical field vector, and the treatment of both the anomalous optical Zeeman effect and of the optical Paschen-Back effect becomes the same as conventional theory, leading to the same physical considerations.[22] This is because in the semiclassical representation, $\mathbf{B}_\Pi$ is akin to a magnetostatic field, albeit generated by a laser.[1]

Finally, it is straightforward to derive order of magnitude estimates of the splitting from any of the equations (13), (17), (27), and (30), given the relation[1-5]

$$|\mathbf{B}_\Pi| = B_0 = \left(\frac{2I_0}{\varepsilon_0 c^3}\right)^{1/2} \sim 10^{-7}I_0^{1/2} \tag{45}$$

between $|\mathbf{B}_\Pi|$ and the intensity of the laser in watts per square meter. Here, ε_0 is the permittivity in vacuo in S.I. units:

$$\varepsilon_0 = 8.854 \times 10^{-12}J^{-1}C^2m^{-1} \tag{46}$$

For an intensity of 100 W cm^{-2} (10^6 W m^{-2}) we expect that the novel property $\hat{B}_\Pi$ will shift the original $^2P_{1/2} \rightarrow {}^2D_{3/2}$ transition typically by of the order 1.5×10^6 Hz. This is 5×10^{-5} cm^{-1} (inverse centimeters), and the splitting is expected to be proportional to the square root of the laser's intensity. There should be no splitting if the laser has no degree of circular polarity. These features should help in identifying the effect of the new fundamental photon property $\hat{B}_\Pi$ in which we are interested, and which has recently been proposed theoretically.[1-5]

Acknowledgments

Support is acknowledged from the Leverhulme Trust, Cornell Theory Center, and the Materials Research Laboratory of Penn State University.

References

1. M. W. Evans, *Physica B*, **182**, 227 (1992).
2. M. W. Evans, ibid, **182**, 237 (1992).
3. S. Borman, *Chem. Eng. News*, 6 April 1992, p. 31.
4. M. W. Evans, *Physica B*, **183**, 103 (1993).
5. M. W. Evans, *Physica B*, in press.
6. P. S. Pershan, *Phys. Rev.* **130**, 919 (1963).
7. Y. R. Shen, *Phys. Rev.* **134**, 661 (1964).
8. J. P. van der Ziel, P. S. Pershan, and L. D. Malmstrom, *Phys. Rev. Lett.* **15**, 190 (1965).
9. P. S. Pershan, J. P. van der Ziel, and L. D. Malmstrom, *Phys. Rev.* **143**, 574 (1966).
10. M. W. Evans, *J. Phys. Chem.* **95**, 2256 (1991).
11. W. S. Warren, S. Mayr, D. Goswami, and A. P. West, Jr., *Science* **255**, 1683 (1992).
12. N. L. Manakov, L. D. Ovsiannikov, and S. Kielich, *Acta Phys. Pol.* **A53**, 595 (1978).
13. P. W. Atkins and M. H. Miller, *Mol. Phys.* **15**, 503 (1968).
14. M. W. Evans, *Phys. Rev. Lett.* **64**, 2909 (1990).
15. M. W. Evans, *Opt. Lett.* **15**, 863 (1990).

16. M. W. Evans, *Physica B* **168**, 9 (1991).
17. M. W. Evans, in I. Prigogine and S. A. Rice (Eds.), Vol. 81, *Advances in Chemical Physics*, Wiley Interscience, New York, 1992.
18. M. W. Evans, *Chem. Phys.* **157**, 1 (1991).
19. M. W. Evans, *Int. J. Mod. Phys. B* **5**, 1963 (1991) (review).
20. M. W. Evans, *Chem. Phys.* **150**, 197 (1990).
21. K. A. Earle, Department of Chemistry, Cornell University, personal communications.
22. P. W. Atkins, *Molecular Quantum Mechanics*, 2d ed., Oxford University Press, Oxford, UK, 1983.
23. B. W. Shore, *The Theory of Coherent Atomic Excitation*, Vols. 1 and 2, Wiley, New York, 1991.
24. L. D. Barron, *Molecular Light Scattering and Optical Activity*, Cambridge University Press, Cambridge, UK 1982.
25. B. L. Silver, *Irreducible Tensorial Methods*, Academic, New York, 1976.
26. R. F. Curl and J. L. Kinsey, *J. Chem. Phys.*, **35**, 1758 (1961).

THE OPTICAL FARADAY EFFECT AND OPTICAL MCD

I. INTRODUCTION

This paper continues a series of articles in which the consequences are developed of the recent deduction[1-6] that the photon generates a magnetic flux quantum, an operator

$$\hat{B}_{\Pi} = B_0 \frac{\hat{J}}{\hbar} \tag{1}$$

Here B_0 is the scalar magnetic flux density amplitude of a beam of N photons (a circularly polarized generator laser), $\hat{J}$ is the photon's angular momentum operator[7] whose eigenvalues are $\pm M_J \hbar$, where M_J is plus or minus one, and where $\hbar$ is the reduced Planck constant. The operator $\hat{B}_{\Pi}$ changes sign with the circular polarity of the generator laser, is unlocalized in space, and has eigenvalues $\pm M_J B_0$, where B_0 is the laser's scalar flux density amplitude. Its classical equivalent is the axial vector $\mathbf{B}_{\Pi}$, a novel magnetostatic flux density generated by circularly polarized electromagnetic plane waves.[1-6] The theoretical existence of $\hat{B}_{\Pi}$ is supported by the experimental evidence for light-induced magnetic effects. The first to be described (in the 1960s) was the inverse Faraday effect,[8-14] and recently it has been shown[15, 16] that NMR resonances are shifted in new and useful ways by the magnetizing effect of circularly polarized argon ion radiation at frequencies far from optical resonance (i.e., where the sample is transparent to the argon ion radiation and does not absorb it). Both these effects can be described in terms of the operator $\hat{B}_\Pi$, or its classical equivalent $\mathbf{B}_\Pi$ (Refs. 1–6, 17–23). It has also been proposed[4-6] that there exist an optical Zeeman effect, in which $\hat{B}_\Pi$ splits singlet electric dipole transition frequencies in atoms; an anomalous optical Zeeman effect for atomic triplet states; and an optical Paschen-Back effect. It has also been proposed theoretically[2, 3] that $\hat{B}_\Pi$ can be detected by examining spectrally a circularly polarized laser at visible frequencies reflected from an electron beam.

The existence of $\hat{B}_\Pi$ also implies that of an optical Faraday effect, in which it rotates the plane of a linearly polarized probe, an effect that is frequency dependent and gives rise, therefore, to optical magnetic circular dichroism (optical MCD spectroscopy). These effects are developed theoretically in this paper for atoms.

Mason[24] has given an interesting discussion of cause and effect in chirality that includes some pertinent historical analysis of interest here. In 1846, Faraday showed experimentally[25] that a magnetostatic field induces optical activity in flint glass and other isotropic transparent media. In 1884, Pasteur[26] proposed on the basis of this result that the magnetic field represented a source of chirality. It is now known[27] that the magnetically induced optical activity in the Faraday effect has a fundamentally different symmetry[28-31] from that of natural optical activity. Nonetheless, it is interesting for our purpose, following Mason,[24] that le Bel[32] in 1874 had independently proposed that circularly polarized radiation also provides a "chiral force" of the type envisioned by Pasteur emanating from the magnetostatic field used by Faraday. Enantio-differentiating photoreactions were indeed reported by Kuhn and Braun[33] in 1929, and it has been proposed repeatedly (e.g., Bonner[34] that circularly polarized solar irradiation may be responsible for the preponderance in nature of one enantiomer over another. However, Mason[24] favors the universal and parity-violating mechanism of the electroweak force as the origin of this dissymmetry because the electroweak force does not depend on time and location on the earth's surface. (Natural solar radiation is only 0.1% circularly polarized, and equally and oppositely so at dawn and dusk.[24]

It is interesting for our purpose to note that both circular polarity in light and the magnetostatic field have been proposed independently (by le Bel and Pasteur, respectively) as sources of chirality. It had thus been sensed more than one hundred years ago that these two concepts have something in common in their effect on material. Pershan et al.,[11] in their first paper on the experimental demonstration of the magnetizing effect of circularly polarized giant ruby laser radiation, edged toward the concept of

$\hat{B}_\Pi$ by describing the effect of the laser as being due to an effective magnetostatic field. In view of these indications, those by Kielich and coworkers,[12, 13] and those by Atkins and Miller,[14] the present author appears to have demonstrated conclusively[1-6] the fact that circularly polarized radiation generates the classical magnetostatic field B_Π whose equivalent in quantum field theory is the photon's flux quantum $\hat{B}_\Pi$, a novel fundamental property of quantum field theory. As noted elsewhere, the concept of $\hat{B}_\Pi$ must be clearly and carefully distinguished from the (standard IUPAC) oscillating B field of the electromagnetic plane wave, because the two are quite different.[1-6] We have therefore resolved[1-6] the conjectures of le Bel (1874) and Pasteur (1884) insofar as to show that circularly polarized light can indeed act in the same way as a magneto-static field, a finding that implies the existence of several novel types of spectroscopy in circumstances where a conventionally applied magnet would be replaced by a circularly polarized laser.

In Section II, the contemporary quantum theory of Faraday's effect of 1846 is developed succinctly for atoms, whereby it becomes relatively straightforward to show that the flux quantum $\hat{B}_\Pi$ must also generate Faraday's observation of optical activity in all material, inherently (struct-urally) chiral or otherwise. In our case, the magnet used by Faraday is replaced by a circularly polarized laser, which generates $\hat{B}_\Pi$ and is therefore referred to as the generator laser. Section III develops the frequency dependence of the optical Faraday effect in atoms through the properties of the magnetically (i.e., $\hat{B}_\Pi$) perturbed antisymmetric polariz-ability of conventional contemporary Faraday effect theory,[35] and there-fore arrives at expressions in atoms for optical MCD. Finally, a discussion is given of possible experimental configurations and order of magnitudes of the expected optical Faraday effect in terms of the intensity in watts per unit area of the generator laser.

II. OPTICALLY INDUCED FARADAY ROTATION IN ATOMS

The contemporary quantum theory of the Faraday effect is based on the work by Serber,[36] which was the precursor for the A, B, and C terms.[37] This section develops the Serber theory for use with a flux quantum $\hat{B}_\Pi$ from the generator laser and shows that the analogy between the optical and conventional Faraday effects is easily forged by using B_Π in place of the conventional magnetic field B_0 from a magnet.

The starting point for the theory of the conventional Faraday effect in quantum mechanics is an equation for the rotation of the plane of linearly polarized radiation. This is derived[37] by a consideration of the effect of a magnetostatic field on the antisymmetric part of an atomic or molecular property tensor called the antisymmetric polarizability α''_{ij}:

$$\alpha''_{ij}(B_{0k}) = \alpha''_{ij}(\mathbf{O}) + \alpha''_{ijk}B_{0k} \qquad (2)$$

where α''_{ijk} is a perturbation tensor of order three. Before embarking on detailed theoretical development it is instructive to consider the fundamental symmetries of these atomic property tensors (we restrict consideration in this paper to atoms) and to recall that the complex electronic electric polarizability, of which α''_{ij} is the antisymmetric, imaginary component, is derived from time-dependent perturbation theory within whose framework[37] α''_{ij} is a product of two transition electric dipole moment operators. The perturbation tensor α''_{ijk} in this context involves two of these and one transition magnetic dipole moment matrix element. This structure introduces frequency dependence into the conventional Faraday effect, leading to MCD. Similarly, frequency dependence occurs in the optical Faraday effect, leading to optical MCD.

The fundamental symmetries considered are parity inversion (represented by the operator $\hat{P}$) and motion reversal (by the operator $\hat{T}$). In this context α''_{ij} is negative to $\hat{T}$ and positive to $\hat{P}$, while α''_{ijk} is positive to both $\hat{P}$ and $\hat{T}$. Therefore α''_{ij} is finite only in the presence of a $\hat{T}$-negative influence, such as $\mathbf{B}_0$ or $\hat{B}_{\Pi}$, and this influence is mediated by α''_{ijk}, which is finite for all atoms and molecules and is described by the ubiquitous B term.[37] The contemporary theory of Faraday's effect depends on a perturbation of a quantity α''_{ij}, which is itself the result of semiclassical, time-dependent, second-order perturbation theory. The basic reason for this is that the observable in the Faraday effect is an angle of rotation ($\Delta\theta$) (or alternatively a change in ellipticity $\Delta\eta$), which must be calculated from Maxwell's equations or Rayleigh refringent scattering theory.[37] Although the A term is closely related to the Zeeman effect,[38] the observables of the two effects are quite different, being traditionally an angle of rotation (Faraday's effect) and a frequency shift in the visible frequency region (Zeeman's effect). The latter can be described by an energy perturbation, while Faraday's effect needs perturbation of the antisymmetric polarizability, because energy does not appear directly in Maxwell's equations, which are needed to calculate refractive indices and therefrom $\Delta\theta$ and $\Delta\eta$.

With these considerations, the starting point for our development of the optical Faraday effect and optical MCD is the equation[37] for angle of rotation in the conventional quantum theory of the Faraday effect:

$$\Delta\theta \doteq \frac{1}{4}\omega\mu_0 cl \left(\frac{N}{d_n}B\right)_{0z} \sum_n \left[\alpha''_{XYZ} - \alpha''_{YXZ} + \frac{m_{nZ}}{kT}(\alpha''_{XY} - \alpha''_{YX})\right] \qquad (3)$$

Here, $\Delta\theta$ is the angle of rotation of plane polarized probe radiation of angular frequency ω parallel to the conventionally generated magnetostatic flux density B_z. The quantity N is the total number of molecules per unit volume in a set of degenerate quantum states of the atom, individually designated[37] ψ_n, where d_n is the degeneracy and the sum is over all components of the degenerate set, with $N_d = Nd_n$. In Eq. (3), μ_0 is the vacuum permeability, c is the speed of light, l is the sample length, m_{nZ} is the Z component of the atomic magnetic dipole moment in state n, and kT is the thermal energy per atom. Recall that the atomic property tensors α''_{ij} and α''_{ijk} are derived from semiclassical, time-dependent, perturbation theory and are frequency dependent in general, so that $\Delta\theta$ mapped over a frequency range has the appearance of a spectrum—the conventional MCD spectrum.

Our task here is to incorporate the novel flux quantum $\hat{B}_{\Pi}$ (Refs. 1–6) into Eq. (3), and thus generate the optical Faraday effect and optical MCD. The terms α''_{ij} and α''_{ijk} as used in Eq. (3) are expectation values of the respective quantum-mechanical operators $\hat{\alpha}''_{ij}$ and $\hat{\alpha}''_{ijk}$; in the same way that m_{ni} is an expectation value of the magnetic electronic dipole moment operator $\hat{m}_n$. The quantity B_{0Z} is a classical magnetostatic vector component, and $\Delta\theta$ is an expectation value of the operator $\Delta\hat{\theta}$. It is convenient to transform the appropriate Cartesian components of the operators $\hat{\alpha}''_{ij}$ and $\hat{\alpha}''_{ijk}$ into spherical form,[37,39] using the Condon/Shortley phase convention:

$$\hat{\alpha}''_{XY} = -\frac{i}{2}\left[\sqrt{2}\,\hat{\alpha}_0^{1\prime\prime} + \left(\hat{\alpha}_2^{2\prime\prime} - \hat{\alpha}_{-2}^{2\prime\prime}\right)\right]$$

$$\hat{\alpha}''_{YX} = \frac{i}{2}\left[\sqrt{2}\,\hat{\alpha}_0^{1\prime\prime} - \left(\hat{\alpha}_2^{2\prime\prime} - \hat{\alpha}_{-2}^{2\prime\prime}\right)\right]$$

$$\hat{\alpha}''_{XYZ} = -\frac{i}{2}\left[\hat{\alpha}_0^{2\prime\prime} + \frac{1}{\sqrt{6}}\left(\hat{\alpha}_2^{2\prime\prime} + \hat{\alpha}_{-2}^{2\prime\prime}\right) + \frac{1}{\sqrt{3}}\left(\hat{\alpha}_2^{3\prime\prime} - \hat{\alpha}_{-2}^{3\prime\prime}\right)\right] \qquad (4)$$

$$\hat{\alpha}''_{YXZ} = \frac{i}{2}\left[\hat{\alpha}_0^{2\prime\prime} - \frac{1}{\sqrt{6}}\left(\hat{\alpha}_2^{2\prime\prime} + \hat{\alpha}_{-2}^{2\prime\prime}\right) - \frac{1}{\sqrt{3}}\left(\hat{\alpha}_2^{3\prime\prime} - \hat{\alpha}_{-2}^{3\prime\prime}\right)\right]$$

from which

$$\hat{\alpha}''_{XY} - \hat{\alpha}''_{YX} = -\sqrt{2}\,i\hat{\alpha}_0^{1\prime\prime}$$

$$\hat{\alpha}''_{XYZ} - \hat{\alpha}''_{YXZ} = -i\hat{\alpha}_0^{2\prime\prime}. \qquad (5)$$

Both $\hat{\alpha}''_{ij}$ and $\hat{\alpha}''_{ijk}$ are purely imaginary in the appropriate spherical representation, indicating that the operators $i\hat{\alpha}^1_0$ and $i\hat{\alpha}^2_0$ are anti-Hermitian, with purely imaginary eigenvalues.[37] (However, the $\hat{T}$ symmetry of $i\hat{\alpha}^1_0$ is negative, while that of $i\hat{\alpha}^2_0$ is positive. Both have positive $\hat{P}$ symmetry.)

The next step in our development for atoms is to replace the vector component B_{0Z} of the conventional quantum theory by the quantum-mechanical operator defined by Eq. (1), which has associated with it the angular momentum quantum number J. An immediate consequence of this replacement is the necessity to consider the magnetic field flux quantum using irreducible tensorial methods of angular momentum coupling theory in quantum mechanics.[39-41] In other words we are now considering a quantized photon angular momentum interacting with an atom, which contains, as usual, quantized orbital and spin electronic angular momentum. Without loss of generality we restrict consideration to atomic singlet states, in which there is a net orbital electronic angular momentum $\hat{L}$, but no net spin angular momentum $\hat{S}$.

With these considerations, Eq. (3) becomes, for the optical Faraday effect

$$\Delta\theta = \langle JM_J; LM_L|\Delta\theta|J'M'_J; L'M'_L\rangle = -\tfrac{1}{4}\omega\mu_0 clN_n i\langle JM_J|\hat{B}_\Pi|J'M'_J\rangle$$
$$\times\left(\frac{\langle LM_L|\hat{m}^1_{n0}|L'M'_L\rangle\langle LM_L|\hat{\alpha}^1_0|L'M'_L\rangle}{kT} + \langle LM_L|\hat{\alpha}^2_0|L'M'_L\rangle\right) \quad (6)$$

where we have used an uncoupled representation[39-41] to describe the net angular momentum generated during the interaction of photon and atom. This is justified because the azimuthal components of the operators $\hat{m}_n$ and $\hat{B}_\Pi$ are both well defined in the uncoupled representation,[39-41] whereas in the coupled representation of the same problem the total angular momentum is defined but the individual azimuthal components are not. With these considerations, the expectation value of the angle of rotation in the optical Faraday effect is

$$\Delta\theta = \mp\frac{1}{4}\omega\mu_0 clN_n\, iM_J B_0\left\{\frac{M_L^2\langle L\|\hat{m}^1_0\|L\rangle\langle L\|\hat{\alpha}^1_0\|L\rangle}{L(L+1)(2L+1)kT}\right.$$
$$\left. +\frac{[3M_L^2 - L(L+1)]\langle L\|\hat{\alpha}^2_0\|L\rangle}{[L(L+1)(2L+1)(2L+3)(2L-1)]^{1/2}}\right\} \quad (7)$$

where we have used the fact that the expectation value of the photon's $\hat{B}_\Pi$

operator is $\pm M_J B_0$, positive for left circularly polarized radiation and negative for right circularly polarized radiation from the generator laser. This is an expression for the angle of rotation induced in plane polarized probe radiation in a sample of atoms by a circularly polarized laser generating the flux quantum $\hat{B}_{\Pi}$. The observation of such a rotation would provide a test for the existence of $\hat{B}_{\Pi}$. Equation (7) is written out in terms of reduced matrix elements of dipole moment and atomic polarizability operators, matrix elements that are products of electric and magnetic dipole transition dipole moment matrix elements from time-dependent perturbation theory.[37] These introduce frequency dependence into the angle of rotation of the optical Faraday effect. The selection rules governing the various atomic property tensors are as follows:

$$\hat{\alpha}_0^1: \Delta L = 0, \quad \Delta M_L = 0$$
$$\hat{\alpha}_0^2: \Delta L = 0, \quad \pm 2; \quad \Delta M_L = 0 \tag{8}$$

where the $\Delta L = \pm 1$ part is parity forbidden, as in magnetic dipole transitions.

III.   OPTICAL MCD: FREQUENCY DEPENDENCE OF $\Delta\theta$

The origin of frequency dependence in the optical Faraday effect can be traced to semiclassical time-dependent perturbation theory, which produces expressions for the polarizability components as given in the conventional theory of magnetic electronic optical activity.[37] These can be further developed as usual in terms of reduced matrix elements of electric and magnetic transition dipole moment operators. For a given generator laser intensity and frequency, therefore, the optical MCD spectrum is a plot of the $\hat{B}_{\Pi}$ induced angle of rotation $\Delta\theta$ against the frequency of the linearly polarized probe. Experimentally, this is built up by replacing the conventional magnet of MCD apparatus by the circularly polarized generator laser.

IV.   DISCUSSION

Using the concept of $\hat{B}_{\Pi}$ the theory of the optical Faraday effect can also be developed and understood simply by replacing the magnetic flux density vector component B_{0Z} wherever it occurs in the conventional theory of MCD by the quantity $\pm B_0 M_J$, where B_0 is the magnetic flux density amplitude of the generator laser, and $\pm M_J$ are the two possible azimuthal quantum numbers of the photon. Thus, the optical Faraday effect can be developed along the lines of the conventional counterpart in Serber's A, B, and C terms, a convenient description of which is given by Barron[37] in his Eqs. (6.2.2) and (6.2.3). It follows that the optical MCD spectrum, which would test for the existence of $\hat{B}_\Pi$, would have the same characteristics as the conventional spectrum. If confirmed experimentally, this would be strong evidence for the photon's fundamental flux quantum $\hat{B}_\Pi$ introduced in Refs. 1–6. If the optical MCD spectrum were found to differ from the conventional MCD spectrum, it would indicate the presence of other mechanisms of magnetization by the generator laser, such as the induction of a magnetic dipole moment through[42–45]

$$m_i = {}^m\beta_{ijk}^{ee}(E_j E_k^* - E_k E_j^*) \equiv {}^m\beta_{ijk}^{ee}\Pi_{jk}^{(A)} \tag{9}$$

where ${}^m\beta_{ijk}^{ee}$ is a hyperpolarizability, and $\Pi_{jk}^{(A)}$ is the antisymmetric conjugate product of the generator laser.[42–45]

Finally, for an order of magnitude estimation of the expected angle of rotation in a linearly polarized probe due to a generator laser of intensity $I_0 = 10^6$ W m^{-2}, we use antisymmetric polarizabilities computed ab initio by Manakov et al.[13] in atomic $S = \frac{1}{2}$ ground states as a guide to orders of magnitude. For example, in Cs at 9440 cm^{-1} we have $\hat{\alpha}_0^1 = 3.4 \times 10^{-39}$ C^{2}m^{2}J^{-1}. Focusing attention on the term in $\hat{\alpha}_0^1$ in Eq. (7), and using $N = 6 \times 10^{26}$ molecules m^{-3} and the Bohr magneton for $\hat{m}_0^1$, we obtain for a generator laser delivering at 300 K

$$B_0 \sim 10^{-7}I_0^{1/2} = 10^{-4} \text{ T} \tag{10}$$

an angle of rotation of 0.8 rad m^{-1}, easily measurable with a spectropolarimeter. This result should be proportional to the square root of the intensity I_0 of the generator laser and inversely proportional to temperature. There is also a contribution from the rank three perturbating tensor of Eq. (7). These features would add to the evidence for the existence of $\hat{B}_\Pi$ already available from the inverse Faraday effect[8–14] and light-induced NMR shifts.[15, 16]

Acknowledgments

The Leverhulme Trust, Cornell Theory Center, and the MRL at Penn State are thanked for research support.

References

1. M. W. Evans, *Physica B*, **182**, 227 (1992).
2. M. W. Evans, *Physica B*, **182**, 237 (1992).
3. M. W. Evans, *Physica B*, **183**, 103 (1993).

4. S. Borman, *Chem. Eng. News*, 6 April 1992.

5. M. W. Evans, *Physica B*, in press (1993).

6. M. W. Evans, submitted for publication.

7. P. W. Atkins, *Molecular Quantum Mechanics*, 2d ed., Oxford University Press, Oxford, UK, 1983.

8. P. S. Pershan, *Phys. Rev.* **130**, 919 (1963).

9. Y. R. Shen, *Phys. Rev.* **134**, 661 (1964).

10. J. P. van der Ziel, P. S. Pershan, and L. D. Malmstrom, *Phys. Rev. Lett.* **15**, 190 (1965).

11. P. S. Pershan, J. P. van der Ziel, and L. D. Malmstrom, *Phys. Rev.* **143**, 574 (1966).

12. S. Kielich, *Proc. Phys. Soc.* **86**, 709 (1965).

13. N. L. Manakov, V. D. Ovsiannikov, and S. Kielich, *Acta Phys. Pol.* **A53**, 581, 595 (1978).

14. P. W. Atkins and M. H. Miller, *Mol. Phys.* **15**, 503 (1968).

15. M. W. Evans, *J. Phys. Chem.* **95**, 2256 (1991).

16. W. S. Warren, S. Mayr, D. Goswami, and A. P. West., Jr., *Science*, **255**, 1683 (1992).

17. M. W. Evans, *Int. J. Mod. Phys. B*, **5**, 1963 (1991) (review).

18. M. W. Evans, *Chem. Phys.* **157**, 1 (1991).

19. M. W. Evans, S. Woźniak, and G. Wagnière, *Physica B* **173**, 357 (1991); 416 (1991).

20. S. Woźniak, M. W. Evans, and G. Wagnière, *Mol. Phys.* **75**, 81 (1992).

21. S. Woźniak, M. W. Evans, and G. Wagnière, *Mol. Phys.* **75**, 99 (1992).

22. M. W. Evans, *Physica B* **168**, 9 (1991).

23. M. W. Evans, *J. Mol. Spectrosc.* **146**, 351 (1991).

24. S. F. Mason, *Bio Systems* **20**, 27 (1987).

25. M. Faraday, *Philos. Mag.* **28**, 294 (1846).

26. L. Pasteur, *Rev. Sci.* **7**, 2 (1884).

27. L. D. Barron, *Chem. Soc. Rev.* **15**, 189 (1986).

28. M. W. Evans, in I. Prigogine and S. A. Rice, *Advances in Chemical Physics*, Vol. 81, Wiley Interscience, New York, 1992.

29. S. F. Mason, *Molecular Optical Activity and the Chiral Discriminations*, Cambridge University Press, Cambridge, UK, 1982.

30. M. W. Evans, *Phys. Rev. Lett.* **50**, 371 (1983).

31. P. G. de Gennes, *C. R. Hebd. Seances Acad. Sci., Ser. B* **270**, 891 (1970).

32. J. A. le Bel, *Bull. Soc. Chim. France* **22**, 337 (1874).

33. W. Kuhn and F. Braun, *Naturwissenchaften* **17**, 227 (1929).

34. W. A. Bonner, *Origins Life* **14**, 383 (1984).

35. S. B. Piepho and P. N. Schatz, *Group Theory in Spectroscopy with Applications to Magnetic Circular Dichroism*, Wiley, New York, 1983.

36. R. Serber, *Phys. Rev.* **41**, 489 (1932).

37. L. D. Barron, *Molecular Light Scattering and Optical Activity*, Cambridge University Press, Cambridge, UK, 1982.

38. B. W. Shore, *The Theory of Coherent Atomic Excitation*, Vol. 2, Wiley, New York, 1990.

39. A. R. Edmonds, *Angular Momentum in Quantum Mechanics*, 2d ed., Princeton University Press, Princeton, NJ, 1960.

40. B. L. Silver, *Irreducible Tensor Methods*, Academic, New York, 1976.

41. R. Zare, *Angular Momentum*, Wiley, New York, 1988.

42. S. Kielich, *Acta Phys. Pol.* **32**, 385 (1966).

43. S. Kielich, *Acta Phys. Pol.* **29**, 875 (1966).

44. S. Kielich and A. Piekara, *Acta Phys. Pol.* **18**, 1297 (1958).

45. A. Piekara and S. Kielich, *J. Chem. Phys.* **29**, 1297 (1958).

THE PHOTON'S MAGNETOSTATIC FLUX DENSITY $\hat{B}_\Pi$:* THE INVERSE FARADAY EFFECT REVISITED

I. INTRODUCTION

Intense, circularly polarized laser pulses produce a net magnetization M_Z (A m^{-1}) in atomic, molecular, and other condensed material such as dilute magnetic semiconductors.[1, 2] This magneto-optic property was first proposed by Piekara and Kielich,[3-6] and was demonstrated experimentally in the early 1960s by Pershan et al.[7, 9, 10] and Shen.[8] Since then, no further experimental work appears to have been reported on the effect. The theory of the inverse Faraday effect rests on the foundations built by Piekara and Kielich,[3-6] and was developed by Pershan et al.[7, 9, 10] in terms of the antisymmetric part of the tensor $E_i E_j^*$, where E_i is the electric field strength of a circularly polarized laser pulse in V m^{-1} and E_i^* is its own complex conjugate. This antisymmetric intensity is conveniently expressed in vector notation as the cross product $\mathbf{E} \times \mathbf{E}^*$, which is negative[11, 12] to motion reversal ($\hat{T}$) and positive to parity inversion ($\hat{P}$). It therefore has the necessary $\hat{P}$ and $\hat{T}$ symmetries of magnetic flux density, which is the qualitative explanation for the ability of a circularly polarized laser to magnetize.

Further development of the theory is due to Kielich,[13-15] Atkins and Miller,[16] Wagnière,[17] Woźniak et al.,[18, 19] and Evans et al.[20-22] with computer simulation of the magnetization. These theories all rely on the property $\mathbf{E} \times \mathbf{E}^*$ of the laser pulse. However, it has been shown recently[23-27] that this property, $\mathbf{E} \times \mathbf{E}^*$, is directly proportional to a novel, fundamental, magnetic flux density vector, $\mathbf{B}_\Pi$, of the classical electromagnetic field. In quantum-field theory[24] this becomes the novel,

*Printed by permission, based on Physica B, **183**, 103 (1993).

fundamental, and ubiquitous magnetic flux density operator, $\hat{B}_\Pi$, of the photon itself. In this paper it is argued that the inverse Faraday effect must be described semiclassically as a combination of terms in all positive integral powers of the classical $\mathbf{B}_\Pi$. The original theory, which relies on $\mathbf{E} \times \mathbf{E}^*$, is shown to be equivalent to considering only the term in $|\mathbf{B}_\Pi|^2$.

II.   DESCRIPTION OF $\mathbf{B}_\Pi$

It is straightforward to show that[11, 12, 17, 23-27]

$$\mathbf{E} \times \mathbf{E}^* = 2E_0^2\,\mathbf{ik} \tag{1}$$

where $\mathbf{k}$ is a unit axial vector, negative to $\hat{T}$ and positive to $\hat{P}$. This is purely imaginary and proportional to the square of E_0, the scalar electric field strength amplitude of a circularly polarized laser. In free space $E_0 = cB_0$ and

$$\mathbf{E} \times \mathbf{E}^* = 2E_0 c\,iB_0\mathbf{k} \equiv 2E_0 c\,i\mathbf{B}_\Pi \tag{2}$$

where B_0 is the scalar magnetic flux density amplitude (tesla) and c is the speed of light. The vector $\mathbf{B}_\Pi$ is the product $B_0\mathbf{k}$, which is in units of tesla. From these simple considerations,

$$\mathbf{B}_\Pi = \frac{\mathbf{E} \times \mathbf{E}^*}{2E_0 c\,i} = B_0\mathbf{k} = \frac{E_0}{c}\mathbf{k} = \left(\frac{I_0}{\varepsilon_0 c^3}\right)^{1/2}\mathbf{k} \sim 10^{-7}I_0^{1/2}\mathbf{k}$$

$$= \left(\frac{|\mathbf{N}|}{2\varepsilon_0 c^3}\right)^{1/2}\mathbf{k} \tag{3}$$

Here I_0 is the scalar intensity in W m^{-2}, which in free space is

$$I_0 = \varepsilon_0 cE_0^2 \tag{4}$$

where ε_0 is the free space permittivity. In Eq. (3), $|\mathbf{N}|$ is the scalar magnitude of Poynting's vector

$$\mathbf{N} = \frac{1}{\mu_0}\mathbf{E} \times \mathbf{B}^* \tag{5}$$

where μ_0 is the free space permeability. The vector $\mathbf{N}$ (Ref. 28) is a flux of energy density, and the novel vector $\mathbf{B}_{\Pi}$ is a flux of magnetic density. Although $\mathbf{N}$ and $\mathbf{B}_{\Pi}$ have the same negative $\hat{T}$ symmetry, the former is negative to $\hat{P}$ and the latter, as we have seen, is positive to $\hat{P}$. We note that $\hat{N}$ is nonzero in linear polarization, but $\mathbf{B}_{\Pi}$ vanishes, because the latter reverses sign with circular polarity, whereas the former does not. Furthermore, $\mathbf{N}$ can be expressed as

$$\mathbf{N} = 2I_0 \mathbf{n} \tag{6}$$

where $\mathbf{n}$ is the vector whose scalar magnitude is the real refractive index in the direction of propagation, and which is well known[28, 29] to be proportional to the $\hat{P}$- and $\hat{T}$-negative polar wave vector κ. This must be carefully distinguished from the $\hat{P}$-positive, $\hat{T}$-negative unit axial vector $\mathbf{k}$, which is multiplied by B_0 to form the novel $\mathbf{B}_{\Pi}$. Note that both $\mathbf{N}$ and $\mathbf{B}_{\Pi}$ are independent of the phase of the laser, and therefore of its angular frequency ω. In other words, the time averages over many cycles of both $\mathbf{N}$ and $\mathbf{B}_{\Pi}$ are nonzero, and it follows that $\mathbf{B}_{\Pi}$ is quite different from the usual oscillating $\mathbf{B}$ field of the electromagnetic plane wave, which vanishes when averaged over time. $\mathbf{B}$ is a complex quantity, with components mutually orthogonal (i.e., in X and Y) to the propagation direction (Z) of the wave, but none in that direction itself. In contrast, $\mathbf{B}_{\Pi}$ is a purely real quantity,[23-27] and is directed exclusively in Z, with no components in X and Y. Remarkably, its existence appears to have gone unrecognized in the long and illustrious history of the theory of electromagnetic fields.

III. THE ROLE OF $\mathbf{B}_{\Pi}$ IN THE INVERSE FARADAY EFFECT: SEMICLASSICAL TREATMENT

Using the vector $\mathbf{B}_{\Pi}$ it becomes straightforward to develop any magnetic effect of the circularly polarized electromagnetic plane wave, because we can now say that such a wave can magnetize material with which it interacts. There exists in nature an optical magnet, which delivers a magnetic flux density in tesla of

$$|\mathbf{B}_{\Pi}| \sim 10^{-7} I_0^{1/2} \tag{7}$$

Thus, for a circularly polarized laser of intensity $I_0 = 10\,000$ W m^{-2} (1.0 W cm^{-2}) the $\mathbf{B}_{\Pi}$ field is 10^{-5} T, or 0.1 G, about a tenth of the earth's mean magnetic field.

When $\mathbf{B}_{\Pi}$ is used, the theory of magnetization by circularly polarized light becomes standard and straightforward, because we have only to adapt the existing semiclassical theory[28-30] of magnetization by an ordinary magnetostatic field, $\mathbf{B}_S$, and replace $\mathbf{B}_S$ everywhere by $\mathbf{B}_\Pi$. Thus, the magnetization is given by

$$M_Z = \frac{1}{\mu}\left(\frac{\kappa}{1+\kappa}\right)B_{\Pi Z} = N\langle m_Z\rangle_U \tag{8}$$

where κ is the volume susceptibility, μ_0 is the vacuum permeability, N is the number density, and $\langle m_Z\rangle$ is the mean magnetic dipole moment. It is assumed here that the total magnetic dipole moment is a sum of permanent and induced components:

$$m_Z = m_Z^{(0)} + m_Z^{(\text{ind})}$$
$$m_Z^{(\text{ind})} = \tfrac{1}{2}\xi_{ZZ}B_{\Pi Z}/\mu_0 \tag{9}$$

where ξ_{ZZ} is the molecular magnetizability.[28-30] The ensemble average appearing in Eq. (8) is assumed to be the usual thermodynamic average[28-30]

$$\langle m_Z\rangle_U = \frac{\int\left(m_Z^{(0)} + m_Z^{(\text{ind})}\right)e^{-U/kT}\,d\Omega}{\int e^{-U/kT}\,d\Omega} \tag{10}$$

where the energy of interaction U is defined by

$$U = -m_Z^{(0)}B_{\Pi Z} - \tfrac{1}{2}\frac{\xi_{ZZ}}{\mu_0}B_{\Pi Z}^2 \tag{11}$$

The approximation[28-30]

$$\langle m_Z\rangle_U = \langle m_Z\rangle_0 - \frac{1}{kT}(\langle m_Z U\rangle_0 - \langle m_Z\rangle_0\langle U\rangle_0) + \cdots \tag{12}$$

is used for the thermodynamic average. Here $\langle\ \rangle_0$ denotes ensemble averaging in the absence of U, and $\langle\ \rangle_U$ denotes ensemble in averaging the presence of U. Since $\langle m_Z\rangle_0$ is zero in an initially isotropic material such as a gas or liquid, we have

$$\langle m_Z\rangle_U = -\frac{1}{kT}\langle m_Z U\rangle_0 + \cdots \tag{13}$$

This entirely standard semiclassical approach gives the following result for the temperature-dependent part of the effect we seek to describe:

$$\langle m_Z \rangle_U = \frac{1}{kT} \left(\langle m_Z^{(0)2} \rangle_0 B_{\Pi Z} + \frac{1}{\mu_0} \langle \xi_{ZZ} m_Z^{(0)} \rangle_0 B_{\Pi Z}^2 + \frac{1}{4\mu_0^2} \langle \xi_{ZZ}^2 \rangle_0 B_{\Pi Z}^3 \right)$$

$$+ \cdots \tag{14}$$

which shows that the magnetization (A/m) is described by terms in the first three powers of $B_{\Pi Z}$ within the first approximation (12) of the thermodynamic average (10). The term in $B_{\Pi Z}^2$ in Eq. (14) vanishes, however, because the theory of tensor invariants[28-30] shows that the ensemble average $\langle \xi_{ZZ} m_Z^{(0)} \rangle_0$ must vanish in isotropic media (but not in certain crystals). So we are left with terms in $B_{\Pi Z}$ and $B_{\Pi Z}^3$ for which the premultiplying ensemble averages do not vanish in liquids or gases.

In addition to these temperature-dependent terms, there exists the temperature-independent term considered in the usual theory of the inverse Faraday effect,[3-22] which has recently been put in the following simple form by Woźniak et al.[18]:

$$\langle m_Z \rangle_U = \frac{E_0^2}{3} \left({}^m \gamma_{123}^{ee} + {}^m \gamma_{231}^{ee} + {}^m \gamma_{312}^{ee} \right) \tag{15}$$

Here ${}^m \gamma_{ijk}^{ee}$ are molecule fixed-frame components of the appropriate[18, 19] molecular hyperpolarizability tensor. Using

$$E_0 = c |\mathbf{B}_\Pi| \tag{16}$$

it is immediately clear that this term is proportional to $|\mathbf{B}_\Pi|^2$. The complete expression for the inverse Faraday effect within the approximations we have made here is therefore

$$M_Z \doteq \frac{Nc^2}{3} \left({}^m \gamma_{123}^{ee} + {}^m \gamma_{231}^{ee} + {}^m \gamma_{312}^{ee} \right) B_{\Pi Z}^2$$

$$+ \frac{N}{kT} \left(\langle m_Z^{(0)2} \rangle_0 B_{\Pi Z} + \frac{\langle \xi_{ZZ}^2 \rangle_0}{4\mu_0^2} B_{\Pi Z}^3 \right) \tag{17}$$

IV. ORDER OF MAGNITUDE ESTIMATES

To estimate the various orders of magnitude of the contributing terms in Eq. (17), the magnetic dipole moment is estimated roughly as a tenth of the electronic Bohr magneton, i.e., as 10^{-24} J T^{-1}. A rough order of magnitude approximation to the volume magnetic susceptibility κ is obtained from a model calculation given by Atkins,[28] which gives κ of about 10^{-5}. From this, the magnetizability can be obtained using $\kappa = N\xi_{ZZ}$ where N is the number density.[28] The order of magnitude of the hyperpolarizability $^{m}\gamma_{ijk}^{ee}$ is obtained from the Faraday effect theory of Woźniak et al.[18, 31] as about 10^{-45} A m^{4}V^{-2} for a typical diamagnetic. For a paramagnetic with a permanent magnetic dipole moment it is assumed that the hyperpolarizability is roughly 100 times bigger, i.e., 10^{-43} A m^{4}V^{-2}.

In Eq. (17) N is set at 10^{28} molecules for the molar volume in meters cubed (Ref. 18) and kT at 4×10^{-21} J molecule^{-1}, equivalent to 300 K. The order of magnitude of $B_{\Pi Z}$ is set at 1.0 T, corresponding to a pulse of intensity about 3×10^{15} W m^{-2}, available from a contemporary mode-locked laser, which must be accurately circularly polarized.

These rough estimates give an order of magnitude of magnetization (A m^{-1}) of about 2.5 A m^{-1} for the term in $B_{\Pi Z}$, about 2.0 A m^{-1} for the term in $B_{\Pi Z}^3$, and about 30 A m^{-1} for the temperature-independent term proportional to $B_{\Pi Z}^2$. Clearly these figures depend on the estimates we have used for $m_Z^{(0)}$, ξ_{ZZ}, and $^{m}\gamma_{ijk}^{ee}$ but all three terms contribute to the total magnetization. In our estimate, the term in $B_{\Pi Z}^2$ happens to be dominant, but at very low T and with less intense laser pulses, the term in $B_{\Pi Z}$ dominates, provided there is a permanent magnetic dipole moment. (If the latter is zero, there are terms in $B_{\Pi Z}^2$ and $B_{\Pi Z}^3$, but not in $B_{\Pi Z}$.)

Note that the magnetization changes sign with the circular polarity of the laser. The term in $B_{\Pi Z}$ changes sign because the vector $\mathbf{B}_\Pi$ is switched from positive (left) to negative (right). The conjugate product $\mathbf{E} \times \mathbf{E}^*$ changes sign with circular polarization,[18] and the product $\mathbf{E} \times \mathbf{E}^*$ is proportional to $|\mathbf{B}_\Pi|^2\mathbf{k}$, where $\mathbf{k}$ is a unit axial vector.

V. CONCLUSION

The inverse Faraday effect is characterized by a laser-induced magnetization that is proportional to all positive integral powers of $B_{\Pi Z}$. The conventional theory[18] is based solely on a consideration of $\mathbf{E} \times \mathbf{E}^*$, and produces a magnetization proportional to $B_{\Pi Z}^2$ only, from $\mathbf{E} \times \mathbf{E}^*$ multiplied by the sample's hyperpolarizability. We have argued that there is also a magnetization produced by a product of $B_{\Pi Z}$ and the sample's magnetizability.

172   M. W. EVANS

Acknowledgments

The Leverhulme Trust is thanked for a Fellowship, and the Cornell Theory Center and Penn State Materials Research Laboratory are thanked for additional support.

References

1. J. Frey, R. Frey, C. Flytzanis, and R. Triboulet, *Opt. Commun.* **84**, 76 (1991).
2. J. K. Furdyna, *J. Appl. Phys.* **64**, R29 (1988).
3. A. Piekara and S. Kielich, *Archives des Sciences*, II, fasc. special, 7e Colloque Ampère, 1958, p. 304.
4. S. Kielich and A. Piekara, *Acta Phys. Pol.* **17**, 209 (1958); **18**, 439 (1959).
5. A. Piekara and S. Kielich, *J. Chem. Phys.* **29**, 1292 (1958).
6. A. Piekara and S. Kielich, *J. Phys. Rad.* **18**, 490 (1957).
7. P. S. Pershan, *Phys. Rev.* **130**, 919 (1963).
8. Y. R. Shen, *Phys. Rev.* **134**, 661 (1964).
9. J. P. van der Ziel, P. S. Pershan, and L. D. Malmstrom, *Phys. Rev. Lett.* **15**, 190 (1965).
10. P. S. Pershan, J. P. van der Ziel, and L. D. Malmstrom, *Phys. Rev.* **143**, 574 (1966).
11. M. W. Evans, *Physica B* **168**, 9 (1991).
12. M. W. Evans, in I. Prigogine and S. A. Rice (Eds.), *Advances in Chemical Physics*, Vol. 81, Wiley Interscience, New York, 1992.
13. S. Kielich, *Acta Phys. Pol.* **32**, 405 (1967).
14. S. Kielich, *Bulletin de la Societe des Amis des Sciences et des Lettres de Poznán* **21B**, 35, (1968/1969).
15. S. Kielich, *J. Colloid Interface Sci.* **30**, 159 (1969).
16. P. W. Atkins and M. H. Miller, *Mol. Phys.* **15**, 503 (1968).
17. G. Wagnière, *Phys. Rev. A* **40**, 2437 (1989).
18. S. Woźniak, M. W. Evans, and G. Wagnière, *Mol. Phys.* **75**, 81 (1992).
19. S. Woźniak, M. W. Evans, and G. Wagnière, *Mol. Phys.* **75**, 99 (1992).
20. M. W. Evans, S. Woźniak, and G. Wagnière, *Physica B* **173**, 357 (1991).
21. M. W. Evans, S. Woźniak, and G. Wagnière, *Physica B* **175**, 416 (1991).
22. M. W. Evans, S. Woźniak, and G. Wagnière, *Physica B*, **176**, 33 (1992).
23. M. W. Evans, *Physica B*, **182**, 227 (1992).
24. M. W. Evans, *Physica B*, **182**, 237 (1992).
25. M. W. Evans, *Physica B*, in press (1993).
26. M. W. Evans, *Physica B*, **179**, 157 (1992).
27. M. W. Evans, in S. Kielich and M. W. Evans (Eds.), *Modern Nonlinear Optics*, Topical Issue of I. Prigogine and S. A. Rice (Series Eds.), *Advances in Chemical Physics*, Wiley Interscience, New York, 1993, 3 Vols.
28. P. W. Atkins, *Molecular Quantum Mechanics*, 2d ed., Oxford University Press, Oxford, UK 1983.
29. S. Kielich, *Nonlinear Molecular Optics*, Nauka, Moscow, 1981.
30. L. D. Barron, *Molecular Light Scattering and Optical Activity*, Cambridge University Press, Cambridge, UK, 1982.
31. S. Woźniak, B. Linder, and R. Zawodny, *J. Phys. Paris* **44**, 403 (1983).

THE PHOTON'S MAGNETOSTATIC FLUX QUANTUM: THE OPTICAL COTTON-MOUTON EFFECT

I. INTRODUCTION

The ability of circularly polarized electromagnetic radiation to produce anisotropy in magnetic permeability was first proposed by Piekara and Kielich,[1,2] who systematically described light-induced anisotropy in material electric permittivity ($\Delta\varepsilon$), magnetic permeability ($\Delta\mu$), and refractive index (Δn). In Ref. 1 for example, formulated in the pre-laser era, it was proposed that "On observe alors des changements de ε, μ, ou n, dus à l'action du champ polarisant." We are concerned in this paper with the formulation of an optical Cotton-Mouton effect, a relative of the optical Kerr effect first proposed by Buckingham[3] and classified by Piekara and Kielich in their references. We define the novel optical Cotton-Mouton effect as a change in refractive index (linear dichroism) due to the novel, recently proposed, static magnetic field ($\mathbf{B}_\Pi$) of a circularly polarized electromagnetic plane wave.[4-8] Piekara and Kielich[1,2] described "saturation optique dans un champ optique." This effect later became known as the optical Kerr effect, or Buckingham effect.[3]

This paper is developed from the recent deduction[4-8] that the photon carries a magnetostatic flux quantum, $\hat{B}_\Pi$, whose classical equivalent is a phase-independent magnetic field $\mathbf{B}_\Pi$ generated in a circularly polarized light beam, an axial vector with the symmetry characteristics of a static magnetic flux density (tesla). The latter must be an axial vector positive to the parity inversion operator $\hat{P}$, and negative to the motion reversal operator $\hat{T}$ (Ref. 9). The classical field $\mathbf{B}_\Pi$ of the circularly polarized electromagnetic plane wave is a purely real quantity that is proportional to the antisymmetric (purely imaginary) part of the tensor $E_i E_j^*$, where E_i is the electric field strength of the wave in volts per meter. The scalar part of the tensor $E_i E_j^*$ is proportional to the phase-independent intensity of the plane wave in watts per meter squared, and we have shown elsewhere[4-8] that the vector part of $E_i E_j^*$ (i.e., its antisymmetric part) is proportional to the phase-independent magnetic flux density $\mathbf{B}_\Pi$ and vanishes if there is no degree of circular polarity. Furthermore, we have shown[8] that $\mathbf{B}_\Pi$ can be expressed in terms of the ubiquitous third Stokes parameter S_3 (Ref. 10) and therefore that phenomena such as circular dichroism and ellipticity are fundamentally magnetic.

The definition[4-8] of the $\hat{B}_\Pi$ operator per photon allows a wide range of novel optical/photonic phenomena to be forecasted straightforwardly, on the grounds that circularly polarized electromagnetic radiation can magnetize. This conclusion is independent of the phase of the plane wave, and therefore independent of its angular frequency, ω (rad/s). It follows that the time average of the classical vector $\mathbf{B}_\Pi$ is nonzero. It is emphasized that $\mathbf{B}_\Pi$ is fundamentally different from the usual oscillating $\mathbf{B}$ field of the plane wave[10]: $\mathbf{B}$ vanishes when time averaged, because it is phase dependent, and has no component in the propagation axis Z of the wave. The vector $\mathbf{B}_\Pi$ is directed exclusively in Z, and has no components in X and Y. By expressing the antisymmetric part of the tensor $E_i E_j^*$ as a vector product, $\mathbf{E} \times \mathbf{E}^*$ (refs. 4–8), it becomes clear that $\mathbf{B}_\Pi$ is a relative of the Poynting vector,[8-10] $\mathbf{N} = (\mathbf{E} \times \mathbf{B}^*)/\mu_0$, where μ_0 is the free space permeability. However, the polar vector $\mathbf{N}$ is $\hat{T}$- and $\hat{P}$-negative, whereas the axial vector $\mathbf{B}_\Pi$ is $\hat{T}$-negative, $\hat{P}$-positive,[4-8,11] a critically important symmetry difference. Accordingly, $\mathbf{N}$ is interpreted physically as a flux of energy density, and $\mathbf{B}_\Pi$ as a flux of magnetic density. Remarkably, $\mathbf{N}$ has been well known for many years, and $\mathbf{B}_\Pi$ appears to be entirely novel.

The flux density vector $\mathbf{B}_\Pi$ is clearly generated in vacuo (i.e., in free space), in direct analogy with $\mathbf{N}$. Both vectors $\mathbf{N}$ and $\mathbf{B}_\Pi$ are generated from solutions of Maxwell's classical equations through vector cross products of the usual, oscillating, phase-dependent $\mathbf{E}$ and $\mathbf{B}$ components of the electromagnetic plane wave solutions, cross products that multiply a vector with a complex conjugate vector, thus removing the phase dependence. For example, the complex conjugate $(\mathbf{E}^*)$ of $\mathbf{E}$, a plane wave solution of Maxwell's equations, is also an allowed solution of Maxwell's equations, and the vector product of $\mathbf{E}$ and $\mathbf{E}^*$, two allowed solutions, generates the purely imaginary conjugate product $\Pi^{(A)}$, which is proportional[4-8] to $\mathbf{B}_\Pi$. Similarly, the vector product of $\mathbf{E}$ and $\mathbf{B}^*$ is proportional to $\mathbf{N}$. Therefore, although Maxwell's equations allow no direct, phase-independent solutions in free space, vector products of allowed solutions, such as $\mathbf{N}$ and $\mathbf{B}_\Pi$, are physically meaningful phase-independent quantities whose time averages are nonzero.

It follows that $\mathbf{B}_\Pi$ can interact with material to produce observable effects, again in direct analogy with $\mathbf{N}$. The scalar part of $\mathbf{N}$ is the intensity I_0, and the intensity (for example, of a laser) is clearly a free space quantity that affects and interacts with material. (For example, a sample is heated by intense light, light that travels through a vacuum.) Similarly, $\mathbf{B}_\Pi$ is a free space magnetic flux density that can also affect material. For example, $\mathbf{B}_\Pi$ forms a vector dot product with an electronic or nuclear magnetic dipole moment to give an interaction Hamiltonian (whose expectation value is an observable and measurable energy). This leads to the recently observed phenomenon of optical NMR[12] in which a circularly polarized laser shifts NMR resonances in new and unexpected ways, leading to useful new fingerprints for the analytical laboratory.[13] These shifts were found experimentally to vanish in the uncertainty of measurement when the laser's circular polarization was removed, strong evidence that they depend in an as yet incompletely understood manner on $\mathbf{B}_\Pi$. (There are as many, if not more, mechanisms involving $\mathbf{B}_\Pi$ in optical NMR as there are involving the ordinary magnetostatic field in conventional NMR.)

It is easy to see that if circularly polarized light is simply regarded as an "optical magnet," there should be observable in one way or another all the well-known phenomena of conventionally produced magnetism,[14] such as the Faraday, magnetic circular dichroic, Zeeman, Cotton-Mouton, Gerlach-Stern, Aharonov-Bohm, NMR, and ESR phenomena. Thus far, the $\mathbf{B}_\Pi$ concept has been developed for the optical Zeeman effect,[5] anomalous optical Zeeman effect,[6] the optical Faraday effect,[7] optical effects in Compton scattering,[4] and the inverse Faraday effect,[8] which is bulk magnetization by $\mathbf{B}_\Pi$ of a circularly polarized laser. In general, whenever a magnetic can be used in physics, so can a circularly polarized laser, which generates $\mathbf{B}_\Pi$. The magnitude of $\mathbf{B}_\Pi$ is approximately 10^{-7} $I_0^{1/2}$ in tesla,[4-8] so that an accurately circularly polarized laser of intensity 1.0 W cm^{-2} generates 10^{-5} T, about a tenth of the earth's mean magnetic field. Clearly, pulses of laser radiation of say, up to 10^{16} W m^{-2}, available in principle,[15] generate a substantial 10 T over the duration of the laser pulse. (Normally incoherent radiation, such as daylight, produces no $\mathbf{B}_\Pi$, because there is no mean circular polarization; a linearly polarized laser, however, intense, produces no $\mathbf{B}_\Pi$, because such a laser always contains equal and opposite amounts of right and left circularly polarized light—right and left photons.)

In this paper, an example is given of the straightforward way in which the $\mathbf{B}_\Pi$ vector can be used to anticipate the existence of a novel optical phenomenon—the optical Cotton-Mouton effect. Section II defines $\mathbf{B}_\Pi$ in its classical limit in terms of fundamental constants, and brings out the precise analogy between $\mathbf{B}_\Pi$ and an ordinary magnetostatic flux density, $\mathbf{B}_0$. This allows the semiclassical theory[10] of the Cotton-Mouton effect to be developed straightforwardly in terms of $\mathbf{B}_\Pi$ in section III. The order of magnitude of the linear dichroism (or ellipticity) induced by $\mathbf{B}_\Pi$ is estimated in Section IV.

II. DEFINITION OF THE CLASSICAL $\mathbf{B}_\Pi$ OF A CIRCULARLY POLARIZED LASER

The classical vector $\mathbf{B}_\Pi$ of a circularly polarized laser in free space is obtained straightforwardly[4-8] by a consideration of the $\hat{T}$ and $\hat{P}$ symmetries of the conjugate product—the vector part of $E_i E_j^*$:

$$\mathbf{\Pi}^{(A)} = \mathbf{E} \times \mathbf{E}^* = 2E_0^2 i\mathbf{k} = 2E_0 c i\mathbf{B}_\Pi \tag{1}$$

Here the axial vector $\mathbf{B}_\Pi$ is in tesla, is $\hat{T}$-negative and $\hat{P}$-positive, and is directed in the propagation axis of the laser. Thus, $\mathbf{B}_\Pi$ has the necessary and sufficient characteristics to define a magnetic flux density vector. This simple derivation shows that circularly polarized radiation magnetizes material with which it comes into contact from free space. A relation between $\mathbf{B}_\Pi$ and the Poynting vector $\mathbf{N}$ is obtained straightforwardly from a consideration of[10]

$$\mathbf{B} = \frac{1}{c}\mathbf{n} \times \mathbf{E} \qquad \mathbf{E} = -\frac{c}{n^2}\mathbf{n} \times \mathbf{B} \qquad (2)$$

Here, $\mathbf{n}$ is a $\hat{T}$- and $\hat{P}$-negative polar vector, whose scalar magnitude is the refractive index, and which is related to the classical wave vector of the laser by

$$\boldsymbol{\kappa} = \frac{\omega}{c}\mathbf{n} \qquad (3)$$

In free space, the scalar magnitude of $\mathbf{n}$ is unity, and it follows that

$$\mathbf{N} = 2I_0\mathbf{n} \qquad (4)$$

where the magnitude of $\mathbf{N}$ is defined through the scalar intensity of the laser in W m^{-2}:

$$I_0 = \varepsilon_0 c E_0^2 \qquad (5)$$

Here ε_0 is the vacuum permittivity in S.I. units and c is the speed of light. It follows that $\mathbf{B}_\Pi$ is related to the square root of the Poynting vector:

$$\mathbf{B}_\Pi = B_0\mathbf{k} = \frac{E_0}{c}\mathbf{k} = \left(\frac{I_0}{\varepsilon_0 c^3}\right)^{1/2}\mathbf{k} = \left(\frac{|\mathbf{N}|}{2\varepsilon_0 c^3}\right)^{1/2}\mathbf{k} \qquad (6)$$

From these simple derivations it follows that the scalar part (or trace) of the tensor $E_i E_j^*$ is responsible for the Poynting vector's magnitude, and that the antisymmetric (vector) part of the tensor $E_i E_j^*$ is responsible for the novel phase-independent magnetic flux density $\mathbf{B}_\Pi$. In quantum field theory it has been shown elsewhere[4-9] that $\mathbf{B}_\Pi$ becomes a novel elementary magnetic field of the photon itself—an operator $\hat{B}_\Pi$.

III. APPLICATION TO THE OPTICAL COTTON-MOUTON EFFECT

Since $\mathbf{B}_{\Pi}$ has all the characteristics of a magnetostatic flux density, it can be used to describe a variety of novel magneto-optic effects, an example of which is an optical Cotton-Mouton effect, developed in this section with a standard semiclassical approach. The optical Cotton-Mouton effect is the development of linear birefringence in a probe light beam propagating in axis Z through a suitable sample and linearly polarized at 45° to the direction of an applied pump laser in the X axis. The latter is circularly polarized and generates $B_{\Pi X}$. Elliptical polarization in the probe is produced by $B_{\Pi X}$ of the pump, which plays the role of the ordinary magnet of the original effect discovered by Cotton and Mouton[16] in 1907. The pump's $B_{\Pi X}$ produces a phase difference in the two coherent resolved components of the probe, linearly polarized parallel and perpendicular, respectively, to the direction X of $B_{\Pi X}$ of the pump. This phase difference is[10]

$$\delta = \frac{\omega}{c} l(n_{\parallel} - n_{\perp}) \tag{7}$$

where $n_{\parallel}$ and $n_{\perp}$ are the refractive indices for light linearly polarized parallel and perpendicular to X. The resulting ellipticity is $\delta/2$.

At absorbing wavelengths, the two components $n_{\parallel}$ and $n_{\perp}$ are accompanied by two different absorption coefficients, signaling the presence of linear dichroism due to $B_{\Pi X}$ of the circularly polarized pump laser. There is a rotation of the major axis of the polarization ellipse of the probe laser because a difference in amplitude develops between two orthogonal resolved components for which no phase difference exists.[10]

Kielich and Piekara[2] have summarized the various theories of the standard Cotton-Mouton effect, under their classification scheme denoted "optical saturation in a magnetic field." In our case this magnetic field is $B_{\Pi X}$ of the circularly polarized pump laser in direction X. The novel $\mathbf{B}_{\Pi}$ concept allows these theories to be adapted directly for the optical Cotton-Mouton effect suggested here. We have simply replaced an ordinary magnet with an optical magnet, which is an intense circularly polarized pump laser, operable at any electromagnetic frequency, from infrared to X-ray regions. Notable theories include those of Raman and Krishnan,[17] Piekara,[18, 19] Peterlin and Stuart,[20] Snelman,[21] and the semiclassical approach at Buckingham and Pople.[22, 23] Kielich has developed the conventional Cotton-Mouton and related effects in several directions, for example, (1) the theory of the inverse Cotton-Mouton effect,[24] which he described as the induction of magnetic anisotropy by an intense laser beam; (2) the theories in colloids of the inverse Cotton-Mouton effect[25] and the important but neglected Majorana effect[26] in colloids, liquid crystals, and polymers; and (3) general theories of magneto-optics.[27]

These theories can now be recast to great advantage, in principle, using the $\mathbf{B}_\Pi$ concept, or its equivalent for magneto-photonics, the operator $\hat{B}_\Pi$ (Ref. 15) of the photon itself. As an example we take the semiclassical theory of Buckingham and Pople[23] given originally for the ordinary Kerr effect, and adapt it straightforwardly for the optical Cotton-Mouton effect by substituting $\mathbf{B}_\Pi$ for the ordinary static electric field $\mathbf{E}_0$ of the Kerr effect, or the ordinary static $\mathbf{B}_0$ field of the standard Cotton-Mouton effect.[10] In so doing it is convenient to follow the summary given by Barron[10] for the Kerr effect and indicate the simple changes needed for the optical Cotton-Mouton effect along the way.

The starting point is the expression for probe ellipticity in Rayleigh refringent scattering theory[10]:

$$\eta = -\tfrac{1}{4}N\omega\mu_0 cl\big[\alpha'_{XX}(f) - \alpha'_{YY}(f)\big] \tag{8}$$

in terms of laboratory frame components of the real parts of the polarization tensor $\alpha'_{\alpha\beta}$ of a molecules of the sample. Here N is the number of molecules in the sample, ω is the angular frequency of the probe laser, μ_0 is the vacuum magnetic permeability, and l is the sample length in meters through which the probe passes. The $\mathbf{B}_\Pi$ vector of the circularly polarized pump laser generates anisotropy in the sample because $\mathbf{B}_\Pi$ interacts with the permanent and induced magnetic dipole moments in each molecule (or atom). The total magnetic dipole moment per molecule is, accordingly,

$$m_\alpha = m_{0\alpha} + \chi'_{\alpha\beta}B_{\Pi\beta} + \cdots \tag{9}$$

where $m_{0\alpha}$ is the permanent molecular electronic magnetic dipole moment (if nonzero), and $\chi'_{\alpha\beta}$ is the real static susceptibility, a symmetric second-rank property tensor.[10] The dynamic polarizability is perturbed by $\mathbf{B}_\Pi$ of the pump laser as follows:

$$\alpha'_{\alpha\beta}(\mathbf{B}_\Pi) = \alpha'_{\alpha\beta}(0) + \alpha'_{\alpha\beta\gamma} + \alpha'_{\alpha\beta\gamma\delta}B_{\Pi\gamma}B_{\Pi\delta} + \cdots \tag{10}$$

and in the evaluation of η in Eq. (8) an ensemble average is taken of the polarizability tensor components perturbed by $\mathbf{B}_\Pi$ of the pump. In forming this ensemble average, an interaction potential energy is used of the type

$$V(\Omega) = -m_{0X}B_{\Pi X} - \tfrac{1}{2}\chi'_{XX}B_{\Pi X}^2 + \cdots \tag{11}$$

From tensor invariant theory[10] the ellipticity is finally obtained, in precise parallel with the theory of the Kerr effect, as

$$
\eta = -\frac{1}{120}\omega\mu_0 clNB_{\Pi X}^2\Bigg[\alpha'_{\alpha\beta\alpha\beta}(f) - \alpha'_{\alpha\alpha\beta\beta}(f)
$$

$$
+ \frac{2}{kT}\left(3\alpha'_{\alpha\beta\alpha}(f)m_{0\beta} - \alpha'_{\alpha\alpha\beta}(f)m_{0\beta}\right)
$$

$$
+ \frac{1}{kT}\left(3\alpha'_{\alpha\beta}(f)\chi'_{\alpha\beta} - \alpha'_{\alpha\alpha}(f)\chi'_{\beta\beta}\right) \tag{12}
$$

$$
+ \frac{1}{k^2T^2}\left(3\alpha'_{\alpha\beta}(f)m_{0\alpha}m_{0\beta} - \alpha'_{\alpha\alpha}(f)m_{0\beta}m_{0\beta}\right)\Bigg]
$$

which is valid rigorously at transparent frequencies only.

IV. DISCUSSION

For simplicity we consider a sample that has no permanent magnetic dipole moment. For this sample the probably dominant term in Eq. (12) involves a product of the molecular polarizability and molecular suscepti-bility. The ellipticity developed in the probe is second order in $B_{\Pi X}$, or first order in the intensity I_0 of the pump laser. Accordingly, the sign of η should not be changed by switching the circular polarization of the pump from left to right, thus reversing $\mathbf{B}_{\Pi}$ (Refs. 4–8). However, if the pump is linearly polarized, $B_{\Pi X}$ and thus η should be zero for all I_0 of the pump.

With these overall considerations and taking a sample molecular elec-tric polarizability[27] of the order $10^{-40}C^2m^2J^{-1}$, a static molecular suscep-tibility of the order $10^{-24}C^2m^{-4}J^{-1}S^{-1}$, N of the order 10^{26} molecules m^{-3}, l of 1 m, ω about 10^{15} rad s^{-1}, and kT of 4.14×10^{-21} J, corresponding to 300 K, we obtain

$$
\eta \doteq 10^{-2}B_{\Pi X}^2 \tag{13}
$$

Therefore, for a pump laser delivering a $B_{\Pi X}$ pulse of 1.0 T, the ellipticity change is 0.01 rad, or 0.6° m^{-1}. As first discussed by Kielich,[25] this could be enhanced by up to six orders of magnitude in colloidal solution, or in suitable liquid crystals just above the isotropic to mesophase transition, i.e., in a state where the sample is still transparent to pump and probe lasers. The effect of the pump's $\mathbf{B}_{\Pi}$ pulse can be picked up by a probe using highly developed contemporary timing technology, as in work on the rotation of the elliptical polarization ellipse by a circularly polarized, giant ruby laser pump pulse.[28-30] It therefore appears possible to observe the optical Cotton-Mouton effect as proposed in this work in terms of the novel $\mathbf{B}_\Pi$ vector, whose photon equivalent is the $\hat{B}_\Pi$ operator, the photon's magnetostatic flux density.

Acknowledgments

The Leverhulme Trust, Cornell Theory Center, and the Materials Research laboratory at Penn State are thanked for research support.

References

1. A. Piekara and S. Kielich, *Archives des Sciences*, II, fasc. special, 7e Colloque Ampère, 1958, p. 304.

2. S. Kielich and A. Piekara, *Acta Phys. Pol.* **18** (1959) 439.

3. A. D. Buckingham, *Proc. Phys. Soc* **68A**, 910 (1955); **69B**, 344 (1956); **70B**, 753 (1957).

4. M. W. Evans, *Physica B*, **182**, 227 (1992).

5. M. W. Evans, *Physica B*, **182**, 237 (1992).

6. M. W. Evans, *Physica B*, **182**, 103 (1993).

7. M. W. Evans, *Physica B*, in press (1993).

8. M. W. Evans, *Physica B*, in press (1993).

9. M. W. Evans, in I. Prigogine and S. A. Rice (Eds.) *Advances in Chemical Physics*, Vol. 81, Wiley Interscience, New York, 1992.

10. L. D. Barron, *Molecular Light Scattering and Optical Activity*, Cambridge University Press, Cambridge, UK, 1982.

11. S. Kielich and M. W. Evans (Eds.), *Modern Non-Linear Optics*, special topical issue of I. Prigogine and S. A. Rice (Series Eds.), *Advances in Chemical Physics*, Wiley Interscience, New York, 1993, 3 Vols.

12. W. S. Warren, S. Mayr, D. Goswami, and A. P. West, Jr., *Science* **255**, 1683 (1992).

13. S. Borman, *Chem. Eng. News*, April 1992, p. 32.

14. P. W. Atkins, *Molecular Quantum Mechanics*, 2d ed., Oxford University Press, Oxford, UK, 1983.

15. B. W. Shore, *The Theory of Coherent Atomic Excitation*, Wiley Interscience, New York, 1990, Vols. 1 and 2.

16. A. Cotton and H. Mouton, *Ann. Chim. Phys.* **11**, 145, 289 (1907).

17. V. Raman and K. Krishnan, *Proc. R. Soc. London Ser. A* **117**, (1927).

18. A. Piekara, *Proc. R. Soc. London Ser. A* **172**, 360 (1939).

19. A. Piekara, *Acta Phys. Pol.* **10**, 37, 107 (1950).

20. A. Peterlin and H. Stuart, *Z. Phys.* **113**, 663 (1939).

21. O. Snelman, *Philos. Mag.* **40**, 983 (1949).

22. A. D. Buckingham and J. A. Pople. Proc. Phys. Soc. **69B**, 1133 (1956).

23. A. D. Buckingham and J. A. Pople, *Proc. Phys. Soc.* **68A**, 905 (1955).

24. S. Kielich, *Acta Phys. Pol* **32** 405 (1967).

25. S. Kielich, *J. Colloid Interface Sci.* **30**, 159 (1969).

26. S. Kielich, *Bulletin de la Societé des Amis des Sciences et des Lettres de Poznań*, **21B**, 35 (1968/69).

27. S. Kielich, in M. Davies (Senior Reporter), *Dielectric and Related Molecular Processes*, Chem. Soc. Specialist Periodical Report, London, 1972, Vol. 1.

28. P. D. Maker and R. W. Terhune, *Phys. Rev.* **137**, A801 (1965).

29. C. C. Wang, *Phys. Rev.* **152**, 149 (1966).

30. F. Shimizu, *J. Phys. Soc. Japan* **22**, 1070 (1967).

THE PHOTON'S MAGNETOSTATIC FLUX QUANTUM: FORWARD–BACKWARD BIREFRINGENCE INDUCED BY A LASER

I. INTRODUCTION

When a magnetostatic flux density $\mathbf{B_S}$ is applied to an initially isotropic chiral liquid, that liquid develops forward–backward birefringence, otherwise known as magneto-chiral birefringence and magneto-spatial dispersion. The refractive index in the direction $(+Z)$ of forward propagation of a probe laser becomes different from that in the backward direction $(-Z)$. The Kramers-Kronig theorem implies that the same happens to the power absorption coefficient. The presence of this effect in liquids has been proposed several times theoretically, but has never been detected experimentally. The effect appears to have been first proposed by Portigal and Burstein[1] in magnetic crystal symmetries, and was measured by Mankelov et al.[2] The theory was extended by Kielich and Zawodny[3] to crystals with magnetic ordering. Working with liquids, Baranova and Zel'dovich[4] described the refractive index change in circularly polarized probe radiation in terms of the dot product $\mathbf{B_S} \cdot \mathbf{\kappa}$, where $\mathbf{\kappa}$ is the classical wave vector of the probe, and implied the presence of forward–backward birefringence. The first detailed papers on the subject in chiral liquids are due to Woźniak and Zawodny,[5,6] who defined the molecular point groups able to support the effect, and developed a theory based on electronic distortion and reorientation of the permanent molecular magnetic dipole moment, if nonvanishing. Wagnière and coworkers[7-9] developed the theory of the effect for power absorption as well as refractive index, and Barron and Vrbancich[10] contributed a comprehensive paper on forward–backward birefringence and dichroism in chiral liquids based on time-odd, complex, molecular property tensors. In this work, an unsuccessful attempt to measure the effect experimentally was reported briefly. Woźniak later developed the semiclassical theory of the effect in diamagnetic molecules in the presence of electric[11,12] and optical[13] fields, and later[14] for para-magnetic molecules. Wagnière[15] considered the consequences of using the effect to measure parity violation in atoms, and has also proposed a similar, but much weaker, forward–backward effect, which he named "inverse magneto-chiral birefringence."[16] The latter has recently been developed in the context of frequency dependence by Woźniak et al.[17,18] Barron[19] has reviewed the effect in the context of motion reversal ($\hat{T}$) and parity inversion symmetry ($\hat{P}$).

Recently[20-25] the present author has proposed an optically induced forward–backward birefringence, in which the magnetostatic field is replaced by a circularly polarized pump laser, wherein the antisymmetric component ($\mathbf{\Pi}^{(A)}$) of the intensity tensor of the pump laser is the optical property responsible for the development of forward–backward birefringence in a probe laser directed parallel to the circularly polarized pump. The component $\mathbf{\Pi}^{(A)}$ can be expressed as a $\hat{T}$-negative, $\hat{P}$-positive axial vector directed in the propagation axis Z of the circularly polarized pump. This effect, known as spin chiral dichroism,[20-25] is proportional to the scalar magnitude I_0 of the pump laser intensity, and vanishes if there is no degree of pump circular polarization.

In this paper, we use the recently proposed[26-31] magnetostatic flux density operator $\hat{\mathbf{B}}_{\Pi}$ of the photon to demonstrate straightforwardly that there exists a first-order optically induced forward–backward birefringence in which the traditional $\mathbf{B}_S$ field produced by a strong magnet is replaced by the $\hat{\mathbf{B}}_{\Pi}$ operator of a circularly polarized pump laser pulse. In the classical theory of fields, $\hat{B}_{\Pi}$ becomes the axial flux density vector $\mathbf{B}_{\Pi}$, directed along the propagation axis Z of the pump laser. This new effect is proportional to the square root of the pump laser scalar intensity, i.e., $I_0^{1/2}$. We refer hereafter to this effect as $\hat{B}_{\Pi}$-induced forward–backward anisotropy (BFBA), an effect that is one of a large number of new magneto-photonic phenomena based on the existence of $\hat{B}_{\Pi}$, or its classical equivalent $\mathbf{B}_{\Pi}$. Among these are optical NMR, recently detected experimentally,[32] in which $\hat{B}_{\Pi}$ causes unexpected and characteristic shifts in conventional NMR resonance patterns in N dimensions, optical ESR and the optical Zeeman effects,[27] the optical Faraday effect,[28] the optical Cotton-Mouton effect,[29] and other $\hat{B}_{\Pi}$-induced effects. It has also been shown[30,31] that $\hat{B}_{\Pi}$ is ubiquitous in physical optics, being interpretable in terms of the third Stokes parameter S_3, and is therefore the origin of such well-known phenomena as circular dichroism and ellipticity. The operator $\hat{B}_{\Pi}$ also provides a new fundamental explanation for the de Broglie wave particle duality in photons,[33] in that it can be defined simultaneously in terms of the angular and linear momenta of the photon. The same conclusion applies for other massless leptons that propagate at the speed of light in any reference frame, for example, the neutrino. In Section II we summarize briefly the pertinent properties of the $\mathbf{B}_\Pi$ vector of a circularly polarized pump laser, and relate it to such well-known quantities as the Poynting vector $\mathbf{N}$ (Ref. 34) and the usual electric field strength amplitude (E_0) and magnetic flux density amplitude (B_0) of the classical electromagnetic plane wave. A simple equation relates the scalar magnitude (tesla) B_0 of $\mathbf{B}_\Pi$ to the square root of I_0 of the pump laser (watts per square meter). Section III is a straightforward application of the $\mathbf{B}_\Pi$ vector to the various theories proposed for forward–backward birefringence due to an ordinary magnetostatic field, $\mathbf{B}_S$. An order of magnitude of the BFBA is given in terms of the intensity I_0 of the pump laser, thus anticipating the feasibility of an experimental investigation of the effect.

II. SUMMARY OF $\mathbf{B}_\Pi$ PROPERTIES

The $\mathbf{B}_\Pi$ vector is the classical equivalent of the quantum-field operator $\hat{B}_\Pi$ (Ref. 26), and can be expressed in terms of the vector cross product $\mathbf{E} \times \mathbf{E}^*$ of the free space electromagnetic plane wave. Here $\mathbf{E}$ is the usual, oscillating, electric field strength vector of the wave, and $\mathbf{E}^*$ is its complex conjugate. We have

$$\mathbf{B}_\Pi = \frac{\mathbf{E} \times \mathbf{E}^*}{2E_0 c i} = B_0 \mathbf{k} = \frac{E_0}{c}\mathbf{k} = \left(\frac{I_0}{\varepsilon_0 c^3}\right)^{1/2}\mathbf{k} = \left(\frac{|\mathbf{N}|}{2\varepsilon_0 c^3}\right)^{1/2}\mathbf{k} \quad (1)$$

Here $\mathbf{k}$ is an axial unit vector in the propagation axis of the wave, ε_0 is the vacuum permittivity in S.I. units, and c is the speed of light. From these definitions[26-33] it is clear that the novel $\mathbf{B}_\Pi$ is a member of the same class of optical properties as the Poynting vector $\mathbf{N}$:

$$\mathbf{N} = \frac{\mathbf{E} \times \mathbf{B}^*}{\mu_0} \quad (2)$$

where μ_0 is the vacuum permeability in S.I. units and $\mathbf{B}$ is the usual oscillating electromagnetic flux density vector of the plane wave in free space. (Note that the novel $\mathbf{B}_\Pi$ is independent of the phase of the plane wave, is directed exclusively in its propagation axis Z and is purely real, reversing its sign with a left to right-switch in circular polarization and vanishing, therefore, in linearly polarized or incoherent radiation such as normal daylight. On the other hand, $\mathbf{B}$ of the plane wave is known[34] to be a complex quantity, phase dependent, with components in X and Y, mutually orthogonal to the propagation axis.) The novel $\mathbf{B}_\Pi$ and the well-known $\mathbf{N}$ are both derived by forming vector cross products from oscillating plane wave solutions of the Maxwell equations, but are themselves independent of the phase of the plane wave, and thus of its angular frequency ω. The vector $\mathbf{N}$ is a flux of energy density, and $\mathbf{B}_\Pi$ is a flux of magnetic density in tesla (i.e., magnetic flux density). However, the $\hat{P}$ symmetries of $\mathbf{B}_\Pi$ and $\mathbf{N}$ are opposite: The former is $\hat{P}$-positive and the latter is $\hat{P}$-negative, implying that the $\mathbf{B}_\Pi$ is an axial vector and that $\mathbf{N}$ is a polar vector. The $\hat{T}$ symmetries of both vectors are identical, both are $\hat{T}$-negative,[35-38] a conclusion arrived at after a careful consideration of the $\hat{T}$ symmetries of $\mathbf{E} \times \mathbf{E}^*$ and $\mathbf{E} \times \mathbf{B}^*$. The scalar magnitude of $\mathbf{N}$ is the quantity $2I_0$, which is the trace (or scalar part) of the tensor product $E_i E_j^*$ in tensor subscript notation.[34] The magnitude of $\mathbf{B}_\Pi$ is derived from the antisymmetric part of $E_i E_j^*$, which is its vector part.[39]

The following approximate relation is useful for assessing the magnitude of $\mathbf{B}_\Pi$ in terms of I_0:

$$|\mathbf{B}_\Pi| = B_0 \doteq 10^{-7} I_0^{1/2} \tag{3}$$

so that a circularly polarized laser delivering one watt per square centimeter ($10\,000$ W m^{-2}) generates a $\mathbf{B}_\Pi$ field of 10^{-5} T, which is 0.1 G or about one-tenth of the earth's mean magnetic field. A laser pulse of 10^{16} generates 10.0 T, equivalent to a contemporary superconducting magnet for the duration of the pulse.

III. EXPRESSIONS FOR FORWARD–BACKWARD ANISOTROPY DUE TO $\mathbf{B}_\Pi$

Using the novel $\mathbf{B}_\Pi$ vector, it is possible to adapt immediately the key results of previous work on forward–backward birefringence. Adapting the results of Woźniak and Zawodny[5,6] shows, for example, that the magneto-spatial change in the refractive index is proportional to the scalar product $\mathbf{B}_\Pi \cdot \boldsymbol{\kappa}$, where $\boldsymbol{\kappa}$ is the wave vector of a circularly polarized probe laser parallel to the pump laser generating $\mathbf{B}_\Pi$. By substituting $\mathbf{B}_\Pi$ for $\mathbf{B}_S$ in the development of these authors, this change in refractive index can be interpreted through the same electronic distortion mechanism and through the same reorientational process mediated by the permanent magnetic dipole moment. Furthermore, the molecular symmetries (magnetic point groups) mediating the magneto-spatial effect of $\mathbf{B}_\Pi$ are the same as those derived by Woźniak and Zawodny for the magneto-spatial effect of an ordinary magnetic field, $\mathbf{B}_S$. The key semiclassical expression of magneto-spatial dispersion due to $\mathbf{B}_{\Pi}$ is the precise analogue of the last term of Eq. (52) of Woźniak and Zawodny[6]:

$$n_{\pm} = -(V + V^{\mathrm{T}})(\mathbf{B}_{\Pi} \cdot \mathbf{S}) + \cdots \qquad (4)$$

where $n_{\pm}$ are the real refractive indices measured by the probe laser in the $\pm Z$ directions, the quantities V and V^{T} are defined in terms of molecular property tensors by Eqs. (58) and (59) of Ref. 6, and $\mathbf{s}$ is a unit vector in Z. In an addendum to their original paper,[6] these authors discuss other contributions to their original Eq. (52). They have provided extensive tabulation of the symmetry of the molecular property tensors making up V and V^{T}; but no order of magnitude estimate was made.

Similarly, it is possible to adapt straightforwardly the development of magneto-chiral birefringence due to Barron and Vrbancich.[10] These authors make extensive use of tensor algebra and invariant ensemble averages to describe the phenomenon semiclassically. They provide an approximate estimate of the anticipated order of magnitude of the effect, and also report an attempted experimental observation with a modified Rayleigh interferometer. Our purpose here is to adopt their main results for use with $\mathbf{B}_{\Pi}$ of a circularly polarized pump laser, thus providing immediately a theory of BFBA in chiral liquids. This is achieved by replacing the $\mathbf{B}$ field of the permanent magnet, wherever it occurs, by the $\mathbf{B}_{\Pi}$ field of a circularly polarized pump laser. The symbol $\mathbf{B}$ in the development by Barron and Vrbancich[10] is replaced wherever it occurs in their paper by $\mathbf{B}_{\Pi}$. It is therefore unnecessary to repeat the complicated tensor algebra here and we focus on a result such as their Eq. (3.17a), which gives the magneto-chiral birefringence in terms of appropriate molecular property tensor elements, averaged with the principles of tensor invariants.[10, 34] This result is later approximated[10] to base an estimate upon

$$n^{1\uparrow} - n^{1\downarrow} \doteq \frac{\frac{1}{6}\mu_0 c N B_{\Pi Z} \varepsilon_{\alpha\beta\gamma} m_{0\gamma} G_{\alpha\beta}(f)}{kT} + \cdots$$

where $n^{1\uparrow} - n^{1\downarrow}$ is the forward–backward refractive index difference due to a pump laser generating $+B_{\Pi Z}$ and $-B_{\Pi Z}$ (left and right circular polarizations, respectively) to a probe laser in the Z axis. Here $\varepsilon_{\alpha\beta\gamma}$ is the Levi-Civita symbol and $G_{\alpha\beta}(f)$ the appropriate[10] molecular property tensor elements in semiclassical approximation. The tensor $G_{\alpha\beta}(f)$ is supported only by chiral ensembles. The property $m_{0\gamma}$ is an appropriate component of the permanent magnetic dipole moment, and kT is the thermal energy per molecule. Under the conditions discussed on page 728 of Ref. 10:

$$n^{1\uparrow} - n^{1\downarrow} \sim 10^{-7} \qquad (6)$$

for a $\mathbf{B}_\Pi$ field of 1.0 T delivered by a pump laser pulse. To generate optically a $\mathbf{B}_\Pi$ of 1.0 T requires a pulse of intensity about 10^{14} W m^{-2} (10^{10} W cm^{-2}). Such an intensity is a practical possibility with a picosecond pulse from an instrument such as a mode-locked dye laser,[35] which is circularly polarized and if necessary, focused. To detect the change $n^{1\uparrow} - n^{1\downarrow}$ experimentally requires the most sensitive type of Rayleigh refractometer,[10] in which a right circularly polarized pump pulse is delivered in one arm of the refractometer and a left circularly polarized pump pulse is delivered simultaneously in the other arm. The sample is chosen optimally as described by Barron and Vbrancich[10] in the context of a conventional magnetostatic $\mathbf{B}$ from a superconducting magnet.

Woźniak[11-14] and Wagnière[15] have made some interesting developments of the theory of magneto-spatial dispersion. Woźniak has considered the effect of additional electric and optical fields, and Wagnière the possibility of detecting $\hat{p}$ violation with forward–backward effects in atoms. It is straightforward to adapt these theories for use with an optical magnet by replacing the conventional $\mathbf{B}$ by $\mathbf{B}_\Pi$, in precisely the same manner as considered already. Additionally, it is possible to use a combination of $\mathbf{B}$ and $\mathbf{B}_\Pi$. These considerations point toward a range of new optically induced forward–backward anisotropy, whose variations with the probe frequency are novel spectroscopic signatures.

The disucssion has been restricted thus far to BFBA to first order in $\mathbf{B}_\Pi$, and in this context, the various phenomena of spin chiral dichroism[20-25] suggested by the present author are effects that are mediated by the antisymmetric conjugate product $\mathbf{E} \times \mathbf{E}^*$, which from Eq. (1) is seen to be proportional to the magnitude of $\mathbf{B}_\Pi$ squared, and to the axial unit vector $\mathbf{k}$. These terms therefore also change sign with circular polarity of the laser and mediate forward–backward birefringence. In general, spin chiral anisotropy is present in addition to BFBA, the former being proportional to I_0 and the latter to $I_0^{1/2}$. The molecular property tensors involved in spin chiral anisotropy and BFBA are clearly different properties. This can best be seen from the fact that $\mathbf{B}_\Pi$ forms an interaction energy with a magnetic dipole moment, but $\mathbf{E} \times \mathbf{E}^*$ must form an interaction energy with the antisymmetric part of the electric polarizability.[20-25]

Note that there is no forward–backward birefringence proportional to an ordinary magnetostatic flux density squared because such a quantity is positive to the $\hat{T}$ operator. Forward–backward birefringence needs an influence that is $\hat{T}$-negative, such as $\mathbf{B}_\Pi$, $\mathbf{E} \times \mathbf{E}^*$, or the ordinary $\mathbf{B}$. This is a direct result of Wigner's theorem on motion reversal in the complete experiment, as discussed by Barron[10, 34] and the present author.[35]

Acknowledgments

The Cornell Theory Center and Penn State University are thanked for research support, and the Leverhulme Trust is thanked for a Fellowship.

References

1. D. L. Portigal and E. Burstein, *J. Phys. Chem. Solids* **32**, 603 (1971).

2. V. A. Mankelov, M. A. Novikov, and A. A. Turkin, *JETP Lett.* **25**, 378 (1977).

3. S. Kielich and R. Zawodny, *Physica B* **89**, 122 (1977).

4. N. B. Baranova and B. Ya Zel'dovich, *Mol. Phys.* **38**, 1085 (1979).

5. S. Woźniak and R. Zawodny, *Phys. Lett. A* **85**, 111 (1981).

6. S. Woźniak and R. Zawodny, *Acta Phys. Pol.* **A61**, 175 (1982); **A68**, 675 (1985).

7. G. Wagnière and A. Meier, *Chem. Phys. Lett.* **93**, 78 (1982).

8. G. Wagnière and A. Meier, *Experientia* **39**, 1090 (1983).

9. G. Wagnière, *Z. Naturforsch, Teil A* **39**, 254 (1984).

10. L. D. Barron and J. Vrbancich, *Mol. Phys.* **51**, 715 (1984).

11. S. Woźniak, *Mol. Phys.* **59**, 421 (1986).

12. S. Woźniak, *Phys. Lett.* **119**, 256 (1986).

13. S. Woźniak, *J. Chem. Phys.* **85**, 4217 (1986).

14. S. Woźniak, *Acta Phys. Pol* **A72**, 779 (1987).

15. G. Wagnière, *Z. Phys. D* **8**, 229 (1988).

16. G. Wagnière, *Phys. Rev. A* **40**, 2437 (1989).

17. S. Woźniak, M. W. Evans, and G. Wagnière, *Mol. Phys.* **75**, 81 (1992).

18. S. Woźniak, M. W. Evans, and G. Wagnière, *Mol. Phys.* **75**, 99 (1992).

19. L. D. Barron, *Chem. Soc. Rev.* **15**, 189 (1986).

20. M. W. Evans, *Chem. Phys. Lett.* **152**, 33 (1988).

21. M. W. Evans, *Phys. Rev. Lett.* **64**, 2909 (1990).

22. M. W. Evans, *Opt. Lett.* **15**, 863 (1990).

23. M. W. Evans, *Phys. Lett. A* **146**, 475 (1990); **147**, 364 (1990).

24. M. W. Evans, *Physica B* **168**, 9 (1991).

25. M. W. Evans, *J. Mol. Spectrosc.* **143**, 327 (1990).

26. M. W. Evans, *Physica B*, **179**, 237 (1992).

27. M. W. Evans, *Physica B*, **179**, 157 (1992).

28. M. W. Evans, *Physica B*, **182**, 227 (1992).

29. M. W. Evans, *Physica B*, **182**, 237 (1992).

30. M. W. Evans, *Physica B*, **183**, 103 (1993).

31. M. W. Evans, *Physica B*, in press (1993).

32. W. S. Warren, S. Mayr, D. Goswami, and A. P. West, Jr., *Science* **255**, 1683 (1992).

33. M. W. Evans, *Physica B*, **179**, 342 (1992).

34. L. D. Barron, *Molecular Light Scattering and Optical Activity*, Cambridge University Press, Cambridge, UK, 1982.

35. M. W. Evans, in I. Prigogine and S. A. Rice (Eds.), *Advances in Chemical Physics*, Vol. 81, Wiley Interscience, New York, 1992.

36. M. W. Evans, *J. Mol. Spectrosc.* **146**, 351 (1991).

37. M. W. Evans, in S. Kielich and M. W. Evans (Eds.), *Modern Nonlinear Optics*, a Special Issue of *Advances in Chemical Physics*, I. Prigogine and S. A. Rice (Series Eds.), Wiley Interscience, New York, 1993, 3 Vols.

38. M. W. Evans, S. Woźniak, and G. Wagnière, *Physica B* **176**, 33 (1992); **175** 412 (1991); **173**, 357 (1991).

39. S. Kielich, *Nonlinear Molecular Optics*, Nauka, USSR, 1981.

THE PHOTON'S MAGNETIC FIELD B_{Π}: THE MAGNETIC NATURE OF ANTISYMMETRIC LIGHT SCATTERING

I. INTRODUCTION

The classical intensity of electromagnetic radiation is a tensor (I_{ij}) proportional to the tensor product $E_i E_j^*$ (Refs. 1–3). Here E_i is a component of the electric field strength and E_i^* denotes its complex conjugate.[4] In free space, the scalar part of the intensity is

$$I_0 = \varepsilon_0 c E_0^2 \tag{1}$$

where ε_0 is the free space permittivity, c the speed of light and E_0 the scalar amplitude of the electric field strength of the electromagnetic plane wave.[5,6] The vector part of $E_i E_j^*$ is conveniently expressed as the conjugate vector cross product:

$$\mathbf{\Pi}^{(A)} = \mathbf{E} \times \mathbf{E}^* = 2E_0^2 i \mathbf{k} \tag{2}$$

which is purely imaginary as a consequence of the fact that $E_i E_j^*$ is a Hermitian tensor of rank two.[7,8] It has recently been shown[9–12] that the conjugate product $\mathbf{\Pi}^{(A)}$ is directly proportional to a novel magnetostatic flux density vector $\mathbf{B}_{\Pi}$ of the classical electromagnetic plane wave:

$$\mathbf{B}_{\Pi} = \frac{\mathbf{\Pi}^{(A)}}{2E_0 c i} = B_0 \mathbf{k} = \frac{E_0}{c} \mathbf{k} = \left(\frac{I_0}{\varepsilon_0 c^3}\right)^{1/2} \mathbf{k} = \left(\frac{|\mathbf{N}|}{2\varepsilon_0 c^3}\right)^{1/2} \mathbf{k} \tag{3}$$

Here **k** is a unit axial vector in the propagation axis of the plane wave, B_0 is the plane wave's scalar magnetic flux density amplitude, and **N** is the Poynting vector:

$$\mathbf{N} = \frac{1}{\mu_0}\mathbf{E} \times \mathbf{B}^* \tag{4}$$

Remarkably, the vector $\mathbf{B}_\pi$, a flux of magnetic density (in tesla) independent of the phase of the plane wave, has been overlooked in the long and illustrious history of the classical theory of fields,[8] whereas its close relative **N**, a flux of energy density, has been well known for many years. Furthermore, the interpretation of $\mathbf{B}_\Pi$ in the quantum theory of fields leads straightforwardly[10] to the conclusion that the photon generates on the most fundamental level a magnetic flux density operator

$$\hat{B}_\Pi = B_0\frac{\hat{J}}{\hbar} \tag{5}$$

directly proportional to its angular momentum $\hat{J}$, a well-known boson operator.[13] Here $\hbar$ is the unit of angular momentum in quantum mechanics, the reduced Planck constant.

Quite generally, therefore, the ubiquitous antisymmetric part of the electromagnetic intensity (denoted by I_{ij}^-) can be rewritten in terms of the novel magnetic vector $\mathbf{B}_\Pi$, leading immediately to novel insights about all processes in physical optics that depend on I_{ij}^-.

In this paper we illustrate this conclusion with reference to the antisymmetric part of Rayleigh scattering from molecular liquids, a process first considered by Placzek[14] in 1934. Section II defines the antisymmetric part of the scattered intensity in terms of the scattered $\mathbf{B}_\Pi$ vector, adapting the arguments of Knast and Kielich.[7] Section III continues the development in terms of Rayleigh refringent scattering theory, used to relate the incoming and scattered magnetic vectors $\mathbf{B}_\Pi$. Section IV is a discussion of these results, leading to the conclusions that antisymmetric light scattering in general can be reinterpreted fundamentally as a purely magnetic process, whereby the incoming $\mathbf{B}_\Pi$ (or $\hat{B}_\Pi$ of quantum field theory) interacts with the molecular ensemble forming the scattering volume, and is scattered as the vector $\mathbf{B}_{\Pi S}$. The two magnetic vectors $\mathbf{B}_\Pi$ and $\mathbf{B}_{\Pi S}$ are related through the molecular property tensors of the scattering volume.

190   M. W. EVANS

II. DEFINITION OF THE SCATTERED $B_{\Pi S}$ VECTOR

It is convenient to define the scattered $B_{\Pi S}$ vector in terms of the development of Knast and Kielich,[7] who have also tabulated extensively the magnetic point group symmetries of the antisymmetric part of the molecular and atomic polarizability. In so doing we define the intensity tensor of the incoming light as

$$I_{ij} = \varepsilon_0 c E_i E_j^* \tag{6}$$

from which it follows that the vector part of the intensity, the antisymmetric component I_{ij}^- of Knast and Kielich[7] can be expressed simply as the purely imaginary axial vector:

$$\mathbf{I}^- \equiv \tfrac{1}{2}\varepsilon_{ijk}\big(I_{ij} - I_{ji}\big) = iI_0 k$$
$$= \varepsilon_0 c^2 E_0 i\mathbf{B}_\Pi \tag{7}$$

where $\mathbf{k}$ is a unit axial vector in the propagation direction. From these definitions we deduce immediately that the incoming $\mathbf{B}_\Pi$ vector is proportional to the square root of the incoming I_0:

$$\mathbf{B}_\Pi = \left(\frac{I_0}{\varepsilon_0 c^2 E_0}\right)\mathbf{k} = \left(\frac{I_0}{\varepsilon_0 c^3}\right)^{1/2}\mathbf{k} \tag{8}$$

The antisymmetric part of the scattered light intensity tensor is defined by Knast and Kielich to be:

$$I_{ijs}^-(t) = \left(\frac{\omega_0}{c}\right)^4 \big\langle M_i(\mathbf{r}, t_0) M_j^*(\mathbf{r}, t)\big\rangle_{t_0}^- \tag{9}$$

where

$$M_i(\mathbf{r}, t_0) = \mu_i^{(p)}(t_0)\exp\big[i(\Delta\kappa \cdot \mathbf{r}(t_0))\big] \tag{10}$$

where $\mu_i^{(p)}$ is the ith component of the electric dipole moment induced in a molecule p by the light's electric field strength vector, as usual in the theory of scattering.[15, 16] The quantity $\Delta\kappa$ is a difference in wave vectors of incident and scattered light, as usual, and the summation extends over all the N molecules of the volume. The angular brackets in Eq. (9) denote a time-correlation function[17] of the fluctuating quantity M_i. It is well known

from the theory of nonequilibrium statistical mechanics[17] that the normalized time-correlation function starts at unity and evolves to zero with time
t. Its Fourier transform is a spectral function of angular frequency ω. In
Rayleigh scattering[15-17] the spectrum is the experimental Rayleigh band-
shape, which is Fourier transformed to give the time-correlation function.
The scattered antisymmetric intensity $I_{ijk}^-(k)$ is therefore frequency dependent, and forms the antisymmetric part of the Rayleigh scattering spec-
trum. However, $I_{ijs}^-(t)$ is directly proportional to the scattered $\mathbf{B}_{\Pi S}$ vector,
(or in quantum-field theory[10] the scattered flux density operator $\hat{B}_{\Pi S}$) and
we reach the significant conclusion that the spectrum of antisymmetric
Rayleigh scattering is a graph of the scattered $\beta_{\Pi S}(\omega)$ plotted against the
change in frequency $(\omega - \omega_0)$, where ω_0 is the incoming laser frequency.
Antisymmetric Rayleigh scattering is therefore a process that can be
described entirely in terms of the vector $\mathbf{B}_{\Pi}$.

This conclusion can be underlined by expressing the fundamental
equation (9) in terms of a mean magnetic dipole moment $\langle \mathbf{m}_S(t) \rangle$, formed
at time t from the antisymmetric part of the time correlation function
$\langle M_i(\mathbf{r}, t_0) M_j^*(\mathbf{r}, t) \rangle_{t_0}$. This development is based on the relation

$$\langle \mathbf{m}_S(t) \rangle = \xi \langle \mathbf{M}(\mathbf{r}, t_0) \times \mathbf{M}^*(\mathbf{r}, t) \rangle_{t_0} \tag{11}$$

where ξ is a proportionality coefficient; i.e., the vector cross product of
two nonidentical electric dipole moment vectors is proportional to a
magnetic dipole moment. This conclusion can be illustrated by the following simple model.

Express the electric dipole moments $\boldsymbol{\mu}$ and $\boldsymbol{\mu}^*$ as products of electronic
charge e and position vectors $\mathbf{r}$ and $\mathbf{r}^*$. Then the cross product can be
expressed as an area

$$\boldsymbol{\mu} \times \boldsymbol{\mu}^* = e^2 \mathbf{r} \times \mathbf{r}^* = e^2 A \mathbf{k} \tag{12}$$

Considering the simple model of the motion of charge with instantaneous
linear velocity v around a circle of radius r, the magnetic dipole moment
is known[18] to be proportional to the product IA, where A is the area of
the circle, and I is the quantity $ev/(2\pi r)$. It follows that, in general, a
magnetic dipole moment is proportional to area, and therefore to the
cross product of $\boldsymbol{\mu}$ and $\boldsymbol{\mu}^*$. The same conclusion is derived for the cross
product of transition electric dipole moments by Atkins and Miller.[19] It is
also a consequence of the fact that antisymmetric electric polarizability is
proportional to the cross product of transition electric dipole moments,
and therefore has the same symmetry[20] as a magnetic dipole moment.

From these considerations, we reach the equation

$$\mathbf{m}_S(t) = \frac{1}{\xi}\left(\frac{c}{\omega_0}\right)^4 \varepsilon_0 c^2 E_{0s} \mathbf{B}_{\Pi S}(t) \tag{13}$$

showing that the magnetic dipole moment $\mathbf{m}_S(t)$ is proportional to the scattered $\mathbf{B}_{\Pi S}(t)$ vector at time t. Fourier transformation leads immediately to the conclusion that the magnetic dipole moment at frequency ω is proportional to the scattered $\mathbf{B}_{\Pi S}(\omega)$ vector at the frequency ω of the antisymmetric Rayleigh scattering spectrum.

III. REFRINGENT SCATTERING APPROACH

In this section we adapt straightforwardly Rayleigh refringent scattering theory to provide an expression linking the incoming $\mathbf{B}_\Pi$ and the scattering $\mathbf{B}_{\Pi S}$ in terms of a parameter Ξ_Z, which is defined in terms of the molecular property tensors of the scattering volume. The scattered light intensity tensor is defined in semiclassical Rayleigh refringent scattering theory by [20]

$$I_{\alpha\beta}^{(s)} = \varepsilon_0 c E_\alpha^{(S)} E_\beta^{*(S)}$$
$$= \varepsilon_0 c \left(\frac{\omega^2 \mu_0}{4\pi R}\right)^2 a_{\alpha\gamma} a_{\beta\delta}^* E_\gamma^{(0)} E_\delta^{*(0)} \tag{14}$$

where

$$E_\alpha^{(s)} = \frac{\omega^2 \mu_0}{4\pi R} \exp\left[i\omega\left(\frac{R}{c} - t\right)\right] a_{\alpha\beta} E_\beta^{(0)} \tag{15}$$

is the scattered electric field detected in the wave zone at a point d at a distance R from the molecular origin. The origin of scattered light is considered to be the characteristic radiation field generated by the oscillating electric and magnetic multipole moments induced in a molecule by the electromagnetic fields of the incident light wave. Here $a_{\alpha\beta}$ is the scattering tensor, a molecular property of the scattering volume for particular incident and scattered directions given by unit vectors $\mathbf{n}^{(0)}$ and $\mathbf{n}^{(S)}$. In Eq. (14) ω is the angular frequency of the incoming wave, whose electric field strength vector is denoted $\mathbf{E}^{(0)}$, so that the scattered intensity tensor $I_{\alpha\beta}^{(S)}$ is expressed in terms of the incident intensity tensor $I_{\gamma\delta}^{(0)}$ by

$$I_{\alpha\beta}^{(S)} = \left(\frac{\omega^2\mu_0}{4\pi R}\right)^2 a_{\alpha\gamma}a_{\beta\delta}^* I_{\gamma\delta}^{(0)} \tag{16}$$

It is immediately clear, therefore, that the incident $\mathbf{B}_\Pi$ in antisymmetric scattering can be expressed in terms of the scattered $\mathbf{B}_{\Pi S}$ in a similar way. Thus, we arrive at the conclusion that antisymmetric light scattering, in general, is a process whereby the incoming $\mathbf{B}_\Pi$ is transformed into a scattered $\mathbf{B}_{\Pi S}$; i.e., antisymmetric Rayleigh scattering is a purely magneto-optic process. This argument is developed by consideration of the antisymmetric (vector) part of the scattered intensity tensor

$$I_\varepsilon^{(AS)} = \varepsilon_0 c\Pi_\varepsilon^{(AS)}$$

$$\Pi_\varepsilon^{(AS)} = \frac{1}{2}\left(\frac{\omega^2\mu_0}{4\pi R}\right)^2 \varepsilon_{\alpha\beta\varepsilon}\left(E_\alpha^{(S)}E_\beta^{*(S)} - E_\beta^{(S)}E_\alpha^{*(S)}\right) \tag{17}$$

For the Z component

$$\Pi_Z^{(AS)} = 2E_0^{(0)2}i\Xi_Z \tag{18}$$

where

$$\Xi_Z = \frac{\omega^4\mu_0^2}{32\pi^2 R^2 i}\Big[(a_{XX}a_{YX}^* - a_{YX}a_{XX}^*) + (a_{XY}a_{YY}^* - a_{YY}a_{XY}^*) \tag{19}$$

$$+i(a_{XX}a_{YY}^* - a_{YX}a_{XY}^*) - i(a_{XY}a_{YX}^* - a_{YY}a_{XX}^*)\Big]$$

is in general a complex quantity. Using the result

$$\Pi_Z^{(A)} = 2E_0^{(0)2}i \tag{20}$$

it follows that

$$\Pi_Z^{(AS)} = \Pi_Z^{(A)}\Xi_Z \tag{21}$$

and that

$$B_{\Pi S Z} = \Xi_Z B_{\Pi Z} \tag{22}$$

For forward scattering, there is no component of $\mathbf{B}_{\Pi S}$ other than $B_{\Pi S Z}$; but there are components of $\mathbf{B}_{\Pi S}$ in X, Y, and Z, depending on the scattering angle. These are all generated from the incoming $B_{\Pi Z}$ by tensor multiplication with $\Xi_{\alpha\beta}$ in its second-rank tensor form.

Without loss of generality we concentrate on forward scattering in the rest of this section, so that[20]

$$\text{Re}\left(a_{\alpha\beta}^{f}\right) = \alpha'_{\alpha\beta}(f) + \zeta'_{\alpha\beta\gamma}(f)n'_{\gamma} + \alpha''_{\alpha\beta}(g) + \zeta''_{\alpha\beta\gamma}(g)n \quad (23)$$

$$\text{Im}\left(a_{\alpha\beta}^{f}\right) = -\alpha''_{\alpha\beta}(f) - \zeta''_{\alpha\beta\gamma}(f)n'_{\gamma} + \alpha'_{\alpha\beta}(g) + \zeta'_{\alpha\beta\gamma}(g)n_{\gamma} + \cdots \quad (24)$$

where $\alpha'_{\alpha\beta}$ and $\alpha''_{\alpha\beta}$ are respectively the real and imaginary parts of the molecular polarizability tensor[20] and $\zeta'_{\alpha\beta\gamma}$ and $\zeta''_{\alpha\beta\gamma}$ are those of the zeta tensor defined by Barron.[20] In the forward direction, the process becomes one of antisymmetric spectral absorption, in which the incoming and outgoing $B_{\Pi Z}$ and $B_{\Pi S Z}$ define the absorption coefficient:

$$A^{(A)} \propto \frac{I_Z^{(s)}}{I_Z^{(0)}} = \frac{B_{\Pi S Z}}{B_{\Pi Z}} = \Xi_Z \quad (25)$$

After ensemble averaging[20]

$$\langle \Xi_Z \rangle = \frac{\omega^4 \mu_0^2}{192\pi^2 R^2}\left(a_{\alpha\alpha} a_{\beta\beta}^* - \alpha_{\alpha\beta} a_{\alpha\beta}^* \right) \quad (26)$$

and

$$B_{\Pi S Z} = \langle \Xi_Z \rangle B_{\Pi Z} \quad (27)$$

describes the process in terms of tensor invariants[20] of the molecular ensemble constituting the scattering volume.

IV. DISCUSSION

The historical development and experimental evidence for antisymmetric Rayleigh scattering has been reviewed in detail by Barron.[20] In this work and that of Knast and Kielich,[7] the magnetic nature of the phenomenon is implied indirectly, for example, through the fact that the process is described with the antisymmetric part of molecular property tensors such as the electric polarizability. Using the relation

$$\alpha''_k = \tfrac{1}{2}\varepsilon_{ijk}(\alpha''_{ij} - \alpha''_{ji}) \quad (28)$$

this tensor can be described as an axial vector, α''_k (Refs. 21–25), which has the same symmetry as a magnetic dipole moment, and whose irreducible representations in various molecular point groups are the same as those of the magnetic dipole moment or angular momentum. Knast and Kielich,[7] in their Eq. (36), also point out that the vector α''_k is negative to the motion reversal operator $\hat{T}$, and it follows[21-25] that it can form an interaction energy only with another $\hat{T}$-negative property, the antisymmetric conjugate product $\mathbf{E} \times \mathbf{E}^*$ of Eq. (2) of this paper. This argument shows that $\mathbf{E} \times \mathbf{E}^*$ must have the symmetry of a magnetic field, since α''_k has the symmetry of a magnetic dipole moment. In fact, as discussed in the introduction, $\mathbf{E} \times \mathbf{E}^*$ is proportional to the novel magnetic field $\mathbf{B}_\Pi$, the classical equivalent of the operator $\hat{B}_\pi$ for each individual photon. Equation (13) of this paper now shows that antisymmetric scattering can be thought of as the induction by the scattered magnetic field $\mathbf{B}_{\Pi S}$ of the magnetic dipole moment $\langle \mathbf{m}_S(t) \rangle$ (which is the value at t of a time-correlation function), and this result can also be generalized in quantum field theory, or the theory of magneto-photonics. The Fourier transform of $\langle m_S(t) \rangle$ is a point on the spectrum of scattered light at the frequency ω. The magnetic dipole moment $\langle \mathbf{m}_S \rangle$ has the same irreducible representations in the appropriate point groups[7] as the antisymmetric polarizability considered by Knast and Kielich,[7] and $\mathbf{B}_{\Pi S}(t)$ has the same symmetry as the antisymmetric part of the scattered intensity, denoted in tensor notation by I^-_{ij} by these authors.[7] It follows that the same conclusions arrived at by Knast and Kielich[7] for the properties of the antisymmetric polarizability hold for the novel magnetic dipole moment $\langle \mathbf{m}_S(t) \rangle$. For example, $\langle \mathbf{m}_S(t) \rangle$ is nonzero only in the presence of a $\hat{T}$-negative influence, which in Eq. (13) is the magnetic field $\mathbf{B}_{\Pi S}(t)$.

Another conclusion that becomes immediately obvious in our magnetic interpretation of forward antisymmetric Rayleigh light scattering is that it involves circular polarization. The magnetic fields $\mathbf{B}_\Pi$ and $\mathbf{B}_{\Pi S}$ vanish if there is no degree of circular polarization, respectively in the incoming and scattered radiation. These findings are reinforced by the arguments, summarized in Section 3.5.3 of Ref. 20, for Rayleigh scattering in the near forward direction from refringent scattering theory. In this case, the degree of circular polarization is directly proportional to the pseudoscalar magnitude[12] of the scattered $\mathbf{B}_{\Pi S}$, which is the third Stokes parameter of the scattered radiation. Thus, in purely antisymmetric, near forward Rayleigh scattering, if the incident beam is completely circularly polarized, so is the scattered beam. This is summarized in our terms by Eq. (27), in which $B_{\Pi Z}$ and $B_{\Pi S Z}$ are both well defined, and in which the coefficient $\langle \Xi_Z \rangle$ is a finite ensemble average. Clearly, if $B_{\Pi Z}$ is zero (no degree of circular polarization in the incoming beam), then $B_{\Pi S Z}$ is also zero, because the molecular ensemble average $\langle \Xi_Z \rangle$ is nonzero in general. Therefore, there is no near forward scattering.

V. CONCLUSION

The phenomenon of antisymmetric light scattering has been interpreted in terms of the novel incident and scattered magnetostatic flux density vectors $\mathbf{B}_\Pi$ and $\mathbf{B}_{\Pi S}$, respectively. This shows that antisymmetric scattering is a purely magneto-optic phenomenon, giving information on the nature of the scattered $\mathbf{B}_{\Pi S}$ vector. In magneto-photonics, the vector $\mathbf{B}_\Pi$ is replaced by the operator $\hat{B}_\Pi$, and the appropriate quantum theory must be employed.

Acknowledgments

The Leverhulme Trust is thanked for a Fellowship and additional research support is acknowledged from the Cornell Theory Center and the Materials Research Laboratory of Penn State University.

References

1. S. Kielich, *Molecular Nonlinear Optics*, PWN, Warszawa-Poznań, 1977.

2. L. D. Landau and E. M. Lifshitz, *The Classical Theory of Fields*, 4th ed., Pergamon, Oxford, UK, 1975.

3. M. Born and E. Wolf, *Principles of Optics*, 6th ed., Pergamon, Oxford, UK, 1975.

4. M. W. Evans, in I. Prigogine and S. A. Rice (Eds.), *Advances in Chemical Physics*, Vol. 81, Wiley, New York, 1992.

5. S. Kielich, *J. Chem. Phys.* **46**, 4090 (1967).

6. S. Kielich, *Chem. Phys. Lett.* **10**, 516 (1971).

7. K. Knast and S. Kielich, *Acta Phys. Pol.* **A55**, 319 (1979).

8. M. W. Evans, *Physica B* **168**, 9 (1991).

9. M. W. Evans, *Physica B*, **182**, 227 (1992).

10. M. W. Evans, *Physica B*, **182**, 237 (1992).

11. M. W. Evans, *Physica B*, **183**, 103 (1993).

12. M. W. Evans, *Physica B*, in press (1993).

13. P. A. M. Dirac, *The Principles of Quantum Mechanics*, 4th ed., Oxford University Press, Oxford, UK, 1958.

14. G. Placzek, in E. Marx (Ed.), *Handbuch der Radiologie*, Akad. Verlag., Leipzig, 1934.

15. B. J. Berne and R. Pecora, *Dynamical Light Scattering with Applications in Physics, Chemistry and Biology*, Wiley Interscience, New York, 1976.

16. I. L. Fabelinskii, *Molecular Scattering of Light*, Plenum, New York, 1968.

17. M. W. Evans, G. J. Evans, W. T. Coffey, and P. Grigolini, *Molecular Dynamics and the Theory of Broad Band Spectroscopy*, Wiley, New York, 1982.

18. P. W. Atkins, *Molecular Quantum Mechanics*, 2d ed., Oxford University Press, Oxford, UK, 1983.

19. P. W. Atkins and M. H. Miller, *Mol. Phys.* **15**, 503, Eqs. (75) ff (1968).

20. L. D. Barron, *Molecular Light Scattering and Optical Activity*, Cambridge University Press, Cambridge, UK, 1982.

21. M. W. Evans, *Phys. Rev. Lett.* **64** 2909 (1990).

22. M. W. Evans, *Opt. Lett.* **15**, 863 (1990).

23. M. W. Evans, *J. Mol. Spectrosc.* **146**, 351 (1991).

24. M. W. Evans, *Int. J. Mod. Phys. B* **5**, 1963 (1991) (review).

25. M. W. Evans, *J. Phys. Chem.* **95**, 2256 (1991).

MANIFESTLY COVARIANT THEORY OF THE ELECTROMAGNETIC FIELD IN FREE SPACETIME, PART 1: ELECTRIC AND MAGNETIC FIELDS AND MAXWELL'S EQUATIONS

I. INTRODUCTION

It has recently been shown[1-5] that there exist longitudinal solutions of Maxwell's equations in free spacetime which are independent of the phase ϕ of the traveling plane wave. These longitudinal electric and magnetic fields, denoted $\mathbf{E}^{(3)}$ and $\mathbf{B}^{(3)}$, respectively, are consistent with the conclusion of quantum electrodynamics that there exist four photon polarizations in free spacetime, one timelike ((0)), two transverse spacelike ((1) and (2)), and one longitudinal spacelike ((3)).[6,7] However, the existence of four photon polarizations has to date been regarded[7] as being in conflict with the deduction that the photon can have only two helicities, $+1$ and -1. This in turn has led to the arbitrary assertion that only the two transverse spacelike polarizations (1) and (2) can be "physically meaningful" in free spacetime. The timelike ((0)) and longitudinal spacelike ((3)) are conventionally discarded as physically meaningless. This implies that the theory of the electromagnetic field in free spacetime loses manifest covariance.[7] This fundamental difficulty is well described by Ryder,[7] from whose Chapter 4 we quote the following: "the electromagnetic field, like any massless field, possesses only two independent components, but is covariantly described by a (potential) four vector A_μ. In choosing two of these components as the physical ones, and thence quantizing them, we lose manifest covariance. Alternatively, if we wish to keep covariance, we have two redundant components."

Clearly, if the theory of the electromagnetic field in free spacetime is to be made manifestly covariant and therefore rigorously consistent with special relativity, then all four photon polarizations must be physically meaningful. This implies that electric and magnetic fields in vacuo must be four vectors, E_μ and B_μ, respectively, in spacetime. The difficulty with this notion to date appears to have been the preconception that any longitudinal solution of Maxwell's equations in free spacetime must necessarily be phase dependent, so that the longitudinal spacelike solution cannot be solenoidal. This means that Gauss's theorem in differential form is violated by such a solution.[8-15] However, with the recent discovery[1-5] that the longitudinal solutions to Maxwell's equations in vacuo are not phase dependent, the conflict with Gauss's theorem disappears, and one of the most intractable difficulties of electromagnetic field theory is removed. In so doing, the very basis of electrodynamics is changed profoundly, because at present the subject is based on the existence in vacuo of a potential four vector A_μ, whose four-curl gives the antisymmetric electromagnetic field tensor $F_{\mu\nu} = -F_{\nu\mu}$ in spacetime. The components of $F_{\mu\nu}$ contain no explicit reference to the timelike component of the four vectors E_μ and B_μ, and the longitudinal components that appear in $F_{\mu\nu}$ are evidently discarded as unphysical. To maintain manifest covariance the timelike and longitudinal components must be retained, and must have physical meaning. In other words, the electric and magnetic parts of the electromagnetic plane wave in free space are treated conventionally as three vectors in Euclidean space, and not as manifestly covariant four vectors in pseudo-Euclidean spacetime. This reveals an internal inconsistency in electrodynamics in vacuo, in that the d'Alembert equation

$$\Box A_\mu = 0 \tag{1}$$

allows four photon polarizations, but the Maxwell equations

$$\frac{\partial F_{\mu\nu}}{\partial x_\nu} = 0 \qquad \frac{\partial \tilde{F}_{\mu\nu}}{\partial x_\nu} = 0 \tag{2}$$

$$(x \equiv (X, Y, Z, ict))$$

link only the spacelike components of E_μ and B_μ. They make no explicit reference to their timelike components $E^{(0)}$ and $B^{(0)}$. (In Eq. (2), the Maxwell equations are stated in terms of the four divergence of $F_{\mu\nu}$ and of its dual, $\tilde{F}_{\mu\nu}$ (Ref. 7).) A consistent, manifestly covariant, and rigorous theory of electrodynamics in vacuo must link E_μ and B_μ to the tensor $F_{\mu\nu}$, which is the four-curl of A_μ.

In Section II of this paper, a brief review is given of the phase-independent longitudinal components $E^{(3)}$ and $B^{(3)}$ of the electromagnetic plane wave in vacuo. These are identified with the conclusions of quantum field theory[7] that physical photon states in a manifestly covariant gauge such as the Lorentz gauge are described as admixtures of operators of the field, namely the creation and annihilation operators.

Section III links the four vectors E_μ and B_μ to the four tensor $F_{\mu\nu}$, and shows that E_μ and B_μ take the form of a Pauli Lubanski vector and pseudovector, respectively, in spacetime. These are well defined[7] within the inhomogeneous Lorentz group (or Poincaré group). This leads in turn to the conclusion that the two photon helicities, $+1$ and -1, can be reconciled rigorously with four physically meaningful photon polarizations, because the helicities can be described equally well in terms either of (1) and (2) polarizations or of (0) and (3) polarizations. This is consistent with our earlier[1,2] conclusion that one photon generates the longitudinal magnetic field component:

$$\mathbf{B}^{(3)} = \langle \psi | \hat{B}^{(3)} | \psi \rangle = \frac{B^{(0)}}{\hbar} \langle \psi | \hat{J} | \psi \rangle \tag{3}$$

where $|\psi\rangle$ is an eigenstate of the photon and where the eigenvalues of the operator $\hat{J}$ are $M_J \hbar$; $M_J = +1$ and -1, the photon helicities. The result (3) is generalized in Section III through the definition of E_μ and B_μ as Pauli-Lubanski types in spacetime.

Section IV deals with some consequences in vacuum electrodynamics of the existence of manifestly covariant E_μ and B_μ, with four physically meaningful components. The Maxwell equations, in particular the differential form of Gauss's theorem, are developed covariantly in terms of E_μ and B_μ. Specifically, Gauss's theorem in differential form becomes

$$\frac{\partial E_\mu}{\partial x_\mu} = 0 \quad \text{or} \quad \nabla \cdot \mathbf{E} + \frac{1}{c} \frac{\partial E^{(0)}}{\partial t} = 0 \tag{4a}$$

and

$$\frac{\partial B_\mu}{\partial x_\mu} = 0 \quad \text{or} \quad \nabla \cdot \mathbf{B} + \frac{1}{c} \frac{\partial B^{(0)}}{\partial t} = 0 \tag{4b}$$

The electromagnetic energy and energy flux densities in vacuo are expressed in terms of products of E_μ and B_μ, showing that the (3) and (0) polarizations do not make explicit contributions to either on a time-averaged basis. The four Stokes parameters, however, are profoundly affected by the manifest covariance of E_μ and B_μ, in that it is no longer sufficient to describe S_0, S_1, S_2, and S_3 in terms of Pauli matrices.[17] It is shown that the covariant description of the Stokes parameters in vacuo can be obtained through the use of Dirac matrices.[18] This description maintains the fundamental relation

$$S_0^2 = S_1^2 + S_2^2 + S_3^2 \tag{5}$$

on the Poincaré sphere, while at the same time showing that S_1 and S_2 become different in a description based on E_μ and B_μ rather than on the usual transverse spacelike E and B. A new term appears in both S_1 and S_2 due to the existence of physically meaningful (0) and (3) states of the electromagnetic field. The parameters S_0 and S_3, on the other hand, are unaffected.

Section V is a discussion of the available experimental evidence for $\mathbf{B}^{(3)}$, and suggests several experimental tests of its physical existence when the electromagnetic field interacts with matter.

II. THE LONGITUDINAL SOLUTIONS OF MAXWELL'S EQUATIONS IN FREE SPACETIME: (0) AND (3) POLARIZATIONS

Longitudinal solutions of Maxwell's equations in vacuo appear not to have been considered as physically meaningful in the great majority of standard texts. Jackson[8] simply states that the differential form of Gauss's theorem demands that phase-dependent solutions are transverse. The possibility of phase-independent solutions appears not to be considered. It is frequently considered[8-15] that a plane, monochromatic, electromagnetic wave traveling in Z (the propagation axis) in vacuo is simply the sum of two coherent waves linearly polarized in the orthogonal axes X and Y. Atkins[11] and Landau and Lifshitz[12] similarly consider only transverse fields, and thus transverse polarizations, in a Cartesian or circular basis. Whitner,[9] however, mentions briefly and without further development that "Plane waves are an important example but they do constitute a special case; we must not conclude that all electromagnetic waves are transverse." Similarly, other authors[8-15] in classical and quantum electrodynamics in vacuo make little or no mention of longitudinal solutions.

Recently, however, Evans[1-4] and Farahi and Evans[5] have systematically considered the theory of phase-independent longitudinal electric and magnetic fields, which are solutions to the free spacetime Maxwell equations and thus obey the differential form of Gauss's theorem in free spacetime. This work has developed rapidly from the observation[1] that spacelike components of the plane wave in vacuo are interrelated by

$$\mathbf{B}^{(3)} = \frac{\mathbf{E}^{(1)} \times \mathbf{E}^{(2)}}{icE_0} \tag{6}$$

where

$$\mathbf{E}^{(1)} \equiv E_0 \hat{\mathbf{e}}^{(1)} e^{i\phi} \tag{7a}$$

$$\mathbf{E}^{(2)} \equiv E_0 \hat{\mathbf{e}}^{(2)} e^{-i\phi} \tag{7b}$$

are the transverse electric field components. Here

$$\hat{\mathbf{e}}^{(1)} = \frac{\mathbf{i} - i\mathbf{j}}{\sqrt{2}}$$

$$\hat{\mathbf{e}}^{(2)} = \frac{\mathbf{i} + i\mathbf{j}}{\sqrt{2}}$$

are unit vectors in the circular basis,[19] where $\mathbf{i}$ and $\mathbf{j}$ are unit cartesian vectors in X and Y, orthogonal to the propagation axis Z. Here ϕ is the phase of the traveling monochromatic plane wave, defined by

$$\phi = \omega t - \boldsymbol{\kappa} \cdot \mathbf{r}$$

where ω is its angular frequency at instant t and $\boldsymbol{\kappa}$ its wave vector at position $\mathbf{r}$ in Euclidean space. We have from Eqs. (6) and (7)

$$\mathbf{B}^{(3)} = B_0 \hat{\mathbf{e}}^{(3)} \tag{8}$$

with

$$\hat{\mathbf{e}}^{(1)} \times \hat{\mathbf{e}}^{(2)} = i\hat{\mathbf{e}}^{(3)} \tag{9}$$

and the well-known free spacetime relation[8-15]

$$E_0 = cB_0 \tag{10}$$

With $B_0 \equiv B^{(0)}$ we write Eq. (8) as

$$B_0 - |\mathbf{B}^{(3)}| = 0 \tag{11}$$

where $|\mathbf{B}^{(3)}|$ denotes the scalar magnitude of the vector $\mathbf{B}^{(3)}$.

From Eqs. (6), (7), and (11) it is clear that field polarizations (0) and (3) are not independent of polarizations (1) and (2). Furthermore, by considering the results of quantum field theory,[7] polarization (0) can be identified as being timelike in a manifestly covariant description, and (3) as longitudinal spacelike. The field $\mathbf{B}^{(3)}$ is consistent with Maxwell's equations in vacuo and therefore with Gauss's theorem. The key to this result is that the phase ϕ has been removed in the conjugate product (6). The electric counterpart of Eq. (11) is, in general:

$$E^{(0)} - |\mathbf{E}^{(3)}| = 0; \; E^{(0)} \propto E_0 \tag{12a}$$

$$\mathbf{E}^{(3)} = E^{(0)}\hat{\mathbf{e}}^{(3)} \tag{12b}$$

Equations (12) also represent solutions of Maxwell's equations in vacuo.

Equations (11) and (12) are, furthermore, related[5] through conservation of electromagnetic energy by the Euclidean space equation:

$$\mathbf{E}^{(3)} \times \mathbf{B}^{(2)} = \mathbf{B}^{(3)} \times \mathbf{E}^{(2)} \tag{13}$$

showing that if $\mathbf{B}^{(3)}$ is real, as in Eq. (6), $\mathbf{E}^{(3)}$ is imaginary. Relations such as (6) and (11) to (13) show that there are only two independent states for $\mathbf{E}$ and $\mathbf{B}$ because from Eq. (6) either of states (1) and (2) can be expressed in terms of (3); and the latter can be expressed in terms of (0) through Eq. (11) for $\mathbf{B}^{(3)}$ and Eq. (12) for $\mathbf{E}^{(3)}$. Finally, states (0) for $\mathbf{E}$ and $\mathbf{B}$ are related by Eq. (10) and states (3) by Eq. (13). This result, that there are only two independent states out of the four possible, (0) to (3), is evidently the classical expression of the fact[7] that the massless gauge field possesses only two independent components, but is at the same time covariantly described by a four vector, A_μ made up of four physically meaningful polarization states (0), (1), (2), and (3).

Thus far, we have used a conventional, classical description in terms of spacelike vectors $\mathbf{E}$ and $\mathbf{B}$, but have introduced the novel $\mathbf{E}^{(3)}$ and $\mathbf{B}^{(3)}$. By using the relativistic quantum description of the electromagnetic field[7] we now introduce the concept of electric and magnetic field four vectors E_μ and B_μ, respectively.

It is well known[7] that the quantization of Eq. (1) in the Lorentz gauge proceeds through a condition derived by Gupta and Bleuler in the early days[6] of quantum field theory:

$$\frac{\partial \hat{A}_\mu^{(+)}}{\partial x_\mu} |\psi\rangle = 0 \tag{14}$$

where $\hat{A}_{\mu}^{(+)}$ is the operator equivalent of A_{μ} and acts on a photon eigenstate $|\psi\rangle$. Equation (14) leads to the result[7] that physical photon states are admixtures of (0) and (3) photon polarizations in such a way that

$$(\hat{a}^{(0)} - \hat{a}^{(3)})|\psi\rangle = 0 \tag{15a}$$

$$\langle\psi|\hat{a}^{(0)+}\hat{a}^{(0)}|\psi\rangle = \langle\psi|\hat{a}^{(3)+}\hat{a}^{(3)}|\psi\rangle \tag{15b}$$

where $\hat{a}$ and $\hat{a}^{+}$ are annihilation and creation operators, respectively. Furthermore, the energy density of the quantized field is proportional[7] to the sum

$$\sum_{\lambda=0}^{3} (\hat{a}^{(\lambda)+}\hat{a}^{(\lambda)} - \hat{a}^{(0)+}\hat{a}^{(0)}) \tag{16}$$

and from Eq. (15b)[7] the contribution of the longitudinal ((3)) and timelike ((0)) states cancel. It will be shown in this section that the classical but manifestly covariant equivalents of Eqs. (15) and (16) are obtained from the four vectors $\mathbf{E}_{\mu}$ and $\mathbf{B}_{\mu}$.

We use the well-known relations[19, 20] (in S.I. units),

$$\hat{E}^{(0)} = \left(\frac{2\hbar\omega}{\varepsilon_0 V}\right)^{1/2} \hat{a}^{(0)} \qquad \hat{E}^{(3)} = \left(\frac{2\hbar\omega}{\varepsilon_0 V}\right)^{1/2} \hat{a}^{(3)}$$

$$\hat{B}^{(0)} = \left(\frac{2\mu_0\hbar\omega}{V}\right)^{1/2} \hat{a}^{(0)} \qquad \hat{B}^{(3)} = \left(\frac{2\mu_0\hbar\omega}{V}\right)^{1/2} \hat{a}^{(3)} \tag{17}$$

to link the annihilation operators in states (0) and (3) to the equivalent field operators. Here ε_0 is the permittivity and μ_0 the permeability of the vacuum state, $\hbar$ is the reduced Planck constant, and V the quantization volume. From Eqs. (15a) and (17),

$$(\hat{E}^{(0)} - \hat{E}^{(3)})|\psi\rangle = 0 \qquad (\hat{B}^{(0)} - \hat{B}^{(3)})|\psi\rangle = 0 \tag{18}$$

and from Eq. (15b),

$$\langle\psi|\hat{E}^{(0)+}\hat{E}^{(0)}|\psi\rangle = \langle\psi|\hat{E}^{(3)+}\hat{E}^{(3)}|\psi\rangle$$

$$\langle\psi|\hat{B}^{(0)+}\hat{B}^{(0)}|\psi\rangle = \langle\psi|\hat{B}^{(3)+}\hat{B}^{(3)}|\psi\rangle \tag{19}$$

The classical equivalent of Eq. (18) is

$$E^{(0)} - |\mathbf{E}^{(3)}| = 0 \qquad B^{(0)} - |\mathbf{B}^{(3)}| = 0 \qquad (20a)$$

and that of Eqs. (19) is

$$\mathbf{E}^{(0)2} = \mathbf{E}^{(3)} \cdot \mathbf{E}^{(3)}$$
$$\mathbf{B}^{(0)2} = \mathbf{B}^{(3)} \cdot \mathbf{B}^{(3)} \qquad (20b)$$

where $\mathbf{E}^{(3)} = \langle \psi | \hat{\mathbf{E}}^{(3)} | \psi \rangle$, etc. Equation (20a) is identical with Eq. (11), and Eq. (20b) is consistent with Eqs. (11) and (12a). However, Eq. (19) was derived from a quantized counterpart, Eq. (15a), which is manifestly covariant in that physical photon states are admixtures of states (0) and (3) of the quantum field.[7] It follows that classical field states in vacuo are also admixtures of the classical (0) and (3) polarizations as defined by Eqs. (19) and (20). From this we arrive at two fundamentally important conclusions:

1. The electric and magnetic components of the electromagnetic field in vacuo are manifestly covariant four vectors in spacetime, E_μ and B_μ, respectively, all of whose four components must be physically meaningful.

2. From Eq. (19) the fields $\mathbf{E}^{(3)} = E^{(0)}\hat{\mathbf{e}}^{(3)}$ and $\mathbf{B}^{(3)} = B^{(0)}\hat{\mathbf{e}}^{(3)}$ are independent of the phase of the traveling plane wave, which is consistent with Eq. (6) and the development thereof.

The four physical states of the classical, manifestly covariant, electromagnetic field are formed from the (0) and (3) admixtures $E^{(0)} - |\mathbf{E}^{(3)}|$ and $B^{(0)} - |\mathbf{B}^{(3)}|$ and from the well-known transverse ((1) and (2)) components.[8-15] Although Maxwell's phenomenological equations of the 1860s are conventionally accepted as being consistent with special relativity, the electric and magnetic fields that they relate in vacuo are purely spacelike. The field potentials in terms of which $\mathbf{E}$ and $\mathbf{B}$ are described in the conventional theory[8-15] are, on the other hand, taken to be components of the potential four vector in spacetime:

$$A_\mu \equiv (\mathbf{A}, +i\phi) \qquad (21)$$

where $\mathbf{A}$ is the spacelike (vector) potential and ϕ is the timelike (scalar) potential. The four-curl of A_μ conventionally produces the electromagnetic field tensor in spacetime (see Appendix A):

$$F_{\mu\nu} = -F_{\nu\mu} = \varepsilon_0 \begin{bmatrix} 0 & cB_Z & -cB_Y & -iE_x \\ -cB_Z & 0 & cB_X & -iE_Y \\ cB_Y & -cB_X & 0 & -iE_Z \\ iE_X & iE_Y & iE_Z & 0 \end{bmatrix} \qquad (22)$$

The difficulty with $F_{\mu\nu}$ and with the conventional theory is that $F_{\mu\nu}$ contains no explicit reference to the timelike components of E_μ and B_μ. These are removed by the mathematical nature of the four-curl:

$$F_{\mu\nu} = \frac{\partial A_\nu}{\partial x_\mu} - \frac{\partial A_\mu}{\partial x_\nu} \qquad (23)$$

A method must be found to relate E_μ and B_μ to the four potential A_μ, thus making the theory rigorously self-consistent and manifestly covariant.

It is reasonable to base this method in contemporary quantum field theory,[7] in which the electromagnetic field is an example of a massless gauge field (as opposed to a spinor field), described in general by the Poincaré group. The latter incorporates three Lorentz rotation generators (J_i), three Lorentz boost generators (K_i), and four generators of spacetime translation (P_μ). Details are well summarized in Ref. 7. The difference between the Poincaré and Lorentz groups is that the former incorporates the generator of spacetime translations, defined by

$$P_\mu \equiv i \frac{\partial}{\partial x_\mu} \qquad (24)$$

and which is proportional through a factor $\hbar$ to the momentum energy four vector operator. The Pauli-Lubansky pseudovector in spacetime, W_μ, characterizes the Poincaré group by forming its second (spin) Casimir invariant $W_\mu W_\mu$. (The first (mass) Casimir invariant is formed from $P_\mu P_\mu$.) The Pauli Lubansky pseudovector is defined by

$$W_\mu = -\tfrac{1}{2}\varepsilon_{\mu\nu\rho\sigma} J_{\nu\rho} P_\sigma \qquad (25)$$

where $\varepsilon_{\mu\nu\rho\sigma}$ is the totally antisymmetric spacetime tensor of rank four, and where the four tensor $J_{\nu\rho}$ is

$$J_{\mu\sigma}(\mu, \sigma = 0, \ldots, 3) \begin{bmatrix} J_{ij} = -J_{ji} = \varepsilon_{ijk}J_k \\ J_{i0} = -J_{0i} = -K_i \end{bmatrix}$$

$$i, j, k = 1, 2, 3 \tag{26}$$

III. THE LINK BETWEEN E_μ, B_μ, $F_{\mu\nu}$, AND A_μ

The photon helicity is defined in the lightlike condition applied to Eq. (25), in which condition W_μ becomes proportional to P_μ, so that the helicity is a number, $+1$ (Ref. 7). The opposite value, -1, is given by considerations[7] of parity inversion. We show in this section that E_μ and B_μ can be defined in terms of $F_{\mu\nu}$, and therefore of the four curl of A_μ, by an equation whose structure is the same as that of Eq. (25). Thus, E_μ and B_μ are identified as a Pauli Lubansky vector and pseudovector, respectively. This procedure succeeds in expressing electric and magnetic four vectors in terms of a single potential four vector, and covariantly describes the conventional[8-15] relations between **E**, **B**, **A**, and ϕ in vacuo.

The primary basis of the derivation is the observation that $J_{\mu\nu}$ has the same structure as $F_{\mu\nu}$, both being antisymmetric four tensors of the type (26). The $ij = -ji$ components of $F_{\mu\nu}$ are therefore identified as being proportional to rotation generators of the Poincaré group.[7] With this observation, it becomes obvious that the spacelike electric components in Eq. (22) are proportional to boost generators of the Poincaré group.[7] Pure boost Lorentz transformations[7] are those connecting two inertial frames moving at a relative speed v. A Lorentz rotation is a four vector rotation in spacetime. Therefore, the conventional assertion that $cB^{(3)}$ and $E^{(3)}$ in $F_{\mu\nu}$ are physically meaningless is tantamount to asserting that one out of three rotation generators and one out of three boost generators are physically meaningless. This is a reductio ad absurdum, and a vivid demonstration of the fact that the conventional assertion that $E^{(3)}$ and $B^{(3)}$ are unphysical is flawed fundamentally, i.e.,is geometrically unsound.

Secondly, Eq. (3), which we have derived elsewhere[2] using independent considerations in the quantum field, shows that $\hat{B}^{(3)}$ is directly proportional to the photon's quantized angular momentum boson operator $\hat{\mathbf{J}}$. Classically, the rotation generators $\mathbf{J}_X$, $\mathbf{J}_Y$, and $\mathbf{J}_Z$ of the Poincaré group are matrices of numbers which obey the commutation relations

$$[\mathbf{J}_X, \mathbf{J}_Y] = i\mathbf{J}_Z \tag{27}$$

and cyclic permutations thereof. These are immediately recognizable to be the commutators of quantized angular momentum within the factor $\hbar$. This suggests that $F_{\mu\nu}$ is proportional to $J_{\mu\nu}$ through a four scalar invariant of spacetime. For example, components 12 and 21 of $F_{\mu\nu}$ are respectively $cB^{(3)}$ and $-cB^{(3)}$, and all other off-diagonal components of $F_{\mu\nu}$ have the same dimensions as the 12 and 21 components. The 12 and 21 components of $J_{\mu\nu}$ are the angular momenta $J^{(3)}$ and $-J^{(3)}$ within a factor $\hbar$. The 21 and 12 component proportionality is therefore embodied in Eq. (3).

Thirdly, the contemporary quantum field description of photon helicity in terms of W_μ and P_μ clearly involves the concept of spacetime translation within the Poincaré group, introduced by Wigner[21] in 1939, and whose generator, as we have seen, is P_μ. This is missing from the Lorentz group.[7] The concept of spacetime translation is also missing from the Maxwell equations, which do not deal explicitly in the timelike field polarization (0). Spacetime translation is implied in d'Alembert's equation (1),[7] but if and only if all four field polarizations are taken to be physically meaningful. To see this, recall (1) that the four-curl (23) removes the (0) polarization, and (2) that the Maxwell equations (2) are equations[7] in $F_{\mu\nu}$ and its dual $\tilde{F}_{\mu\nu}$. From the proportionality of $F_{\mu\nu}$ to $J_{\mu\nu}$ it becomes clear, however, that the components of $F_{\mu\nu}$ must be either rotation or boost generators of the Poincaré group, and there is no reference within $F_{\mu\nu}$ to spacetime translation. Photon helicity, on the other hand, is described in terms of the proportionality and orthogonality in spacetime of W_μ to P_μ (Ref. 7) in the lightlike condition. Therefore, the description of electric and magnetic components in vacuo in terms of $F_{\mu\nu}$ is inconsistent with the contemporary description of helicity. This inconsistency can be remedied if and only if electric and magnetic components of the electromagnetic field in vacuo are manifestly covariant four vectors E_μ and B_μ.

Fourthly, defining a photon state $|k\rangle$ in the lightlike condition, the photon helicity (λ) in contemporary thought[7] is given by the condition

$$(W_\mu - \lambda P_\mu)|k\rangle = 0 \qquad (28)$$

so that for the massless photon, λ is a number ($+1$), which is the ratio of W_μ to P_μ and which has the dimensions of angular momentum,[7] provided that P_μ has the dimensions of linear momentum/energy by multiplication by $\hbar$. For lightlike particles with no mass, such as the photon[7],

$$k_\mu \equiv (0, 0, k, -ik) \qquad (29)$$

which can be regarded as a unit four vector

$$\delta_\mu \equiv (0, 0, 1, -i) \tag{30}$$

describing a massless particle moving at the speed of light in the Z spacelike axis (the propagation axis of the electromagnetic wave). Equation (30) can be incorporated into Eq. (25) by dividing the left and right sides of Eq. (25) by $\hbar$, so that W_μ, $J_{\nu\rho}$, and P_σ become numbers. This is consistent with the definition of rotation and boost generators as matrices of numbers (Eqs. (2.65)–(2.67) of Ref. 7).

With these considerations, we are led to the following fundamental definitions of B_μ and E_μ in terms of $F_{\nu\rho}$ and δ_σ:

$$cB_\mu = -\frac{i}{2\varepsilon_0} \varepsilon_{\mu\nu\rho\sigma} F_{\nu\rho} \delta_\sigma \tag{31a}$$

$$E_\mu = \frac{1}{2\varepsilon_0} \varepsilon_{\mu\nu\rho\sigma} \tilde{F}_{\nu\rho} \delta_\sigma \tag{31b}$$

In these equations, we recall that if $\varepsilon_{0123} = 1$, then its other nonzero elements are $+1$ and -1, according as to whether ε_{0123} can be generated by an even or odd number of subscript pair permutations. Thus, for example,

$$
\begin{array}{lll}
\varepsilon_{3120} = -1 & \varepsilon_{3210} = 1 & \varepsilon_{2310} = -1 \\
\varepsilon_{3120} = -1 & \varepsilon_{1320} = 1 & \varepsilon_{1230} = -1
\end{array} \tag{32}
$$

and so on. All elements of $\varepsilon_{\mu\nu\rho\sigma}$ are zero in which two or more subscripts are equal. The elements of $F_{\nu\rho}$ are labeled explicitly as

$$
F_{\nu\rho}(\nu, \rho = 0, 1, 2, 3) =
\begin{bmatrix}
11 & 12 & 13 & 10 \\
21 & 22 & 23 & 20 \\
31 & 32 & 33 & 30 \\
01 & 02 & 03 & 00
\end{bmatrix} \tag{33}
$$

With these definitions it is verified by tensor algebra (Appendix B) that the real elements (labeled (1), (2), (3), and (0)) of the magnetic and electric field four vectors

$$
\begin{aligned}
E_\mu &\equiv (E^{(1)}, E^{(2)}, E^{(3)}, -iE^{(0)}) \\
B_\mu &\equiv (B^{(1)}, B^{(2)}, B^{(3)}, -iB^{(0)})
\end{aligned} \tag{34}
$$

are given by Eqs. (31). (The dual $\tilde{F}_{\mu\nu}$ of $F_{\rho\sigma}$ is obtained by the well-known[7] dual transformation $\tilde{F}_{\mu\nu} = \frac{1}{2}\varepsilon_{\mu\nu\rho\sigma}F_{\rho\sigma}$.)

Equation (31a) covariantly defines B_μ as a Pauli-Lubansky pseudovector, and Eq. (31b) covariantly defines E_μ as a Pauli-Lubansky vector. These definitions imply several properties of both B_μ and E_μ:

1. Since $F_{\mu\nu}$ is the four-curl of A_ν (Eq. (23)), Eqs (31) covariantly relate B_μ and E_μ to A_μ in spacetime.

2. Since E_μ and B_μ are four vectors in Minkowski spacetime, it follows in pseudo-Euclidean geometry that $E_\mu E_\mu$ and $B_\mu B_\mu$ are constants in spacetime, and that

$$\frac{\partial E_\mu}{\partial x_\mu} = \text{constant} \qquad \frac{\partial B_\mu}{\partial x_\mu} = \text{constant} \qquad (35)$$

3. Equation (31a) is dual with Eq. (31b), because under the dual transformation of fields

$$F_{\rho\sigma} \to \tilde{F}_{\mu\nu} \qquad E_\mu \to -icB_\mu \qquad (36)$$

4. The parity inversion $\hat{P}$ and motion reversal $\hat{T}$ symmetries of B_μ are consistent with those of $F_{\nu\rho}$ and δ_σ, bearing in mind that the latter is the unit generator of spacetime translation. Since Eq. (31b) is dual with Eq. (31a), its symmetries are consistent with those of Eq. (31a). B_μ is a pseudovector because its spacelike component **B** is positive to $\hat{P}$ and negative to $\hat{T}$. E_μ is a vector because **E** is negative to $\hat{P}$ and positive to $\hat{T}$.

5. From the properties of the Pauli-Lubansky pseudovectors and vectors,[7] both E_μ and B_μ are orthogonal to δ_μ in spacetime:

$$B_\mu\delta_\mu = 0 \qquad E_\mu\delta_\mu = 0 \qquad (37)$$

6. Both E_μ and B_μ are defined in Eqs. (31) in terms of the unit generator of spacetime translations δ_σ, allowing E_μ and B_μ to be covariantly and consistently interpreted in terms of helicity in the lightlike condition.[7]

7. Since E_μ and B_μ are defined covariantly, the timelike components $E^{(0)}$ and $B^{(0)}$, respectively, are both explicitly and implicitly stated to be physically meaningful in spacetime.

8. The products $B_\mu B_\mu$ and $E_\mu E_\mu$ are both Casimir invariants[7] of the Poincaré group, specifically Casimir invariants of the second kind, or

"spin" invariants. The product $\delta_\mu \delta_\mu$ is a Casimir invariant of the first kind ("mass" invariant). This deduction follows from the definition (31) of B_μ and E_μ as Pauli-Lubansky types. In the lightlike condition, i.e., for the massless electromagnetic gauge field,

$$\delta_\mu \delta_\mu = 0 \qquad E_\mu E_\mu = 0 \qquad B_\mu B_\mu = 0 \qquad (38)$$

Equation (38) is a classical statement of the fact that the photon in the quantum field is massless and possesses spin.

9. From Eqs. (37) and (38), E_μ and B_μ are both orthogonal and proportional to δ_μ in spacetime. The proportionality constant (a scalar in spacetime) expresses the helicity of the electromagnetic gauge field.

It can be verified explicitly that the fundamental conditions (37) and (38) are satisfied by the circularly polarized transverse components of Eq. (7) in combination with the longitudinal components of Eqs. (8) and (12b). For example,

$$
\begin{aligned}
E_\mu E_\mu &\equiv E^{(1)2} + E^{(2)2} + E^{(3)2} - E^{(0)2} \\
&= E^{(0)2}\left(\hat{\mathbf{e}}^{(1)} \cdot \hat{\mathbf{e}}^{(1)} e^{2i\phi} + \hat{\mathbf{e}}^{(2)} \cdot \hat{\mathbf{e}}^{(2)} e^{-2i\phi} + \hat{\mathbf{e}}^{(3)} \cdot \hat{\mathbf{e}}^{(3)} - 1\right) \\
&= E^{(0)2}\left(\hat{\mathbf{e}}^{(1)} \cdot \hat{\mathbf{e}}^{(1)} e^{2i\phi} + \hat{\mathbf{e}}^{(2)} \cdot \hat{\mathbf{e}}^{(2)} e^{-2i\phi}\right) \qquad (39) \\
&= \frac{E^{(0)2}}{2}\left((\mathbf{i} - \mathbf{ij}) \cdot (\mathbf{i} - \mathbf{ij}) e^{2i\phi} + (\mathbf{i} + \mathbf{ij}) \cdot (\mathbf{i} - + \mathbf{ij}) e^{-2i\phi}\right) \\
&= 0
\end{aligned}
$$

IV. CONSEQUENCES FOR VACUUM ELECTRODYNAMICS

Equations (31) covariantly define the four vectors E_μ and B_μ in spacetime. This means that the fundamentals of vacuum electrodynamics are changed. One immediate consequence is that Eqs. (35) restate the Gauss theory in covariant form. Using the polarizations defined in Eqs. (7), (8), and (12b) it is clear that the constant in Eqs. (35) is zero and that the Gauss theorem in differential form is covariantly written as

$$\frac{\partial E_\mu}{\partial x_\mu} = 0 \quad \text{or} \quad \nabla \cdot \mathbf{E} + \frac{1}{c}\frac{\partial E^{(0)}}{\partial t} = 0 \qquad (40)$$

and

$$\frac{\partial B_\mu}{\partial x_\mu} = 0 \quad \text{or} \quad \nabla \cdot \mathbf{B} + \frac{1}{c}\frac{\partial B^{(0)}}{\partial t} = 0 \tag{41}$$

Equations (40) and (41) replace the conventional spacelike statements of the Gauss theorem in differential form in vacuo:

$$\nabla \cdot \mathbf{E} = 0 \qquad \nabla \cdot \mathbf{B} = 0 \tag{42}$$

Therefore, it becomes clear that the covariant definitions (30) lead to the following covariant statements of the Maxwell equations in vacuo:

$$\frac{\partial F_{\mu\nu}}{\partial x_\mu} = 0 \qquad \frac{\partial \tilde{B}_\mu}{\partial x_\mu} = 0$$

$$\frac{\partial \tilde{F}_{\mu\nu}}{\partial x_\mu} = 0 \qquad \frac{\partial B_\mu}{\partial x_\mu} = 0 \tag{43}$$

in which $\tilde{B}_\mu$ is the dual of $\mathbf{B}_\mu$, and $\tilde{F}_{\mu\nu}$ that of $\mathbf{F}_{\rho\sigma}$. The Maxwell equations in the form (43) are covariantly consistent with the d'Alembert equation (1), which was the starting point of our development.

The vacuum electromagnetic energy density is, from Eqs. (31), covariantly defined in S.I. units as (see Appendix C)

$$U = \frac{1}{2}\left(\varepsilon_0 E_\mu E_\mu + \frac{1}{\mu_0} B_\mu B_\mu\right) \tag{44}$$

where μ_0 is the permeability in vacuo. (Note that it is not consistent to refer to the vacuum as "free space"; it is covariantly described as "free spacetime.") From the example of Eqs. (39), it is clear that field polarizations (0) and (3), although physically meaningful, do not contribute to U, so that Eq. (44) happens to reduce to the conventional[8-15] spacelike definition of U:

$$U(\text{conventional}) \equiv \frac{1}{2}\left(\varepsilon_0 \mathbf{E} \cdot \mathbf{E} + \frac{1}{\mu_0}\mathbf{B} \cdot \mathbf{B}\right) \tag{45}$$

This deduction is consistent with the quantum field theory leading[7] to Eq. (16).

Similarly, the vacuum electromagnetic flux density (the conventional, spacelike, Poynting vector[8-15]) is covariantly defined from Eqs. (30) as the four vector product of E_μ and B_μ, the four tensor, in S.I. units:

$$S_{\mu\nu} = \frac{1}{\mu_0}(E_\mu B_\nu - B_\mu E_\nu) \qquad (46)$$

The conventional statement of the law of conservation of electromagnetic energy in vacuo is the Poynting theorem,[8-15] expressed through the continuity equation:

$$\nabla \cdot \mathbf{S} + \frac{1}{c^2}\frac{\partial U}{\partial t} = 0 \qquad (47)$$

This is already Lorentz covariant in structure, because it is an equation in the four divergence of the Poynting four vector $S_\mu = (S, -iS^{(0)})$, i.e.,

$$\frac{\partial S_\mu}{\partial x_\mu} = 0 \qquad (48)$$

However, the conventional definition (47) implies that the two spacelike components of the Poynting four vector S_μ orthogonal to the propagation direction of the electromagnetic wave in vacuo must vanish. The definition (47), although Lorentz covariant, is not necessarily manifestly covariant, because it is based on the conventional[8-15] assumption that $\mathbf{E}$ and $\mathbf{B}$ are spacelike and transverse.

In a manifestly covariant description it is necessary to relate the Poynting four vector of Eq. (48) to the Poynting four tensor $S_{\mu\nu}$ formed from the vector product in spacetime of the novel four vectors E_μ and B_μ. It is reasonable to propose that this relation is

$$S_\mu = \frac{i}{2}\varepsilon_{\mu\nu\rho\sigma}S_{\nu\rho}\delta_\sigma \qquad (49)$$

where $\varepsilon_{\mu\nu\rho\sigma}$ and δ_σ have the same meaning as in Eq. (31). Explicitly,

$$S^{(1)} \equiv S_1 = \frac{i}{2}(\varepsilon_{1230}S_{23}\delta_0 + \varepsilon_{1320}S_{32}\delta_0 + \varepsilon_{1203}S_{20}\delta_3 + \varepsilon_{1023}S_{02}\delta_3)$$
$$(50a)$$

$$S^{(2)} \equiv S_2 = \frac{i}{2}(\varepsilon_{2310}S_{31}\delta_0 + \varepsilon_{2130}S_{13}\delta_0 + \varepsilon_{2013}S_{01}\delta_3 + \varepsilon_{2103}S_{10}\delta_3)$$
$$(50b)$$

$$S^{(3)} \equiv S_3 = \frac{i}{2}(\varepsilon_{3210}S_{21}\delta_0 + \varepsilon_{3120}S_{12}\delta_0)$$

$$-iS^{(0)} \equiv -iS_0 = \frac{i}{2}(\varepsilon_{0123}S_{12}\delta_3 + \varepsilon_{0213}S_{21}\delta_3) \qquad (50c)$$

$$\text{with} \quad \delta_3 = 1 \quad \delta_0 = -i \quad (\delta_1 = \delta_2 = 0)$$

$$\varepsilon_{0123} = 1 \quad \varepsilon_{0213} = -1 \quad \varepsilon_{3120} = -1 \quad \varepsilon_{3210} = -1$$
$$\varepsilon_{2130} = 1 \quad \varepsilon_{2310} = -1 \quad \varepsilon_{2013} = 1 \quad \varepsilon_{2103} = -1 \quad (50d)$$
$$\varepsilon_{1230} = -1 \quad \varepsilon_{1320} = 1 \quad \varepsilon_{1203} = 1 \quad \varepsilon_{1023} = -1$$

Equations (50a) and (50b) show that in this definition, the Poynting four vector in spacetime develops components in the spacelike axes orthogonal to the propagation axis (3).

The definition (49) of the manifestly covariant Poynting vector introduces the unit generator of spacetime translations, δ_σ, for an electromagnetic wave traveling in vacuo in the spacelike axis (3). In direct analogy with our fundamental definitions, Eqs. (31), of E_μ and B_μ, S_μ is thereby defined within the Poincaré group rather than the Lorentz group, and spacetime translation is included explicitly in the definition. This means that the manifestly covariant Poynting vector is also a Pauli-Lubansky vector within the Poincaré group in spacetime. Note that from Eqs. (50c) and (50d),

$$S^{(3)} - |S^{(0)}| = 0$$
$$S^{(3)} \cdot S^{(3)} - |S^{(0)}|^2 = 0 \qquad (51)$$

i.e., the conventional, spacelike Poynting vector, which has only one spacelike component, (3), and no timelike component, (0), becomes within the structure of Eq. (49) a physical state that is an admixture of (3) and (0) polarizations. The other spacelike components of S_μ, i.e., (1) and (2), also become physically meaningful through Eqs. (50a) and (50b). At an instant in spacetime, these components (1) and (2) are experimental observables. However, observations (Appendix 3) of the electromagnetic energy flux density, known as the Poynting vector,[8-15] are made by the observer with an instrument such as a power meter, which gives only the time-averaged value of the Poynting vector. The components $S_{23}, S_{32}, S_{20}, S_{02}, S_{31}$, S_{13}, S_{01}, S_{10} disappear upon time averaging (Appendix 3) because they are made up of products of one phase-dependent component and one that is phase independent. The components S_{21} and S_{12}, on the other hand, are products of two phase-dependent components, one of which is the complex conjugate of the other, so that the phase disappears in the product, which is thereby nonzero after time averaging. For these reasons, the conventional Poynting theorem (47), which is not manifestly covariant, happens to be an adequate description of the law of conservation of electromagnetic energy, but only on a time-averaged basis. If it were experimentally possible to observe electromagnetic energy flux density in an instant in spacetime, then the components S_{23} and so on would contribute explicitly to the law of conservation of energy. Clearly, if S_μ is a Pauli-Lubansky vector in the Poincaré group, then the product is a Casimir invariant of type two of the Poincaré group, and S_μ is orthogonal to δ_μ in spacetime.

The description of the electromagnetic field polarization in vacuo through the four Stokes parameters in terms of E_μ and B_μ requires a modification[22] of the conventional description[23] based on Pauli matrices:

$$S_0 = [E_X E_Y] \begin{bmatrix} 1 & 0 \\ 0 & 1 \end{bmatrix} \begin{bmatrix} E_X^* \\ E_Y^* \end{bmatrix} \tag{52a}$$

$$S_1 = [E_X E_Y] \begin{bmatrix} 1 & 0 \\ 0 & -1 \end{bmatrix} \begin{bmatrix} E_X^* \\ E_Y^* \end{bmatrix} \tag{52b}$$

$$S_2 = [E_X E_Y] \begin{bmatrix} 0 & 1 \\ 1 & 0 \end{bmatrix} \begin{bmatrix} E_X^* \\ E_Y^* \end{bmatrix} \tag{52c}$$

$$S_3 = [E_X E_Y] \begin{bmatrix} 0 & -i \\ i & 0 \end{bmatrix} \begin{bmatrix} E_X^* \\ E_Y^* \end{bmatrix} \tag{52d}$$

so that

$$S_0^2 = S_1^2 + S_2^2 + S_3^2 \tag{53a}$$

$$[S_1, S_2] = iS_3 \tag{53b}$$

Within a factor $\hbar$, the Stokes parameters obey the commutation rules of quantized angular momentum, and form a four vector $S_\mu \equiv (\mathbf{S}, iS_0)$. (The Stokes vector S_μ should not be confused with the Poynting vector, unfortunately also denoted $\mathbf{S}$ in the conventional literature.) It follows from Eq. (53b) that the Pauli matrices also obey the angular momentum commutation rules. Equations (52) omit the longitudinal and timelike polarizations of the four vector E_μ, and for a manifestly covariant description, these must be included in the basic definition of the four Stokes parameters (real numbers in the conventional theory[8-15]). The generalization of S_0 to S_3 must conform with Eqs. (53), and S_0, which is proportional to the electromagnetic energy density in vacuo,[8-15] must conform to our earlier results (16) and (39). It is natural to propose the replacement in Eqs. (52) of the field vectors $[E_X, E_Y]$ and $\begin{bmatrix} E_X^* \\ E_Y^* \end{bmatrix}$ by their manifestly covariant equivalents (four vectors), and to replace the Pauli matrices by Dirac matrices (angular momentum operators[24]). The latter obey the same commutation rules and their structure is that of a "doubled" (4×4) Pauli matrix. The following generalization is manifestly covariant and conforms with Eq. (53):

$$S_0 = \begin{bmatrix} E_X & E_Y & \dfrac{E_Z}{2} & \dfrac{-iE^{(0)}}{2} \end{bmatrix} \begin{bmatrix} 1 & 0 & 0 & 0 \\ 0 & 1 & 0 & 0 \\ 0 & 0 & 1 & 0 \\ 0 & 0 & 0 & 1 \end{bmatrix} \begin{bmatrix} E_X^* \\ E_Y^* \\ E_Z/2 \\ -iE^{(0)}/2 \end{bmatrix} \tag{54a}$$

$$S_1 = \begin{bmatrix} E_X & E_Y & \dfrac{E_Z}{2} & \dfrac{-iE^{(0)}}{2} \end{bmatrix} \begin{bmatrix} 1 & 0 & 0 & 0 \\ 0 & -1 & 0 & 0 \\ 0 & 0 & 1 & 0 \\ 0 & 0 & 0 & -1 \end{bmatrix} \begin{bmatrix} E_X^* \\ E_Y^* \\ E_Z/2 \\ -iE^{(0)}/2 \end{bmatrix} \tag{54b}$$

$$S_2 = \begin{bmatrix} E_X & E_Y & \dfrac{E_Z}{2} & \dfrac{-iE^{(0)}}{2} \end{bmatrix} \begin{bmatrix} 0 & 1 & 0 & 0 \\ 1 & 0 & 0 & 0 \\ 0 & 0 & 0 & 1 \\ 0 & 0 & 1 & 0 \end{bmatrix} \begin{bmatrix} E_X^* \\ E_Y^* \\ E_Z/2 \\ -iE^{(0)}/2 \end{bmatrix} \tag{54c}$$

$$S_3 = \begin{bmatrix} E_X & E_Y & \dfrac{E_Z}{2} & \dfrac{-iE^{(0)}}{2} \end{bmatrix} \begin{bmatrix} 0 & -i & 0 & 0 \\ i & 0 & 0 & 0 \\ 0 & 0 & 0 & -i \\ 0 & 0 & i & 0 \end{bmatrix} \begin{bmatrix} E_X^* \\ E_Y^* \\ E_Z/2 \\ -iE^{(0)}/2 \end{bmatrix} \tag{54d}$$

(The factor $\frac{1}{2}$ follows from the definitions, Eqs. (31) and Appendix B, Eq. (B.7).) Explicitly written out, the covariant Stokes parameters for one sense of circular polarization become

$$S_0 = E_X E_X^* + E_Y E_Y^* = E^{(0)2} \tag{55a}$$

$$S_1 = E_X E_X^* - E_Y E_Y^* + \tfrac{1}{4}\left(E_Z^2 + E^{(0)2}\right) = \tfrac{1}{2} E^{(0)2} \tag{55b}$$

$$S_2 = E_X E_Y^* + E_Y E_X^* - \frac{i}{4}\left(E_Z E^{(0)} + E^{(0)} E_Z\right) = -\frac{1}{2} i E^{(0)2} \tag{55c}$$

$$S_3 = -i\left(E_X E_Y^* - E_Y E_X^*\right) = E^{(0)2} \tag{55d}$$

We find that the conventional result $S_1 = S_2 = 0$ in circular polarization[8-15] is replaced by

$$S_1 = iS_2 = \tfrac{1}{2} E^{(0)2} = \tfrac{1}{2} \mathbf{E}^{(3)} \cdot \mathbf{E}^{(3)} \tag{56}$$

Our covariant theory leaves the value of S_0 unchanged, as required, and finally, S_3 is also unchanged. Significantly, the results (55) can be expressed entirely in terms of $\mathbf{E}^{(3)}$ (or $\mathbf{B}^{(3)}$):

$$S_0 = |S_3| = \mathbf{E}^{(3)} \cdot \mathbf{E}^{(3)} \tag{57}$$

together with Eq. (56). Since $E^{(0)} = cB^{(0)}$ in free spacetime, Eqs. (56) and (57) can be represented in terms of $\mathbf{B}^{(3)}$. In particular,

$$|S_3| = c^2 \mathbf{B}^{(3)} \cdot \mathbf{B}^{(3)} = c^2 B^{(0)} |\mathbf{B}^{(3)}| \tag{58}$$

a result derived previously[1-5] through the relation

$$\mathbf{B}^{(3)} = \frac{\mathbf{E}^{(1)} \times \mathbf{E}^{(2)}}{icE^{(0)}} \tag{59}$$

It is interesting to note that the following eigenvalue (operator type but classical) equation consistently reconciles the existence of only one photon helicity for one sense of circular polarization:

$$\begin{bmatrix} 0 & -i & 0 & 0 \\ i & 0 & 0 & 0 \\ 0 & 0 & 0 & -i \\ 0 & 0 & i & 0 \end{bmatrix} \begin{bmatrix} E_X \\ E_Y \\ E_Z/2 \\ -iE^{(0)}/2 \end{bmatrix} = \lambda^{(i)} \begin{bmatrix} E_X \\ E_Y \\ E_Z/2 \\ -iE^{(0)}/2 \end{bmatrix} \tag{60}$$

The eigenvalues are

$$
\lambda^{(1)} = \frac{-iE_Y}{E_X} = -1 \qquad \lambda^{(2)} = \frac{iE_X}{E_Y} = -1
$$

$$
\lambda^{(3)} = \frac{-E^{(0)}}{E_Z} = -1 \qquad \lambda^{(4)} = \frac{-E_Z}{E^{(0)}} = -1
$$

(61)

For the opposite sense of circular polarization, E_Y and E_Z change sign and the four eigenvalues $\lambda^{(i)}$ becomes $+1$. Equation (60) therefore provides the result

$$
\lambda^{(1)} = \lambda^{(2)} = \lambda^{(3)} = \lambda^{(4)} = \mp 1
$$

(62)

for different senses of circular polarization, and reconciles the existence of four different field polarizations with only two different field helicities. (The four polarizations are right and left circular spacelike, longitudinal spacelike, and timelike. The two helicities are $+1$ and -1.) In the conventional theory,[8-15] Eq. (60) becomes

$$
\begin{bmatrix} 0 & -i \\ i & 0 \end{bmatrix} \begin{bmatrix} E_X \\ E_Y \end{bmatrix} = \lambda^{(i)} \begin{bmatrix} E_X \\ E_Y \end{bmatrix}
$$

(63)

i.e., helicities $\lambda^{(3)}$ and $\lambda^{(4)}$ are missing, and the remaining two "transverse" helicities are generated by a Pauli matrix rather than a Dirac matrix.

It is therefore concluded that the structures of the Stokes parameters are changed in the manifestly covariant description of electrodynamics in vacuo, and therefore so is the fundamental specification of the polarization characteristics of light: the Hermitian polarization density, or coherency matrix of Born and Wolf[13] and the polarization tensor of Landau and Lifshitz.[12] Specifically, the Stokes parameters S_1 and S_2 no longer vanish in circular polarization, and this is a direct consequence of the covariant nature of E_μ, in that its longitudinal and timelike components now contribute to a purely real, nonzero, S_1, and a purely imaginary S_2 with the opposite sign. Conventionally,[8-15] S_1 and S_2 are nonzero only in elliptical polarization. They can be described in terms of excess of linear polarization, and conventionally it is considered that there is no excess of linear polarization when the beam is fully right or left circularly polarized. However, in the covariant description, there is an additioinal longitudinal component in the propagation axis of the beam, even in a completely circularly polarized beam. The longitudinal components $\mathbf{E}^{(3)}$ and $\mathbf{B}^{(3)}$

vanish, however, if the beam has an equal amount of transverse right and transverse left circularly polarized components. In this state of transverse linear polarization, the imaginary contribution to S_2 vanishes, but the real contribution to S_1 doubles. This can be interpreted to mean that although $\mathbf{E}^{(3)}$ changes sign between right and left transverse circular polarization, its square $E^{(3)2}$ evidently does not. It is $E^{(3)2}$ that contributes to S_1. This emphasizes the fact that the Stokes parameters are quadratic in the electric part of the electromagnetic field.

S_3 is unchanged in the covariant description, because S_3 is defined in this description by

$$\mathbf{B}^{(3)} = \frac{\mathbf{E}^{(1)} \times \mathbf{E}^{(2)}}{icE^{(0)}} = B^{(0)}\mathbf{k} \tag{64a}$$

$$B^{(0)} = \left(\frac{\varepsilon_0}{I_0 c}\right)^{1/2} |S_3| \tag{64b}$$

V. DISCUSSION

The covariant description of the electromagnetic field in vacuo shows that there are physically meaningful fields $\mathbf{B}^{(3)}$ and $\mathbf{E}^{(3)}$ that satisfy Maxwell's equations. These fields do not appear explicitly in the conventional theory,[8-15] and are assumed to be physically meaningless. It is therefore necessary to identify experiments that can distinguish between the conventional theory and the manifestly covariant theory of the electromagnetic field. One immediately obvious consequence of $\mathbf{B}^{(3)}$ is that circularly polarized electromagnetic radiation can magnetize matter. Before embarking on a development of these properties, however, we show that effects such as natural optical activity, the electrical Kerr effect, and the development of ellipticity in an initially circularly polarized light beam can be explained in terms of changes in $\mathbf{B}^{(3)}$ as they traverse a sample. The essential reason for this is that whenever the Stokes parameter S_3 appears in physical optics, it signals (vide supra) the existence of $\mathbf{B}^{(3)}$, to whose magnitude it is directly proportional:

$$|\mathbf{B}^{(3)}| = \left(\frac{\varepsilon_0}{I_0 c}\right)^{1/2} |S_3| = \frac{|S_3|}{c^2 B^{(0)}} \tag{65}$$

Therefore, S_3 can be replaced whenever it occurs by the scalar quantity

$$\pm c^2 B_0 | \mathbf{B}^{(3)} | \equiv \pm c^2 \mathbf{B}^{(3)} \cdot \mathbf{B}^{(3)} \tag{66}$$

In material media, as opposed to free space, Kielich[25] has shown that linear and nonlinear optical activity depends on S_3, and in the Rayleigh theory[26] of natural optical activity in chiral media, it is well known that whatever the nature of the several molecular property tensors participating in the polarization and magnetization of the material, the observable of circular dichroism has pseudoscalar symmetry and is proportional to the third Stokes parameter. For different enantiomers for a given sense of transverse circular polarization, or for one enantiomer for different sense of transverse circular polarization,

$$\frac{I_R - I_L}{I_R + I_L} = \pm \frac{S_3}{S_0} \tag{67}$$

where I_R and I_L are the intensities of right and left components transmitted by structurally chiral material, with

$$I_0 = I_R + I_L \tag{68}$$

for the transmitted total beam intensity. From Eqs. (67) and (68) we derive the result (with $S_0 = c^2 B^{(0)2}$),

$$\pm \frac{S_3}{S_0} = \pm \frac{| \mathbf{B}^{(3)2} |}{B^{(0)2}} = \frac{I_R - I_L}{I_R + I_L} \tag{69}$$

which reveals the fundamental origin of the phenomenon of circular dichroism at all electromagnetic frequencies, because it shows that the observable $(I_R - I_L)$ is proportional to $| \mathbf{B}^{(3)2} |$.

The origin of circular dichroism, therefore, resides in the photon's longitudinal magnetostatic flux quantum $\hat{B}^{(3)}$, whose expectation value is $\mathbf{B}^{(3)}$.

The observable $I_R - I_L$ is therefore a spectral consequence of the interaction of $\mathbf{B}^{(3)}$ with structurally chiral material. From Eq. (69), $I_R - I_L$ is proportional to the real pseudoscalar $\pm | \mathbf{B}^{(3)} |$ after they emerge from the chiral material through which the beam has passed, i.e., after interaction has occurred between the flux quantum $\hat{B}^{(3)}$ and the appropriate molecular property tensors.[26] For one photon, the observable $I_R - I_L$ provides an experimental measure of the transmitted elementary $\mathbf{B}^{(3)}$ at each frequency. Although $\mathbf{B}^{(3)}$ itself is independent of frequency, the interacting molecular property tensor is not. Semiclassical perturbation theory[26] gives, for linear optical activity,

$$\frac{S_3}{S_0} = \frac{|\mathbf{B}^{(3)2}|}{B^{(0)2}}\tanh\left[\omega\mu_0 clN\zeta''_{XYZ}(g)\right] \tag{70}$$

where μ_0 is the permeability in vacuo, ω the angular frequency of the beam, l the sample path length, and ζ''_{XYZ} a combination[26] of molecular property tensors, which may be electric and/or magnetic in nature. For nonlinear optical activity, Eq. (70), as shown by Kielich,[25] contains additional terms.

Therefore, every time natural optical activity is observed with $I_R - I_L$, as in circular dichroism, the quantity $\mathbf{B}^{(3)}$, has been measured. In this context, a covariant description of the electromagnetic field is one that identifies the phenomenon of circular dichroism with the longitudinal field $\mathbf{B}^{(3)}$, showing that the latter is physically meaningful and is, indeed, well measured in the literature although not explicitly recognized as a magnetic field. In the conventional description on the other hand, natural optical activity is measured by S_3/S_0, which is given by

$$\frac{S_3}{S_0} = \frac{-i(E_X E_Y^* - E_Y E_X^*)}{E_X E_X^* + E_Y E_Y^*} \tag{71}$$

and $\mathbf{E}^{(3)}$ and $\mathbf{B}^{(3)}$ are conventionally supposed to be physically meaningless. However, S_3/S_0 is, of course, also expressible in the covariant description by Eq. (71), showing that the covariant description is both simpler and more complete than the conventional one. The conventional assertion that $\mathbf{B}^{(3)}$ be physically meaningless conflicts with Eq. (6), and becomes unsustainable, because $\mathbf{E}^{(1)} \times \mathbf{E}^{(2)}$ is a physically meaningful quantity directly proportional to S_3. It is more complete, more revealing, and more "natural" to describe optical activity as changes in $\mathbf{B}^{(3)}$ as a medium is traversed by a light beam. In other words, the phenomenon of natural optical activity is definitive experimental evidence for the existence of $\mathbf{B}^{(3)}$ in physical optics.

More generally, it can be shown that any phenomenon in optics that involves S_3 must involve $\mathbf{B}^{(3)}$ in its quantized or classical forms, whichever is the more appropriate to a given situation. Throughout the contemporary literature[27] that there are many of these optical phenomena, one commonplace example being the development of ellipticity in an initially circularly polarized light beam. For example, in the electric Kerr effect,[28] beam ellipticity (η) is expressed in terms of S_3, and is induced with an electric field in a probe laser. The electric Kerr effect is therefore

$$\frac{|\mathbf{B}^{(3)2}|}{B^{(0)2}} = \sin(2\eta) \qquad (72)$$

where η is the ellipticity developed in the transmitted probe as a result of the application of an electric field to a sample. This is experimental evidence for the existence of the longitudinal field $\mathbf{B}^{(3)}$. Note that for an initially linearly polarized beam, $\mathbf{B}^{(3)}$ is zero, and so Eq. (72) shows that the development of ellipticity in the Kerr effect is a direct consequence of the interaction of $\mathbf{B}^{(3)}$ with the medium through which the probe laser has passed.

Rayleigh refringent scattering theory[26] shows that the third Stokes parameter S_3 is associated with a change $d\eta/dZ$ in ellipticity in a beam passing through a sample of thickness Z. Therefore, $d\eta/dZ$ measures changes in $\mathbf{B}^{(3)}$ as it traverses the sample thickness Z. We arrive at the generally valid conclusion that ellipticity in the electromagnetic plane wave is directly related to $\mathbf{B}^{(3)}$, and that the development of ellipticity can be expressed in terms of these fields. The scalar magnitude of $\mathbf{B}^{(3)}$ is $|\mathbf{B}^{(3)}|$, respectively, associated with the timelike polarization (0) of the electromagnetic field. The timelike polarization always appears as an admixture with the longitudinal polarization, and both are physically meaningful because they are observed in fundamental optical phenomena. Equation (70), for example, shows that circular dichroism is related to the molecular property tensor sum represented by ζ'', which is made up of the Rosenfeld tensor and the electric quadrupole tensor. Note carefully, however, that ζ'' is a material property, while $\mathbf{B}^{(3)}$ is a property of free spacetime, which interacts with matter. The definition of S_3 in free spacetime in terms of $\mathbf{B}^{(3)}$ is obviously unaffected by any material property because $\mathbf{B}^{(3)}$ is associated with fundamental photon polarizations. In Eq. (70) we have used the result that $|\mathbf{B}^{(3)2}|$ is directly proportional to the Stokes parameter S_3 in free spacetime, and have replaced the Stokes parameter by a term proportional to $|\mathbf{B}^{(3)2}|$.

In summary, there is copious experimental evidence for the existence of $\mathbf{B}^{(3)}$, which is a physically meaningful magnetic field in free spacetime. Through Eq. (6), the conventional description is supplanted in physical optics by the more complete and more rigorous covariant description, i.e., by a description that is fully compatible with the theory of special relativity. Although the conventional description is self-consistent up to a point, the key equation (6) of this paper shows that it lacks the polarizations (3) and (0), which are present in the quantum field[7] but usually wrongly asserted to be physically meaningless. Note that Eq. (6) is invariant to the fundamental symmetries of physics[29]: charge conjugation $\hat{C}$, parity inversion $\hat{P}$, and motion reversal $\hat{T}$ and is therefore a rigorously self-consistent equation of electrodynamics in free spacetime.

It is also interesting to note[29] that in the field of high-energy particle physics, experimental evidence exists for timelike photons that can be produced in electron positron annihilation processes at extremely high energy. This presumably means that in such processes the concomitant magnetic and electric field amplitudes $B^{(0)}$ and $E^{(0)}$ exist independently, and are therefore physically meaningful.

Having established with available data the existence of $\mathbf{B}^{(3)}$, it is now possible to reinterpret known optical phenomena and to predict with some degree of confidence the existence of hitherto unmeasured optical phenomena based on $\mathbf{B}^{(3)}$. The everyday phenomenon of optical absorption is described by the Beer–Lambert law:

$$\alpha(\bar{\nu}) = \frac{1}{d} \log_e \frac{I_0}{I} \qquad (73)$$

Here I_0 is the incident beam intensity, I the transmitted beam intensity, and d the sample length. α is the power absorption coefficient[30] in neper m^{-1}, and this quantity can be reinterpreted in terms of $\mathbf{B}^{(3)}$, because the zeroth Stokes parameter S_0 is proportional to beam intensity. Therefore, simple optical absorption is a process that can be interpreted in terms of the longitudinal electric and/or magnetic fields of the electromagnetic plane wave, an interpretation that is just as valid as the usual one[30] in terms of the transverse components $\mathbf{E}^{(1)}$ or $\mathbf{E}^{(2)}$. In general, since all four Stokes parameters in covariant electrodynamics can be expressed in terms of $\mathbf{B}^{(3)}$, all optical phenomena involving beam polarization or optical coherence processes in linear physical optics can also be described in terms of these longitudinal fields.

In nonlinear optics,[31] the light beam is used to induce phenomena in material media (e.g., molecular matter such as liquids), phenomena that depend nonlinearly on the electric and magnetic components of the intense laser beam. A large number of such phenomena have been observed in the past thirty years,[31] and the theory of such processes has been systematically developed by Kielich and coworkers,[32] following early inroads by Piekara and Kielich,[33] who were among the first to consider systematically statistical molecular theories of optically induced phenomena in isotropic dielectric and diamagnetic media. These earlier theories are, of course, formulated in terms of the transverse spacelike components of our covariant description, and should be modified to take into account the existence of $\mathbf{B}^{(3)}$ in free spacetime. These fields are expected to produce observable magnetization and polarization when they interact with matter. For laser beams that are intense enough, various optical saturation phenomena due to $\mathbf{B}^{(3)}$ should occur. A classic work such as the early paper by Kielich[34] on frequency and spatially variable electric and magnetic polarization induced in nonlinear media by electromagnetic fields should be covariantly developed, so that the Born-Infeld electrodynamics[35] to which it refers can be extended to include $\mathbf{B}^{(3)}$ within a manifestly covariant structure. Terms such as $\mathbf{E} \times \mathbf{E}^*$ in the work by Kielich[34] can be replaced by $\mathbf{B}^{(3)}$, for example, thus predicting birefringence effects proportional to the square root of intensity, in addition to the traditional effects proportional to intensity, such as the inverse Faraday effect.[36]

In another classic paper by Kielich,[37] on nonlinear processes resulting from multipole interaction between molecules and electromagnetic fields, it would be interesting to explore the role played by $\mathbf{B}^{(3)}$ in the various nonlinear optical phenomenon proposed in this work, for example, (1) a covariant reformulation of the Dirac theory to describe the absorption of a flux quantum $\hat{B}^{(3)}$; (2) a covariant scattering theory for $\mathbf{B}^{(3)}$; (3) the role of $\mathbf{B}^{(3)}$ in the nonlinear optical processes where linear superposition is lost; (4) investigations of the probability of an n photon process with magnetic transitions involving an incoming $\hat{\mathbf{B}}^{(3)}$ flux quantum; (5) scattering theory involving the classical $\mathbf{B}^{(3)}$. Again, in the theory of nonlinear light scattering from colloidal media,[38] $\mathbf{B}^{(3)}$ is expected to play a basic part in defining the depolarization ratio, since, as we have seen, $\mathbf{B}^{(3)}$ is proportional to $I_R - I_L$. In general, in Rayleigh refringent scattering theory, the Stokes parameters in our covariant description enter in terms of $\mathbf{B}^{(3)}$, so that the longitudinal field is fundamental to any description. The role of $\mathbf{B}^{(3)}$ in the Majorana effect,[39] and intensity dependent optical circular birefringence[40] is also fundamental. The interesting phenomenon of ellipse self-rotation by a circularly polarized laser[41] is also fundamentally dependent on the longitudinal field $\mathbf{B}^{(3)}$.

More recently, the phenomena associated with light squeezing in quantum electrodynamics have become prevalent in the literature[42] and in this context Tanaś and Kielich[19] have systematically investigated the effect of squeezing on a large number of optical phenomena, including the effect on the four Stokes operators, the quantum equivalent of the four Stokes parameters.[43] It was deduced that the parameters S_1 and S_2 are in general affected by squeezing, and it would be interesting to develop this result in a manifestly covariant description, where classically, as we have seen, the four Stokes parameters are affected in basic structure, and new terms are added to S_1 and S_2. The field $\mathbf{B}^{(3)}$ also plays a role in light self-squeezing in Kerr media, discovered by Kielich et al.[44] and in general in all nonlinear quantum electrodynamics, fundamentally changing the structure of the theory.

For example, Frey et al.[45] have recently observed azimuth rotation due to an intense laser beam (the optical Faraday effect), and this has been shown by Farahi and Evans[5] to be a linear function of the square root of laser intensity, i.e., to be linearly dependent on the magnitude of $\mathbf{B}^{(3)}$. This is the first experimental evidence for the ability of $\mathbf{B}^{(3)}$ to magnetize a material, in this case a magnetic semiconductor.[45] Magnetization by a circularly polarized light beam has been observed as the inverse Faraday effect,[36] and recently, as laser-induced shifts in NMR spectra.[46] Light shifts in atomic spectra have also been observed experimentally[47] and can be reinterpreted in terms of $\mathbf{B}^{(3)}$. In general, a large number of phenomena can be reinterpreted in terms of longitudinal[48] fields in vacuo phenomena that are at present attributed solely to the transverse fields $\mathbf{E}^{(1)}$ and $\mathbf{E}^{(2)}$. In theory, optical effects due to $\mathbf{B}^{(3)}$ can be identified and separated from the concomitant effects due to $\mathbf{E}^{(1)}$ and $\mathbf{E}^{(2)}$, or $\mathbf{B}^{(1)}$ and $\mathbf{B}^{(2)}$, because the former are expected to be proportional to the square root of laser intensity (and integral powers thereof), and the latter to even powers only of laser intensity.

APPENDIX A: CARTESIAN AND CIRCULAR REPRESENTATIONS

The subscripts in the matrix in Eq. (22) are conventionally[8-15] given in the Cartesian basis, (X, Y, Z), while circular polarization is described in the circular basis $((1), (2), (3))$. Any physical phenomenon should be independent of the basis (i.e., laboratory frame of reference) used in its description, and in this paper the link between the two representations is given in terms of the following unit vector equations. Superscripts (1), (2), and (3) refer respectively to the first and second sense of transverse circular polarization, and the longitudinal polarization:

$$\hat{e}^{(1)} \equiv \frac{1}{\sqrt{2}}(\mathbf{i} - i\mathbf{j}) \tag{A.1}$$

$$\hat{e}^{(2)} = \frac{1}{\sqrt{2}}(\mathbf{i} + i\mathbf{j}) \tag{A.2}$$

$$\hat{e}^{(3)} \equiv \mathbf{k} \tag{A.3}$$

where **i**, **j**, and **k** are Cartesian unit vectors in X, Y, and Z, respectively. Thus,

$$\hat{e}^{(1)} \times \hat{e}^{(2)} = i\hat{e}^{(3)} = i\mathbf{k} \qquad (A.4)$$

The circular basis is used in Eq. (34) to define E_μ and B_μ in terms of polarizations (0), (1), (2), and (3), which are respectively timelike, transverse circular spacelike (1) and (2), and longitudinal spacelike. In Eq. (33), $F_{\nu\rho}$ is accordingly defined in the circular basis. In Appendix B, however, the explicit demonstration of Eqs. (31) is carried out in the Cartesian basis. Equations (31) are, of course, valid in any frame of reference fixed in the laboratory. The longitudinal spacelike and timelike components are the same in the Cartesian and circular basis, while the transverse components can be interrelated with Eqs. (A.1) and (A.2). Equation (22) has been obtained from a Cartesian representation of A_μ, the four potential, using a four curl, Eq. (23), in the Cartesian frame for the spacelike components.

APPENDIX B: EXPLICIT DEMONSTRATION OF EQUATIONS (31)

In this Appendix we provide an explicit demonstration of the self-consistency of Eqs. (31), both for the B_μ and E_μ vectors, because these equations form the basis of our manifestly covariant theory of vacuum electrodynamics. From Eqs. (31a) and the definition of δ_σ in Eq. (30),

$$cB_1 = -\frac{i}{2\varepsilon_0}(\varepsilon_{1230}F_{23}\delta_0 + \varepsilon_{1320}F_{32}\delta_0 + \varepsilon_{1203}F_{20}\delta_3 + \varepsilon_{1023}F_{02}\delta_3) \quad (B.1)$$

$$cB_2 = -\frac{i}{2\varepsilon_0}(\varepsilon_{2310}F_{31}\delta_0 + \varepsilon_{2130}F_{13}\delta_0 + \varepsilon_{2013}F_{01}\delta_3 + \varepsilon_{2103}F_{10}\delta_3) \quad (B.2)$$

$$cB_3 = -\frac{i}{2\varepsilon_0}(\varepsilon_{3210}F_{21}\delta_0 + \varepsilon_{3120}F_{12}\delta_0) \qquad (B.3)$$

$$-icB_0 = -\frac{i}{2\varepsilon_0}(\varepsilon_{0123}F_{12}\delta_3 + \varepsilon_{0213}F_{21}\delta_3) \qquad (B.4)$$

with

$$\delta_1 = \delta_2 = 0 \qquad \delta_3 = 1 \qquad \delta_0 = -i$$

$$\varepsilon_{0123} = 1 \qquad \varepsilon_{0213} = -1 \qquad \varepsilon_{3120} = -1 \qquad \varepsilon_{3210} = 1$$

$$\varepsilon_{2130} = 1 \qquad \varepsilon_{2310} = -1 \qquad \varepsilon_{2013} = 1 \qquad \varepsilon_{2103} = -1$$

$$\varepsilon_{1230} = -1 \quad \varepsilon_{1320} = 1 \qquad \varepsilon_{1203} = 1 \qquad \varepsilon_{1023} = -1$$

$$F_{23} = c\varepsilon_0 B_X = -F_{32} \qquad F_{20} = -i\varepsilon_0 E_Y = -F_{02}$$

$$F_{31} = c\varepsilon_0 B_Y = -F_{13} \qquad F_{01} = i\varepsilon_0 E_X = -F_{10}$$

$$F_{21} = -c\varepsilon_0 B_Z = -F_{12}$$

For one sense of circular polarization, we have

$$E_X = \frac{E_0}{\sqrt{2}} \qquad E_Y = -\frac{iE_0}{\sqrt{2}} \qquad B_X = \frac{iB_0}{\sqrt{2}} \qquad B_Y = \frac{B_0}{\sqrt{2}} \qquad \text{(B.5)}$$

$$cB_Y = E_X \qquad cB_X = -E_Y \qquad \text{(B.6)}$$

so that in Eqs. (B.1) and (B.2),

$$cB_1 = cB_X - E_Y = 2cB_X$$
$$cB_2 = cB_Y + E_X = 2cB_Y \qquad \text{(B.7)}$$

and so the left sides become magnetic components in vacuo with the vacuum relation $E_0 = cB_0$.

Similarly, the dual of $F_{\mu\nu}$ is the four tensor[8-15],

$$\tilde{F}_{\mu\nu} \equiv -\varepsilon_0 \begin{bmatrix} 0 & iE_Z & -iE_Y & -cB_X \\ -iE_Z & 0 & iE_X & -cB_Y \\ iE_Y & -iE_X & 0 & -cB_Z \\ cB_X & cB_Y & cB_Z & 0 \end{bmatrix} \qquad \text{(B.8)}$$

and in Eq. (31b),

$$E_1 = \tfrac{1}{2}\left(\varepsilon_{1230}\tilde{F}_{23}\delta_0 + \varepsilon_{1320}\tilde{F}_{32}\delta_0 + \varepsilon_{1203}\tilde{F}_{20}\delta_3 + \varepsilon_{1023}\tilde{F}_{02}\delta_3\right) \qquad \text{(B.9)}$$

$$E_2 = \tfrac{1}{2}\left(\varepsilon_{2310}\tilde{F}_{31}\delta_0 + \varepsilon_{2130}\tilde{F}_{13}\delta_0 + \varepsilon_{2013}\tilde{F}_{01}\delta_3 + \varepsilon_{2103}\tilde{F}_{10}\delta_3\right) \qquad \text{(B.10)}$$

$$E_3 = +\tfrac{1}{2}\left(\varepsilon_{3210}\tilde{F}_{21}\delta_0 + \varepsilon_{3120}\tilde{F}_{12}\delta_0\right) \qquad \text{(B.11)}$$

$$-iE_0 = -\tfrac{1}{2}\left(\varepsilon_{0123}\tilde{F}_{12}\delta_3 + \varepsilon_{0213}\tilde{F}_{21}\delta_3\right) \qquad \text{(B.12)}$$

With the relations (B.5) and (B.6) it can be shown that the components in Eqs. (B.9) and (B.10) are the electric components $2E_X$ and $2E_Y$, with

$$\tilde{F}_{12} = -\tilde{F}_{21} = i\varepsilon_0 E_Z \qquad \tilde{F}_{13} = -\tilde{F}_{31} = -i\varepsilon_0 E_Y$$

$$\tilde{F}_{10} = -\tilde{F}_{01} = -c\varepsilon_0 B_X \qquad \tilde{F}_{23} = -\tilde{F}_{32} = i\varepsilon_0 E_X$$

Similarly, it may be checked explicitly that

$$
\begin{aligned}
B_\mu \delta_\mu &\equiv |B_1|\,|\delta_1| + |B_2|\,|\delta_2| + |B_3|\,|\delta_3| - |B_0|\,|\delta_0| \\
&= \quad 0 \quad + \quad 0 \quad + \quad B_Z \quad - \quad B_Z \\
&= \quad 0
\end{aligned}
$$

APPENDIX C: THE ELECTRODYNAMICAL ENERGY DENSITY AND TIME-AVERAGED ENERGY DENSITY, OR INTENSITY, I_0

Equation (38) produces the free spacetime result

$$E_\mu E_\mu = 0 \qquad\qquad (C.1)$$

This is interpreted to mean that the scalar product of the two four vectors E_μ and E_μ is zero in the lightlike condition. In the conventional theory[8-15] the equivalent of Eq. (C.1) is

$$\mathbf{E} \cdot \mathbf{E} = 0 \qquad\qquad (C.2)$$

Equations (C.1) and (C.2) do not mean, however, that the time-averaged electromagnetic energy density I_0 is zero in vacuo. The quantity I_0 (W m^{-2}) is defined covariantly by

$$
\begin{aligned}
I_0 &= \varepsilon_0 c E^{(0)2} \\
&\equiv \tfrac{1}{2}\varepsilon_0 c E_\mu E_\mu^*
\end{aligned}
\qquad\qquad (C.3)
$$

where E_μ^* is the complex conjugate of E_μ in vacuo. Explicitly,

$$
\begin{aligned}
E_\mu &\equiv (E^{(1)}, E^{(2)}, E^{(3)}, -iE^{(0)}) \\
E_\mu^* &\equiv (E^{(1)*}, E^{(2)*}, E^{(3)}, -iE^{(0)})
\end{aligned}
\qquad\qquad (C.4)
$$

and

$$E_\mu E_\mu^* = \frac{E^{(0)2}}{2}((\mathbf{i} - \mathbf{ij}) \cdot (\mathbf{i} + \mathbf{ij}) + (\mathbf{i} + \mathbf{ij}) \cdot (\mathbf{i} - \mathbf{ij})) \quad (C.5)$$

$$= 2E^{(0)2}$$

This result shows that $E_\mu E_\mu^*$ is covariantly described because it is a constant in free spacetime. Equation (C.3) is known as the time-averaged energy density[8-15] or beam *intensity*. This is invariant to Lorentz transformation and is a scalar quantity. Note that although E_μ^* is defined in Eq. (C.4) as the complex conjugate of E_μ, the sign of the timelike component $-iE^{(0)}$ does not change, because the operation $E_\mu \to E_\mu^*$ takes place in a fixed frame of reference $(X, Y, Z, -ict)$ in pseudo-Euclidean spacetime. Finally, $E^{(3)}$ is defined as having no imaginary part, and is invariant under $E_\mu \to E_\mu^*$. Thus, $E^{(3)}$ and $-iE^{(0)}$ do not contribute to I_0.

APPENDIX D: SIMPLE LORENTZ TRANSFORMATION OF E_μ AND B_μ

The simple Lorentz transformation of the four vector E_μ is given covariantly by

$$E'^{(0)} = E^{(0)}\xi; \qquad E'^{(3)} = E^{(3)}\xi \qquad E^{(2)} = E^{(2)'} \qquad E^{(1)} = E^{(1)'} \quad (D.1)$$

and, similarly, the transformation of B_μ is

$$B'^{(0)} = B^{(0)}\xi \qquad B'^{(3)} = B^{(3)}\xi \qquad B^{(2)} = B^{(2)'} \qquad B^{(1)} = B^{(1)'} \quad (D.2)$$

The transform is from the covariantly defined frame $(X, Y, Z, -ict)$ to $(X', Y', Z', -ict')$; which translates along Z at speed v relative to the former. In Eqs. (D.1) and (D.2),

$$\xi = \frac{1 - v/c}{\left(1 - v^2/c^2\right)^{1/2}} \quad (D.3)$$

This is referred to as a simple Lorentz transformation because there is no rotation and no translation generator considered. In other words, the origin of frame $(X, Y, Z, -ict)$ does not translate, and no rotations are considered in spacetime. For the electromagnetic plane wave in vacuo,

$v = c$, and Eqs. (D.1) and (D.2) give

$$E^{(0)} = E^{(0)'} \qquad B^{(0)} = B^{(0)'}$$
$$E^{(3)} = E^{(3)'} \qquad B^{(3)} = B^{(3)'} \qquad \text{(D.4)}$$

which confirm that the equations

$$|\mathbf{E}^{(3)}| = E^{(0)} \qquad |\mathbf{B}^{(3)}| = B^{(0)} \qquad \text{(D.5)}$$

are invariant to the simple Lorentz transformation. The results (D.1) to (D.4) confirm that the E_μ and B_μ fields are the same for all v, because $v = c$ in vacuo, and c is the universal constant of special relativity. Therefore, all four components of both E_μ and B_μ are formally invariant to the simple Lorentz transformation.

It is important to note that this result is fully consistent with, but contains additional information compared with, the standard approach,[12] which applies the simple Lorentz transformation to the four potential vector A_μ and to the second rank tensor $F_{\mu\nu}$. In S.I. units the standard approach gives the well-known result

$$E'_X = \frac{E_X - vB_Y}{\left(1 - v^2/c^2\right)^{1/2}}$$
$$E'_Y = \frac{E_Y + vB_X}{\left(1 - v^2/c^2\right)^{1/2}} \qquad \text{(D.6)}$$
$$E'_Z = E_Z$$

and using the free space relations

$$cB_Y = E_X \qquad cB_X = -E_Y \qquad \text{(D.7)}$$

we obtain

$$E'_X = \xi E_X \qquad E'_Y = \xi E_Y \qquad E_Z = E'_Z \qquad \text{(D.8)}$$

For $v = 0$, these equations show that the three spacelike components of E_μ (and of B_μ) are separately invariant to Lorentz transformation, but say nothing about the timelike component $\mathbf{E}^{(0)}$ or its relation to $\mathbf{E}^{(3)}$. For this, a more complete theory, as in this paper, is needed. Since the simple Lorentz transformation does not involve the generator of translations, it is an incomplete description of the properties of the electromagnetic field.

The photon is never at rest, but being massless, always moves at the velocity of light c, implying that the origin of frame $(X, Y, Z, -ict)$ also moves at c. The generator of spacetime translations is automatically required, therefore, for a description of the photon, since the latter always translates at c in any frame of reference. Since c is a universal constant, the assumption that there is a frame $(X', Y', Z', -ict')$ which moves at v relative to $(X, Y, Z, -ict)$ conflicts with Einstein's second principle. In other words it is not possible for the photon to define a frame moving at a speed v relative to one that is moving at speed c.

Acknowledgments

Many interesting letters were received from S. Kielich of Adam Mickiewicz University, Poznań, Poland, during the course of this work, and are gratefully acknowledged. Many interesting e-mail discussions are acknowledged with K. A. Earle of Cornell University.

References

1. M. W. Evans, *Mod. Phys. Lett.* **6**, 1237 (1992).

2. M. W. Evans, *Physica B* **182**, 227, 237 (1992).

3. M. W. Evans, *Physica B*, **183**, 103 (1993).

4. M. W. Evans, *The Photon's Magnetic Field*, World Scientific, Singapore, 1993.

5. F. Farahi and M. W. Evans, *Phys. Rev. E*, in press.

6. W. Heitler, *The Quantum Theory of Radiation*, 3d ed., Clarendon, Oxford, UK, 1954.

7. L. S. Ryder, *Quantum Field Theory*, 2d ed., Cambridge University Press, Cambridge, UK, 1987.

8. J. D. Jackson, *Classical Electrodynamics*, Wiley, New York, 1962.

9. R. M. Whitner, *Electromagnetics*, Prentice Hall, Englewood Cliffs, NJ, 1962.

10. A. F. Kip, *Fundamentals of Electricity and Magnetism*, McGraw Hill, New York, 1962.

11. P. W. Atkins, *Molecular Quantum Mechanics*, 2d ed., Oxford University Press, Oxford, UK, 1983.

12. L. D. Landau and E. M. Lifshitz, *The Classical Theory of Fields*, 4th ed., Pergamon, Oxford, UK, 1975.

13. M. Born and E. Wolf, *Principles of Optics*, 6th ed., Pergamon, Oxford, UK, 1975.

14. C. Cohen-Tannoudji, J. Dupont-Roc, and G. Grynberg, *Photons and Atoms: Introduction to Quantum Electrodynamics*, Wiley, New York, 1989.

15. W. M. Schwarz, *Intermediate Electromagnetic Theory*, Wiley, New York, 1964.

16. F. I. Fedorov, *J. Appl. Spectrosc.* **2**, 244 (1965).

17. B. W. Shore, *The Theory of Coherent Atomic Excitation*, Vols. 1 and 2, Wiley, New York, 1990.

18. P. A. M. Dirac, *The Principles of Quantum Mechanics*, 4th ed., Clarendon, Oxford, UK, 1958.

19. R. Tanaś and S. Kielich, *J. Mod. Opt.* **37**, 1935 (1990).

20. R. Tanaś, S. Kielich, and R. Zawodny, in S. Kielich and M. W. Evans (Eds.), *Modern Nonlinear Optics*, a topical issue of I. Prigogine and S. A. Rice (Eds.), *Advances in Chemical Physics*, Vols. 85A and 85B, Wiley, New York, 1993.

21. E. P. Wigner, *Ann. Math.* **40**, 149 (1939).

22. M. W. Evans, *Phys. Lett. A*, in press.

23. For a good description, see Ref. 17, Vol. 1.

24. B. L. Silver, *Irreducible Tensor Methods*, Academic, New York, 1976.

25. S. Kielich, *Proc. Phys. Soc.* **90**, 847 (1967); *Optoelectronics* **1**, 75 (1969).

26. L. Nafie and D. Che, in Ref. 20, Vol. 85B; L. D. Barron, *Molecular Light Scattering and Optical Activity*, Cambridge University Press, Cambridge, UK, 1982.

27. S. Kielich, in M. Davies, (senior reporter), *Dielectric and Related Molecular Processes*, Vol. 1, Chemical Society, London, 1972.

28. M. W. Evans, S. Woźniak, and G. Wagnière, *Physica B* **173**, 357 (1991).

29. L. S. Ryder, *Elementary Particles and Symmetries*, Gordon & Breach, 1986 (enlarged ed.).

30. M. W. Evans, G. J. Evans, W. T. Coffey, and P. Grigolini, *Molecular Dynamics and the Theory of Broad Band Spectroscopy*, Wiley, New York, 1982.

31. S. Kielich, *Nonlinear Molecular Optics*, Nauka, Moscow, 1981.

32. Reviewed in Ref. 20, Vol. 85(2).

33. A. Piekara and S. Kielich, *J. Chem. Phys.* **29**, 1292 (1958); *Acta Phys. Pol.* **17**, 209 (1958); *J. Phys. Rad.* **18**, 490 (1957).

34. S. Kielich, *Acta Phys. Pol.* **29**, 875 (1966).

35. M. Born and L. Infeld, *Proc. R. Soc. London Ser. A* **144**, 425 (1934); **147**, 522 (1934); **150**, 141 (1935).

36. P. S. Pershan, *Phys. Rev.* **130**, 919 (1963); J. P. van der Ziel, P. S. Pershan, and L. D. Malmstrom, *Phys. Rev. Lett.* **15**, 574 (1965); *Phys. Rev.* **143**, 574 (1966).

37. S. Kielich, *Proc. Phys. Soc.* **86**, 709 (1965).

38. S. Kielich, *J. Colloid Interface Sci.* **30**, 159 (1969); **27**, 432 (1968).

39. S. Kielich, *Bull. Soc. Amis Sci. Lett. Poznań* **21B**, 35 (1968/1969).

40. S. Kielich and R. Zawodny, *Opt. Commun.* **15**, 267 (1975).

41. P. D. Maker and R. W. Terhune, *Phys. Rev.* **137**, A801 (1965); C. C. Wang, *Phys. Rev.* **152**, 149 (1966); F. Shimizu, *J. Phys. Soc. Japan* **22**, 1070 (1967).

42. Special Issues: *J. Mod. Opt.* **34**(6/7) (1987) *J. Opt. Soc. Am. B* **4**(10) (1987); also reviewed extensively in Ref. 20.

43. J. M. Jauch and F. Rohrlich, *The Theory of Photons and Electrons*, Addison Wesley, London, 1959.

44. R. Tanaś and S. Kielich, *Opt. Commun.* **45**, 351 (1983); *Opt. Acta* **31**, 81 (1984); S. Kielich, R. Tanaś, and R. Zawodny, *Phys. Rev. A* **36**, 5670 (1987); *J. Opt. Soc. Am. B* **4**, 1627 (1987); *Appl. Phys. B* **45**, 249 (1988).

45. J. Frey, R. Frey, C. Flytzanis, and R. Triboulet, *Opt. Commun.* **84**, 76 (1991).

46. W. S. Warren, D. Goswami, S. Mayr, and A. P. West, Jr., *Science* **255**, 1683 (1992).

47. Reviewed in Ref. 14.

48. J. Jortner, R. D. Levine, I. Prigogine, and S. A. Rice (Eds.), *Advances in Chemical Physics*, Wiley, New York, 1981; T. Kobayashi, in Ref. 20, Vol. 85B.

MANIFESTLY COVARIANT THEORY OF THE ELECTROMAGNETIC FIELD IN FREE SPACETIME, PART 2: THE LORENTZ FORCE EQUATION

I. INTRODUCTION

In Part 1 of this series a manifestly covariant theory was developed for the electromagnetic field in free spacetime, in which[1] the electric and magnetic fields are treated as four vectors E_μ and B_μ, all of whose components are physically meaningful. This is a departure from the conventional approach suggested by the recent discovery[2-5] that the longitudinal ((3)) and transverse ((1) and (2)) components of the electromagnetic field are linked by[2]

$$\mathbf{B}^{(3)} = \frac{\mathbf{E}^{(1)} \times \mathbf{E}^{(2)}}{icE_0} \qquad (1)$$

Here $\mathbf{B}^{(3)}$ is the longitudinal magnetic field of the electromagnetic plane wave, and $\mathbf{E}^{(1)}$ and $\mathbf{E}^{(2)}$ define the transverse electric fields through

$$\mathbf{E}^{(1)} = E_0\hat{\mathbf{e}}^{(1)} e^{i\phi} \qquad (2a)$$

$$\mathbf{E}^{(2)} = E_0\hat{\mathbf{e}}^{(2)} e^{-i\phi} \qquad (2b)$$

where E_0 (proportional to the timelike polarization $E^{(0)}$ is the scalar amplitude of the plane wave, and c is the speed of light in free spacetime, the universal constant of special relativity. The unit vectors $\hat{\mathbf{e}}^{(1)}$ and $\hat{\mathbf{e}}^{(2)}$ are defined in the circular basis[1] by

$$\hat{\mathbf{e}}^{(1)} = \frac{\mathbf{i} - i\mathbf{j}}{\sqrt{2}} \qquad (3a)$$

$$\hat{\mathbf{e}}^{(2)} = \frac{\mathbf{i} + i\mathbf{j}}{\sqrt{2}} \qquad (3b)$$

$$\hat{\mathbf{e}}^{(1)} \times \hat{\mathbf{e}}^{(2)} = i\hat{\mathbf{e}}^{(3)} \qquad (3c)$$

where $\mathbf{i}$ and $\mathbf{j}$ are unit vectors in X and Y, mutually orthogonal to the propagation axis Z of the electromagnetic plane wave. The phase ϕ is defined as usual[6-15] by

$$\phi = \omega t - \boldsymbol{\kappa} \cdot \mathbf{r} \qquad (4)$$

where κ is the wave vector at position $\mathbf{r}$, and ω is the angular frequency at instant t in free spacetime.

Equation (1) is the key to the theory of covariant electrodynamics, essentially because the novel longitudinal field $\mathbf{B}^{(3)}$ is independent of the phase of the plane wave, and thus satisfies the conventionally defined Gauss theorem in differential form. Equation (1) is invariant to the fundamental symmetries, charge conjugation $\hat{C}$, parity inversion $\hat{P}$, and motion reversal $\hat{T}$, i.e., the right and left sides have the same $\hat{C}$, $\hat{P}$, and $\hat{T}$ symmetries.[1-5] Furthermore, the numerator on its right side is proportional[16-18] to the antisymmetric part of the light intensity tensor I_{ij}. The latter is known to be proportional to the third Stokes parameter S_3 and to mediate experimentally observable phenomena, such as antisymmetric light scattering[16-18] and the inverse Faraday effect,[19] and can therefore be considered a nonzero property of a circularly polarized electromagnetic wave in free space. Inter alia, $\mathbf{B}^{(3)}$ from Eq. (1) is similarly nonzero in free space, because the denominator on the right side of Eq. (1) is nonzero for finite $E^{(0)}$. It is more logical to state that the right side of Eq. (1) is nonzero because $\mathbf{B}^{(3)}$ is nonzero rather than the other way around, because $\mathbf{B}^{(3)}$ is a fundamental solution of Maxwell's equations in free spacetime. The quantity $\mathbf{E}^{(1)} \times \mathbf{E}^{(2)}$ is, on the other hand, built up from a cross product of fundamental transverse electric fields. It is clear, however, that if the antisymmetric part of the light intensity is nonzero, then $\mathbf{B}^{(3)}$ is nonzero. In other words, $\mathbf{B}^{(3)}$ is the source of the antisymmetric part of light intensity and all concomitant experimental phenomena. It is worth noting in the context of $\hat{C}$ symmetry[19] that

$$\hat{C}(A_\mu) = -A_\mu \tag{5}$$

where A_μ is the well-known potential four vector in free spacetime. The $\hat{C}$ symmetries of A_μ, E_μ, and B_μ are all negative, so that the concomitant fields of the photon change sign under $\hat{C}$. Although the photon is stated[19] to be its own antiparticle, the antiphoton, generated by $\hat{C}$ from the photon, is associated with electric and magnetic fields of the opposite sign. For this reason, the antiphoton is a distinct entity from the photon. Furthermore, all four components of A_μ, E_μ, and B_μ must change sign under $\hat{C}$; i.e., all four polarizations (0), (1), (2), and (3) change sign. On the other hand, spacelike quantities, such as the propagation vector κ, by definition are unaffected by $\hat{C}$, so that the $\hat{C}$ operator produces an electromagnetic wave propagating in the same direction, but with all four polarizations reversed. The electromagnetic wave produced in vacuo by $\hat{C}$ defines the antiphoton in the quantum field, a distinct entity from the photon. The fact that the concommitant fields are reversed in sign does not mean that $\mathbf{B}^{(3)}$ of Eq. (1) violates $\hat{C}$ symmetry. In the same way, Eq. (5) does not mean that A_μ violates $\hat{C}$ symmetry in vacuo. We conclude that Eq. (1) satisfies $\hat{C}$, $\hat{P}$, and $\hat{T}$ invariance in vacuo, and is a legitimate equation of electrodynamics.

In Part 1 of this series the vectors E_μ and B_μ were defined in terms of the electromagnetic field four tensor[6-15] $F_{\mu\nu}$, and its dual, $\tilde{F}_{\mu\nu}$. It was shown[1] that both E_μ and B_μ are Pauli-Lubanski types within the Poincaré group (the inhomogeneous Lorentz group), and that the products $E_\mu E_\mu$ and $B_\mu B_\mu$ form Casimir invariants of the Poincaré group. The Maxwell equations, Poynting theorem, and Stokes parameters were derived in manifestly covariant form, and it was shown that phenomena such as natural optical activity, ellipticity, and the electric Kerr effect can be expressed in terms of changes in B_μ (or its electric counterpart E_μ). It was shown that optical absorption can be defined in terms of B_μ and E_μ, and suggestions were made for experiments to detect the magnetizing effect of B_μ and the polarizing effect of E_μ as an electromagnetic wave interacts with matter. In this paper (part 2), the Lorentz force equation is investigated in manifestly covariant form; i.e., a manifestly covariant theory is given of the interaction of an electromagnetic wave with the electron.

In Section II, the Lorentz force equation is derived from the covariant definitions of E_μ and B_μ, and expressed in terms of its magnetic and electric components. Section III examines the individual terms in the equation and shows that in manifestly covariant form, the Lorentz equation contains extra terms that, in principle, produce experimentally observable effects on the electron. There follows a discussion that suggests possible experiments for the detection of the extra manifestly covariant forces on the electron.

II. DERIVATION OF THE MANIFESTLY COVARIANT LORENTZ FORCE EQUATION

Our aim is to derive the equation describing the interaction of E_μ and B_μ with an electron, this being a manifestly covariant description of the interaction of an electromagnetic wave with particulate matter. In Part 1, the four vectors E_μ and B_μ were defined as[1]

$$E_\mu = \tfrac{1}{2}\varepsilon_{\mu\nu\rho\sigma}\tilde{F}_{\nu\rho}\delta_\sigma \qquad (6a)$$

$$B_\mu = -\frac{i}{2c}\varepsilon_{\mu\nu\rho\sigma}F_{\nu\rho}\delta_\sigma \qquad (6b)$$

where $F_{\nu\rho}$ is the four curl of A_ν in free spacetime, and $\tilde{F}_{\nu\rho}$ is its dual. The unit tensors $\varepsilon_{\mu\nu\rho\sigma}$ and δ_σ are respectively the totally antisymmetric unit tensor in four dimensions and the unit generator of spacetime translations.[1, 20] These quantities are written out for reference as follows:

$$E_\mu \equiv (E^{(1)}, E^{(2)}, E^{(3)}, -iE^{(0)}) \tag{7a}$$

$$B_\mu \equiv (B^{(1)}, B^{(2)}, B^{(3)}, -iB^{(0)}) \tag{7b}$$

$$\delta_\mu \equiv (0, 0, 1, -i) \tag{7c}$$

$$F_{\mu\nu} \equiv \begin{bmatrix} 0 & cB^{(3)} & -cB^{(2)} & -iE^{(1)} \\ -cB^{(3)} & 0 & cB^{(1)} & -iE^{(2)} \\ cB^{(2)} & -cB^{(1)} & 0 & -iE^{(3)} \\ iE^{(1)} & iE^{(2)} & iE^{(3)} & 0 \end{bmatrix} \tag{7d}$$

$$\tilde{F}_{\mu\nu} \equiv \begin{bmatrix} 0 & -iE^{(3)} & iE^{(2)} & cB^{(1)} \\ iE^{(3)} & 0 & -iE^{(1)} & cB^{(2)} \\ -iE^{(2)} & iE^{(1)} & 0 & cB^{(3)} \\ -cB^{(1)} & -cB^{(2)} & -cB^{(3)} & 0 \end{bmatrix} \tag{7e}$$

The need to define E_μ and B_μ as four vectors in spacetime is a direct consequence of Eq. (1), because the latter implies that there is a relation between the transverse and longitudinal spacelike components of the electromagnetic wave in vacuo. The conventional assertion that longitudinal components be "unphysical"[6-15] is no longer tenable in view of Eq. (1), because if $\mathbf{E}^{(1)}$ and $\mathbf{E}^{(2)}$ be physically meaningful, then so must $\mathbf{B}^{(3)}$. It has been demonstrated[1-5] that the existence of $\mathbf{B}^{(3)}$ implies the existence of $\mathbf{E}^{(3)}$, and quantum field theory[20] leads to

$$|\mathbf{B}^{(3)}| - B^{(0)} = 0 \qquad |\mathbf{E}^{(3)}| - E^{(0)} = 0,$$
$$\mathbf{B}^{(3)} \cdot \mathbf{B}^{(3)} = B^{(0)2} \qquad \mathbf{E}^{(3)} \cdot \mathbf{E}^{(3)} = E^{(0)2} \tag{8}$$

i.e., that physical states are admixtures of polarizations (3) and (0). Therefore, all four polarizations are physically meaningful. This is consistent with the fact that A_μ has four components.

The Lorentz force equation can be expressed covariantly by

$$f_\mu = F_{\mu\nu} J_\nu \tag{9}$$

where f_μ is a force four vector[6] and J_μ is the charge current four vector. In the conventional theory this is taken to be an adequate description of the interaction of the electric and magnetic components of the electro-magnetic field with an electron. The conventional approach, however, assumes that the longitudinal and timelike components of these fields are unphysical, which means essentially that the longitudinal component is set to zero. In view of Eqs. (8) this is an illogical procedure, because if $E^{(3)}$ or $B^{(3)}$ be zero, then so must $E^{(0)}$ and $B^{(0)}$, but the latter are also propor-tional to the amplitudes of transverse components such as $\mathbf{E}^{(1)}$ and $\mathbf{E}^{(2)}$, and in defining these, $E^{(0)}$ is obviously not zero. The conventional ap-proach is therefore logically inconsistent. In the manifestly covariant theory,[1] on the other hand, this inconsistency is remedied. The inconsis-tency of the conventional approach is "hidden" by the mathematical nature of the four curl, which defines $F_{\mu\nu}$ as

$$F_{\mu\nu} \equiv \frac{\partial A_\nu}{\partial x_\mu} - \frac{\partial A_\mu}{\partial x_\nu} \qquad (10)$$

since this four curl leaves the timelike components of E_μ and B_μ unde-fined, i.e., the matrix $F_{\mu\nu}$ contains only the spacelike components on its off diagonals. The conventional antisymmetric tensor $F_{\mu\nu}$ contains no refer-ence to the timelike polarizations $E^{(0)}$ and $B^{(0)}$. It follows, therefore, that the Lorentz force equation in covariant form (9) cannot be manifestly covariant, because $F_{\mu\nu}$ is used to define the Lorentz force vector f_μ. Manifest covariance means that the physically meaningful polarizations (0) and (3) must be taken into consideration when calculating the force on the electron.

This is achieved by solving Eqs. (6a), (6b), and (9) simultaneously as follows. We note firstly the definitions of J_μ and f_μ:

$$J_\mu \equiv \left(\rho\frac{v^{(1)}}{c}, \rho\frac{v^{(2)}}{c}, \rho\frac{v^{(3)}}{c}, i\rho \right) \qquad (11a)$$

$$f_\mu \equiv \left(f^{(1)}, f^{(2)}, f^{(3)}, f^{(0)} \right) \qquad (11b)$$

where ρ is the charge density, and $v^{(1)}$, $v^{(2)}$, and $v^{(3)}$ are the spacelike velocity components of the electron. The inverse of J_μ is defined so that

$$J_\mu J_\mu^{-1} = 1 \qquad (12)$$

Multiplying both sides of Eq. (9) from the right by J_ν^{-1} we obtain

$$F_{\mu\nu} = f_\mu J_\nu^{-1} \qquad (13)$$

so that in Eq. (6b)

$$cB_\mu = -\frac{i}{2}\varepsilon_{\mu\nu\rho\sigma}f_\nu J_\rho^{-1}\delta_\sigma \tag{14}$$

Multiplying both sides from the right by δ_σ^{-1} yields

$$cB_\mu\delta_\sigma^{-1} = -\frac{i}{2}\varepsilon_{\mu\nu\rho\sigma}f_\nu J_\rho^{-1} \tag{15}$$

and multiplying both sides of this equation from the right by J_ρ gives

$$cB_\mu J_\rho\delta_\sigma^{-1} = -\frac{i}{2}\varepsilon_{\mu\nu\rho\sigma}f_\nu \tag{16}$$

Here we have used the fact that

$$J_\rho\delta_\sigma^{-1} = \delta_\sigma^{-1}J_\rho \tag{17}$$

Finally, multiplying both sides of Eq. (16) from the right by δ_σ gives the magnetic part of the Lorentz force equation in manifestly covariant form:

$$cB_\mu J_\rho = -\frac{i}{2}\varepsilon_{\mu\nu\rho\sigma}f_\nu\delta_\sigma \tag{18}$$

Similarly, the electric part of the Lorentz force equation is

$$E_\mu J_\rho = \tfrac{1}{2}\varepsilon_{\mu\nu\rho\sigma}g_\nu\delta_\sigma \tag{19}$$

where g_μ is defined through the dual $\tilde{F}_{\mu\nu}$ of the electromagnetic field tensor $F_{\mu\nu}$:

$$g_\mu \equiv \tilde{F}_{\mu\nu}J_\nu \tag{20}$$

III. COMPONENTS OF THE MANIFESTLY COVARIANT LORENTZ EQUATION

In this section the structure of the tensor Eqs. (18) and (19) is investigated for individual terms, and the result is compared with the conventional Lorentz force equation for an electromagnetic plane wave interacting with

TABLE I
Summary: Generalized Lorentz Equation

Force	Components
f_0	$E_1J_1i,\ E_2J_2i$
	New E_3J_3i
f_1	$-2cB_2J_3$
	$2cB_2\rho = 2E_1\rho$
	New $2cB^{(0)}J_2 = 2cB^{(3)}J_2$
f_2	$2cB_1J_3$
	$-2cB_1\rho = 2E_2\rho$
	New $-2cB^{(3)}J_1 = -2cB^{(0)}J_1$
f_3	$c(B_2J_1 - B_1J_2)$
	New $-E_3\rho$

an electron. Note that in the conventional approach, only the two transverse components (1) and (2) exist, the components (0) and (3) are discarded as unphysical. In our manifestly covariant approach, components (0), (1), (2), and (3) are physically meaningful. Explicit calculations are given, and the individual results from Eq. (18) are presented in Table I.

A. Components f_1 and f_2

The Lorentz force on the electron due to B_0 is given as follows:

$$cB_0J_2 = -\frac{i}{2}\varepsilon_{0123}f_1\delta_3$$

$$cB_0J_1 = -\frac{i}{2}\varepsilon_{0213}f_2\delta_3 \tag{21}$$

and with the definitions $B_0 = -iB^{(0)}$ and $\delta_3 \equiv 1$ we obtain the 1 and 2 components of f_μ

$$f_1 = 2cB^{(0)}J_2 = 2\rho v_2 B^{(0)}$$

$$f_2 = -2cB^{(0)}J_1 = -2\rho v_1 B^{(0)} \tag{22}$$

Similarly, it may be shown that

$$f_1 = 2cB^{(3)}J_2$$

$$f_2 = -2cB^{(3)}J_1 \tag{23}$$

so that

$$f_1 = 2cB^{(0)}J_2 = 2cB^{(3)}J_2$$
$$f_2 = -2cB^{(0)}J_2 = -2cB^{(3)}J_2 \tag{24}$$

which is consistent with Eq. (8), i.e., with the quantum theoretical result that physical photon states are admixtures of (0) and (3) polarizations such that Eq. (8) holds.

Clearly, the forces in Eq. (24) are absent from the conventional Lorentz equation. It is seen by inspection of Eqs. (22) and (23) that they have precisely the same form as the equations of motion[8] of a charge in a static magnetic field $\mathbf{B}^{(3)}$, whose magnitude is equal to $B^{(0)}$. This is consistent with the phase-independent definition of $\mathbf{B}^{(3)}$, Eq. (1), although $\mathbf{B}^{(3)}$ is generated by a photon traveling at the speed of light and cannot be regarded as a conventional magnetostatic field. It is a longitudinal magnetic field which travels with the photon at the speed of light. In principle, an experiment can be devised for measuring these extra forces on the electron in the manifestly covariant theory. This possibility is discussed further below.

Additional contributions to f_1 and f_2 arise from the timelike component of the charge current four vector J_μ, but unlike the contributions from Eq. (24), these are also present in the conventional theory. They arise as follows:

$$cB_1 J_0 = -\frac{i}{2}\varepsilon_{1203}f_2$$
$$f_2 = -2cB_1\rho \tag{25}$$

and using the relation in circular polarization,

$$cB_1 = -E_2 \tag{26}$$

we obtain

$$f_2 = 2E_2\rho \tag{27}$$

Similarly,

$$f_1 = 2cB_2\rho \tag{28}$$

and using

$$cB_2 = E_1 \tag{29}$$

we obtain

$$f_1 = 2E_1\rho \tag{30}$$

B. The f_0 and f_3 Forces

The timelike f_0 components from Eq. (18) are obtained by

$$
\begin{aligned}
f_0 &= -2cB_1J_2i = E_2J_2i \\
f_0 &= 2cB_2J_1i = E_1J_1i
\end{aligned} \tag{31}
$$

and correspond to the well-known[8] $\mathbf{E} \cdot \mathbf{J}$ force from the conventional Lorentz force equation in covariant from. Therefore, the manifestly covariant Eq. (18) provides no new terms in $\mathbf{E} \cdot \mathbf{J}$.

The f_3 component is obtained from

$$f_0 + if_3 = \frac{2}{i}cB_1J_2 \tag{32}$$

together with

$$f_0 + if_3 = -\frac{2}{i}cB_2J_1 \tag{33}$$

Solving Eqs. (32) and (33) simultaneously gives

$$
\begin{aligned}
f_0 &= 0 \\
f_3 &= c(B_2J_1 - B_1J_2)
\end{aligned} \tag{34}
$$

i.e., there is no contribution to f_0 from the B_2 and B_1 components interacting with J_1 and J_2 respectively, and the overall f_3 component is the same as that in the conventional Lorentz force equation. Finally, there are manifestly covariant forces:

$$
\begin{aligned}
f_0 &= F_{03}J_3 = iE_3J_3 \\
f_3 &= F_{30}J_0 = -E_3\rho
\end{aligned} \tag{35}
$$

direct from Eq. (9), the equation from which (18) is derived using (6a) and (6b).

These results are summarized in Table I, which shows that there are additions to the conventional f_0, f_1, f_2, and f_3 in the manifestly covariant description. From Table I, Eq. (18) reduces to the conventional Lorentz equation if $\mathbf{B}^{(3)} = \mathbf{0}$. However, this assumption of the conventional approach is illogical, because it conflicts with Eq. (1).

IV. DISCUSSION

The question arises immediately as to whether the extra terms in Table I marked "new" are observable experimentally. This type of observation might be able to distinguish between the conventional theory and the manifestly covariant approach based on Eq. (1). It is probably difficult to isolate a single electron in a vacuum in order to test the theory directly, but the use of electron beams may be feasible. The resultant force on a single electron due to an electromagnetic plane wave is given in the manifestly covariant approach by a combination of Eqs. (9) (the conventional equation) and (18), taking into account thereby the existence of Eqs. (1) and (8). In the conventional approach, the force is described by the spacelike components of Eq. (9) alone, and Eqs. (1), (8), and (18) are not considered.

The conventional calculation of the trajectory of an electron in a monochromatic, circularly polarized plane wave is a standard problem (e.g., Landau and Lifshitz,[10] p. 118), in which there are no linear forces such as $-E_3\rho$ of the manifestly covariant theory. Presumably, such a force would cause the linear deflection of an electron beam when the latter is acted upon by a circularly polarized electromagnetic beam, such as an X-ray beam. However, this assumption does not consider statistical effects in either the electromagnetic or electron beam. Extra Lorentz precession terms due to f_1 and f_2 are expected in the manifestly covariant theory. If extra forces can be observed unequivocally, this would add to the considerable experimental evidence for the manifestly covariant theory reviewed in Part 1 of this series, evidence from sources such as absorption of circularly polarized light, circular dichroism, ellipticity, and the electric Kerr effect.

Acknowledgments

The author acknowledges many interesting letters from Stanislaw Kielich during the course of this work, and many interesting discussions with F. Farahi and other members of the physics department at UNCC.

References

1. M. W. Evans, submitted.
2. M. W. Evans, *Mod. Phys. Lett.* **20**, 1237 (1992).
3. M. W. Evans, *Physica B* **182**, 227, 237 (1992).
4. M. W. Evans, *Physica B*, **183**, 103 (1993).
5. F. Farahi and M. W. Evans, *Phys. Rev. E*, in press.

6. W. Heitler, *The Quantum Theory of Radiation*, 3d ed., Clarendon, Oxford, UK, 1954.

7. J. D. Jackson, *Classical Electrodynamics*, Wiley, New York, 1962.

8. R. M. Whitner, *Electrodynamics*, Prentice Hall, Englewood Cliffs, NJ, 1962.

9. A. F. Kip, *Fundamentals of Electricity and Magnetism*, McGraw Hill, New York, 1962.

10. L. D. Landau and E. M. Lifshitz, *The Classical Theory of Fields*, 4th ed., Pergamon, Oxford, UK, 1975.

11. M. Born and E. Wolf, *Principles of Optics*, 6th ed., Pergamon, Oxford, UK, 1975.

12. W. M. Schwarz, *Intermediate Electromagnetic Theory*, Wiley, New York, 1964.

13. B. W. Shore, *The Theory of Coherent Atomic Excitation*, Wiley, New York, 1990.

14. S. Kielich, *Nonlinear Molecular Optics*, Nauka, Moscow, 1981.

15. S. Kielich, in M. Davies (Senior Rep.), *Dielectric and Related Molecular Processes*, Vol. 1, Chem. Soc., London, 1972.

16. K. Knast and S. Kielich, *Acta Phys. Pol.* **A55**, 319 (1979).

17. G. Placzek, in E. Marx (Ed.), *Handbuch der Radiologie*, Akad. Verlag., Leipzig, 1934.

18. M. W. Evans, *The Photon's Magnetic Field*, World Scientific, Singapore, 1993.

19. P. S. Pershan, J. P. van der Ziel, and L. D. Malmstrom, *Phys. Rev.* **143**, 574 (1966).

20. E. P. Wigner, *Ann. Math.* **40**, 149 (1939).

MANIFESTLY COVARIANT THEORY OF THE ELECTROMAGNETIC FIELD IN FREE SPACETIME, PART 3: Ĉ, P̂, AND T̂ SYMMETRIES

I. INTRODUCTION

It has recently been observed[1-5] that there exists an equation of electrodynamics in vacuo that defines a longitudinal magnetic field, $B^{(3)}$, which is independent of the phase of the electromagnetic plane wave, thus showing for the first time that there exist physically meaningful longitudinal solutions to Maxwell's equations in vacuo. Parts 1 and 2 of this series[1,2] developed the theory of manifestly covariant electrodynamics from this basic observation, and recent work by Farahi and Evans[4] has shown that the existence of $B^{(3)}$ implies the existence of its longitudinal electric counterpart $iE^{(3)}$. In Part 1[1] it was shown that $iE^{(3)}$ and $B^{(3)}$ do not contribute to the electromagnetic energy density, and that Poynting's theorem can be expressed in terms of four, rather than two, polarizations. The existence of four photon polarizations, (0), (1), (2), and (3), was reconciled with two photon helicities, $+1$ and -1, by noting[1] that the helicities can be defined in terms either of (0) and (3) or of (1) and (2). Here (0) denotes the timelike photon polarization, (1) and (2) the transverse spacelike, and (3) the longitudinal spacelike. In Part (2), the Lorentz force equation was expressed in manifestly covariant form.

In this paper (Part 3), the fundamental symmetries of physics are applied to the basic equation

$$\mathbf{B}^{(3)} = \frac{\mathbf{E}^{(1)} \times \mathbf{E}^{(2)}}{\mathrm{i}E_0 c} \qquad (1)$$

of manifestly covariant electrodynamics (MCE). Here $\mathbf{B}^{(3)}$ is linked[1-5] to the transverse, oscillating, electric fields $\mathbf{E}^{(1)}$ and $\mathbf{E}^{(2)}$ of the plane wave in vacuo, where c is the speed of light. Here $\mathbf{E}^{(1)}$ is the complex conjugate of $\mathbf{E}^{(2)}$:

$$\mathbf{E}^{(1)} \equiv E_0 \hat{\mathbf{e}}^{(1)} \, \mathrm{e}^{\mathrm{i}\phi} \qquad (2a)$$

$$\mathbf{E}^{(2)} \equiv E_0 \hat{\mathbf{e}}^{(2)} \, \mathrm{e}^{-\mathrm{i}\phi} \qquad (2b)$$

where

$$\phi = \omega t - \boldsymbol{\kappa} \cdot \mathbf{r} \qquad (3)$$

is the phase of the plane wave, with, as usual, ω as the angular frequency at instant t, $\boldsymbol{\kappa}$ the wave vector at position $\mathbf{r}$. The circular basis[6,7] is used to define the unit vectors $\hat{\mathbf{e}}^{(1)}$ and $\hat{\mathbf{e}}^{(2)}$:

$$\hat{\mathbf{e}}^{(1)} = \frac{1}{\sqrt{2}}(\mathbf{i} - \mathrm{i}\mathbf{j}) \qquad (4a)$$

$$\hat{\mathbf{e}}^{(2)} = \frac{1}{\sqrt{2}}(\mathbf{i} + \mathrm{i}\mathbf{j}) \qquad (4b)$$

where $\mathbf{i}$ and $\mathbf{j}$ are unit vectors in axes X and Y of the Cartesian frame (X, Y, Z).

In Section II, it is shown that Eq. (1) is invariant under the following conditions:

1. The charge conjugation operator $\hat{C}$, which changes the sign of charge in classical electrodynamics, and in particle physics produces the antiparticle from the original particle
2. The parity inversion operator $\hat{P}$
3. The motion reversal operator $\hat{T}$

In other words, the left and right sides of Eq. (1) remain balanced after application of $\hat{C}$, $\hat{P}$, and $\hat{T}$ to each variable on both sides. Equation (1) is therefore a legitimate equation of electrodynamics, and $\mathbf{B}^{(3)}$ has the $\hat{C}$, $\hat{P}$, and $\hat{T}$ symmetries, and units, of magnetic flux density. $\mathbf{B}^{(3)}$ is also a solution of Maxwell's equations[1-5] in vacuo, and is therefore a real, physically meaningful, longitudinal magnetic field with polarization (3). It has been shown in Parts 1 and 2 of this series that as a direct consequence, electrodynamics (both classical and quantum) must be made manifestly covariant in nature.

In Section II, the fundamental symmetries $\hat{C}$, $\hat{P}$, and $\hat{T}$ are applied to electromagnetic radiation in vacuo, represented by the helicity λ and the potential four vector A_μ. These are the two fundamental elements of the electromagnetic plane wave in vacuo. The helicity λ is negative to $\hat{P}$, and is a number, $+1$ or -1. In contemporary quantum field theory[8] λ is defined for the massless electromagnetic gauge field as the ratio of the Pauli-Lubansky pseudovector W_μ to the generator of spacetime translations P_μ. It is related in the lightlike condition to the second (spin) Casimir invariant of the inhomogeneous Lorentz group (or Poincaré group). The first (mass) Casimir invariant is zero for the electromagnetic field, and so λ is the only nonzero quantity that is invariant to the most general type of Lorentz transformation in the theory of special relativity. The Lorentz invariant spacetime character of the electromagnetic wave is described therefore in terms of λ. In the quantum field the photon is described by two helicities, $+1$ and -1. On the other hand, the concomitant electrodynamic properties of the electromagnetic field in vacuo are described by d'Alembert's equation:

$$\Box A_\mu = 0 \qquad (5)$$

where $\Box$ is the d'Alembertian and A_μ the potential four vector. The electric and magnetic parts of the electromagnetic field can be described in terms of A_μ (Refs. 9–14). It is therefore necessary and sufficient to describe electromagnetism in vacuo in terms of the fundamental spacetime quantity λ, and the fundamental electrodynamic quantity A_μ. Section III therefore considers $\hat{C}$, $\hat{P}$, and $\hat{T}$ symmetry applied to λ and A_μ, and defines the response of the electromagnetic field to $\hat{C}$, $\hat{P}$, and $\hat{T}$ in terms of λ and A_μ. Specifically, it is shown that nonzero longitudinal solutions of Maxwell's equations are consistent with $\hat{C}$, $\hat{P}$, and $\hat{T}$ in vacuo. Finally, a detailed discussion is given of the correct way in which to apply $\hat{C}$, $\hat{P}$, and $\hat{T}$ to manifestly covariant electrodynamics, addressing some misconceptions in the recent literature.[15]

II. THE $\hat{C}$, $\hat{P}$, AND $\hat{T}$ SYMMETRIES OF THE FUNDAMENTAL EQUATION (1) OF MCE

We first note that the numerator on the right side of Eq. (1) is the antisymmetric part of the light intensity tensor[16-21] of the standard literature. It is a nonzero quantity in vacuo whose absolute magnitude is the same as the absolute magnitude of the Stokes parameter S_3 of circularly polarized light.[6] The denominator in Eq. (1) is the product of the scalar amplitude, E_0, of the electric component of the radiation with the speed of light c, and is also nonzero. The quantity $\mathbf{B}^{(3)}$ is therefore nonzero in general, provided that there is some element of circular polarity. $\mathbf{B}^{(3)}$ changes sign with the sense of circular polarity, as does $\mathbf{E}^{(1)} \times \mathbf{E}^{(2)}$ (Refs. 1–5). Since $\mathbf{B}^{(3)}$ is a magnetic field, and is physically meaningful, conventional electrodynamics becomes untenable, because it is no longer sufficient to consider just two polarizations ((1) and (2)) and to arbitrarily discard[10-14] polarization (3) as being "physically meaningless."

It is fundamentally important, therefore, to show that Eq. (1) conserves the symmetries of physics, $\hat{C}$, $\hat{P}$, and $\hat{T}$, and is therefore legitimate in all respects as an equation of electrodynamics in vacuo, because Eq. (1) must mean that conventional electrodynamics is an incomplete description, both in the classical and quantum fields.

A. $\hat{C}$ Symmetry

The charge conjugation operator $\hat{C}$ is defined as[22]

$$\hat{C}(A_\mu) = -A_\mu \tag{6}$$

and in particle physics, the photon, represented by A_μ, is negative to $\hat{C}$, being changed to the antiphoton. By definition, all spacetime quantities are unaffected by $\hat{C}$. Therefore,

$$\hat{C}(\lambda) = \lambda \tag{7}$$

From this, it follows that $\hat{C}$ changes the sign of the scalar amplitudes E_0 and B_0 of the plane wave in vacuo, and therefore changes the sign of the timelike and all spacelike components of the manifestly covariant four-vectors[1, 2] E_μ and B_μ. Thus,

$$\hat{C}(\mathbf{B}^{(3)}) = -\mathbf{B}^{(3)} \qquad \hat{C}(\mathbf{E}^{(1)} \times \mathbf{E}^{(2)}) = \mathbf{E}^{(1)} \times \mathbf{E}^{(2)}$$
$$\hat{C}(E_0) = -E_0 \qquad \hat{C}(\mathrm{i}c) = \mathrm{i}c \tag{8}$$

so that it is clear that Eq. (1) conserves $\hat{C}$ symmetry. (The $\hat{C}$ symmetry of both sides of Eq. (1) is negative.)

B. $\hat{P}$ Symmetry

The $\hat{P}$ operator,[23] parity inversion, is defined as $\hat{P}(\mathbf{r}) = -\mathbf{r}$; $\hat{P}(\mathbf{v}) = -\mathbf{v}$; where $\mathbf{r}$ and $\mathbf{v} = \dot{\mathbf{r}}$ are position and velocity, respectively. It has been shown[23] that

$$\hat{P}(\mathbf{E}^{(1)} \times \mathbf{E}^{(2)}) = \mathbf{E}^{(1)} \times \mathbf{E}^{(2)} \qquad (9)$$

The $\hat{P}$ symmetry of magnetic flux density is positive,[23] and since c and $E^{(0)}$ are scalars, Eq. (1) conserves $\hat{P}$ symmetry (both sides are positive).

C. $\hat{T}$ Symmetry

The $\hat{T}$ operator,[23] motion reversal, is defined as $\hat{T}(\mathbf{r}) = \mathbf{r}$; $\hat{T}(\mathbf{v} = \dot{\mathbf{r}}) = -\mathbf{v}$; and reverses all motions in the same frame of reference. It has been shown[23] that

$$\hat{T}(\mathbf{E}^{(1)} \times \mathbf{E}^{(2)}) = -\mathbf{E}^{(1)} \times \mathbf{E}^{(2)} \qquad (10)$$

The $\hat{T}$ symmetry of magnetic flux density is negative,[23] and since c and $E^{(0)}$ are scalars, they are positive to $\hat{T}$. Therefore, Eq. (1) conserves $\hat{T}$ symmetry (both sides are negative).

Therefore, Eq. (1) conserves $\hat{C}$, $\hat{P}$, and $\hat{T}$, the fundamental symmetries of physics, and is a legitimate equation of electrodynamics in vacuo. $\mathbf{B}^{(3)}$ has the $\hat{C}$, $\hat{P}$, $\hat{T}$ symmetries, and units, of magnetic flux density in vacuo, and is a solution of Maxwell's equations in vacuo. It is therefore physically meaningful, longitudinal, and phase independent.

III.   THE FUNDAMENTAL SYMMETRIES OF THE ELECTROMAGNETIC PLANE WAVE

Since $\mathbf{B}^{(3)}$ (and its electric counterpart $\mathbf{E}^{(3)}$) are physically meaningful solutions of Maxwell's equations, they must be invariant to $\hat{C}$, $\hat{P}$, and $\hat{T}$, in the same way that the well-accepted, oscillating, transverse solutions (1) and (2) are invariant to $\hat{C}$, $\hat{P}$, and $\hat{T}$. The invariance of Eq. (1) is already sufficient proof that $\mathbf{B}^{(3)}$ satisfies these basic symmetry constraints in vacuo. However, a set of self-consistent rules is necessary by which the symmetries of electromagnetic radiation can be identified in terms of its most fundamental variables. We take these to be the helicity λ and the potential four vector A_μ for reasons given already.

The symmetry properties of the electromagnetic wave in vacuo can now be defined as follows:

$$[\lambda, A_\mu] \xrightarrow{\hat{C}} [\lambda, -A_\mu]$$

$$[\lambda, A, \phi] \xrightarrow{\hat{P}} [-\lambda, -A, \phi]$$

$$[\lambda, A, \phi] \xrightarrow{\hat{T}} [\lambda, -A, \phi] \tag{11}$$

$$A_\mu \equiv (A, i\phi)$$

Therefore, the $\hat{C}$ operator leaves the spacetime quantity λ unchanged by definition, while changing the sign of A_μ by definition. $\hat{C}$ thus produces a distinct entity which we identify classically as the antiwave and quantum mechanically as the antiphoton, since, by definition, $\hat{C}$ produces the antiparticle from the original particle. The antiwave is defined as the classical electromagnetic entity with the same λ as the original wave but with reversed A_μ and therefore with concomitant electric and magnetic fields of the opposite sign. The spacetime parameter λ of the antiwave is the same as that of the original wave, while the electrodynamic parameter A_μ of the antiwave is opposite in sign. This emphasizes that the antiwave is a distinct entity from the wave.

The $\hat{P}$ operator reverses the sign of λ by definition. The $\hat{P}$ symmetry of the spacelike part of A_μ (the vector potential) is negative, and that of the timelike part (the scalar potential) is positive. $\hat{P}$ again produces a distinct entity, classically the wave with opposite helicity, and quantum mechanically the photon with opposite helicity. The $\hat{T}$ operator does not change the sign of λ, and the $\hat{T}$ symmetry of the spacelike part of A_μ is negative, while that of the timelike part is positive. $\hat{T}$ again produces a distinct entity from the original wave or photon.

Therefore, distinct entities are produced by the application of all three symmetries, $\hat{C}$, $\hat{P}$, and $\hat{T}$, to the two fundamental properties, λ and A_μ of the classical electromagnetic wave or quantized photon. We denote λ and A_μ as symmetry elements of electromagnetic radiation in vacuo. Note that in the above, we have implicitly assumed that the scalar potential is nonzero, and have thus worked in a gauge such as the Lorentz gauge that allows this. In the Coulomb gauge it is assumed that the scalar potential is zero, but this loses manifest covariance unless a zero scalar potential be regarded as physically meaningful. The $\hat{C}$, $\hat{P}$, and $\hat{T}$ symmetries of a zero scalar potential are, however, the same as a nonzero scalar potential, respectively, negative, positive, and positive. In the Coulomb gauge, therefore, the $\hat{C}$, $\hat{P}$, and $\hat{T}$ symmetries are no different from a manifestly covariant gauge such as the Lorentz gauge. Conventionally, gauge invariance means that electric and magnetic fields obtained are invariant to gauge transformation. In MCE, however, it is necessary to regard the scalar potential as physically meaningful, because Eq. (1) implies the existence of four physically meaningful polarizations. Equation (1) automatically satisfies the principle of gauge invariance, because the longitudinal magnetic field $\mathbf{B}^{(3)}$ is formed from the vector product of two transverse electric fields $\mathbf{E}^{(1)}$ and $\mathbf{E}^{(2)}$ which are separately gauge invariant.

IV. DISCUSSION

In Sections II and III it has been demonstrated that Eq. (1), the key equation of manifestly covariant electrodynamics, is invariant to the fundamental symmetries of physics, and defines a quantity $\mathbf{B}^{(3)}$, which is a physically meaningful, gauge invariant, magnetic field. Equation (1) defines $\mathbf{B}^{(3)}$, a solution in vacuo of Maxwell's equations, in terms of other components of the electromagnetic plane wave in vacuo, thus showing that if the transverse components are physically meaningful, then so must be the longitudinal, making the conventional view of electrodynamics untenable, both in the classical and quantum fields.

The fundamental symmetries $\hat{C}$, $\hat{P}$, and $\hat{T}$ have been applied to the symmetry elements λ and A_μ of the electromagnetic field, and it has been shown that $\hat{C}$, $\hat{P}$, and $\hat{T}$ all produce distinct entities, or "distinct situations" by operating on the original entity or situation. The symmetry elements have been defined as λ and A_μ because these parameters are necessary and sufficient to define the spacetime and electrodynamic properties, respectively, of electromagnetic radiation in vacuo. The spacetime symmetry element λ has been chosen because it is the only nonzero Casimir invariant of the Poincaré (inhomogeneous Lorentz) group for electromagnetic radiation; and the symmetry element A_μ has been chosen because it is the only electrodynamic element that appears in d'Alembert's equation.

It is important to find a reasonable (i.e., objective) basis such as this for the definition of symmetry elements for electromagnetic radiation in vacuo. Other, arbitrary, choices of symmetry elements can (i.e., may or may not) lead to erroneous conclusions that conflict with the symmetry invariance of Eq. (1). For example, Barron[15] has recently examined the symmetry of electromagnetic radiation in vacuo using three symmetry elements, which appear to have been chosen subjectively. Since λ and A_μ are sufficient to describe the spacetime and electrodynamic properties of the radiation, Barron has one superfluous element in the three chosen, these being[15] the wave vector κ, the sense of rotation, and the axial magnetic field $\mathbf{B}^{(3)}$. It is clear that the first two of these elements can be combined into one, the helicity, which can be regarded as a product of the linear and angular momenta of the electromagnetic radiation, and that the third, $\mathbf{B}^{(3)}$, is related to the potential four vector A_μ, and can be expressed in terms of A_μ. Barron asserts that since κ is unchanged by $\hat{C}$, and $\mathbf{B}^{(3)}$ is reversed in sign by $\hat{C}$, then $\mathbf{B}^{(3)}$ must be zero. In coming to this conclusion, he asserts that "the photon is its own antiphoton." However, Barron's result conflicts with our explicitly demonstrated symmetry invariance of Eq. (1), and with the fact that the numerator and denominator on the right side of Eq. (1) are both nonzero in general. His assertion that the photon is its own antiphoton conflicts with the symmetry equation

$$\left[\lambda, A_\mu\right] \xrightarrow{\hat{C}} \left[\lambda, -A_\mu\right] \tag{12}$$

i.e., $\hat{C}$ changes the sign of A_μ while leaving λ unchanged, and thus produces the antiwave (or antiphoton) from the original wave or photon. Barron's choice of three symmetry elements has therefore led to the incorrect conclusion that $\mathbf{B}^{(3)}$ is zero.

In the context of $\hat{T}$ symmetry, Barron,[15] on the other hand, concludes that $\hat{T}$ applied to his three elements does not rule out $\mathbf{B}^{(3)}$. Barron argues that $\hat{T}$ does not produce a distinct situation because the three symmetry elements he uses are all changed in sign by $\hat{T}$, and therefore $\hat{T}$ does not produce a distinct situation. However, we have seen in Section III that the use of the fundamental symmetry elements λ and A_μ produces a distinct situation when operated upon by $\hat{T}$. For this reason, $\hat{T}$ does not rule out the existence of the longitudinal field $\mathbf{B}^{(3)}$. Similarly, $\hat{P}$ acting on λ and A_μ produces a distinct situation that does not rule out $\mathbf{B}^{(3)}$. The choice of symmetry elements is therefore critically important to any argument based on $\hat{C}$, $\hat{P}$, and $\hat{T}$ symmetry that purports to show the existence or nonexistence of electric and/or magnetic fields in vacuo.

Several other arguments may be used to demonstrate why Barron's choice of symmetry elements has led to the erroneous conclusion that $\mathbf{B}^{(3)}$ is zero. These are discussed in detail as follows.

The $\hat{C}$ symmetry of all components of A_μ is negative, so that it follows by Barron's argument that oscillating transverse components such as $\mathbf{E}^{(1)}$ and $\mathbf{E}^{(2)}$ are also zero in the electromagnetic plane wave in vacuo, an erroneous conclusion. Barron's argument also implies, incorrectly, that the scalar amplitude, E_0, of the plane wave is zero for the following reason. $\hat{C}$ is a symmetry that acts on a scalar, such as charge, reversing its sign, and by definition $\hat{C}$ leaves all spacetime quantities unchanged. Any electric or magnetic field can be expressed as the product of scalar amplitude with a

J

vector, e.g.,

$$\mathbf{E} = E_0 \zeta \tag{13}$$

where ζ is a vector, a spacetime quantity. Regardless of whether ζ is transverse or longitudinal, it is unchanged by $\hat{C}$ by definition, and E_0 reverses sign by definition when operated upon by $\hat{C}$. The product

$$\hat{C}(E_0)\hat{C}(\zeta) \tag{14}$$

is therefore always negative, and it cannot be deduced on the grounds of $\hat{C}$ symmetry that in one direction the field is zero, and in orthogonal (or any other) directions nonzero, since direction, by definition, is a spacetime quantity invariant to $\hat{C}$. In manifestly covariant electrodynamics, E_0 is the timelike component of the electric field,[1, 2] a nonzero quantity.

Maxwell's equations in vacuo are invariant to $\hat{C}$, $\hat{P}$, and $\hat{T}$. It follows that all legitimate solutions of Maxwell's equations in vacuo are also invariant to $\hat{C}$, $\hat{P}$, and $\hat{T}$, and Eq. (1) shows that the novel longitudinal solution $\mathbf{B}^{(3)}$ is so. The solution $\mathbf{B}^{(3)}$ cannot violate $\hat{C}$ because the equation to which it is a solution does not violate $\hat{C}$. Similarly, transverse solutions to Maxwell's equations, such as $\mathbf{E}^{(1)}$ and $\mathbf{E}^{(2)}$, conserve $\hat{C}$, $\hat{P}$, and $\hat{T}$ in vacuo. Underpinning Barron's argument is a subjective choice of three symmetry elements, described already, and the assumption that the photon and antiphoton are the same in all respects, i.e., one is not "distinct" from the other. We have argued that the photon and antiphoton are different entities, i.e.:

$$[\lambda, A_\mu] \quad \xrightarrow{\hat{C}} \quad [\lambda, -A_\mu] \tag{15}$$
$$\text{Photon} \qquad\qquad \text{Antiphoton}$$

and that the choice of λ and A_μ as symmetry elements is rooted in contemporary theory of electromagnetic radiation. Only two elements are needed to define the symmetry of electromagnetic radiation in vacuo, one being an invariant of the Poincaré group, the other being the single variable of the d'Alembert equation.

Barron, therefore, bases his argument[15] on the assumption that the photon and antiphoton are indistinct, so that in an indistinct situation all variables must be relatively the same, so that $\mathbf{B}^{(3)}$ relative to κ must not change when both are acted upon by $\hat{C}$. We argue that $\hat{C}$ operates to produce a distinct situation, embodied in the antiwave, or antiphoton, and in a distinct situation, it is no longer reasonable to expect that all variables must be relatively unchanged, so that $\mathbf{B}^{(3)}$ may change sign with respect to

κ, and so may E_0, $\mathbf{E}^{(1)}$, and $\mathbf{E}^{(2)}$. Even within the framework of his own argument, Barron has shown only that *either* $\mathbf{B}^{(3)}$ or κ must be zero, so that on his grounds κ may be zero and $\mathbf{B}^{(3)}$ nonzero. It is therefore not possible to assert unequivocally, even within his own argument, that $\mathbf{B}^{(3)}$ is zero.

Barron proceeds to argue, on the basis of his three symmetry elements and on the basis of his assertion that the photon and the antiphoton are distinct, that $\hat{C}$ symmetry does not imply that the inverse Faraday effect,[24] magnetization by a circularly polarized laser, cannot exist. Before commenting on Barron's viewpoint in this context, we note that the inverse Faraday effect is accommodated straightforwardly by Eq. (1). This is because $\mathbf{E}^{(1)} \times \mathbf{E}^{(2)}$ interacts with a $\hat{C}$-negative rank three property tensor to induce a magnetic dipole moment.[25] Similarly, $\mathbf{B}^{(3)}$ interacts with a $\hat{C}$-positive rank two molecular property tensor (the susceptibility) to induce a magnetic dipole moment.[26]

In Barron's viewpoint, on the other hand, $\mathbf{B}^{(3)}$ is zero, so that by Eq. (1), $\mathbf{E}^{(1)} \times \mathbf{E}^{(2)}$ is zero. Despite this, his argument asserts that there is a nonzero inverse Faraday effect, showing conclusively that his viewpoint is illogical. The root error in Barron's approach is that under $\hat{C}$, both $\mathbf{E}^{(1)}$ and $\mathbf{E}^{(2)}$ are separately negative, because they are electric fields of the antiwave, which is distinct from the wave. His assertion that the wave and antiwave are in all respects identical (i.e., "indistinct") implies in his view that $\mathbf{E}^{(1)}$ and $\mathbf{E}^{(2)}$ must be separately zero, and that the product $\mathbf{E}^{(1)} \times \mathbf{E}^{(2)}$ is zero. This means that there is no inverse Faraday effect, in direct conflict with experimental data.[24] It may be argued in Barron's favor[27] that $\mathbf{E}^{(1)} \times \mathbf{E}^{(2)}$, and quantities such as $\mathbf{E}^{(1)} \times \mathbf{B}^{(2)}$, do not change sign with $\hat{C}$, but this is spurious, because neither is an electric or a magnetic field. On these albeit spurious grounds it can be similarly asserted that $\mathbf{B}^{(3)}$ is nonzero because quantities such as $\mathbf{B}^{(3)} \times \mathbf{E}^{(1)}$ and $\mathbf{B}^{(3)} \times \mathbf{B}^{(1)}$ do not change sign under $\hat{C}$, so that $\mathbf{B}^{(3)}$ is nonzero.

It is clear that energy is invariant to $\hat{C}$ and also that, in our viewpoint, the interaction energy of the antiwave with antimatter is identical with the interaction energy of the wave and matter, for example:

$$
\begin{aligned}
\hat{C}\left(\boldsymbol{\mu} \cdot \mathbf{E}^{(1)}\right) &= (-\boldsymbol{\mu}) \cdot \left(-\mathbf{E}^{(1)}\right) = \boldsymbol{\mu} \cdot \mathbf{E}^{(1)} \\
\hat{C}\left(\mathbf{m} \cdot \mathbf{B}^{(3)}\right) &= (-\mathbf{m}) \cdot \left(-\mathbf{B}^{(3)}\right) = \mathbf{m} \cdot \mathbf{B}^{(3)}
\end{aligned}
\tag{16}
$$

where $\boldsymbol{\mu}$ is an electric and $\mathbf{m}$ a magnetic dipole moment. It is therefore quite natural in our argument that the inverse Faraday effect and similar effects may exist in nature, because they are governed by an interaction energy that is invariant under the basic symmetries of physics, $\hat{C}$, $\hat{P}$, and

$\hat{T}$. Thus, in the inverse Faraday effect, for example, $\mathbf{B}^{(3)}$ forms an interaction energy with an electronic magnetic dipole moment, and $\mathbf{E}^{(1)} \times \mathbf{E}^{(2)}$ with an electronic antisymmetric polarizability, both types of interaction energy being invariant under $\hat{C}$, $\hat{P}$, and $\hat{T}$.

In Barron's[15] view, however, the argument is convoluted and obscure, because if the wave be indistinct from the antiwave, as in his view, the interaction energy of the antiwave with antimatter, produced by $\hat{C}$ from matter, is no longer invariant under $\hat{C}$. Obviously, matter must be distinct from antimatter, but if wave be indistinct from antiwave, it follows that the interaction energy of antiwave with antimatter is opposite in sign to the interaction energy of wave with matter, an insupportable conclusion because all forms of energy must be indistinct under the basic symmetries of physics. Thus, if matter be distinct from antimatter, wave must be distinct from antiwave, as in our argument. Barron does not consider interaction energy in his paper,[15] but uses the fact that the induced magnetic dipole moment is $\hat{C}$ negative, which is also naturally accommodated within our argument, and also in that of Woźniak,[25] which Barron does not dispute.

There are several considerations of classical electrodynamics, for example, in the classic text by Jackson,[10] that appear to conflict with the assertion by Barron, that all forms of longitudinal solutions to Maxwell's equations in vacuo are zero. The following examples are mentioned briefly, but there are several more available.[10]

The Maxwell equations in vacuo have spherical solutions,[10] which in general are not transverse, as in plane wave solutions. The most general form of these solutions is given by Jackson's equation (16.35):

$$B = \sum_{l,m} \left[A_{lm}^{(1)}h_l^{(1)}(kr) + A_{lm}^{(2)}h_l^{(2)}(kr) \right] Y_{lm}(\theta,\phi) \qquad (17)$$

where A_{lm} are arbitrary constant vectors; where $h_l^{(1)}$ and $h_l^{(2)}$ are Hankel functions, and where Y_{lm} are spherical harmonics. The longitudinal component is found for $\theta = 0$:

$$B_{\theta=0} = \sum_l \left[A_l^{(1)}h_l^{(1)}(kr) + A_l^{(2)}h_l^{(2)}(kr) \right] \left(\frac{2l+l}{4\pi} \right)^{1/2} \qquad (18)$$

and this is clearly not zero in vacuo, being a special case of the general spherical solution (17) of the vacuum Maxwell equations:

$$(\nabla^2 + \kappa^2)\mathbf{B} = 0$$
$$\nabla \cdot \mathbf{B} = 0 \qquad (19)$$

Transverse solutions are also special cases of Eq. (17), cases that are described, for example, in Jackson's equation (16.42).[10] It is impossible to assert on the grounds of $\hat{C}$ symmetry that the fields defined in Eq. (18) are zero, while those in Eq. (16.42) of Ref. 10 are nonzero, because both Eqs. (18) and (16.42) of Ref. 10 are special cases of Eq. (17).

These are longitudinal solutions of Maxwell's equations in conducting media, where they are augmented by Ohm's law:

$$\mathbf{J} = \sigma E \tag{20}$$

where $\mathbf{J}$ is a current and σ the conductivity. The effect of $\hat{C}$ on Ohm's law is as follows:

$$-\mathbf{J} = \sigma(-E) \tag{21}$$

i.e., the conductivity is $\hat{C}$-positive. Similarly, electric permittivity and magnetic permeability in a conductor are both $\hat{C}$-positive, so changing conducting matter to conducting antimatter by operating with $\hat{C}$ does not change conductivity, permittivity, and permeability. Thus, Maxwell's equations and Ohm's law in conducting anti matter are the same as in conducting matter, and longitudinal solutions exist in both situations. The interaction of the antiwave with conducting antimatter is therefore the same as that of wave and conducting matter, as in our argument given already for the inverse Faraday effect. In Barron's view there is no antiwave, and the interaction is distinct, an insupportable conclusion.

Longitudinal, but phase dependent, solutions of Maxwell's equations exist in waveguide theory,[10] through equations such as

$$B_Z = B_0 = B_0 \cos\left(\pi\frac{x}{a}\right)e^{i\phi} \tag{22}$$

which are $\hat{C}$-invariant. Here a is a waveguide dimension and x a coordinate in this dimension. The only quantity on the right side of this equation that changes sign with $\hat{C}$ is the magnetic flux density amplitude B_0. All others are spacetime quantities which are invariant to $\hat{C}$ by definition. If we take $a \to \infty$, then for finite x,

$$B_Z \xrightarrow[a \to \infty]{} B_0 e^{i\phi} \tag{23}$$

i.e., at a point x inside a waveguide of infinite dimension a, the longitudinal magnetic field B_Z is nonzero. Since all parameters in Eq. (22) are $\hat{C}$-invariant except B_0, then according to Barron's view $\hat{C}$ operating on B_Z

of the waveguide does not produce a distinct situation, and so B_Z must vanish, an erroneous conclusion that conflicts with waveguide theory. Furthermore, if $a \to \infty$ in the waveguide, the wave B_Z is effectively propagating in a container of infinite volume, i.e., free space, and so in this situation B_Z remains nonzero in the free space limit, obtained by setting $a \to \infty$. According to Barron's view, B_Z is zero.

A closely related situation is that of resonant cavities, which again support longitudinal, phase-dependent solutions of Maxwell's equations, as in the example of a right cylindrical cavity, in which the longitudinal electric field is

$$E_Z = E_0 J_0 \left(2.405 \frac{\rho}{R} \right) e^{-i\omega t} \qquad (24)$$

Here,[10] J_0 is a Bessel function, ρ is a point on the radius R of the cylinder, and E_0 is a scalar electric field strength amplitude. Again, the only quantity on the right side of Eq. (24) that changes sign with $\hat{C}$ is E_0, so that all properties of the cavity are invariant to $\hat{C}$. Therefore, applying $\hat{C}$ in Barron's view produces an indistinct situation in which the wave vector of the field E_Z has not changed, no cavity property has changed, but E_Z has changed. So in Barron's view E_Z is zero, an incorrect conclusion which conflicts with the theory of resonant cavities.

APPENDIX: $\hat{C}\hat{P}\hat{T}$ THEOREM

In the text of the paper, it has been shown that $\mathbf{B}^{(3)}$ does not violate any of the three discrete symmetries. $\hat{C}$, $\hat{P}$, and $\hat{T}$, and by Eq. (1), is nonzero if $\mathbf{E}^{(1)} \times \mathbf{E}^{(2)}$ is nonzero. It follows that the nonobservation of $\mathbf{B}^{(3)}$ would violate $\hat{C}\hat{P}\hat{T}$, striking at the roots of quantum theory. $\hat{C}\hat{P}\hat{T}$ theorem implies that any quantum (and by implication classical) theory of fields that is compatible with special relativity and assumes only local interactions does not violate $\hat{C}\hat{P}\hat{T}$.[8] Therefore, if a physically meaningful magnetic flux density, $\mathbf{B}^{(3)}$, is inviolate of $\hat{C}$, $\hat{P}$, and $\hat{T}$ separately, and is nonzero, it must be an observable by the $\hat{C}\hat{P}\hat{T}$ theorem. If it is a nonobservable, the $\hat{C}\hat{P}\hat{T}$ theorem is violated. If $\mathbf{B}^{(3)}$ is observable experimentally, on the other hand, it provides evidence for the manifest covariance of electrodynamics and conservation of $\hat{C}\hat{P}\hat{T}$.

In this context, we note that if $\mathbf{B}^{(3)}$ is an observable, and if it is inviolate of $\hat{P}$ and $\hat{T}$, and therefore of $\hat{P}\hat{T}$, then the $\hat{C}\hat{P}\hat{T}$ theorem shows that it cannot violate $\hat{C}$. Barron's argument[15] is therefore shown to be incorrect, because he has assumed that $\mathbf{B}^{(3)}$ is an observable, and has himself concluded (albeit in a subjective argument) that in consequence $\mathbf{B}^{(3)}$ does not violate $\hat{P}\hat{T}$. It follows that if $\hat{P}\hat{T}$ is conserved, and $\mathbf{B}^{(3)}$ is an observable, as assumed by Barron[15]; then it cannot violate $\hat{C}$. Thus, if $\mathbf{B}^{(3)}$ is an observable, it must conserve $\hat{C}\hat{P}\hat{T}$. Conversely, conservation of $\hat{C}\hat{P}\hat{T}$ means that $\mathbf{B}^{(3)}$ must be an observable.

It follows that the conventional electrodynamical notion that the longitudinal solutions of Maxwell's equations $\mathbf{B}^{(3)}$ and $i\mathbf{E}^{(3)}$ (Ref. 4), be "unphysical" implies $\hat{C}\hat{P}\hat{T}$ violation, thus putting in doubt the fundamentals of quantum field theory applied to the electromagnetic field. Either $\mathbf{B}^{(3)}$ is an observable and $\hat{C}\hat{P}\hat{T}$ is conserved, or $\mathbf{B}^{(3)}$ is a nonobservable and $\hat{C}\hat{P}\hat{T}$ is violated. In other words, the only possible reason why the left side of Eq. (1) is not equal to the right side is if $\mathbf{B}^{(3)}$ violated $\hat{C}\hat{P}\hat{T}$. Therefore, quantum and classical electromagnetic field theory implies that the left and right sides of Eq. (1) must be equal, and in this field theory, since $\mathbf{E}^{(1)} \times \mathbf{E}^{(2)}$ is physically meaningful, then $\mathbf{B}^{(3)}$ must be physically meaningful. Otherwise, electromagnetic field theory is fundamentally flawed.

Acknowledgments

Many interesting discussions are acknowledged with S. Kielich, K. A. Earle, and F. Farahi. L. D. Barron is thanked for a preprint of Ref. 15 and correspondence.

References

1. M. W. Evans, submitted for publication.

2. M. W. Evans, submitted for publication.

3. M. W. Evans, *Physica B* **182**, 227, 237 (1992).

4. F. Farahi and M. W. Evans, *Phys. Rev. E*, in press.

5. M. W. Evans, *The Photon's Magnetic Field*, World Scientific, Singapore, 1993.

6. R. Tanáś and S. Kielich, *J. Mod. Opt.* **37**, 1935 (1990).

7. S. Kielich, in M. Davies (Senior Reporter), *Dielectric and Related Molecular Processes*, Vol. 1, Chem. Soc., London, 1972.

8. L. S. Ryder, *Quantum Field Theory*, Cambridge University Press, Cambridge, UK, 1987.

9. S. Kielich, *Nonlinear Molecular Optics*, Nauka, Moscow, 1981.

10. J. D. Jackson, *Classical Electrodynamics*, Wiley, New York, 1962.

11. L. D. Landau and E. M. Lifshitz, *The Classical Theory of Fields*, 4th ed., Pergamon, Oxford, UK, 1975.

12. M. Born and E. Wolf, *Principles of Optics*, 6th ed., Pergamon, Oxford, UK 1975.

13. C. Cohen-Tannoudji, J. Dupont-Roc, and G. Grynberg, *Photons and Atoms: Introduction to Quantum Electrodynamics*, Wiley, New York, 1989.

14. W. M. Schwartz, *Intermediate Electromagnetic Theory*, Wiley, New York, 1964.

15. L. D. Barron, in press, *Physica B*.

16. G. Placzek, in E. Marx (Ed.), *Handbuch der Radiologie*, Akad. Verlag, Leipzig, 1934.

17. K. Knast and S. Kielich, *Acta. Phys. Pol.* **A55**, 319 (1979).

256   M. W. EVANS

18. S. Kielich, *Acta Phys. Pol.* **29**, 875 (1966).
19. S. Kielich, *Acta Phys. Pol.* **31**, 929 (1967).
20. S. Kielich, *Proc. Phys. Soc.* **86**, 709 (1965).
21. S. Kielich and M. W. Evans (Eds.), *Modern Nonlinear Optics*, special issue of I. Prigogine and S. A. Rice (Eds.), *Advances in Chemical Physics*, Vols. 85(A) and 85(B), Wiley, New York, 1993.
22. L. S. Ryder, *Elementary Particles and Symmetries*, Gordon & Breach, London, 1986.
23. M. W. Evans, in I. Prigogine and S. A. Rice (Eds.), *Advances in Chemical Physics*, Vol. 81, Wiley, New York, 1992.
24. P. S. Pershan, J. P. van der Ziel, and L. D. Malmstrom, *Phys. Rev.* **143**, 574 (1966).
25. S. Woźniak, M. W. Evans, and G. Wagnière, *Mol. Phys.* **75**, 81 (1992).
26. M. W. Evans, Physica B, in press.
27. L. D. Barron, personal communication to the author, Dec. 1992, from Department of Chemistry, University of Glasgow, Scotland.

THE ELECTROSTATIC AND MAGNETOSTATIC FIELDS GENERATED BY LIGHT IN FREE SPACE*

I. INTRODUCTION

The phenomenological equations of J. C. Maxwell form the basis of the classical understanding of light. The equations were formulated in the mid nineteenth century, before relativity was fully developed, and before the quantum theory came into existence. They were later put on a microscopic basis by H. A. Lorentz in his theory of the electron, and have become the starting point of a vast number of contemporary papers on the nature of light in free space and in materials. In this paper we show that there exist novel electro and magnetostatic fields in the propagation axis of the classical electromagnetic plane wave, fields that propagate in free space and conserve the structure of the well-defined Poynting vector, and therefore do not affect the law of conservation of electromagnetic energy in free space. It is usually assumed that the following are solutions to the free space Maxwell equations for a completely circularly polarized plane wave:

$$\mathbf{E}(\mathbf{r}, t) = \frac{1}{\sqrt{2}} E_0(\mathbf{i} + i\mathbf{j})e^{i\phi} \tag{1}$$

$$\mathbf{B}(\mathbf{r}, t) = \frac{1}{\sqrt{2}} B_0(\mathbf{j} - i\mathbf{i})e^{i\phi} \tag{2}$$

*R. Gauthier and F. Farahi of UNCC are thanked for suggesting the possibility of $\mathbf{E}_{\text{II}}$.

Here E_0 is the scalar electric field strength amplitude, and B_0 the scalar magnetic flux density amplitude, **i** and **j** are unit vectors in X and Y of the laboratory frame, and ϕ is the phase of the plane wave. These solutions are oscillatory and time and space dependent through the phase

$$\phi = \omega t - \boldsymbol{\kappa} \cdot \mathbf{r} \tag{3}$$

where ω is the angular frequency of the wave, t the time, $\boldsymbol{\kappa}$ the wave vector, and **r** a position vector as usual. A whole literature is available concerning their properties.

However, the equations

$$\mathbf{E}^{G} = \mathbf{E}(\mathbf{r}, t) + \mathbf{E}_{\Pi} \tag{4}$$

$$\mathbf{B}^{G} = \mathbf{B}(\mathbf{r}, t) + \mathbf{B}_{\Pi} \tag{5}$$

are also valid solutions to the free space Maxwell equations. Here $\mathbf{E}_{\Pi}$ and $\mathbf{B}_{\Pi}$ are uniform, time-independent, electric and magnetic fields directed in the propagation axis Z of the plane wave. It appears always to have been implicitly assumed that $\mathbf{E}_{\Pi}$ and $\mathbf{B}_{\Pi}$ are both zero in free space, and that there is no component in Z of the plane wave in vacuo. There is no mathematical reason for this supposition, however, and as we shall see, the vectors $\mathbf{E}_{\Pi}$ and $\mathbf{B}_{\Pi}$ can be related to the well-known $\mathbf{E}(\mathbf{r}, t)$ and $\mathbf{B}(\mathbf{r}, t)$. The source of $\mathbf{E}_{\Pi}$ and $\mathbf{B}_{\Pi}$ is therefore the same as the source of $\mathbf{E}(\mathbf{r}, t)$ and $\mathbf{B}(\mathbf{r}, t)$. If the latter are nonzero, then so are both $\mathbf{E}_{\Pi}$ and $\mathbf{B}_{\Pi}$, in general.

Section II introduces $\mathbf{B}_{\Pi}$ using the imaginary conjugate product,

$$\boldsymbol{\Pi}^{(A)} \equiv E_0 c \, \mathrm{Im}(\mathbf{B}_{\Pi}) = \mathbf{E}(\mathbf{r}, t) \times \mathbf{E}^{*}(\mathbf{r}, t) = -iE_0^2 \mathbf{k} \tag{6}$$

of the electromagnetic plane wave,[1-8] where $\mathbf{E}^{*}(\mathbf{r}, t)$ is the complex conjugate of $\mathbf{E}(\mathbf{r}, t)$, i.e.,

$$\mathbf{E}^{*}(\mathbf{r}, t) = \frac{1}{\sqrt{2}} E_0(\mathbf{i} - i\mathbf{j}) e^{-i\phi} \tag{7}$$

We see in Appendix A that the real and imaginary parts of $\mathbf{B}_{\Pi}$ are the same.

The law of conservation of energy for a plane wave in free space can be expressed through the continuity equation:

$$\nabla \cdot \mathbf{N} = -\frac{\partial U}{\partial t} \tag{8}$$

where **N** is the Poynting vector:

$$N = \frac{1}{\mu_0} E(r, t) \times B(r, t) \qquad (9)$$

and U a scalar field. Here μ_0 is the magnetic permeability of free space. The vector **N** is the flux of electromagnetic energy of the plane wave, and the scalar U is the electromagnetic field's energy density. Therefore, **N** is electromagnetic power per unit area, and U is power per unit volume. The scalar amplitude of the Poynting vector is the light intensity I_0. Therefore, Eq. (8) expresses, in classical electrodynamics, the law of conservation of electromagnetic energy in free space. This idea of field energy has no meaning[9] unless the wave interacts with matter (e.g., an electron). In Section IV, it is shown that the continuity equation (8) is unchanged for nonzero E_Π and B_Π provided they are both complex and

$$B_\Pi \times E(r, t) = E_\Pi \times B(r, t) \qquad (10)$$

In other words, Eq. (10) is the condition for conservation of free space electromagnetic energy given the general solutions (4) and (5) of the Maxwell equations. Equation (10) shows that if B_Π is real, and defined through the conjugate product (6), then E_Π is imaginary. Finally, a discussion is given of the physical meaning of the novel vectors E_Π and B_Π, with order of magnitude estimates, and experimental consequences.

II. THE DEFINITION OF B_Π THROUGH THE CONJUGATE PRODUCT

The conjugate product $E \times E^*$ appears in the antisymmetric part of Maxwell's stress tensor[10] and is a well-defined property of light. It is an axial vector with magnetic symmetry,[11, 12] i.e., that of angular momentum: positive to parity inversion $\hat{P}$, and negative to motion reversal $\hat{T}$. The vector notation $E \times E^*$ is equivalent to the tensor notation

$$\Pi_i^{(A)} = \tfrac{1}{2}\varepsilon_{ijk}\left(E_j E_k^* - E_k E_j^*\right) \qquad (11)$$

where ε_{ijk} is the Levi-Civita symbol. This shows that the axial vector $E \times E^*$ is equivalent to a polar rank two tensor:

$$\Pi_{jk}^{(A)} = \tfrac{1}{2}\left(E_j E_k^* - E_k E_j^*\right) \qquad (12)$$

which is the antisymmetric part of the tensor $E_j E_k^*$. Therefore, $\mathbf{E} \times \mathbf{E}^*$ is the vector part of light intensity.

The quantity

$$\text{Im}(\mathbf{B}_{\Pi}) = \frac{i\mathbf{E} \times \mathbf{E}^*}{E_0 c} \tag{13}$$

is a uniform, divergentless, time-independent, magnetic flux density vector with the required symmetry and units. The magnetic field $\mathbf{B}_{\Pi}$ exists in free space because $\mathbf{E} \times \mathbf{E}^*$ exists in free space, and is defined in the Z axis:

$$\text{Im}(\mathbf{B}_{\Pi}) = + \frac{E_0}{c}\mathbf{k} = +B_0\mathbf{k} \tag{14}$$

where $\mathbf{k}$ in axial unit vector. The magnitude of $\mathbf{B}_{\Pi}$, i.e., $|\mathbf{B}_{\Pi}|$, is the scalar amplitude B_0 defined in the introduction. A real interaction Hamiltonian is produced from $\mathbf{E} \times \mathbf{E}^*/(E_0 c)$ when it forms a scalar product with the usual imaginary magnetic dipole moment operator, $i\hat{m}''$, in quantum mechanics.[13-15] Similarly, the imaginary $\mathbf{E} \times \mathbf{E}^*$ produces a well-defined[1-8] real interaction Hamiltonian when it multiplies the imaginary part of molecular electric polarizability operator, $i\hat{\alpha}''$. The latter is the vectorial polarizability,[16, 17] which vanishes at zero frequency from time-dependent perturbation theory. Both $\hat{m}''$ and $\hat{\alpha}''$ are directly proportional (using the Wigner-Eckart theorem, for example,[16, 17] to the net molecular electronic angular momentum operator $\hat{J}$:

$$\hat{m}'' = \gamma_e \hat{J} \tag{15}$$

$$\hat{\alpha}'' = \gamma_\Pi \hat{J} \tag{16}$$

where γ_e is the gyromagnetic ratio[13-15] and γ_Π is the gyroptic ratio.[18-20] Consequently,

$$\hat{\alpha}'' = \frac{\gamma_\Pi}{\gamma_e}\hat{m}'' \tag{17}$$

showing that $\hat{m}''$ and $\hat{\alpha}''$ have the same $\hat{T}$-negative, $\hat{P}$-positive symmetry, and are both axial vector operators. The conjugate product $\mathbf{E} \times \mathbf{E}^*$ forms a real Hamiltonian operator when it multiplies $i\hat{\alpha}''$, and because $\hat{\alpha}''$ is directly proportional to $\hat{J}$ and thus to $\hat{m}''$, it follows that $\mathbf{E} \times \mathbf{E}^*$ must be proportional to a magnetic field, which we have identified as $\mathbf{B}_{\Pi}$ in Eq. (13). Clearly, $\hat{m}''$ can form a real Hamiltonian operator only when multiplied by a magnetic field. The root of Eq. (13) is therefore found in the fact that the molecular property tensors $\hat{\alpha}''$ *and* $\hat{m}''$ are both axial vectors with magnetic symmetry. This point can be emphasized by assuming that the real part of $\hat{m}'' \cdot \mathbf{B}_\Pi$ is an interaction Hamiltonian and investigating the logical consequences. To do this, it is convenient to write the interaction Hamiltonian[16, 17] between $i\hat{\alpha}''$ and $\mathbf{E} \times \mathbf{E}^*$ as

$$\Delta \hat{H} = -i\hat{\alpha}'' \cdot \mathbf{E} \times \mathbf{E}^* = -iE_0c\hat{\alpha}'' \cdot \frac{\mathbf{E} \times \mathbf{E}^*}{E_0c} \equiv -E_0c\hat{\alpha}'' \mathrm{Im}(\mathbf{B}_\Pi) \quad (18)$$

where we have defined $\mathbf{B}_\Pi$ in terms of $\mathbf{E} \times \mathbf{E}^*$ as in Eq. (13). Using the proportionality (17) between the magnetic dipole moment and the vectorial polarizability, Eq. (18) becomes

$$\Delta \hat{H} = -E_0c\frac{\gamma_\Pi}{\gamma_e}\hat{m}'' \cdot \mathrm{Im}(\mathbf{B}_\Pi) = -i\hat{\alpha}'' \cdot \mathbf{E} \times \mathbf{E}^* \quad (19)$$

showing that the product $\hat{m} \cdot \mathbf{B}_\Pi$ is directly proportional to the product $i\hat{\alpha}'' \cdot \mathbf{E} \times \mathbf{E}^*$ through a nonzero proportionality constant. Therefore, if the energy $i\hat{\alpha}'' \cdot \mathbf{E} \times \mathbf{E}^*$ is nonzero, then so must the energy $\hat{m}'' \cdot \mathbf{B}_\Pi$ be nonzero.

Using the Wigner-Eckart theorem, the gyromagnetic and gyroptic ratios can be defined as follows in an atom with net electronic angular momentum, $\hat{J}$, showing that in this case γ_e and γ_Π are nonzero in general:

$$m_0^{1''} = \frac{\langle J \| \hat{m}^{1''} \| J \rangle}{\langle J \| \hat{J}^1 \| J \rangle} J_0^1 = \gamma_e J_0^1 \quad (20)$$

$$\alpha_0^{1''} = \frac{\langle J \| \hat{\alpha}^{1''} \| J \rangle}{\langle J \| \hat{J}^1 \| J \rangle} J_0^1 = \gamma_\Pi J_0^1 \quad (21)$$

$$\langle J \| \hat{J}^1 \| J \rangle = [J(J + 1)(2J + 1)]^{1/2} \quad (22)$$

III. THE CONSERVATION OF ELECTROMAGNETIC ENERGY

We have assumed that Eqs. (4) and (5) are solutions of the free space Maxwell equations:

$$\nabla \times \mathbf{E}^G = -\frac{\partial \mathbf{E}^G}{\partial t} \quad (23)$$

$$\nabla \times \mathbf{B}^G = \frac{1}{c^2}\frac{\partial \mathbf{E}^G}{\partial t} \quad (24)$$

in SI units. It is clear that if $\mathbf{E}_\Pi$ and $\mathbf{B}_\Pi$ are defined in the Z axis of the plane wave, then

$$\nabla \times \mathbf{E}_\Pi = \frac{\partial \mathbf{B}_\Pi}{\partial t} = 0 \tag{25}$$

$$\nabla \times \mathbf{B}_\Pi = \frac{\partial \mathbf{E}_\Pi}{\partial t} = 0 \tag{26}$$

because these fields are time independent and have no X and Y components. Consider the divergence of $\mathbf{E}^G(\mathbf{r}, t)$ and $\mathbf{B}^G(r, t)$. Using the vector identity,

$$\nabla \cdot (\mathbf{A} \times \mathbf{B}) = \mathbf{B} \cdot (\nabla \times \mathbf{A}) - \mathbf{A} \cdot (\nabla \times \mathbf{B}) \tag{27}$$

it follows that we may expand:

$$\begin{aligned} \nabla \cdot (\mathbf{E}^G \times \mathbf{B}^G) = \nabla \cdot (\mathbf{E} \times \mathbf{B}) + \nabla \cdot (\mathbf{E} \times \mathbf{B}_\Pi) \\ + \nabla \cdot (\mathbf{E}_\Pi \times \mathbf{B}) + \nabla \cdot (\mathbf{E}_\Pi \times \mathbf{B}_\Pi) \end{aligned} \tag{28}$$

where $\mathbf{E} \times \mathbf{B}$ is proportional to the Poynting vector of the law of conservation of energy, Eq. (8). From the relations

$$\begin{aligned} \nabla \cdot (\mathbf{E} \times \mathbf{B}_\Pi) = \mathbf{B}_\Pi \cdot (\nabla \times \mathbf{E}) - \mathbf{E} \cdot (\nabla \times \mathbf{B}_\Pi) \\ \nabla \times \mathbf{B}_\Pi = 0 \end{aligned} \tag{29}$$

and

$$\mathbf{B}_\Pi \cdot (\nabla \times \mathbf{E}) = -\mathbf{B}_\Pi \cdot \frac{\partial \mathbf{B}}{\partial t} = 0 \tag{30}$$

it follows that

$$\nabla \cdot (\mathbf{E} \times \mathbf{B}_\Pi) = 0 \tag{31}$$

and similarly

$$\nabla \cdot (\mathbf{E}_\Pi \times \mathbf{B}) = 0 \tag{32}$$

Also, the last term in Eq. (28) vanishes because $\mathbf{E}_\Pi$ is parallel to $\mathbf{B}_\Pi$ in Z. It follows, therefore, that

$$\nabla \cdot (\mathbf{E}^G \times \mathbf{B}^G) = \nabla \cdot (\mathbf{E} \times \mathbf{B}) \tag{33}$$

i.e., the continuity equation (8) is unaffected by the presence of $\mathbf{E}_\Pi$ and

$\mathbf{B}_\Pi$, and the relation (Eq. (8)) of the field energy flux density ($\mathbf{N}$) to the electromagnetic field energy density (U) is unchanged in the free space electromagnetic plane wave. In other words, the electromagnetic powers per unit area generated by $\nabla \cdot (\mathbf{E} \times \mathbf{B}_\Pi)$ and by $\nabla \cdot (\mathbf{E}_\Pi \times \mathbf{B})$ are both zero, and therefore so are the associated electromagnetic powers per unit volume.

This result is true only if $\mathbf{E}_\Pi$ and $\mathbf{B}_\Pi$ are both in the propagation axis of the plane wave. The argument so far shows that $\mathbf{E}_\Pi$ and $\mathbf{B}_\Pi$ may be separately non zero, or that $\mathbf{E}_\Pi$ may be zero and $\mathbf{B}_\Pi$ nonzero, as defined in Eq. (13).

To obtain a relation between $\mathbf{E}_\Pi$ and $\mathbf{B}_\Pi$ we use the result, from Eq. (8):

$$\nabla \cdot (\mathbf{E}^G \times \mathbf{B}^G) = \nabla \cdot (\mathbf{E} \times \mathbf{B}) = -\frac{\partial U}{\partial t} \tag{34}$$

which implies that the divergence of the product $\mathbf{E}^G \times \mathbf{B}^G$ is nonzero and identical with the divergence of the product $\mathbf{E} \times \mathbf{B}$. This implies that

$$\mathbf{E}^G \times \mathbf{B}^G = \mathbf{E} \times \mathbf{B} + \text{constant} \tag{35}$$

However, we know that

$$\mathbf{E}^G \times \mathbf{B}^G = \mathbf{E} \times \mathbf{B} + \mathbf{E}_\Pi \times \mathbf{B} + \mathbf{E} \times \mathbf{B}_\Pi \tag{36}$$

and from Eqs. (35) and (36) we derive the key result:

$$\mathbf{E}_\Pi \times \mathbf{B} = \mathbf{B}_\Pi \times \mathbf{E} \tag{37}$$

assuming that the constant of integration in Eq. (35) is zero (see Appendix D) and demonstrating that if $\mathbf{B}_\Pi$ is real, then $\mathbf{E}_\Pi$ must be imaginary.

In precise analogy with $\mathbf{E}(\mathbf{r}, t)$ and $\mathbf{B}(\mathbf{r}, t)$, the fields $\mathbf{E}_\Pi$ and $\mathbf{B}_\Pi$ take meaning only when there is wave–particle or wave–matter interaction, but these fields propagate through free space (i.e., vacuum). Clearly, electromagnetic waves can be detected only when there is particulate matter with which the waves can interact, otherwise there would be no experimental evidence at all for the existence of electromagnetic fields. The source of $\mathbf{E}_\Pi$ and $\mathbf{B}_\Pi$ is the same as the source of the oscillating fields $\mathbf{E}(\mathbf{r}, t)$ and $\mathbf{B}(\mathbf{r}, t)$, because both the static and oscillating components are needed for the complete solution of the free space Maxwell equations and the presence of oscillating components implies through Eqs. (13) and (37) the presence of static components. The static and oscillating components are both relativistic in nature, because the plane wave propagates at the speed of light. In quantum field theory, there are operator equivalents[18-21] of $\mathbf{E}_\Pi$ and $\mathbf{B}_\Pi$. A fundamentally important difference between the oscillating and static components of the plane wave is that the former vanish upon time averaging and the latter do not. This is the source of several novel physical phenomena when there is wave–matter interaction. Equation (37) conserves $\hat{P}$ and $\hat{T}$ symmetry, and the static components of the solution are related through Eqs. (13) and (37) to the oscillating components with, as we have seen, conservation of electromagnetic energy. The components are therefore completely defined and the definition is self-consistent.

From the properties of the dual transform of special relativity (see Appendix A) the Maxwell equations are invariant to

$$
\begin{aligned}
c\mathbf{B} &\rightarrow -i\mathbf{E} \\
c\mathbf{B}_\Pi &\rightarrow -i\mathbf{E}_\Pi
\end{aligned}
\tag{38}
$$

The dual transform implies immediately that

$$
\mathbf{B}_\Pi \times \mathbf{E} = -i\frac{\mathbf{E}_\Pi}{c} \times \left(-c\frac{\mathbf{B}}{i}\right) = \mathbf{E}_\Pi \times \mathbf{B}
\tag{39}
$$

which confirms that the sum

$$
\mathbf{E}_\Pi \times \mathbf{B} + \mathbf{E} \times \mathbf{B}_\Pi = 0
\tag{40}
$$

and that the general solution of Maxwell's equations must be of the form (see Appendix A)

$$
\mathbf{E}^G = \mathbf{E}(\mathbf{r},t) \pm E_0(i-1)\mathbf{k}
\tag{41}
$$

$$
\mathbf{B}^G = \mathbf{B}(\mathbf{r},t) \pm B_0(i+1)\mathbf{k}
\tag{42}
$$

to be consistent with the theory of special relativity applied to the Maxwell equations.

It is easily checked that Eq. (39) is consistent with Eqs. (1) and (2), with

$$
\mathbf{B}_\Pi = \pm B_0(i+1)\mathbf{k}
\tag{43}
$$

$$
\mathbf{E}_\Pi = \pm E_0(i-1)\mathbf{k}
\tag{44}
$$

Equation (39) is also consistent with the generalized continuity equation, and with the fact (see appendix A) that

$$
\mathbf{F}_\Pi = \mathbf{E}_\Pi + ic\mathbf{B}_\Pi
\tag{45}
$$

264 M. W. EVANS

and

$$F_\Pi^2 = E_\Pi^2 - c^2 B_\Pi^2 + 2ic E_\Pi \cdot B_\Pi \qquad (46)$$

are invariants of the Lorentz transform. From Eq. (40), the net contribution of E_Π and B_Π to free space electromagnetic energy is zero.

IV. DISCUSSION

The orders of magnitude of E_Π and B_Π can be estimated directly from the intensity I_0 of the light beam in W m^{-2}, through the free space relations

$$|B_\Pi| = B_0 = \left(\frac{I_0}{\varepsilon_0 c^3}\right)^{1/2} \qquad |E_\Pi| = E_0 = \left(\frac{I_0}{\varepsilon_0 c}\right)^{1/2} \qquad (47)$$

where ε_0 is the electric permittivity in vacuo (8.854×10^{-12} J^{-1} C^{2}m^{-1} in S.I. units). Thus, for a beam of $10\,000$ W m^{-2}, (1.0 W cm^{-2}), B_0 is about 10^{-8} T and E_0 about 20 V m^{-1}. These are also the scalar amplitudes B_0 and E_0 of the oscillating part of the solution to Maxwell's equations, and the scalar intensity I_0 of the beam is unaffected by the presence of E_Π and B_Π because I_0 is the magnitude of the Poynting vector. However, B_Π is nonzero after time averaging because it is independent of time, and forms a real, nonzero, interaction Hamiltonian with particulate matter. This Hamiltonian leads, therefore, to the prediction of novel physical phenomena, which can be measured as a function of I_0 and of the polarization state of the light beam. If the latter is linearly or incoherently polarized, $E \times E^*$ is zero and in consequence, so are B_Π and E_Π; otherwise E_Π and B_Π are proportional to the square root of I_0. Because E_Π and B_Π are electrostatic and magnetostatic fields that form part of the general solution of Maxwell's equations in free space, they have the properties of such fields when light interacts with matter. This is the main conclusion of this paper.

On the basis of this conclusion it is easy to see that the various theories of the interaction of conventionally generated electric and magnetic fields can be applied directly to the real field B_Π, and examples of these applications have been given elsewhere for B_Π (Refs. 22–26). These include the inverse Faraday effect, the optical Faraday and Zeeman effects, optically induced shifts in NMR resonances ("optical NMR", recently observed experimentally,[27] the optical Cotton Mouton effect, optical ESR, optical forward backward birefringence, and a reinterpretation of antisymmetric light scattering and related phenomena in terms of

$\mathbf{B}_\Pi$. It has also been deduced[18-21] that the quantum field equivalent of $\mathbf{B}_\Pi$ is the operator

$$\hat{B}_\Pi = B_0 \frac{\hat{J}}{\hbar} \qquad (48)$$

where $\hat{J}$ is the quantized photon angular momentum, and $\hbar$ the reduced Planck constant. It has also been shown,[25] using the properties of the classical Lorentz transformation, that there can be no Faraday induction in free space due to a time derivative of the type $d\mathbf{B}_\Pi/dt$, produced, for example, by modulating a laser beam. (Note however, that Faraday induction occurs via the inverse Faraday effect[28] when a circularly polarized laser interacts with matter inside an induction coil.) The reason for this is that the Lorentz transformations do not allow free space X and Y components either of $\mathbf{B}_\Pi$ or of $\mathbf{E}_\Pi$, and also show that the Z components $\mathbf{E}_\Pi$ and $\mathbf{B}_\Pi$ must be relativistically invariant.[25]

One of the simplest consequences of the presence of $\mathbf{B}_\Pi$ is an optical Zeeman effect, whose semiclassical theory regards $\mathbf{B}_\Pi$ as a classical vector.[24] In this approximation the theory of the optical Zeeman effect is the same as that of the conventional Zeeman effect,[29] with the conventional magnetostatic $\mathbf{B}_s$ replaced by $\mathbf{B}_\Pi$. In the simplest case, the Zeeman shift is proportional to

$$\Delta f = \hat{m} \cdot \frac{\mathbf{B}_\Pi}{h} \qquad (49)$$

and therefore to the square root of the laser intensity $I_0^{1/2}$. This occurs in addition to an optical Zeeman shift caused[30] by the interaction of $\mathbf{E} \times \mathbf{E}^*$ with $\hat{\alpha}''$, a mechanism that is proportional to intensity I_0. There appear to be no experimental investigations to date of the optical Zeeman effect, which requires only a minor modification of optical Stark effect apparatus to circularly polarize the pump laser.

Because $\mathbf{E}_\Pi$ is imaginary when $\mathbf{B}_\Pi$ is real, no simple physical effects are expected due to $\mathbf{E}_\Pi$, and significantly, none appears to have been reported in the literature.

The experimental evidence for the presence of uniform and time-independent components in the free space solutions of Maxwell's equations is available in at least three forms: (1) the inverse Faraday effect (IFE),[28] the optical Faraday effect (OFE),[30] and optically induced frequency shifts in NMR (ONMR).[27] In the IFE, bulk magnetization has been observed when a circularly polarized giant ruby laser pulse was passed through a sample in an inductance coil, thereby producing a measurable voltage that was not present in linear polarization and that changed sign with the sense of circular polarization. These are characteristic properties of $\mathbf{B}_\Pi$. In ONMR, a continuous wave argon ion laser was used[27] to shift NMR resonances and the shifts were much larger in circular than in linear polarization of the laser, too large to be explained by mechanisms based on the oscillating $\mathbf{E}$ and $\mathbf{B}$, which time average to zero. For example it may be conjectured that the oscillating $\mathbf{B}(\mathbf{r}, t)$ induces in semiclassical theory a magnetic dipole moment in the electrons of a molecule in ONMR, a dipole moment that sets up a magnetic field at the resonating nucleus. However, this induced magnetic field would produce shifts much smaller than those observed,[27] and would time average to zero and no ONMR shift would be observable. The novel field $\mathbf{B}_\Pi$ does not time average to zero, and in principle sets up an interaction Hamiltonian $\hat{\mathbf{m}}_N \cdot \mathrm{Im}(\mathbf{B}_\Pi)$, where $\hat{\mathbf{m}}_N$ is the nuclear magnetic dipole moment, causing ONMR shifts in circular polarization, as observed,[27] but not in linear polarization. In general, $\hat{\mathbf{B}}_\Pi$ is an operator in quantum field theory, and its interaction with the operator $\hat{\mathbf{m}}_N$ must be described properly in terms of quantization both in the applied laser field and in the nucleus. ONMR provides information about the nature of the interaction between $\hat{\mathbf{B}}_\Pi$ and the nucleus. However, there are several competing mechanisms in ONMR, and the data cannot yet be interpreted unequivocally. Recently, Frey et al.[30] have reported the optical Faraday effect in magnetic semiconductors, in which the polarization direction and phase of a laser beam are modified by interaction with the material. The authors interpret these changes in terms of nonlinear Faraday processes, but it is interesting to note that a plot of their experimentally observed rotation of the polarization of the laser by the sample against the square root of the laser intensity is a straight line within the experimental uncertainty (Fig. 1). This is the result expected from a mechanism of self-rotation based on the presence of the vector $\mathbf{B}_\Pi$ in the laser beam. These results are suggestive, but not unequivocal, because the straight line does not go through the origin in Fig. 1, and because it is not clear from the experimental arrangement of Frey et al.[30] whether the laser they used was circularly polarized. The output beam from the sample was analyzed by these authors with a Soleil compensator to determine its state of polarization, and the fact that a rotation of polarization was observed[30] (Fig. 1) suggests an excess of circular polarization in the output beam. A more critical test for the presence of $\mathbf{B}_\Pi$ would consist of a completely circularly polarized pump laser incident on a magnetic semiconductor (showing a giant Zeeman effect) together with a linearly polarized probe laser. The plane of polarization of the latter would be rotated by the $\mathbf{B}_\Pi$ vector of the pump, and this rotation should be proportional to the square root of the pump laser's intensity if there are no competing mechanisms.

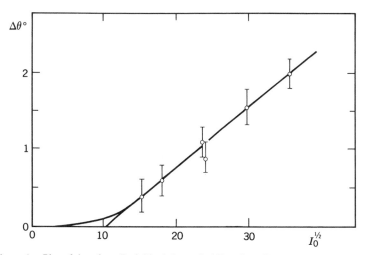

Figure 1. Plot of data from Ref. 30 of the optical Faraday effect, angle of rotation versus the square root of laser intensity. The points and uncertainty bars are those of Ref. 30. —, Best fit line of Ref. 30; - - -, linear extrapolation (this work).

The sense of rotation should be reversed on reversing the sense of circular polarization of the pump, and should vanish when the pump is linearly polarized. The semiclassical theory of this effect is the same as that[31] of the conventional Faraday effect of 1846, with the magnetostatic field of the latter replaced by $\mathbf{B}_{\Pi}$ of the pump laser. The quantum field theory would treat $\hat{\mathbf{B}}_{\Pi}$ as a quantum operator, Eq. (48), and there is no reason to suppose that the results in quantum field theory would be the same as that in semiclassical theory, i.e., there are nonclassical effects, in general, due to $\mathbf{B}_{\Pi}$ treated as quantum field operators.

The available experimental evidence for the existence of $\mathbf{B}_{\Pi}$ is suggestive, but not unequivocal; therefore, the challenge is to separate the particular influence of $\mathbf{B}_{\Pi}$ from the simultaneous influence of $\mathbf{E}(\mathbf{r}, t)$ and $\mathbf{B}(\mathbf{r}, t)$. It appears that one of the clearest ways of demonstrating the existence of $\mathbf{B}_{\Pi}$ would be through its characteristic square root intensity dependence, and through the fact that both vectors change sign with the sense of circular polarization, vanishing in linear polarization. If no experimental evidence for $\mathbf{B}_{\Pi}$ were found, such an eventuality would in itself be a major challenge to contemporary understanding of the nature of light and electromagnetic radiation in general. The reason for this is that the vector $\mathbf{B}_{\Pi}$ is directly proportional (Eq. (13)) to the conjugate product $\mathbf{E} \times \mathbf{E}^*$, in free space, and if the notion of conjugate product is accepted, as is the contemporary practice, then $\mathbf{B}_{\Pi}$ must be accepted, and vice versa.

If $\mathbf{B}_\Pi$ is real then E_Π is imaginary, through Eq. (37), and it has been demonstrated in this work that $\mathbf{E}_\Pi$ and $\mathbf{B}_\Pi$ do not affect the law of conservation of electromagnetic energy, the widely accepted continuity equation (8) of the classical theory of fields. The notions of $\mathbf{E} \times \mathbf{E}^*$, $\mathbf{B}_\Pi$, and $\mathbf{E}_\Pi$ are inextricably and ineluctably interrelated, therefore, and experimental evidence for the presence of any one is evidence for all. Conversely, if there is no apparent evidence for one, then all must not exist. The inverse Faraday effect has been interpreted through the notion of $\mathbf{E} \times \mathbf{E}^*$ (Refs. 7 and 8), but this provides an explanation in terms of only one mechanism, proportional to intensity. It has been argued here that there must be another mechanism present, proportional to the square root of intensity (the $\mathbf{B}_\Pi$ mechanism). If these are found experimentally, contemporary understanding would be strengthened. But if evidence for one mechanism (e.g., $\mathbf{E} \times \mathbf{E}^*$) is found and evidence for another (e.g., $\mathbf{B}_\Pi$) is not found, then the theory of electromagnetic fields would be challenged at the most fundamental level.

Clearly, the notion of $\mathbf{E} \times \mathbf{E}^*$ implies that this object is transmitted through free space in an electromagnetic plane wave, and when this wave meets particulate matter, an interaction Hamiltonian is formed between $\mathbf{E} \times \mathbf{E}^*$ and a material property. In atoms and molecules with net electronic angular momentum, this property is the vectorial polarizability vector $\hat{\alpha}''$, well defined and accepted in semiclassical time-dependent perturbation theory, based on the time-dependent Schrödinger equation.[29] Since $\mathbf{B}_\Pi$ is directly proportional to $\mathbf{E} \times \mathbf{E}^*$, it cannot be argued that $\mathbf{E} \times \mathbf{E}^*$ exists and that $\mathbf{B}_\Pi$ does not. The source of $\mathbf{E}_\Pi$ and $\mathbf{B}_\Pi$ is clearly the same as that of $\mathbf{E}(\mathbf{r}, t)$ and $\mathbf{B}(\mathbf{r}, t)$. Furthermore, it has been shown that $\mathbf{B}_\Pi$ is part of the general solution of the equations of Maxwell, and is therefore phenomenologically indistinguishable from uniform, magnetostatic flux density, whose symmetry and units it possesses. It cannot therefore be argued that $\mathbf{B}_\Pi$ cannot form an interaction Hamiltonian with the appropriate material property (a magnetic dipole moment), and it has been shown in Eq. (19) that if $i\hat{\alpha}'' \cdot \mathbf{E} \times \mathbf{E}^*$ is accepted as an interaction energy, then $\hat{\mathbf{m}} \cdot \mathrm{Im}(\mathbf{B}_\Pi)$ must also be accepted. Finally, the classical presence of $\mathbf{E}_\Pi$ and $\mathbf{B}_\Pi$ must also have meaning in quantum field theory, where these vector fields become operators (Appendix B).

APPENDIX A: LORENTZ COVARIANCE

The theory of the Lorentz covariance of Maxwell's equations show[10] that the complex quantity

$$\mathbf{F}_\Pi = \mathbf{E}_\Pi + ic\mathbf{B}_\Pi \tag{A.1}$$

is an invariant of the Lorentz transformation of special relativity. Therefore,

$$F_{\Pi}^2 = E_{\Pi}^2 - c^2 B_{\Pi}^2 + 2ic\mathbf{E}_{\Pi} \cdot \mathbf{B}_{\Pi} \qquad (A.2)$$

is an invariant of $\mathbf{F}_{\Pi}$ with respect to rotation in $(\mathbf{Z}, t)$ in Minkowski four space. This is equivalent to a rotation in the X, Y plane through an imaginary angle in three dimensions. Thus, $E_{\Pi}^2 - c^2 B_{\Pi}^2$ and $\mathbf{E}_{\Pi} \cdot \mathbf{B}_{\Pi}$ are the only two independent invariants of the antisymmetric four tensor of the electromagnetic field in the four-dimensional representation. This is the tensor F^{ik}, with

$$F_{ik}F^{ik} = \text{inv.}$$
$$e^{iklm}F_{ik}F_{lm} = \text{inv.} \qquad (A.3)$$

and where e^{iklm} denotes the completely antisymmetric unit tensor of rank four.

It is important to note that these invariants are zero only when

$$E_{\Pi}^2 = c^2 B_{\Pi}^2 \qquad (A.4)$$

and

$$\mathbf{E}_{\Pi} \cdot \mathbf{B}_{\Pi} = 0 \qquad (A.5)$$

Because, for complex $\mathbf{B}_{\Pi}$ and $\mathbf{E}_{\Pi}$,

$$|\mathbf{E}_{\Pi}| = \sqrt{2}\, E_0 \quad and \quad |\mathbf{B}_{\Pi}| = \sqrt{2}\, E_0 \qquad E_0 = cB_0,$$

we obtain

$$E_{\Pi}^2 - c^2 B_{\Pi}^2 = 0 \qquad (A.6)$$

However, because $\mathbf{E}_{\Pi}$ and $\mathbf{B}_{\Pi}$ are parallel in Z, the propagation axis,

$$\mathbf{E}_{\Pi} \cdot \mathbf{B}_{\Pi} \neq 0 \qquad (A.7)$$

If $F_{\Pi}^2 \neq 0$, it is known that $\mathbf{F}_{\Pi} = a\mathbf{n}$ (Ref. 10), where $\mathbf{n}$ is a complex unit vector ($nn^* = 1$). Using a complex rotation it is always possible to direct $\mathbf{n}$ along a coordinate axis,[10] and $\mathbf{n}$ becomes real, determining the directions of $\mathbf{E}_{\Pi}$ and $\mathbf{B}_{\Pi}$:

$$\mathbf{F}_{\Pi} = (E_{\Pi} + icB_{\Pi})\mathbf{n} \qquad (A.8)$$

so E_Π is parallel to B_Π. Therefore, in any frame of reference, E_Π must be parallel to B_Π; and there can be no components of either E_Π and B_Π perpendicular to the direction of the plane wave.

It is concluded that E_Π and B_Π form a nonzero invariant, Eq. (A.7), of the Lorentz transformation of special relativity. The oscillating components $E(\mathbf{r}, t)$ and $B(\mathbf{r}, t)$ form zero invariants, because they are mutually perpendicular and $E_0 = cB_0$.

Furthermore, the dual transformation of special relativity corresponds to[10]

$$cB^G \rightarrow -iE^G$$
$$-iE^G \rightarrow cB^G \qquad (A.9)$$

which leaves the Maxwell equations invariant in vacuo. The dual transformation therefore corresponds to

$$c(B(\mathbf{r}, t) + iB_\Pi) \rightarrow -iE(\mathbf{r}, t) + E_\Pi$$
$$-i(E(\mathbf{r}, t) + E_\Pi) \rightarrow c(B(\mathbf{r}, t) + B_\Pi) \qquad (A.10)$$

Therefore, the Maxwell equations are invariant in vacuo to the transformations $cB_\Pi \rightarrow -iE_\Pi$ and $-iE_\Pi \rightarrow cB_\Pi$. Exchanging cB_Π and $-iE_\Pi$ everywhere, or vice versa, the Maxwell equations are the same in any frame of reference, which is consistent with the fact that E_Π is parallel to B_Π in Z, and if B_Π is real, then E_Π is imaginary. A gauge transformation leaves the Lorentz relation unchanged, and leaves E_Π and B_Π unaltered. Thus, gauge invariance means that E_Π and B_Π are unaltered in any valid gauge.

It is important to note that the invariant in Eq. (A.1) is a complex quantity, and that E_Π and B_Π also appear in Eqs. (4) and (5) as complex. Otherwise, the invariant in Eq. (A.6) would not vanish, and E_Π and B_Π would contribute to the electromagnetic energy density and flux density. This is essentially the result of the special theory of relativity, which uses four-dimensional Minkowski space:

$$m = (X, Y, Z, ict)$$

The Lorentz transformation describes rotations in this four space, and it follows that E_Π and B_Π are unaffected by rotations in spacetime. The fourth coordinate of Minkowski space is ict, i.e., imaginary. The reason is that this gives a space in which the four-dimensional Pythagorean theorem has the same form as the usual three-dimensional theorem. This is taken as a fundamental criterion of a Cartesian system. Maxwell's equations can be written in tensor form in Minkowski space, leading to the Lorentz transformation.

It is concluded that E_Π and B_Π are entirely consistent with the Lorentz covariance of the Maxwell equations, and form invariants of the Lorentz transformation. The fields E_Π and B_Π are therefore physically meaningful in the classical theory of electromagnetic radiation. The free space vector

$$F_\Pi = E_\Pi + i c B_\Pi \qquad (A.11)$$

is a nonzero invariant of the Lorentz transformation in the special theory of relativity. The existence of B_Π implies that Maxwell's equations in free space support the nonlinear solution:

$$B^G(r, t) = B(r, t) + \frac{1}{iE_0 c} E(r, t) \times E^*(r, t) \qquad (A.12)$$

Finally, the dual transformations show that the Maxwell equations are invariant to

$$
\begin{aligned}
E_0(i + ij)e^{i\phi} &\to B_0(i + ij)e^{i\phi} \\
B_0(j - ii)e^{i\phi} &\to E_0(j - ii)e^{i\phi} \\
- E_0 k &\to i E_0 k \\
i B_0 k &\to B_0 k
\end{aligned}
\qquad (A.13)
$$

so Eq. (40) is also satisfied by

$$
\begin{aligned}
B^G &= B - i B_0 k \\
E^G &= E - E_0 k
\end{aligned}
\qquad (A.14)
$$

Therefore (A.14) is also a valid solution of Maxwell's equations, and it is therefore possible to obtain valid solutions of Maxwell's equations in which E_Π is real and B_Π is imaginary, or in which E_Π is imaginary and B_Π is real. Since B_Π from eqn. (18) is real, E_Π is imaginary.

It may also be verified that solutions of the type

$$
\begin{aligned}
E^G &= E(r, t) \pm E_0(i - 1)k \\
B^G &= B(r, t) \pm B_0(i + 1)k
\end{aligned}
\qquad (A.15)
$$

satisfy Maxwell's equations in vacuo and also relation (40) of the text. The most general solution of Maxwell's equations is therefore in this case:

$$
\mathbf{E}^G = \frac{E_0}{\sqrt{2}} \left[(\mathbf{i} \pm \mathbf{ij})e^{i\phi} \pm \sqrt{2}\,(\mathbf{i} - 1)\mathbf{k} \right]
$$
$$
\mathbf{B}^G = \frac{B_0}{\sqrt{2}} \left[(\mathbf{j} \mp \mathbf{ii})e^{i\phi} \pm \sqrt{2}\,(\mathbf{i} + 1)\mathbf{k} \right]
$$

(A.16)

where the normalisation factor $1/\sqrt{2}$ has been used, as is the standard practice.

Solutions (A.16) show that, in general, Maxwell's equations in vacuo support components in $\mathbf{i}, \mathbf{j}$, and $\mathbf{k}$ unit vectors in X, Y, and Z, respectively. In this case, the fields $\mathbf{E}_{\Pi}$ and $\mathbf{B}_{\Pi}$ are both complex:

$$
\mathbf{B}_{\Pi} = B_0(1 + i)\mathbf{k}
$$
$$
\mathbf{E}_{\Pi} = E_0(-1 + i)\mathbf{k}
$$

(A.17)

APPENDIX B: QUANTIZATION OF $\mathbf{B}_{\Pi}$ AND $\mathbf{E}_{\Pi}$

Defining the annihilation and creation operators

$$
\hat{a}(t) = \hat{a}(0)e^{-i\omega t}
$$
$$
\hat{a}^+(t) = \hat{a}^+(0)e^{i\omega t}
$$

(B.1)

the quantized equivalent of the oscillating electric field

$$
\mathbf{E}(t) = \frac{E_0}{\sqrt{2}} e^{i\boldsymbol{\kappa} \cdot \mathbf{r}} \left(\mathbf{i}e^{-i\omega t} + \mathbf{ij}e^{-i\omega t} \right)
$$

(B.2)

becomes the operator

$$
\hat{\mathbf{E}}(t) = \frac{E_0}{\sqrt{2}} e^{i\boldsymbol{\kappa} \cdot r} \left(\hat{a}_X(t)\mathbf{i} + i\hat{a}_Y(t)\mathbf{j} \right)
$$

(B.3)

Similarly,

$$
\hat{\mathbf{E}}^*(t) = \frac{E_0}{\sqrt{2}} e^{-i\boldsymbol{\kappa} \cdot \mathbf{r}} \left(\hat{a}_X^+(t)\mathbf{i} - i\hat{a}_Y^+(t)\mathbf{j} \right)
$$

(B.4)

and

$$\mathbf{E} \times \mathbf{E}^* = -\mathrm{i}\frac{E_0^2}{2}\left(\hat{a}_X \hat{a}_Y^+ - \hat{a}_Y \hat{a}_X^+\right)\mathbf{k} \qquad (\text{B}.5)$$

where

$$\hat{a}_X \hat{a}_Y^+ \equiv \hat{a}_X(t)\hat{a}_Y^+(t) = \hat{a}_X(0)\hat{a}_Y^+(0) \qquad (\text{B}.6)$$

and

$$E_0^2 = \frac{2\hbar\omega}{\varepsilon_0 L_0^3}$$

Here L_0^3 is the volume of quantization[9] and ε_0 the vacuum permittivity.

The expectation value of the operator in Eq. (B.5) is the classical conjugate product $\mathbf{E} \times \mathbf{E}^*$:

$$\langle n|\hat{\mathbf{E}} \times \hat{\mathbf{E}}^*|n\rangle = -\mathrm{i}E_0^2\mathbf{k} \qquad (\text{B}.7)$$

implying that

$$\langle n|\hat{a}_X \hat{a}_Y^+ - \hat{a}_Y \hat{a}_X^+|n\rangle = 2 \qquad (\text{B}.8)$$

It follows from Eq. (B.5) that

$$\hat{B}_\Pi = \tfrac{1}{2}B_0\left(\hat{a}_X \hat{a}_Y^+ - \hat{a}_Y \hat{a}_X^+\right)\mathbf{k} \qquad (\text{B}.9)$$

where

$$B_0 = \left(\frac{2\mu_0\omega}{L_0^3}\right)^{1/2} \qquad (\text{B}.10)$$

with μ_0 denoting the vacuum permeability. As is the standard practice in quantum field theory, $|n\rangle$ is an eigenfunction whose eigenvalues are nonnegative integers n, i.e.,

$$\hat{N}|n\rangle = n|n\rangle \qquad (\text{B}.11)$$

where $\hat{N}$ is the photon number operator, whose eigenstates are photon number states of the quantized field. The photon is a quantum of the field, with energy $\hbar\omega$. Thus, $\hat{a}^+$ operates on $|n\rangle$ to produce a field increment of energy $\hbar\omega$, i.e., to produce a photon, and $\hat{a}$ operates conversely:

$$\hat{a}^+ |n\rangle = (n + 1)^{1/2}|n + 1\rangle$$
$$\hat{a}^- |n\rangle = n^{1/2}|n - 1\rangle \tag{B.12}$$

the normalization factor being chosen so that[9]

$$\langle n|n'\rangle = \delta_{nn'} \tag{B.13}$$

which is the orthonormality condition.

It is instructive to note that Eq. (B.9) for the operator $\hat{B}_\Pi$ can be derived independently of the conjugate operator $\hat{E} \times \hat{E}^*$, showing, inter alia, that $\hat{B}_\Pi$ *and* $\hat{E} \times \hat{E}^*$ are rigorously proportional to each other in quantum field theory. The independent derivation proceeds from a direct quantization of the classical magnetostatic field

$$\mathbf{B}_\Pi = B_0\mathbf{k} \tag{B.14}$$

where we have made no assumptions concerning the origin of this field. Here $\mathbf{k}$ is a axial unit vector in the Z axis of the frame (X, Y, Z):

$$\mathbf{k} = \mathbf{i} \times \mathbf{j} \tag{B.15}$$

where $\mathbf{i}$ is a polar unit vector in X and $\mathbf{j}$ a polar unit vector in Y. Thus, B_0 is a scalar magnetic flux density amplitude (tesla).

It is always possible to write Eq. (B.14) as

$$\begin{aligned}
\mathbf{B}_\Pi = B_0\mathbf{k} &\equiv \tfrac{1}{2}B_0 e^{i\kappa \cdot \mathbf{r}} e^{-i\kappa \cdot \mathbf{r}} \\
&\times (e^{-i\omega t}\mathbf{i} \times e^{i\omega t}\mathbf{j} - e^{-i\omega t}\mathbf{j} \times e^{i\omega t}\mathbf{i}) \\
&= \tfrac{1}{2}B_0 e^{-i\omega t} e^{i\omega t} \\
&\times \left(\begin{vmatrix} \mathbf{i} & \mathbf{j} & \mathbf{k} \\ 1 & 0 & 0 \\ 0 & 1 & 0 \end{vmatrix} - \begin{vmatrix} \mathbf{i} & \mathbf{j} & \mathbf{k} \\ 0 & 1 & 0 \\ 1 & 0 & 0 \end{vmatrix} \right)
\end{aligned} \tag{B.16}$$

a purely mathematical identity, which leads to the well-known conclusion of tensor algebra that an axial vector is equivalent to a second-rank polar tensor.

If we now assume that ω is the angular frequency of the classical electromagnetic plane wave, then the quantized form of any magnetic field operator of the form (B.14) carried by such a wave pattern in quantum field theory must be

$$\hat{\mathbf{B}}_{\Pi} = \frac{B_0}{2}\left(\hat{a}_X\hat{a}_Y^+ - \hat{a}_Y\hat{a}_X^+\right)\mathbf{k} \tag{B.17}$$

which, except for sign, is the same as Eq. (B.9). The sign of $\hat{\mathbf{B}}_{\Pi}$ is switched[1-7] by switching from left to right circular polarization, and in consequence $\hat{\mathbf{B}}_{\Pi}$ can always be defined as plus or minus. Equation (B.17) has been derived directly from Eq. (B.16) using the definitions (B.1), and using no other assumption. This shows that the magnetic field operator $\hat{\mathbf{B}}_{\Pi}$ and the conjugate product operator $\hat{\mathbf{E}} \times \hat{\mathbf{E}}^*$ are rigorously proportional in quantum field theory. Both are described by the operator $(\hat{a}_X\hat{a}_Y^+ - \hat{a}_Y\hat{a}_X^+)$ whose expectation values between states is always 2.

Further insight to the physical interpretation of $\hat{\mathbf{B}}_{\Pi}$ *and* $\hat{\mathbf{E}} \times \hat{\mathbf{E}}^*$ can be gained by using the quantum field definition[9] of the Stokes parameter S_3:

$$\hat{S}_3 = -\frac{E_0^2}{2}\left(\hat{a}_X\hat{a}_Y^+ - \hat{a}_Y\hat{a}_X^+\right) \tag{B.18}$$

so that

$$\hat{\mathbf{B}}_{\Pi} = -\frac{\hat{S}_3}{E_0 c}\mathbf{k} \tag{B.19}$$

showing that $\hat{\mathbf{B}}_{\Pi}$ *and* $\hat{\mathbf{E}} \times \hat{\mathbf{E}}^*$ are both directly proportional to the scalar operator $\hat{S}_3$, the third Stokes operator of quantum field theory.

Furthermore, the Stokes operators $\hat{S}_1$, $\hat{S}_2$, *and* $\hat{S}_3$ obey the commutator equations of angular momentum in quantum field theory,[9] showing that $\hat{\mathbf{B}}_{\Pi}$ has the properties of quantized angular momentum of the electromagnetic wave. Standard theory[9] shows that

$$\langle n|\hat{S}_3|n\rangle = \langle\hat{\sigma}_Z\rangle = \frac{\langle\hat{J}_Z\rangle}{\hbar} \tag{B.20}$$

where $\hat{\sigma}_Z$ is the Pauli matrix operator:

$$\hat{\sigma}_Z \equiv \begin{pmatrix} 0 & -i \\ i & 0 \end{pmatrix} \tag{B.21}$$

so that

$$\mathbf{\psi} = \frac{E_0}{\sqrt{2}}\left(\begin{pmatrix} 1 \\ 0 \end{pmatrix} + \begin{pmatrix} 0 \\ 1 \end{pmatrix}\right) \tag{B.22}$$

is a spinor. The operator $\hat{J}_Z$ is defined by

$$\langle n|\hat{J}_Z|n\rangle = \hbar S_3 \tag{B.23}$$

Using Eq. (B.21), we obtain

$$\hat{\mathbf{B}}_\Pi = B_0 \frac{\hat{J}_Z}{\hbar}\mathbf{k} = B_0 \frac{\hat{J}}{\hbar} \tag{B.24}$$

with B_0 given by Eq. (B.10) in SI units.

It is seen from Eq. (B.23) that the expectation value of $\hat{J}_Z$ between eigenfunctions $|n\rangle$ is S_3, the classical third Stokes parameter, multiplied by $\hbar$, the unit angular momentum in quantum theory. The expectation value of $\hat{J}_Z$ is S_3 units of quantized angular momentum for any eigenstate $|n\rangle$ of the quantized field. Classically, the third Stokes operator is

$$S_3 = -\frac{i}{2}(E_X i E_Y^* + i E_Y E_X^*) = \tfrac{1}{2}(E_X E_Y^* + E_Y E_X^*) \equiv E_L^2 \tag{B.25}$$

a real quantity. The intensity of the beam is classically

$$I_0 = \varepsilon_0 c E_L^2 \tag{B.26}$$

showing that the expectation value of the angular momentum operator $\hat{\mathbf{J}}$ is

$$\frac{1}{\hbar}\langle n|\hat{\mathbf{J}}|n\rangle = \frac{I_L}{\varepsilon_0 c} \tag{B.27}$$

which is directly proportional to the intensity of a beam that is fully left circularly polarized. In a beam that has elements of both left and right circular polarization,

$$\frac{1}{\hbar}\langle n|\hat{\mathbf{J}}|n\rangle = \frac{1}{\varepsilon_0 c}(I_L - I_R) \tag{B.28}$$

If the beam were to consist of a wave pattern corresponding to one photon of energy $\hbar\omega$, then $|n\rangle = |1\rangle$ and $\langle 1|\hat{\mathbf{J}}|1\rangle$ is the expectation value of $\hat{\mathbf{J}}$ for one photon. In this case, $\hat{\mathbf{B}}_\Pi$ is the magnetic flux density operator of one photon, whose scalar magnitude we denote $B_0^{(1)}$:

$$B_0^{(1)} = \left(2\mu_0 \frac{\hbar\omega}{L_0^3}\right)^{1/2} \tag{B.29}$$

which is proportional to the square root of $\hbar\omega$, the energy of the single photon under consideration. The energy of n photons is describable by the expectation value[9]

$$\langle n|\hat{H}^R|n\rangle = \left(n + \tfrac{1}{2}\right)\hbar\omega \qquad (B.30)$$

i.e., by an integer number n of energy quanta $\hbar\omega$, plus a "background" $\hbar\omega/2$ independent of n. Therefore,

$$nB_0^{(1)2} = \frac{2n\omega_0}{L_0^3}\hbar\omega \qquad (B.31)$$

and the energy $n\hbar\omega$ is proportional to $nB_0^{(1)2}$. Using $E_0 = cB_0$ and Eq. (B.26), it becomes clear that the intensity of a beam of n photons is proportional to $nB_0^{(1)2}$, so that $B_0^{(1)}$ is an elementary quantum of magneto-static flux density associated with one photon. This is in analogy with the fact that the quantum of energy associated with the photon is $\hbar\omega$. From Eq. (B.29), $B_0^{(1)}$ vanishes if $\omega = 0$, i.e., if the frequency of the wave is zero. In this case the energy $\hbar\omega$ is zero, and there are no photons. Alternatively, if $L_0^3 \to \infty$, i.e., if the quantization volume tends to infinity, then $B_0^{(1)}$ tends to zero, even for finite ω.

Under all other conditions, $B_0^{(1)}$ is nonzero, and produces finite and measurable physical effects, as described in the text.

The quantization of the imaginary $\mathbf{E}_\Pi$ proceeds similarly using the classical dual transformation

$$\mathbf{B}_\Pi \to -\frac{i}{c}\mathbf{E}_\Pi \qquad (B.32)$$

And Eq. (B.9)

$$i\hat{\mathbf{E}}_\Pi = \tfrac{1}{2}E_0 i\left(\hat{a}_X\hat{a}_Y^+ - \hat{a}_Y\hat{a}_X^+\right)\mathbf{k} \qquad (B.33)$$

whose expectation value is

$$\mathbf{E}_\Pi = \langle n|\hat{\mathbf{E}}_\Pi|n\rangle = -E_0\mathbf{k} \qquad (B.34)$$

is an acceptable definition (see Appendix A) of $\mathbf{E}_\Pi$. Here $\mathbf{k}$ is of course a polar unit vector.

It is seen that both $\hat{\mathbf{B}}_\Pi$ and $i\hat{\mathbf{E}}_\Pi$ are defined in terms of the operator $\hat{a}_X\hat{a}_Y^+ - \hat{a}_Y\hat{a}_X^+$, which operates on any number state $|n\rangle$ to give the constant expectation value of 2. This expectation value is independent of the number state $|n\rangle$ of the photons, and generates the third Stokes parameter S_3 of the classical field.

Classically, the $\mathbf{B}_\Pi$ is an axial vector, $\hat{P}$-positive and $\hat{T}$-negative, and proportional to angular momentum. The field $\mathbf{E}_\Pi$ is a polar vector, $\hat{P}$-negative and $\hat{T}$-positive, and cannot therefore be proportional to an angular momentum. It is essential to note, therefore, that $\mathbf{k}$ in Eq. (B.9) is an axial unit vector ($\hat{T}$-negative and $\hat{P}$-positive) and that $\mathbf{k}$ in Eq. (B.33) is a polar unit vector ($\hat{T}$-positive and $\hat{P}$-negative).

Finally, we note that $\hat{a}_X \hat{a}_Y^* - \hat{a}_Y \hat{a}_X^*$ operates to give an eigenvalue of 2, and this does not change the energy $n\hbar\omega$ of n photons. This is in agreement with the classical theory (see text), which shows that $\mathbf{B}_\Pi$ and $\mathbf{E}_\Pi$ do not contribute to the field energy.

APPENDIX C: DEFINITION OF $\mathbf{B}_\Pi$ AND $\mathbf{E}_\Pi$ IN TERMS OF THE VECTOR POTENTIAL IN FREE SPACE

In free space, the oscillating fields $\mathbf{E}$ and $\mathbf{B}$ of the plane wave are defined in terms of the vector potential $\mathbf{A}$. Using the Coulomb gauge:

$$\mathbf{E} = -\frac{\partial \mathbf{A}}{\partial t} - \nabla\phi; \quad \mathbf{B} = \nabla \times \mathbf{A} \qquad (C.1)$$

In free space the scalar part, $\nabla\phi$ is zero and the Lorentz condition and Coulomb gauge are both describable by

$$\nabla \cdot \mathbf{A} = 0 \qquad (C.2)$$

From the definition in the text,

$$\mathbf{B}_\Pi = \frac{1}{E_0 c}\left(\frac{\partial \mathbf{A}}{\partial t} \times \frac{\partial \mathbf{A}^*}{\partial t}\right) \equiv B_0 \mathbf{k} \qquad (C.3)$$

where

$$\mathbf{E}^* = -\frac{\partial \mathbf{A}^*}{\partial t} - \nabla\phi^* \qquad (C.4)$$

Using the condition for conservation of energy,

$$\mathbf{E}_\Pi \times \mathbf{B} = \mathbf{B}_\Pi \times \mathbf{E} \qquad (C.5)$$

with the definition (C.3) implies

$$\mathbf{E}_\Pi = i E_0 \mathbf{k} \tag{C.6}$$

and

$$\mathbf{B} \times \mathbf{E}_\Pi = i B_0 \mathbf{E} \tag{C.7}$$

From (C.1) in (C.7),

$$(\nabla \times \mathbf{A}) \times \mathbf{E}_\Pi = i B_0 \mathbf{E} = -i B_0 \left(\frac{\partial \mathbf{A}}{\partial t} - \nabla \phi \right) \tag{C.8}$$

which defines $\mathbf{E}_\Pi$ in terms of $\mathbf{A}$.

Equation (C.8) can be simplified to

$$\nabla (\mathbf{E}_\Pi \cdot \mathbf{A}) = i B_0 \left(\frac{\partial \mathbf{A}}{\partial t} - \nabla \phi \right) \tag{C.9}$$

using

$$(\nabla \times \mathbf{A}) \times \mathbf{E}_\Pi = \mathbf{A}(\mathbf{E}_\Pi \cdot \nabla) - \nabla(\mathbf{E}_\Pi \cdot \mathbf{A}) \tag{C.10}$$

and

$$\mathbf{E}_\Pi \cdot \nabla = \nabla \cdot \mathbf{E}_\Pi = 0 \tag{C.11}$$

Note that (C.9) is a type of continuity equation which defines the imaginary $\mathbf{E}_\Pi$ in terms of $\mathbf{A}$ of the oscillating components $\mathbf{E}$ and $\mathbf{B}$ of the plane wave.

In texts on the electromagnetic plane wave it is usually asserted that $\mathbf{E}$ and $\mathbf{B}$ are transverse plane waves, with no components in the direction of propagation. Equations (C.1) and (C.2) are usually taken as justification for this conclusion. Most texts assert that electromagnetic plane waves in vacuo are necessarily time-varying, because the solutions for constant E and B from Maxwell's equations in the absence of charge and current are zero. While this is true for linear solutions, we can form a nonlinear solution, Eq. (C.3), for $\mathbf{B}_\Pi$, which is well defined as in this paper, and which is a product of time-varying solutions. With the condition (C.5), derived in the text of this paper, the field $\mathbf{E}_\Pi$ is also well defined in terms of $\mathbf{A}$ as in Eq. (C.9), a novel continuity equation.

Nonlinear solutions of Maxwell's equations therefore support the existence of $\mathbf{E}_\Pi$ and $\mathbf{B}_\Pi$ in the axis of propagation of the plane wave in vacuo.

APPENDIX D: THE CONSTANT OF INTEGRATION IN EQUATION (35)

Most generally, from Eq. (35),

$$\mathbf{E}_\Pi \times \mathbf{B} = \mathbf{B}_\Pi \times \mathbf{E} + \text{constant} \qquad (D.1)$$

The dual transformation of special relativity means that $\mathbf{E}_\Pi$ and $-(c/i)\mathbf{B}_\Pi$, for example, are indistinguishable solutions of Maxwell's equations; i.e., it is possible to replace $\mathbf{E}_\Pi$ everywhere by $-(c/i)\mathbf{B}_\Pi$ without changing the Maxwell equations, and therefore without changing the solutions to the equations. The dual transformation, however, does not affect the constant in Eq. (D.1), which is independent of $\mathbf{E}$, $\mathbf{B}$, $\mathbf{E}_\Pi$, and $\mathbf{B}_\Pi$. Thus, applying the dual transform,

$$\mathbf{B}_\Pi \times \mathbf{E} = \mathbf{E}_\Pi \times \mathbf{B} + \text{constant} \qquad (D.2)$$

Adding Eqs. (D.1) and (D.2) yields

$$\text{Constant} = 0$$

Acknowledgments

Support for this work is acknowledged from Cornell Theory Center, the Materials Research Laboratory of Penn State University, and Optical Ventures Inc. Many interesting discussions are acknowledged with colleagues at UNCC and elsewhere.

References

1. A. Piekara and S. Kielich, *Archives des Science*, II, fasc. special, 7è Colloque Ampère, 1958, p. 304.

2. S. Kielich and A. Piekara, *Acta Phys. Pol.* **18**, 439 (1959).

3. S. Kielich, *Proc. Phys. Soc.* **86**, 709 (1965).

4. S. Kielich, in M. Davies (Ed.), *Dielectric and Related Molecular Processes*, Vol. 1, Chem. Soc., London, 1972.

5. S. Kielich and M. W. Evans (Eds.), *Modern Nonlinear Optics*, a special topical issue of *Advances in Chemical Physics*, Vols. 85(1) and 85(2) (I. Prigogine and S. A. Rice, Series Eds.), Wiley, New York, 1993.

6. P. W. Atkins and M. H. Miller, *Mol. Phys.* **15**, 503 (1968).

7. S. Woźniak, M. W. Evans, and G. Wagnière, *Mol. Phys.* **75**, 81 (1992).

8. S. Woźniak, M. W. Evans, and G. Wagnière, *Mol. Phys.* **75**, 99 (1992).

9. B. W. Shore, *The Theory of Coherent Atomic Excitation*, Wiley, New York, 1990, Chapter 9.

10. L. D. Landau and E. M. Lifshitz, *The Classical Theory of Fields*, Pergamon, Oxford, UK, 1974.

11. M. W. Evans, *Phys. Rev. Lett.* **64**, 2909 (1990).

12. M. W. Evans, *Opt. Lett.* **15**, 863 (1990).

13. A. R. Edmonds, *Angular Momentum in Quantum Mechanics*, Princeton University Press, Princeton, NJ, 1960.

14. B. L. Silver, *Irreducible Tensor Methods*, Academic, New York, 1976.

15. S. B. Piepho and P. N. Schatz, *Group Theory in Spectroscopy with Applications to Magnetic Circular Dichroism*, Wiley, New York, 1983.

16. N. L. Manakov, V. D. Ovsiannikov, and S. Kielich, *Acta Phys. Pol.* **A53**, 581, 595 (1978).

17. R. Zawodny, in Ref. 5, Vol. 85(1), a review with ca. 150 references.

18. M. W. Evans, *J. Phys. Chem.* **95**, 2256 (1991).

19. M. W. Evans, *Int. J. Mod. Phys. B* **5**, 1963 (1991) (review).

20. M. W. Evans, *J. Mol. Spectrosc.* **146**, 351 (1991).

21. M. W. Evans, *The Photon's Magnetic Field*, World Scientific, Singapore, 1992, in press.

22. M. W. Evans, *Physica B*, **182**, 227 (1992).

23. M. W. Evans, *Physica B*, **182**, 237 (1992).

24. M. W. Evans, *Physica B*, **183**, 103 (1993).

25. M. W. Evans, *Physica B*, in press (1993).

26. M. W. Evans, submitted for publication.

27. W. S. Warren, S. Mayr, D. Goswami, and A. P. West, Jr., **255**, 1683 (1992).

28. P. S. Pershan, J. P. van der Ziel, and L. D. Malmstrom, *Phys. Rev.* **143**, 574 (1966).

29. P. W. Atkins, *Molecular Quantum Mechanics*, Oxford University Press, Oxford, UK, 1983.

30. J. Frey, R. Frey, C. Flytzanis, and R. Triboulet, *Opt. Commun.* **84** 76 (1991).

31. L. D. Barron, *Molecular Light Scattering and Optical Activity*, Cambridge University Press, Cambridge, UK, 1982.

ON LONGITUDINAL FREE SPACETIME ELECTRIC AND MAGNETIC FIELDS IN THE EINSTEIN-DE BROGLIE THEORY OF LIGHT

I. INTRODUCTION

It is usually concluded in electrodynamical literature[1-16] that the photon is massless and that the range of the electromagnetic field is infinite. This conclusion is not, however, supported by experimental data. To the contrary, Vigier[17] has recently reviewed a substantial amount of evidence that leads to the conclusion of finite photon rest mass. These data include, to take two of many examples, the direction-dependent anisotropy of the frequency of light in cosmology and frequently observed anomalous red shifts.

In papers and correspondence circa 1916 to 1919,[17] Einstein[18] proposed a photon rest mass[19] that can be estimated from the Hubble constant to be about 10^{-68} kg. An immediate consequence is that the d'Alembert equation is replaced by the Einstein-de Broglie-Proca (EBP) equation, which can be expressed[20, 21] in the form

$$\Box A_\mu = -\xi^2 A_\mu \qquad (1)$$

where

$$\xi = \frac{m_0 c}{\hbar}$$

Here m_0 is the photon rest mass, c the speed of light, the universal constant of special relativity, and $\hbar$ the reduced Planck constant. The potential four vector A_μ of the de Broglie-Proca field is manifestly covariant, and has four, physically meaningful, components, one timelike ((0)) and three spacelike, of which two are transverse ((1) and (2)) and one is longitudinal ((3)). From Eq. (2), the range ξ^{-1} of the field becomes 10^{26} m, cosmic in dimensions, but finite. Equation (1) is an expression of the Einstein-de Broglie theory of light[17] and implies that gauge transformations of the first and second kind[20, 21] can no longer be interpreted as implying zero photon rest mass. It is well known that the EBP equation implies mathematically[20, 21] the Lorentz condition

$$\frac{\partial A_\mu}{\partial x_\mu} = 0 \qquad (2)$$

for the massive boson. If the photon has rest mass, it is always described by the Lorentz condition. Experimental evidence[17] for finite photon rest mass implies that gauge invariance must be reinterpreted fundamentally, and this is part of the purpose of this paper, in which it is shown that finite m_0 is consistent with gauge invariance of the first and second kind if and only if

$$A_\mu A_\mu = 0$$
$$m_0 \neq 0 \qquad (3a)$$

a condition that implies

$$\phi = c|\mathbf{A}| \qquad (3b)$$

where

$$A_\mu = \left(\mathbf{A}, \frac{i}{c}\phi\right) \qquad (3c)$$

and ϕ is the scalar potential and $\mathbf{A}$ the vector potential of the de Broglie-Proca field. Condition (3a) is consistent with the Lorentz condition (2), but is inconsistent with a massless gauge such as the traditional Coulomb gauge.[1-16]

Furthermore, the notion of zero photon rest mass leads to considerable physical obscurity, for example, in the quantization of the Maxwellian electromagnetic field.[20, 21] The traditional theory abandons the longitudinal and timelike field polarizations as being "unphysical," and in so doing inevitably loses manifest covariance. Another traditional difficulty[20] is that the little group of the Poincaré group[20] for the massless photon becomes the Euclidean E(2), which is physically obscure. The Lie algebra for the Maxwellian electromagnetic field on the other hand is that of the Lorentz group. These difficulties are accepted because it is traditionally thought that special relativity implies zero photon mass, and that gauge invariance of the first and second kind can be interpreted only in terms of zero photon rest mass. In this paper it is shown that both of these traditional viewpoints are flawed, and that in consequence, the Einstein-de Broglie theory of light is consistent with both special relativity and gauge transformation. We recall for reference that the massless electromagnetic field is summarized in the d'Alembert equation:

$$\Box \, A_\mu = 0 \qquad (4)$$

Quantization[20] of Eq. (1) is straightforward, but that of Eq. (4) is beset with considerable difficulty. From quantization of Eq. (1), for the massive boson, the conclusion is reached that the massive boson is a particle (the photon) with finite mass and three physically meaningful spacelike polarizations, (1), (2), and (3). Quantization[20] of Eq. (4) traditionally proceeds in the Coulomb or Lorentz gauge. To quote from Ryder,[20] "Quantisation of the electromagnetic field suffers from difficulties posed by gauge invariance. The quantisation procedure is outlined in both the radiation (Coulomb) gauge, in which there appear only the two physical (transverse) polarisation states, and in the Lorentz gauge, in which all four polarisation states appear, the formalism being Lorentz covariant. The resulting difficulties are resolved by the method of Gupta and Bleuler." The reader is referred to Ryder[20] for an excellent account of these difficulties. The Coulomb gauge is inconsistent, furthermore, with a nonzero photon rest mass, so that, conversely, finite m_0 implies immediately that the notion of there being only two physically meaningful photon polarization states must be abandoned. One is led ineluctably to the conclusion that there are four physically meaningful photon polarizations ((0) to (3)).

Lorentz gauge quantization[20] in the limit $m_0 \to 0$ is possible only with the Gupta-Bleuler condition,[22] which leads to the conclusion that admixtures of timelike and longitudinal spacelike photon polarizations are physical states.[20] In a diametrically self-contradictory procedure, the traditional theory abandons these physical states as unphysical.

This procedure is logically untenable, and recently[23-28] this has become clear through the discovery of a simple relation between longitudinal and transverse solutions of Maxwell's equations in vacuo:

$$\mathbf{B}^{(3)} = \frac{\mathbf{E}^{(1)} \times \mathbf{E}^{(2)}}{E_0 c i} = \frac{\mathbf{B}^{(1)} \times \mathbf{B}^{(2)}}{B_0 i} = B_0 \mathbf{k} \qquad (5)$$

Equation (5) comes directly from the original Maxwell equations, without the introduction of scalar and vector potentials, and is an entirely novel relation between physically meaningful electric and magnetic components of the electromagnetic field in vacuo. It can be derived without reference to gauge theory, but is consistent with gauge invariance. Here $\mathbf{E}^{(1)}$ and $\mathbf{E}^{(2)}$ are the oscillating transverse components of the electric field, taken to be a plane wave in vacuo. The vector product in Eq. (5) is defined by the Stokes parameter S_3:

$$\mathbf{E}^{(1)} \times \mathbf{E}^{(2)} = -S_3 \mathbf{k} \qquad (6)$$

In a light beam in which there is some degree of circular polarization, therefore, S_3 is always nonzero, implying that the longitudinal magnetic field $\mathbf{B}^{(3)}$ is nonzero in vacuo. The transverse components in Eq. (5) are the usual vacuum plane wave solutions of Maxwell's equations:

$$\mathbf{E}^{(1)} = \frac{E_0}{\sqrt{2}}(\mathbf{i} - \mathbf{ij})e^{i\phi} \qquad \mathbf{E}^{(2)} = \frac{E_0}{\sqrt{2}}(\mathbf{i} + \mathbf{ij})e^{-i\phi}$$

$$\mathbf{B}^{(1)} = \frac{B_0}{\sqrt{2}}(\mathbf{ii} + \mathbf{j})e^{i\phi} \qquad \mathbf{B}^{(2)} = \frac{B_0}{\sqrt{2}}(-\mathbf{ii} + \mathbf{j})e^{-i\phi} \qquad (7)$$

where the phase is

$$\phi = \omega t - \boldsymbol{\kappa} \cdot \mathbf{r}$$

Here ω is the angular frequency at an instant t, and $\boldsymbol{\kappa}$ is the wave vector at a point **r**. It can also be shown[23] that the concommitant longitudinal electric field $\mathbf{E}^{(3)}$ exists in vacuo, and is related to $\mathbf{B}^{(3)}$ by

$$\mathbf{E}^{(3)} \times \mathbf{B}^{(2)} = \mathbf{B}^{(3)} \times \mathbf{E}^{(2)} \tag{8}$$

so that $\mathbf{E}^{(3)}$ is nonzero if $\mathbf{B}^{(3)}$ is nonzero. The imaginary $i\mathbf{E}^{(3)}$ is expressible as:

$$i\mathbf{E}^{(3)} \propto iE_0 \mathbf{k} \tag{9}$$

It is worth demonstrating explicitly that $\mathbf{B}^{(3)}$ and $\mathbf{E}^{(3)}$ are solutions in vacuo of the Maxwell equations, because

$$
\begin{array}{ll}
\nabla \times \mathbf{E}^{(3)} = 0 & -\dfrac{\partial \mathbf{B}^{(3)}}{\partial t} = 0 \\[2ex]
\nabla \times \mathbf{B}^{(3)} = 0 & \dfrac{1}{c^2}\dfrac{\partial \mathbf{E}^{(3)}}{\partial t} = 0 \\[2ex]
\nabla \cdot \mathbf{E}^{(3)} = 0 & \nabla \cdot \mathbf{B}^{(3)} = 0
\end{array}
\tag{10}
$$

These relations follow from Eqs. (5) and (9); i.e., $\mathbf{B}^{(3)}$ and $\mathbf{E}^{(3)}$ are solenoidal and phase independent.

In this paper we show that $\mathbf{B}^{(3)}$ and $\mathbf{E}^{(3)}$ are natural consequences of the Einstein-de Broglie theory of light, and are physically meaningful magnetic and electric fields. Experiments to detect them would support the theory of Einstein and de Broglie. Equations (5) and (9) are therefore relations between longitudinal and transverse field components in the massless limit of the Einstein-de Broglie theory. This conclusion is consistent with the recent development[24] by the present author of manifestly covariant electrodynamics, using electric and magnetic four vectors. This development is equivalent to the Einstein-de Broglie theory in the massless limit ($m_0 \to 0$), and is a direct consequence of the existence of $\mathbf{B}^{(3)}$ and $\mathbf{E}^{(3)}$ defined by Eqs. (5) and (9), respectively. It is impossible to reconcile the existence of Eqs. (5) and (9) with traditional thinking, in which $\mathbf{B}^{(3)}$ and $\mathbf{E}^{(3)}$ are abandoned as unphysical. Clearly, $\mathbf{B}^{(3)}$ and $\mathbf{E}^{(3)}$ are formed from physical quantities such as the Stokes parameter S_3. In the Einstein-de Broglie theory, on the other hand, $\mathbf{B}^{(3)}$ and $\mathbf{E}^{(3)}$ are physical fields, components of the four vectors E_μ and B_μ in vacuo. A longitudinal solution of Eq. (1) for $\mathbf{B}^{(3)}$ is given in Section II, where it is shown that $\mathbf{B}^{(3)}$ is an exponentially decaying function of ξ in the propagation axis Z of the light beam. The divergence of $\mathbf{B}^{(3)}$ is nonzero for finite m_0, and is given by $-\xi B^{(0)}$, a magnetic monopole in vacuo. The numerical value of ξ (10^{-26} m^{-1}) is so small that for all practical purposes, and for laboratory dimensions, $\mathbf{B}^{(3)}$ is a constant magnetic field, independent of distance and time. Section III derives general solutions of Eq. (1) for the transverse and longitudinal fields of the electromagnetic plane wave in vacuo. A discussion follows of the role of $\mathbf{B}^{(3)}$ and $\mathbf{E}^{(3)}$ in various experimental tests of the Einstein-de Broglie theory of light, taking into account experimental evidence[17] for finite photon mass.

II. LONGITUDINAL SOLUTIONS OF THE EBP EQUATION IN VACUO

In quantum optics interpreted by Einstein and de Broglie[17] light is constituted by real Maxwellian waves which coexist in spacetime with moving particles—photons. In the Copenhagen interpretation of Bohr, Schrödinger, Pauli, Glauber, and others, on the other hand, light is made up of waves of probability, which cannot coexist in spacetime with photons. In the interpretation of the Einstein-de Broglie school, the photon is massive; in that of the Copenhagen school, it is not necessarily so. The basic electrodynamical equations are therefore (1) and (4), respectively. Although it is frequently asserted[1–16, 20, 21] that the photon is massless in its rest frame, there is no supporting experimental evidence. Indeed, it appears to be impossible to test the hypothesis of zero m_0, because it is impossible to test the implication that the range of electromagnetic radiation is infinite. On the other hand, finite m_0 leads[17] to such observable implications as anomalous red shifts, reported on numerous occasions, and tired light phenomena. Einstein,[18] some years after his theory of special relativity (1905), and during his development of general relativity, proposed that the photon's rest mass is finite, i.e., that the mass of the photon is finite in a frame of reference moving at the speed of light. This leads[17–19] to Eq. (1). It is clear therefore that Einstein saw no contradiction with special relativity in his proposal; i.e., Eq. (1) is Lorentz covariant, even though the photon rest mass, m_0, is nonzero. Several conclusions flow immediately from this proposal.

Firstly, the notion that the photon is massless in the frame of the observer (laboratory frame) because it travels at the speed of light is incorrect if the photon rest mass m_0 is nonzero. In the contemporary description[20] of special relativity, the reason for this is that the quantity

$$C = P_\mu P_\mu \tag{11}$$

is the first (or "mass") Casimir invariant of the Poincaré (inhomogeneous Lorentz) group. Here P_μ is the generator of spacetime translations, first introduced by Wigner in 1939.[29] A spacetime translation is defined by the operation

$$x'_\mu = x_\mu + a_\mu \tag{12}$$

where x_μ is the distance/time four vector of Minkowski spacetime. P_μ does not appear in the homogeneous Lorentz group,[20] i.e., in a group made up only of boost transformations and Lorentz rotations. The quantity m_0^2 (the square of the rest mass) is therefore invariant to Lorentz transformations, i.e., is the same in the rest frame of the photon (which travels at the speed of light) and in the observer frame. The invariant m_0^2 appears in Eq. (1), which is Lorentz covariant, i.e., fully consistent with special relativity. The latter theory does not imply, therefore, that the photon rest mass is zero. It is clear that Einstein himself[17, 18] saw no inconsistency with special relativity in his proposal of finite m_0, and contemporary theory also shows that m_0^2 is an invariant of the Poincaré group. The Einstein-de Broglie theory of light is therefore consistent with special relativity. This means that the rest frame momentum of the photon (a massive boson) is timelike, not lightlike, and that the photon has rest energy $m_0 c^2$, i.e., that the energy of the photon in its own frame of reference, which moves at the speed of light, is $m_0 c^2$, about 10^{-57} J. The spacelike momentum of the photon in its own rest frame is zero, because it does not move relative to this rest frame. In its rest frame, the photon is thus described by a four vector:

$$
\begin{aligned}
q_\mu &= (0,0,0,im_0c) \\
&= \left(0,0,0,i\frac{En_0}{c}\right)
\end{aligned}
\tag{13}
$$

in Minkowski spacetime. In the laboratory frame of the observer, however, the photon's momentum is finite, and the vector (13) is transformed into

$$p_\mu = L_{\mu\nu}q_\nu \tag{14}$$

where $L_{\mu\nu}$ denotes a Lorentz transformation[20] which transforms q_ν into p_μ. Clearly, the latter is observed in the laboratory. Wigner[29] showed that this transformation can be described from a knowledge of the rotation group, and that the little group for q_μ is a rotation group.

As discussed by Vigier[17] the consequences are that photons slow down in the laboratory frame of an observer, although the rest frame must move at the speed of light, which is a universal constant of special relativity.

Photons in the frame of the observer behave like relativistic nonzero mass particles, with rest mass $m_0 \doteq 10^{-68}$ kg. The energy momentum four vector in the observer frame is p_μ, with components[17]

$$p_\mu \equiv \left(\mathbf{p}, \mathrm{i}\frac{En}{c} \right)$$

$$En = h\nu = m_0 c^2 \left(1 - \frac{v^2}{c^2} \right)^{-1/2} \tag{15}$$

$$|\mathbf{p}| \doteq \frac{h\nu}{v} \doteq \frac{h\nu}{c}$$

The velocity of the photon in the observer frame is therefore not c, but v, defined from the Guiding Theorem of de Broglie, the basis of wave mechanics:

$$En_0 = h\nu_0 = m_0 c^2 \tag{16}$$

In other words, the energy of the photon in the rest frame is

$$En_0 = m_0 c^2 = h\nu_0 \tag{17}$$

and its energy in the observer frame is

$$En = m_0 c^2 \left(1 - \frac{v^2}{c^2} \right)^{-1/2} \tag{18}$$
$$= h\nu$$

so that there is a change in the frequency of light from one frame to the other. This is the origin of *observed* distance proportional shifts,[17] the "tired light" of Hubble and Tolman. There are photons, therefore, that move at low velocities and contribute to the mass of the universe. Clearly, this is a consequence of the fact that the field has a finite range, of about 10^{26} m, as discussed in the introduction. This conclusion does not contradict the principle of conservation of energy, because in special relativity, the quantity $P_\mu P_\mu$ is invariant to Lorentz transformation. Therefore, special relativity does not imply that the rest mass of the photon is zero, as in the traditional interpretations.[1-16, 20, 21]

Secondly, if m_0 is not zero, the traditional interpretation[20, 21] of gauge transformations must be revised fundamentally, because it leads to the conclusion that the photon rest mass m_0 is zero and therefore contradicts the Einstein-de Broglie theory and experimental evidence[17] for finite photon mass. Traditional considerations of gauge transformations also lead to the principle of gauge invariance (eicheninvarianz prinzip), which holds if and only if the photon mass is identically zero. For these reasons, we consider carefully the basic Lagrangian formalism of gauge theory, and modify its interpretation to make it consistent with finite m_0. The result of our considerations is Eq. (3a) of the introduction.

Geometrically, a gauge transformation of the first kind[20, 21] is a rotation in the $(1, 2)$ plane of the "vector" field

$$\phi = \phi_1 \mathbf{i} + \phi_2 \mathbf{j} \tag{19}$$

through an angle Λ. Under such a rotation, Noether's theorem leads to conserved charge Q in a volume V

$$Q = i \int \left(\phi^* \frac{\partial \phi}{\partial t} - \phi \frac{\partial \phi^*}{\partial t} \right) dV \tag{20}$$

and a conserved current

$$J_\mu = i \left(\phi^* \frac{\partial \phi}{\partial x_\mu} - \phi \frac{\partial \phi^*}{\partial x_\mu} \right) \tag{21}$$

The existence of Q and J_μ is based on the invariance of action. When the action is real, the Lagrangian is[20]

$$\mathscr{L} = \left(\frac{\partial \phi}{\partial x_\mu} \right) \left(\frac{\partial \phi^*}{\partial x_\mu} \right) - m^2 \phi^* \phi \tag{22}$$

where m is a mass associated with the complex field ϕ, defined by

$$\phi = \frac{\phi_1 + i\phi_2}{\sqrt{2}}$$
$$\phi^* = \frac{\phi_1 - i\phi_2}{\sqrt{2}} \tag{23}$$

Since Λ is a constant (an angle in $(1, 2)$) the gauge transformation of the first kind, which can be expressed[20] as

$$\phi \rightarrow e^{-i\Lambda}\phi \qquad \phi^* \rightarrow e^{i\Lambda}\phi \qquad (24)$$

is the same at all points in spacetime, so that at an instant t the same rotation occurs for all points in space. This contradicts special relativity[20] whose universal constant is the speed of light, and which implies that action at a distance is impossible. Electrodynamics cannot, therefore, be invariant to a gauge transformation of the first kind. To comply with special relativity, Λ is made an arbitrary function of spacetime:

$$\Lambda \equiv \Lambda(x_\mu) \qquad (25)$$

so defining a gauge transformation of the second kind. For $\Lambda \ll 1$, electrodynamics is invariant to the gauge transformation of the second kind:

$$\phi \rightarrow \phi - i\Lambda(x_\mu)\phi \qquad (26)$$

Condition (25) implies,[20] however, that $\partial\phi/\partial x_\mu$ does not transform in the same way as ϕ, i.e., does not transform covariantly, so that the action is no longer invariant[20, 21]:

$$\delta\mathscr{L} = J_\mu \frac{\partial\Lambda}{\partial x_\mu} \neq 0 \qquad (27)$$

with $\mathscr{L}$ defined by Eq. (22). To preserve the invariance of action under (26), the potential four vector A_μ is introduced through

$$\mathscr{L}_1 = -eJ_\mu A_\mu \qquad (28)$$

This implies the need for two more conditions[20]:

$$A_\mu \rightarrow A_\mu + \frac{1}{e}\frac{\partial\Lambda}{\partial x_\mu} \qquad (29)$$

and

$$\mathscr{L}_2 \equiv e^2 A_\mu A_\mu \phi^* \phi \qquad (30)$$

Equations (28) to (30) imply[20]

$$\delta\mathscr{L} + \delta\mathscr{L}_1 + \delta\mathscr{L}_2 = 0 \tag{31}$$

In fundamental gauge theory, therefore, A_μ of the conventional d'Alembert equation is introduced to produce Eq. (31) in association with the extra term (30). So far, nothing has been said about the need for zero mass. We note that if $A_\mu A_\mu = 0$, $\mathscr{L}_2$ is automatically zero.

The field A_μ itself makes a contribution to the Lagrangian, implying the need for an additional $\mathscr{L}_3$ to maintain a zero overall action[20]:

$$\mathscr{L}_3 \equiv -\tfrac{1}{4}F_{\mu\nu}F_{\mu\nu} \tag{32}$$

where

$$F_{\mu\nu} = \frac{\partial A_\nu}{\partial x_\mu} - \frac{\partial A_\mu}{\partial x_\nu} \tag{33}$$

the four curl of A_μ, is the electromagnetic field four tensor,[20] an invariant under (29). The complete Lagrangian is therefore

$$\mathscr{L}_{\text{tot}} = \mathscr{L} + \mathscr{L}_1 + \mathscr{L}_2 + \mathscr{L}_3 \tag{34}$$

If the mass, m_0, associated with the electromagnetic field is not zero, then the form of the Lagrangian is changed from (32) to

$$\mathscr{L}_4 = -\tfrac{1}{4}F_{\mu\nu}F_{\mu\nu} + \tfrac{1}{2}m_0^2 A_\mu A_\mu \tag{35}$$

and this is invariant to Eq. (29) if and only if

$$m_0^2 A_\mu A_\mu = 0 \tag{36}$$

If $m_0 \neq 0$, then

$$A_\mu A_\mu = 0 \qquad A_\mu \neq 0 \tag{37}$$

is the only alternative possible, as described in the introduction. Conventionally, it is asserted[20, 21] that the invariance of $\mathscr{L}_4$ under (29) means that $m_0 = 0$. However, in the Einstein-de Broglie theory, Eq. (37) is consistent with Eq. (29), and m_0 is quantized as the photon rest mass. Equation (37) is also consistent with Eq. (31) of fundamental gauge theory, because[20] $\mathscr{L}_2 = 0$ if $A_\mu A_\mu = 0$. This implies that

$$\delta\mathscr{L}_2 = 2eA_\mu\left(\frac{\partial\Lambda}{\partial x_\mu}\right)\phi^*\phi = 0 \tag{38}$$

so that

$$\delta\mathscr{L} + \delta\mathscr{L}_1 = -\delta\mathscr{L}_2$$

$$= -2eA_\mu\left(\frac{\partial\Lambda}{\partial x_\mu}\right)\phi^*\phi \tag{39}$$

$$= 0$$

i.e., the action is conserved as in Eq. (31). The condition (37) for finite m_0 is one in which the EBP equation is invariant to the gauge transformation (29), which is implied by the need to conserve action under the gauge transformation of the second kind, Eq. (26). We therefore conclude that gauge theory does not imply that photon rest mass is zero.

If $m_0 \not= 0$, the quantity $A_\mu A_\mu$ vanishes, implying that $\phi = c|A|$ where $A_\mu \equiv (A, i\phi/c)$. This condition is furthermore consistent with Eqs. (1) and (2), which is

$$\nabla\cdot A + \frac{1}{c^2}\frac{\partial\phi}{\partial t} = 0 \tag{40}$$

that is,

$$\nabla\cdot A = -\frac{1}{c}\frac{\partial A}{\partial t} \tag{41}$$

Additionally, using the Lorentz condition, Eq. (40),

$$\left(\nabla^2 - \frac{1}{c^2}\frac{\partial^2}{\partial t^2}\right)A = 0 \tag{42a}$$

$$\left(\nabla^2 - \frac{1}{c^2}\frac{\partial^2}{\partial t^2}\right)\phi = 0 \tag{42b}$$

whose solutions are the Liénard-Wiechert potentials.[1-16, 20] Clearly, Eqs. (42a) and (42b) become the same if $\phi = c|A|$ in S.I. units.

Fundamental gauge theory does not imply that the photon rest mass must be zero, contrary to much of the current literature.[1-16, 20, 21] Secondly, special relativity, as pointed out by Einstein[17, 18] also does not imply zero photon rest mass. Thirdly, there is experimental evidence,[17] for nonzero m_0, and none for zero photon rest mass. Fourthly, the transverse, radiation, or Coulomb gauge[1-16, 20] is inconsistent with $\phi = c|\mathbf{A}|$, because in that gauge $\phi = 0$, $\mathbf{A} \neq 0$. The Lorentz gauge and Dirac gauge[17] are, on the other hand, consistent with $m_0 \neq 0$.

Having argued in some detail in this way, it becomes easy to see that much of the obscurity in the current thought on electromagnetism is due to the notion that the photon is massless and travels at the speed of light. Both statements contradict experimental evidence.[17] These notions result in "too much gauge freedom," in that Eqs. (26) and (29) can be satisfied with $m_0 = 0$ by the Coulomb, Lorentz, and other gauges. For $m_0 \neq 0$, as in the Einstein-de Broglie theory, the Coulomb gauge is invalidated, but the Lorentz gauge is a direct mathematical consequence of the EBP equation (1). The excess gauge freedom for $m_0 = 0$ results in severe[20, 21] difficulties of quantization of the electromagnetic field, whereas quantization of the EBP equation (a wave equation) is straightforward,[20] leading to longitudinal, physically meaningful, spacelike photon polarization, as well as the two transverse spacelike polarizations. It is natural to expect that a particle, the photon, should have three spacelike polarizations in three physical dimensions, X, Y, and Z.

The $m_0 = 0$ assertion is conventionally associated with the notion that the electromagnetic field is a massless gauge field with two independent components, customarily identified with left and right circular polarization. However, even in the limit $m_0 = 0$, the same Maxwellian field is covariantly described by the four components of A_μ. The Bohm-Aharonov effect[20] shows that A_μ is physically meaningful. Recent work,[23-28] leading to Eq. (5), shows conclusively that there is a well-defined relation between the transverse ((1) and (2)) and longitudinal ((3)) components of solutions of Maxwell's field equations in vacuo. It is straightforward to show that the three magnetic field components form a classical cyclic permutation in the circular basis, (1), (2), and (3): with $B^{(0)} = B_0$

$$\mathbf{B}^{(1)} \times \mathbf{B}^{(2)} = i B^{(0)} \mathbf{B}^{(3)*} \qquad \mathbf{B}^{(3)*} = \mathbf{B}^{(3)} \qquad (43a)$$

$$\mathbf{B}^{(2)} \times \mathbf{B}^{(3)} = i B^{(0)} \mathbf{B}^{(1)*} \qquad \mathbf{B}^{(1)*} = \mathbf{B}^{(2)} \qquad (43b)$$

$$\mathbf{B}^{(3)} \times \mathbf{B}^{(1)} = i B^{(0)} \mathbf{B}^{(2)*} \qquad \mathbf{B}^{(2)*} = \mathbf{B}^{(1)} \qquad (43c)$$

Furthermore, there exist classical permutations involving $\mathbf{E}^{(3)}$. If we assert $\mathbf{E}^{(3)} \equiv E^{(0)}\mathbf{k}$, these are, algebraically,

$$\mathbf{E}^{(1)} \times \mathbf{E}^{(2)} = iE^{(0)}c\mathbf{B}^{(3)*} \tag{44}$$

$$\mathbf{E}^{(2)} \times \mathbf{E}^{(3)} = -E^{(0)}c\mathbf{B}^{(1)*}$$
$$\mathbf{E}^{(3)} \times \mathbf{E}^{(1)} = E^{(0)}c\mathbf{B}^{(2)*}$$
$$\mathbf{E}^{(1)} \times \mathbf{B}^{(2)} = B^{(0)}\mathbf{E}^{(3)*} \tag{45}$$
$$\mathbf{E}^{(2)} \times \mathbf{B}^{(3)} = iB^{(0)}\mathbf{E}^{(1)*}$$
$$\mathbf{E}^{(3)} \times \mathbf{B}^{(1)} = -B^{(0)}\mathbf{E}^{(2)*}$$

$$\mathbf{E}^{(1)} \times \mathbf{B}^{(1)} = 0$$
$$\mathbf{E}^{(2)} \times \mathbf{B}^{(2)} = 0 \tag{46}$$
$$\mathbf{E}^{(3)} \times \mathbf{B}^{(3)} = 0$$

and are reminiscent of the Lie algebra of the Lorentz group,[20] a classical commutator algebra that is built up with boost and rotation generators defined in Minkowski spacetime. However, all the eqns. (45) violate $\hat{T}$ symmetry, which is a consequence of the fact that $\mathbf{E}^{(3)}$ is imaginary and cannot be derived from transverse solutions of Maxwell's equations. Eqns. (45) are not valid equations of electrodynamics while eqn. (44) is valid and identical with eqn. (43a). This does not mean that $\mathbf{E}^{(3)}$ itself violates $\hat{T}$ symmetry.

Before proceeding to the derivation of $\mathbf{B}^{(3)}$ for nonzero m_0, the purpose of this section, it is demonstrated that Lie algebra also applies to the electric and magnetic components of electromagnetic radiation in vacuo (the Maxwellian field) provided that these components are defined as classical field operators directly proportional respectively to the boost and rotation generators of the Lorentz transformation. This is a mathematical demonstration of the fact that if the longitudinal spacelike components of these fields are unphysical (i.e., zero), then the Lie algebraic structure of the Lorentz group is contradicted. This means that the Lorentz transformation itself is incorrectly defined, in that the longitudinal (Z) boost and rotation generator components are incorrectly asserted to be zero. This is equivalent to destroying the geometrical structure of Minkowski spacetime. Even in the Maxwellian limit $m_0 \rightarrow 0$, therefore, the assertion that $\mathbf{B}^{(3)}$ and $\mathbf{E}^{(3)}$ are zero results in a mathematical reductio ad absurdum.

That the Maxwell equations in vacuo are the Lorentz covariant equations[20, 21]:

$$\frac{\partial F_{\mu\nu}}{\partial x_\mu} = 0 \qquad \frac{\partial \tilde{F}_{\mu\nu}}{\partial x_\mu} = 0 \tag{47}$$

where $\tilde{F}_{\mu\nu}$ is the dual of $F_{\mu\nu}$, the electromagnetic field four tensor. The latter is antisymmetric under Lorentz transformation and its structure can be displayed as

$$F_{\mu\nu} \equiv \begin{bmatrix} 0 & -E_1 & -E_2 & -E_3 \\ E_1 & 0 & -cB_3 & cB_2 \\ E_2 & cB_3 & 0 & -cB_1 \\ E_3 & -cB_2 & cB_1 & 0 \end{bmatrix} \tag{48}$$

We note that this structure is identical with that of the Lie algebra of the Lorentz group, defined[20] by the dimensionless, boost generator $\hat{K}_i$, an operator, and the rotation generator $\hat{J}_i$, also a dimensionless operator. The Lie algebra of the Lorentz group can be displayed as

$$\hat{J}_{\mu\nu} = \begin{bmatrix} 0 & \hat{K}_1 & \hat{K}_2 & \hat{K}_3 \\ -\hat{K}_1 & 0 & \hat{J}_3 & -\hat{J}_2 \\ -\hat{K}_2 & -\hat{J}_3 & 0 & \hat{J}_1 \\ -\hat{K}_3 & \hat{J}_2 & -\hat{J}_1 & 0 \end{bmatrix} \tag{49}$$

i.e., as

$$\hat{J}_{\mu\nu}(\mu, \nu = 0, \dots, 3) \begin{bmatrix} \hat{J}_{ij} = -\hat{J}_{ji} = \varepsilon_{ijk}, \hat{J}_k \\ \hat{J}_{i0} = -\hat{J}_{0i} = -\hat{K}_i \end{bmatrix} \tag{50}$$

$$(i, j, k = 1, 2, 3)$$

Equations (49) and (50) are condensed representations of the classical commutator (Lie) algebra of the Lorentz group[20]:

$$\left[\hat{J}_X, \hat{J}_Y \right] = i\hat{J}_Z \text{ and cyclic permutations} \tag{51a}$$

$$\left[\hat{K}_X, \hat{K}_Y \right] = -i\hat{J}_Z \text{ and cyclic permutations} \tag{51b}$$

$$\left[\hat{K}_X, \hat{J}_Y \right] = i\hat{K}_Z \text{ and cyclic permutations} \tag{51c}$$

$$\left[\hat{K}_X, \hat{J}_X \right] = 0 \text{ etc.} \tag{51d}$$

The geometrical equivalence of (48) and (49) means that

$$\begin{aligned} \hat{E}_i &= E_0 \hat{K}_i \\ \hat{B}_i &= B_0 \hat{J}_i \end{aligned} \tag{52}$$

where $\hat{E}_i$ and $\hat{B}_i$ are classical electric and magnetic field operators, a result that is implied by the proportionality of the classical operator matrices $\tilde{F}_{\mu\nu}$ and $\hat{J}_{\mu\nu}$. In the Cartesian basis (X, Y, Z),

$$\left[\hat{B}_X, \hat{B}_Y\right] = iB_0\hat{B}_Z \text{ and cyclic permutations} \tag{53a}$$

$$\left[\hat{E}_X, \hat{E}_Y\right] = -icB_0\hat{B}_Z \text{ and cyclic permutations} \tag{53b}$$

$$\left[\hat{E}_X, \hat{B}_Y\right] = iB_0\hat{E}_Z \text{ and cyclic permutations} \tag{53c}$$

$$\left[\hat{E}_X, \hat{B}_X\right] = 0 \text{ etc.} \tag{53d}$$

Equations (53) represent a classical operator equivalent of the vector products in Eqs. (36), where the Maxwellian fields are vectors in space, and not operators defined in spacetime.

The ansatz (52) is based on the fundamental Lie algebraic structure of Minkowski spacetime, and implies the following:

1. The classical electric field operator $\hat{E}_i$ is proportional to a boost generator, and the classical magnetic field operator $\hat{B}_i$ is proportional to a rotation generator in Minkowski spacetime.

2. If the longitudinal component operators $\hat{B}_Z$ and $\hat{E}_Z$ are asserted to be zero, or unphysical, the structure of the Lie algebra is destroyed in Eqs. (53). For example, if $\hat{B}_Z = \hat{E}_Z = \hat{0}$, $[\hat{B}_X, \hat{B}_Y] \doteq \hat{0}$ and from the structure of $\hat{J}_i$ in Eq. (52), this is mathematically incorrect. Explicitly,

$$\left[\hat{B}_X, \hat{B}_Y\right] = B_0^2\left[\hat{J}_X, \hat{J}_Y\right] = B_0^2\hat{J}_Z \neq 0 \tag{54}$$

because[20]

$$\hat{J}_X \equiv -i\begin{pmatrix} 0 & 0 & 0 & 0 \\ 0 & 0 & 0 & 0 \\ 0 & 0 & 0 & 1 \\ 0 & 0 & -1 & 0 \end{pmatrix} \tag{55a}$$

$$\hat{J}_Y \equiv -i\begin{pmatrix} 0 & 0 & 0 & 0 \\ 0 & 0 & 0 & -1 \\ 0 & 0 & 0 & 0 \\ 0 & 1 & 0 & 0 \end{pmatrix} \tag{55b}$$

$$\hat{J}_Z \equiv -i\begin{pmatrix} 0 & 0 & 0 & 0 \\ 0 & 0 & 1 & 0 \\ 0 & -1 & 0 & 0 \\ 0 & 0 & 0 & 0 \end{pmatrix} \tag{55c}$$

3. The Maxwell equations (47) are seen to be relations between boost and rotation generators defined in spacetime:

$$\frac{\partial J_{\mu\nu}}{\partial x_\mu} = 0 \qquad \frac{\partial \tilde{J}_{\mu\nu}}{\partial x_\mu} = 0 \qquad (56)$$

and are thus given a precise geometrical interpretation. In this light, it is seen that the d'Alembert equation (4) is also geometrical in nature:

$$\Box \, \hat{L}_\mu = 0 \qquad (57)$$

where $\hat{J}_{\mu\nu}$ is the four curl of $\hat{L}_\mu$:

$$\hat{J}_{\mu\nu} = \frac{\partial \hat{L}_\nu}{\partial x_\mu} - \frac{\partial \hat{L}_\mu}{\partial x_\nu} \qquad (58)$$

4. It may be seen precisely that the conventional notion that the Maxwellian $\hat{B}_Z$ (and $\hat{E}_Z$) is unphysical is equivalent to the geometrically incorrect assertion

$$\hat{J}_Z = \begin{pmatrix} 0 & 0 & 0 & 0 \\ 0 & 0 & 0 & 0 \\ 0 & 0 & 0 & 0 \\ 0 & 0 & 0 & 0 \end{pmatrix} \qquad (59)$$

which by implication habitually[1-16, 20, 21] replaces the correct rotation generator (55c).

There is of course no experimental evidence for Eq. (59), even in the massless limit $m_0 \to 0$ conventionally associated with the Maxwellian field.

Our geometrical interpretation of the Maxwell field equations is a direct logical consequence of the geometry of Minkowski spacetime itself and of the theory of special relativity. This is consistent with the fact that Einstein's considerations of the Maxwell equations led to his formulation of special relativity. If it is asserted that longitudinal solutions of Maxwell's equations be unphysical, special relativity is contradicted and the structure of the Lorentz group and its associated Lie algebra is destroyed. There is no experimental evidence whatsoever that the longitudinal solutions of Maxwell's equations in vacuo are unphysical, and there is no evidence for $m_0 = 0$.

The commutator relations (51a) and (53a) lead to a method of quantization of the Maxwellian field simply by noting the ordinary angular momentum commutator relations of quantum mechanics. In Cartesian terms,

$$\left[\hat{J}_X, \hat{J}_Y\right] = i\hbar \hat{J}_Z \tag{60}$$

are structurally identical with Eq. (51a) except for $\hbar$ (which has the units of angular momentum). In quantum mechanics, the $\hat{J}$ operators in Eq. (60) are angular momentum operators. Quantized angular momentum is therefore a consequence of the classical rotation generator,[20] as is well known. The quantized equivalent of Eq. (53a) must therefore be

$$\left[\hat{B}_X, \hat{B}_Y\right] = i\hbar \left(\frac{B_0}{\hbar}\right) \hat{B}_Z \tag{61}$$

to balance units, symmetries, and dimensions on the left and right sides. This implies

$$\hat{B} = B^{(0)} \frac{\hat{J}}{\hbar} \tag{62}$$

which is identical with the result obtained recently by the present author[25] using an independent method of derivation. Therefore,

$$\hat{B}_Z = B^{(0)} \frac{\hat{J}_Z}{\hbar} \tag{63}$$

is the elementary longitudinal component of the quantized Maxwellian magnetic field in vacuo. In the same way that $\hbar$ is the archetypical elementary quantum of angular momentum, $B^{(0)}$ is the elementary quantum of magnetic flux density of the Maxwellian field in vacuo.

The eigenvalues of $\hat{J}_Z$ in Eq. (63) may be identified with those of a massless boson (the "conventional" photon), i.e., $\hbar M_J$, where $M_J = \pm 1$, so that the classical limit of Eq. (63) is

$$\mathbf{B}_Z = B^{(0)} \mathbf{k} \tag{64}$$

which is Eq. (5) in Cartesian terms instead of a circular basis. Equation (5) is therefore geometrically consistent with the Lie algebra of the Lorentz group. The generalization of our development to $m_0 \not= 0$ is now straightforward.

Having considered in some detail the geometrical structure of the Lorentz group, we revert to a simpler development of the EBP equation (1), solving it as a classical eigenvalue equation with the differential operator:

$$\Box \equiv -\nabla^2 + \frac{1}{c^2}\frac{\partial^2}{\partial t^2} \tag{65}$$

The order of magnitude of ξ is such that

$$\Box A_\mu \doteq 10^{-52} A_\mu \tag{66}$$

which closely approximates the d'Alembert equation (4). It is clear, therefore, that the classical interpretation of the EBP field closely approximates the Maxwellian field. However, in the EBP field, the Coulomb gauge is inconsistent with Eq. (66), which must be written in terms of the spacelike **A** as

$$\left(\nabla^2 - \frac{1}{c^2}\frac{\partial^2}{\partial t^2}\right)\mathbf{A} = \xi^2\mathbf{A} \tag{67}$$

with $\phi = c|\mathbf{A}|$. In the Galilean limit this equation becomes

$$\nabla^2\mathbf{A} = \xi^2\mathbf{A} \tag{68}$$

Using the relation

$$\mathbf{B} = \nabla \times \mathbf{A} \tag{69}$$

it can be seen that the equation

$$\nabla^2\mathbf{B} = \xi^2\mathbf{B} \tag{70}$$

is the same as Eq. (68), because

$$\nabla^2(\nabla \times \mathbf{A}) = \xi^2\nabla \times \mathbf{A} \tag{71a}$$

$$\nabla \times \nabla^2\mathbf{A} = \nabla \times \xi^2\mathbf{A} \tag{71b}$$

In considering the Galilean limit, we have removed the time dependence in the solution for **B** of Eq. (70). Furthermore, since

$$\nabla^2\mathbf{B} = 10^{-52}\mathbf{B} \sim 0 \tag{72}$$

describes the magnetic component in vacuo of an electromagnetic field closely resembling the Maxwellian field, we know that the time-independent solution to Eq. (70) must be the longitudinal component, defined in the propagation axis Z. The solution to Eq. (70) in Cartesian terms is therefore

$$\mathbf{B} = B^{(0)} \exp(-\xi Z)\mathbf{k}$$
$$|\mathbf{B}| = B_Z \tag{73}$$

and since $\xi \sim 10^{-26}$ m^{-1}, this is for all practical purposes identical with Eq. (64) of the Maxwellian field. Several physical consequences follow from Eqs. (64) and (73):

1. The longitudinal solution for $\mathbf{B}$ of the EBP field, Eq. (73), is for all practical purposes identical with the corresponding Maxwellian solution, Eq. (64). By the caveat "for all practical purposes" we imply laboratory dimensions and time scales. On a cosmic scale, in which $Z \sim 1/\xi$, Eq. (73) is different from Eq. (64) in general. In the "tired light" terminology of Hubble,[17] $\mathbf{B}$ becomes a "tired field" if Z is big enough (ca. 10^{26} m).

2. Physically meaningful, practically identical, and longitudinal solutions exist for $\mathbf{B}$ from the EBP and Maxwell equations, the former being considered as a classical wave equation. To assert $\mathbf{B} = \mathbf{0}$ in Eq. (64) is mathematically incorrect in the Maxwellian field, because it corresponds to the assertion (59) in spacetime. For all practical purposes, therefore, this assertion is incorrect in the EBP field. Quantization of the EBP field[20] confirms this conclusion, leading to a physically meaningful longitudinal photon polarization.

3. Since the EBP and Maxwellian fields are practically (i.e., in the laboratory) identical, the EBP field obeys the various commutator relations of this paper for all practical purposes, and the transverse EBP solutions are practically those of Eqs. (7). In the cosmology of light from distant sources, however, this simple classical interpretation is no longer tenable.

Quantization of the EBP field is straightforward,[20] whereas that of the Maxwellian field is obscure. Although the rest mass m_0 of the photon is very small, it is essential that it be rigorously nonzero to maintain a logical and self-consistent, physically meaningful, structure for the quantized electromagnetic field in vacuo. If this is done, quantization results in a consistent particle interpretation[20] in terms of a massive boson, with eigenvalues $M_J \hbar$, $M_J = -1, 0, +1$. The three polarization vectors of the quantized EBP field are orthonormal and spacelike; i.e., there are physically meaningful longitudinal and transverse components. The little group of Wigner[29] is a physically meaningful rotation group, utilizing the three dimensions of space. If $m_0 = 0$, on the other hand, the constraint $A_\mu A_\mu = 0$ is conventionally lost, resulting in "too much gauge freedom." The two Casimir invariants[20] of the Poincaré group vanish for $m_0 = 0$, meaning that physical quantities that are invariant under the most general type of Lorentz transformation must vanish identically for the massless gauge field. This implies $A_\mu A_\mu = 0$, if $A_\mu A_\mu$ is to be an invariant of the Poincaré group, diametrically contradicting the conventional use of gauge freedom for a massless particle, i.e., contradicting the conventional assertion that $A_\mu A_\mu \neq 0$ for $m_0 = 0$. Thus, the conventional assertion $A_\mu A_\mu \neq 0$ for $m_0 = 0$ is geometrically unsound, i.e., contradicts the geometry of Minkowski spacetime, a geometry that requires $A_\mu A_\mu$ to be an invariant of the Poincaré (inhomogeneous Lorentz) group. We are forced to conclude that the widespread use of the Coulomb gauge, in which $A_\mu A_\mu = 0$, is relativistically incorrect. The conventional assertion that m_0 must be zero because $A_\mu A_\mu$ is nonzero is also basically incorrect, because $A_\mu A_\mu$ is always zero in vacuo.

It is the habitual use of the Coulomb (or "transverse") gauge that more than any other factor leads to the conventional assertion that the electromagnetic field can have no longitudinal solution that is physically meaningful. The Coulomb gauge is relativistically incorrect, and is inconsistent with finite photon rest mass, for which there is experimental evidence.[17] The widespread use of the Coulomb gauge[1-16, 20, 21] should therefore be viewed with caution. It is obvious that quantization in the Coulomb gauge cannot be consistent with special relativity, because its use is equivalent to the incorrect assertion (59). These difficulties are frequently compounded in the literature by a series of misstatements, traceable to the relativistically incorrect assertion $A_\mu A_\mu \neq 0$. For example, it is frequently asserted that the Lorentz gauge does not define A_μ uniquely. This is true if and only if $A_\mu A_\mu \neq 0$. If $A_\mu A_\mu = 0$, then the Lorentz condition defines A_μ uniquely. Quantization in the Coulomb gauge is therefore a mathematically incorrect procedure, and we discard its results as meaningless. In other words it is meaningless to assert that the Maxwellian field has only two transverse polarizations.

Quantization of the Maxwellian field in the Lorentz gauge[20] retains manifest covariance, but is physically obscure. It also relies on the notion that the gauge field is massless, so that quantization of the field must lead to a massless photon. In consequence, the internally inconsistent notion $A_\mu A_\mu \neq 0$ is habitually retained in the Lorentz gauge. This immediately leads to the difficulty that the Lagrangian has to be modified with a gauge fixing term, a procedure that leads to a non-Maxwellian equation of motion.[20] Even with this artifice, the conjugate momentum field $\boldsymbol{\Pi}^0$ vanishes,[20] and the traditional method is forced to assert that the Lorentz condition, within whose framework the method is developed, cannot hold as an operator identity. This difficulty is habitually resolved by the method of Gupta and Bleuler,[20] a method that results in the conclusion that admixtures of timelike and longitudinal spacelike photon polarizations are physical states.[20] Despite this conclusion, these states are abandoned as unphysical in order to comply with the results of Coulomb gauge quantization, which, as we have just seen, are incorrect. Quantization of the Maxwellian field, regarded as a massless gauge field, is therefore inconsistent and physically obscure.

In considerations of the Poincaré group, the notion of a massless gauge field, habitually associated with the Maxwellian field, leads to the little group[20, 29] E(2), the Euclidean group of rotations and translations in a plane. The physical significance of this little group is obscure.[20] Its Lie algebra does not correspond to that of a rotation group, but it is the group that is needed to maintain a lightlike vector invariant under the most general Lorentz transformation. This suggests that the notion of a massless field is physically meaningless. The traditional line of reasoning, however, considers a massless particle traveling in the propagation axis (Z) described by a lightlike four vector k_μ. Invariance of k_μ under the most general type of Lorentz transformation leads to the Lie algebra:

$$\left[\hat{L}_1, \hat{L}_2\right] = 0$$

$$\left[\hat{J}_3, \hat{J}_1\right] = i\hat{L}_3 \tag{74}$$

$$\left[\hat{L}_2, \hat{J}_3\right] = i\hat{L}_1$$

where

$$\hat{L}_1 \equiv \hat{K}_1 - \hat{J}_2$$

$$\hat{L}_2 \equiv \hat{K}_2 + \hat{J}_1$$

Thus,

$$\left[\hat{L}_1, \hat{L}_2\right] = \left[\hat{K}_1, \hat{K}_2\right] + \left[\hat{K}_1, \hat{J}_1\right]$$

$$- \left[\hat{J}_2, \hat{K}_2\right] - \left[\hat{J}_2, \hat{J}_1\right] \tag{75}$$

$$= 0$$

In Cartesian terms, $X = 1$, $Y = 2$, $Z = 3$ and if we attempt to apply to Eq. (75) the Lorentz group algebra of Eqs. (51),

$$\left[\hat{K}_1, \hat{K}_2\right] = \left[\hat{K}_X, \hat{K}_Y\right] = -i J_Z \tag{76a}$$

$$\left[\hat{J}_2, \hat{J}_1\right] = -\left[\hat{J}_X, \hat{J}_Y\right] = -i\hat{J}_Z \tag{76b}$$

$$\left[\hat{K}_1, \hat{J}_1\right] = \left[\hat{K}_X, \hat{J}_X\right] = 0 \tag{76c}$$

$$\left[\hat{J}_2, \hat{K}_2\right] = \left[\hat{J}_Y, \hat{K}_Y\right] = 0 \tag{76d}$$

we obtain

$$\left[\hat{L}_1, \hat{L}_2\right] = 2i\hat{J}_Z \neq \hat{0} \tag{77}$$

Since in the Lorentz group

$$\hat{J}_Z \neq \hat{0} \tag{78}$$

in general, Eq. (77) contradicts Eq. (75).

Therefore, the most general Lorentz transformation that leaves the lightlike momentum vector $\mathbf{k}_\mu$ invariant cannot be described by the Lie algebra of the Lorentz group. This implies that the notion of lightlike momentum (a massless particle traveling at the speed of light), is not relativistically self-consistent. This is another way of demonstrating that the quantization of a massless field into a massless particle is beset with obscurity; i.e., we are led to the conclusion that the Maxwellian field has no meaning in quantum theory. Attempts to impose a meaning lead into physical obscurity as we have described. In the Einstein-de Broglie theory of light, the quantization of the EBP field leads directly and without difficulty[20] to a particle interpretation of light in terms of a massive boson. Quantum/classical equivalence in the EBP field is therefore clear. The only physically meaningful and consistent interpretation is to accept the photon as a massive boson whose classical field is described by the classical limit of the EBP equation. The mathematical limit of this field for zero mass is the Maxwellian field. Direct quantization of the Maxwellian field, regarded as a classical massless gauge field, is physically obscure. The quantized Maxwellian field must therefore be defined as being for all practical purposes the quantized EBP field, with which it is practically identical because photon mass is numerically very small.

III. TRANSVERSE SOLUTIONS IN VACUO FOR FINITE PHOTON MASS

The EBP equation can be written in terms of the tensor $F_{\mu\nu}$ defined in Eq. (43) as

$$\frac{\partial F_{\mu\nu}}{\partial x_\mu} = -\xi^2 A_\nu \sim 0 \tag{79}$$

and, as we have seen, the Lie algebra associated with $F_{\mu\nu}$ is given by Eqs. (43)–(46). Therefore, transverse solutions of the EBP equation in its classical limit obey the classical cross products in Eqs. (43)–(46). Using Eq. (43a) with the longitudinal solution of the EBP equation (73), we obtain

$$\mathbf{B}^{(1)} = \frac{B_0}{\sqrt{2}}(\mathbf{i}\mathbf{i} + \mathbf{j})e^{i\phi}\,e^{-\xi Z/2}$$

$$\mathbf{B}^{(2)} = \mathbf{B}^{(1)*} \tag{80}$$

and its complex conjugate. The difference between this solution and the equivalent Eq. (7c) for the Maxwell equations can be expressed by replacing the wave vector of the Maxwell equations by

$$\kappa_Z \rightarrow \kappa_Z - \frac{i\xi}{2} \quad \text{for polarization 1}$$

$$\kappa_Z \rightarrow \kappa_Z + \frac{i\xi}{2} \quad \text{for polarization 2}$$

At visible frequencies, the order of magnitude of the Maxwellian κ in vacuo is given by

$$\kappa_Z = \frac{\omega}{c} \sim \frac{10^{15}}{10^8} \sim 10^7 \text{ m}^{-1} \tag{81}$$

so that at these frequencies κ_Z is about 33 orders of magnitude greater than ξ. For all practical purposes, therefore, the transverse solutions of the classical limit of the EBP equation are identical with those of the Maxwell equations.

This is a simple demonstration in the classical limit that the fields associated with the EBP and Maxwell equations contain physically meaningful longitudinal as well as transverse components in vacuo. In the next

section we discuss several experimental consequences of physically meaningful longitudinal fields when electromagnetic radiation interacts with
matter. Firstly, however, we review the available experimental evidence for
finite photon mass, following a recent account by Vigier.[17]

IV. DISCUSSION

There is available an increasing amount of evidence for finite photon rest
mass, upon which is based the theory of Einstein and de Broglie. A recent
experiment by Mizobuchi and Ohtake[17] has demonstrated for single
photons the simultaneity of classical wave and particle behavior in light.
This has demonstrated for the first time that the Copenhagen interpretation cannot be valid, but supports the Einstein-de Broglie interpretation as
reviewed recently by Vigier,[17] an interpretation that implies, for example,
that photons are emitted from a source in quanta of energy with well-
defined directionality. The wave associated with a single photon has a
physical reality. Light is constituted by massive bosons (photons) controlled or piloted[17] by real surrounding spin one fields. The motion of the
photon is thus controlled by a quantum potential. The photons are the
only directly observable elements of light and behave in Minkowski spacetime as relativistic particles with finite mass. Light is also constituted in the
Einstein-de Broglie theory by physically meaningful fields (waves), which,
as we have seen, obey Maxwell's equations for all practical purposes,
essentially because the photon rest mass is finite (10^{-68} kg) but small.
These fields are described by complex vector waves, which also describe
photon motion. Thus, if there is a longitudinal photon polarization, there
must be a longitudinal field polarization, as described already. Longitudinal field solutions of the EBP equation were first derived by Schrödinger
and de Broglie and, in general, the EBP equation has longitudinal and
transverse WAVE solutions.[17] Since these are also wave solutions of
Maxwell's equations for all practical purposes, it becomes clear that
Maxwell's equations must have physically meaningful longitudinal solutions. The relation of these to the transverse solutions has only recently
become clear,[23-28] as described in Sections II and III of this paper.

Following Vigier's recent description[17] there are several consequences
of finite photon mass. The r dependence of the Coulomb potential is
replaced by that of the Yukawa potential:

$$V_Y \propto \frac{\exp(-\xi Z)}{Z} \tag{82}$$

There exist low velocity photons (i.e., photons traveling at considerably less than the speed of light c), whose small but finite mass contributes to that of the universe. There is, thirdly, a red shift proportional to $\exp(-\xi Z)$, which can be applied to explain recent astronomical observations of anomalous red shifts from several distant sources, such as quasars. These "tired light" phenomena originate in the EBP equation and may account for observed anomalies in double-star motions, galaxy clusters, observed variations of the Hubble "constant," and other evidence reviewed in the literature.[17] A photon with finite rest mass behaves relativistically in the frame of observation, leading to the expectation[17] of a direction dependent anisotropy in the frequency of light in the observer frame. Such an anisotropy has been observed experimentally by Hall et al.[30] in the direction of the apex of the 2.7 K background of microwave radiation. These Boulder experiments are currently being repeated in Copenhagen by Poulsen and coworkers.[17] Experimental evidence for the Einstein-de Broglie theory of light has also been reviewed by Vigier[17] in the following areas:

1. Super-luminal action at a distance, a facet of Einstein's interpretation of light
2. The question of locality or nonlocality of the quantum potential
3. Direct experimental testing of Heisenberg's uncertainty principle using single photons
4. Experimental testing for the existence of particle trajectories in light (einweg/welcherweg)
5. Testing the existence of physically meaningful waves without the presence of particles, for example, the recent experimental observation by Bartlett and Corle[31] of the Maxwell displacement current in vacuo
6. Testing directly the existence of the quantum potential with intersecting laser beams and laser-induced fringe patterns

There is, therefore, a considerable amount of experimentation in progress concerning the existence of finite photon mass, and it is no longer tenable to assert[1-16, 20, 21] that the photon mass is zero.

Similarly, it is not reasonable to assert that $\mathbf{B}^{(3)}$ and $\mathbf{E}^{(3)}$ must be zero, "irrelevant," "unphysical," or similar, as in much of the contemporary literature. It is in fact implied, but not specifically stated, in the work of de Broglie and Schrödinger[17] that $\mathbf{B}^{(3)}$ and $\mathbf{E}^{(3)}$ must exist. They exist, as we have seen, both for finite photon mass and in the Maxwellian limit, but finite photon rest mass is essential for a natural quantization of the electromagnetic field. For all intents and purposes, therefore, evidence for $\mathbf{B}^{(3)}$ and $\mathbf{E}^{(3)}$ is evidence for finite photon mass, and corroboration for other sources of evidence quoted already. The present author has proposed a number of different magneto-optic experiments[23-28] that would test for $\mathbf{B}^{(3)}$ through its interaction with matter, using its characteristic square root dependence on light intensity I_0 (W m^{-2}). In free space, fundamental electrodynamics leads to[23-28]

$$|\mathbf{B}^{(3)}| \sim 10^{-7} I_0^{1/2} \qquad (83)$$

and assuming that $\mathbf{B}^{(3)}$ acts as a magnetic field whose time average is nonzero, it is to be expected[23-28] that there exist the following effects (collected details in Ref. 26) proportional to the square root of laser intensity, provided that the laser is circularly polarized: (1) inverse Faraday effect (magnetization due to $\mathbf{B}^{(3)}$), (2) optical Faraday effect (azimuth rotation due to $\mathbf{B}^{(3)}$), (3) effects of $\mathbf{B}^{(3)}$ in NMR (preliminary observations reported in Ref. 32) and ESR spectroscopy, (4) Cotton-Mouton effect due to $\mathbf{B}^{(3)}$, (5) forward–backward birefringence due to $\mathbf{B}^{(3)}$, and (6) reinterpretation of antisymmetric light scattering and similar phenomena in terms of $\mathbf{B}^{(3)}$.

Finally, we propose the Bohm-Aharonov effect due to $\mathbf{B}^{(3)}$ of a circularly polarized laser, which replaces the solenoid, or iron whisker[20] of the conventional Bohm–Aharonov effect. The Bohm-Aharonov effect[20] indicates that the vector potential in quantum mechanics is physically meaningful, and that the vacuum has a nontrivial topology. It is therefore one of the most incisive effects in contemporary electrodynamics. The experiment has been repeated independently several times and consists of placing a small solenoid between two slits, which are used to generate interference fringes due to electron beams. The magnetic flux density $\mathbf{B}$ (tesla) is confined within the solenoid, and is inaccessible to the interfering electrons passing through the two slits. Despite this, the solenoid is observed experimentally[20] to produce a shift in the interference pattern (or fringes) set up by the electrons. This shift is due to the curl of the vector potential $\mathbf{A}$ set up outside the solenoid. Essentially, $\mathbf{A}$ changes the electron wave function

$$\psi = |\psi| \exp\left(i \frac{\mathbf{p} \cdot \mathbf{r}}{\hbar}\right) \qquad (84)$$

because $\mathbf{p}$, the electron momentum, is changed to $\mathbf{p} - e\mathbf{A}$, where e is the electronic charge. This does not occur in classical mechanics, but in quantum theory, the electronic wave function, and thus the electron, is influenced by $\mathbf{A}$ even though it travels in regions where magnetic flux density $\mathbf{B}$ is zero. This means that there is nonlocality in the integral

$\oint \mathbf{A} \cdot \mathbf{dr}$ (Ref. 20). The Bohm-Aharonov effect is therefore evidence for this type of nonlocality.
The shift is given in meters by

$$\Delta x = \frac{L\lambda}{d} \frac{e}{h} \Phi \tag{85}$$

where λ is the wavelength of the electron beam entering the two slits, L is the distance between the screen containing the two slits and the detector plane, d is the distance between the two slits, and

$$\Phi = \int \mathbf{B} \cdot \mathbf{dS} = \oint \mathbf{A} \cdot \mathbf{dr} \tag{86}$$

is a surface integral.

It is clear that if the solenoid is replaced by a thin, circularly polarized, laser beam, there should be a Bohm-Aharonov effect due to $\mathbf{B}^{(3)}$ in which this field shifts the interference pattern of the electrons, with $\mathbf{B}$ of Eq. (85) replaced by $\mathbf{B}^{(3)}$. This shift should be proportional to the square root of the laser intensity, reverse with the sense of circular polarization of the laser (because $\mathbf{B}^{(3)}$ changes sign), and disappear if the laser is linearly polarized or incoherently polarized. This laser-induced fringe displacement would be a particularly interesting investigation of the nature of $\mathbf{B}^{(3)}$, and of its concommitant $\mathbf{A}^{(3)}$. Presumably $\mathbf{B}^{(3)}$ is confined to the radius of the laser beam, and $\mathbf{A}^{(3)}$ exists outside this beam, as in a solenoid generating a conventional, longitudinal, magnetostatic field. The experiment would prove both the existence and the nonlocality of $\mathbf{A}^{(3)}$.

Acknowledgments

J. P. Vigier is thanked for a preprint of Ref. 17, and for his suggestion in a letter of 4 January 1993 that $\mathbf{B}^{(3)}$ and $\mathbf{E}^{(3)}$ can be accommodated naturally within the Einstein-de Broglie theory of light.

References

1. J. D. Jackson, *Classical Electrodynamics*, Wiley, New York, 1962.
2. R. M. Whitner, *Electromagnetics*, Prentice Hall, Englewood Cliffs, NJ, 1962.
3. A. F. Kip, *Fundamentals of Electricity and Magnetism*, McGraw Hill, New York, 1962.
4. L. D. Landau and E. M. Lifshitz, The Classical Theory of Fields, 4th ed., Pergamon, Oxford, UK, 1975.
5. M. Born and E. Wolf, *Principles of Optics*, 6th ed., Pergamon, Oxford, UK, 1975.
6. W. M. Schwartz, *Intermediate Electromagnetic Theory*, 2d ed., Wiley, New York, 1964.
7. P. W. Atkins, *Molecular Quantum Mechanics*, Oxford University Press, Oxford, UK, 1983.

8. C. Cohen-Tannoudji, J. Dupont-Roc, and G. Grynberg, *Photons and Atoms: Introduction to Quantum Electrodynamics*, Wiley, New York, 1989.

9. L. D. Barron, *Molecular Light Scattering and Optical Activity*, Cambridge University Press, Cambridge, UK, 1982.

10. B. W. Shore, *The Theory of Coherent Atomic Excitation*, Vols. 1 and 2, Wiley, New York, 1990.

11. D. E. Soper, *Classical Field Theory*, Wiley, New York, 1976.

12. E. L. Hill, *Rev. Mod. Phys.* **23**, 253 (1951).

13. S. S. Schweber, *An Introduction to Relativistic Quantum Field Theory*, Harper & Row, New York, 1962.

14. N. N. Bogoliubov and D. V. Shirkov, *Introduction to the Theory of Quantised Fields*, 3d ed., Wiley Interscience, New York, 1980.

15. D. Lurie, *Particles and Fields*, Wiley Interscience, New York, 1968.

16. R. Jost, in M. Fierz and V. F. Wisskopf (Eds.), *Theoretical Physics in the Twentieth Century*, Wiley Interscience, New York, 1960.

17. J. P. Vigier, *Present Experimental Status of the Einstein/de Broglie Theory of Light*, Conference Reprint, communication to the author of 4 Jan. 1993, from Université Pierre et Marie Curie, Paris.

18. A. Einstein, for example, *Werk. Deutsch. Phys. Ges.* **18**, 318 (1916); *Mitt. Phys. Ges. Zurich* **16**, 47 (1916); *Phys. Zeit.* **18**, 121 (1917); Letters to Besso, 8 Aug., 6 Sep. (1916).

19. L. de Broglie, *La Mecanique Ondulatoire du Photon*, Gauthier Villars, Paris, 1936.

20. L. S. Ryder, *Quantum Field Theory*, 2d ed., Cambridge University Press, Cambridge, UK, 1987.

21. L. S. Ryder, *Elementary Particles and Symmetries*, Gordon & Breach, London, 1986.

22. See for example, W. Heitler, *The Quantum Theory of Radiation*, 3d ed., Clarendon, Oxford, UK, 1954.

23. M. W. Evans, *Mod. Phys. Lett.* **6**, 1237 (1992).

24. M. W. Evans, *Physica B* **182**, 237 (1992).

25. M. W. Evans, *Physica B* **182**, 227 (1992).

26. M. W. Evans, *The Photon's Magnetic Field*, World Scientific, Singapore, 1993.

27. F. Farahi and M. W. Evans, *Phys. Rev. E*, in press.

28. M. W. Evans, *Physica B*, **182**, 103 (1993).

29. E. P. Wigner, *Ann. Math.* **40**, 149 (1939).

30. Hall et al., *Phys. Rev. Lett.* **60**, 81 (1988).

31. Bartlett and Corle, *Phys. Rev. Lett.* **55**, 59 (1985).

32. W. S. Warren, D. Goswami, S. Mayr, and A. P. West, Jr., *Science* **255**, 1681 (1992).

FREQUENCY-DEPENDENT CONTINUUM ELECTROMAGNETIC PROPERTIES OF A GAS OF SCATTERING CENTERS

AKHLESH LAKHTAKIA

Department of Engineering Science and Mechanics, The Pennsylvania State University, University Park, PA

CONTENTS

Modern Nonlinear Optics, Part 2, Edited by Myron Evans and Stanisław Kielich. Advances in Chemical Physics Series, Vol. LXXXV.
ISBN 0-471-57546-1 © 1993 John Wiley & Sons, Inc.

I. INTRODUCTION

Lasers are extensively utilized to examine optical scattering phenomena—associated with the names of Rayleigh, Brillouin, Raman, and others —to arrive at an understanding of the interaction of electromagnetic fields with matter at the microscopic level.[1-3] An excellent introduction by Delone and Krainov[4] has recently become available, though, for sheer lucidity, Baldwin[5] is still hard to beat.

The four postulates of Maxwell are perfectly linear and, often contrary to commonly received wisdom, given in the time domain[6-8]. Therefore, nonlinear response can only be built into the constitutive equations (which also must be specified in the time domain, but rarely are) that have become widely used in the Heaviside-Lorentz conceptualization, which came after the discovery of the Hall effect.[9]

Current research on nonlinear optics is focused chiefly on dielectrics.[10, 11] Nonlinear dielectrics can be phenomenologically described by polarizability tensor (of ranks two and higher) operators in the time domain.[12, 13] When interpreted in the frequency domain, polarizability tensors give rise to phenomenological descriptions of elastic as well as inelastic scattering. Because the nonlinear effects at off-resonance frequencies are generally small and the frequency-domain constitutive relations in electromagnetic theory are generally a small-signal concept, common (but not necessarily always correct) usage is to treat nonlinear response as a perturbation of the monochromatic linear theory[5, 14]; see also Ref. 15 on semiconductor clusters.

Matter is discrete, and electromagnetism has acquired a microphysical basis, which sets it apart from elastodynamics, on which classical as well as quantum-mechanical treatments can be borne upon. Yet, handling phenomenally large numbers of discrete entities entails the use of statistical techniques (e.g., Ref. 16), which converts the microphysical basis into a continuum approach.[9]

The quantities measured in most experiments conducted at optical and suboptical frequencies can be interpreted directly only through the continuum approach and the microphysical properties inferred therefrom. Such estimation can be quantitative (e.g., Refs. 10, 17, 18), but is more likely to be qualitative (e.g., Refs. 2, 3, 19). Even the well-known isotropic chiral materials, for which elaborate quantum-mechanical models have been devised,[20] are experimentally investigated only through continuum approaches; see, for example, Emeis et al.[21] and Urry and Krivacic.[22] Sophistry aside, the dichotomy is conspicuous, in the exception, by the confusion in terminology. For instance, the term Rayleigh scattering was originally used for electrically small particles,[23] but is often (mis-)applied to molecular scattering as well.[19]

With these issues in mind, the author has planned this chapter on the continuum properties of a gas of scattering centers as follows: The Heaviside-Lorentz conceptualization of continua is reviewed in Section II, and is then symmetrized to include bianisotropy in the time-domain. Then, we enter the frequency domain to consider a gas of scattering centers that can be either electrically small particles or molecules. Particles, being small pieces of continua, can never become molecular, regardless of the minuteness of their sizes. A material continuum can always be subdivided into small pieces, each piece retaining the same properties as the bigger original. Next, using coupled volume integral equations, in Section III we obtain the polarizability dyadics of electrically small bianisotropic particles of convex shapes. In Section IV the nature of the polarizability dyadics of molecules is briefly discussed. The developments of Sections III and IV are then used in Section V to formulate the Maxwell Garnett continuum that is macroscopically equivalent to the gas of scattering centers. It is expected that the formalism will apply easily to polymers, gases, and dilute suspensions and colloids.[24]

Field-theoretic relations pertaining to the Maxwell Garnett continuum are given in Section VI. Section VII directly relates to commonly performed experiments on optical rotation and dichroism. The Maxwell curl postulates in the equivalent continuum are converted into a matrix differential equation when the electromagnetic fields are allowed to vary only in a given direction. A formulation for the reflection and the transmission of normally incident plane waves by a continuum slab is obtained, followed by an examination of the theoretical results pertaining to optical rotation and dichroism. As a parting shot, unidirectionally varying fields in inhomogeneous slabs are investigated.

II. THE MAXWELL POSTULATES: MICROPHYSICS AND CONTINUA

In modern notation, the Maxwell postulates for the electromagnetic field in vacuum are usually stated as[25]

$$\nabla \cdot \mathbf{E}(\mathbf{x}, t) = \frac{\rho_e(\mathbf{x}, t)}{\varepsilon_0} \tag{1a}$$

$$\nabla \cdot \mathbf{B}(\mathbf{x}, t) = 0 \tag{1b}$$

$$\nabla \times \mathbf{E}(\mathbf{x}, t) = -\left\{\frac{\partial}{\partial t}\right\}\mathbf{B}(\mathbf{x}, t) \tag{1c}$$

$$\nabla \times \mathbf{H}(\mathbf{x}, t) = \varepsilon_0\left\{\frac{\partial}{\partial t}\right\}\mathbf{E}(\mathbf{x}, t) + \mathbf{J}(\mathbf{x}, t) \tag{1d}$$

Here, $\mathbf{E}$ is the electric field intensity (volt meter^{-1}), $\mathbf{B}$ is the magnetic induction (tesla), $\mathbf{H}$ is the magnetic field intensity (ampere meter^{-1}), ρ_e is the volumetric electric charge density (C m^{-3}), while $\mathbf{J}$ is the volumetric electric current density (A m^{-3}); $\varepsilon_0 = 8.854 \times 10^{-12}$ F m^{-1} and $\mu_0 = 4\pi \times 10^{-7}$ H m^{-1} are the usual free space constants in SI units. Of these, (1a) can be inferred from Articles 69, 71, and 77 of Maxwell's *Treatise*,[26] (1b) is as per Articles 400 and 403, (1c) follows from Articles 590 and 616, and (1d) follows from Article 607. A linear relationship between $\mathbf{B}$ and $\mathbf{H}$ was proposed on heuristic grounds in Article 614 on the same lines as the linear relationship between $\mathbf{E}$ and $\mathbf{D}$, the electric displacement (C m^{-2}) in Article 68; specifically, in modern notation, $\mathbf{D} = \varepsilon_0 \mathbf{E}$ and $\mathbf{B} = \mu_0 \mathbf{H}$ hold in vacuum. In the Heaviside-Lorentz formulation, (1d) is generally stated as

$$\nabla \times \mathbf{B} = \mu_0 \left(\varepsilon_0 \frac{\partial \mathbf{E}}{\partial t} + \mathbf{J} \right) \qquad (1d')$$

which is also in accordance with Article 616 of the *Treatise*, but only after the linear relationship between $\mathbf{B}$ and $\mathbf{H}$ had been hypothesized in Article 614. Finally, the continuity of the current $\mathbf{J}$ and the charge is expressed by

$$\nabla \cdot \mathbf{J}(\mathbf{x}, t) + \left\{ \frac{\partial}{\partial t} \right\} \rho_e(\mathbf{x}, t) = 0 \qquad (1e)$$

which is of hydrodynamic origin, as implied in Article 295 of the *Treatise*. Maxwell's own contribution is the $\partial \mathbf{E}/\partial t$ term on the right sides of (1d) and (1d'), which led to the development of electromagnetic communication systems.

A. The Heaviside-Lorentz View of Material Continua

From a microphysical basis, the nature of matter is built into the Maxwell postulates by modeling matter to be composed of point charges and currents that are suspended in vacuum. Thus, the electric current density $\mathbf{J}$ in (1d') can be broken up into free-standing components

$$\mathbf{J} = \mathbf{J}' + \mathbf{J}_{pol} + \mathbf{J}_{mag} \qquad (2a)$$

where the subscripts of pol and mag, respectively, stand for polarization and magnetization, while $\mathbf{J}' = \mathbf{J} - \mathbf{J}_{pol} - \mathbf{J}_{mag}$ is due to all other agents. Due to the continuity condition (1e), the electric charge density should also be broken up similarly into free-standing components as

$$\rho_e = \rho_e' + \rho_{e,\text{pol}} + \rho_{e,\text{mag}} \tag{2b}$$

The polarization components due to the dielectric nature of the continuum are usually modeled as

$$\mathbf{J}_{\text{pol}} = \left\{\frac{\partial}{\partial t}\right\}\mathbf{P}(\mathbf{x}, t) \tag{3a}$$

$$\rho_{e,\text{pol}} = -\boldsymbol{\nabla} \cdot \mathbf{P}(\mathbf{x}, t) \tag{3b}$$

where $\mathbf{P}$ is the polarization vector, and give rise to the electric displacement

$$\mathbf{D}(\mathbf{r}, t) = \varepsilon_0\mathbf{E}(\mathbf{x}, t) + \mathbf{P}(\mathbf{x}, t) \tag{3c}$$

when (3a) and (3b) are substituted into (1d').

On the other hand, the magnetization components due to the magnetic nature of the continuum are postulated in the Heaviside-Lorentz view as

$$\mathbf{J}_{\text{mag}} = \boldsymbol{\nabla} \times \mathbf{M}(\mathbf{x}, t) \tag{4a}$$

$$\left\{\frac{\partial}{\partial t}\right\}\rho_{e,\text{mag}} \equiv 0 \tag{4b}$$

where $\mathbf{M}$ is the magnetization field. Thus, $\rho_{e,\text{mag}}(\mathbf{x}, t)$ is independent of time, but is taken to be identically zero as per the widespread belief that magnetic monopoles do not exist. In any case, the use of (4a) and (4b) in (1d') yields the relation

$$\mathbf{H}(\mathbf{x}, t) = \mu_0^{-1}\mathbf{B}(\mathbf{x}, t) - \mathbf{M}(\mathbf{x}, t) \tag{4c}$$

Using (3c) and (4c), the macroscopic Maxwell postulates can be set down as

$$\boldsymbol{\nabla} \cdot \mathbf{D}(\mathbf{x}, t) = \rho_e'(\mathbf{x}, t) \tag{5a}$$

$$\boldsymbol{\nabla} \cdot \mathbf{B}(\mathbf{x}, t) = 0 \tag{5b}$$

$$\boldsymbol{\nabla} \times \mathbf{E}(\mathbf{x}, t) = -\left\{\frac{\partial}{\partial t}\right\}\mathbf{B}(\mathbf{x}, t) \tag{5c}$$

$$\boldsymbol{\nabla} \times \mathbf{H}(\mathbf{x}, t) = \left\{\frac{\partial}{\partial t}\right\}\mathbf{D}(\mathbf{x}, t) + \mathbf{J}'(\mathbf{x}, t) \tag{5d}$$

along with the continuity condition

$$\nabla \cdot \mathbf{J}'(\mathbf{x}, t) + \left\{ \frac{\partial}{\partial t} \right\} \rho_e'(\mathbf{x}, t) = 0 \tag{5e}$$

For homogeneous material continua, linear constitutive equations

$$\mathbf{P}(\mathbf{x}, t) = \varepsilon_0 \int_0^\infty dt\, \underline{\underline{\chi}}_e(\tau) \mathbf{E}(\mathbf{x}, t - \tau) = \varepsilon_0 \underline{\underline{\chi}}_e(t) \# \mathbf{E}(\mathbf{x}, t) \tag{6a}$$

$$\mathbf{M}(\mathbf{x}, t) = \mu_0^{-1} \int_0^\infty dt\, \underline{\underline{\chi}}_m(\tau) \mathbf{B}(\mathbf{x}, t - \tau) = \mu_0^{-1} \underline{\underline{\chi}}_m(t) \# \mathbf{B}(\mathbf{x}, t) \tag{6b}$$

are used, where the tensor $\underline{\underline{\chi}}_e(t)$ is the dielectric susceptibility, and $\underline{\underline{\chi}}_m(t)$ is the magnetic susceptibility; the symbol # denotes the temporal convolution operation.[27] Equations (6a) and (6b) represent the most general spatially local and linear relationships. Finally, $\underline{\underline{\chi}}_e(t)$ and $\underline{\underline{\chi}}_m(t)$ must be causal and satisfy the Kramers-Kronig relationships.[28]

B. Magnetic Charges in Vacuum

Into the current density $\mathbf{J}'$ and the charge density ρ_e' can be incorporated all other forms, such as those due to convection, conduction, or impressed sources. These need not be considered in detail here. What is interesting is the unequal treatment of the polarization $\mathbf{P}$ and the magnetization $\mathbf{M}$ in the preceding schema for continua. Specifically, $\mathbf{M}$ is incorporated in such a way that it gives rise to an electric current density but not to an electric charge density. Thus, magnetic dipoles are taken into account, but magnetic monopoles are excluded. Although magnetic monopoles have never been unambiguously observed, there is no sufficiently logical reason for their exclusion either.[29]

To include magnetic monopoles, however, an autonomous source pair, made of a magnetic current density $\mathbf{K}$ and a magnetic charge density ρ_m, must be prescribed. The Maxwell postulates (1a)–(1d), thus symmetrized, in vacuum read as

$$\nabla \cdot \mathbf{E}(\mathbf{x}, t) = \frac{\rho_e(\mathbf{r}, t)}{\varepsilon_0} \tag{7a}$$

$$\nabla \cdot \mathbf{B}(\mathbf{x}, t) = \rho_m(\mathbf{x}, t) \tag{7b}$$

$$\nabla \times \mathbf{E}(\mathbf{x}, t) = -\left\{ \frac{\partial}{\partial t} \right\} \mathbf{B}(\mathbf{x}, t) - \mathbf{K}(\mathbf{x}, t) \tag{7c}$$

$$\nabla \times \mathbf{H}(\mathbf{x}, t) = \varepsilon_0 \left\{ \frac{\partial}{\partial t} \right\} \mathbf{E}(\mathbf{x}, t) + \mathbf{J}(\mathbf{x}, t) \tag{7d}$$

along with the continuity conditions

$$\nabla \cdot \mathbf{J}(\mathbf{x}, t) + \left\{ \frac{\partial}{\partial t} \right\} \rho_e(\mathbf{x}, t) = 0 \tag{7e}$$

$$\nabla \cdot \mathbf{K}(\mathbf{x}, t) + \left\{ \frac{\partial}{\partial t} \right\} \rho_m(\mathbf{x}, t) = 0 \tag{7f}$$

Equations (7a)–(7f) constitute a non-Abelian theory (noncommutative), as opposed to (1a)–(1e), which are Abelian (commutative),[30] as has been pointed out by Barrett.[8] Certainly, (7a)–(7f) are more general than (1a)–(1e); but, even more importantly, Harmuth[6, 31] has shown the necessity of using free-standing source pairs $\{\mathbf{J}, \rho_e\}$ and $\{\mathbf{K}, \rho_m\}$ for electromagnetic transient analysis, as opposed to $\{\mathbf{J}, \rho_e\}$ alone.

To convert (7a)–(7f) for application to matter on a macroscopic basis, the procedure going from (2) to (5) is repeated. Thus,

$$\mathbf{J} = \mathbf{J}' + \mathbf{J}_{pol} \tag{8a}$$

$$\rho_e = \rho_e' + \rho_{e, pol} \tag{8b}$$

$$\mathbf{J}_{pol} = \frac{\partial \mathbf{P}}{\partial t} \tag{8c}$$

$$\rho_{e, pol} = -\nabla \cdot \mathbf{P} \tag{8d}$$

$$\mathbf{K} = \mathbf{K}' + \mathbf{K}_{mag} \tag{8e}$$

$$\rho_m = \rho_m' + \rho_{m, mag} \tag{8f}$$

$$\mathbf{K}_{mag} = -\mu_0 \frac{\partial \mathbf{M}}{\partial t} \tag{8g}$$

$$\rho_{m, mag} = \mu_0 \nabla \cdot \mathbf{M} \tag{8h}$$

are substituted into (7a)–(7d), leading to

$$\nabla \cdot \mathbf{D}(\mathbf{x}, t) = \rho_e'(\mathbf{x}, t) \tag{9a}$$

$$\nabla \cdot \mathbf{H}(\mathbf{x}, t) = \frac{\rho_m'(\mathbf{x}, t)}{\mu_0} \tag{9b}$$

$$\nabla \times \mathbf{E}(\mathbf{x}, t) = -\mu_0 \left\{ \frac{\partial}{\partial t} \right\} \mathbf{H}(\mathbf{x}, t) - \mathbf{K}'(\mathbf{x}, t) \tag{9c}$$

$$\nabla \times \mathbf{H}(\mathbf{x}, t) = \left\{ \frac{\partial}{\partial t} \right\} \mathbf{D}(\mathbf{x}, t) + \mathbf{J}'(\mathbf{x}, t) \tag{9d}$$

along with the definitions $\mathbf{D} = \varepsilon_0 \mathbf{E} + \mathbf{P}$ and $\mathbf{H} = \mathbf{B}/\mu_0 - \mathbf{M}$.

C. The Symmetrized Heaviside-Lorentz View
of Material Continua

The preceding subsection can now be regularized for operation either in vacuum or in a material continuum. It is necessary to break up the source charges and currents into two parts each: one part due to the microscopic nature of matter, and the other due to externally impressed sources. Thus,

$$\nabla \cdot \mathbf{D}(\mathbf{x}, t) = \rho_{e,\text{imp}}(\mathbf{r}, t) \tag{10a}$$

$$\nabla \cdot \mathbf{H}(\mathbf{x}, t) = \rho_{m,\text{imp}}(\mathbf{x}, t)/\mu_0 \tag{10b}$$

$$\nabla \times \mathbf{E}(\mathbf{x}, t) = -\mu_0 \left\{ \frac{\partial}{\partial t} \right\} \mathbf{H}(\mathbf{x}, t) - \mathbf{K}_{\text{imp}}(\mathbf{x}, t) \tag{10c}$$

$$\nabla \times \mathbf{H}(\mathbf{x}, t) = \left\{ \frac{\partial}{\partial t} \right\} \mathbf{D}(\mathbf{x}, t) + \mathbf{J}_{\text{imp}}(\mathbf{x}, t) \tag{10d}$$

where the externally impressed sources obey the continuity conditions

$$\nabla \cdot \mathbf{J}_{\text{imp}}(\mathbf{x}, t) + \left\{ \frac{\partial}{\partial t} \right\} \rho_{e,\text{imp}}(\mathbf{x}, t) = 0 \tag{11a}$$

$$\nabla \cdot \mathbf{K}_{\text{imp}}(\mathbf{x}, t) + \left\{ \frac{\partial}{\partial t} \right\} \rho_{m,\text{imp}}(\mathbf{x}, t) = 0 \tag{11b}$$

The linear constitutive equations are of the forms

$$\mathbf{D}(\mathbf{x}, t) = \varepsilon_0 \mathbf{E}(\mathbf{x}, t) + \underline{\underline{\chi}}_{ee}(\mathbf{x}, t) @\#\mathbf{E}(\mathbf{x}, t) + \underline{\underline{\chi}}_{em}(\mathbf{x}, t) @\#\mathbf{B}(\mathbf{x}, t) \tag{12a}$$

$$\mathbf{H}(\mathbf{x}, t) = \frac{\mathbf{B}(\mathbf{x}, t)}{\mu_0} + \underline{\underline{\chi}}_{me}(\mathbf{x}, t) @\#\mathbf{E}(\mathbf{x}, t) + \underline{\underline{\chi}}_{mm}(\mathbf{x}, t) @\#\mathbf{B}(\mathbf{x}, t) \tag{12b}$$

If (12a) and (12b) are to be used in vacuum, only the first terms on the right sides are necessary. The susceptibilities $\chi_{ee}(\mathbf{r}, t)$, etc. in (12a) and (12b), which reflect the nature of matter on a macroscopic basis, should be interpreted as integro-differential operators, and can be sums of tensors of any order. The operation @ is a spatial convolution.[32] It is assumed here that space and time are not interdependent so that the order of appearance of spatial and temporal operations is of no consequence; hence, $@\# \equiv \#@$.

Emphasis has been placed in (12a) and (12b) on linearity simply because electromagnetic fields in nonlinear materials are, though widely studied for many years, as yet only poorly understood. However, there is nothing to prevent χ_{ee}, etc. from being explicit functions of $\mathbf{E}, \mathbf{B}, \mathbf{E} \cdot \mathbf{B}, \mathbf{E} \cdot \mathbf{E}, \mathbf{EE}, \mathbf{EB}$, and so on.[13, 33] Indeed, the inclusion of such nonlinear dependences is necessary for nonlinear electromagnetic responses; see, for example, Ducuing[12] and Eaton[10] on nonlinear dielectrics.

III. POLARIZABILITY DYADICS OF SMALL HOMOGENEOUS PARTICLES

Returning to our objective of finding the frequency-dependent macroscopic properties of the gas of scattering centers, we now attend to the scattering response of an electrically small, convex, bianisotropic particle immersed in free space. In the analysis to follow, the particle is replaced by an equivalent distribution of volume electric and magnetic current densities in order to obtain two coupled integral equations. Because the particle is electrically small, a long-wavelength approximation[33, 34] is utilized to convert the integral equations into algebraic equations. Solution of these algebraic equations yields the desired scattering response in terms of four polarizability dyadics. The dyadic functions employed throughout can be interpreted in terms of 3×3 matrices,[35, 36] and are therefore convenient to handle. Dyadics can also be interpreted very easily in terms of second-rank Cartesian tensors.[37]

From now on, we will use $\mathbf{x} \equiv (x_1, x_2, x_3)$ for local coordinate systems, and $\mathbf{X} \equiv (X_1, X_2, X_3)$ for the laboratory fame or the global coordinate system.

A. The Equivalent Volume Current Densities

Let all space be made of a region V completely enclosed by the region V_0 that extends out to infinity, as shown in Fig. 1, and S is the simply-connected closed surface that separates V_0 from V; in other words, S does not intersect with itself and a unique normal can be prescribed at all points on

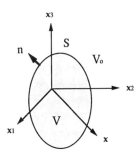

Figure 1. Relevant to the scattering characteristics of electrically small bianisotropic particles.

S. The region V_0 is vacuous whose constitutive equations are given as

$$\mathbf{D}(\mathbf{x}) = \varepsilon_0 \mathbf{E}(\mathbf{x}) \qquad \mathbf{x} \in V_0 \tag{13a}$$

$$\mathbf{B}(\mathbf{x}) = \mu_0 \mathbf{H}(\mathbf{x}) \qquad \mathbf{x} \in V_0 \tag{13b}$$

It must be borne in mind that (13a) and (13b) are frequency-independent and yield a Lorentz-invariant characterization of vacuum.[38]

We begin by assuming an $\exp(-i\omega t)$ harmonic time dependence, where ω is the circular frequency of the small-signal field exciting the particle. The origin $\mathbf{x} = \mathbf{0}$ of the local coordinate system is assumed to be located at the center of the scattering volume V that is filled with a general, linear, homogeneous, nondiffusive bianisotropic continuum with frequency-dependent constitutive equations

$$\mathbf{D}(\mathbf{x}) = \varepsilon_0 \left[\underline{\underline{\varepsilon}}_r \cdot \mathbf{E}(\mathbf{x}) + \underline{\underline{\xi}}_r \cdot \mathbf{H}(\mathbf{x}) \right], \qquad \mathbf{x} \in V \tag{14a}$$

$$\mathbf{B}(\mathbf{x}) = \mu_0 \left[\underline{\underline{\zeta}}_r \cdot \mathbf{E}(\mathbf{x}) + \underline{\underline{\mu}}_r \cdot \mathbf{H}(\mathbf{x}) \right] \qquad \mathbf{x} \in V \tag{14b}$$

where $\underline{\underline{\varepsilon}}_r$ is the relative permittivity dyadic, $\underline{\underline{\mu}}_r$ is the relative permeability dyadic, and $\underline{\underline{\xi}}_r$ and $\underline{\underline{\zeta}}_r$ represent the magnetoelectric dyadics. Note should be taken of the differences between (12a) and (12b) on one hand, and (14a) and (14b) on the other. For homogeneous linear materials without memory, the temporal Fourier transforms of (12a) and (12b) can be algebraically manipulated to yield (14a) and (14b). Furthermore, all electromagnetic fields from this section on are the temporal Fourier transforms of the corresponding fields of Section II, the $\exp(-i\omega t)$ dependence being implicit in what follows. Finally, the magnetization has been redefined as $\mathbf{M} = \mathbf{B} - \mu_0 \mathbf{H}$, conforming thus to the older and the more symmetric convention.[39]

Bianisotropic media occur readily in nature. Materials with dyadic permittivity $(\underline{\underline{\mu}}_r = \mu_r \underline{\underline{I}}, \underline{\underline{\zeta}}_r = \underline{\underline{\xi}}_r = \underline{\underline{0}})$ abound as crystals[40] and magnetoplasmas,[36] while ferrites[36] have dyadic permeability $(\underline{\underline{\varepsilon}}_r = \varepsilon_r \underline{\underline{I}}, \underline{\underline{\xi}}_r = \underline{\underline{\zeta}}_r = \underline{\underline{0}})$, $\underline{\underline{I}}$ being the identity dyadic and $\underline{\underline{0}}$ being the null dyadic. Natural optically active materials $(\underline{\underline{\varepsilon}}_r = \varepsilon_r \underline{\underline{I}}, \underline{\underline{\mu}}_r \mu_r = \underline{\underline{I}}, \underline{\underline{\zeta}}_r = \zeta_r \underline{\underline{I}}, \underline{\underline{\xi}}_r = \xi_r \underline{\underline{I}})$ are well known to organic and physical chemists.[20, 41] Magnetoelectric materials $(\underline{\underline{\zeta}}_r \neq \underline{\underline{0}}$ and/ or $\underline{\underline{\xi}}_r \neq \underline{\underline{0}})$ were theoretically predicted by Dzyaloshinskii[42] while studying the antiferromagnetic Cr_2O_3, and experimentally confirmed by Astrov.[43] Indenbom,[44] Birss,[45] and Rado[46] put the magnetoelectric behavior of both ferromagnetics and antiferromagnetics on a surer footing. On a slightly different note, in the Minkowski formulation, a simply moving dielectric body appears to have magnetization when interrogated by a stationary observer.[47] Physically realizable forms of the constitutive dyadics in (14a) and (14b) have been discussed at length by Post[38] within the framework of Lorentz covariance and causality. Incidentally, the medium of (14a, b) is called bianisotropic since both $\mathbf{D}$ and $\mathbf{B}$ are connected to both $\mathbf{E}$ and $\mathbf{H}$ through constitutive dyadics; it is nondiffusive, since we ignore the drift and diffusion of charge carriers.[48, 49]

In the absence of any externally impressed sources, the Maxwell curl postulates can be expressed in V_0 as

$$\nabla \times \mathbf{E}(\mathbf{x}) - i\omega\mu_0\mathbf{H}(\mathbf{x}) = \mathbf{0} \qquad \mathbf{x} \in V_0 \qquad (15a)$$

$$\nabla \times \mathbf{H}(\mathbf{x}) + i\omega\varepsilon_0\mathbf{E}(\mathbf{x}) = \mathbf{0} \qquad \mathbf{x} \in V_0 \qquad (15b)$$

Similarly, we have in V,

$$\nabla \times \mathbf{E}(\mathbf{x}) - i\omega\mu_0\Big[\underline{\underline{\xi}}_r \cdot \mathbf{E}(\mathbf{x}) + \underline{\underline{\mu}}_r \cdot \mathbf{H}(\mathbf{x})\Big] = \mathbf{0} \qquad \mathbf{x} \in V \quad (16a)$$

$$\nabla \times \mathbf{H}(\mathbf{x}) + i\omega\varepsilon_0\Big[\underline{\underline{\varepsilon}}_r \cdot \mathbf{E}(\mathbf{x}) + \underline{\underline{\zeta}}_r \cdot \mathbf{H}(\mathbf{x})\Big] = \mathbf{0} \qquad \mathbf{x} \in V \quad (16b)$$

with $\mathbf{0}$ being the null vector. Somewhat disingenuously at this stage, (16a) and (16b) can be rewritten respectively as

$$\nabla \times \mathbf{E}(\mathbf{x}) = i\omega\mu_0\mathbf{H}(\mathbf{x}) + \Big\{ i\omega\mu_0\Big[\underline{\underline{\xi}}_r \cdot \mathbf{E}(\mathbf{x}) + \underline{\underline{\mu}}_r \cdot \mathbf{H}(\mathbf{x})\Big] - i\omega\mu_0\mathbf{H}(\mathbf{x}) \Big\}$$
$$\mathbf{x} \in V \quad (17a)$$

$$\nabla \times \mathbf{H}(\mathbf{x}) = -i\omega\varepsilon_0\mathbf{E}(\mathbf{x}) + \Big\{ -i\omega\varepsilon_0\Big[\underline{\underline{\varepsilon}}_r \cdot \mathbf{E}(\mathbf{x}) + \underline{\underline{\zeta}}_r \cdot \mathbf{H}(\mathbf{x})\Big] + i\omega\varepsilon_0\mathbf{E}(\mathbf{x}) \Big\},$$
$$\mathbf{x} \in V \quad (17b)$$

The aim in writing (17a) and (17b) is to recast the scattering problem in terms of a radiation problem. In other words, the scattering particle is to be replaced by volume electric and magnetic current densities, $\mathbf{J}(\mathbf{x})$ and $\mathbf{K}(\mathbf{x})$, radiating into vacuum. This is in keeping with the microscopic nature of matter,[50] but it can also be (unpejoratively) regarded as mathematical skulduggery.

Therefore, let $\mathbf{J}$ and $\mathbf{K}$ replace the bianisotropic particle; then, the Maxwell curl postulates everywhere can be rewritten as[51]

$$\nabla \times \mathbf{E}(\mathbf{x}) - i\omega\mu_0\mathbf{H}(\mathbf{x}) = -\mathbf{K}(\mathbf{x}) \qquad \mathbf{x} \in V_0 + V \qquad (18a)$$

$$\nabla \times \mathbf{H}(\mathbf{x}) + i\omega\varepsilon_0\mathbf{E}(\mathbf{x}) = \mathbf{J}(\mathbf{x}) \qquad \mathbf{x} \in V_0 + V \qquad (18b)$$

On comparing (15) and (18), it is clear that

$$\mathbf{J}(\mathbf{x}) = \mathbf{0} \qquad \mathbf{K}(\mathbf{x}) = \mathbf{0} \qquad \mathbf{x} \in V_0 \qquad (19)$$

and the comparison of (17) and (18) yields

$$\mathbf{J}(\mathbf{x}) = i\omega\varepsilon_0\left[\left(\underline{\underline{I}} - \underline{\underline{\varepsilon}}_r\right) \cdot \mathbf{E}(\mathbf{x}) - \underline{\underline{\xi}}_r \cdot \mathbf{H}(\mathbf{x})\right] \qquad \mathbf{x} \in V \qquad (20a)$$

$$\mathbf{K}(\mathbf{x}) = i\omega\mu_0\left[-\underline{\underline{\zeta}}_r \cdot \mathbf{E}(\mathbf{x}) + \left(\underline{\underline{I}} - \underline{\underline{\mu}}_r\right) \cdot \mathbf{H}(\mathbf{x})\right] \qquad \mathbf{x} \in V \qquad (20b)$$

Thus, the influence of the bianisotropic particle can be treated as that due to certain volume distributions of the electric and the magnetic current densities in vacuum.

The solutions of (18a) and (18b) are given as the coupled volume integral equations[52]

$$\mathbf{E}(\mathbf{x}) - \mathbf{E}_{\text{inc}}(\mathbf{x}) = i\omega\mu_0\iiint_V d^3x'\left\{\underline{\underline{G}}(\mathbf{x},\mathbf{x}') \cdot \mathbf{J}(\mathbf{x}')\right\}$$

$$-\iiint_V d^3x'\left\{\left[\nabla \times \underline{\underline{G}}(\mathbf{x},\mathbf{x}')\right] \cdot \mathbf{K}(\mathbf{x}')\right\} \qquad \mathbf{x} \in V_0 + V$$

$$(21a)$$

$$\mathbf{H}(\mathbf{x}) - \mathbf{H}_{\text{inc}}(\mathbf{x}) = i\omega\varepsilon_0\iiint_V d^3x'\left\{\underline{\underline{G}}(\mathbf{x},\mathbf{x}') \cdot \mathbf{K}(\mathbf{x}')\right\}$$

$$+\iiint_V d^3x'\left\{\left[\nabla \times \underline{\underline{G}}(\mathbf{x},\mathbf{x}')\right] \cdot \mathbf{J}(\mathbf{x}')\right\} \qquad \mathbf{x} \in V_0 + V$$

$$(21b)$$

where

$$\underline{\underline{G}}(\mathbf{x},\mathbf{x}') = \frac{1}{4\pi}\left[\underline{\underline{I}} + \left(\frac{1}{k_0^2}\right)\nabla\nabla\right]\frac{\exp(ik_0|\mathbf{x} - \mathbf{x}'|)}{|\mathbf{x} - \mathbf{x}'|} \qquad (22)$$

is the free space Green's dyadic function and $k_0 = \omega\sqrt{\mu_0\varepsilon_0}$ is the free space wave number. We recall that every differential equation has two kinds of solutions: the complementary function which is the solution in the absence of a source term, and the particular solution, which depends on the source. The fields $\mathbf{E}_{\text{inc}}(\mathbf{x})$ and $\mathbf{H}_{\text{inc}}(\mathbf{x})$ are the respective complementary functions of the homogeneous (i.e., right side = 0) counterparts of the differential equations (18a) and (18b), and represent the electromagnetic field actually incident on the particle. The right sides of (21a) and (21b) are the particular solutions.

There is no requirement of homogeneity in V thus far, the integral equations (21a) and (21b) being perfectly general.[53] However, homogeneity is required in the ensuing developments.

B. The Long-Wavelength Approximation

Since the particle is electrically small, it may be assumed that the electromagnetic field inside V is spatially constant; i.e., $E(x) \cong E(0)$ and $H(x) \cong H(0)$ for all $x \in V$,[30, 51, 52] provided the maximum linear cross-sectional extent of the particle is small with respect to all wave numbers possible in $V + V_0$ at the given frequency ω. With this long-wavelength approximation, (21a) and (21b), respectively, yield

$$i\omega\varepsilon_0\left[E(0) - E_{inc}(0)\right] = \underline{\underline{L}} \cdot J(0) \qquad (23a)$$

$$i\omega\mu_0\left[H(0) - H_{inc}(0)\right] = \underline{\underline{L}} \cdot K(0) \qquad (23b)$$

Here $\underline{\underline{L}}$ is a depolarization dyadic[54, 55] dependent on the shape of the particle, and is a consequence of the singularity of the $\nabla\nabla[\exp(ik_0|x - x'|)/|x - x'|]$ term in the free space Green's function at $x = x'$ in (21a) and (21b).

Elimination of the fields $E(0)$ and $H(0)$ from (20) and (23) yields

$$J(0)/i\omega\varepsilon_0 = \left[\left(\underline{\underline{\varepsilon}}_r - \underline{\underline{I}}\right) \cdot \underline{\underline{A}}_e^{-1} \cdot \underline{\underline{\zeta}}_r^{-1} \cdot \left(\underline{\underline{L}}^{-1} - \underline{\underline{I}} + \underline{\underline{\mu}}_r\right) - \underline{\underline{\xi}}_r \cdot \underline{\underline{A}}_h^{-1}\right]$$

$$\cdot \underline{\underline{\xi}}_r^{-1} \cdot \underline{\underline{L}}^{-1} \cdot E_{inc}(0)$$

$$+ \left[-\left(\underline{\underline{\varepsilon}}_r - \underline{\underline{I}}\right) \cdot \underline{\underline{A}}_e^{-1} + \underline{\underline{\xi}}_r \cdot \underline{\underline{A}}_h^{-1} \cdot \underline{\underline{\xi}}_r^{-1}\right] \qquad (24)$$

$$\cdot \left(\underline{\underline{L}}^{-1} - \underline{\underline{I}} + \underline{\underline{\varepsilon}}_r\right)\right] \cdot \underline{\underline{\zeta}}_r^{-1} \cdot \underline{\underline{L}}^{-1} \cdot H_{inc}(0)$$

$$K(0)/i\omega\mu_0 = \left[\left(\underline{\underline{\mu}}_r - \underline{\underline{I}}\right) \cdot \underline{\underline{A}}_h^{-1} \cdot \underline{\underline{\xi}}_r^{-1} \cdot \left(\underline{\underline{L}}^{-1} - \underline{\underline{I}} + \underline{\underline{\varepsilon}}_r\right) - \underline{\underline{\zeta}}_r \cdot \underline{\underline{A}}_e^{-1}\right]$$

$$\cdot \underline{\underline{\zeta}}_r^{-1} \cdot \underline{\underline{L}}^{-1} \cdot H_{inc}(0)$$

$$+ \left[-\left(\underline{\underline{\mu}}_r - \underline{\underline{I}}\right) \cdot \underline{\underline{A}}_h^{-1} + \underline{\underline{\zeta}}_r \cdot \underline{\underline{A}}_e^{-1} \cdot \underline{\underline{\zeta}}_r^{-1}\right] \qquad (25)$$

$$\cdot \left(\underline{\underline{L}}^{-1} - \underline{\underline{I}} + \underline{\underline{\mu}}_r\right)\right] \cdot \underline{\underline{\xi}}_r^{-1} \cdot \underline{\underline{L}}^{-1} \cdot E_{inc}(0)$$

where $\underline{\zeta}_r^{-1}$ is the inverse of $\underline{\zeta}_r$, etc., while

$$\underline{\underline{A}}_h = \underline{\underline{I}} - \underline{\underline{\xi}}_r^{-1} \cdot \left(\underline{\underline{L}}^{-1} - \underline{\underline{I}} + \underline{\underline{\varepsilon}}_r \right) \cdot \underline{\underline{\zeta}}_r^{-1} \cdot \left(\underline{\underline{L}}^{-1} - \underline{\underline{I}} + \underline{\underline{\mu}}_r \right) \quad (26a)$$

$$\underline{\underline{A}}_e = \underline{\underline{I}} - \underline{\underline{\zeta}}_r^{-1} \cdot \left(\underline{\underline{L}}^{-1} - \underline{\underline{I}} + \underline{\underline{\mu}}_r \right) \cdot \underline{\underline{\xi}}_r^{-1} \cdot \left(\underline{\underline{L}}^{-1} - \underline{\underline{I}} + \underline{\underline{\varepsilon}}_r \right) \quad (26b)$$

Since dyadics can be interpreted in terms of matrices, computations of the dyadics and their inverses in (24)–(26) can be carried out rather easily.[36]

C. The Polarizability Dyadics

A long-wavelength approximation can also be made for the scattered fields.[52] This leads to the prescription of an equivalent electric dipole moment $\mathbf{p}$ and an equivalent magnetic dipole moment $\mathbf{m}$ such that the fields scattered by the bianisotropic particle can be asymptotically expressed as[56]

$$\mathbf{E}_{sc}(\mathbf{x}) \cong \left\{ \omega^2 \mu_0 \left\{ \underline{\underline{I}} - x^{-2} \mathbf{xx} \right\} \cdot \mathbf{p} - \omega k_0 x^{-1} \mathbf{x} \times \mathbf{m} \right\} \frac{\exp(ik_0 x)}{4\pi x} \qquad \mathbf{x} \in V_0$$

$$(27a)$$

$$\mathbf{H}_{sc}(\mathbf{x}) \cong \left\{ \omega^2 \varepsilon_0 \left\{ \underline{\underline{I}} - x^{-2} \mathbf{xx} \right\} \cdot \mathbf{m} + \omega k_0 x^{-1} \mathbf{x} \times \mathbf{p} \right\} \frac{\exp(ik_0 x)}{4\pi x} \qquad \mathbf{x} \in V_0$$

$$(27b)$$

in the limit $k_0 x \to \infty$; these dipole moments can be computed from[52]

$$\mathbf{p} = (i/\omega) \iiint_V d^3 \mathbf{x} \, \mathbf{J}(\mathbf{x}) \qquad (28a)$$

$$\mathbf{m} = (i/\omega) \iiint_V d^3 \mathbf{x} \, \mathbf{K}(\mathbf{x}) \qquad (28b)$$

We observe here the treatment of $\mathbf{p}$ and $\mathbf{m}$ on an equal footing, which is in accord with the symmetric definition of magnetization as $\mathbf{M} = \mathbf{B} - \mu_0 \mathbf{H}$, and which will be utilized in Section V.C. Furthermore, it is clear from (27) that the Sommerfeld radiation conditions have been satisfied.[51]

Using (24) and (25), and in view of the long-wavelength approximations, $\mathbf{p} = (i/\omega)\nu \mathbf{J}(\mathbf{0})$ and $\mathbf{m} = (i/\omega)\nu \mathbf{K}(\mathbf{0})$, that come from (28a) and (28b), we have

$$\mathbf{p} = \underline{\underline{a}}_{ee} \cdot \mathbf{E}_{inc}(\mathbf{0}) + \underline{\underline{a}}_{em} \cdot \mathbf{H}_{inc}(\mathbf{0}) \tag{29a}$$

$$\mathbf{m} = \underline{\underline{a}}_{me} \cdot \mathbf{E}_{inc}(\mathbf{0}) + \underline{\underline{a}}_{mm} \cdot \mathbf{H}_{inc}(\mathbf{0}) \tag{29b}$$

where

$$\begin{aligned}
\underline{\underline{a}}_{ee} = -\nu\varepsilon_0 &\left[\left(\underline{\underline{\varepsilon}}_r - \underline{\underline{I}} \right) \cdot \underline{\underline{\Delta}}_e^{-1} \cdot \underline{\underline{\zeta}}_r^{-1} \cdot \left(\underline{\underline{L}}^{-1} - \underline{\underline{I}} + \underline{\underline{\mu}}_r \right) - \underline{\underline{\xi}}_r \cdot \underline{\underline{\Delta}}_h^{-1} \right] \\
&\cdot \underline{\underline{\xi}}_r^{-1} \cdot \underline{\underline{L}}^{-1}
\end{aligned} \tag{30a}$$

$$\begin{aligned}
\underline{\underline{a}}_{em} = -\nu\varepsilon_0 &\left[-\left(\underline{\underline{\varepsilon}}_r - \underline{\underline{I}} \right) \cdot \underline{\underline{\Delta}}_e^{-1} + \underline{\underline{\xi}}_r \cdot \underline{\underline{\Delta}}_h^{-1} \cdot \underline{\underline{\xi}}_r^{-1} \cdot \left(\underline{\underline{L}}^{-1} - \underline{\underline{I}} + \underline{\underline{\varepsilon}}_r \right) \right] \\
&\cdot \underline{\underline{\zeta}}_r^{-1} \cdot \underline{\underline{L}}^{-1}
\end{aligned} \tag{30b}$$

$$\begin{aligned}
\underline{\underline{a}}_{me} = -\nu\mu_0 &\left[-\left(\underline{\underline{\mu}}_r - \underline{\underline{I}} \right) \cdot \underline{\underline{\Delta}}_h^{-1} + \underline{\underline{\zeta}}_r \cdot \underline{\underline{\Delta}}_e^{-1} \cdot \underline{\underline{\zeta}}_r \cdot \left(\underline{\underline{L}}^{-1} - \underline{\underline{I}} + \underline{\underline{\mu}}_r \right) \right] \\
&\cdot \underline{\underline{\xi}}_r^{-1} \cdot \underline{\underline{L}}^{-1}
\end{aligned} \tag{30c}$$

$$\begin{aligned}
\underline{\underline{a}}_{mm} = -\nu\mu_0 &\left[\left(\underline{\underline{\mu}}_r - \underline{\underline{I}} \right) \cdot \underline{\underline{\Delta}}_h^{-1} \cdot \underline{\underline{\xi}}_r^{-1} \cdot \left(\underline{\underline{L}}^{-1} - \underline{\underline{I}} + \underline{\underline{\varepsilon}}_r \right) - \underline{\underline{\zeta}}_r \cdot \underline{\underline{\Delta}}_e^{-1} \right] \\
&\cdot \underline{\underline{\zeta}}_r^{-1} \cdot \underline{\underline{L}}^{-1}
\end{aligned} \tag{30d}$$

Expressions (30a)–(30d) can also be specialized to the cases of perfectly conducting particles by setting $\underline{\underline{\mu}}_r = \bar{\omega}\underline{\underline{I}}$, $\underline{\underline{\varepsilon}}_r = \underline{\underline{I}}/\bar{\omega}$, and $\underline{\underline{\zeta}}_r = \underline{\underline{\xi}}_r = \underline{\underline{0}}$.[57] In the limit $\bar{\omega} \to 0$, the particle becomes a perfect electric conductor, and one gets $\underline{\underline{a}}_{ee} = \nu\varepsilon_0\underline{\underline{L}}^{-1}$ and $\underline{\underline{a}}_{mm} = \nu\mu_0(\underline{\underline{L}} - \underline{\underline{I}})^{-1}$. Conversely, in the limit $\bar{\omega} \to \infty$, the particle becomes a perfect magnetic conductor, and we have $\underline{\underline{a}}_{ee} = \nu\varepsilon_0(\underline{\underline{L}} - \underline{\underline{I}})^{-1}$ and $\underline{\underline{a}}_{mm} = \nu\mu_0\underline{\underline{L}}^{-1}$. In either case, the cross dyadics $\underline{\underline{a}}_{em} = \underline{\underline{a}}_{me} = \underline{\underline{0}}$.

The polarizability dyadics (30a)–(30d) can be derived from surface integral equation procedures if the scatterer is spherical in shape and isotropic in its constitution. Reference is made in this connection to Ma et al.,[58] Lakhtakia,[59] and Dungey and Bohren.[60]

D. The Depolarization Dyadic

The dyadics $\underline{\underline{a}}_{ee}$, etc. are the polarizability dyadics of a small bianisotropic particle in vacuum, and contain not only the time-harmonic continuum properties of the matter making up the scatterer, but also the geometry of the scatterer. The quantity

$$\nu = \iiint_V d^3\mathbf{x} \tag{31a}$$

is the volume of the particle. The depolarization dyadic, given by Yaghjian[55]
as

$$\underline{\underline{L}} = \frac{1}{4\pi} \iint_S \frac{d^2\mathbf{x}\,\mathbf{n}\mathbf{x}}{x^3} \tag{31b}$$

contains information on the shape of the particle. Here $\mathbf{n}$ is the unit
outward normal at the point $\mathbf{x} \in S$ on the surface of the particle. It should
be noted that $\underline{\underline{L}}$ is always real symmetric and nonsingular; consequently, it
can always be expressed, at its most complicated, in the biaxial form.[36]
Furthermore, $\underline{\underline{L}}$ has unit trace.

From Yaghjian's exhaustive review of the singularity of $\underline{\underline{G}}(\mathbf{x}, \mathbf{x}')$, it
appears that (31b) may also be applicable for particles that are not convex
in shape; however, V must be a simply connected volume. Though it seems
to have been left unstated in earlier literature, it is clear that the function
$s(\mathbf{x}) = 0$ that defines the surface S must be at least once-differentiable
with finite partial derivatives: This requirement ensures that $\mathbf{n} = \nabla s(\mathbf{x})/
|\nabla s(\mathbf{x})|$ can be determined unambiguously at every point $\mathbf{x} \in S$. Tongue-
in-cheek, it is pointed out that the presented formalism, as well as others,
may not be applicable to particles with truly fractal surfaces,[61] though the
procedures may be admissible when $s(\mathbf{x})$ is a prefractal function.[62]

By far the particles most commonly investigated are ellipsoids or
reductions thereof. Let the semiprincipal axes of the ellipsoidal scatterer
coincide with the coordinate axes $(a\|x_1, b\|x_2, c\|x_3)$. The depolarization
dyadic $\underline{\underline{L}}$ is then given by[54]

$$\underline{\underline{L}} = L_1\mathbf{u}_1\mathbf{u}_1 + L_2\mathbf{u}_2\mathbf{u}_2 + L_3\mathbf{u}_3\mathbf{u}_3 \tag{32}$$

where $\mathbf{u}_m$ is the unit vector corresponding to the coordinate axis x_m
$(m = 1, 2, 3)$,

$$L_1 = (abc/2)\int_0^\infty d\bar\omega\left[(\bar\omega + a^2)^{-3/2}(\bar\omega + b^2)^{-1/2}(\bar\omega + c^2)^{-1/2}\right] \tag{33a}$$

$$L_2 = (abc/2)\int_0^\infty d\bar\omega\left[(\bar\omega + a^2)^{-1/2}(\bar\omega + b^2)^{-3/2}(\bar\omega + c^2)^{-1/2}\right] \tag{33b}$$

$$L_3 = 1 - L_1 - L_2 \tag{33c}$$

and $\nu = 4\pi abc/3$. For a sphere, $\underline{\underline{L}} = \underline{\underline{I}}/3$; for a spheroid $(a = b)$, $L_1 = L_2$;

for needle-shaped objects ($c \gg a$, $c \gg b$), $L_3 \cong 0$; and for disk-shaped objects ($c \ll a$, $c \ll b$), $L_3 \cong 1$. The analysis contained in this subsection can always be applied to the quasi-two-dimensional cases of infinitely long cylinders of electrically small cross sections. Although those cases can be treated as reductions of the preceding analysis, care must be taken to ensure the proper limiting conditions along the x_3 axis; reference is made to Lakhtakia[63] in this connection.

IV. POLARIZABILITY DYADICS OF MOLECULES

That matter is discrete has long been established. Therefore, the polarizability dyadics developed in Section III really do not apply at the molecular (or the atomic) level.[64] To appreciate this, consider the fact that the frequency-dependent constitutive equations (14) are given for a bianisotropic continuum. True, the continuum material is actually made of molecules, but in the process of obtaining the macroscopic relations (14), the small-scale fluctuations of the electromagnetic field were smoothed out.[9, 56, 65-67] Furthermore, though one can use statistics to go from the microscopic to the macroscopic, the reverse transition is impossible: However small the bianisotropic particles, they can never become molecular. The situation has some parallel with ecology and sociology: Whereas the functions of a multicellular entity are determined by those of its constituent cells, the functions of a constituent cell cannot be determined with certitude from those of the entire entity.[68]

Molecular polarizabilities, therefore, cannot come from macroscopic analyses and may only be obtainable as ab initio estimates; see, for example, Andre and Champagne,[69] and Meyers and Bredas.[70] However, a vast literature in chemical physics testifies to the difficulties underlying such investigations, as exemplified by Glover and Weinhold[71] on two-electron atoms; see also the recent book by Anastasovski,[72] but that work must be read with some caution as to notation. Consider the fact that an isolated molecule, although a scattering center, has such a small scattering efficiency at nonresonant frequencies as to be almost unidentifiable from the background; this argument is buttressed by reports that extremely small dielectric spheres also have minute scattering efficiencies.[73] Indeed, at very small (but supermolecular) length scales, (low-frequency) scattering is observable simply because there are lots of molecules present to scatter.[23, 73] Therefore, estimating molecular polarizabilities should be considered a never-ending interactive game in which these microscopic quantities are interpreted via some model from observations on continua.[10, 12, 17, 74]

At optical and suboptical frequencies, the atom can be regarded as a point-polarizable entity, and can be modeled as an electric dipole with its dielectric polarizability being possibly dyadic. Except at much lower frequencies, this quasielectrostatic model cannot hold true for molecules, in general, but is often so held.[75] Modeling a multiatomic molecule, particularly a macromolecule, as solely a point electric dipole is tantamount to ignoring its conformation, as well as the interatomic influences. This statement should be interpreted in light of the free space Green's dyadic function $\underline{G}(x, x')$: The source-influence felt at a field point is dependent on the vectorial displacement $(x - x')$ of the field point x from the source point x', and was apparently first appreciated by Gray[76] during his studies on the optical activity of liquids and gases. A graphic illustration was experimentally provided by Lindman[77, 78] while reporting his pioneering 1914 studies wherein he compared the electromagnetic responses of optically active molecules and miniature metallic helices, which has been subsequently confirmed by a host of computer studies.[41]

Without going into any details regarding the current state of modeling molecular electromagnetic responses, from either the classical or the quantum viewpoints, it will suffice to state here that the particulate forms,

$$\mathbf{p} = \underline{\underline{a}}_{ee} \cdot \mathbf{E}_{inc}(0) + \underline{\underline{a}}_{em} \cdot \mathbf{H}_{inc}(0) \tag{34a}$$

$$\mathbf{m} = \underline{\underline{a}}_{me} \cdot \mathbf{E}_{inc}(0) + \underline{\underline{a}}_{mm} \cdot \mathbf{H}_{inc}(0) \tag{34b}$$

should serve equally well for molecules at low frequencies.[79-82] The frequency-dependent polarizability dyadics, now, are functions of (1) The constitution of the molecule,[83] (2) the conformation of the molecule with respect to the ambient conditions,[84-86] and (3) the electromagnetic fields present at some other frequency (including electrostatic and magnetostatic fields), thereby giving rise to nonlinear responses.[87-92] In addition, if the coherent fields are highly energetic, these fields can themselves distort the molecular conformations.[93] Last, but not the least, observational subjectivity,[94] the overall environment, and the history may not be ignored either.[80, 95, 96] That the polarizability dyadics in (34a) and (34b) are purely phenomenological cannot be exaggerated: It is most clearly evident in studies on second-harmonic generation at optical frequencies.[10, 97-99]

V. THE EQUIVALENT CONTINUUM

Whether the scattering centers are considered to be particulate or molecular, the propagation of an electromagnetic wave in a rarefield gas of such centers should properly be thought of as a many-body problem. The trouble with many-body problems is indicated in their name: There are just too many bodies to keep track of. It is, of course, possible to deal with a few elementary scattering centers—to wit, the coupled-dipole method[53] or the Purcell-Pennypacker scheme.[100] Even there, computer memories are quickly swamped by a relatively small number of electric dipolar scattering centers,[60] so that the consideration of more detailed interaction models (e.g., Refs. 101–103) has become feasible only with the advent of supercomputers. Although considerable insight is provided by such few-body studies, to deal with a large number necessitates the use of statistical techniques. It is these techniques that allow one to go from the microscopic to the continuum, but, in the process, the reverse transformation becomes impossible to any reasonable degree.

The electromagnetic field scattered by any scatterer (in free space) can be decomposed into multipoles.[56] In particular, at low enough frequencies, the scattering response of a homogenous dielectric sphere is isomorphic with the radiation characteristics of a point electric dipole; hence, an electrically small dielectric sphere may be adequately characterized by an electric polarizability.[104] This fact, along with the Clausius-Mossotti relation, was utilized by Maxwell Garnett[105] to fashion a theory for the macroscopic properties of a composite medium constructed by randomly dispersing small dielectric spherical inclusions in a dielectric host material. The resulting approach has been considerably augmented for different cases (e.g., Refs. 23, 106, 107); it also has its competitors in the Bruggeman model[108] and others (e.g., Ref. 109, 110), as well as in the more rigorous multiple scattering theories.[111–113] But the elegant simplicity of the Maxwell Garnett approach has lead to its extensive usage, despite its many limitations.[23, 107] The Maxwell Garnett approach is developed in this section.

A. Orientational Averaging

Before creating the Maxwell Garnett continuum model of the gas, it is necessary to explore some ancillary aspects. The scattering centers are not stationary, even though the volume filled with the gas may be stationary with respect to an observer. However, the motions of the scattering centers are random and bounded, in the absence of any externally imposed bias, so that the centers can be treated as being stationary in the characteristic Nyquist time intervals $\pi/2\omega$.[27]

An externally imposed bias may create an orientational distribution. As mentioned at the beginning of Section 3, let $\mathbf{x} \equiv (x_1, x_2, x_3)$ be a Cartesian coordinate system specific to one particular scattering center, and let $\mathbf{X} \equiv (X_1, X_2, X_3)$ be the laboratory Cartesian coordinate frame. To transform from the (x_1, x_2, x_3) to the (X_1, X_2, X_3) system, the Euler angles (ψ_1, ψ_2, ψ_3) have to be specified;[103] in matrix notation,

$$
\begin{bmatrix} X_1 \\ X_2 \\ X_3 \end{bmatrix} = \begin{bmatrix} \cos\psi_3 & \sin\psi_3 & 0 \\ -\sin\psi_3 & \cos\psi_3 & 0 \\ 0 & 0 & 1 \end{bmatrix} \begin{bmatrix} \cos\psi_2 & 0 & -\sin\psi_2 \\ 0 & 1 & 0 \\ \sin\psi_2 & 0 & \cos\psi_2 \end{bmatrix}
$$
$$
\times \begin{bmatrix} \cos\psi_1 & \sin\psi_1 & 0 \\ -\sin\psi_1 & \cos\psi_1 & 0 \\ 0 & 0 & 1 \end{bmatrix} \begin{bmatrix} x_1 \\ x_2 \\ x_3 \end{bmatrix} \tag{35}
$$

Let now $n(\psi_1, \psi_2\,\psi_3)$ be the volumetric number density of otherwise identical scattering centers which require the rotation (ψ_1, ψ_2, ψ_3) in order to be described in the laboratory frame. For particulate scatterers, a weak mean-field approach can be used to obtain an orientation-averaged depolarization dyadic

$$
\langle \underline{\underline{L}} \rangle = \left[\int_0^{2\pi} d\psi_3 \int_0^{\pi} d\psi_2 \sin\psi_2 \int_0^{2\pi} d\psi_1 \left\{ \underline{\underline{L}}(\psi_1, \psi_2\,\psi_3) n(\psi_1, \psi_2\,\psi_3) \right\} \right] \Big/
$$
$$
\left[\int_0^{2\pi} d\psi_3 \int_0^{\pi} d\psi_2 \sin\psi_2 \int_0^{2\pi} d\psi_1\, n(\psi_1, \psi_2, \psi_3) \right\} \right]; \tag{36}
$$

then $\langle \underline{\underline{L}} \rangle$ can be used in place of $\underline{\underline{L}}(\psi_1, \psi_2, \psi_3)$ in (26) and (30).

Averaging, however, should be done at the very highest level possible. Therefore, in a more direct approach, the polarizability dyadics should be orientation-averaged as

$$
\langle \underline{\underline{a}}_{ee} \rangle = \left[\int_0^{2\pi} d\psi_3 \int_0^{\pi} d\psi_2 \sin\psi_2 \int_0^{2\pi} d\psi_1 \left\{ \underline{\underline{a}}_{ee}(\psi_1, \psi_2, \psi_3) n(\psi_1, \psi_2, \psi_3) \right\} \right] \Big/
$$
$$
\left[\int_0^{2\pi} d\psi_3 \int_0^{\pi} d\psi_2 \sin\psi_2 \int_0^{2\pi} d\psi_1\, n(\psi_1, \psi_2, \psi_3) \right\} \right], \tag{37}
$$

etc. If all scattering centers are identically-aligned, there is no need for orientation averaging; and if the distribution $n(\psi_1, \psi_2, \psi_3)$ is isotropic, then

$$
\langle \underline{\underline{a}}_{ee} \rangle = \left[\int_0^{2\pi} d\psi_3 \int_0^{\pi} d\psi_2 \sin\psi_2 \int_0^{2\pi} d\psi_1 \left\{ \underline{\underline{a}}_{ee}(\psi_1, \psi_2, \psi_3) \right\} \right] \Big/ 8\pi^2, \tag{38}
$$

etc. In the following, the angular brackets are dropped, it being under-stood that the polarizability dyadics have been orientation-averaged.

B. Multiphase Mixtures

The next form of averaging that may be needed is for mixtures of different types of scattering centers.[114, 115] Consider a gas containing $j \in \{1, 2, \ldots, J\}$ species, where N_j is the volumetric number density of the jth species such that

$$N_1 + N_2 + \cdots + N_J = N \tag{39}$$

where N is total number of scattering centers of all types per unit volume. Let $\underline{a}_{ee, j}$, etc., be the orientation-averaged polarizability dyadics of the jth species. Provided the gas is sufficiently sparse, i.e., the total "volume per unit volume" occupied by all types of scattering centers is small, one can use the phase-averaged polarizability dyadics

$$\langle\langle \underline{a}_{ee} \rangle\rangle = \frac{N_1 \underline{a}_{ee, 1} + N_2 \underline{a}_{ee, 2} + \cdots + N_J \underline{a}_{ee, J}}{N} \tag{40}$$

etc., in constructing the equivalent continuum. If the various species chemically react with each other,[116] all reaction rate time constants must be $\gg \pi/2\omega$.

Again, we drop the angle brackets to denote the phase-averaging process, it being assumed in what follows that the gas is composed of N orientation-averaged, phase-averaged scattering centers per unit volume.

C. The Maxwell Garnett Continuum Model

If the gas composed of the electrically small scattering centers is to be viewed as being effectively homogeneous, its frequency-dependent consti-tutive relations must be of the forms

$$\mathbf{D} = \varepsilon_0 \left[\underline{\varepsilon} \cdot \mathbf{E} + \underline{\xi} \cdot \mathbf{H} \right] \tag{41a}$$

$$\mathbf{B} = \mu_0 \left[\underline{\zeta} \cdot \mathbf{E} + \underline{\mu} \cdot \mathbf{H} \right] \tag{41b}$$

The concept of flux densities $\mathbf{D}$ and $\mathbf{B}$ implies a polarization field $\mathbf{P} = (\mathbf{D} - \varepsilon_0 \mathbf{E})$ and a magnetization field $\mathbf{M} = (\mathbf{B} - \mu_0 \mathbf{H})$. The polarization field $\mathbf{P}$ is defined as the electric dipole moment per unit volume, while the magnetization field $\mathbf{M}$ is the magnetic dipole moment per unit volume; thus,

$$\mathbf{P} = N\mathbf{p} \qquad \mathbf{M} = N\mathbf{m} \tag{42}$$

The equivalent moments, **p** and **m**, of a single scattering center are proportional to the local electric and magnetic fields exciting it.[56, 117] Therefore, either from (29) for particulate scatterers or from (34) for molecular scatterers, and after implementing the averaging procedures discussed above, it follows that

$$\mathbf{p} = \underline{\underline{a}}_{ee} \cdot \mathbf{E}_L + \underline{\underline{a}}_{em} \cdot \mathbf{H}_L \tag{43a}$$

$$\mathbf{m} = \underline{\underline{a}}_{me} \cdot \mathbf{E}_L + \underline{\underline{a}}_{mm} \cdot \mathbf{H}_L \tag{43b}$$

can be used in (42), where the subscript L stands for local. In a cubic Lorentzian model of the effective continuum, the usual[117] prescription for the Lorentz field can be followed; thus,

$$\mathbf{E}_L = \mathbf{E} + \frac{\mathbf{P}}{3\varepsilon_0} \tag{44a}$$

$$\mathbf{H}_L = \mathbf{H} + \frac{\mathbf{M}}{3\mu_0} \tag{44b}$$

Faxén[118] and Lundblad[119] gave a similar procedure for isotropic dielectric media in 1920 to show that the Lorenz-Lorentz equation[120] could be derived from the Maxwell postulates. Some extensions of this equation are reviewed in Ref. 103.

Simultaneous solution of (42)–(44) then yields

$$
\begin{aligned}
\mathbf{P} = &\left\{ \left(\frac{3\mu_0}{N} \right) \underline{\underline{a}}_{em}^{-1} \cdot \left[\underline{\underline{I}} - \left(\frac{N}{3\varepsilon_0} \right) \underline{\underline{a}}_{ee} \right] - \left[\underline{\underline{I}} - \left(\frac{N}{3\mu_0} \right) \underline{\underline{a}}_{mm} \right]^{-1} \cdot \left(\frac{N}{3\varepsilon_0} \right) \underline{\underline{a}}_{me} \right\}^{-1} \\
&\cdot \left\{ \left[\underline{\underline{I}} - \left(\frac{N}{3\mu_0} \right) \underline{\underline{a}}_{mm} \right]^{-1} \cdot N\underline{\underline{a}}_{me} + 3\mu_0 \underline{\underline{a}}_{em}^{-1} \cdot \underline{\underline{a}}_{ee} \right\} \cdot \mathbf{E} \\
&+ \left\{ \left(\frac{3\mu_0}{N} \right) \underline{\underline{a}}_{em}^{-1} \cdot \left[\underline{\underline{I}} - \left(\frac{N}{3\varepsilon_0} \right) \underline{\underline{a}}_{ee} \right] - \left[\underline{\underline{I}} - \left(\frac{N}{3\mu_0} \right) \underline{\underline{a}}_{mm} \right]^{-1} \right. \\
&\left. \cdot \left(\frac{N}{3\varepsilon_0} \right) \underline{\underline{a}}_{me} \right\}^{-1} \\
&\cdot \left\{ \left[\underline{\underline{I}} - \left(\frac{N}{3\mu_0} \right) \underline{\underline{a}}_{mm} \right]^{-1} \cdot N\underline{\underline{a}}_{mm} + 3\mu_0 \underline{\underline{I}} \right\} \cdot \mathbf{H}
\end{aligned}
\tag{45}
$$

and

$$
\begin{aligned}
\mathbf{M} = &\left\{ \left[\left(\frac{3\varepsilon_0}{N} \right) \underline{\underline{a}}_{me}^{-1} \cdot \left[\underline{\underline{I}} - \left(\frac{N}{3\mu_0} \right) \underline{\underline{a}}_{mm} \right] - \left[\underline{\underline{I}} - \left(\frac{N}{3\varepsilon_0} \right) \underline{\underline{a}}_{ee} \right]^{-1} \right. \right. \\
&\left. \cdot \left(\frac{N}{3\mu_0} \right) \underline{\underline{a}}_{em} \right\}^{-1} \\
&\cdot \left\{ \left[\underline{\underline{I}} - \left(\frac{N}{3\varepsilon_0} \right) \underline{\underline{a}}_{ee} \right]^{-1} \cdot N\underline{\underline{a}}_{em} + 3\varepsilon_0 \underline{\underline{a}}_{me}^{-1} \cdot \underline{\underline{a}}_{mm} \right\} \cdot \mathbf{H} \\
&+ \left\{ \left[\left(\frac{3\varepsilon_0}{N} \right) \underline{\underline{a}}_{me}^{-1} \cdot \left[\underline{\underline{I}} - \left(\frac{N}{3\mu_0} \right) \underline{\underline{a}}_{mm} \right] - \left[\underline{\underline{I}} - \left(\frac{N}{3\varepsilon_0} \right) \underline{\underline{a}}_{ee} \right]^{-1} \right. \right. \\
&\left. \cdot \left(\frac{N}{3\mu_0} \right) \underline{\underline{a}}_{em} \right\}^{-1} \\
&\cdot \left\{ \left[\underline{\underline{I}} - \left(\frac{N}{3\varepsilon_0} \right) \underline{\underline{a}}_{ee} \right]^{-1} \cdot N\underline{\underline{a}}_{ee} + 3\varepsilon_0 \underline{\underline{I}} \right\} \cdot \mathbf{E}
\end{aligned} \tag{46}
$$

It follows from (41), as well as from the definitions of **P** and **M**, that the constitutive parameters of the effective medium can now be estimated as

$$
\begin{aligned}
\underline{\underline{\varepsilon}} - \underline{\underline{I}} = &\left\{ \left(\frac{3\mu_0}{N} \right) \underline{\underline{a}}_{em}^{-1} \cdot \left[\underline{\underline{I}} - \left(\frac{N}{3\varepsilon_0} \right) \underline{\underline{a}}_{ee} \right] \right. \\
&\left. - \left[\underline{\underline{I}} - \left(\frac{N}{3\mu_0} \right) \underline{\underline{a}}_{mm} \right]^{-1} \cdot \left(\frac{N}{3\varepsilon_0} \right) \underline{\underline{a}}_{me} \right\}^{-1} \\
&\cdot \frac{1}{\varepsilon_0} \left\{ \left[\underline{\underline{I}} - \left(\frac{N}{3\mu_0} \right) \underline{\underline{a}}_{mm} \right]^{-1} \cdot N\underline{\underline{a}}_{me} + 3\mu_0 \underline{\underline{a}}_{em}^{-1} \cdot \underline{\underline{a}}_{ee} \right\}
\end{aligned} \tag{47}
$$

$$
\begin{aligned}
\underline{\underline{\xi}} = &\left\{ \left(\frac{3\mu_0}{N} \right) \underline{\underline{a}}_{em}^{-1} \cdot \left[\underline{\underline{I}} - \left(\frac{N}{3\varepsilon_0} \right) \underline{\underline{a}}_{ee} \right] - \left[\underline{\underline{I}} - \left(\frac{N}{3\mu_0} \right) \underline{\underline{a}}_{mm} \right]^{-1} \cdot \left(\frac{N}{3\varepsilon_0} \right) \underline{\underline{a}}_{me} \right\}^{-1} \\
&\cdot \frac{1}{\varepsilon_0} \left\{ \left[\underline{\underline{I}} - \left(\frac{N}{3\mu_0} \right) \underline{\underline{a}}_{mm} \right]^{-1} \cdot N\underline{\underline{a}}_{mm} + 3\mu_0 \underline{\underline{I}} \right\}
\end{aligned} \tag{48}
$$

$$
\underline{\underline{\mu}} - \underline{\underline{I}} = \left\{ \left(\frac{3\varepsilon_0}{N} \right) \underline{\underline{a}}_{me}^{-1} \cdot \left[\underline{\underline{I}} - \left(\frac{N}{3\mu_0} \right) \underline{\underline{a}}_{mm} \right] \right.
$$
$$
- \left[\underline{\underline{I}} - \left(\frac{N}{3\varepsilon_0} \right) \underline{\underline{a}}_{ee} \right]^{-1} \cdot \left(\frac{N}{3\mu_0} \right) \underline{\underline{a}}_{em} \right\}^{-1} \tag{49}
$$
$$
\cdot \frac{1}{\mu_0} \left\{ \left[\underline{\underline{I}} - \left(\frac{N}{3\varepsilon_0} \right) \underline{\underline{a}}_{ee} \right]^{-1} \cdot N \underline{\underline{a}}_{em} + 3\varepsilon_0 \underline{\underline{a}}_{me}^{-1} \cdot \underline{\underline{a}}_{mm} \right\}
$$

and

$$
\underline{\underline{\zeta}} = \left\{ \left(\frac{3\varepsilon_0}{N} \right) \underline{\underline{a}}_{me}^{-1} \cdot \left[\underline{\underline{I}} - \left(\frac{N}{3\mu_0} \right) \underline{\underline{a}}_{mm} \right] - \left[\underline{\underline{I}} - \left(\frac{N}{3\varepsilon_0} \right) \underline{\underline{a}}_{ee} \right]^{-1} \right.
$$
$$
\left. \cdot \left(\frac{N}{3\mu_0} \right) \underline{\underline{a}}_{em} \right\}^{-1} \cdot \frac{1}{\mu_0} \left\{ \left[\underline{\underline{I}} - \left(\frac{N}{3\varepsilon_0} \right) \underline{\underline{a}}_{ee} \right]^{-1} \cdot N \underline{\underline{a}}_{ee} + 3\varepsilon_0 \underline{\underline{I}} \right\} \tag{50}
$$

these four expressions constituting an extensive generalization of the original Maxwell Garnett formula of 1904 if the particulate polarizability dyadics (30a)–(30d) are substituted therein.

To be observed in (47)–(50) are the complicated dependences of the Maxwell Garnett estimates of $\underline{\underline{\varepsilon}}$, $\underline{\underline{\xi}}$, $\underline{\underline{\mu}}$, and $\underline{\underline{\zeta}}$ on the number density N. It may be of use to consider the limiting case of an extremely dilute gas. In such a dilute composite, correct to the first order in the number density N, one gets from (47)–(50) that

$$
\underline{\underline{\varepsilon}} \cong \underline{\underline{I}} + \left(\frac{N}{\varepsilon_0} \right) \underline{\underline{a}}_{ee} + \text{higher order terms in } N \tag{51a}
$$

$$
\underline{\underline{\xi}} \cong \left(\frac{N}{\varepsilon_0} \right) \underline{\underline{a}}_{em} + \text{higher order terms in } N \tag{51b}
$$

$$
\underline{\underline{\mu}} \cong \underline{\underline{I}} + \left(\frac{N}{\mu_0} \right) \underline{\underline{a}}_{mm} + \text{higher order terms in } N \tag{51c}
$$

$$
\underline{\underline{\zeta}} \cong \left(\frac{N}{\mu_0} \right) \underline{\underline{a}}_{me} + \text{higher order terms in } N \tag{51d}
$$

This dilute-limit result could also have been obtained by replacing the local fields $\{\mathbf{E}_L, \mathbf{H}_L\}$ by $\{\mathbf{E}, \mathbf{H}\}$ in (43), because sparsely distributed scattering centers will not interact significantly with each other. Higher order terms in N have been deduced for far-infrared collision-induced spectroscopy, for example, in Chapter 11 of Ref. 103.

D. Assessment of the Maxwell Garnett Continuum Model

Equations (47)–(50) tell us that the gas of scattering centers should be thought of as a bianisotropic continuum at macroscopic scales, unless the simplest cases are being dealt with; this conclusion is reinforced by the small-N expressions (51a)–(51d).

It has been shown that in metals $E_L = E$, because of strongly delocalized nearly-free electrons;[12] furthermore, in other situations, more complicated interactions may exist.[121] So the more general equivalents of (44a) and (44b) should be $E_L = E + \underline{\underline{Q}} \cdot P/\varepsilon_0$ and $H_L = H + \underline{\underline{R}} \cdot M/\mu_0$, $\underline{\underline{Q}}$ and $\underline{\underline{R}}$ being the interaction dyadics. For the gaseous continuum considered here, however, $\underline{\underline{Q}} = \underline{\underline{R}} = \underline{\underline{I}}/3$ is a reasonable approximation, as can be inferred also from Evans.[122]

What is interesting about the present Maxwell Garnett model is that no simpler model can possibly be devised that retains the bianisotropic flavor of the scattering centers. This is because of the complicated long-wavelength response, (43a) and (43b), of the scattering centers. Several simpler expressions are available[123–132]—due to Arago and Biot; Oster; Landau, Lifshitz and Looyenga; Böttcher; Rayleigh; Bruggeman and Hanai; and Lichtenecker; among others—that work for isotropic dielectric particles, but may not be easily extended to the bianisotropic scattering centers. Partly, this is due to the empirical natures of some of the formulae, as in the case of that of Arago and Biot; partly, as in the case of the Bruggeman formula, this is due to the requirement of the knowledge of at least the singularities of the infinite-medium Green's dyadic function for a bianisotropic medium: These are not yet known.

The presented Maxwell Garnett continuum model is also the most complete known to this author. More sophisticated models based on quantifying the response of the particle in terms of a T matrix are in use.[111, 113] But these models can only yield, at best, wave numbers for a plane wave moving along the coherent direction in the equivalent continuum. Thus, these models result in an incomplete specification of the equivalent continuum.

VI. ELECTROMAGNETIC FIELDS IN THE CONTINUUM

At the frequency of interest, the equivalent Maxwell Garnett continuum possesses the frequency-dependent relations

$$D = \varepsilon_0 \left[\underline{\underline{\varepsilon}} \cdot E + \underline{\underline{\xi}} \cdot H \right] \tag{52a}$$

$$B = \mu_0 \left[\underline{\underline{\zeta}} \cdot E + \underline{\underline{\mu}} \cdot H \right] \tag{52b}$$

Assuming that these are adequate to describe electromagnetic phenomena in the continuum—and they may not be; for discussion, see Barrett[7]—we are in a position to make several explorations now. The laboratory frame $\mathbf{X} \equiv (X_1, X_2, X_3)$ is used exclusively from here on.

A. Field–Theoretic Relations

Substitution of (52a) and (52b) into the usual time-harmonic Maxwell curl postulates

$$\nabla \times \mathbf{E}(\mathbf{X}) = i\omega \mathbf{B}(\mathbf{X}) - \mathbf{K}(\mathbf{X}) \tag{53a}$$

$$-\nabla \times \mathbf{H}(\mathbf{X}) = i\omega \mathbf{D}(\mathbf{X}) - \mathbf{J}(\mathbf{X}) \tag{53b}$$

yields

$$\nabla \times \mathbf{E} = i\omega \mu_0 \left[\underline{\underline{\zeta}} \cdot \mathbf{E} + \underline{\underline{\mu}} \cdot \mathbf{H} \right] - \mathbf{K} \tag{53c}$$

$$-\nabla \times \mathbf{H} = i\omega \varepsilon_0 \left[\underline{\underline{\varepsilon}} \cdot \mathbf{E} + \underline{\underline{\xi}} \cdot \mathbf{H} \right] - \mathbf{J} \tag{53d}$$

with $\mathbf{J}$ and $\mathbf{K}$ being the impressed source current densities. Likewise, the substitution of (51) into the usual time-harmonic Maxwell divergence postulates

$$\nabla \cdot \mathbf{D}(\mathbf{X}) = \rho_e(\mathbf{X}) \tag{54a}$$

$$\nabla \cdot \mathbf{B}(\mathbf{X}) = \rho_m(\mathbf{X}) \tag{54b}$$

gives

$$\varepsilon_0 \nabla \cdot \left[\underline{\underline{\varepsilon}} \cdot \mathbf{E} + \underline{\underline{\xi}} \cdot \mathbf{H} \right] = \rho_e \tag{54c}$$

$$\mu_0 \nabla \cdot \left[\underline{\underline{\zeta}} \cdot \mathbf{E} + \underline{\underline{\mu}} \cdot \mathbf{H} \right] = \rho_m \tag{54d}$$

respectively, where ρ_e and ρ_m are the impressed source charge densities. The duality[56] inherent in the Maxwell postulates gives rise to the duality transform of the bianisotropic continuum:

$$\mathbf{E} \to \mathbf{H} \quad \mathbf{H} \to -\mathbf{E} \quad \mathbf{J} \to \mathbf{K} \quad \mathbf{K} \to -\mathbf{J} \quad \rho_e \to \rho_m \quad \rho_m \to -\rho_e$$

$$\varepsilon_0 \to \mu_0 \quad \mu_0 \to \varepsilon_0 \quad \underline{\underline{\varepsilon}} \to \underline{\underline{\mu}} \quad \underline{\underline{\mu}} \to \underline{\underline{\varepsilon}}, \quad \underline{\underline{\zeta}} \to -\underline{\underline{\xi}} \quad \underline{\underline{\xi}} \to -\underline{\underline{\zeta}} \tag{55}$$

It can be easily verified by making these interchanges that (53c), (53d), (54c), and (54d) are invariant with respect to the duality transform (55).

Some other symmetry conditions can also be deduced without explicitly solving (53c), (53d), (54c), and (54d). First, the bianisotropic continuum is Lorentz reciprocal provided[118]

$$\underline{\underline{\varepsilon}} = \underline{\underline{\varepsilon}}^{tr} \quad \underline{\underline{\mu}} = \underline{\underline{\mu}}^{tr} \quad \underline{\underline{\zeta}} = -\underline{\underline{\xi}}^{tr} \tag{56}$$

where the superscript tr denotes the transpose. Now, Rumsey's reaction theorem suggests that the continuum complementary to (52a) and (52b) will have its constitutive equations given by[133]

$$\mathbf{D} = \varepsilon_0 \left[\underline{\underline{\varepsilon}}^{tr} \cdot \mathbf{E} - \underline{\underline{\zeta}}^{tr} \cdot \mathbf{H} \right] \tag{57a}$$

$$\mathbf{B} = \mu_0 \left[-\underline{\underline{\xi}}^{tr} \cdot \mathbf{E} + \underline{\underline{\mu}}^{tr} \cdot \mathbf{H} \right] \tag{57b}$$

In view of (56), our bianisotropic continuum is, therefore, self-complementary.

Second, the complex Poynting theorem yields

$$\nabla \cdot (\mathbf{E} \times \mathbf{H}^*) = i\omega(\mathbf{B} \cdot \mathbf{H}^* - \mathbf{E} \cdot \mathbf{D}^*) - (\mathbf{K} \cdot \mathbf{H}^* + \mathbf{E} \cdot \mathbf{J}^*) \tag{58}$$

where the asterisk denotes the complex conjugate. In the absence of externally impressed sources (i.e., $\mathbf{J} = \mathbf{K} = \mathbf{0}$), the left side of (58) must be identically zero if the continuum has no intrinsic loss. Therefore, by substituting (52a) and (52b) on the right side of (58), the conditions for losslessness can be obtained as[36]

$$\underline{\underline{\varepsilon}} = \underline{\underline{\varepsilon}}^{*tr} \quad \underline{\underline{\mu}} = \underline{\underline{\mu}}^{*tr} \quad \underline{\underline{\zeta}} = \underline{\underline{\xi}}^{*tr} \tag{59}$$

On comparing (56) and (59), one cannot help observing that reciprocal lossless continua are really special!

B.   Vector–Dyadic Helmholtz Equations

We note that Eqs. (53c) and (53d) can be, respectively, rewritten as

$$\left[\nabla \times \underline{\underline{I}} - i\omega\mu_0\underline{\underline{\zeta}} \right] \cdot \mathbf{E}(\mathbf{X}) = i\omega\mu_0\underline{\underline{\mu}} \cdot \mathbf{H}(\mathbf{X}) - \mathbf{K}(\mathbf{X}) \tag{60a}$$

$$\left[\nabla \times \underline{\underline{I}} + i\omega\varepsilon_0\underline{\underline{\xi}} \right] \cdot \mathbf{H}(\mathbf{X}) = -i\omega\varepsilon_0\underline{\underline{\varepsilon}} \cdot \mathbf{E}(\mathbf{X}) + \mathbf{J}(\mathbf{X}) \tag{60b}$$

using vector-dyadic algebra. Elimination of $\mathbf{H}(\mathbf{X})$ from (60a) and (60b) yields the source-incorporated Helmholtz equation for the electric field as

$$\underline{\underline{W}}_e(\nabla) \cdot \mathbf{E}(\mathbf{X}) = i\omega\mu_0 \mathbf{J}(\mathbf{X}) - \left[\nabla \times \underline{\underline{I}} + i\omega\varepsilon_0\underline{\underline{\xi}}\right] \cdot \underline{\underline{\mu}}^{-1} \cdot \mathbf{K}(\mathbf{X}) \quad (61a)$$

where the dyadic differential operator

$$\underline{\underline{W}}_e(\nabla) = \left[\nabla \times \underline{\underline{I}} + i\omega\varepsilon_0\underline{\underline{\xi}}\right] \cdot \underline{\underline{\mu}}^{-1} \cdot \left[\nabla \times \underline{\underline{I}} - i\omega\mu_0\underline{\underline{\zeta}}\right] - k_0^2\underline{\underline{\varepsilon}} \quad (61b)$$

The similar elimination of $\mathbf{E}(\mathbf{X})$ from (60a) and (60b) yields the source-incorporated Helmholtz equation for the magnetic field as

$$\underline{\underline{W}}_m(\nabla) \cdot \mathbf{H}(\mathbf{X}) = i\omega\varepsilon_0 \mathbf{K}(\mathbf{X}) + \left[\nabla \times \underline{\underline{I}} - i\omega\mu_0\underline{\underline{\zeta}}\right] \cdot \underline{\underline{\varepsilon}}^{-1} \cdot \mathbf{J}(\mathbf{X}) \quad (62a)$$

where the dyadic differential operator

$$\underline{\underline{W}}_m(\nabla) = \left[\nabla \times \underline{\underline{I}} - i\omega\mu_0\underline{\underline{\zeta}}\right] \cdot \underline{\underline{\varepsilon}}^{-1} \cdot \left[\nabla \times \underline{\underline{I}} + i\omega\varepsilon_0\underline{\underline{\xi}}\right] - k_0^2\underline{\underline{\mu}} \quad (62b)$$

The solutions of (61a) and (62a) must be linear with respect to source distributions; it is an easy matter to verify by direct substitution into (60a) and (60b) that these solutions must be

$$\mathbf{E}(\mathbf{X}) = i\omega\mu_0 \iiint d^3\mathbf{X}' \underline{\underline{G}}_e(\mathbf{X};\mathbf{X}') \cdot \mathbf{J}(\mathbf{X}')$$
$$- \underline{\underline{\varepsilon}}^{-1} \cdot \left[\nabla \times \underline{\underline{I}} + i\omega\varepsilon_0\underline{\underline{\xi}}\right] \cdot \iiint d^3\mathbf{X}' \underline{\underline{G}}_m(\mathbf{X};\mathbf{X}') \cdot \mathbf{K}(\mathbf{X}') \quad (63)$$

and

$$\mathbf{H}(\mathbf{X}) = i\omega\varepsilon_0 \iiint d^3\mathbf{X}' \underline{\underline{G}}_m(\mathbf{X};\mathbf{X}') \cdot \mathbf{K}(\mathbf{X}')$$
$$+ \underline{\underline{\mu}}^{-1} \cdot \left[\nabla \times \underline{\underline{I}} - i\omega\mu_0\underline{\underline{\zeta}}\right] \cdot \iiint d^3\mathbf{X}' \underline{\underline{G}}_e(\mathbf{X};\mathbf{X}') \cdot \mathbf{J}(\mathbf{X}') \quad (64)$$

provided the bianisotropic Green's dyadics $\underline{\underline{G}}_e(\mathbf{X};\mathbf{X}')$ and $\underline{\underline{G}}_m(\mathbf{X};\mathbf{X}')$, respectively, obey the dyadic differential equations

$$\underline{\underline{W}}_e(\nabla) \cdot \underline{\underline{G}}_e(\mathbf{X};\mathbf{X}') = \underline{\underline{I}}\delta(\mathbf{X} - \mathbf{X}') \quad (65a)$$

$$\underline{\underline{W}}_m(\nabla) \cdot \underline{\underline{G}}_m(\mathbf{X};\mathbf{X}') = \underline{\underline{I}}\delta(\mathbf{X} - \mathbf{X}') \quad (65b)$$

with $\delta(\mathbf{X} - \mathbf{X}')$ being the three-dimensional Dirac delta function.

The inversions of (65a) and (65b) are exceedingly difficult to find, in general, and are known only for a few cases. For an ordinary dielectric-magnetic substance ($\underline{\varepsilon} = \varepsilon\underline{I}$, $\underline{\mu} = \mu\underline{I}$, $\underline{\zeta} = \underline{\xi} = \underline{0}$), the results are isomorphic with those for the free space case. The inversions are also available for natural optically active materials,[134-136] biisotropic materials,[137, 138] uniaxial dielectrics,[139-142] and uniaxial dielectric-magnetics.[143, 144]

For the general case, all that can be said is that the spectral solutions

$$\underline{G}_e(\mathbf{X}; \mathbf{X}') = \left(\frac{1}{8\pi^3}\right)\iiint_{-\infty}^{\infty} d^3\mathbf{q}\left[\underline{W}_e(i\mathbf{q})\right]^{-1}\exp\{i\mathbf{q}\cdot(\mathbf{X} - \mathbf{X}')\} \quad (66a)$$

$$\underline{G}_m(\mathbf{X}; \mathbf{X}') = \left(\frac{1}{8\pi^3}\right)\iiint_{-\infty}^{\infty} d^3\mathbf{q}\left[\underline{W}_m(i\mathbf{q})\right]^{-1}\exp\{i\mathbf{q}\cdot(\mathbf{X} - \mathbf{X}')\} \quad (66b)$$

can be thought of. While $[\underline{W}_e(i\mathbf{q})]^{-1}$ and $[\underline{W}_m(i\mathbf{q})]^{-1}$ can be derived using dyadic algebra,[35, 36] the three-dimensional integrals in (66a) and (66b) are analytically intractable in general, and can only be estimated numerically. Use of such spectral forms is generally made when dealing with stratified planar media,[145-149] and scalar treatments are also on the horizon.[150]

VII. ONE-DIMENSIONAL FIELD VARIATIONS IN THE EQUIVALENT CONTINUUM

Experiments conducted in physical chemistry on the optical properties of substances often yield measurements of the optical rotation and the dichroism betrayed by the transmitted plane wave when a plane wave is normally incident on a slab of the material being evaluated; this has been the case at least since Biot formalized the science of saccharimetry.[41, 151] For such investigations it is not necessary that the solutions (66a) and (66b) be known in closed form. (This feature was probably responsible for the inordinate delay in the understanding of classical electromagnetic field behavior in isotropic chiral media.[134, 152]) The developments of this section may also be useful for analysis of dichroic superconductors.[153]

A. The Matrix Differential Equation

Without any loss of generality, therefore, we seek solutions of the homogeneous counterparts of (53c) and (53d) in the form

$$\mathbf{E}(\mathbf{X}) = \mathbf{E}(X_3) \qquad \mathbf{H}(\mathbf{X}) = \mathbf{H}(X_3) \quad (67)$$

Furthermore, let

$$\mathbf{E}(X_3) = E_1(X_3)\mathbf{U}_1 + E_2(X_3)\mathbf{U}_2 + E_3(X_3)\mathbf{U}_3 = \sum_{m=1,2,3} E_m(X_3)\mathbf{U}_m,$$
(68)

and similarly for $\mathbf{H}(X_3)$, while the continuum constitutive dyadics be represented as

$$\underline{\underline{\varepsilon}} = \sum_{m=1,2,3} \sum_{n=1,2,3} \varepsilon_{mn}\mathbf{U}_m\mathbf{U}_n$$
(69)

etc.; the $\mathbf{U}_m$'s are the Cartesian unit vectors in the laboratory frame $\mathbf{X} \equiv (X_1, X_2, X_3)$.

Substitution of the above representations into (53c) and (53d), with $\mathbf{J} = 0$ and $\mathbf{K} = 0$, leads to the two algebraic equations

$$(\zeta_{33}E_3 + \mu_{33}H_3) = -(\zeta_{31}E_1 + \mu_{31}H_1 + \zeta_{32}E_2 + \mu_{32}H_2) \quad (70a)$$

$$(\varepsilon_{33}E_3 + \xi_{33}H_3) = -(\varepsilon_{31}E_1 + \xi_{31}H_1 + \varepsilon_{32}E_2 + \xi_{32}H_2) \quad (70b)$$

and the four first-order differential equations

$$\left\{\frac{d}{dX_3}\right\}E_1 = i\omega\mu_0\mathbf{U}_2 \cdot \left[\underline{\underline{\zeta}} \cdot \mathbf{E} + \underline{\underline{\mu}} \cdot \mathbf{H}\right]$$
(71a)

$$\left\{\frac{d}{dX_3}\right\}E_2 = -i\omega\mu_0\mathbf{U}_1 \cdot \left[\underline{\underline{\zeta}} \cdot \mathbf{E} + \underline{\underline{\mu}} \cdot \mathbf{H}\right]$$
(71b)

$$\left\{\frac{d}{dX_3}\right\}H_1 = i\omega\varepsilon_0\mathbf{U}_2 \cdot \left[\underline{\underline{\varepsilon}} \cdot \mathbf{E} + \underline{\underline{\xi}} \cdot \mathbf{H}\right]$$
(71c)

$$\left\{\frac{d}{dX_3}\right\}H_2 = -i\omega\varepsilon_0\mathbf{U}_1 \cdot \left[\underline{\underline{\varepsilon}} \cdot \mathbf{E} + \underline{\underline{\xi}} \cdot \mathbf{H}\right]$$
(71d)

We assume that (70a) and (70b) are linearly independent when solved for E_3 and H_3; in other words,

$$\varepsilon_{33}\mu_{33} \neq \zeta_{33}\xi_{33}$$
(72)

Then, by solving (70a) and (70b), the eliminating E_3 and H_3 from (71a)–(71d), the matrix differential equation

$$\left\{\frac{d}{dX_3}\right\}[f(X_3)] = i[\Lambda][f(X_3)] \tag{73}$$

is obtained, where

$$[f(X_3)] = \text{col}[E_1(X_3); E_2(X_3); H_1(X_3); H_2(X_3)] \tag{74}$$

is a column 4-vector, while $[\Lambda]$ is a 4×4 matrix whose exact form is too cumbersome to be reproduced here but is easily derivable. Once the solution of (73) is available, $E_3(X_3)$ and $H_3(X_3)$ can be computed using (70a) and (70b).

The solution of (73) is also in a matrix form,[154] viz.,

$$[f(X_3)] = [M(X_3)][f(0)] \tag{75}$$

where

$$[M(X_3)] = \exp(i[\Lambda]X_3) \tag{76}$$

Although the right side of (76) can be computed by appropriately truncating the infinite sum

$$[M(X_3)] = \sum_{n=0,1,\ldots,\infty} \frac{(i[\Lambda]X_3)^n}{n!} \tag{77}$$

more efficient procedures are possible if need be.

B. The Cayley-Hamilton Procedure

The Cayley-Hamilton theorem permits the evaluation of the matrix $[M(X_3)]$ very easily, provided the eigenvalues of $[\Lambda]$ are all distinct; let the four eigenvalues of $[\Lambda]$ be denoted by κ_m, $m = 1-4$. In the present instance, since $[\Lambda]$ is a 4×4 matrix, it follows that[155]

$$[M(X_3)] = \sigma_1(X_3)[I] + \sigma_2(X_3)[\Lambda] + \sigma_3(X_3)[\Lambda]^2 + \sigma_4(X_3)[\Lambda]^3 \tag{78}$$

where $[I]$ is the identity matrix and the four coefficients $\sigma_1(X_3)$–$\sigma_4(X_3)$

can be calculated from the four algebraic equations[156]

$$\exp(i\kappa_m X_3) = \sigma_1(X_3) + \sigma_2(X_3)\kappa_m$$
$$+\sigma_3(X_3)\kappa_m^2 + \sigma_4(X_3)\kappa_m^3 \qquad m = 1, 2, 3, 4 \qquad (79)$$

In matrix form, (79) can be solved to yield

$$
\begin{bmatrix} \sigma_1(X_3) \\ \sigma_2(X_3) \\ \sigma_3(X_3) \\ \sigma_4(X_3) \end{bmatrix} =
\begin{bmatrix} 1 & \kappa_1 & \kappa_1^2 & \kappa_1^3 \\ 1 & \kappa_2 & \kappa_2^2 & \kappa_2^3 \\ 1 & \kappa_3 & \kappa_3^2 & \kappa_3^3 \\ 1 & \kappa_4 & \kappa_4^2 & \kappa_4^3 \end{bmatrix}^{-1}
\begin{bmatrix} \exp(i\kappa_1 X_3) \\ \exp(i\kappa_2 X_3) \\ \exp(i\kappa_3 X_3) \\ \exp(i\kappa_4 X_3) \end{bmatrix} \qquad (80)
$$

Several numerical methods are available to find the eigenvalues of $[\Lambda]$ and algorithms can be found in the books by Carnahan et al.[157] and Strang.[158] The EISPACK and IMSL libraries can be utilized with generally little effort for FORTRAN programs. For matrix inversion the LU decomposition method[158] is recommended, which is available as the FORTRAN subroutine LEQT1C in the IMSL library. European software packages, such as the NAG, are suggested for those who do not have access to American packages. NAG, incidently, has become available recently in the American market, and IBM's package ESSL may also be useful. Finally, computer mathematics software, such as MACSYMA and MATHEMATICA, should be exploited to the hilt to obtain analytical expressions for the eigenvalues.[156, 159]

C. The Diagonalization Procedure

The representation (78) is certainly continuum-specific, but it also depends on X_3 in a complicated fashion. Diagonalization of the matrix $[\Lambda]$ provides a way of computing $[M(X_3)]$ that is far more physically meaningful than (78) by yielding the dependence of $[f(X_3)]$ on X_3 explicitly, and can be easily done if the four eigenvalues of $[\Lambda]$ are distinct.

Let the column 4-vector

$$[t_m] = \operatorname{col}[t_{m1}; t_{m2}; t_{m3}; t_{m4}] \qquad (81)$$

be the eigenvector of $[\Lambda]$ corresponding to the eigenvalue of κ_m, $m = 1\text{--}4$. Then, it can be shown that[154]

$$[M(X_3)] = [T][P(X_3)][T]^{-1} \qquad (82)$$

where the constant matrix

$$[T] = \text{matrix}[[t_1]; [t_2]:[t_3]; [t_4]] \tag{83}$$

and the diagonal matrix

$$[P(X_3)] = \text{diag}[\exp(i\kappa_1 X_3); \exp(i\kappa_2 X_3); \exp(i\kappa_3 X_3); \exp(i\kappa_4 X_3)] \tag{84}$$

It should be noted that $[T]^{-1}[f(X_3)]$ are the elemental plane wave field compositions whose propagator matrix is $[P(X_3)]$, while standard software packages, discussed in Section VII.B can be used to determine $[T]$ and $[P(X_3)]$.

There is no guarantee that $[\Lambda]$ will have distinct eigenvalues; however, all complex continua investigated thus far[156, 159–164] have been found amenable to this procedure. This is because, except for the simplest continua $[\underline{\underline{\varepsilon}} = \varepsilon \underline{\underline{I}}, \underline{\underline{\mu}} = \mu \underline{\underline{I}}, \underline{\underline{\zeta}} = \underline{\underline{\xi}} = \underline{\underline{0}}]$, bianisotropic continua are generally birefringent. The cases of pathological and incidental unirefringence, which lead to nondistinct eigenvalues of $[\Lambda]$, are discussed in Section VII.F.

D.   Fresnel Reflection and Transmission Coefficients for a Slab

Despite the case of normal plane wave incidence being interpretable as pathological,[165] as mentioned at the beginning of this section experimentalists commonly study the transmission of a normally incident plane wave through slabs of materials. Whichever way the matrix is $[M(X_3)]$ evaluated, it can be used for analysis of this situation.

Consider the slab $0 \le X_3 \le d$ made of the bianisotropic continuum being discussed; the zones $X_3 \le 0$ and $X_3 \ge d$ are vacuous. Let the plane wave specified by

$$\mathbf{E}_{\text{inc}}(\mathbf{X}) = [a_1 \mathbf{U}_2 - a_2 \mathbf{U}_1]\exp(ik_0 X_3) \qquad X_3 \le 0 \quad (85a)$$

$$\mathbf{H}_{\text{inc}}(\mathbf{X}) = -\sqrt{\frac{\varepsilon_0}{\mu_0}}\,[a_2 \mathbf{U}_2 + a_1 \mathbf{U}_1]\exp(ik_0 X_3) \qquad X_3 \le 0 \quad (85b)$$

be incident on the slab with known coefficients a_1 and a_2; then, the reflected plane wave is given by

$$\mathbf{E}_{\text{ref}}(\mathbf{X}) = [r_1\mathbf{U}_2 + r_2\mathbf{U}_1]\exp(ik_0X_3) \qquad X_3 \le 0 \quad (86\text{a})$$

$$\mathbf{H}_{\text{ref}}(\mathbf{X}) = \sqrt{\frac{\varepsilon_0}{\mu_0}}\,[-r_2\mathbf{U}_2 + r_1\mathbf{U}_1]\exp(ik_0X_3) \quad X_3 \le 0 \quad (86\text{b})$$

with the as yet unknown coefficients r_1 and r_2. The plane wave transmitted into the zone $X_3 \ge d$ can be expressed as

$$\mathbf{E}_{\text{trans}}(\mathbf{X}) = [t_1\mathbf{U}_2 - t_2\mathbf{U}_1]\exp(ik_0X_3)\exp(-ik_0d) \qquad X_3 \le d$$
$$(87\text{a})$$

$$\mathbf{H}_{\text{trans}}(\mathbf{X}) = -\sqrt{\frac{\varepsilon_0}{\mu_0}}\,[t_2\mathbf{U}_2 + t_1\mathbf{U}_1]\exp(ik_0X_3)\exp(-ik_0d) \quad X_3 \le d$$
$$(87\text{b})$$

where the coefficients t_1 and t_2 are not known either. We observe here that representations (85)–(87) are consistent with Snel's laws.

From (74) and these representations, it follows that

$$[f(0)] = \text{col}\left[r_2 - a_2; r_1 + a_1; (r_1 - a_1)\sqrt{\varepsilon_0/\mu_0}\,;-(r_2 + a_2)\sqrt{\frac{\varepsilon_0}{\mu_0}}\right]$$
$$(88\text{a})$$

$$[f(d)] = \text{col}\left[-t_2; t_1; -t_1\sqrt{\frac{\varepsilon_0}{\mu_0}}\,;\,-t_2\sqrt{\frac{\varepsilon_0}{\mu_0}}\right] \qquad (88\text{b})$$

Per (75), if we define the 4×4 matrix

$$[S] = [M(d)] \qquad (88\text{c})$$

note the continuity of $[f(X_3)]$ across any X_3-constant plane as an intrinsic consequence of the Maxwell postulates, and solve the matrix equation

$$[f(d)] = [S][f(0)] \qquad (88\text{d})$$

we can ascertain the transmission coefficients t_1 and t_2 and the reflection coefficients r_1 and r_2 for given a_1 and a_2.

It stands to reason that the polarization states of the incident, the reflected and the transmitted plane waves are all different, in general. To study the vibration ellipses and the Stokes vectors of the reflected and the transmitted plane waves with respect to those of the incident plane wave, one can utilize standard textbooks (e.g., Refs. 23, 36, 56), for which reason we will not discuss them here.

E. Optical Rotation and Dichroism

Of greater interest here are the eigenvalues κ_m, $m = 1$–4, of the matrix $[\Lambda]$, for which the relations

$$\kappa_1 \kappa_2 \kappa_3 \kappa_4 = \text{determinant}\{[\Lambda]\} \tag{89}$$

and

$$\kappa_1 + \kappa_2 + \kappa_3 + \kappa_4 = \text{trace}\{[\Lambda]\} = \Lambda_{11} + \Lambda_{22} + \Lambda_{33} + \Lambda_{44} \tag{90}$$

can be easily verified from simple matrix algebra. In particular, we look at (90) to arrange the eigenvalues in such a way so that

$$\text{Re}\{\kappa_1\} > 0 \quad \text{Re}\{\kappa_2\} > 0 \quad \text{Re}\{\kappa_3\} < 0 \quad \text{Re}\{\kappa_4\} < 0 \tag{91}$$

If the continuum is electromagnetically lossless at the particular frequency of interest, then

$$\text{Im}\{\kappa_m\} = 0 \quad m = 1\text{–}4 \tag{92}$$

A lossy continuum will have

$$\text{Im}\{\kappa_1\} > 0 \quad \text{Im}\{\kappa_2\} > 0 \quad \text{Im}\{\kappa_3\} < 0 \quad \text{Im}\{\kappa_4\} < 0 \tag{93}$$

while the inequalities for the imaginary parts will be reversed for an active continuum. If $[\Lambda]$ is Hermitian, all of its eigenvalues are real; in other words, $[\Lambda]^{*\,\text{tr}} = [\Lambda]$ implies that $\text{Im}\{\kappa_m\} \equiv 0$.

Optical activity in the $+X_3$ direction is specified by the difference $\kappa_1 - \kappa_2$, and in the $-X_3$ direction by $\kappa_3 - \kappa_4$. For propagation in the $+X_3$ direction, the optical rotation per unit distance will be proportional to $\text{Re}\{\kappa_1 - \kappa_2\}$ and the dichroism per unit distance will be in proportion in $\text{Im}\{\kappa_1 - \kappa_2\}$; likewise, for propagation in the $-X_3$ direction, the optical rotation and the dichroism, per unit distance, will be respectively proportional to $\text{Re}\{\kappa_3 - \kappa_4\}$ and $\text{Im}\{\kappa_3 - \kappa_4\}$.

If $\kappa_3 = -\kappa_1$ and $\kappa_4 = -\kappa_2$, the optical activity in the $+X_3$ directional is equal and opposite to that in the $-X_3$ direction. A word of caution: It is not necessary for a continuum to be reciprocal for optical activity to be reciprocal. This confusion was responsible for the error made by both

AKHLESH LAKHTAKIA

Drude and Born when they wrote down constitutive equations for natural optically materials.[134] Indeed, though both obtained reciprocal optical activity as per their desires, their constitutive equations were for nonreciprocal materials! Several decades later, Fedorov corrected them; see the introductory remarks in the reprint collection compiled by Lakhtakia.[41]

As a simple example, consider the case when the orientation- and phase-averaged polarizability dyadics $\underline{a}_{ee}$ and $\underline{a}_{mm}$ are scalars, while $\underline{a}_{em}$ and $\underline{a}_{me}$ are pseudoscalars;[56] that is, $\underline{a}_{ee} = a_{ee}\underline{I}, \underline{a}_{mm} = \underline{a}_{mm}\underline{I}, \underline{a}_{em} = a_{em}\underline{I}$ and $\underline{a}_{me} = a_{me}\underline{I}$. It follows then that the constitutive parameters in (52a) and (52b) are also scalars or pseudoscalars given as $\underline{\varepsilon} = \varepsilon\underline{I}, \underline{\xi} = \xi\underline{I}, \underline{\mu} = \mu\underline{I}$ and $\underline{\zeta Y} = \zeta\underline{I}$. The eigenvalues in this eventuality turn out to be

$$\kappa_1 = -\kappa_3 = \frac{\omega}{2}\left[i(\mu_0\zeta - \varepsilon_0\xi) + \sqrt{4\mu_0\varepsilon_0(\varepsilon\mu - \zeta\xi) - (\mu_0\zeta - \varepsilon_0\xi)^2}\right]$$

(94a)

$$\kappa_2 = -\kappa_4 = \frac{\omega}{2}\left[-i(\mu_0\zeta - \varepsilon_0\xi) + \sqrt{4\mu_0\varepsilon_0(\varepsilon\mu - \zeta\xi) - (\mu_0\zeta - \varepsilon_0\xi)^2}\right]$$

(94b)

whence

$$\kappa_1 - \kappa_2 = i\omega(\mu_0\zeta - \varepsilon_0\xi)$$

(94c)

shows that the dependence of the Maxwell Garnett estimate of the optical activity on N is quite complicated. Using the small-N estimates (51a)–(51d), however, it is easy to see that

$$\kappa_1 - \kappa_2 \cong i\omega N(a_{me} - a_{em}) + \text{higher order terms in } N \quad (95)$$

In other words, for low number densities, the optical activity is proportional to N, and depends solely on the cross-polarizabilities.

F. Pathological and Incidental Unirefringence

The methods described in Section VII.B and VII.C fail if (1) $\kappa_1 = \kappa_2$, if (2) $\kappa_3 = \kappa_4$, or if (3) $\kappa_1 = \kappa_2$ and $\kappa_3 = \kappa_4$. The simple continuum $[\underline{\varepsilon} = \varepsilon\underline{I}, \underline{\mu} = \mu\underline{I}, \underline{\zeta} = \underline{\xi} = \underline{0}]$ is always unirefringent, satisfying condition (3) along with $\kappa_1 = -\kappa_3$: in this case, combinations of the components of $[f(X_3)]$ can be judiciously constructed and $[\Lambda]$ can be decomposed into two 2×2 diagonalizable matrices; see, for example, Lakhtakia et al.[166] Uniaxial gyroelectromagnetic media show both incidental and pathological unirefringence.[167] In such instances preprocessing using dyadic analysis is recommended. Even otherwise, dyadic analysis can lead to considerable simplification.[48]

The use of dyadic analysis for situations with nondistinct eigenvalues is suggested for the following reason. It is necessary and sufficient that $[\Lambda]$ has four linearly independent eigenvectors for it to be diagonalizable.[168] Of course, having all distinct eigenvalues certainly makes $[\Lambda]$ diagonalizable, but that is not a necessary condition. Provided the eigenvectors $[t_m]$, $m = 1$–4, of $[\Lambda]$ can be ascertained, the matrix $[T]$ of (83) can be determined. It follows then from (76) and (82) that

$$[P_a] = [T]^{-1}[\Lambda][T] \tag{96}$$

is the 4×4 diagonal matrix

$$[P_a] = \text{diag}[\kappa_1; \kappa_2; \kappa_3; \kappa_4] \tag{97}$$

whose elements may not be all distinct; finally, the matrix of (84) can be synthesized as

$$[P(X_3)] = \exp(i[P_a]X_3) = \text{diag}[\exp(i\kappa_1 X_3); \exp(i\kappa_2 X_3); \\ \exp(i\kappa_3 X_3); \exp(i\kappa_4 X_3)] \tag{98}$$

Whereas determining the eigenvectors $[t_m]$ without determining the eigenvalues κ_m is problematic numerically, dyadic analysis can overcome the obstacles at times.[167] Dyadic analysis, however, can itself become overwhelmingly cumbersome, for which reason an alternative approach may be desirable.

Within the framework of a matrix approach, we can utilize a general theorem for the solution of (73) when $[\Lambda]$ has eigenvalues with multiplicities greater than unity. First, we remember that $[\Lambda]$ can have no more than four eigenvalues, all distinct or not; second, since propagational behavior in the $\pm X_3$ directions is contained in (73), no eigenvalue can have a multiplicity greater than two. Let the mth eigenvalue κ_m ($m = 1, 2, \ldots, k$; $2 \le k \le 4$) have a multiplicity n_m ($1 \le n_m \le 2$). Then,[154]

$$[f(X_3)] = \sum_{m=1}^{k} \sum_{p1}^{n_m} (X_3)^{p-1} \exp(i\kappa_m X_3)[\vartheta^{(p,m)}] \tag{99}$$

where $[\vartheta^{(p,m)}]$ are constant 4-vectors, while $n_1 + n_2 + \cdots + n_k = 4$. Substitution of (99) into (73), and equating like powers of X_3 on both side of the resultant will determine the $[\vartheta^{(p,m)}]$ subject to some nonuniqueness involving four unknowns. That nonuniqueness can then be removed by using the boundary condition

$$[f(0)] = \sum_{m=1}^{k} [\vartheta^{(1,m)}] \tag{100}$$

that follows from (99), if $[f(0)]$ is known.

For the problem of Section VII.D we have $[f(0)]$ from (88a), and $[f(d)]$ from (88b). Therefore, (88a), (88b), (99), and (100) can be used to find the unknown r_1, r_2, t_1, and t_2 in terms of the known a_1 and a_2.

A. Extension to Inhomogeneous Slabs

The formalisms developed in the previous subsections can also be extended to some cases when the continuum slab, $0 \le X_3 \le d$, is inhomogeneous along the X_3 direction, the potential utility of such extensions being affirmed by several earlier studies (e.g., Refs. 169 and 170). To that end, we consider the continuum properties specified as

$$\mathbf{D}(X_3) = \varepsilon_0 \Big[\underline{\underline{\varepsilon}}(X_3) \cdot \mathbf{E}(X_3) + \underline{\underline{\xi}}(X_3) \cdot \mathbf{H}(X_3) \Big] \qquad 0 \le X_3 \le d$$

$$\tag{101a}$$

$$\mathbf{B}(X_3) = \mu_0 \Big[\underline{\underline{\zeta}}(X_3) \cdot \mathbf{E}(X_3) + \underline{\underline{\mu}}(X_3) \cdot \mathbf{H}(X_3) \Big] \qquad 0 \le X_3 \le d \tag{101b}$$

while the counterpart of the matrix differential equation (73) becomes

$$\{d/dX_3\}[f(X_3)] = i[\Lambda(X_3)][f(X_3)] \qquad 0 \le X_3 \le d \tag{102}$$

with the inhomogeneous matrix $[\Lambda(X_3)]$. It can be shown that (102) has unique solutions, provided all 16 elements of the matrix $[\Lambda(X_3)]$ are locally integrable.[171] It is reiterated that $[f(X_3)]$ is a continuous function of X_3, this property coming directly from the Maxwell curl postulates;[25] symbolically,

$$\lim_{\delta \to 0} [f(X_3 + \delta)] = \lim_{\delta \to 0} [f(X_3 - \delta)] \qquad |X_3| \le \infty \tag{103}$$

We begin with the case of $[\Lambda(X_3)]$ being piecewise constant, i.e.,

$$[\Lambda(X_3)] = [\Lambda]_{(n)} \qquad X_{3,(n-1)} \le X_3 \le X_{3,(n)} \qquad n = 1, 2, \ldots, N \tag{104}$$

with $X_{3,(0)} = 0$ and $X_{3,(N)} = d$. Then, using the diagonalization procedure of Section VII.C, the solution of (102) can be synthesized from

$$[f(X_3)] = [T]_{(n)}[K(X_3)]_{(n)}[T]_{(n)}^{-1}[f(X_{3,(n-1)})]$$

$$X_{3,(n-1)} \leq X_3 \leq X_{3,(n)} \tag{105}$$

where the diagonal matrix

$$[K(X_3)]_{(n)} = \mathrm{diag}\big[\exp\big(i\kappa_{1,(n)}\{X_3 - X_{3,(n-1)}\}\big);$$

$$\exp\big(i\kappa_{2,(n)}\{X_3 - X_{3,(n-1)}\}\big); \tag{106}$$

$$\times \exp\big(i\kappa_{3,(n)}\{X_3 - X_{3,(n-1)}\}\big); \exp\big(i\kappa_{4,(n)}\{X_3 - X_{3,(n-1)}\}\big)\big]$$

where $\kappa_{m,(n)}$, $m = 1$–4 are the four eigenvalues of $[\Lambda]_{(n)}$, while $[T]_{(n)}$ is the matrix of the corresponding eigenvectors. From (106), it follows that

$$[f(d)] = \big\{[T]_{(N)}[K(X_{3,(N)})]_{(N)}[T]_{(N)}^{-1}\big\}$$

$$\times \big\{[T]_{(N-1)}[K(X_{3,(N-1)})]_{(N-1)}[T]_{(N-1)}^{-1}\big\} \tag{107}$$

$$\times \cdots \big\{[T]_{(1)}[K(X_{3,(1)})]_{(1)}[T]_{(1)}^{-1}\big\}[f(0)]$$

which characterizes the electromagnetic response of the piecewise constant slab.[156, 159, 162] If a particular segment turns out not to have all distinct eigenvalues, the techniques of Section VII.F may be utilized to incorporate the response of that segment into this scheme. We note that (107) can be recast as

$$[f(d)] = [T]_{(N)}[K(X_{3,(N)})]_{(N)}\big\{[T]_{(N)}^{-1}[T]_{(N-1)}\big\}[K(X_{3,(N-1)})]_{(N-1)}$$

$$\times \big\{[T]_{(N-1)}^{-1}[T]_{(N-2)}\big\}[K(X_{3,(N-2)})]_{(N-2)}$$

$$\cdots [K(X_{3,(2)})]_{(2)} \tag{108}$$

$$\times \big\{[T]_{(2)}^{-1}[T]_{(1)}\big\}[K(X_{3,(1)})]_{(1)}[T]_{(1)}^{-1}[f(0)]$$

where the product $[T]_{(n)}^{-1}[T]_{(n-1)}$ indicates the satisfaction of (103) at the plane $X_3 = X_{3,(n-1)}$, while $[K(X_{3,(n)})]_{(n)}$ denotes the field propagation from the plane $X_3 = X_{3,(n-1)}$ to the plane $X_3 = X_{3,(n)}$.

Equation (107) is a more adequate solution of (102) than that possible from a finite-difference procedure.[172, 173] Indeed, the finite-difference method yields

$$\left[f(X_{3,(n)})\right] \cong \left\{i[\Lambda]_{(n)}(X_{3,(n)} - X_{3,(n-1)}) + [I]\right\}\left[f(X_{3,(n-1)})\right] \quad (109)$$

which is less accurate than the expression

$$\left[f(X_{3,(n)})\right] = \exp\left\{i[\Lambda]_{(n)}(X_{3,(n)} - X_{3,(n-1)})\right\}\left[f(X_{3,(n-1)})\right] \quad (110)$$

that is obtained from (105). This is because the matrix on the right side of (109) is the two-term expansion of the matrix on the right side of (110).

Next, if the matrix $[\Lambda(X_3)]$ of (102) has the convergent polynomial expansion

$$[\Lambda(X_3)] = \sum_{k=0,1,\ldots,\infty} [\Lambda]_{(k)}(X_3)^k \quad |X_3| < R \quad R > d \quad (111)$$

where $[\Lambda]_k$ are constant 4×4 matrices, then (102) has the solution[154]

$$[f(X_3)] = \sum_{k=0,1,\ldots,\infty} [g]_{(k)}(X_3)^k \quad |X_3| < R \quad R > d \quad (112)$$

that is uniquely determined by $[f(0)]$. As per the method of Frobenius, the constant 4-vectors $[g]_{(k)}$ are determined by substituting (110) and (112) into (102), and equating the coefficients of $(X_3)^k$ on both sides of the resultant. This leads to the recurrence relation

$$[g]_{(k+1)} = (k+1)^{-1} \sum_{j=0,1,\ldots,k} [\Lambda]_{(k-j)}[g]_{(j)} \quad k = 0,1,2,\ldots$$

$$(113a)$$

along with

$$[g]_{(0)} = [f(0)] \quad (113b)$$

It follows from (112) that

$$[f(d)] = \sum_{k=0,1,\ldots,\infty} [g]_{(k)} d^k \quad (114)$$

For an application of this approach to chiral slabs, see Lakhtakia et al.[174]

Finally, when the property inhomogeneity is periodic, it can be proved that (102) with a given inhomogeneity profile can be reduced to a linear homogeneous system with constant coefficients by a specific transformation of the components of $[f(X_3)]$; it is quite another matter to be able to find that transformation.[175] The general solution of

$$\left\{\frac{d}{dX_3}\right\}[f(X_3)] = i[\Lambda(X_3)][f(X_3)] \qquad (115a)$$

given the periodicity

$$[\Lambda(X_3 + \Omega)] = [\Lambda(X_3)] \qquad \Omega \neq 0 \qquad (115b)$$

is not periodic. Indeed, as per the Floquet-Lyapunov theory,[176, 177]

$$[f(X_3)] = [F(X_3)]\exp(i[K]X_3)[f(0)] \qquad (116)$$

where $[K]$ is a constant matrix, while $[F(X_3)]$ is a periodic matrix such that

$$[F(X_3 + \Omega)] = [F(X_3)] \qquad (117a)$$
$$[F(0)] = [F(\Omega)] = [I] \qquad (117b)$$

$[I]$ being the unit matrix. Thus,

$$[M(X_3)] = [F(X_3)]\exp(i[K]X_3) \qquad (118)$$

The numerical implementation of (116), and ipso facto (118), for an actual problem is difficult at best,[175] for which reason we turn our attention to a perturbational procedure.[178]

For some experimental realizations of the Maxwell Garnett continuum at hand, it is anticipated that a perturbational solution of (115) may indeed be sufficient. For this purpose, we recast that equation into the form

$$\left\{\frac{d}{dX_3}\right\}[f(X_3)] = [\Gamma(X_3)][f(X_3)] \qquad (119)$$

where $[\Gamma(X_3)] = i[\Lambda(X_3)]$. Let

$$[\Gamma(X_3)] \cong [\Gamma_0] + \alpha[\Gamma_1(X_3)] + \alpha^2[\Gamma_2(X_3)] + \text{higher order terms in } \alpha \qquad (120)$$

in which α is a small parameter. Since $[\Gamma(X_3)]$ is periodic, $[\Gamma_n(X_3)]$ are also periodic; hence, Fourier expansions of the type

$$[\Gamma_n(X_3)] = \sum_{m = \pm 1, \pm 2, \ldots} \exp(2\pi i m X_3/\Omega)[\Gamma_{n,m}] \quad n > 0 \quad (121)$$

can be made in which $[\Gamma_{n,m}]$ are constant matrices. The perturbational solution of (115) can be obtained as[176]

$$[f(X_3)] \cong \{[I] + \alpha[F_1(X_3)] + \alpha^2[F_2(X_3)]\}$$
$$\times \exp\{([K_0] + \alpha[K_1] + \alpha^2[K_2])X_3\}[f(0)] \quad (122)$$

whence

$$[M(X_3)] \cong \{[I] + \alpha[F_1(X_3)] + \alpha^2[F_2(X_3)]\}$$
$$\times \exp\{([K_0] + \alpha[K_1] + \alpha^2[K_2])X_3\} \quad (123)$$

Here

$$[K_0] = [\Gamma_0] \quad (124a)$$

$$[K_1] = \frac{1}{\Omega} \int_0^\Omega d\tau [\Gamma_1(\tau)] \quad (124b)$$

$$[K_2] = \frac{1}{\Omega} \int_0^\Omega d\tau [\Gamma_2(\tau)] + \sum_{m \neq 0} [\Gamma_{1,-m}]\Re_m\{[\Gamma_{1,m}]\} \quad (124c)$$

$$[F_1(X_3)] = \sum_{m \neq 0} \exp\left(\frac{2\pi i m X_3}{\Omega}\right)\Re_m\{[\Gamma_{1,m}]\} \quad (124d)$$

$$[F_2(X_3)] = \sum_{m \neq 0} \exp\left(\frac{2\pi i m X_3}{\Omega}\right)\Re_m\{[\Phi_m]\} \quad (124e)$$

while

$$[\Phi_m] = [\Gamma_{2,m}] - \Re_m\{[\Gamma_{1,m}]\}[K_1] + \sum_{j+k=m} [\Gamma_{1,k}]\Re_j\{[\Gamma_{1,j}]\} \quad (125)$$

and the matrix function $\Re_m\{[G]\}$ of a matrix $[G]$ is the solution of the matrix equation

$$[K_0]\Re_m\{[G]\} - \Re_m\{[G]\}[K_0] - (2\pi i m/\Omega)\Re_m\{[G]\} = -[G] \quad (126)$$

Once the matrices appearing in (123) have been determined, one can easily obtain $[f(0)]$ and $[f(d)]$ for calculating the reflection and the transmission characteristics of the periodically inhomogeneous slab. However, solution (123) does become singular at certain frequencies, as has been noted, for example, by Lakhtakia et al.[162, 166] Finally, it can be seen that if $[\Gamma(X_3)] \equiv [\Gamma_0]$, a constant matrix, then (123) reduces to

$$[M(X_3)] = \exp([\Gamma_0]X_3) \tag{127}$$

which is the same as (75).

VIII.   CLOSING REMARKS

In this chapter we began with a discussion of the symmetrization of the Maxwell postulates for general macroscopic continua. We went on to consider a gas of scattering centers, the scattering centers being either particulate or molecular, and N being their volumetric number density. We obtained explicit expressions for the frequency-dependent polarizability dyadics of electrically small bianisotropic particles using volume integral equations, and we discussed the nature of the frequency-dependent polarizability dyadics for molecules. Assuming a cubic Lorentzian interaction, we constructed a Maxwell Garnett continuum that is macroscopically equivalent to the gaseous distribution at the frequency of interest for small-signal analysis. Finally, we explored the symmetries of the equivalent continuum, studied plane wave propagation therein, looked at mathematical techniques to determine the rotational and the dichroic properties, and took stabs at homogeneous and inhomogeneous slabs.

An intriguing use of the formalism presented may be in the exploration of optical nuclear magnetic resonance (ONMR) and optical electron spin resonance (OESR). For the production of these phenomena, Evans[179-181] has suggested the use of a circularly polarized laser. The optical conjugate product of the laser may induce an electronic magnetic dipole moment, which couples with the nuclear magnetic dipole moment for the ONMR and the electron spin for the OESR spectroscopy. The feasibility of these theoretically predicted techniques is currently being evaluated experimentally.[182]

We began with a discussion of the Maxwell postulates (in the time domain), and it seems proper to end with the time domain. Free space is nondispersive; therefore, time-domain analyses can be performed using temporal Fourier transforms and frequency-domain analyses. Though Gibbs' phenomenon bedevils the inverse Fourier transform, one can usually make do. On the other hand, all known (material) continua are dispersive, and have to be.[38] Performing frequency-domain analyses, followed by inverse temporal Fourier transforms, would be not entirely inadequate, were the continuum to be spread homogeneously all around from here to infinity.

But continua are confined to bounded volumes, while all signals are launched and evaluated in the surrounding free space. Thus, there can be no true boundary value problem in the time domain; there can be only initial boundary value problems.[183, 184] Furthermore, time-harmonic waves, being analytic functions of time, contain no information, which can be carried only by the nonanalytic signals.[6, 28] Therefore, note must be made of the inadequacy in interpreting the inverse temporal Fourier transforms of the results obtained in Section VII.

Acknowledgments

No funding agency sponsored this work, nor was any requested to do so. The author thanks Myron Wyn Evans for inviting him to write his review article to felicitate Professor S. Kielich (Poznan, Poland) on an illustrious and productive scientific career that has spanned a score and ten years and more: May his tribe increase! Gratitude is also due to M. W. Evans for educating the author on several aspects of the chemical physics literature.

References

1. B. Chu, *Laser Light Scattering*, Academic, New York, 1974.

2. N. L. Manakov, V. D. Osiannikov, and S. Kielich, *Acta Phys. Pol. A* **53**, 581 (1978).

3. N. L. Manakov, V. D. Osiannikov, and S. Kielich, *Acta Phys. Pol. A* **53**, 595 (1978).

4. N. B. Delone and V. P. Krainov, *Fundamentals of Nonlinear Optics*, Wiley, New York, 1988.

5. G. C. Baldwin, *An Introduction to Nonlinear Optics*, Plenum, New York, 1969.

6. H. F. Harmuth, *Propagation of Nonsinusoidal Electromagnetic Waves*, Academic, New York, 1986.

7. T. W. Barrett, *Annales de la Fondation Louis de Broglie* **15**, 143 (1990).

8. T. W. Barrett, *Annales de la Fondation Louis de Broglie* **15**, 253 (1990).

9. J. Z. Buchwald, *From Maxwell to Microphysics*, University of Chicago Press, Chicago, 1985.

10. D. F. Eaton, *Science* **253**, 281 (1991).

11. H. E. Brandt (Ed.), *Selected Papers on Nonlinear Optics*, SPIE Opt. Engg. Press, Bellingham, WA, 1991.

12. J. Ducuing, in P. G. Harper and B. S. Wherret (Eds.), *Nonlinear Optics*, Academic, London, 1977.

13. M. W. Evans, *Phys. Lett. A* **147**, 364 (1990).

14. J. F. Ward, *Rev. Mod. Phys.* **37**, 1 (1965).

15. Y. Wang, *Acc. Chem. Res.* **24**, 133 (1991).

16. A. A. Abrikosov, L. P. Gorkov, and I. E. Dzyaloshinski, *Methods of Quantum Field Theory in Statistical Physics*, Dover, New York, 1975.

17. Q. Gong, Z. Xia, Y. H. Zou, X. Meng, L. Wei, and F. Li, *Appl. Phys. Lett.* **59**, 381 (1991).

18. S. M. Saltiel, S. Y. Goldberg, and D. Huppert, *Opt. Commun.* **84**, 189 (1991).

19. K. Knast and S. Kielich, *Acta Phys. Pol. A* **55**, 319 (1979).

20. E. Charney, *The Molecular Basis of Optical Activity*, Krieger, Malabar, FL, 1985.

21. C. A. Emeis, L. J. Oosterhoff, and G. de Vries, *Proc. R. Soc. London Ser. A* **297**, 54 (1967).

22. D. W. Urry and J. Krivacic, *Proc. Nat. Acad. Sci. USA* **65**, 845 (1970).

23. C. F. Bohren and D. R. Huffman, *Absorption and Scattering of Light by Small Particles*, Wiley, New York, 1983.

24. I. Ledoux and J. Zyss, in J. Messier, F. Kajzar, and P. Prasad (Eds.), *Organic Molecules for Nonlinear Optics and Photonics*, Kluwer, Dordrecht, 1991.

25. C. H. Durney and C. C. Johnson, *Introduction to Modern Electromagnetics*, McGraw-Hill, New York, 1969, Chapter 3.

26. J. C. Maxwell, *A Treatise on Electricity and Magnetism*, 3rd ed., Clarendon, Oxford, UK, 1891; reprinted in 1954 by Dover, New York.

27. A. Papoulis, *Signal Analysis*, McGraw-Hill, New York, 1977.

28. S. R. De Groot and P. Mazur, *Non-equilibrium Thermodynamics*, Dover, New York, 1984.

29. A. S. Goldhaber and W. P. Trower, *Am. J. Phys.* **58**, 429 (1990).

30. W. Magnus, A. Karrass, and D. Solitar, *Combinatorial Group Theory*, Dover, New York, 1976.

31. H. F. Harmuth, *Radiation of Nonsinusoidal Electromagnetic Waves*, Academic, New York, 1990.

32. J. W. Goodman, *Introduction to Fourier Optics*, McGraw-Hill, New York, 1968.

33. A. Lakhtakia, *Ber. Bunsenges. Phys. Chem.* **94**, 1504 (1990).

34. A. Lakhtakia, *Ber. Bunsenges. Phys. Chem.* **95**, 574 (1991).

35. F. I. Fedorov, *Theory of Gyrotropy*, Nauka i Teknika, Minsk, 1976 (in Russian).

36. H. C. Chen, *Theory of Electromagnetic Waves*, McGraw-Hill, New York, 1983.

37. P. M. Morse and H. Feshbach, *Methods of Theoretical Physics*, Vol. 1, McGraw-Hill, New York, 1953.

38. E. J. Post, *Formal Structure of Electromagnetics*, North-Holland, Amsterdam, 1962.

39. A. Kovetz, *The Principles of Electromagnetic Theory*, Cambridge University Press, Cambridge, UK, 1990.

40. M. Born and E. Wolf, *Principles of Optics*, Pergamon, Oxford, UK, 1987.

41. A. Lakhtakia (Ed.), *Selected Papers on Natural Optical Activity*, SPIE Opt. Engg. Press, Bellingham, WA, 1990.

42. I. E. Dzyaloshinskii, *Sov. Phys. JETP* **10**, 628 (1960).

43. D. N. Astrov, *Sov. Phys. JETP* **11**, 708 (1960).

44. V. L. Indenbom, *Sov. Phys. Crystallogr.* **5**, 493 (1960).

45. R. R. Birss, *Rep. Prog. Phys.* **26**, 307 (1963).

46. G. T. Rado, *Phys. Rev. Lett.* **13**, 335 (1964).

47. B. R. Chawla and H. Unz, *Electromagnetic Waves in Moving Magneto-Plasmas*, Kansas University Press, Lawrence, 1969.

48. A. Lakhtakia, V. K. Varadan, and V. V. Varadan, *Int. J. Infrared Millimeter Waves* **12**, (1990).

49. J. C. Monzon and A. Lakhtakia, *Ind. J. Pure Appl. Phys.* **29**, 541 (1991).

50. R. D. Graglia and P. L. E. Uslenghi, *IEEE Trans. Antennas Propagat.* **32**, 867 (1984).

51. D. S. Jones, *J. Inst. Math. Applic.* **23**, 421 (1979).

52. A. Lakhtakia, *J. Phys. France* **51**, 2235 (1990).

53. A. Lakhtakia, *Opt. Commun.* **79**, 1 (1990).

54. H. C. Van de Hulst, *Light Scattering by Small Particles*, Wiley, New York, 1957.

55. A. D. Yaghjian, *Proc. IEEE* **68**, 248 (1980).

56. J. D. Jackson, *Classical Electrodynamics*, Wiley, New York, 1975.

57. A. Lakhtakia, *Mod. Phys. Lett. B* **5**, 1439 (1991).

58. Y. Ma, V. V. Varadan, and V. K. Varadan, *J. Wave-Mater. Interact.* **4**, 345 (1989).

59. A. Lakhtakia, in *Millimetre Wave and Microwave*, Tata McGraw-Hill, New Delhi, 1990.

60. C. E. Dungey and C. F. Bohren, *J. Opt. Soc. Am. A* **8**, 81 (1991).

61. A. Lakhtakia and R. S. Andrulis, Jr., *Opt. Photon. News* (*Opt. Soc. Am.*) **2**(6), (1991); corrections: **2**(8), 79 (1991).

62. J. Feder, *Fractals*, Plenum, New York, 1988.

63. A. Lakhtakia, *Opt. Commun.* **80**, 303 (1991).

64. A. Lakhtakia, *Chem. Phys. Lett.* **174**, 583 (1990).

65. D. Bedeaux and P. Mazur, *Physica* **67**, 23 (1973).

66. S. K. Kim and D. J. Lee, *J. Chem. Phys.* **74**, 3591 (1981).

67. V. L. Kuzmin and V. P. Romanov, *Opt. Spectrosc.* (*USSR*) **69**, 390 (1990).

68. T. F. H. Allen and T. B. Starr, *Hierarchy*, University of Chicago Press, Chicago, 1982.

69. J. M. Andre and B. Champagne, in J. Messier, F. Kajzar, and P. Prasad (Eds.), *Organic Molecules for Nonlinear Optics and Photonics*, Kluwer, Dordrecht, 1991.

70. F. Meyers and J. L. Bredas, in, J. Messier, F. Kajzar, and P. Prasad (Eds.), *Organic Molecules for Nonlinear Optics and Photonics*, Kluwer, Dordrecht, 1991.

71. R. M. Glover and F. Weinhold, *J. Chem. Phys.* **65**, 4913 (1976).

72. P. K. Anastasovski, *Theory of Magnetic and Electric Susceptibilities for Optical Frequencies*, Nova Science, Commack, NY, 1990.

73. K. B. Nahm and W. L. Wolfe, *Appl. Opt.* **26**, 2995 (1987).

74. J. Applequist, *J. Phys. Chem.* **94**, 6564 (1990).

75. P. N. Prasad, in J. Messier, F. Kajzar, and P. Prasad, (Eds.), *Organic Molecules for Nonlinear Optics and Photonics*, Kluwer, Dordrecht, 1991.

76. F. Gray, *Phys. Rev.* **7**, 472 (1916).

77. K. F. Lindman, *Ann. Phys.* (*Leipzig*) **63**, 621 (1920).

78. K. F. Lindman, *Ann. Phys.* (*Leipzig*) **69**, 270 (1922).

79. D. Zwanziger, *Phys. Rev.* **176**, 1489 (1968).

80. A. C. Eringen, *J. Math. Phys.* **25**, 3235 (1984).

81. V. K. Varadan, A. Lakhtakia, and V. V. Varadan, *J. Wave-Mater. Interact.* **2**, 153 (1987).

82. V. V. Varadan, A. Lakhtakia, and V. K. Varadan, *J. Appl. Phys.* **63**, 280 (1988); errata **66**, 1504 (1989).

83. J. Maruani (Ed.), *Molecules in Physics, Chemistry, and Biology*, Vols. 2 and 3, Kluwer, Dordrecht, 1989.

84. M. Barrón, J. A. Medrano, M. Ferraro, and A. H. Buep, *J. Mol. Structure* **172**, 355 (1988).

85. D. K. Kondepudi, R. J. Kaufmann, and N. Singh, *Science* **250**, 975 (1990).

86. P. G. Higgs and E. Raphael, *J. Phys. I France* **1**, 1 (1991).

87. S. Kielich, *Acta Phys. Pol.* **31**, 929 (1967).

88. S. Kielich, *Acta Phys. Pol.* **32**, 405 (1967).

89. S. Kielich, N. L. Manakov, and V. D. Osiannikov, *Acta Phys. Pol. A* **53**, 737 (1978).

90. M. W. Evans, *Phys. Lett. A* **146**, 475 (1990).

91. M. W. Evans, *J. Mol. Liq.* **48**, 61 (1991).

92. R. Cameron, G. Cantatore, A. C. Melissinos, Y. Semertzidis, H. Halama, D. Lazarus, A. Prodell, F. Nezrick, P. Micossi, C. Rizzo, G. Ruoso, and E. Zavattini, *Phys. Lett. A* **157**, 125 (1991).

93. A. C. Tam and W. Happer, *Phys. Rev. Lett.* **38**, 278 (1977).

94. M. P. Silverman, *Phys. Lett. A* **146**, 175 (1990).

95. J. Skupinski, J. Buchiert, S. Kielich, J. R. Lalanne, and B. Pouligny, *Acta Phys. Pol. A* **67**, 719 (1985).

96. J. Stankowska and T. Jasinski, *Acta Phys. Pol. A* **67**, 1059 (1985).

97. S. Radhakrishna and B. C. Tan (Eds.), *Laser Spectroscopy and Nonlinear Optics of Solids*, Narosa, New Delhi, 1990.

98. P. Chmela and P. Dub, *Czech. J. Phys.* **41**, 258 (1991).

99. U. Hohm, *Chem. Phys. Lett.* **183**, 304 (1991).

100. E. M. Purcell and C. R. Pennypacker, *Astrophys. J.* **186**, 705 (1973).

101. H. Margenau and N. R. Kestner, *Theory of Intermolecular Forces*, Pergamon, Oxford, UK, 1969.

102. J. Jortner, R. D. Levine and S. A. Rice (Eds.), *Advances in Chemical Physics XLVII: Photoselective Chemistry*, Wiley, New York, 1981.

103. M. W. Evans, G. J. Evans, W. T. Coffey, and P. Grigolini, *Molecular Dynamics and Theory of Broad Band Spectroscopy*, Wiley, New York, 1982.

104. B. R. Palmer, P. Stamatakis, C. F. Bohren, and G. C. Salzman, *J. Coat. Technol.* **61**, 41 (1989).

105. J. C. Maxwell Garnett, *Philos. Trans. R. Soc. London Ser. A* **203**, 385 (1904).

106. C. T. O'Konski and S. Krause, in C. T. O'Konski (Ed.), *Molecular Electro-Optics*, Dekker, New York, 1976.

107. L. Ward, *The Optical Constants of Bulk Materials and Films*, Adam Hilger, Bristol, UK, 1988.

108. D. A. G. Bruggemann, *Ann. Phys. (Leipzig)* **24**, 636 (1935).

109. A. Wachniewski and H. B. McClung, *Phys. Rev. B* **33**, 8053 (1986).

110. C. A. Grimes, *The Permittivity and Permeability of Solid, Granular, Electromagnetic Materials*, Thesis, University of Texas, Austin.

111. V. K. Varadan, V. N. Bringi, V. V. Varadan, and A. Ishimaru, *Radio Sci.* **18**, 321 (1983).

112. M. Milgrom and S. Shtrikman, *J. Appl. Phys.* **66**, 3429 (1989).

113. K. H. Ding and L. Tsang, in J. A. Kong (Ed.), *Progress in Electromagnetics Research I*, Elsevier, New York, 1989.

114. R. Piazza, V. Degiorgio, and T. Bellini, *J. Opt. Soc. Am. B* **3**, 1642 (1986).

115. J. Philip and T. A. Prasada Rao, *J. Mol. Liq.* **48**, 85 (1991).

116. M. A. Kalinichenko and V. A. Trofimov, *Sov. J. Chem. Phys.* **8**, 1232 (1991).

117. C. Kittel, *Introduction to Solid State Physics*, Wiley Eastern, New Delhi, 1973.

118. H. Faxén, *Z. Phys.* **2**, 219 (1920).

119. R. Lundblad, *Untersuchungen über die Optik der dispergierenden Medien*, Thesis, University of Uppsala, 1920.

120. H. A. Lorentz, *The Theory of Electrons*, Dover, New York, 1952.

121 E. Fatuzzo and P. R. Mason, *Proc. Phys. Soc. London* **90**, 729 (1967).

122. M. W. Evans, *Physica B* **162**, 293 (1990).

123. L. S. Taylor, *IEEE Trans. Antennas Propagat.* **13**, 943 (1965).

124. P. S. Neelakantaswamy, R. I. Turkman, and T. K. Sarkar, *Electron. Lett.* **21**, 270 (1985).

125. K. Subramaniam, P. S. Neelakanta, and V. Ungvichian, *Electron. Lett.* **27**, 1534 (1991).

126. H. Looyenga, *Physica* **31**, 401 (1965).

127. P. T. Cummings and L. Blum, *J. Chem. Phys.* **85**, 6658 (1986).

128. A. H. Harvey and J. M. Prausnitz, *J. Sol. Chem.* **16**, 857 (1987).

129. J. B. Hasted, *Aqueous Dielectrics*, Chapman & Hall, London, 1973.

130. C. J. F. Böttcher and P. Bordewijk, *Theory of Electric Polarization*, Elsevier, Amsterdam, 1978.

131. R. Deul and E. U. Franck, *Ber. Bunsenges. Phys. Chem.* **95**, 847 (1991).

132. S. Nelson, A. Kraszewski, and T. You, *J. Microwave Power Electromagn. Energy*, **26**, 45 (1991).

133. C. M. Krowne, *IEEE Trans. Antennas Propagat.* **32**, 1224 (1984).

134. A. Lakhtakia, V. K. Varadan, and V. V. Varadan, *Time-Harmonic Electromagnetic Fields in Isotropic Chiral Media*, Springer, Berlin, 1989.

135. A. Lakhtakia, *Arch. Elektron. Über.* **45**, 323 (1991).

136. V. V. Gvozdez and A. N. Serdyukov, *Opt. Spectrosc. (USSR)* **47**, 301 (1979).

137. J. C. Monzon, *IEEE Trans. Antennas Propagat.* **38**, 227 (1990).

138. W. S. Weiglhofer, *J. Electromagn. Waves Applic.* **5**, 953 (1991).

139. A. Lakhtakia, V. K. Varadan, and V. V. Varadan, *Int. J. Electron.* **65**, 1171 (1988).

140. A. Lakhtakia, V. V. Varadan, and V. K. Varadan, *J. Wave-Mater. Interact.* **3**, 1 (1988).

141. J. C. Monzon and A. Lakhtakia, *Int. J. Electron.* **67**, 243 (1989).

142. A. Lakhtakia, V. V. Varadan, and V. K. Varadan, *J. Wave-Mater. Interact.* **4**, 339 (1989).

143. A. Lakhtakia, V. V. Varadan, and V. K. Vardan, *Appl. Opt.* **28**, 1049 (1989).

144. W. S. Weiglhofer, *Proc. IEE-Part H* **137**, 5 (1990).

145. T. M. Roberts, H. A. Sabbagh, and L. D. Sabbagh, *J. Math. Phys.* **29**, 2675 (1988).

146. T. M. Roberts, H. A. Sabbagh, and L. D. Sabbagh, *IEEE Trans. Magn.* **24**, 3193 (1988).

147. A. Lakhtakia, *Arch. Elektron. Über.* **47**, 1 (1993).

148. C. M. Krowne, *Int. J. Electron.* **59**, 315 (1985).

149. S.-K. Yang, V. V. Varadan, A. Lakhtakia, and V. K. Varadan, *J. Phys. D: Appl. Phys.* **24**, 1601 (1991).

150. W. Weiglhofer, *Proc. IEE-Part H* **134**, 357 (1987).
151. T. M. Lowry, *Optical Rotatory Power*, Dover, New York, 1964.
152. A. Lakhtakia, *Speculat. Sci. Technol.* **14**, 2 (1991).
153. Q. P. Li and R. Joynt, *Phys. Rev. B* **44**, 4720 (1991).
154. H. Hochstadt, *Differential Equations: A Modern Approach*, Dover, New York, 1975.
155. R. A. Gabel and R. A. Roberts, *Signals and Linear Systems*, Wiley, New York, 1987.
156. P. S. Reese and A. Lakhtakia, *Optik* **86**, 47 (1990).
157. B. Carnahan, H. A. Luther, and J. O. Wilkes, *Applied Numerical Methods*, Wiley, New York, 1969.
158. G. Strang, *Introduction to Applied Mathematics*, Wellesley-Cambridge, Wellesley, MA, 1986.
159. P. S. Reese and A. Lakhtakia, *Z. Naturforsch. A* **46**, 384 (1991).
160. A. Lakhtakia, *Optik* **84**, 160 (1990).
161. A. Lakhtakia, V. K. Varadan, and V. V. Varadan, *J. Mater. Res.* **5**, 1511 (1989).
162. A. Lakhtakia, V. K. Varadan, and V. V. Varadan, *Z. Naturforsch. A* **45**, 639 (1990).
163. V. V. Varadan, A. Lakhtakia, and V. K. Varadan, *Optik* **83**, 26 (1989); erratum **83**, 181 (1989).
164. S. Visnovsky, *Czech. J. Phys.* **41**, 663 (1991).
165. A. Lakhtakia, *Int. J. Infrared Millimeter Waves* **11**, 1407 (1990).
166. A. Lakhtakia, V. K. Varadan, and V. V. Varadan, *Int. J. Engg. Sci.* **27**, 1267 (1989).
167. A. Lakhtakia, V. K. Varadan, and V. V. Varadan, *Int. J. Electron.* **71**, 853 (1991).
168. D. Y. Finkbeiner II, *Introduction to Matrices and Linear Transformations*, Freeman, San Francisco, 1978.
169. E. Hild and A. Grofcsik, *Infrared Phys.* **18**, 23 (1978).
170. M. W. Evans, *J. Mol. Liq.* **33**, 127 (1987).
171. D. L. Lukes, *Differential Equations: Classical to Controlled*, Academic, New York, 1982.
172. M. Abramowitz and I. A. Stegun, *Handbook of Mathematical Functions*, Dover, New York, 1972.
173. E. Kreyszig, *Advanced Engineering Mathematics*, Wiley, New York, 1962.
174. A. Lakhtakia, V. K. Varadan, and V. V. Varadan, *Optik* **87**, 77 (1991).
175. S. Lefschetz, *Differential Equations: Geometric Theory*, Wiley, New York, 1963.
176. V. A. Yakubovich and V. M. Starzhinskii, *Linear Differential Equations with Periodic Coefficients*, Wiley, New York, 1975.
177. A. M. Samoilenko and N. I. Ronto, *Numerical-Analytical Methods of Investigating Periodic Solutions*, Mir, Moscow, 1979.
178. A. Lakhtakia, V. K. Varadan, and V. V. Varadan, *Optik* **88**, 63 (1991).
179. M. W. Evans, *J. Phys. Chem.* **95**, 2256 (1991).
180. M. W. Evans, *Int. J. Mod. Phys. B* **5**, 1263 (1991).
181. M. W. Evans, *Int. J. Mod. Phys. B* **5**, 1963 (1991).
182. D. Goswami, C. Hillegas, Q. He, J. Tull, and W. S. Warren, *Experimental NMR Conference*, 8–12 April, St. Louis, MO, 1991.
183. P. Hillion, *Rev. Math. Phys.* **2**, 177 (1990).
184. P. Hillion, *Proc. 2nd. Progr. Elecromagnetics Res. Symp.*, 1–5 July, Boston, MA, 1991.

ASPECTS OF THE OPTICAL KERR EFFECT AND COTTON-MOUTON EFFECT OF SOLUTIONS

JEFFREY HUW WILLIAMS

Institut Max von Laue-Paul Langevin, Grenoble, France

CONTENTS

I. INTRODUCTION

I have long held an opinion, almost amounting to conviction, in common I believe with many other lovers of natural knowledge, that the various forms under which the forces of matter are made manifest have one common origin; or, in other words, are so directly related and mutually dependent, that they are convertible, as it were, one into another.[1]

It was with this splendid intuition that in the third decade of the last century, Michael Faraday began a series of experiments that were to create the field of electro-optics. At the time, Faraday had been investigating the laws of electromagnetic induction and was interested in studying

Modern Nonlinear Optics, Part 2, Edited by Myron Evans and Stanisław Kielich. Advances in Chemical Physics Series, Vol. LXXXV.
ISBN 0-471-57546-1 © 1993 John Wiley & Sons, Inc.

the effect on light of an applied magnetic field. By necessity he was also studying the behavior of the magnetic field on the material through which the light propagated. He found the plane of polarization of light to be rotated after passing through a piece of soft lead borosilicate glass mounted between the poles of an electromagnet.[2] Later investigations showed this result to be a general property of all matter and the magnitude of the induced effect to be linear in the strength of the applied magnetic field.

Today we should say that the medium had become optically active; that is, its refractive indices for right and left circularly polarized light, n_+ and n_-, respectively, had become unequal due to the application of the magnetic field, left- and right-handed motions about a magnetic field being inequivalent. Only longitudinal fields were found to cause optical rotation, whose magnitude is expressed as

$$\theta = VH_z l \tag{1}$$

where θ is the induced rotation of the plane of polarization, l is the path length, H_z is the magnetic field in the direction of propagation of the light, and V is termed the Verdet constant, which is material specific and a function of frequency.[3]

Since its discovery, the Faraday effect, or magneto-optic rotation, has become a standard investigative tool for condensed matter science. With the advent of the laser and the possibility of its use in communications, the Faraday effect has found a use as a means of modulating the polarization properties of the light beam and thus allowing the information of the transmitted light to be greatly increased.[4] At the time of its discovery it was held as important evidence for the electromagnetic nature of light. Not surprisingly, led by the view that the Faraday effect marked a profound physical change in the nature of the medium, many other scientists began searching for the electric analogue.

While he was a divinity student at the University of Glasgow in the mid 1840s John Kerr undertook scientific research in the first purpose-built research laboratory in Great Britain. His director of research, William Thomson, was to write later that among the divinity students of the time were some of the best researchers, and they all became better clergymen for having seen something of scientific methods and handled scientific instruments.[5]

Kerr began investigating the effect of electric fields on insulators, his initial experiments were on the same material that Faraday had first studied, soft glass. He found that a birefringence developed in the glass and in other materials when placed in an intense electric field. The maximum effect was produced when the wave front of the incident light was arranged perpendicular to the applied field and to have its plane of polarization inclined to the field direction at an angle of 45°. This effect is called the electro-optic Kerr effect, or more commonly the Kerr effect, to distinguish it from the other electro-optic effect that bears his name, viz. the magneto-optic Kerr effect.[6]

The first observations by Kerr on glass[7] were followed by measurements on a variety of liquids and he showed that the induced effect was proportional to the square of the applied electric field[8] (remember that the Faraday effect is linear in the applied magnetic field). Kerr showed that the effect arises as a direct consequence of the applied field and that it is not due to an anisotropy setup by mechanical deformation, that is, electrostriction, which was thought to have been present in the case of solid samples.

In 1901, Kerr observed that finely divided Fe_3O_4 suspended in water became birefringent when light traversed the medium normal to the lines of force of an applied magnetic field.[9] At about this time, Majoranna independently discovered this phenomenon in various colloidal solutions of iron.[10] Shortly afterwards, Cotton and Mouton initiated a study of this magnetic field-induced birefringence, which resulted in the discovery of the effect that bears their name.[11]

Cotton and Mouton observed that many pure liquids became doubly refracting or birefringent when placed in a magnetic field with the beam of light perpendicular to the lines of force.[12, 13] As in the Kerr effect, an isotropic fluid when placed in the electromagnetic field behaved optically like a uniaxial crystal. The effect was seen to be quite general and small. The size of the magnetic birefringence signal led to confusion with the larger Faraday effect, a rotation of the plane of polarization of light passing through the material parallel to the lines of force of the magnetic field. To separate these two magnetic field effects care has to be taken in setting up the experiment. Cotton and Mouton showed that their effect was analogous to the Kerr effect and followed the same type of relations.

If $n_{\parallel}$ and $n_{\perp}$ are the refractive indices for the components of light vibrating parallel and perpendicular, respectively, to the magnetic field direction in the substance. Their phase difference, ϕ (which is a measure of the induced birefringence) after the light has passed a distance l through the medium in the field $\mathbf{H}$ is

$$\phi = \frac{2\pi l(n_{\parallel} - n_{\perp})}{\lambda_0} = 2\pi Cl\mathbf{H}^2 \qquad (2)$$

where λ_0 is the wavelength of the probing light beam and C is the Cotton-Mouton constant.

For the Kerr effect one exchanges an applied electric field **E** for the magnetic field **H**, and a Kerr constant B for the Cotton-Mouton constant. Thus,

$$\phi = \frac{2\pi l(n_{\parallel} - n_{\perp})}{\lambda_0} = 2\pi Bl\mathbf{E}^2 \tag{3}$$

Both B and C may be positive or negative and vary with substance, temperature, and the probing wavelength. Indeed, it is these "constants" that are of interest. They are determined for a known applied field strength, pathlength, probing wavelength, molecule number density, and temperature. The Kerr constant B is independent of field strength for fields that are not strong enough to cause nonlinear dielectric polarization. The Kerr effect has yielded much valuable information about molecular moments and polarizabilities and has been successfully used to investigate the dynamical properties of large molecules, polymers, and biological macromolecules in solution when the applied orienting voltage is modulated at low frequencies.[14-16] For these dynamic experiments one observes the time dependence of the onset and loss of the induced biregringence as the applied field, for convenience of a square waveform, is switched on and off. This has been reviewed on a number of occasions.[17-24] Similarly, the Cotton-Mouton constant C has been used as a probe of the electromagnetic properties of molecules,[25-28] yielding magnetic susceptibilities, and of the dynamics of large polymer and biological molecules in solution or suspension.[29]

From Eqs. (2) and (3) we see that the Kerr effect is quadratic in the applied electric field and that it has a direct magnetic analogue, the Cotton-Mouton effect. But what of the electric analogue of the Faraday effect for which Kerr, among others, had been searching?

It is possible to show by symmetry arguments[30] that there is no simple electric analogue of the Faraday effect. That is, optical rotation cannot be induced in a linearly polarized light beam propagating through an isotropic achiral (nonintrinsically optically active) medium by application of a static electric field in the direction of propagation. Any physical process that involves the interaction of a molecule with an electromagnetic field must, to be allowed, conserve parity and be invariant under time reversal. That is, the parity operator **P**, which inverts the system coordinates from i(x, y, z) to i'($-x$, $-y$, $-z$), must not affect the experiment, that is, must not affect the total system, which must be considered as the sample plus the applied field together with the probing light beam. However, for the possible electric analogue of the Faraday effect, the direction of the electric field and the direction of propagation of the light beam are reversed when one applies the parity operator. Within the medium in the absence of the field all directions are equivalent and any optical rotation induced by odd powers of E would violate parity, unlike Faraday rotation, which conserves parity. Similarly, with time reversal symmetry, the operator T, which reverses the direction of motion of all physical entities in the system (the time coordinate going from t to $-t$), does not affect the electric field, the medium, or the sense of optical rotation. It does, however, reverse the direction of propagation of the light beam relative to the electric field direction. Thus, any optical rotation induced by odd (but not even) powers of E would violate reversibility. Faraday magneto-optic rotation conserves reversibility. It is also readily seen that Faraday rotation must depend on odd powers of the applied magnetic field, since these changes sign under T, whereas even powers do not.

In the optical Kerr effect the orienting electric field is that associated with a powerful light beam. However, its existence was predicted before the advent of the laser.[31-32] Buckingham suggested that the high-frequency electric field associated with light would give rise to an optical anisotropy in a fluid. Since the electric field of a light beam is proportional to the square root of its intensity, we see from Eq. (3) that the induced birefringence in the optical Kerr effect will be proportional to the intensity of the applied light beam.

With the arrival of the laser the optical Kerr effect has become a standard investigative tool in molecular electro-optics.[33] Recently the optical Kerr effect has been observed with continuous-wave lasers,[34] which because of their much lower light intensities have smaller associated electric fields and consequently induce smaller birefringences.

II. MEASUREMENT OF INDUCED ELLIPTICITIES

In such experiments one measures the birefringence induced in a fluid by application of an electromagnetic field. Thus, one must define the starting polarization state of a probing light beam and have a means of observing any changes in this polarization state after the light beam has propagated through the now, optically anisotropic, that is, oriented medium. Such experimental setups are relatively simple and are represented in Fig. 1a.

In Fig. 1a the light source is normally a laser yielding polarized, continuous, and monochromatic radiation. Even through the light from the laser may be plane polarized it is necessary to improve the polarization quality of the beam by use of a polarizer. Both the polarizer and the analyzer should be of good quality. Glan-Thompson polarizing prisms are

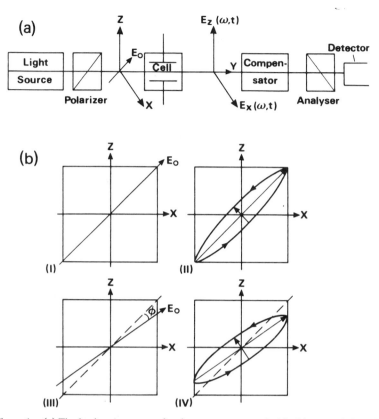

Figure 1. (a) The basic components for the measurement of a birefringence induced in a fluid in the cell. (b) Definition of polarization states in electro-optic experiments: (I) linearly polarized light propagating away from the observer with its electric vector at 45° or 135° to the x and z axes; (II) elliptically polarized light produced by a positive value of $(n_\parallel - n_\perp)$; (III) positive rotation, of magnitude ϕ, of the plane of polarization produced by a positive value of $(k_\parallel - k_\perp)$; (IV) elliptically polarized light, with its major axis rotated, produced by positive values of $(n_\parallel - n_\perp)$ and $(k_\parallel - k_\perp)$.

particularly suitable for visible wavelengths, since they are efficient and do not alter the path of the light beam. When crossed, such polarizers give, an extinction coefficient of 10^{-6} or better.

The polarizer's orientation yields light linearly polarized at 45° to the applied uniform electromagnetic field, **E** or **H**, in the cell. The electric vector of the light wave makes equal angles with the z and the x axes for light propagating along the y direction, as defined in Fig. 1b(i).

The cell, as seen from Eqs. (2) and (3), should be as long as possible, thereby increasing the induced phase difference and reducing the relative significance of end effects due to nonuniformities in the field at the extremities. For Kerr effect experiments, both static and optical, a length of order 1 m is practicable; however, for the Cotton-Mouton effect path lengths tend to be tens of centimeters.

A multipass arrangement in which the light beam traverses the cell a number of times would have obvious advantages, but the loss of polarization at each reflection needs to be considered. A possible solution to this problem may be to place the Kerr or Cotton-Mouton cell inside the cavity of a continuous-wave dye laser. Here, as a consequence of the reflectivity of the output coupler, a photon may well oscillate back and forth along the cavity many hundreds of times before being emitted from the cavity. In such an intracavity electro-optic experiment, polarization selection is ensured by the polarization state of the pumping laser (for example, argon or krypton ion), i.e., the relative orientation of the dye jet to the polarization of the incoming, pumping, laser beam. In this way it may well be possible to improve currently available experimental sensitivities.

Accurate temperature control of the cell is essential and a wide range of temperatures should be available. Low temperatures are particularly advantageous since one is attempting to orient molecules, whereas the thermal energy of the system is trying to randomize this induced molecular orientation. One therefore sees a birefringence signal increasing at constant density as T^{-1}.

The compensator is used to quantify the induced birefringence. Various types of compensators may be used; several types have been discussed by Born and Wolf[35] and by LeFèvre and LeFèvre.[21] A particularly convenient form of compensator is an additional Kerr cell containing a standard reference fluid (generally taken to be CS_2). Another form of compensator includes a quarter-wave plate and a Faraday cell. The former converts the induced phase difference ϕ into a rotation of the plane of polarization $\phi/2$, and this is compensated by an equal but opposite rotation induced in the Faraday cell by the applied magnetic field. The detector in such experiments is either a photomultiplier tube or a photodiode linked to a phase-sensitive detector. In Kerr experiments a modulated electric field is applied to both the fluid under investigation and the reference cell. For a compensating Kerr cell whose signal is out of phase with that of the Kerr cell containing the sample of interest, the nulling potential yields the birefringence of the fluid being studied. Although the Kerr effect arises through the square of the applied electric field, one many chose the phase of the Kerr signal because both static, E_0, and dynamic (the frequency of modulation being, typically, a few hundred hertz), $E(\omega)$, voltages are applied to the Kerr cell and so the total field E_{total} is $E_0 + E(\omega)$. Since the observed signal is proportional to E_{total}^2, with a phase-sensitive detector one may detect the cross term $E_0 E(\omega)$, whose phase is dependent on that of E_0.

For well-crossed polarizers the light level at the detector is given by

$$I = I_0 \sin^2\left(\frac{\phi}{2}\right) \qquad (4)$$

where I_0 is the light intensity incident upon the analyzer. If the applied field is alternating at a frequency, ω, the detector will see a light signal alternating at 2ω (see Eq. (3)). For small ϕ, I will be proportional to $\phi^2/4$.

Typically, ϕ is about 10^{-6} rad. Because of the presence of stray birefringences of about 10^{-4} rad, which depolarize the light beam and thus contribute to the background noise of the measurement, one normally adopts a heterodyne method for measuring the small induced birefringences of interest.[36] Here, both a small modulated birefringence, $\phi^{(\omega)} \sin \omega t$, and a large static birefringence, ϕ_0, are placed between the crossed polarizers. Then, if their axes are parallel, a number of such sources of birefringence act as a single source of birefringence. That is,

$$I = I_0 \left(\sum \phi_i \right)^2 \qquad (5)$$

and we have at the detector[34]

$$I = \frac{I_0}{4}\left[\sin^2\left(\frac{\phi_0}{2}\right) + \frac{1}{2}\sin\phi_0\phi^{(\omega)}\sin\omega t + O(\phi^{(\omega)})^2\right] \qquad (6)$$

A phase-sensitive detector measures the component of I oscillating at the frequency ω. This term may be made larger by increasing the magnitude of ϕ_0. However, the noise level may be determined by the total light level transmitted by the analyzer and hence is strongly dependent on ϕ_0^2 (Ref. 36). In practice ϕ_0 is introduced by a crystal compensator or a dc Kerr cell and is about 0.2 rad.

At the detector, the average number of photons recorded during a particular time interval δt is given by $N = IQ\,\delta t$, where Q is the quantum efficiency of the detector. The number of photons detected in this time interval will be randomly distributed about the mean with standard deviation $\delta N = (N)^{1/2}$. These statistical fluctuations are called shot noise and determine the sensitivity of the detection system for a particular measurement. For a typical electric birefringence apparatus, $\phi(\omega) \ll \phi_0$ and the signal-to-noise ratio is[34]

$$\frac{S}{N} = \frac{\frac{1}{2}QI_0 \, \delta t \, \sin \phi_0 \phi^{(\omega)}}{\sqrt{QI_0 \, \delta t} \, \sin(\phi_0/2)}$$

$$= (QI_0 \, \delta t)^{1/2} \cos\left(\frac{\phi_0}{2}\right) \phi^{(\omega)}$$

(7)

Hence, ϕ_0 should be chosen large in relation to stray birefringences but small compared to 1 rad; the signal-to-noise ratio is then independent of ϕ_0. For example, with a He-Ne laser emitting 10^{16} photons per second at 632.8 nm, assuming $Q = 0.1$ and $\delta t = 1$ s, we would have $\phi^{(\omega)} = 3 \times 10^{-8}$ rad. as the birefringence giving a signal-to-noise ratio of unity.

For all types of field-induced birefringence experiments it is advisable to reduce stray birefringences. A major source of such stray depolarization is the windows of the cell, which may be tightly clamped into place. Consequently, it is useful to construct the windows of the cell from zero-stress birefringence or Pockels glass (manufactured by Chance-Pilkington). This material has a much smaller (10^{-2}) birefringence than silica.[37]

For measurement of the optical Kerr effect one again modulates the applied orienting field; however, a mechanical chopper operating at about 1 kHz is used here. See Fig. 2a for a schematic of the apparatus used for such measurements with a continuous wave laser,[34] and Fig. 3 for an apparatus that is similar to that used for measurement of the optical Kerr effect with pulsed lasers.[38] In the case of continuous-wave optical Kerr effect measurements, the output from the photomultiplier tube or photodiode is again fed to a phase-sensitive detector for demodulation.

Care must be taken in continuous-wave optical Kerr effect experiments because of problems of thermal lensing. For liquids, the refractive index usually increases with increasing applied light intensity due to the molecular reorientation that arises through the optical Kerr effect. Because of the nonuniform intensity distribution inherent with a light beam of finite cross section, the intensity-dependent refractive index causes different parts of the beam to propagate with different phase velocities. A lense-like effect is produced whereby rays move toward the region of higher intensity and thus further increase the intensity there. This increase in intensity is accompanied by a reduction in the effective beam diameter, which continues until limited by other factors. The change of laser beam waist on propagating through a fluid that is polarizable enough to be oriented and give rise to thermal lensing is seen in Fig. 4. Here we see the change of

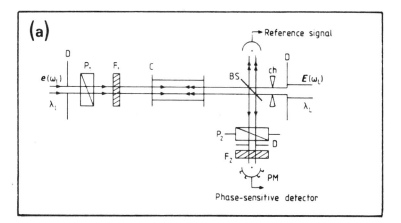

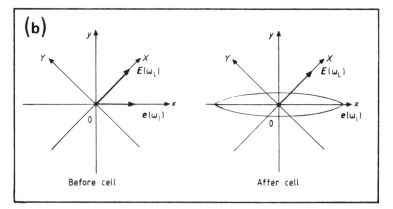

Figure 2. (a) General scheme for the two-laser apparatus for observing the optical Kerr effect with continuous-wave lasers. The more powerful orienting laser, electric field $\mathbf{E}(\omega_L)$, is an Ar^+ laser, $\lambda = 514.5$ nm. The weaker probe laser, electric field, $\mathbf{e}(\omega_l)$, is a He-Ne laser, $\lambda = 632.8$ nm. The polarizers P1 and P2 are Glan polarizers; F1 and F2 are interference filters for 632.8 nm. The cell C is 10 cm in length and BS, a beamsplitter, is a fused quartz disk of optical quality. D is a diaphragm, ch a mechanical chopper, and PM a red-sensitive photomultiplier tube. (Reproduced by kind permission of *Journal of Physics E: Scientific Instruments*. (b) The state of polarization of the two laser beams before and after the cell. The electric vectors are defined in Fig. 2(a). As the weaker probe beam propogates through the medium rendered anisotropic by the orienting effect of the more powerful optical field $\mathbf{E}(\omega_L)$, it becomes elliptically polarized. (Reproduced by kind permission of *Journal of Physics E: Scientific Instruments*.)

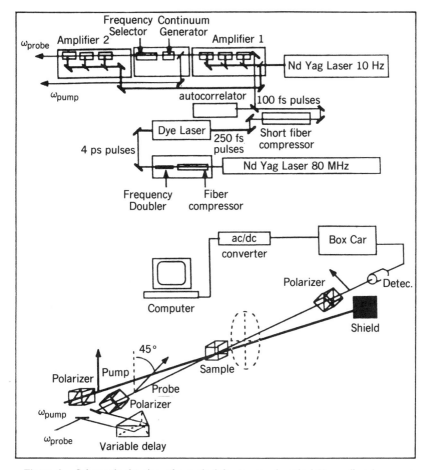

Figure 3. Schematic drawing of a typical femtosecond optical Kerr effect laser system (the TROKE experimental setup at the European Laboratory for Nonlinear Spectroscopy, Florence, Italy). The upper part of the figure gives the details of the laser pulse generating and amplifying system and the lower part of the figure shows the relative polarizations of the pump and probe laser beams (see Fig. 2b for a comparison with the apparatus used for measurements of the optical Kerr effect with a continuous-wave laser). (Reproduced by kind permission of the *Journal of Chemical Physics*.)

JEFFREY HUW WILLIAMS

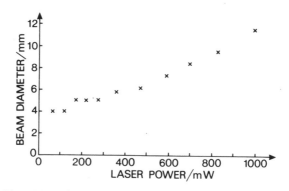

Figure 4. Plot of laser (argon-ion at 514.5 nm) beam diameter versus laser power, in milliwatts, after the beam has propagated through a Kerr cell (at zero applied field) with a 10-cm pathlength of liquid CS_2. The initial diameter of the beam is 3 mm and the measurements were made 2 m after the exit of the liquid CS_2 cell.

diameter of an argon-ion laser beam $\lambda = 514.5$ nm and of 3-mm initial beam diameter, after it has propagated 2 m beyond a Kerr cell containing a 10 cm path length of liquid CS_2 as a function of laser power. There is little or no absorption of this wavelength by CS_2 and the effect arises by molecular orientation.

To remove these effects, which may dominate or even completely obscure (by distortion of the laser beam) the optical Kerr measurements, one rotates the cell at a few hertz around its longitudinal axis.[34] Or one may apply a static electric field to the fluid through which the laser beam is passing and being distorted. Figure 5 shows the effect of such an applied static field on the same sample described in Fig. 4. This applied field is arranged such that its orienting influence on the molecules will be out of phase with and thereby counteract that induced by the optical field of the argon-ion laser. The data are measured at the highest laser power (1000

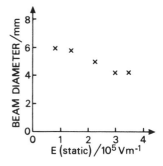

Figure 5. Consequence of applying a static electric field to the CS_2 cell described in Fig. 4. That is, the effect of an electric field on the diameter of the distorted laser beam, measured with a laser power of 1000 mW.

mW at 514.5 nm). Thus, at the zero field position the laser beam is seen, from Fig. 4, to have a diameter of 11 mm and application of the static field substantially reduced the beam cross section. The applied fields are of the same order of magnitude as the optical fields. The fact that one does not completely counteract the orienting effect of the light field may be due to some weak absorption of the laser radiation at these power densities.

This thermal lensing of the laser beam is quite a slow process. Indeed, by chopping the laser beam one may see the time evolution of the distortion. For the development of such focusing or defocusing molecules must move. Thus, the time required depends on the actual beam diameter and the effect will propagate across this beam with the velocity of sound. It can be seen, therefore, that such distortions are likely to be a problem only in experiments with continuous-wave or long pulsed lasers.

For a $2W$ argon-ion laser beam weakly focused to give a power density $\overline{W}$, of about 8×10^6 W m^{-2} in the cell of length 0.1 m, the oscillating electric field $E(\omega)$ is given by

$$E(\omega) = \left(\frac{\overline{W}}{c\varepsilon_0} \right)^{1/2} \quad (8)$$

and is seen to be 55×10^3 V m^{-1}. From the known molar Kerr constant of liquid CS_2, $_mB = 339 \times 10^{-16}$ V^{-2} m (Ref. 23), the induced retardation for the argon-ion laser beam would be 6×10^{-5} rad, a small induced retardation, but measurable with some form of modulation technology. With optical Kerr experiments performed with powerful pulsed lasers it is not possible to modulate the orienting light beam. However, because of the much larger electric fields present in such lasers, the induced retardations are very much greater. With a pulsed laser giving 40×10^{10} W m^{-2} (Ref. 38), the applied electric field is 1.2×10^7 V m^{-1}, an electric field larger than it is possible to apply in a static laboratory experiment. With such a field the induced birefringence is of order 0.3 rad,[38] being now sufficiently large as not to require any modulation techniques for its observation. Also of interest with large pulsed lasers is the magnetic field associated with the intense light beam. This is a factor c (2.998×10^8 ms^{-1}) weaker than the electric field; however, it would be a source of high-frequency magnetic birefringence. If the optical electric field were 3×10^7 V m^{-1}, then the associated magnetic field would be 0.1 T. An electro-optic effect arising from the high-frequency magnetic field of a light beam was discussed by Evans,[39] and its observation was recently reported.[40]

Unfortunately, it is not possible to modulate large laboratory magnetic field in the same manner. Consequently, a different strategy has to be adopted to observe an induced magnetic birefringence signal. A typical experimental setup for a magnetic birefringence experiment is given in Fig. 6 (Ref. 41), which is very similar to that used in Ref. 25.

A photoelastic modulator aligned with its optical axis parallel to the magnetic field direction produces a sinusoidal variation of the phase difference ϕ between the horizontally and vertically polarized light components at a frequency $\omega \approx 50$ kHz and an amplitude of $\Delta \psi \approx 0.2\pi$, $\phi = \Delta \psi \sin \omega t$. If the polarizers are well crossed, then the light level at the detector is given by Eq. (4). With a number of birefringent elements between the polarizer and the analyzer, the light signal at the detector is given by Eq. (5). Thus, in addition to the alternating component arising from the photoelastic modulator, there will be a steady phase difference ϕ_0 made up of the induced static Cotton-Mouton effect ϕ_{CM}, the birefringence arising from the Babinet-Soleil compensator ϕ_{BS}, and any stray birefringence induced by optical components or windows, ϕ_{ext}. The total dc birefringence signal is proportional to

$$\phi_0 = \phi_{CM} + \phi_{BS} + \phi_{ext} \tag{9}$$

Thus, the detector signal is

$$I = \frac{I_0}{4} \left(\Delta \psi^2 \sin^2 \omega t + 2 \, \Delta \psi \, \phi_0 \sin \omega t + \phi_0^2 \right) \tag{10}$$

where the different alternating components can be separated with a phase-sensitive detector.

The noise level of such an experiment is dependent on the total light level passed by the analyzer, and thus strongly dependent on the total static birefringence present in the system. This dc output may be used as an error signal in a feedback loop involving a parity-converting integrator, a high-voltage amplifier, and a longitudinal Pockels cell. Because the birefringence ϕ_P of the Pockels cell is linear in the applied voltage, ϕ_0 can be continuously compensated by ϕ_P; i.e., $\phi_0 = -\phi_P$. We therefore have a means of determining the magnitude and sign of the induced magnetic birefringence, provided the sources of static birefringence are minimized.

Because of the very different forms of detection system traditionally used for such electro-optic experiments, the measured sensitivities vary greatly. In the static Kerr effect, induced nanoradian retardations are measurable.[34, 42] It is straightforward to modulate the applied electric field and use a phase-sensitive detector to demodulate the signal coming from

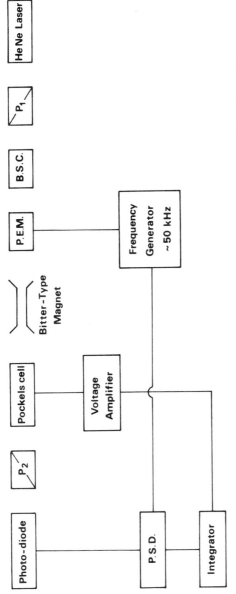

Figure 6. Diagrammatic representation of an apparatus for the measurement of the Cotton-Mouton effect. P1 and P2 are Glan polarizers, BSC is a Babinet-Soleil compensator, PEM is a photoelastic modulator, and PSD is the phase-sensitive detector.

375

the detector. Applied field gradients may also be modulated and the birefringence arising from the orientation of the molecular quadrupole moment may be measured with similar precession.[43] With the optical Kerr effect, the powerful orienting laser beam may be mechanically modulated for continuous-wave experiments, and if thermal lensing problems are overcome one may measure induced retardations of order 10^{-6}–10^{-7} rad. However, very powerful pulsed lasers have such an intense electric field that no modulation is required for the detection of the induced optical Kerr effect.

With the Cotton-Mouton effect, the applied magnetic field is static, which severely limits the experimental sensitivity. The apparatus described above and displayed in Fig. 6 has a limiting measurable birefringence of 10^{-4} rad. This is fairly typical. Buckingham et al.[44] quote 10^{-5} rad for their experimental setup, which is essentially unchanged in the work of Battaglia and Ritchie.[25] Because the interaction of interest, the molecule–applied magnetic field interaction, is not modulated directly, the sensitivity is limited by the total stray light arriving at the detector ϕ_0^2 in Eq. (10).

Recently, however, Scuri et al.[45] have shown the enormous improvements in the sensitivity of measurement of the Cotton-Mouton effect that may be obtained by modulating the applied magnetic field. Scuri et al. spatially modulated their magnetic field by rotating the magnet. The field was quite modest, 0.5 T, and the frequency of rotation is very low, 1.3125 Hz. However, they have achieved a sensitivity of 3×10^{-8} rad in the measurements of the Cotton-Mouton effect in gases, thereby bringing the sensitivity of magnetic birefringence experiments into line with what is possible in electric birefringence experiments. Perhaps this is the way forward for such measurements, to use smaller magnetic fields, but fields that may be modulated. With water-cooled coils it may well be possible to have a magnetic field of near a tesla modulated at a frequency of a few tens of hertz. What would be lost in orienting power would be more than offset by large improvements in the sensitivity given by newer detection technology.

III. THEORY

A. The Kerr and Cotton-Mouton Effects

If a beam of linearly polarized light propagates through an anisotropic medium, it becomes elliptically polarized, and if there is absorption there will also be a rotation of the plane of polarization. Figure 1a illustrates the arrangement in a typical birefringence experiment. The electric vector of a wave of angular frequency $\omega = 2\pi c/\lambda_0$ propagates along the y axis and, initially linearly polarized, can be resolved into its z and x components:

$$
\begin{aligned}
E_z(\omega, t) &= E_z^{(0)} \exp\left\{\frac{-k_z \omega y}{c}\right\} \cos\left\{\omega t - \frac{n_z \omega y}{c}\right\} \\
E_x(\omega, t) &= E_x^{(0)} \exp\left\{\frac{-k_x \omega y}{c}\right\} \cos\left\{\omega t - \frac{n_x \omega y}{c}\right\}
\end{aligned}
\tag{11}
$$

where n_z, n_x and k_z, k_x are the refractive indices and absorption coefficients for radiation whose electric vector is polarized in the z and x directions, respectively. If the polarization is at $45°$ to the z and x directions, as in Fig. 1b(I), the amplitudes $E_z^{(0)} = E_x^{(0)} = E^{(0)}/(2)^{1/2}$. In a medium that has become birefringent by application of an external perturbation, $n_z \neq n_x$ and the wave is elliptically polarized with its electric vector rotating about the direction of propogation, see Fig. 1b(II). In a dichroic medium, $k_z \neq k_x$ and the wave is anisotropically absorbed, producing an optical rotation, see Fig. 1b(III). The effect of birefringence and of dichroism together is to produce two out-of-phase components of different amplitudes, see Fig. 1b(IV), giving elliptically polarized light whose major axis is rotated from the original $45°$ position.

In what follows we are concerned with pure birefringence; that is, the fluids are assumed to be transparent at the frequencies of the probing laser radiation. In such a fluid rendered birefringent the phase difference ϕ between the electric vectors E_z and E_x may be written as

$$
\phi = \frac{2\pi(n_z - n_x)l}{\lambda_0}
\tag{12}
$$

where the terms are as defined earlier. If complex refractive indices are defined by

$$
\begin{aligned}
n_z^* &= n_z + i k_z \\
n_x^* &= n_x + i k_x
\end{aligned}
$$

then a complex Kerr constant is given by

$$
B^* = (n_z^* - n_x^*)\lambda_0^{-1}E^{-2} = \frac{\phi^*}{2\pi l E^2} = B + i B'
\tag{13}
$$

where

$$B' = (k_z - k_x)\lambda_0^{-1}E^{-2} = \frac{\phi'}{2\pi l E^2} \tag{14}$$

The dichroism ϕ' is related to the angle of rotation of the plane of polarization ϕ in Figs. 1b(III) and 1b(IV) by[23]

$$\tan 2\phi = \frac{\sinh \phi'}{\cos \phi}$$

where for small ϕ and small ϕ'

$$\phi = \frac{\phi'}{2} = \frac{\pi(k_z - k_x)l}{\lambda_0}$$

In Eq. (11), $\mathbf{n}$ determines the phase and $\mathbf{k}$ the intensity of the wave at any point in the medium. In the electric dipole approximation, $\mathbf{n}^*$ is related to the mean differential polarizability Π, that is, the extra electric moment induced in a molecule by application of, for example, an applied electric field.

Consider a molecule in a real field,

$$E_\alpha = E_\alpha^{(0)} \cos \omega \left(t - \frac{y}{c} \right) \tag{15}$$

with a time derivative given by

$$\dot{E}_\alpha = -E\alpha^{(0)}\omega \sin \omega \left(t - \frac{y}{c} \right) \tag{16}$$

Then the dipole proportional to $E^{(0)}$ induced in the molecule by the applied field is

$$\mu_\alpha = \Pi_{\alpha\beta}E_\beta + \frac{\dot{\Pi}'_{\alpha\beta}E_\beta}{\omega} \tag{17}$$

Thus, $\Pi_{\alpha\beta}$ gives the α component of the dipole in phase with $\mathbf{E}$ induced in the molecule by the β component of $\mathbf{E}$; $\Pi'_{\alpha\beta}$ gives the α component of the dipole out of phase with $\mathbf{E}$ induced by the β component of the time derivative of $\mathbf{E}$, with the symmetric part of Π' being responsible for absorption and Π for refraction of the wave.

The refractive indices n_z and n_x for light with electric vector in the z and x directions are given within the electric dipole approximation by the Lorentz-Lorenz equation,

$$\frac{n_z^2 - 1}{n_z^2 + 2} = \left(\frac{4\pi N}{12\pi\varepsilon_0}\right)\overline{\Pi}_{zz} \tag{18}$$

and

$$\frac{n_x^2 - 1}{n_x^2 + 2} = \left(\frac{4\pi N}{12\pi\varepsilon_0}\right)\overline{\Pi}_{xx} \tag{19}$$

where ε_0 is the permicivity of free space, N is the number density of molecules, and $\overline{\Pi}_{zz}$ and $\overline{\Pi}_{xx}$ are mean polarizabilities averaged over the molecular translational and rotational motion and over internal vibrational and electronic states. In most samples it is sufficient to treat the translations and rotations classically and at the temperatures of most experiments only the ground vibrational and ground electronic states are populated. If the difference between n_z and n_x is small, which it is away from regions of dispersion, as in pure birefringence, we may write

$$n_z - n_x = \left(\frac{2\pi N}{36\pi\varepsilon_0}\right)\left\{\frac{(n^2 + 2)^2}{n}\right\}\{\overline{\Pi}_{zz} - \overline{\Pi}_{xx}\} \tag{20}$$

where

$$\frac{\mathrm{d}}{\mathrm{d}n}\left\{\frac{n^2 - 1}{n^2 + 2}\right\} = \frac{(n^2 + 2)^2}{n}$$

The applied electromagnetic field causes a torque to be exerted on the molecules in the sample. In fluids this orienting effect is an even function of the field $\mathbf{E}$ or $\mathbf{H}$, since it must be unaffected by reversal of the field direction.

In an applied uniform electric field $\mathbf{E}$, we may write the the total dipole moment of a molecule as[46, 47]

$$\mu_\alpha = \mu_\alpha^{(0)} + \alpha_{\alpha\beta}E_\beta + \tfrac{1}{2}\beta_{\alpha\beta\gamma}E_\beta E_\gamma + \tfrac{1}{6}\gamma_{\alpha\beta\gamma\delta}E_\beta E_\gamma E_\delta + \cdots \tag{21}$$

where $\mu_\alpha^{(0)}$ is the permanent electric dipole moment of the molecule, α is the familiar molecular polarizability, and the first (β) and second (γ) hyperpolarizabilities describe the distorting effect of the applied field on the polarizability. The first hyperpolarizability is responsible for frequency doubling and the second is responsible for frequency tripling; that is, they describe the dipole moments induced by a light field that oscillates at twice and three times the incident frequency. The differential polarizability is then found by differentiating the total dipole moment with respect to the applied field, i.e.,

$$\Pi_{\alpha\beta} = \frac{\partial \mu_\alpha}{\partial E_\beta}$$

$$= \alpha_{\alpha\beta} + \beta_{\alpha\beta\gamma} E_\gamma + \tfrac{1}{2}\gamma_{\alpha\beta\gamma\delta} E_\gamma E_\delta + \cdots \qquad (22)$$

For a diamagnetic molecule in an applied magnetic field $\mathbf{H}$, we may write for the differential polarizability[48]

$$\Pi_{\alpha\beta} = \alpha_{\alpha\beta} + \eta_{\alpha\beta,\gamma\delta} H_\gamma H_\delta + \cdots \qquad (23)$$

where $\eta_{\alpha\beta,\gamma\delta}$ is a measure of the distortion of the molecular polarizability by the applied magnetic field.

In these applied fields the potential energy of the molecule U is a continuous function of the orientation of the molecule and we must now average the differential polarizability of a molecule over all possible configurations τ in the presence of the external field. Thus, for an applied field[46]

$$\Pi_{\alpha\beta}(\tau, E_0) = \frac{\int \Pi_{\alpha\beta}(\tau, E_0)\exp\{-U(\tau, E_0)/kT\}\,d\tau}{\int \exp\{-U(\tau, E_0)/kT\}\,d\tau} \qquad (24)$$

where $U(\tau, E_0)$ is the potential energy of the system in the presence of E_0 and in the configuration τ. For an applied static electric field, the potential energy for a dipolar molecule may be written as

$$U = U_0 - \mu_\alpha^{(0)} E_\alpha - \tfrac{1}{2}\alpha_{\alpha\beta}^{(0)} E_\alpha E_\beta - \tfrac{1}{6}\beta_{\alpha\beta\gamma} E_\alpha E_\beta E_\gamma - \cdots \qquad (25)$$

Here $\alpha^{(0)}$ is the static polarizability of the molecule, that is, the polarizability capable of responding to a very low-frequency electric field. For a diamagnetic molecule in an applied magnetic field we may write for the potential energy[48]

$$U = U_0 - \tfrac{1}{2}\chi_{\alpha\beta} H_\alpha H_\beta - \eta_{\alpha\beta,\gamma\delta} H_\alpha H_\beta H_\gamma H_\delta - \cdots \qquad (26)$$

where χ is the diamagnetic magnetic susceptibility or magnetizability.

It is now necessary to expand Π as a power series in the field $\mathbf{E}$, taking note that it depends on $\mathbf{E}$ both through $\Pi(\tau, E)$ and through the potential energy U. For convenience we will introduce the notation $\langle \Phi \rangle$ for the average of the quantity $\Phi(\tau, E)$ in the fluid, with $E = 0$; that is,

$$\langle \Phi \rangle = \frac{\int \Phi(\tau, 0) \exp\{-U^{(0)}/kT\} \, d\tau}{\int \exp\{-U^{(0)}/kT\} \, d\tau} \tag{27}$$

The coefficients of E in the expansion can be obtained by differentiating Eq. (24) with respect to E and then putting $E = 0$. As expected, the leading term in the expansion of Π is in E^2, the coefficients being[46]

$$\frac{1}{2} \left\{ \frac{\partial^2 \overline{\Pi}}{\partial E^2} \right\}_{E=0} = \frac{1}{2} \left\langle \frac{\partial^2 \Pi}{\partial E^2} \right\rangle - \left(\frac{1}{2kT} \right) \left\langle 2 \left(\frac{\partial \Pi}{\partial E} \right) \left(\frac{\partial U}{\partial E} \right) + \Pi \left(\frac{\partial^2 U}{\partial E^2} \right) \right\rangle$$
$$+ \left(\frac{1}{2k^2 T^2} \right) \left\langle \Pi \left(\frac{\partial U}{\partial E} \right)^2 \right\rangle \tag{28}$$

The angle brackets denote an averaged quantity. This expression is applicable to the derivation of an expression for either the Kerr effect or the Cotton-Mouton effect. The various terms that appear in Eq. (28) can be determined from the above equations. Thus, for the Kerr effect one substitutes the appropriate derivatives of Eqs. (22) and (25) into Eq. (28), and for the Cotton-Mouton effect one substitutes the appropriate derivatives of Eqs. (23) and (26) into Eq. (28).

Consider the Kerr effect. Making the necessary substitutions into Eq. (28) gives for the induced anisotropy in the differential polarizability[46]

$$\overline{\Pi}_{zz} - \overline{\Pi}_{xx} = \frac{E_z^2}{30} \left\{ (3\gamma_{\alpha\beta\alpha\beta} - \gamma_{\alpha\alpha\beta\beta}) \right.$$
$$+ \frac{2}{kT} \{ 3\beta_{\alpha\beta\alpha} \mu_\beta^{(0)} - \beta_{\alpha\alpha\beta} \mu_\beta^{(0)} \}$$
$$+ \frac{1}{kT} \{ 3\alpha_{\alpha\beta} \alpha_{\alpha\beta}^{(0)} - \alpha_{\alpha\alpha} \alpha_{\beta\beta}^{(0)} \} \tag{29}$$
$$+ \left. \frac{1}{k^2 T^2} \{ 3\alpha_{\alpha\beta} \mu_\alpha^{(0)} \mu_\beta^{(0)} - \alpha_{\alpha\alpha} \mu_\beta^{(0)} \mu_\beta^{(0)} \} \right\}$$

This is a general expression which may be greatly simplified by inclusion of molecular symmetry. However, as it stands it is still a representation of the anisotropy in a molecule fixed axis system. The evaluation of the isotropic averages of the tenser components present in Eq. (29) is straightforward. The problem reduces to the determination of products of direction cosines, between particular pairs of axes in the molecule fixed and the laboratory fixed or space fixed coordinate system, averaged over all possible relative orientations of the two coordinate systems.[23, 49] This done, we may write for molecules that have a threefold or higher rotational axis (symmetric tops where, for example, we may define a polarizability anisotropy, $\Delta \alpha = (\alpha_{\parallel} - \alpha_{\perp})$ and $\alpha_{\parallel} = \alpha_{zz}$ and $\alpha_{\perp} = \alpha_{xx} = \alpha_{yy}$) as

$$\overline{\Pi}_{zz} - \overline{\Pi}_{xx} = E_z^2 \left\{ \frac{\gamma}{3} + \frac{\Delta\alpha \, \Delta\alpha^{(0)}}{15kT} + \frac{2\beta\mu^{(0)}}{9kT} + \frac{\Delta\alpha \, \mu^{2(0)}}{15k^2T^2} \right\} \quad (30)$$

which may be related to the experimental observable ϕ through Eqs. (20) and (12).

For a pure material it is convenient to define a molar Kerr constant, $_mB$. At low densities this is[46, 47]

$$_mB = \left(\frac{2}{27} \right) V_m \lim_{E_z = 0} \left\{ (n_z - n_x)/E_z^2 \right\}$$

$$= \left(\frac{4\pi N_A}{27} \right) \lim_{E_z = 0} \left\{ (\overline{\Pi}_{zz} - \overline{\Pi}_{xx})/E_z^2 \right\}$$

where N_A is Avogadro's number and V_m is the molar volume. Thus, for the static Kerr effect of a pure fluid we may write (in SI units)[23]

$$_mB = \frac{N_A}{81\varepsilon_0} \left\{ (\gamma) + \frac{\Delta\alpha \, \Delta\alpha^{(0)}}{5kT} + \frac{2\beta\mu^{(0)}}{3kT} + \frac{\Delta\alpha \, \mu^{2(0)}}{5k^2T^2} \right\} \quad (31)$$

An investigation of the magnitudes of the various terms that appear in Eq. (31) and contribute to the measured Kerr effect reveals that for polar molecules in static electric fields it is the terms involving the permanent dipole moments that make the biggest contribution to the observed static Kerr effect. However, if this polar molecule is subject to an oscillating electric field, whose frequency of modulation is too high for the permanent dipole moment to be able to respond (a molecule will need 10^{-12} s or longer to rotate in a fluid, that is to orient, but an optical $E(\omega)$ will reverse every 10^{-15} s), then we may rewrite Eq. (30), without its dipolar contribution, as

$$\overline{\Pi}_{zz} - \overline{\Pi}_{xx} = E_z^2 \left\{ \frac{\gamma}{3} + \frac{\Delta\alpha \, \Delta\alpha^{(0)}}{15kT} \right\} \tag{32}$$

Thus for a nonpolar molecule such as CS_2, the optical Kerr effect follows the same expression for the induced optical anisotropy as does the static Kerr effect. Consider Eq. (32). In the static Kerr effect the nonpolar molecule is oriented through the interaction of its static polarizability $\alpha^{(0)}$ with the applied static electric field. One is able to measure the induced birefringence by using a laser that becomes elliptically polarized upon passage through the oriented fluid by interaction with the optical polarizability α, which is a function of frequency. In the optical Kerr effect, on the same molecule, both polarizability terms in the temperature-dependent part of Eq. (32) are frequency-dependent optical polarizabilities, since both the orienting and weaker probe laser are in the visible or near ultraviolet.

If the molecule under investigation is a spherical top, for example, SF_6 or CH_4 or a spherical atom such as argon, then $\alpha_{xx} = \alpha_{yy}$ and $\Delta\alpha = 0$. However, a Kerr effect, static and optical, will still be observable through the temperature-independent hyperpolarizability γ, which exists for all atoms and molecules. This term describes the distortion of the initially spherical atom or molecule by the applied field to give an anisotropic system that has the ability to depolarize light.

The equivalent expression for the molar Cotton-Mouton constant of a pure fluid of molecules with a rotational axis of threefold or greater symmetry, $_mC$, in SI units is[25]

$$_mC = \frac{N_A \mu_0^2}{270\varepsilon_0} \left\{ \eta + \frac{2\Delta\alpha \, \Delta\chi}{3kT} \right\} \tag{33}$$

where μ_0 is the vacuum permeability and $\Delta\chi = \chi_{\parallel} - \chi_{\perp}$, that is, the anisotropy in the magnetic susceptibility. Here we see how the orientation of the molecule arises through the interaction of the applied magnetic field and the magnetic susceptibility and the orientation is observed via the electronic polarizability as in the Kerr effects. Again there is a temperature-independent distortion which will give rise to a magnetic birefringence in spherical atoms and molecules.

An extreme form of this temperature-independent distortion which contributes to the Cotton-Mouton effect may be seen in the work of Bacri and Salin[50] who investigated particle shape in fine dispersions of ferrofluids subjected to applied magnetic fields. It was found that the initially spherical particles, of diameter ≈ 100 Å, were seen to become ellipsoids, with their long axis of rotation along the field direction, in fields of a few tens of gauss.[50] Such systems, which change their shape to become anisotropically polarizable and anisotropically magnetizable upon application of a magnetic field, serve as an example of how the particle of interest, in the absence of a field has no intrinsic anisotropy, but distorts in the applied field and may then be oriented by the applied field. Similar temperature-independent distortions, which then give rise subsequently to induced birefringence, have been observed in the application of magnetic field to initially spherical lipid bilayers.[51]

B. Solutions

The derivation of the molar Kerr or molar Cotton-Mouton constants of a pure fluid, as given in the previous section, is fairly straightforward. However, problems arise when one considers dense fluids.[47] Molecular interactions perturb the intrinsic polarizabilities of the molecules and give rise to a solvent effect. There is therefore the possibility of a difference arising between the electro-optic properties of a molecule as measured in a low-pressure gas and the corresponding property derived from measurements made on the same species as a liquid.

This internal field problem has been much studied,[52, 53] but there is no general solution. For a dense fluid the notion of independent molecules has limitations. What is generally undertaken in analyzing, for example, the results of the optical Kerr effect measurements on liquids is to apply a Lorentz field correction, that is, to correct for the difference in the electric field as applied by the experimenter and as experienced by the molecule under investigation. Then the applied field $\mathbf{E}_{applied}$ is related to the local field $\mathbf{E}_{local}$ seen by the molecule via

$$\mathbf{E}_{local} = \left\{ \frac{\varepsilon + 2}{3} \right\} \mathbf{E}_{applied} = \left\{ \frac{n^2 + 2}{3} \right\} \mathbf{E}_{applied} \qquad (34)$$

where ε is the dielectric constant of the medium and n its refractive index ($\varepsilon = n^2$) for water at ambient temperatures and a wavelength of 589 nm, $n = 1.3333$.

The problem of local field corrections does not arise when the samples under investigation are gases. However, for many experimental situations investigation of liquids and solutions are unavoidable. For example, the Cotton-Mouton effect has a limited sensitivity and the measurement of induced magnetic birefringence in gaseous samples has, in comparison with induced electric birefringence, only recently become routine. In the liquid there are more molecules per unit path length and such experiments are easier to perform. The study of the Kerr effect and the Cotton-Mouton effect in solutions of liquids over wide concentration ranges are of interest in themselves because they permit measurement of the interactions between molecules of different species. For example, the molar Kerr constant $_mB$ was defined by LeFèvre and LeFèvre[20] in terms of specific Kerr constants $_sB$, which represented the properties of the components present in solution. At the time there was interest in comparing electro-optic measurements made on molecules in the gaseous state to those on the same molecule in the liquid state.

It was shown how the molar Kerr or molar Cotton-Mouton constant may be defined as the difference between the molecular refractions parallel and perpendicular to an applied field of unit strength. The molar refractivities may be expanded to give the refractive indices. LeFèvre and LeFèvre showed that, for example, the molar Cotton-Mouton constant is given by[24, 54]

$$_mC = \frac{6\lambda_s Cm}{(n^2 + 2)^2 d} \tag{35}$$

where d is the fluid density, m is the molar mass, and n is the refractive index when the applied field $\mathbf{H} = 0$. To obtain the molar Cotton-Mouton constant of a component in a mixture one extrapolates the measured Cotton-Mouton constants of the solutions for a particular solute of concentration w to infinite dilution in the solvent. Thus, the infinite dilution partial Cotton-Mouton constant is given by[24, 54]

$$_\infty(_sC_{\text{solute}}) = {}_sC_{\text{solvent}} + \lim_{w \to 0} \left\{ \frac{\mathrm{d}(_sC_{\text{solution}})}{\mathrm{d}w} \right\} \tag{36}$$

As the concentration of solute falls one may express the incremental change of solution refractive index n, solution density d, and specific Cotton-Mouton constant $_sC$ in terms of those of the solvent and a weighted (by concentration of solute) contribution for the solute. Then, with $n_{\text{solution}} = n_{\text{solvent}} + wG$, $_sC_{\text{solution}} = {}_sC_{\text{solvent}} + wD$, and $d_{\text{solution}} = d_{\text{solvent}} + wB$ substituted into Eq. (36) and differentiated with respect to w, one obtains an expression for $_\infty(_sC_{\text{solute}})$, an expression for the specific Cotton-Mouton constant of the solute in terms of that for the solvent.

To obtain the molar Cotton-Mouton constant one must multiply the specific Cotton-Mouton constant by the appropriate molar mass. A similar procedure gives the Kerr constant at infinite dilution. Such a model gives, for a particular solute in a particular solvent, one Cotton-Mouton or one Kerr constant. However, if there are strong specific interactions that change with solute concentration, it may be better not to make the extrapolation to infinite dilution.

Such models were introduced to measure and interpret molecular interactions. For this type of experimental program electro-optic experiments were and in many ways still are ideally suited. It was shown earlier how one may measure very small induced birefringences; an induced retardation of $\phi = 10^{-7}$ rad is readily observed. Then from Eq. (2) or (3), with a path length of 0.1 m and at a measuring wavelength of 514.5 nm, one may measure an anisotropy in the refractive index of a fluid of 8×10^{-14}. The ability to measure such small differences in a fluid property that is itself affected by molecular interactions is almost unique. For example, a polar solute in a nonpolar solvent may well give a different result than measurements made on the pure solute. This difference way be interpretable in terms of the various models of molecular interactions. The measurements of the static Kerr constants made on solutions of polar fluorobenzenes in benzene[55] show how stereo-specific information may be derived from solvent dependent investigations.

Kielich has discussed in detail the solvent specific interactions that may arise in electro-optic experiments.[56, 57] Consider, for example, the specific Kerr constant $_sB$ of a multicomponent system:

$$_sB = \sum_i x_i\,_sB^{(i)} + \sum_{i,j} x_i x_j\,_sB^{(ij)} + \cdots \qquad (37)$$

The constants $_sB^{(i)}$ contain only the respective molecular parameters that determine the electro-optic properties of the isolated molecules of the various components. In order that the constant $_sB$ of the solution is a strictly additive quantity, it is necessary that the component parts $_sB^{(1)}, _sB^{(2)}, \ldots$, are independent of concentration. In reality, however, this is rarely the case. There are strong deviations from the additivity rule even for dilute solutions.[58]

In a later section we will discuss electro-optic measurements made on aqueous solutions of simple electrolytes. To analyze these data we have used a modified form of Eq. (35). The systems of interest are considered to be two component mixtures: water (the solvent) and the electrolyte (the solute). The electrolyte will have dissociated upon solution. However, only one ion will have any appreciable effect, i.e., be intrinsically anisotropic.

For the alkali metal nitrates this will be the planar NO_3^- anion. We may then, after Eq. (37), write

$$B_{\text{solution}} = x_1 B_{\text{water}} + x_2 B_{\text{solute}} + x_1 x_2 B_{\text{int}} \qquad (38)$$

where B_{int} is taken to be an interaction-induced contribution to the measured effect. From Eq. (38) the property of interest is B_{solute}, B_{solution} is a measured quantity, as is the effect in the pure solvent, and the mole fractions present in solution are known; for example, $x_1 = $ (Number of moles of water)/(Number of moles water + Number of moles solute).

IV. DIFFERENCES BETWEEN STATIC AND OPTICAL KERR CONSTANTS

It was pointed out above that when the electric field applied to a fluid oscillates at optical frequencies the permanent molecular dipole moments will be unable to follow the field changes and one therefore loses the dipolar contribution to the induced birefringence. Consequently, for a molecule such as CS_2 the two Kerr constants—the static Kerr constant $_mB^{(s)}$ and the optical Kerr constant $_mB^{(o)}$—should be very similar. These two values are found in Table I, where it is seen that they are similar. A difference will arise because of the dispersion properties of the electronic polarizability of CS_2. The light sources used to probe the induced optical anisotropy were of differing wavelengths and consequently the measured

TABLE I
Liquid Kerr Constants

	$_mB^{(s)a}/\text{V}^{-2}\,\text{m}$	$_mB^{(o)b}/\text{V}^{-2}\,\text{m}$	$_mB^{(o)c}/\text{V}^{-2}\,\text{m}$
Water	$294 \times 10^{-16\,e}$	$3.24 \times 10^{-16\,f}$	
CS_2	$339 \times 10^{-16\,d}$	$467 \times 10^{-16\,f}$	$436 \times 10^{-16\,g}$
	$479 \times 10^{-16\,f}$		
CCl_4	$9.2 \times 10^{-16\,d}$	$5.7 \times 10^{-16\,f}$	
C_6H_6	$46 \times 10^{-16\,d}$	$44.7 \times 10^{-16\,f}$	$78 \times 10^{-16\,g}$
	$41 \times 10^{-16\,e}$		
$C_6H_5NO_2$	$44\,000 \times 10^{-16\,d}$	$324 \times 10^{-16\,f}$	$369 \times 10^{-16\,g}$

[a] Measured at a probing wavelength of 589 nm.
[b] Measured at an orienting wavelength of 694.3 nm and a probing wavelength of 488 nm.
[c] Measured at an orienting wavelength of 500 nm.
[d] Taken from Ref. 23.
[e] Taken from Ref. 64.
[f] Taken from Ref. 38.
[g] Taken from Ref. 72.

signal will be different. The electronic polarizability increases as one moves toward the ultraviolet where electronic resonances are found. However, away from resonance (where the polarizability will become very large[59]) one may scale the change of polarizability with wavelength as the ratio of the wavelengths.

Conversely, from Table I, a substantial difference is seen between the static and optical Kerr constants of water. Here there are no electronic resonances in the visible and so the difference between $_mB^{(s)}$ and $_mB^{(o)}$ arises because of the ability of the molecular properties, seen in Eq. (31), to respond differently to the optical and static electric fields.

Consider Eq. (31), and for simplicity consider the water molecule to be a symmetric top. This is an approximation introduced only for an order of magnitude calculation. Then the polarizability contribution to both the static and optical Kerr effects may be represented as a polarizability anisotropy, $\Delta\alpha = (\alpha_\| - \alpha_\perp) = \alpha_{zz} - \{(\alpha_{xx} + \alpha_{yy})/2\}$, where[60] $\alpha_{zz} = 1.634 \times 10^{-40}$ C^2 m^2 J^{-1} (at 514.5 nm), $\alpha_{xx} = 1.699 \times 10^{-40}$ C^2 m^2 J^{-1} (at 514.5 nm), and $\alpha_{yy} = 1.573 \times 10^{-40}$ C^2 m^2 J^{-1} (at 514.5 nm), from which we determine $\Delta\alpha$ to be -247×10^{-45} C^2 m^2 J^{-1}. We wish to make a comparison of the magnitudes of the four terms that contribute to Eq. (31) for water. For this we take the permanent dipole moment to be 1.85 D (Ref. 61) and take literature values for the two hyperpolarizabilities, $\beta = -170 \times 10^{-54}$ C^3 m^3 J^{-2} and $\gamma = 178 \times 10^{-63}$ C^4 m^4 J^{-3} (Refs. 62 and 63). These two values have been determined by a study of the temperature dependence of electric field-induced second-harmonic generation in liquid water. We are therefore able to calculate the ratios of the magnitudes of the terms that occur in Eq. (31) at 20°C. Thus the values $10^{63}\gamma$, $10^{63}\beta\mu(2/3kT)$, $10^{63}\Delta\alpha^2/5kT$, and $10^{63}\mu^2\Delta\alpha/5k^2T^2$ are calculated to be 178, -173, 0.003, and -115 C^4 m^4 J^{-3}. The measured static Kerr constant for water is dominated by the dipolar contribution and, surprisingly, the γ hyperpolarizability. Similarly, the loss of the dipole contribution to the Kerr effect at optical frequencies gives the very large difference between $_mB^{(s)}$ and $_mB^{(o)}$ seen in Table I.

Indeed, we may extend this model and point out that in the optical Kerr effect of liquid water essentially all the signal comes from the hyperpolarizability term. From Table I, the optical Kerr constant for water was measured at 488 nm by Paillette[38] to be 3.24×10^{-16} V^{-2} m, from which, using Eq. (31), we obtain $\gamma = 3.38 \times 10^{-60}$ C^4 m^4 J^{-3}, assuming that the temperature-dependent polarizability term is zero.

The static Kerr constant of water is seen from Eq. (31) to be made up of many more terms; however, it is again reasonable to ignore the term in $\Delta\alpha^2$. Then we estimate $\mu^2\Delta\alpha/5k^2T^2$ to be -115×10^{-63} C^4 m^4 J^{-3} at 20°C, and with $_mB^{(s)} = 294 \times 10^{-16}$ V^{-2} m (Ref. 64) and the value of γ

derived from the optical Kerr effect we find $\beta = -3.9 \times 10^{-51}$ C^3 m^3 J^{-2} for liquid water, where we use the same polarity as given in Refs. 62 and 63.

The value we have derived for β and γ from the measured Kerr constants are larger than one would normally expect for such properties, see above. We have not applied any local field correction and these may be important, particularly in the static Kerr constant. However, it is the lack of optical anisotropy in this molecule that is the main cause for the relative importance of the hyperpolarizabilities.

That these hyperpolarizabilities derived from the Kerr effect are different from those obtained from harmonic frequency generation is not unexpected. Consider the electric field applied to the molecule as consisting of a static part, E_0, and a dynamic part, $E^{(\omega)}\cos \omega t$. Then from Eq. (21), which gives the total molecular dipole moment of the molecule in the presence of the field, we see how in the presence of the static and the dynamic electric field there are now dynamic polarizabilities and dynamic hyperpolarizabilities in addition to their static counterparts. For example, the dipole induced through the polarizability α now has a static part and a dynamic part:

$$\mu = \alpha(0;0)E_0 + \alpha(-\omega;\omega)E(\omega)\cos \omega t \tag{39}$$

Similar expressions may be written for the hyperpolarizability contributions to the molecular dipole moment in the applied static and dynamic fields. Such an analysis[65] gives $\beta(-\omega;0,\omega)$ for the second hyperpolarizability, which contributes to the Kerr effect, and $\beta(-2\omega;\omega,\omega)$ for that which contributes to the second harmonic experiment.[63] Similarly $\gamma(-\omega;0,0,\omega)$ contributes to the Kerr effect and $\gamma(-2\omega;0,\omega,\omega)$ to the harmonic generation experiment. The minus sign comes from the convention that the sum of the frequencies in the brackets is zero. These molecular properties will have different scalar values[62, 63] and it is perhaps not surprising that we see this difference in the above analysis of the different hyperpolarizabilities of water.

Also of importance in an analysis of the various polarizabilities of molecules determined from either gas-phase or condensed-phase measurements, is the presence of a vibrational contribution to the measured values. The electronic contribution to a property ζ is defined as the expectation value $\langle 0|\zeta_{electronic}|0\rangle$ where $\zeta_{electronic}$ is the expression for ζ using the electronic wave functions only. The vibrational contribution to the property ζ is the sum of all terms in the expression for $\langle \zeta \rangle$ excluding electronic contributions.

The experimental and theoretical work of Ward and coworkers[65-67] and Shelton and coworkers[68-70] together with the theoretical calculations of Bishop[71] has given a detailed understanding of the vibrational contributions to the polarizabilities and hyperpolarizabilities that occur in different electro-optic experiments. For example, the vibrational contribution to the isotropic polarizability of a small polyatomic molecule is calculated from the expression[69]

$$\alpha_{\text{vib}}(\omega) = \frac{2\pi}{h} \sum |\mu_{ij}|^2 \left\{ \frac{2\Omega_{ij}}{\Omega_{ij}^2 - \omega^2} \right\} \tag{40}$$

where Ω_{ij} is a vibrational transition frequency, $|\mu_{ij}|$ is a matrix element of the transition dipole moment, and the summation excludes the ground state. At optical frequencies we may approximate $\Omega_{ij}^2 - \omega^2 = -\omega^2$ and obtain, for a second-harmonic generating experiment[69]

$$\Delta\alpha_{\text{vib}}(\omega) = \alpha_{\text{vib}}(2\omega) - \alpha_{\text{vib}}(\omega)$$
$$= \frac{6\pi}{2h}\omega^{-2} \sum |\mu_{ij}|^2 \Omega_{ij} \tag{41}$$

For the evaluation of the vibrational contribution to the polarizability the required matrix elements are found from infrared and Raman intensities. For CF_4 the vibrational contribution is found to be[69]

$$\Delta\alpha_{\text{vib}}(\omega) = (1.36 \times 10^{-34} C^2\ m^2\ J^{-1}\ cm^{-2})\omega^{-2}$$

for ω given in cm^{-1}. We are therefore able to calculate the contribution of each vibrational mode to the molecular polarizability. For the polarizability these corrections are found to be modest, a few percent. However, for the hyperpolarizability γ they may be as much as 30% for the $\gamma(-\omega; 0, 0, \omega)$ measured in the Kerr effect experiment in CF_4 (Ref. 69).

This analysis allows us to rewrite the expression for the anisotropic polarizability measured in the optical Kerr effect, Eq. (32), to include a vibrational hyperpolarizability, γ_{vib}:

$$\overline{\Pi}_{zz} - \overline{\Pi}_{xx} = \frac{E_z^2}{3}\left\{ \gamma_{\text{electronic}} + \gamma_{\text{vib}} + \frac{\Delta\alpha\,\Delta\alpha^{(0)}}{5kT} \right\} \tag{42}$$

where $\gamma_{\text{electronic}}$ is the electronic contribution to the γ hyperpolarizability. We will return to the implications of this separation of vibrational and electronic contributions to the various molecular polarizabilities and hyperpolarizabilities when we consider the time dependence of the optical Kerr effect in the next section.

V. OPTICAL KERR EFFECT STUDIES OF MOLECULAR DYNAMICS

Of considerable current interest is the use of the optical Kerr effect to investigate liquid and solution structure and the localized motions of molecules in condensed phases. When a very short pulse of intense light passes through a fluid it produces a transient birefringence. The rapid onset and relaxation of this birefringence with the arrival and departure of the light pulse, particularly in CS_2, has led to the use of the optical Kerr effect in ultrafast optical switches. See Table I for the magnitude of the optical Kerr constant for CS_2 relative to several other organic liquids. Nitrobenzene has the largest static Kerr constant of any common organic fluid. However, when we go to the optical frequency regime we lose the dipole contribution to the Kerr effect and we are left with an optical anisotropy smaller than that for CS_2 (Refs. 23, 38, 64, and 72). The large static response of nitrobenzene has been put to use in electro-optic switches operating at very different frequencies from the optical region.

The development of powerful femtosecond (10^{-15} s) laser pulses has lead to a new field of experimental investigation of molecular dynamics in both pure liquids and in solutions.[73-76] For example, the reorientational dynamics for an anisotropically polarizable target molecule in a strongly associated electrolyte solution exhibit time scales for relaxation extending into hundreds of picoseconds (10^{-12} s), whereas carbon disulfide molecules in pure CS_2 or in solutions of CS_2 in alkanes are seen to reorient in 1.6 ps (Refs. 74 and 75).

In Fig. 2a we display an apparatus used for making measurements of the optical Kerr effect with continuous-wave lasers, and in Fig. 3 we show an apparatus used for pulsed optical Kerr effect measurements. In the latter experimental setup the pump and probe lasers are generated from the same source to reduce any temporal jitter (which is of some importance when dealing with such short pulses) between coupled laser sources, and have the same wavelength. With the impressive temporal resolution of such equipment (≈ 50 fs is now typical, giving about 25 wavelengths of light per pulse), it has been observed, for example, that the transient induced Kerr effect in liquids of small polyatomics, such as CH_2Br_2 (Ref. 76) and CS_2 (Ref. 74, 75), is structured. To interpret these latter measurements, Kenney-Wallace and coworkers have put forward a model in which the reorientational dynamics of a molecule in such a weakly associated, covalent fluid, as probed by the transient optical Kerr effect, consists of five dynamically distinguishable contributions[74, 75, 77]: (1) a purely electronic hyperpolarizability, that is, the electronic response of the molecules to the intense electric field of the laser (in such experiments the applied laser field is of order 10^7 V m^{-1}); (2) a rapidly dephased, coherently excited by the laser, intermolecular librational motion, the rocking of one molecule within the cage of molecules that surround it, with a reorientational time of < 170 fs and occurring in a frequency regime 25–60 cm^{-1}; (3) intramolecular vibrational contributions, corresponding to vibrational frequencies in the hundreds to few thousand wave numbers; (4) at longer times, local translational fluctuations of the center-of-mass decaying over, typically, 400 fs, which correlates with density fluctuations, hence to the time dependence of the solvent cage, and consequently, to the collision-induced anisotropy in the molecular polarizability; (5) a long-time reorientational anisotropy with a time scale for decay of about 1 ps.

In Fig. 7 we display some of these data. Part (a) is the transient Kerr effect of liquid CHCl$_3$ (Ref. 78) and part (b) is the transient Kerr effect of liquid CBrCl$_3$ (Ref. 78). An instantaneous electronic response is seen together with slower nuclear relaxations. Of interest is the structure seen on these slower components. These are the intramolecular vibrational contributions, the Raman-induced Kerr effect, which arises from a particular third-order susceptibility or hyperpolarizability, $\gamma(-\omega 2; \omega 2, -\omega 1, \omega 1)$; see Ref. 65 for a description of the symmetry properties of the different hyperpolarizabilities. Where $\omega 1$ is the frequency of the strong pump laser beam that gives rise to the virtual birefringence, and $\omega 2$ is the frequency of the weaker probe laser beam. When $|\omega 1 - \omega 2|$ approaches a Raman active frequency in the target molecule, the observed birefringence exhibits a resonance. In the experiment of Fig. 7 the pump and probe lasers are at the same wavelength center; however, the laser pulses are broadened to give a 300 cm^{-1} bandwidth. In this way the complete Raman spectrum over the bandwidth of the laser (with a bandwidth of 300 cm^{-1} only the lowest frequency allowed Raman transition is observed) may be determined from the optical Kerr effect experiment by placing a spectrometer after the analyzing polarizer.

Because of the high frequency of the probing laser field, only the Raman effect can contribute to the induced optical anisotropy[78, 79]; the permanent molecular dipoles are unable to respond. For chloroform a value of 262 cm^{-1} is found for the Raman active molecular vibration contributing to the measurement from the difference between the oscillations seen in Fig. 7 and for CBrCl$_3$ the corresponding frequency is 192 cm^{-1} (Ref. 78). Within the bandwidth of the laser pluses the depolarization ratio of the mode determines the probability of observation,[78, 79] that is, the size of the component induced in the probe beam which is

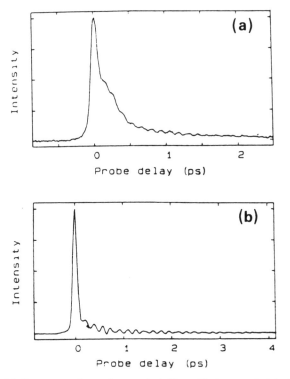

Figure 7. (a) Femtosecond transient optical Kerr effect generated in $CHCl_3$ with a resolution of 8.3 fs per point. (Reproduced by kind permission of *Chemical Physics Letters*.) (b) Femtosecond transient optical Kerr effect generated in $CBrCl_3$ with a resolution of 10 fs per point. (Reproduced by kind permission of *Chemical Physics Letters*.)

linearly polarized perpendicular to the polarization direction of the initial beam, thus resulting in increased transmission through the analyzer.

In Section IV it was pointed out that there may well be a sizeable vibrational contribution to observed electro-optic phenomena in polyatomic molecules. Consider Eq. (42). The optical Kerr effect signal is proportional to a temperature-independent term containing a purely electronic hyperpolarizability component and a temperature-dependent orientational component. This is what is seen in the femtosecond Kerr transient experiments: There is an instantaneous electronic response of the target molecule followed by a much slower nuclear response that is likely to be temperature dependent.

The dependence of the orientational timescales of the nuclear contributions to the transient optical Kerr effect on the physical properties of the liquid (temperature, viscosity, and number density of CS_2 in an inert covalent solvent) has revealed many new aspects of solvation dynamics. The four nuclear responses seen in pure CS_2 retain their identity in CS_2-alkane solutions but indicate that the laser-induced responses and subsequent relaxations reflect both changing microscopic solvation conditions in the intermolecular potential, as well as changes in the macroscopic continuum properties, such as viscosity.[74, 75, 77] The solvation dynamics of CS_2 in alkanes is consistent with the picture of a liquid as an ensemble of continually evolving, van der Waals-like solvation clusters for which femtosecond and picosecond probing times reveal several discrete molecular motions. Molecules are seen to become trapped in locally ordered intermolecular potentials, where they display motions that are characterized by specific frequencies and relaxations times.

By contrast, in strongly associated liquids, for example, resorufin (a large polarizable anion) in aqueous and methanolic solutions, it is seen that rotational reorientation dominates the observed optical Kerr effect measurements with timescales running into hundreds of picoseconds. Pure water is seen to have a reorientational timescale of 75 ps, while an aqueous solution of LiCl has a time scale of 275 ps (Ref. 77). Here it is undoubtedly the small, highly polarizing Li^+ cation that is giving rise to a strong polarization of the water structure. The lithium cation is sometimes termed a structure-making cation, thus making relaxation processes more difficult than in the moderately structured (hydrogen bonded) pure water.

These experiments have just started to give a fascinating insight into intermolecular interactions in both pure liquids and solutions. However, the actual experimental technique is not limited to liquids. For example, Foggi et al.[80] used the optical Kerr effect to investigate the dynamical properties of polarizable organic molecules in disordered solids. They investigated succinonitrile[80, 81] and with the optical Kerr effect induced by femtosecond laser pulses were able to distinguish a number of relaxation processes in this solid and how the timescale for these relaxations changes with sample temperature within the cubic plastic phase that exists between about 230 K and the solid's melting point of 331 K. Again an instantaneous, purely electronic response is seen, followed by a number of slower nuclear contributions to the observed Kerr transient. The fastest of these nuclear responses (subpicosecond) is attributed to relaxation of the damped librational and torsional vibrations of the molecules coherently excited by the laser pulse, not within a solvent cage as in a liquid but within the unit cell, which is partially disordered. The intermediate decay times, ranging from 4 ps at 323 K to 30 ps at 250 K, are interpreted as rotational reorientations of the molecule, particularly the *trans* conformation, within the cubic unit cell. The rotational motion of the *gauche* conformation is responsible of the slowest observed nuclear response, being more strongly hindered. In these experiments the temperature dependence of the dynamics has yielded the activation barriers to reorientation. Eq. (42) shows that in such experiments one is observing the change of the anisotropic polarizability of the molecule or clusters of molecules. The hyperpolarizability contributions are independent of temperature. Perhaps it is by investigation of the temperature dependence of the various components that one may separate them and be able to gain further insight into condensed-phase dynamics.

For larger more complex polyatomic molecules, in pure liquid or solution, the use of optical Kerr effect measurements, induced by femtosecond pulses will enable examination of not only whole molecule dynamics but also the motion of parts of the molecules. Indeed, the investigation of coupling between the overall molecular motion and that of parts of the molecule may be possible; for example, see the transient optically induced birefringence study of biphenyl.[82]

Such femtosecond optical measurements yield results obtainable in the same frequency domain as NMR techniques. However, thereas NMR probes individual nuclei and their immediate surroundings, the optical Kerr effect, like other optical techniques, probes the average long-range evolution of the polarizability. In this way it is possible to observe in succinonitrile that as the temperature falls to the order–disorder transition at about 230 K, the measured relaxation time for the fastest nuclear component increases, indicating an increase of long-range order, i.e, crystallinity.

The observation of the separability of the purely electronic and purely nuclear contributions to the reorientational processes present in the condensed phase has consequences in many fields. For example, the optical properties of strongly bound or ionic glasses—here the particle motions, rotations, and vibrations—may be followed as a function of sample temperature, and when they have become arrested due to the strong intermolecular forces one observes, thereafter, a purely electronic response.

VI.   DISCUSSION

A.   Optical Kerr Measurements on Ionic Solutions

Having discussed the theory of the optical Kerr and Cotton-Mouton effects, together with their different modes of measurement, we shall discuss in this section a set of results, taken from the literature,[38] of the optical Kerr effect made on aqueous solutions of simple electrolytes. In this way we illustrate the type of information that is available from such experiments. The measurements in question were made over a wide range of solute concentrations, up to 20 mol/L, and they show deviations from the expected linear behavior of optical Kerr signal with solute concentration. However, because of the wealth of information they contain and the questions they raise we consider them of note and worthy of investigation.

Earlier we discussed the optical Kerr effect of pure liquid water, where we invoked, for the purpose of an order of magnitude calculation, the approximation that the water molecule is a symmetric top, thus simplifying our analysis. For a true symmetric top the temperature-dependent orientational term that appears in the optical Kerr effect is $\Delta\alpha^2/5kT$; however, for the water molecular (H_2O is of C_2v symmetry) we should write

$$\frac{1}{10kT}\left\{(\alpha_{xx} - \alpha_{yy})(\alpha_{xx}^0 - \alpha_{yy}^0) + (\alpha_{yy} - \alpha_{zz})(\alpha_{yy}^0 - \alpha_{zz}^0)\right.$$

$$\left. + (\alpha_{zz} - \alpha_{xx})(\alpha_{zz}^0 - \alpha_{xx}^0)\right\}$$

Problems of nonsymmetric-top symmetry do not occur when we consider most of the ionic solutions discussed in this section. The nitrate anion NO_3^- is planar and optically anisotropic, and its polarizability can in a birefringence experiment be defined by a single parameter, $\Delta\alpha = \alpha_\parallel - \alpha_\perp$. Table II contains representative results of measurements of the optical Kerr effect for simple nitrates and some other solutions.[38]

These data have been taken from the work of Paillette, in whose experiments the applied oscillating field was that of a pulsed ruby laser, $\lambda = 694.3$ nm. This field had a pulsewidth at half maximum of between 24 and 40×10^{-9} s with a peak mean power W of 40×10^6 W cm^{-2}. We may then calculate the peak applied electric field as $E = \sqrt{W/c\varepsilon_0} = 1.2 \times 10^7$ V m^{-1}. This pulsed laser is the source of the orienting electric field and the degree of induced orientation is measured with a second, lower power laser (in Paillette's work an argon laser at 488.0 nm). The apparatus used for these experiments is similar to that shown in Fig. 3.

Table II contains some of these data. In Fig. 8 we plot the measured optical Kerr effect for the solution, $B_{solution}$, corrected via Eq. (38) for the signal arising from the solvent, water. It is seen to be reasonably well behaved. In all the solutions displayed in Fig. 8, the NO_3^- anion will be the largest contributor to the measured Kerr effect signal. If the interaction term B_{int} were zero for these solutions, then B_{solute} should be independent of solute concentration. It is seen from the second column of data in Table II that this is not strictly the case. Although for any one solution the values are similar at each solute concentration, there appears to be a

TABLE II
Summary of the Results from the Optical Kerr Effect Measurements of Paillette[38]

	mol/L	B_{solute} $(10^9$ esu)	$\overline{\Delta\alpha}^a$ $/10^{-40}$ C^2 m^2 J^{-1}
NaNO$_3$	10.29	18.1	-12.4
NaNO$_3$	5.88	18.9	-16.7
NaNO$_3$	3.35	20.6	-23.2
NaNO$_3$	1.12	35.3	-52.4
NH$_4$NO$_3$	19.53	9.21	-6.41
NH$_4$NO$_3$	13.75	8.14	-7.19
NH$_4$NO$_3$	1.31	10.0	-25.8
NH$_4$Cl	6.54	3.05	6.38
NH$_4$Cl	1.87	0.73	5.83
NaNO$_2$	6.52	13.0	13.2
NaNO$_2$	1.45	14.7	29.7
H$_2$O	55.55	0.584	0.96^b

Note. The first column lists the solute concentration and the second column gives the observed solute optical Kerr constant. The final column lists the average polarizability anisotropy extracted from the data at 488 nm; see text for details.
^{a}Measured at 488 nm.
^{b}See text for definition.

distinct dependence of B_{solute} on solute concentration. We represent these trends in Fig. 9. Here the finite slope for the lines may be taken as a measure of B_{int}. Due to the presence of these strong interactions it is probably more appropriate to analyze each concentration of each solute independently and not attempt to obtain an infinite dilution Kerr constant for each solute.

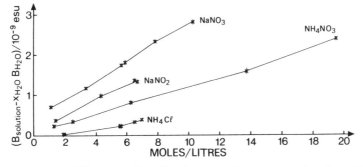

Figure 8. Plot of $\{B_{\text{solution}} - X_{H_2O}B_{H_2O}\}$ versus solute concentration, in mol/L, for different aqueous solutions. The data are taken from the optical Kerr effect measurements of Paillette. The error bars are given by the vertical lines through the data points.

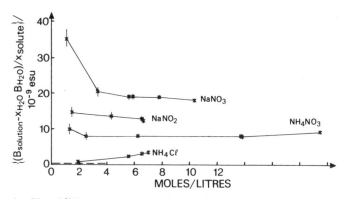

Figure 9. Plot of $\{(B_{solution} - x_{H_2O}B_{H_2O})/x_{solute}\}$ versus solute concentration, in mol/L, for different aqueous solutions. The data are taken from the optical Kerr effect measurements of Paillette and show the presence of intermolecular interactions. The dashed line gives the signal level for pure water. The error bars are given by the vertical lines through the data points.

Using Eq. (32) it is possible to derive from these optical Kerr measurements, the mean polarizability anisotropies, $\overline{\Delta\alpha}$, assuming $\overline{\Delta\alpha^0} = \overline{\Delta\alpha}$. A Lorentz field correction (see Section III.B) has been applied and the temperature-independent contribution to the measurement has been taken to be zero, i.e., $\gamma = 0$ in Eq. (32). The polarizabilities displayed in Table II are with reference to 488 nm. The concentration dependence of $\overline{\Delta\alpha}$ is shown in Fig. 10; also displayed is the signal arising from pure water. The effect in pure water has been calculated assuming that water is a symmetric top. This is an approximation adopted to give an order of magnitude estimate for the contribution of water to the solution data displayed in Fig. 10.

An important point to bear in mind is that the phase of the measured $\Delta\alpha$, Eq. (32), shows that the actual measurable quantity in the optical Kerr effect experiment on axially symmetric molecule is proportional to $\overline{\Delta\alpha^2}$; for $\Delta\alpha \approx \Delta\alpha^{(0)}$. From this we have no knowledge of the sign of $\Delta\alpha$. However, $\Delta\alpha = \alpha_{\parallel} - \alpha_{\perp}$ and the main rotational axis of the planar NO_3^- ion is the C_3 axis passing through the N atom perpendicular to the plane of the molecule, with $\alpha_{\parallel}$ being parallel to this direction and with $\alpha_{\perp}$ containing the contributions from the atoms in the perpendicular plane. This, $\alpha_{\perp} > \alpha_{\parallel}$ and $\Delta\alpha$ is negative as in the case for benzene. The sign of the values of $\Delta\alpha$ for the nitrates listed in Table II have been assigned from this point polarizability argument. It is seen from the measurements displayed in Fig. 10 that as the mean spacing between ions changes; that

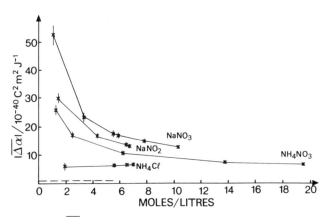

Figure 10. Plot of $|\overline{\Delta\alpha}|$, average polarizability anisotropy versus solute concentration, in mol/L. The dashed line is the measured induced optical anisotropy arising from pure water. The error bars are given by the vertical lines through the data points.

is, as the concentration of ions in solution changes so the measured polarizability anisotropy fluctuates.

Of the solutions whose polarizability anisotropy have been investigated, the nitrates will always have a significant effect. NO_3^- will at all dilutions contribute strongly to the measured optical Kerr effect. However, as in other liquids, there will be a large collision-induced contribution to the measured optical Kerr effect. That is, we may write the measured polarizability anisotropy $\overline{\Delta\alpha}$ as the sum of the isolated ion polarizability anisotropy and a term arising through molecular interactions, which will be a function of the distance between ions in the solution, $\overline{\Delta\alpha} = \Delta\alpha + \Delta\alpha(R)$. This type of model is exemplified by NH_4Cl. Here the two component ions have no intrinsic polarizability anisotropy and at lower concentration should have an optical Kerr effect indistinguishable from that of water. However, at higher concentrations the ions polarize and give rise to a collision-induced polarizability anostropy. Consider a pair of isotropically polarizable groups, NH_4^+ and Cl^-, both solvated. They will have an interaction-induced polarizability, $\alpha(R)$. If R is the distance between the ions and α_0 is the isolated polarizability, we may write for the induced polarizability anisotropy[83]

$$(\alpha_{\parallel} - \alpha_{\perp})(R) = \Delta\alpha(R) = \left\{ \frac{6\alpha_0^2}{R^3 4\pi\varepsilon_0} \right\} + \left\{ \frac{6\alpha_0^3}{R^6(4\pi\varepsilon_0)^2} \right\} \quad (43)$$

and for the mean interaction induced polarizability

$$\alpha(R) = 2\alpha_0^2 + \left\{ \frac{4\alpha_0^2}{R^6(4\pi\varepsilon_0)^2} \right\} \tag{44}$$

For the solutions of intrinsically anisotropic ions we see from Fig. 10 that as the concentration increases the measured polarizability anisotropy falls. This is contrary to what one might expect from Eq. (43), which states that as the solution becomes more concentrated, i.e., as the average interionic spacing falls, $\Delta\alpha(R)$ should increase. However, an explanation may lie in the nature of the hydration process. At low concentrations all the X^+ and NO_3^- ions will be completely solvated. This polarization of the aqueous medium by the field of the ions produces large-scale structures in the solution which may well have substantially larger polarizability anisotropies than the bare NO_3^- ion or water molecule. However, as the solution becomes more concentrated, there are simply not enough water molecules available to maintain these extended solvent sheaths. For example, in 6 M $NaNO_3$ solution with each Na^+ ion requiring 6 water molecules and each NO_3^- ion requiring about 2 molecules for solvation, there is barely enough water present to give a first completed solvent sheath around each ion. The strong dependence of cation hydration and hence anion hydration number on solute concentration may be seen in the dielectric work of Wei et al.[84] Here we observe that for Li^+, in LiCl solution at a concentration of 1 mol/L the cation hydration number is found to be 7.5–8.0 while at a solute concentration of 5 mol/L the hydration number has fallen to 5. We may estimate from the data in Ref. 84 results for hydration of the Na^+ cation: We find a hydration number of 6 at a concentration of 1 mol/L and a hydration number of 4 at 5 mol/L. We therefore have a picture of long-range, polarizable structures established at low concentrations which are disrupted by the strong ionic forces produced by the addition of further solute.

In Fig. 10 we see that all the derived values of $\overline{\Delta\alpha}$ are tending, with increasing solute concentration, to some limiting value. A solution of ammonium nitrate with 19.53 mol/L of solute[38] is likely to be fairly viscous; indeed, it is probably better described as a glass. The same is likely to be the case for all of the stronger solutions. We pointed out above how at these concentrations there are not enough water molecules available to completely solvate all the ions. In the limit of large concentration we would have a completely disordered ionic glass and we have to consider how and why $\Delta\alpha$ tends to some limiting value for such systems.

Equation (42) gives the answer. In the optical Kerr effect there is a temperature-independent term and a temperature-dependent term which contribute to the measurement. The former describes the distortion of the polarizability of the subject molecule NO_3^- by the applied oscillating electric field. The latter describes the orientation of the NO_3^- ion induced by the torque arising through the interaction of the ions polarizability and the applied field. In the case of strongly associated ionic glasses where the planar anisotropic anions are no longer free to rotate in the medium it is only the temperature-independent distortion that will survive. The induced orientational torque will not be large enough to overcome the strong ionic intermolecular interactions present in the disordered material. We are therefore observing the loss of this temperature-dependent orientational contribution, $(\Delta\alpha^2/5kT)$, to the optical Kerr effect when we look at the decrease in $\overline{\Delta\alpha}$ as a function of solute concentration in Fig. 10. At these high concentrations we are left only with the electronic contribution, that is, $\gamma_{electronic}$ in Eq. (42), to the optical susceptibility, which we see from Fig. 10 can be a sizeable fraction of the total polarizability.

This separation of electronic and nuclear contributions to the observed induced optical anisotropy of a medium is similar to that observed in the femtosecond transient optical Kerr effect discussed in Section IV. In the concentration dependence of the optical Kerr effect signal we are observing the "freezing out" of molecular motions by strong intermolecular forces which prevent rotational reorientation, i.e., field-induced alignment, while in the femtosecond experiments we see a separation of electronic and nuclear contributions because of the different timescales required for electronic (essentially instantaneous) and nuclear (slower) relaxation. Both techniques are explicable with Eq. (42) and should yield similar information.

B. Cotton-Mouton Measurements of Ionic Solutions

Measurements of the optical Kerr effect of conducting electrolyte solutions are rare and static Kerr effect measurements on such fluids are not possible, due to ionic conduction. Detailed measurements of the Cotton-Mouton effect for these types of solutions have only recently become available,[41] although some such ionic solutions were investigated very early on in the history of the Cotton-Mouton effect.[85]

Here we discuss some Cotton-Mouton measurements on the same types of solutions as were discussed in the previous section. Consequently, as we are dealing with symmetric tops, we are able to define the nitrate ions magnetizability as an anisotropy, $\Delta\chi = \chi_\parallel - \chi_\perp$, as was possible for its polarizability. Table III contains some data taken from a recent investiga-

TABLE III
Summary of the Results of Measurements of the Cotton-Mouton Effect Made in Aqueous Soluti

	$C_{\text{solution}}/10^{15}$ $(\text{G}^{-2}\,\text{cm}^{-1})$	$C_{\text{solute}}/10^{15}$ $(\text{G}^{-2}\,\text{cm}^{-1})$	$N/10^{26}\,\text{m}^{-3}$	$\overline{\Delta\alpha}/10^{-40}$ $\text{C}^2\,\text{m}^2\,\text{J}^{-1}$	$\overline{\Delta\alpha\,\Delta\chi}/10^{-}$ $\text{C}^2\,\text{m}^2\,\text{T}^{-}$
H_2O	−1.035		333.3		(−0.00284
$ZnCl_2$	0.12	32.25	12.0		2.46
$ZnBr_2$	−0.64	6.35	18.85		0.31
$HgCl_2$	−0.09	308.8	1.02		277
NaCl	1.51	29.79	30.0		0.91
KCl	−1.26	−4.04	27.49		−0.137
HNO_3	9.70	468.3	7.81	(−159)	54.81
HNO_3	5.45	559.95	3.905	(−359)	131.2
$NaNO_3$	23.08	167.48	55.67	−13.25	2.75
$NaNO_3$	4.69	142.39	13.92	−37.0	9.393
NH_4NO_3	17.30	126.96	56.34	−9.0	2.044
NH_4NO_3	5.16	113.39	18.78	−15.75	5.603
$LiNO_3$	10.61	251.1	16.16	(−43.7)	14.22
$LiNO_3$	4.42	257.00	7.21	(−95.6)	32.69
KNO_3	2.38	153.10	4.17	(−53.0)	18.3
$CsNO_3$	0.36	168.60	2.206	(−157)	56.44
$NaNO_2$	24.55	396.2	22.93	−19.0	15.80
$NaNO_2$	37.31	409.0	34.39	−14.2	10.87
$NaNO_2$	27.89	378.6	27.51	−16.0	12.58
Data taken from Landolt and Börnstein					
H_2O	−3.9(a)		333.3		−(0.009)
	−11(b)		333.3		
$NaNO_3$	30.0	276	45.9		4.7
$NaNO_3$	13.2	205	29.6		5.4
NH_4NO_3	33	233	61.6		3.0
NH_4NO_3	15.4	174	40.5		3.4
$NaNO_2$	69	566	48.5		9.2
HNO_3	63	541	46.7		9.7
KNO_2	135	879	62.25		11.1

Note. The final column lists the average values of $\Delta\alpha\,\Delta\chi$ extracted from the measurement, in uni $10^{-66}\,\text{C}^2\,\text{m}^2\,\text{T}^{-2}$. The C_{solution} column contains the measured data. The column labeled C_{solute} the Cotton-Mouton constant corrected for the solvent via Eq. (38); the solute number density is in the next column. To obtain the molar Cotton-Mouton constant for the solute data one multiply the C_{solute} values by the wavelength of the laser used for the measurements (632.8 nm) divide by the number of mol/cm³. This will give C_{solute} in units of $\text{G}^{-2}\,\text{cm}^3\,\text{mol}^{-1}$. To obtain the equivalent $(\text{m}^5\,\text{A}^{-2}\,\text{mol}^{-1})$ one divides by $(10^6/4\pi)^2$. In the column labeled $\overline{\Delta\alpha}$, the num without parentheses are those values obtained directly from the optical Kerr measurements giv Table II. Those in parentheses are derived from the Cotton-Mouton measurements using a concen tion-dependent mean value for the anisotropy of the diamagnetic susceptibility of the NO_3^- (a) Corresponds to Ref. 85 and (b) corresponds to Ref. 86. See the text for details.

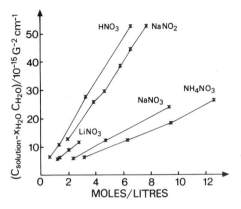
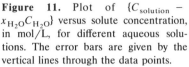

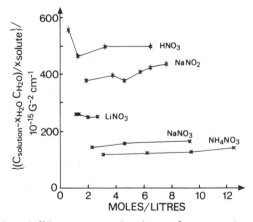

Figure 11. Plot of $\{C_{\text{solution}} - x_{H_2O}C_{H_2O}\}$ versus solute concentration, in mol/L, for different aqueous solutions. The error bars are given by the vertical lines through the data points.

tion of the Cotton-Mouton effect of ionic solutions measured in fields of 13 and at a probing laser wavelength of 632.8 nm (Ref. 41).

In Table III, C_{solute} is the Cotton-Mouton constant for the solution corrected for the signal arising from water. That is, the Cotton-Mouton constant equivalent of Eq. (38) has been used for this analysis. The concentration dependence of these derived values of C_{solute} are displayed in Fig. 11. In Fig. 12 we display the same data corrected for solute concentration. It is clearly seen that they are not independent of concentration and that therefore there is a finite C_{int} interaction contribution to the measured Cotton-Mouton constant of the solution.

Figure 12. Plot of $\{(C_{\text{solution}} - x_{H_2O}C_{H_2O})/x_{\text{solute}}\}$ versus solute concentration, in mol/L, for different aqueous solutions. The curves show that there may well be an interaction-induced contribution to the measurement. The error bars are given by the vertical lines through the data points.

Listed at the bottom of Table III are the results of the only other determination of the Cotton-Mouton effect in aqueous solutions. The numbers are taken from Landolt and Börnstein[85] and were made at a variety of temperatures and concentrations with a white light or resonance lamp optical source. Some of these measurements go back a long way. The solution values date from the original experiments of Cotton and Mouton[13] and from early Indian work.[86, 87]

From Eq. (33) it is seen that in symmetric tops the Cotton-Mouton effect yields $\Delta\alpha\,\Delta\chi$, assuming the temperature-independent hypermagnetizability to be negligible. To proceed further and obtain values of the separate molecular properties, polarizability and magnetizability, from such measurements requires the combination of a number of different experimental techniques to identify the various unknowns. As an example, optical Kerr effect measurements may be used to separate the two terms that are found from the Cotton-Mouton experiment. Hence, use may be made of the values of $\overline{\Delta\alpha}$ for $NaNO_3$ and NH_4NO_3 solutions to obtain $\overline{\Delta\chi}$ from the Cotton-Mouton results at similar concentrations. It is clearly seen that both $\overline{\Delta\chi}$ and $\overline{\Delta\alpha\,\Delta\chi}$ are concentration dependent. For example, with $\overline{\Delta\alpha\,\Delta\chi} = 2.75 \times 10^{-66}$ C^2 m^2 T^{-2} for $NaNO_3$ at a concentration of 9.28 mol/L and $\overline{\Delta\alpha\,\Delta\chi} = 1.74 \times 10^{-66}$ C^2 m^2 T^{-2} for NH_4NO_3 at 12.52 mol/L, we obtain $\overline{\Delta\chi} = -2.07 \times 10^{-27}$ J T^{-2} for the $NaNO_3$ solution and $\overline{\Delta\chi} = -2.32 \times 10^{-27}$ J T^{-2} for the NH_4NO_3 solution when we have extracted the relevant $\overline{\Delta\alpha}$ from Fig. 10 ($\overline{\Delta\alpha}(NaNO_3) = -13.25 \times 10^{-40}$ C^2 m^2 J^{-1} and $\overline{\Delta\alpha}(NH_4NO_3) = -7.5 \times 10^{-40}$ C^2 m^2 J^{-1}). It is possible, therefore, to derive values of $|\overline{\Delta\chi}|$, care being taken of the relative phases of $\Delta\alpha$, $\Delta\chi$, and $\Delta\alpha\,\Delta\chi$ (the concentration dependence of $\Delta\alpha\,\Delta\chi$ is shown in Fig. 13), which may be plotted versus solute concentration as in Fig. 14. These data are not available from any other source. For a comparison, the mean magnetizability $\overline{\chi}$ in concentrated aqueous solution for these materials is $\overline{\chi}(NaNO_3) = -4.38 \times 10^{-28}$ J T^{-2} and $\overline{\chi}(NH_4NO_3) = -5.47 \times 10^{-28}$ J T^{-2}, respectively.[85] We may do this substitution of $\overline{\Delta\alpha}$ to obtain $|\overline{\Delta\chi}|$ provided the two sets of measurements were made with similar concentrations. Thus, in Table III, in the column labeled $\overline{\Delta\alpha}$, the values without parentheses are those values of $\overline{\Delta\alpha}$ calculated from the concentration dependence of the optical Kerr effect. These may be used directly to obtain $|\overline{\Delta\chi}|$ from the measured values of $\overline{\Delta\alpha\,\Delta\chi}$ in the final column in Table III.

From the data in Landolt and Börnstein[85, 88] one may obtain the separate ion contributions to the mean magnetic susceptibility. It is seen that NO_3^- is, in comparison to the cations present in these experiments, by far the biggest contributor. For example, for the sequence Na^+, K^+, NH_4^+, and NO_3^-, the values of the magnetic susceptibility are given as[85, 88]

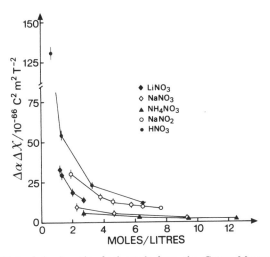

Figure 13. Plot of $\Delta\alpha\,\Delta\chi$, the final result from the Cotton-Mouton measurements, versus solute concentration in mol/L. See text for details.

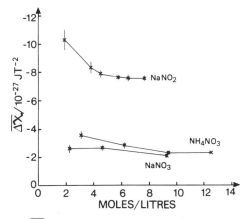

Figure 14. Plot of $|\overline{\Delta\chi}|$, average anisotropy of the diamagnetic susceptibility (derived from Cotton-Mouton and optical Kerr effect data), versus solute concentration, in mol/L, for different aqueous solutions. The error bars are given by the vertical lines through the data points.

$\chi = -8.3 \times 10^{-29}$, -21.6×10^{-29}, -19.1×10^{-29}, and -33.2×10^{-29} J T^{-2}, respectively. With regard to the anisotropy of $\bar{\chi}$, NO_3^-, being intrinsically optically anisotropic, will be the dominant contribution to the Cotton-Mouton effect of aqueous nitrate solutions. Figure 14 shows the similarity between the extracted values of $\overline{\Delta\chi}$ for the different solutions of $NaNO_3$ and NH_4NO_3. It is the nitrate ion which is giving the observable effect; the cation is not really contributing. It is therefore possible to use Fig. 14 to obtain an average $\overline{\Delta\chi}$ for the NO_3^- ion at different concentrations and then extract $\overline{\Delta\alpha}$, at the appropriate concentration, from the Cotton-Mouton measurements. We have in this way obtained the values of $\overline{\Delta\alpha}$ listed in parentheses in Table III. The phase has been assigned on the basis of the argument discussed previously.

For $NaNO_2$, assuming $\alpha_{xx} = \alpha_{yy} = \alpha_\perp$ and $\alpha_{zz} = \alpha_\parallel$ for the NO_2^- ion, one may similarly use the optical Kerr effect to obtain a value of $\overline{\Delta\alpha}$, which can be used in conjunction with the Cotton-Mouton measurement to derive $\overline{\Delta\chi}$. It is curious that the values of $|\overline{\Delta\chi}|$ for NO_3^- and NO_2^- are so different, $|\overline{\Delta\chi}\ NO_2^-| \sim 3|\overline{\Delta\chi}\ NO_3^-|$. This difference may also be seen in the comparison of $\overline{\Delta\alpha\,\Delta\chi}$ measured here for KNO_3 and the data for KNO_2 extracted from Landolt and Börnstein[85] (see also Table III). However, it may well be that here we are observing a paramagnetic contribution to the measurement. The neutral species NO_2 is an open shell molecule; however, one would expect a purely diamagnetic behavior from the NO_2^- ion. Perhaps in the intense applied magnetic field there is a mixing of low-lying π states, which could give rise to an induced paramagnetic current density. Its form would be the same as the current that exists when the orbitals are degenerate and the molecule is in a state of well-defined orbital angular momentum, the difference being one of magnitude.

We may make a similar qualitative analysis with regard to the other Cotton-Mouton results presented in Table III. For those salts that have no intrinsically anisotropically polarizable ions the measured effect arises through molecular interactions, e.g., $NaCl$ and KCl. These ion pairs at infinite separation have anisotropy in neither magnetizability or polarizability. They will, however, still have a Cotton-Mouton effect due to the temperature-independent term seen in Eq. (33). Indeed, this hypermagnetizability for a completely solvated ion may be quite appreciable. However, as these two closed-shell ions approach one another, an induced polarizability anisotropy will arise, as shown above. We may also interpret this difference in the Cotton-Mouton effect of pure water and in aqueous solutions of $NaCl$ and KCl as arising because of the break up of the complex hydrogen bonded structure of water. The statistical average over the angular coordinates of the molecules in the presence of the applied electromagnetic fields which is involved in deriving Eq. (33), that is, the orientational average represented by Eq. (24) together with the isotropic averages calculated on going from Eq. (29) to Eq. (30) are made more complicated in water because of the presence of the strong hydrogen bonds which make up the liquid. When small spherical ions are present, however, they polarize the medium and disrupt these hydrogen bonds, obtaining a solvent sheath in the process. If it is assumed that the ions make no net contribution to the measurement, i.e., they are unobservable electro-optically, then the Cotton-Mouton effect for aqueous solutions of spherical ions such as Na^+ (aq) Cl^- (aq) and K^+ (aq) Cl^- (aq) give a measure of the Cotton-Mouton effect in disordered water.

We see from the second column of Table III that the Cotton-Mouton effect in aqueous NaCl is positive while that for KCl is smaller and negative. However, a difference in the solvation dynamics of Na^+ and K^+ is well known.[89-91] The smaller, polarizing Na^+ cation is termed a structure-making cation, while the K^+ cation, being larger and less able to polarize the medium, is considered a structure-breaking cation.[89, 90]

With regard to these experimental values of the Cotton-Mouton constants of aqueous NaCl solution and aqueous KCl solution it is possible to say something more about the process of solvation. An analysis of the refractive index data for aqueous solutions of NaCl and KCl, as taken from the *Handbook of Chemistry and Physics*[92] using the Lorentz-Lorenz equation gives average polarizabilities for the solvated ions. For NaCl and KCl solutions, of concentration given in Table III (4.5–5 mol/L), it is seen that the derived polarizabilities are very similar, $\alpha_{Na^+ (aq)} + \alpha_{Cl^- (aq)} = 57 \times 10^{-39}$ C^2 m^2 J^{-1} and $\alpha_{K^+ (aq)} + \alpha_{Cl^- (aq)} = 63 \times 10^{-39}$ C^2 m^2 J^{-2}. The question arises, therefore, as to why there should be a difference of magnitude and of phase when one considers the Cotton-Mouton effect on these two solutions. In both cases it should be noted that the ion pairs, Na^+, Cl^- and K^+, Cl^-, are closed-shell spherical ions, i.e., with no intrinsic anisotropic polarizability.

At these ionic concentrations the dielectric measurements of Wei et al.[84] give hydration numbers for these cations of 4.75 $\pm$ 0.25 for Na^+ (aq) and 3.75 $\pm$ $-$ 0.25 for K^+ (aq). Similarly, the neutron scattering measurements of Nielson and Enderby[89] and Skipper and Nielson[90] give hydration number of 5 for Na^+ (aq) and 4 for K^+ (aq). With four water molecules, which from space requirements would be tetrahedrally disposed around the K^+ cation, this solvated aqua ion will have neither anisotropic polarizability nor anisotropic magnetizability, being spherically symmetric. In the case of Na^+, however, solvated with 5 water molecules, the structure will be such as to possess an anisotropic polarizability and an anisotropic magnetizability. Both solvated cations will have a, possibly large, temperature independent hypermagnetizability contribution to the observed Cotton-Mouton effect. If because of tetrahedral symmetry $\Delta\alpha\,\Delta\chi = 0$ for the solvated potassium cation, then it may well be that for this system the hypermagnetizability is negative, as observed in Table III.

Provided we are not at solution concentrations that are so large as to prevent the formation of complete solvent sheaths, it is not unreasonable, considering the nonspherical symmetry of the solvated sodium cation, that the measured Cotton-Mouton constant of aqueous NaCl is very different from that of aqueous KCl, the large positive temperature-dependent contribution to the Cotton-Mouton effect in solvated sodium cations (negative $\Delta\alpha$ and negative $\Delta\chi$) canceling the smaller negative hypermagnetizability contribution, seen in the potassium solution, giving the phase change observed between the two solutions. The Cl^- anion at these concentrations is believed to be solvated with 6 water molecules[91] and, consequently, with such an octahedral disposition of water ligands, will have neither $\Delta\alpha$ or $\Delta\chi$. From such an analysis of the contribution of the solvent shell to this type of electro-optic experiment it may be predicted that aqueous Li^+ solutions will behave like those of K^+. The Li^+ ion takes 6 water molecules into its solvent shell,[91] thereby producing an octahedral solvation complex which will possess neither $\Delta\alpha$ or $\Delta\chi$.

Arguments similar to those just applied to the Cotton-Mouton effect in solvated spherical ions would also apply to the optical Kerr effect on such systems. Neilson and Enderby[89] point out that the water molecules in the solvent shells of solvated metal ions exchange with the bulk water and that for the alkali cations this exchange occurs on a timescale of 10^{-11} to 10^{-10} s. Although there are no optical Kerr effect measurements available for the aqueous alkali metal halides, such data would be of great interest. Values derived from long-pulse laser experiments would yield polarizability anisotropies for the solvated ions, and dynamic data (optical Kerr measurements with a picosecond, or shorter, pulsed laser) would give the change of the total polarizability anisotropy as the water molecules exchange.

The values of $\overline{\Delta\alpha_{NO_3^-}}$ obtained by assuming fixed values of $\overline{\Delta\chi_{NO_3^-}}$ are large. As to the magnitude, in the HNO_3 and $LiNO_3$ ion pairs the cations can approach very close to the anion. For example, the proton will be strongly attached to a water molecule and so the effective inter-ion distance is the nearest-neighbor distance in water. Thus, from Eq. (43) it can be readily seen how we could have a particularly large collision-induced contribution to the measured $\overline{\Delta\alpha}$. This argument applies to a lesser extent with regard to Li^+ and for the other nitrate ion pairs there is likely to be a solvation sheath between them, thereby greatly increasing R and via Eq. (43) reducing the induced $\overline{\Delta\alpha}$. It is also likely that the larger cations, NH_4^+, Cs^+, and K^+, because of their large intrinsic polarizabilities, will have a significant contribution to α_0 and hence to $\overline{\Delta \alpha}$ even as the $\langle R \rangle$ falls. The measurements of the average $\overline{\Delta \chi_{NO_3^-}}$ extend down to 2 mol/L. Below this we must assume a linear extrapolation to derive $\overline{\Delta \alpha}$, and this may not be reasonable.

With $ZnBr_2$, $ZnCl_2$, and $HgCl_2$ it is possible that there are chemical effects present in aqueous solution that contribute to the measurement. It is well known that the zinc halides have a complex chemistry in aqueous solution[93] and that some of the complex cations formed may well have a disproportionately large Cotton-Mouton constant. In the case of $HgCl_2$ there is the possibility of covalent behavior in solution, i.e., very little dissociation into ion pairs. It has a large Cotton-Mouton effect for a comparatively small amount of material. Such observations raise the possibility of using the Cotton-Mouton effect as a probe of aqueous solution chemistry. For example, Raman spectra[93] show that, depending on the concentration, the species present in aqueous solutions of $ZnCl_2$ are $[Zn(H_2O)_6]^{2+}$, $ZnCl^+$, $ZnCl_2$, and $[ZnCl_4(H_2O)_2]^{2-}$. Using the Cotton-Mouton effect to probe the magnitude of the induced optical anisotropy may allow one to place estimates on the relative quantities of these species, of differing intrinsic optical anisotropy, present at a particular concentration.

VII. CONCLUSIONS

The analysis of the electro-optic properties of aqueous solutions made in the previous section is qualitative and makes many approximations. However, it illustrates the type of information that is available from these experiments and points the way for further work with greater precision and more detailed considerations of concentration effects, particularly in the low concentration regime where the observed electro-optic properties are seen to increase rapidly (see Figs. 10 and 13). Such experiments are of interest because these systems are of great importance, but experimental values for their electromagnetic properties are rare. These measurements show, for example, that there is an appreciable effect due to an applied magnetic field on the distribution of ions within aqueous solutions. This observation is of some interest because recent epidemiological experiments[94] suggest that low-frequency electromagnetic fields produced, for example, by power cables and household electrical appliances could be harmful to health. Since these electromagnetic field oscillate at 50 or 60 Hz (depending upon country), it is conceivable that their magnetic field component induces small orientational fluctuations in ionic distributions within tissue, thus influencing in vivo interactions.

Electo-optic experiments, such as the Kerr effect, optical Kerr effect and the Cotton-Mouton effect, have been our major source of information about the electromagnetic properties of molecules. The values of molecular properties obtained from such bulk susceptibility measurements do not possess the precision of, for example, molecular beam Stark measurements of the electric dipole moment and electric polarizability or molecular beam Zeeman determinations of the magnetic susceptibility. They give an averaged picture of molecular interactions in condensed phases, as opposed to the molecular beam methods which give information on a particular interaction of a particular molecular pair. Susceptibility measurements made on solutions are often beset by the uncertainty of the contribution of molecular interactions, which can itself be of interest. They have, however, over the last 75 years contributed enormously to our understanding of condensed-phase chemical physics. For example, much important structural information, particularly molecular conformational analysis, has been established by the experimental work of the LeFèvres and their successors in Australia. Indeed, the use of molecular electro-optic measurements to determine molecular structure and conformation was pioneered by this group.

As an example of the particular advantage of birefringence measurements for the investigation of condensed phase physical phenomena, consider the Cotton-Mouton measurements on spherically polarizable ions, NaCl and KCl (see Table III). The average polarizability of a volume element of solution may be found from the refractive index via the Lorentz-Lorenz equation, see Section 3. However, if the species of interest, whether molecule or ion, can be oriented in an applied laboratory field, magnetic or electric, then one may determine by suitable polarization selection of the probing light beam the distribution of polarizable matter in the oriented sample. That is, one may measure the mean polarizability parallel and perpendicular to the electric vector of the laser beam used to investigate the oriented sample. This difference, the induced birefringence or retardation, may be related, as shown in Section 3, to molecular properties. For the ionic solutions the refractive index gives the mean polarizability of a solvated ion, while the induced birefringence measurement yields the polarizability anisotropy of the solvated ion, which is seen to be different for solvated sodium ions compared, at a similar concentration level, with solvated potassium ions.

In the future, pulsed laser measurements of the optical Kerr effect should give much new information on the dynamics of the solvent sheaths of ions in aqueous solutions. Modern laser technology allows us to vary over many orders of magnitude the pulsing frequency of powerful noncontinuous lasers and this will afford an opportunity to study in detail the stability and behavior of solvent–ion interactions over a wide range of ion types and concentrations.

Acknowledgments

The author gratefully acknowledges the Directors of the Institut Max von Laue-Paul Langevin for allowing him the time to undertake some of the measurements described here and to write this article. He also expresses his thanks to David Buckingham, FRS, for having introduced him to the field of molecular electro-optics as a graduate student.

References

1. M. Faraday, *Philos. Mag.* **19**, 295 (1845).

2. M. Faraday, *Philos. Mag.* **28**, 294 (1846); Phil. Trans. R. Soc. London **1** (1846).

3. A. D. Buckingham and P. J. Stephens, *Ann. Rev. Phys. Chem.* **17**, 399 (1966).

4. A. Yariv, *Quantum Electronics*, Wiley, New York, 1967, p. 195.

5. John Kerr, 1824–1907, Obituary notice of fellows deceased, *Proc. R. Soc. London Ser.* A **82** (1909).

6. J. Kerr, *Philos. Mag.* **3**, 321 (1877).

7. J. Kerr, *Philos. Mag.* **50**, 337 (1875).

8. J. Kerr, *Philos. Mag.* **50**, 446 (1875); Philos. Mag. **9**, 157 (1880).

9. J. Kerr, *Br. Assoc. Rep.* 568 (1901).

10. E. Majoranna, *R. Accad. Lincei* **11**, 374 (1902).

11. A. Cotton and H. Mouton, *C. R. Séance Acad. Sci. Paris*, **141**, 317, 349 (1905); **142**, 203 (1906); **145**, 229 (1907); *Ann. Chem. Phys.* **11**; 145, 289 (1907).

12. A. Cotton, H. Mouton, and C. R. Weiss, *C. R. Séance Acad. Sci. Paris* **145**, 870 (1907).

13. A. Cotton and H. Mouton, *Ann. Chem. Phys.* **19**, 53 (1910); **28**, 209 (1913); **30**, 310 (1913); *C. R. Séance Acad. Sci. Paris* **147**, 51, 193 (1908); **149**, 340 (1909); **150**, 774, 857 (1910); **154**, 818, 930 (1912); **156**, 1456 (1913); A. Cotton and Tsai Belling, C. R. Séance Acad. Sci, Paris, 198, 1989, (1934).

14. C. T. O'Konski, *Encyclopedia of Polymer Science and Technology*, Vol. 9, Interscience, New York, 1968, p. 551.

15. C. T. O'Konski, K. Yoshioka, and W. H. Orttung, *J. Chem. Phys.* **63**, 1558 (1959); S. Krause and C. T. O'Konski, *J. Am. Chem. Soc.* **81**, 5082 (1959); K. Yoshioka and C. T. O'Konski, *Biopolymers* **4**, 499 (1966).

16. M. S. Beevers, *Mol. Cryst. Liq. Cryst.* **31**, 333 (1975); M. Davies, R. Moutran, A. H. Price, M. S. Beevers, and G. Williams, *Trans. Faraday Soc.* 2 **72**, 1447 (1976); M. S. Beevers and G. Williams, *Trans. Faraday Soc.* 2 **72**, 2171 (1976); M. S. Beevers, J. Crossley, D. C. Garrington, and G. Williams, *Trans. Faraday Soc.* 2 **72**, 1482 (1976).

17. J. W. Beams, *Rev. Mod. Phys.* **4**, 133 (1932).

18. H. A. Stuart, *Die Physik der Hochpolymeren*, Vol. 1, Springer, Berlin, 1952, p. 415.

19. J. R. Partington, *An Advanced Treatise on Physical Chemistry*, Vol. 4, Longmans, London, 1953, p. 278.

20. C. G. LeFèvre and R. J. W. LeFèvre, *Rev. Pure Appl. Chem.* **5**, 261 (1955).

21. C. G. LeFèvre and R. J. W. LeFèvre, in A. Weissberger and B. W. Rossiter (Eds.), *Physical Methods of Chemistry*, Part 3c, Wiley-Interscience, New York, 1972, p. 399.

22. A. D. Buckingham and B. J. Orr, *Quart. Rev. (London)* **21**, 195 (1967).

23. A. D. Buckingham, in C. T. O'Konski (Ed.), *Molecular Electro-optics*, Dekker, New York, 1976, p. 27.

24. R. J. W. LeFèvre, P. H. Williams, and J. M. Eckert, *Aust J. Chem.* **18**, 1133 (1965); also see subsequent articles by R. J. W. LeFèvre and coworkers in the same journal (1966–1971).

25. M. R. Battaglia and G. L. D. Ritchie, *Trans. Faraday Soc.* 2 **73**, 209 (1977).

26. M. P. Bogaard, A. D. Buckingham, M. G. Corfield, D. A. Dunmur, and A. H. White, *Chem. Phys. Lett.* **12**, 558 (1972).

27. C. L. Cheng, D. S. N. Murphy, and G. L. D. Ritchie, *Mol. Phys.* **22**, 1137 (1971).

28. H. Kling and W. Huttner, *Chem. Phys.* **90**, 207 (1984).

29. G. Maret and G. Well, *Biopolymers* **22**, 2727 (1983).

30. L. D. Barron, *Nature* **238**, 17 (1972).

31. A. D. Buckingham, *Proc. Phys. Soc.* **B69**, 344 (1956).

32. On being asked by the author how he (A. D. Buckingham) had envisaged making a measurement of the optical Kerr effect in the mid 1950s. The response was given that the light from a World War II searchlight could have been focused down into a suitable polarizable fluid.

33. M. Schubert and B. Wilhelmi, *Nonlinear Optics and Quantum Electronics*, Wiley-Interscience, New York, 1986, Chapter 13.

34. A. D. Buckingham and J. H. Williams, *J. Phys. E.* **22**, 790 (1989).

35. M. Born and E. Wolf, *Principles of Optics*, Pergamon, New York, 1975, p. 690.

36. J. Badoz, *J. Phys. Radium* **17**, 143 *A* (1956).

37. D. E Gray (Ed.), *American Institute of Physics Handbook*, 3d. ed. New York, McGraw-Hill, 1972, p. 6–230.

38. M. Paillette, *Ann. Phys. Paris* **4**, 671 (1969).

39. M. W. Evans, *J. Phys. Chem.* **95**, 2256 (1991).

40. W. S. Warren, S. Mayr, D. Goswami, and A. P. West, Jr., *Science*, **255**, 1683 (1992).

41. J. H. Williams and J. Torbet, submitted. *J. Phys. Chem.* **96**, 10477 (1992).

42. D. P. Shelton and R. E. Cameron, *Rev. Sci. Instrum.* **59**, 430 (1988).

43. A. D. Buckingham, C. G. Graham, and J. H. Williams, *Mol. Phys.* **49**, 703 (1983).

44. A. D. Buckingham, W. H. Prichard, and D. H. Whiffen, *Trans. Faraday Soc.* **63**, 1057 (1967).

45. F. Scuri, G. Stefanini, E. Zavattini, S. Carusotto, E. Iacopini, and E. Polacco, *J. Chem. Phys.* **85**, 1789 (1986).

46. A. D. Buckingham and J. A. Pople, *Proc. Phys. Soc.* **A68**, 905 (1955).

47. A. D. Buckingham, *Proc. Phys. Soc.* **A68**, 910 (1955).

48. A. D. Buckingham and J. A. Pople, *Proc. Phys. Soc.* **B69**, 1133 (1956).

49. L. D. Barron, *Molecular Light Scattering and Optical Activity*, Cambridge University Press, Cambridge, UK, 1982, p. 160.

50. J.-C. Bacri and D. Salin, *J. Phys. Lett.* **43**, L179, L649 (1982).

51. W. Helfrich, *Phys. Lett.* **43A**, 409 (1973).

52. A. D. Buckingham and J. A. Pople, *Disc. Faraday Soc.* **22**, 17 (1956).

53. A. K. Burnham, G. R. Alms, and W. H. Flygare, *J. Chem. Phys.* **62**, 3289 (1975).

54. C. G. Lefèvre and R. J. W. Lefèvre, *J. Chem. Soc.* 4041 (1955).

55. R. J. W. Lefèvre, D. V. Radford, G. L. D. Ritchie, and P. J. Stiles, *J. Chem. Soc.* (*B*) 148 (1968).

56. S. Kielich, *Mol. Phys.* **6**, 49 (1963).

57. S. Kielich, in M. Davies (Ed.), *Dielectrics and Related Molecular Processes*, Vol. 1, Specialist Periodical Reports, The Chemical Society, London, 1972, p. 192.

58. R. Hellworth and N. George, *Optoelectronics* **1**, 213 (1969).

59. A. D. Buckingham, *Proc. R. Soc. London Ser. A* **267**, 271 (1962).

60. W. F. Murphy, *J. Chem. Phys.* **67**, 5877 (1977).

61. W. H. Kirchhoff and D. R. Lide, Jr., *Natl. Std. Ref. Data Ser. Natl. Bur. Std.* **10** (1967).

62. J. F. Ward and C. K. Miller, *Phys. Rev.* **A19**, 826 (1979).

63. B. F. Levine and C. G. Bethea, *J. Chem. Phys.* **65**, 2429 (1976).

64. M. J. Aronay, M. R. Battaglia, R. Ferfoglia, D. Millar, and R. K. Pierens, *Trans. Faraday Soc. 2* **72**, 724 (1976).

65. B. J. Orr and J. F. Ward, *Mol. Phys.* **20**, 513 (1971).

66. D. S. Elliot and J. F. Ward, *Mol. Phys.* **51**, 45 (1984).

67. J. W. Dudley and J. F. Ward, *J. Chem. Phys.* **82**, 4673 (1985).

68. D. P. Shelton, *J. Chem. Phys.* **85**, 4234 (1986).

69. Z. Lu and D. P. Shelton, *J. Chem. Phys.* **87**, 1967 (1987).

70. D. P. Shelton, *Mol. Phys.* **60**, 65 (1987).

71. D. M. Bishop and B. Lam, *Mol. Phys.* **62**, 721 (1987).

72. G. Mayer and F. Gires, *C. R. Séance Acad. Sci. Paris* **258**, 2039 (1964).

73. G. R. Fleming, *Chemical Applications of Ultrafast Spectroscopy*, Oxford University Press, London, 1986, p. 145.

74. C. Kalpouzos, D. McMorrow, W. T. Lotshaw, and G. A. Kenney-Wallace, *Chem. Phys. Lett.* **150**, 138 (1988).

75. C. Kalpouzos, D. McMorrow, W. T. Lotshaw, and G. A. Kenney-Wallace, *Chem. Phys. Lett.* **155**, 240 (1989).

76. S. Ruhman, A. G. Joly, and K. A. Nelson, *J. Chem. Phys.* **86**, 6563 (1987).

77. G. A. Kenney-Wallace, S. Paone, and C. Kalpouzos, *Disc. Faraday Soc.* **85**, 185 (1988).

78. R. Back, G. A. Kenney-Wallace, W. T. Lotshaw and D. McMorrow, *Chem. Phys. Lett.* **191**, 423 (1992).

79. M. D. Levenson, *Introduction to Nonlinear Laser Spectroscopy*, Academic, New York, 1982.

80. P. Foggi, R. Righini, R. Torre, L. Angeloni, and S. Califano, *J. Chem. Phys.* **96**, 110 (1992).

81. G. Cardini, R. Righini, and S. Califano, *J. Chem. Phys.* **95**, 679 (1991).

82. F. W. Deeg, J. J. Stankus, S. R. Greenfield, V. J. Newell, and M. D. Fayer, *J. Chem. Phys.* **90**, 6893 (1989).

83. A. D. Buckingham, P. H. Martin, and R. S. Watts, *Chem. Phys. Lett.* **21**, 186 (1973).

84. Y.-Z. Wei, P. Chiang, and S. Sridhar, *J. Chem. Phys.* **96**, 4569 (1992).

85. Landolt and Börnstein, Springer, Berlin, 1962, 6 *Auflage Zahlenwerte und Funktionen*, II Band, Vol. 8, p. 5–827. In particular, M. A. Haque, *C. R. Séances Acad. Sci. Paris* **190**, 789 (1930).

86. M. Ramanadham, *Indian J. Phys.* **4**, 15 (1929).
87. S. W. Chinchalker, *Indian J. Phys.* **6**, 165 (1931); M. Ramanadham, *Indian J. Phys.* **4**, 109 (1929).
88. Landolt and Börnstein, Springer, New York, 1986, New Series, Vol. 16, p. 402.
89. G. W. Nielson and J. E. Enderby, *Adv. Inorganic Chem.* **34**, 195 (1989).
90. N. T. Skipper and G. W. Neilson, *J. Phys.: Condensed Matter* **1**, 4141 (1989).
91. I. Howell, G. W. Neilson, and P. Chieux, *J. Mol Structure* **250**, 281 (1991).
92. *Handbook of Chemistry and Physics*, 70th ed., CRC Press, Boca Raton, FL, 1989, D-221.
93. F. A. Cotton and G. Wilkinson, *Advanced Inorganic Chemistry*, Wiley, New York, 1972, Chapter 18.
94. R. Pool, *Science* **249**, 1096, 1378 (1990).

AB INITIO COMPUTATIONS OF POLARIZABILITIES AND HYPERPOLARIZABILITIES OF ATOMS AND MOLECULES

AHMED A. HASANEIN

Department of Chemistry, College of Sciences, King Sand University, Riyadh, Saudi Arabia, and Department of Chemistry, Faculty of Science, Alexandria University, Alexandria, Egypt

CONTENTS

Modern Nonlinear Optics, Part 2, Edited by Myron Evans and Stanisław Kielich. Advances in Chemical Physics Series, Vol. LXXXV.
ISBN 0-471-57546-1 © 1993 John Wiley & Sons, Inc.

I. INTRODUCTION

In the last several years there has been a significant advancement in the methods used for the determination of polarizabilities such as atomic and molecular beam methods[1-4] and collision-induced scattering of light.[5-8] A number of nonlinear optical phenomena involving hyperpolarizabilities discovered in experiments employing strong static and dynamic electric and magnetic fields are now also well advanced in the laser era. Kerr effect experiments,[9, 10] electric field-induced second-harmonic generation experiments,[11-18] and third-harmonic generation techniques[19-23] are three electro-optic effects useful in obtaining polarizability and hyperpolarizability data. At the same time there exists an elusive theoretical knowledge of atomic and molecular electrical properties by means of different quantum-mechanical methods and specifically through ab initio calculations. Such a host of theoretical computations has led to a significant advancement in understanding the role of these electrical properties in certain chemical and physical phenomena such as laser technology where development of materials with specified optical properties is required.

We are concerned in this chapter with quantum-mechanical computations of electric dipole polarizabilities and hyperpolarizabilities. We also deal briefly with some phenomena involving these electrical properties and the principal methods by which these properties have been determined. We do not provide a comprehensive coverage of the literature, which is extensive, but rather give an insight into the structure of the field and an awareness of its main points.

II. POLARIZABILITIES AND HYPERPOLARIZABILITIES

Definitions of various electrical properties are available at the textbook level,[24, 25] but a summary of the theory of polarizability is convenient. When an external electric field of moderate intensity is applied on a dielectric it gives rise to a dipole density or, in other words, it polarizes the dielectric. Electronic, atomic, and orientation polarizations take place. The electrons are shifted relative to the positive charges and atoms or atom groups are displaced relative to each other. If the dielectric contains molecules that are permanent dipoles, the field tends also to align these dipoles along its own direction. This is sometimes called a rotation or orientation effect, which is counteracted by the thermal movement of the molecules. On the other hand, fields of high intensities tend to direct an anisotropic particle to an orientation such that its axis of highest polarizability coincides with the direction of the field. Besides, chemical equilibria between components with different permanent dipole moments are shifted in favor of the component with a high permanent dipole moment. To investigate the dependence of the polarization **P** of a dielectric on molecular quantities it is convenient to assume that it can be given as a sum of two terms:

$$\mathbf{P} = \mathbf{P}_d + \mathbf{P}_o \tag{1}$$

where $\mathbf{P}_d$ is the distortion polarization and $\mathbf{P}_o$ is the orientation polarization.

The dependence of the polarization **P** on the field strength **E** can take several forms. The simplest form is the scalar proportionality:

$$\mathbf{P} = \chi\mathbf{E} \tag{2}$$

where the proportionality factor χ is called the scalar electric susceptibility, which must be replaced by a tensor susceptibility for nonisotropic dielectrics. It is given as

$$\chi = \frac{\varepsilon - 1}{4\pi} \tag{3}$$

where ε is the permittivity of the dielectric. Equation (2) is valid only at moderate field intensities. At field intensities of about 10^4 V/cm or higher, deviations from Eq. (2) become noticeable and a series development of **P** with respect to **E** is needed. For isotropic systems we may write

$$\mathbf{P} = \chi\mathbf{E} + \xi E^2 \mathbf{E} \tag{4}$$

and for anisotropic systems we have

$$\mathbf{P} = \chi \cdot \mathbf{E} + \xi \vdots \mathbf{EEE} \tag{5}$$

where ξ is a tensor of the fourth order. The different effects causing nonlinearity of the polarization, namely, normal saturation, anomalous saturation, anisotropic polarizabilities and hyperpolarizabilities effects, have been discussed in detail by Böttcher.[25]

Both parts of the polarization can be calculated from molecular parameters by two methods. The first method starts with a single molecule subject to an electric field and the environment of this molecule is considered to be a continuum with the macroscopic properties of the dielectric. This method does not account for specific molecular interactions. A survey and discussion of the different theories in the framework of this method was given by Buckingham[26] and Böttcher.[25] In the second

AHMED A. HASANEIN

method specific molecular interactions have been accounted for using statistical mechanics, which provide a way of obtaining macroscopic quantities when the properties of the molecules and their interactions are known. Molecular interaction is assumed to follow a hard-sphere or a Lennard-Jones potential with the molecule being represented by an ideal dipole and a scalar polarizability or by an ideal dipole in the center of a dielectric sphere. Even for these simplified models the calculations are often too difficult to perform because the system comprises a very large number of molecules exerting long-range dipole–dipole forces on each other. Therefore, calculations can be performed only if simplifications are made as crude as those made in the continuum approach to the environment of the molecule. Also, Buckingham[26] and Böttcher[25] have reviewed these statistical mechanical theories.

To compute $\mathbf{P}_d$ we must consider the magnitude of the dipole moment induced by the field acting on the particle. In fact, the applied field also results in a change in all higher multipole moments. If the particle has a permanent dipole moment $\boldsymbol{\mu}_0$ in the absence of the external field and its total moment is $\boldsymbol{\mu}$ when the field has been applied, the induced dipole moment $\mathbf{m}$ is simply $\mathbf{m} = \boldsymbol{\mu} - \boldsymbol{\mu}_0$. As long as the gradient of the external field is small, the potential energy of the higher moments will be small and effects due to permanent or induced multipole moments higher than the dipole moment will be small. Also, for small field strengths the displacements of the charges from their equilibrium positions will be a linear function of the applied field strength $\mathbf{E}$ and for isotropic particles the induced moment will have the same direction as the applied field so that, $\mathbf{m} = \alpha \mathbf{E}$ and we may write

$$\boldsymbol{\mu} = \boldsymbol{\mu}_0 + \alpha \mathbf{E} \qquad (6)$$

where α is the scalar or tensorial polarizability, a quantity of importance for many physical and chemical phenomena, such as dielectric, refractive, and light-scattering properties of matter.

It is generally assumed that the polarizabilities of monoatomic ions and undistorted quasi-spherical molecules are independent of the field direction. However, diatomic molecules and the majority of polyatomic molecules are not isotropically polarizable. The degree to which a given molecule displays anisotropy of polarizability is sensitively connected with structure and conformation. For example, in crystalline materials the molecules experience anisotropic interactions with their environments and the potential energy of a polar molecule is dependent on its orientation relative to the crystal axes.

The polarizability α is a tensor of rank 2 and can be characterized in an adequate system of coordinates by the principal polarizabilities α_{xx}, α_{yy}, and α_{zz}. It is customary to use an alternative notation in certain cases. For diatomic molecules having cylindrical symmetry, the polarizabilities are denoted by $\alpha_\| = \alpha_{zz}$ and $\alpha_\perp = \alpha_{xx} = \alpha_{yy}$ where the internuclear axis is taken to coincide with the z axis. The three principal polarizabilities are identical for spherically symmetric systems, while for systems not having spherical symmetry the distortion depends on the orientation of the system. It is then convenient to define an average polarizability $\bar{\alpha}$:

$$\bar{\alpha} = \tfrac{1}{3}(\alpha_{xx} + \alpha_{yy} + \alpha_{zz}) \qquad (7)$$

For an ellipsoidal body of dielectric constant ε and with principal axes $2a$, $2b$, and $2c$ in the direction of the x, y, and z axes, respectively. It has been shown[27-29] that the polarizability tensor is diagonal in this coordinate system, with α_a given as

$$\alpha_a = \frac{\varepsilon - 1}{3[1 + (\varepsilon - 1)A_a]} abc \qquad (8)$$

with similar expressions for α_b and α_c. Extensive tabulations of the factors A_a, A_b, and A_c as a function of a, b, and c are given in Refs. 28 and 29.

Slight changes in the dielectric constant ε have been observed in measurements using fields of considerably high intensities.[30-36] These alterations of ε, although negligibly small, imply a nonrectilinear dependence of induced polarization on field strength. These changes have been attributed to normal and anomalous saturation effects. Quantitative treatments of both effects are given in Refs. 25, 37, and 38.

In calculating the nonlinear part of the polarization, consideration of the effects due to the permanent moments of the molecules is not sufficient. Induced polarization plays also a role. A molecule with an anisotropic polarizability is directed by an external field with its axis of greatest polarizability in the direction of the field, which leads to a nonlinear term in the polarization. Deviations of the linearity of the induced moment as a function of the applied field must also be considered.[25] The contribution of the normal saturation in case of molecules having permanent dipole moments is larger in absolute value than the contribution of the anisotropy of the polarizability.

Following Buckingham[39, 40] and provided that the electric field strength **E** at the molecule is not too large, the electric dipole moment **μ** may be expanded in powers of **E** as[41]

$$\boldsymbol{\mu} = \boldsymbol{\mu}_0 + \alpha\mathbf{E} + K\beta\mathbf{E}^2 + \lambda\gamma\mathbf{E}^3 + \cdots \qquad (9)$$

the polarizability α is a tensor of rank 2 and the hyperpolarizabilities $\beta, \gamma, \ldots$ are tensors of rank $3, 4 \ldots$, respectively. These tensor quantities allow the induced moment to be in the direction other than that of the applied field. The quantities $K, \lambda, \ldots$ are numerical factors which depend on the combination of the electric field frequencies involved in each nonlinear process. There exists in the literature[41, 42] considerable variation in the choice of these numerical factors. Buckingham and Pople[40] have modified the Langevin-Born equations[43-46] for electric birefringence using the concept of hyperpolarizability and an equation of the form of Eq. (9) with $K = \frac{1}{2}$ and $\lambda = \frac{1}{6}$. A prescribed convention[41] which allows a direct comparison of hyperpolarizabilities derived from different nonlinear processes also yields the same values for K and λ for a static electric field.

Therefore, when calculating the induced dipole moment of a molecule we have to consider derivatives of the field at the origin of the molecule,[47] fields of the higher multipoles of other molecules, nonlinear distortion through the hyperpolarizabilities,[48] and distortion due to interactions that are not of an electrostatic nature, for example, dispersion forces[49] and overlap effects.[26, 50-52] Corrections to the polarizabilities and hyperpolarizabilities for molecular vibrations are small for diatomic molecules and increase as the molecule becomes larger, especially for floppy polyatomic molecules containing polar groups.[53, 54] The different theoretical methods used to calculate polarizabilities and hyperpolarizabilities are discussed in Section V.

III. PHENOMENA INVOLVING POLARIZABILITIES AND HYPERPOLARIZABILITIES

Atomic or molecular polarizabilities and hyperpolarizabilities are important in many of the quantitative descriptions of physical and chemical behavior of matter. Polarizability effects are directly involved in many basic principles throughout organic chemistry. However, the relation between chemically inferred polarizabilities and those from physical measurements, such as refractivity data, depends on the extent of our knowledge of the polarizability α and hyperpolarizabilities $\beta, \gamma, \ldots$.

All matter becomes birefringent in the presence of an electric field. This phenomenon is called the electro-optic Kerr effect.[55-57] For nonpolar gases the molar Kerr constant is given by[57, 58]

$$m^{K(T)} = \frac{4\pi N_0 \gamma}{81} + \frac{B_K(T)}{v} \tag{10}$$

where N_0 is Avogadro's number, v is the mole volume, γ is the atomic hyperpolarizability, and B_K is the second virial Kerr coefficient. For

spherically symmetric molecules the hyperpolarizability β is zero. Therefore, for nonpolar spherically symmetric molecules the apparent molar
Kerr constant is simply given[59, 60] by the first term of Eq. (10). This
provides a possible experimental route to the hyperpolarizability γ through
the large amount of experimental data available for molar Kerr
constants.[61-66] The relationship between the static dielectric constant ε
and the atomic polarizability α of nonpolar gases is given by the Clausius-
Mosotti function[67, 68]:

$$\frac{\varepsilon - 1}{\varepsilon + 2} = \frac{4\pi N_0 \alpha}{3v} + \frac{N_0^2 B_\varepsilon}{v^2} + \cdots \tag{11}$$

where B_ε is the second virial dielectric coefficient. At the frequencies of
visible light the dynamic ε is commonly replaced by the square of the
refractive index $\varepsilon = n^2$ and dynamic polarizabilities can thus be determined by what is now called the Lorenz-Lorentz formula.[8, 41, 56, 69, 70] For
ideal or dilute imperfect gases the second term on the right side of Eq.
(11) is zero or vanishingly small. At densities roughly comparable to
atmospheric densities, the second virial dielectric coefficient B_ε becomes
important.[71-75]

Some other phenomena involving polarizabilities are optical rotatory
dispersion, reactivity and reaction kinetics, and different spectral effects
taking place where electromagnetic radiation interacts with matter. Since
fields in the neighborhoods of ions or dipoles must far surpass those
attainable by experiment, hyperpolarizabilities should not be ignored when
applying electrostatic considerations to deal with reactivity and reaction
kinetics. Low-field polarizabilities have usually been used in such approaches.[76, 77]

Rational correlations between experimental atomic polarizabilities and
natural optical rotatory power have been obtained by combining earlier
empirical and theoretical approaches.[78-82] The optical rotatory dispersion
is considered to be a polarizability-based phenomenon. The rotatory
power of a helical macromolecule is seen as an anisotropic property[83-86]
and can be altered by applying external electric fields. The birefringence
of anisotropic macromolecule is often high in consistency with the high
apparent moments attributed to proteins, nucleic acids, and viruses in
aqueous solutions.

The interactions between the applied electric field and anisotropically
polarizable molecules are directly involved in measurements of molecular
quadrupole moments[87] and also in the determination of the absolute signs
of spin–spin coupling constants in NMR spectra of polyatomic molecules,[88]
particularly if a voltage gradient is imposed across the sample where the
emergence of field-induced overtones would constitute a significant ad-

vance in NMR methodology.[89] Anisotropic shielding effects in NMR spectra could be explained through knowledge of electronic polarizabilities, which leads to more information about anisotropies of diamagnetic susceptibility.[90, 91] Theoretical correlations between principal polarizabilities and principal susceptibilities have been obtained.[60, 92, 93] Anisotropic molecular polarizabilities also appear in expressions for the Faraday effect.[60, 94, 95] Polarizabilities are currently invoked in connection with solvent effects on infrared and Raman spectra, particularly frequency shifts and intensity changes.

All methods for determining molecular dipole moments in the first excited singlet states are based on the position change of the spectral band caused by an electric field either external (electrochromism) or internal (solvatochromis). The electro-optical methods used in electrochromism, such as electric polarization of fluorescence, electric dichroism, and Stark splitting of rotational levels of the 0-0 vibrational band, are generally believed to be far more accurate[96] than those based on solvent shift methods.[97-100] The truly accurate methods based on the Stark splitting are restricted to relatively simple molecules. In the solvent-shift methods the spectral maximum $Y(A, s)$ of a compound A in a solvent S is separated into a linear sum of terms. Each of these terms is a product of a solvent factor X_i, which depends mainly on the dielectric constant and refractive index of the solvent and a solute factor B_i, which depends mainly on the dipole moment μ and polarizability α of compound A in its ground and excited states[97]:

$$Y(A, s) = Y^0(A, g) + \sum_{i=1}^{j} B_i(A) X_i(s) \tag{12}$$

$$B_i(A) = B_i(\mu_g, \mu_e, \alpha_g, \alpha_e, a_O, \dots) \tag{13}$$

$$X_i(s) = X_i(\varepsilon, n, \dots) \tag{14}$$

where g and e refers to ground and excited states respectively, and a_O is the Onsager spherical cavity radius. The subscript i describes the type of the solute–solvent interactions, namely dipole–dipole, dipole–induced dipole, dispersion, higher-order, and specific interactions.

As mentioned above, polarizabilities and hyperpolarizabilities become directly concerned whenever electromagnetic radiation interacts with matter. They are directly involved in light scattering by small molecules and by macromolecules. If the system has a uniform distribution of its particles, then coherent scattering will take place, which is encountered in the phenomenon of refraction consisting of what is called forward scattering of an intensity comparable with that of the incident wave. In case of a fluid as an example of a system with random distribution of its particles, not only forward scattering is observed, but random phase relations between the emitted radiation will take place. This will result in an incoherent scattering known as Rayleigh scattering.

Hyperpolarizabilities may be defined for inelastic nonlinear optical processes of which hyper-Raman scattering[101] is a good example. This phenomenon consists of incoherent scattering as a result of excitation by two photons. The transition hyperpolarizability β, which describes such a process, is analogous to the polarizability α, which is responsible for the conventional Raman effect. In general, multiphoton absorption[102-105] is one of the nonlinear optical phenomena that are directly dependent on hyperpolarizabilities. Its origin is in the imaginary part of optical hyperpolarizabilities in resonant situations.

Second-harmonic generation and hyper-Rayleigh scattering are hyperpolarizability phenomena. For molecules lacking a center of inversion the hyperpolarizability component of the electric dipole moment is nonzero. Therefore, under suitable conditions, both coherent scattering (second-harmonic generation) and incoherent scattering (hyper-Rayleigh) may be observed. Second-harmonic generation is now a widely exploited means of producing intense coherent light at shorter wavelengths.[14-16, 106-108] The incident laser beam and the scattered waves differ in frequency and usually propagate with different velocities in the material. Examination of scattering intensities and Raman displacements and other possible effects from plasmas irradiated by intense laser beams was one interesting suggestion,[109, 110] which further underlines the widespread applicability of polarizabilities.

Polarizability is a complex quantity. Its imaginary part, called antisymmetric polarizability, contributes to the laser-enhanced ESR spectra. Optical third-harmonic generation[111] arises from the nonlinear hyperpolarizability γ. However, coherent third-harmonic generation is not restricted to noncentrosymmetric materials[41]; it has also been observed in crystals and isotropic fluids.

IV. SOME RECENT EXPERIMENTAL TECHNIQUES FOR DETERMINING POLARIZABILITIES AND HYPERPOLARIZABILITIES AND THEIR PROPERTIES

As mentioned above, there exists in the literature a host of experimental and theoretical data on the optical dielectric properties of gases and simple fluids.[41, 60, 112-115] Light scattering intensities, dielectric constant, Kerr effect, and refractivity data have been used for obtaining polarizabilities of a wide variety of atomic and molecular systems. Experimental and theoretical means for the determination of polarizabilities and hyperpolarizabilities and also data from various sources have been reviewed.[25, 26, 116-118] The hyperpolarizabilities β and γ have been determined using the Kerr effect,[57, 119-123] second-harmonic generation,[11, 12, 16] electric field-induced second-harmonic generation,[11-13, 16, 124] three-wave mixing,[11, 12, 125] and third-harmonic generation[19] techniques. In recent years there has been significant advancement in the experimental and theoretical means for determining polarizabilities and hyperpolarizabilities. Some of these recent techniques are briefly discussed here.

A. Atomic Beam Experiments

Atomic polarizability is generally a useful parameter when dealing with any atomic process in which the outer part of the electronic wave function plays an important role and a direct experimental determinations of atomic polarizabilities provide a very sensitive check on the accuracy of electronic wave function.[4] Static polarizabilities of ground and excited states of neutral atoms and molecules are directly involved in the calculations of van der Waals interaction, the description of charge transfer, and other collision processes.

Atomic beam methods have been used for measuring polarizabilities.[1-4] Three different methods have been employed. The Strong-field deflection method was first used by Scheffers and Stark[126] to measure the polarizabilities of lithium, potassium, and cesium. The weak-field deflection method has been used[1] to measure the polarizabilities of all the alkali metals. The third method is the electric field–magnetic field gradient balance method.[2, 3, 127] An inhomogeneous magnetic field is established in the same region which contains the inhomogeneous electric field. This method was first used by Salop et al.[2, 3] to measure the polarizabilities of the alkalis and argon. It is expected[4] to fail for states that do not possess magnetic moments. However, for suitably designed electric and magnetic fields, nuclear moments would suffice for an electric field–magnetic field gradient balance rather than the atomic moment.[4]

B. Molecular Beam Electrical Resonance Spectroscopy (MBER)

Electrical moments for ground and first excited vibrational states to remarkably high precision have been obtained using this technique.[128] Use of electric fields has become widespread[54, 129-131] for probing molecular details. The changes of molecular properties by excitation to another quantum state caused by electromagnetic radiation resonant with the transition frequency result in selective deflection of molecules. These molecular beam deflection techniques are important tools for investigating permanent moments, even very small ones.[132-136]

The significant advantage of MBER techniques over spectroscopic electrical resonance[137, 138] is linewidth. Since the lifetime of a rotational state is about the same as the time between collisions, the linewidth of a resonant absorption associated with reorientation in a strong electric field is very large, but this is not the case in a molecular beam experiment.[54] In collision experiments, the nature of the interaction is governed by the potential energy surface, which is a function of the molecular properties of the colliding partners. Usually the potential energy is written in a multipole expansion, whereby the electrical properties are displayed in the long-range terms.[139] The potential that is generated must satisfy[54] simultaneously molecular beam scattering data, virial coefficients, spectral moment data, and so on.[140] For properties incorporated into the long-range term, experiments must reflect sensitivity in that region.[54]

Molecular beam data can be useful in equilibrium and long-range regions. Recent applications include $He \cdot \cdot CO$ and $Ar \cdot \cdot CH_4$ systems,[141-143] where from the derived isotropic part and anisotropic component of the potential, values of the polarizabilities have been extracted.

C. The Inverse Faraday Effect, the Inverse Magnetochiral Birefringence, and Field Applied Molecular Dynamics

Studies of optical effects in gases and liquids exposed to external electric, magnetic, or electromagnetic fields increase our knowledge of molecular properties and intermolecular correlations.[144-150] One of the effects of a static magnetic field acting on a medium in which a linearly polarized beam of monochromatic light propagates in the direction of the magnetic field consists of the rotation of the plane of polarization discovered by Faraday.[151] Extensive reviews on the Faraday effect may be found in the literature.[152-154]

The ability of an intense, circularly polarized pump laser to produce bulk magnetization was first predicted theoretically by Pershan[155] and was demonstrated in a series of novel experiments by Van der Ziel et al.,[156, 157] who called induced static magnetization the inverse Faraday effect. This effect has been developed theoretically.[24, 158, 159] The spectral consequences of the analogy between a static magnetic field and circularly polarized light include the inverse Zeeman effect[160] and possible detection of the magnetization in an NMR or ESR spectrometer; the prototype theory[161] and experiment,[162] which have been initiated recently; and laser-induced forward–backward birefringence[163] as well as laser-induced dynamic polarization,[164] the latter being sustained in chiral paramagnetic molecules.

It has been shown[165-167] that the change in refractive index and in absorption coefficients in the presence of a static magnetic field applied along the direction of light propagation is not a circular differential effect, and that it can occur in optically active media for circularly, linearly, and unpolarized light, leading to magnetochiral birefringence and magnetochiral dichroism.[168, 169] The molecular theory of both effects has been extended to the case of additional external static electric[170] or optical[171] fields for media composed of diamagnetic as well as paramagnetic molecules,[172] and has also been considered in connection with parity violation in atoms.[173]

As was recently shown,[174] a coherent beam of light of arbitrary polarization traveling in a medium composed of randomly oriented chiral molecules induces a static magnetization parallel or antiparallel to the direction of propagation. This nonlinear optical effect in analogy to the inverse Faraday effect is called inverse magnetochiral birefringence.[175] The inverse Faraday effect in the optical resonance regions of molecular systems has been analyzed, and the inverse magnetochiral birefringence has been discussed.[149, 150] Complete quantum-mechanical expressions for the nonlinear polarizabilities involved in both effects are presented and their properties and orders of magnitude discussed.

Field applied molecular dynamics has been extended for use with circularly polarized pump lasers[176] and recently to nonlinear optical effects[177-179] in general. Field applied molecular dynamics has been used to simulate the inverse Faraday effect and inverse magnetochiral birefringence.[148, 150] Ab initio calculated or experimental data on the scalar elements of the polarizability tensors have been used.

V. THEORETICAL APPROACHES TO (HYPER)POLARIZABILITIES

Theoretical knowledge of molecular electrical properties has reached an important threshold and the methods used to calculate polarizabilities have advanced considerably in the last few years. The role of these properties in many physical and chemical phenomena have been significantly clarified through updated experimental techniques and theoretical methods. Before dealing with ab initio calculations of polarizabilities, which are reviewed in Section VII, we discuss the early theoretical methods used for the calculations of (hyper)polarizabilities.

A. Classical Treatment

In the early classical theory of radiation scattering we start with an electron of mass m_0 and a charge e assumed to be elastically held to an equilibrium distance r from the nucleus by a linear restoring force, i.e., a case of charged isotropic harmonic oscillator having a natural frequency of vibration $\omega_0 = 2\pi\nu_0 = (K/m_0)^{1/2}$, where K is the force constant. The action of an electric field of strength $\mathbf{E}$ causes an energy change:

$$\Delta\mathscr{E} = \mathscr{E} - \mathscr{E}^0 = -\int \mu \, d\mathbf{E} \tag{15}$$

Using Eq. (9) with $K = \frac{1}{2}$ and $\lambda = \frac{1}{6}$ we can write

$$\mathscr{E} = \mathscr{E}^0 - \mu_0 E - \tfrac{1}{2}\alpha E^2 - \tfrac{1}{6}\beta E^3 - \tfrac{1}{24}\gamma E^4 \cdots \tag{16}$$

where $\mathscr{E}^0$ is the energy in the absence of the field. Considering energy change due to the polarizability α only and assuming that the system does not have a permanent dipole (μ_0 = zero), the interaction energy may be written as

$$\Delta\mathscr{E} = -\frac{1}{2}E\alpha E = -\frac{1}{2}\sum_u \sum_v E_u \alpha_{uv} E_v \tag{17}$$

where $u, v = x, y, z$, and E_u are the components of the electric field. In a nonuniform electric field the deflection of an S-state atom is an illustration of Eq. (17). Another illustration of quantum-mechanical nature is the quadratic Stark effect observed in atomic and molecular spectra. If damping effects are not considered the classical formula for the polarizability α is given as

$$\alpha = \frac{e^2}{m_0} \frac{1}{\omega_0^2 - \omega^2} \tag{18}$$

However, free oscillations of a radiative system must necessarily be damped since the system loses energy and the polarizability α must have the following expression:

$$\alpha = \frac{e^2}{m_0} \frac{1}{\omega_0^2 - \omega^2 + i\omega\Gamma} \tag{19}$$

The external electric field is assumed to be oscillating with frequency ω, and Γ is the radiation damping constant.

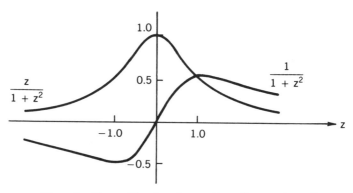

Figure 1. The auxiliary functions $z/(1 + z^2)$ and $1/(1 + z^2)$.

Writing $\alpha = \alpha' + i\alpha''$, the following expressions for the real α' and imaginary α'' parts of the polarizability can be given:

$$\alpha' = -\frac{e^2}{m_0\omega\Gamma}\frac{z}{1 + z^2} \qquad (20)$$

$$\alpha'' = -\frac{e^2}{m_0\omega\Gamma}\frac{1}{1 + z^2} \qquad (21)$$

where $z = (\omega^2 - \omega_0^2)/\omega\Gamma$. If the frequency is close to resonance, we may rewrite $\omega^2 - \omega_0^2 = (\omega - \omega_0)(\omega + \omega_0) \approx 2\omega(\omega - \omega_0)$ and $z \approx 2(\omega - \omega_0)/\Gamma$. The two auxiliary functions $z/(1 + z^2)$ and $1/(1 + z^2)$ are shown in Fig. 1.

The absorption of light depends on the imaginary part given by Eq. (21). Γ is only a function of the frequency and does not depend on the amplitude of the oscillator. It is given as[180]

$$\Gamma = \frac{2e^2\omega_0^2}{3m_0C^3} \qquad (22)$$

Damping also leads to a small shift of the intensity peak of the order Γ^2/ω_0. The damped oscillation no longer represents a monochromatic wave, but contains a whole set of frequencies. Thus, spectral lines are broadened and a half-breadth of the emitted spectral line will be of the order of magnitude of $\Gamma/2$. Quantum-mechanical considerations of the probability of emission of a photon with a frequency in the interval $\omega, \omega + d\omega$ in a transition from a state m to a state n ($\mathscr{E}_m - \mathscr{E}_n = \hbar\omega_0$) led to a linewidth Γ given as[180]

$$\Gamma = \Gamma_m + \Gamma_n = \frac{2e^2\omega_0^2}{m_0c^3}f_{mn} \tag{23}$$

where f_{mn} is the oscillator strength of this transition. Equations (20) and (21) show the way in which variations of the electric field will be followed by the system, particularly at frequencies well below resonance, $\omega \ll \omega_0$. As the frequency is allowed to increase to $\omega < \omega_0$ but of the order of ω_0, a rapidly growing absorption takes place which reaches its maximum at resonance ($\omega = \omega_0$), determined by the value of the damping constant Γ. Beyond resonance $\omega > \omega_0$ the neutral response of the system begins to diminish, leading to steadily decreasing absorption. For sufficiently high frequencies $\omega \gg \omega_0$ the system virtually ceases to respond to the fluctuations of the electric field.

B. Frequency Doubled Optical Stark Effect

The Stark effect consists of the splitting and shifting of atomic levels under the action of an external field. The spectral lines arising from transitions between these levels are shifted or split. Perturbation theory gives the second-order energy change due to the applied electric field as

$$\Delta \mathscr{E}_k^{(2)} = -E^2 \sum_{l \neq k} \frac{|\mu_{kl}|^2}{\mathscr{E}_l^0 - \mathscr{E}_k^0} \tag{24}$$

where μ_{kl} is given by

$$\mu_{kl} = \langle \Psi_k^{(0)} | \mu | \Psi_l^{(0)} \rangle = \langle k | \mu | l \rangle \tag{25}$$

$\Psi_k^{(0)}$ and $\Psi_l^{(0)}$ are the unperturbed eigenfunctions of the states k and l, respectively, with eigenvalues $\mathscr{E}_k^{(0)}$ and $\mathscr{E}_l^{(0)}$, respectively. Comparing Eq. (24) with Eq. (17), the second-order perturbation expression for the polarizability tensor α^k of the state k can be written as[181]

$$\alpha^k = 2e^2 \sum_{l \neq k} \frac{\langle k | \sum_i r_i | l \rangle \langle l | \sum_i r_i | k \rangle}{\mathscr{E}_l^{(0)} - \mathscr{E}_k^{(0)}} \tag{26}$$

This expression gives the polarizability in terms of the transition moment integrals between states l and k and may also be stated in terms of the polarizability tensor components:

$$\alpha_{uv}^{k} = 2e^{2} \sum_{l \neq k} \frac{\langle k | \sum_{i} r_{ui} | l \rangle \langle l | \sum_{i} r_{vi} | k \rangle}{\mathscr{E}_{l}^{(0)} - \mathscr{E}_{k}^{(0)}} \tag{27}$$

where r_{ui} is the u component of the vector r_{i}. The Stark effect has been observed in pure rotational spectra of molecules,[113, 182, 183] and the static polarizabilities of a number of molecules have been obtained. Stark shifts may be obtained from molecular beam resonance spectroscopy[184, 185] and therefore the mean polarizabilities for atoms or molecules may be obtained. Because of its complexity, Eq. (24) is not very suitable for actual calculations. Exceptions are the ground state and the case of a strong interaction with the nearest level where the principal contribution to Eq. (24) is given only by one of the terms in the sum. Generally, polarizability of an atom or a molecule should be strongly dependent on the availability of fairly low-energy unoccupied levels.

If we consider a perturbation depending explicitly on time and assuming that only the leading electric dipole term need be considered, the second-order energy change due to the applied electric field is[180]

$$\Delta \bar{\mathscr{E}}_{k}^{(2)} = \frac{1}{\hbar} \sum_{l \neq k} \frac{\omega_{kl} |\mu_{kl}|^{2}}{\omega_{kl}^{2} - \omega^{2}} \tag{28}$$

where $\mathscr{E}_{k}^{(0)} - \mathscr{E}_{l}^{(0)} = \hbar \omega_{kl}$, and ω is the optical frequency. The polarizability of a state k can therefore be given as

$$\alpha^{k} = -\frac{2}{\hbar} \sum_{l \neq k} \frac{\omega_{kl} |\mu_{kl}|^{2}}{\omega_{kl}^{2} - \omega^{2}} \tag{29}$$

In the limiting case of a static field $\omega \to 0$, Eq. (29) becomes the simple static polarizability given by Eq. (26), and Eq. (28) becomes the usual quadratic Stark effect formula (24).

The optical Stark effect is generally defined through the induction of an electric dipole moment by the following type of interaction between molecular polarizability and the electric field strength E of an electromagnetic plane wave:

$$\mu_{i}^{(\text{ind})} = \alpha_{ij}(-\omega; \omega) E_{j} \tag{30}$$

At the outset the electric field was defined[186] as the plus conjugate of a left circularly polarized plane wave:

$$\mathbf{E}_L^+ = E_0(\mathbf{i} - \mathbf{ij})\exp(\mathrm{i}\phi_L) \qquad (31)$$

where $\mathbf{i}$ and $\mathbf{j}$ are unit vectors in the X and Y axes, respectively, of the laboratory frame (X, Y, Z), and the phase factor is defined by

$$\phi_L = \omega t - \mathbf{\kappa}_L \cdot \mathbf{r} \qquad (32)$$

where ω is the angular frequency in radians per second of the plane wave, t is the time, $\mathbf{K}_L$ is the left-handed propagation vector, and $\mathbf{r}$ is the position vector. For the purposes of computer simulations the phase has been approximated[186] by

$$\phi_L \approx \omega t \qquad (33)$$

Usually in the treatment of the optical Stark effect[187] the potential energy and torque generated by the interaction of $\mathbf{E}$ with the induced electric dipole moment $\mathbf{\mu}^{(\mathrm{ind})}$ are described respectively by the following time-independent expression:

$$H_S' = -\mathbf{\mu}^{(\mathrm{ind})} \cdot \mathbf{E}^* \qquad (34)$$

and

$$\mathbf{T}_S = -\mathbf{\mu}^{(\mathrm{ind})} \times \mathbf{E}^* \qquad (35)$$

The complex conjugate of $\mathbf{E}_L^+$ is $\mathbf{E}_L^{*+}$, i.e.

$$\mathbf{E}_L^{*+} = E_0(\mathbf{i} + \mathbf{ij})\exp(-\mathrm{i}\phi_L) \qquad (36)$$

so that the phase factor ϕ_L disappears in the dot and cross vector products of Eqs. (34) and (35), respectively. The time-independent energy term (34) can then be applied straight-forwardly[186] to the generation of Langevin-Kielich functions[188] which describe ensemble orientational averages of the optical Stark effect. The thermodynamic and molecular dynamic characteristics of a torque of the following type have been considered[186]:

$$\mathbf{T}_S^{(2\omega)} = -\mathbf{\mu}^{(\mathrm{ind})} \times \mathbf{E}_L^+ \qquad (37)$$

The optical Stark effect in this case is accompanied by second-harmonic generation. The energy term corresponding to this torque is

$$H_S'^{(2\omega)} = -\mu^{(ind)} \cdot E_L^+ \tag{38}$$

and disappears when time averaged.

Use is made[186] of field applied molecular dynamics computer simulation to investigate the orientations thermodynamics with rise transients and time correlation functions. A field applied molecular dynamics computer simulation has also been performed for the optical Kerr effect[179] and the inverse Faraday effect.[148] In both the inverse Faraday effect and in the frequency doubled optical Stark effect, the existence of transients signal that of transient birefringence. The latter is observable in principle[186] by adaptations of contemporary optical Kerr effect apparatus with femtosecond time resolution.[189] It has been shown[186] that the frequency doubled optical Stark effect produces second-order rise transients, birefringence, and anisotropy in liquids as well as gases. The birefringence anticipated in the field applied molecular dynamics simulation[186] is a biaxial birefringence; i.e., the refractive index in the pump laser's propagation axis Z becomes different from those in the orthogonal X and Y axes. The antisymmetric part of the polarizability is also invoked in an approach to optically enhanced ESR using a circularly polarized laser.

C. Perturbation Calculations of Polarizabilities and Hyperpolarizabilities

The theory of atomic and molecular polarizability has been formulated with the aid of quantum-mechanical perturbation theory.[191, 192] Time-dependent perturbation theory shows how the solutions of the time-dependent Schrödinger equation $|\Psi(t)\rangle$ for the system evolve in the presence of the perturbation $H'(t) = -\mu E(t) = -\mu E_0 \cos \omega t$ where an appropriate phase choice has made $E(t)$ real. Let the system be in the state K before switching on the perturbation due to a variable field. The wave functions $|\Psi(t)\rangle$ may be expanded in wave functions of the unperturbed system $|\Psi_i^{(0)}(t)\rangle$. As mentioned above, Eqs. (28) and (29) can be obtained for the second-order energy change and polarizability can be calculated. Equation (29) can be expressed in terms of the oscillator strength f_{kl} of the transition $k \rightarrow l$ and the polarizability of an atom may be given as[180]

$$\alpha^k = \frac{e^2}{m_0} \sum_{l \neq k} \frac{f_{kl}}{\omega_{kl}^2 - \omega^2} \tag{39}$$

Generally, let $|\Psi^{(1)}\rangle$ be the first-order correction to the unperturbed wave function $|\Psi^{(0)}\rangle$. The wave equation for $|\Psi^{(1)}\rangle$ is then

$$(\mathscr{H}^0 - \mathscr{E}^0)|\Psi^{(1)}\rangle = -H'|\Psi^{(0)}\rangle \tag{40}$$

If Eq. (40) can be solved for $|\Psi^{(1)}\rangle$, the second-order energy is

$$\mathscr{E}^{(2)} = \langle \Psi^{(1)}|H'|\Psi^{(0)}\rangle \tag{41}$$

Equations (40) and (41) serve as the starting point for all approximate perturbation calculations of the polarizability.

The perturbation theory has been applied within the Hartree-Fock formalism where electron exchange is included. The coupled-Hartree-Fock perturbation theory,[193-195] which introduces into the Fock equations the external perturbations and computes the perturbed wave functions and energies self-consistently and accounts rigorously for the distortion of the single determinental density due to the external one-electron perturbing operator, has also been used. For a detailed discussion of some perturbation methods used, the reader in referred to the review articles of Dalgarno[192] and Langhoff et al.[196] Another approach to the calculation of polarizabilities describes[192, 196] the perturbation by a series of differential equations, which can be solved by numerical, variational, or analytical means and can be applied to both time-dependent and time-independent problems within the Hartree-Fock approximation.

Perturbation methods have been also used to calculate the diatom polarizability. The diatom polarizability, is simply the pair polarizability minus the sum of the polarizabilities of the two separated atoms. Diatom polarizabilities of simple gases are important in their collision-induced Raman spectra. For two-interacting atoms forming a quasimolecular complex in which the two atoms may either be free and in a collisional encounter or be lightly bound together as a so-called dimer or van der Waals molecule, the second-order perturbation theory gives the following expression for the dynamic polarizability tensor components in atomic units[8]:

$$\alpha_{\|, \perp}^{k}(\omega) = \sum_{l \neq k} \frac{(f_{\|, \perp})_{kl}}{(E_l - E_k)^2 - \omega^2} \tag{42}$$

with $\alpha_\|$ and $\alpha_\perp$ being the diatom polarizabilities in directions parallel and perpendicular to the internuclear axis, and $\omega = 2\pi\nu$ where ν is the frequency of the incident photon. The oscillator strength are defined[8] as usual by

$$(f_{\parallel})_{kl} = \frac{2}{3}\left(\frac{g_l}{g_k}\right)(E_l - E_k)\langle k|z|l\rangle\langle l|z|k\rangle \tag{43}$$

$$(f_{\perp})_{kl} = \frac{4}{3}\left(\frac{g_l}{g_k}\right)(E_l - E_k)\langle k|x|l\rangle\langle l|x|k\rangle \tag{44}$$

with E_k and E_l being the diatom energies of the ground and excited states, respectively, and g_l and g_k representing the degeneracies. The computation of $\alpha_{\parallel}$ and $\alpha_{\perp}$ involves determining the first-order correction $\Psi_k^{(1)}$ to the wave function in the presence of the external field perturbation. Calculations fall into two main categories: those based on correlated configuration interaction wave functions and those based on uncorrelated Hartree-Fock schemes. The direct sum-over-states method expands the first-order corrections $\Psi_k^{(1)}$ in terms of the complete set of unperturbed eigenfunctions,[197] and the sum in (42) is formally closed in same way.

Perturbation treatments of first- and higher-order polarizabilities have been reviewed by many authors.[8, 41, 42, 196, 198] A general and convenient approach was introduced by Bogaard and Orr.[41] The electric transition dipole between quantum states k and l of a molecule $\mu_u^{kl} = \langle k|\mu_u|l\rangle$ has been given[41, 42] the following expression:

$$\mu_u^{kl}(t) = \exp(i\omega_{kl}t)\langle k|\mu_u|l\rangle + \frac{1}{2\hbar}\sum_{\pm}\exp[i(\omega_{kl} \pm \omega)t]$$
$$\times \sum_n\left[\frac{\langle k|\mu_u|n\rangle\langle n|\mu_v|l\rangle}{(\Omega_{nl} + \omega)} \pm \frac{\langle k|\mu_v|n\rangle\langle n|\mu_u|l\rangle}{(\Omega_{nk}^* \mp \omega)}\right]E_{0v} + \cdots \tag{45}$$

where $\Omega_{nl} - \omega_{nl} = \frac{1}{2}i\Gamma_{nl}$, with ω_{nl} being the difference in energy in angular velocity units between the unperturbed levels n and l, and Γ_{nl} the radiative damping constant defined in Section V.A. The suffixes $u, v, \ldots$ imply summation over Cartesian x, y, and z components; for example, $A_uB_{uv} = A_xB_{xv} + A_yB_{yv} + A_zB_{zv}$, where v also equals x, y, or z. The symbol $\Sigma_{\pm}$ denotes a summation over ensuing $\pm$ alternatives, and the Ω_{nl} term is complex in order to provide for radiative damping in resonant situations.[41, 42] Various phenomena can be explained using Eq. (45) for example, Rayleigh scattering ($k = l$) and Raman scattering ($k \neq l$) in both resonant and nonresonant cases. The optical polarizability given simply by Eq. (29) for Rayleigh scattering in the nonresonant case can be obtained by setting $k = l$ in Eq. (45) and omitting radiative damping. The same method can be used to obtain the static polarizability of Eq. (26). This can be done by using the following two equations:

$$\mu_u^{(ll)}(t) = \mu_u^{(l)} + \alpha_{uv}^{(l)}(-\omega;\omega)E_{0v}\cos\omega t + \cdots \tag{46}$$

$$\mu_u^{(ll)}(\text{static}) = \mu_u^{(l)} + \alpha_{uv}^{(l)}(0;0)E_v + \cdots \tag{47}$$

where $\mu_u^{(l)} = \langle l|\mu_u|l\rangle$ is the permanent electric dipole moment for the molecule in the quantum state l. If the optical frequency ω approaches a molecular transition frequency ω_{nl}, provision must be made for resonance effects. Retention of radiative damping in Eq. (45) with $k = l$ yields the following result[41]:

$$\mu_u^{(ll)}(t) = \mu_u^{(l)} + \text{Re}\left[\alpha_{uv}^{(l)}(-\omega;\omega)\right]E_{0v}\cos\omega t$$
$$+ \text{Im}\left[\alpha_{uv}^{(l)}(-\omega;\omega)\right]E_{0v}\sin\omega t + \cdots \tag{48}$$

where Re and Im denote real and imaginary parts, respectively. In addition to the contribution due to Re(α), which oscillates in phase with the radiation field $E(t)$, a second contribution, which involves the tensor Im(α) and is out of phase with $E(t)$, appears in the resonant case. The former in-phase contribution is responsible for the so-called anomalous dispersion of refractive index and the latter may be identified with absorption of radiation at resonance. The tensor components of the symmetric and antisymmetric polarizabilities are given[41] as follows:

$$\left(\alpha_{uv}^l\right)' = \sum_n \frac{2\omega_{nl}\left(\omega_{nl}^2 - \omega^2 + \frac{1}{4}\Gamma_{nl}^2\right)}{\hbar\left(\omega_{nl}^2 - \omega^2 + \frac{1}{4}\Gamma_{nl}^2\right)^2 + \hbar\omega^2\Gamma_{nl}^2}\langle l|\mu_u|n\rangle\langle n|\mu_v|l\rangle \tag{49}$$

$$\left(\alpha_{uv}^l\right)'' = \sum_n \frac{2\omega_{nl}\omega\Gamma_{nl}}{\hbar\left(\omega_{nl}^2 - \omega^2 + \frac{1}{4}\Gamma_{nl}^2\right)^2 + \hbar\omega^2\Gamma_{nl}^{j2}}\langle l|\mu_u|n\rangle\langle n|\mu_v|l\rangle \tag{50}$$

Equations (49) and (50) are the quantum-mechanical expressions corresponding to the classical expressions given by Eqs. (20) and (21), respectively. The origin of inelastic (Raman) scattering may be established by considering the real part of Eq. (45) with $k \neq l$ and the Raman processes may be described in terms of the following transition polarizabilities[41]:

$$\alpha_{uv}^{(kl)}(\omega_{kl} \pm \omega; \mp\omega) = \sum_n \left[\frac{\langle k|\mu_u|n\rangle\langle n|\mu_v|l\rangle}{\hbar(\omega_{nl} \pm \omega)} + \frac{\langle k|\mu_v|n\rangle\langle n|\mu_u|l\rangle}{\hbar(\omega_{nl} \mp \omega)}\right] \tag{51}$$

where the upper and lower signs refer to anti-Stokes and Stokes processes, respectively.

The perturbation theory has been extended to higher-order polarizabilities[42] contributing to nonlinear optical phenomena. Expressions for the various hyperpolarizabilities have been obtained[42, 192, 199-201] by extension of the perturbation theory to higher orders where time-dependent[192] and time-independent[201] processes have been considered for both resonant and nonresonant situations. The formulations are much more complicated than the first-order polarizabilities, but relevant formulas have been explicitly given[42, 196] for tensorial and static second- and third-order hyperpolarizabilities β and γ, respectively. We give here only the expression for the static hyperpolarizability β (Ref. 41):

$$\beta_{u,v,w}^{(n)}(0;0;0) = \frac{1}{\hbar^2} \sum_{u,v,w} \sum_{\substack{k,l \\ (\neq n)}} \frac{\langle n|\mu_u|l\rangle \langle \overline{l|\mu_v|k}\rangle \langle k|\mu_w|n\rangle}{\omega_{ln}\omega_{kn}} \tag{52}$$

where $\Sigma_{u,v,w}$ specifies summation over all permutation of Cartesian x, y, and z suffices and $\langle \overline{l|\mu_v|k}\rangle = \langle l|\mu_v|k\rangle - \langle n|\mu_v|n\rangle$. The transition hyperpolarizability $\beta^{(mn)}(\omega_{mn} \pm 2\omega; \mp\omega, \mp\omega)$ describes the hyper-Raman scattering,[101] which consists of incoherent scattering at frequencies $(2\omega \pm \omega_{mn})$ as a result of excitation by two photons of frequency ω. Such a process is analogous to the transition polarizability $\alpha^{(mn)}(\omega_{mn} \pm \omega; \mp\omega)$, which is responsible for the conventional Raman effect. The electric field-induced second-harmonic generation arises from the hyperpolarizability $\gamma(-2\omega; \omega, \omega, 0)$. An explicit formulas has been also given[41] for $\gamma(0;0,0,0)$. The expressions given above for calculating polarizabilities and hyperpolarizabilities are quantitatively useful only when a small number of terms contribute significantly to the finite sums.

The finite-field method originally proposed by Cohen and Roothaan[202] has been extensively used to calculate dipole polarizabilities. The Hamiltonian potential energy term for the interaction between an applied external field of a fixed strength E_z and the ith electron is given simply as $E_z\mu_{zi}$. With the usual dipole moment expansion given by Eq. (9), the polarizability may be obtained as a finite difference

$$\alpha_{z,z}(E_z) = \frac{\mu_z - \mu_z^0}{E_z} \tag{53}$$

where μ_z^0 is the permanent field-free moment. If this is done for several choices of E_z it may be possible to take the limit as $E_z \to 0$ and obtain the dipole polarizability as strictly defined:

$$\alpha_{i,j} = \left. \frac{\partial \mu_i}{\partial E_j} \right|_{V=0} \tag{54}$$

The fields used must be small to ensure that a numerical derivative is accurate. Contaminations of the polarizabilities by the next higher-order term can be removed if fields of the same magnitude but of opposite sign are used. A one-electron operator, which consists of a particular moment-element operator scaled by the choice of the finite field or the finite field gradient and so on, is added to the Hamiltonian. Typical values for finite-field and field gradient strengths employed in these types of calculations range from 0.0001 to 0.01 a.u. The finite-field approaches can be carried out at the SCF level, the CI level, or the many-body perturbation theory level[203-206] with equal ease. Many calculations using the finite-field method have been reported for atoms,[207] simple molecules,[209-212] and small hydrocarbons.[213-215] The use of the finite-field method with semiempirical model Hamiltonian methods is reviewed in the next section and with ab initio calculations in Sections VII.

VI. COMPUTATIONS OF POLARIZABILITIES AND HYPERPOLARIZABILITIES USING SEMIEMPIRICAL MODEL HAMILTONIAN METHODS

The design and synthesis of new molecules having optimized nonlinear optical characteristics currently represents progress in the development of many technologies, such as optical information processing, integrated optics, optical computers, and laser technology. In recent years much attention has been given to organic compounds in view of their unusually large nonlinear response[216-219] and also their application to SHG from the near infrared to the visible. Numerous organic materials have been synthesized and their optical properties have been evaluated. Materials with conjugated π-electron systems have been found to exhibit extremely large optical nonlinear responses[220] and therefore are useful in a variety of electro-optical devices. Also, noncentrosymmetric polar unsaturated molecules in which intramolecular charge transfer transitions occur and in which the dipole moments of the excited states are large appear to exhibit unusually large hyperpolarizabilities. Theoretical and experimental interest has centered on the nature of highly charge-correlated π-electron states in such organic molecules, especially polymeric crystalline structures that comprise the microscopic origin of exceptional second-order nonlinear optical responses, particularly as exhibited in SGH and linear electro-optic effect properties. The lower $\pi \to \pi^*$ excited states in these molecules make the important if not the major contribution to SHG.

To this end an understanding of the relation between molecular structure and polarizabilities may lead to the discovery of materials with improved nonlinear optical characteristics. Theoretical computations of molecular polarizabilities have the potential to help greatly in this field. Several theoretical approaches to calculating molecular nonlinear optical properties have appeared in the literature. They differ first in the manner in which the nonlinear response is calculated in the framework of perturbation theory and details of these different approaches have been given in the preceding section. Another main difference is the form of the molecular wave function used to calculate the nonlinear optical properties. The principal goal in many approaches was to develop and carefully check computational procedures for reasonable accuracy and economic computations in terms of computer time and storage. Owing to their computationally inexpensive characteristics and their ability to handle large molecular systems, the semiempirical model-Hamiltonian methods have been extensively used to calculate molecular polarizabilities and hyperpolarizabilities. It is the aim of the present section to review these semiempirical calculations.

Perturbation theory coupled with various semiempirical Hamiltonians have been used to calculate polarizabilities and hyperpolarizabilities. The simple Hückel method[221] which had wide applications[222, 223] has been used to calculate polarizabilities and hyperpolarizabilities in the pioneering work of Hameka's group and others.[224-231] Owing to its crude approximations its wave functions proved to be unsuitable for practical purposes.

The PPP method,[232-234] π-electron Hamiltonian has been used in a variational perturbation theory approach (VPT-PPP) to calculate polarizabilities and hyperpolarizabilities of various π-electron systems.[229-231, 235-242] This SCF-LCAO model employs a π-type atomic orbital basis set and all monoexcited configurations were included in the CI calculations. The advantage of this model lies in the greater number of π-electron configurations that can be included in the sum over states calculations. Therefore, more elaborate chromophores and chromophore assemblies with different conformations and polarization frequencies have been handled.

The contribution of the π electrons to the SGH tensor in a series of selected molecules has been calculated.[239] At the near IR–visible wavelengths, the π electrons make the dominant contribution to the hyperpolarizability for SGH. Correct orders of magnitude and consistent trends have been obtained[239] within classes of analogous molecules. Reasonable values for the β tensor has been obtained[237] for aniline, nitrobenzene, and p-nitroaniline using all monoexcited configurations within the PPP method. Moreover, PPP-derived vector components for frequency doubling $\beta_{vec}(-2\omega, \omega, \omega)$ were found to be in excellent agreement with experiment over a wide frequency range. Molecular architecture effects on β have been also used[237] to investigate the utility of the PPP model Hamiltonian in designing new elaborate nonlinear chromophores.

Several hypothetical molecules of sequentially varied substituents have been examined. Chromophore–laser compatibility and the hyperpolarizability density, which is simply the actual number of chromophores that can be packed in a given volume of material, have also been considered. It has been concluded that [237] at shorter wavelengths where the excitation of σ electrons comes into play, the application of computational procedures that take into account these σ electrons undoubtedly become necessary. Molecular interactions also play an important role in evaluating the SHG hyperpolarizabilities of molecular crystals. The first attempt to use a Hamiltonian in which σ electrons are treated even partially was made by Hameka's group,[236] where the extended Hückel model Hamiltonian[243] was used in combination with the PPP model. The σ and $\sigma - \pi$ electron contributions were given by the extended Hückel model, while the π contribution was given by the PPP method.

The more elaborate all-valence electron semiempirical MO methods have been used to calculate molecular electrical properties. The CNDO/2 model Hamiltonian[244-246] has been coupled with perturbation theory in different ways. A VPT-CNDO/2 approach has been used to calculate the polarizabilities and hyperpolarizabilities of various systems.[247-250] The polarizability tensor components given by Eq. (2) and the average polarizability $\bar{\alpha}$ given by Eq. (7) were calculated. Hush and Williams[251-253] used finite field perturbation theory and the CNDO method (FPT-CNDO) to calculate the polarizabilities of some simple molecules such as CO, CO_2, CH_4, and C_2H_2 (Refs. 251–252) and also the hyperpolarizabilities of CH_3F, CH_2F_2, and CHF_3 molecules.[253] The calculated values were found to be sensitive to the choice of basis sets where an extended basis (EB) even on the H only has been found[254] to produce good values of the polarizabilities for saturated hydrocarbons, using FPT-EB-CNDO calculations.

A rigorous approach to calculating α, β, and γ using the CNDO/2 Hamiltonian has been developed[255-258] in which the coupled Hartree-Fock perturbation theory has been used together with an extended basis CNDO method (CHF-PT-EB-CNDO). This method has been applied for a host of molecules.[259-272] This approach was based on the fact that within McWeeny et al.[193-195] CHF theory, we are after the polarization of the

zeroth-order charge distribution and not the total energy. Ab initio varia-
tionally obtained wave functions are not only unmanageable from the
computational point of view, but perhaps even unnecessary. Induced
moment calculations are very sensitive to the choice of basis sets, especially for the out-of-plane components.[273, 274] The use of the off-diagonal
one-center dipole moment integrals violates the ZDO approximation;
nevertheless, two-center dipole moment integrals have been considered.[257]
Basis sets including up to f orbitals for carbon and up to d orbitals for
hydrogen were used. Optimization and standardization of these basis sets
for a series of structurally similar compounds have been also done in order
to account for the correlation effects of importance for γ calculations.

It has generally been found that small basis sets optimized with respect
to experimental values are as successful as larger ones in describing
polarization phenomena.[256, 257] These CHF-PT-EB-CNDO semiempirical
calculations bypass the computational bottleneck for large molecules and
have given good results for calculated polarizabilities for a wide variety of
molecules.[256-258] The calculated values for γ were found to be more
sensitive to geometry changes than α values.[258] Moreover, for aromatic
molecules the optimum basis set for calculating α was found to be
different from that used to calculate γ. The average polarizabilities α and
hyperpolarizabilities γ of various polyacetylenes have been reported[265]
using the CHF-PT-EB-CNDO method. The high rate of change of α and
more particularly γ with the chain length has been explained. The effects
due to the intermolecular interactions, delocalization, and charge transfer
processes as key factors in the rationalization of the nonlinear polarization
phenomena have been also established.[259, 263] Furthermore, the zeroth-
order reciprocal HOMO-LUMO energy gap has been found to be an
indicator of the γ values.

In conclusion, the CHF-PT-EB-CNDO method has given reason-
ably good agreement between calculated α, β, and γ values and those
measured experimentally. Various systems have been investigated, for
example, alkanes,[256] polyenes,[257] aromatics,[258, 260] amines,[267] and polyacetylenes.[265]. The correlation effects, which are essential for any meaningful determination[274] of γ, are incorporated to a considerable extent via
the calibration of the wave functions,[275] as the reasonably accurate property values for many molecules have shown.[256-258] The role of the basis set
in the computation of α and γ cannot be overemphasized[265] and therefore
numerous tests on a large number of molecules have been performed.[256-260]
It has been found that basis sets that include polarization and diffuse
functions permit the correct description of the polarized charge
structure.[260, 265]

The CNDO/S-CI method Hamiltonian[276-278] has been used to calculate molecular polarizabilities. A PT-EB-CNDO-CI approach[279, 280] has been used and PT-CNDO/S-CI calculations[281, 282] have been performed for a wide variety of molecules. Many other authors[283-293] have used an FPT-CNDO/S-CI approach. The average polarizabilities $\bar{\alpha}$ for most of the systems studied were found to agree quite well with experiment, while the out-of-plane polarization α_{zz} was found to be too small for all cases studied.[281] This has been ascribed to the insufficiency of the CNDO or the CNDO/S-CI wave functions, which allow for little migration of charge normal to the molecular plane in aromatic molecules. Increasing the number of singly excited configurations to include all possible excitations resulted in a marginal change in the out-of-plane polarizability and a small increase in the in-plane components, but no quantitative relationship between the number ofconfigurations in the CI expansion and the value of α and its components has been found.[281]

The CNDO/S method has been reparameterized[284] by correlating computed and measured dipole moments and transition wavelengths for a wide range of organic conjugated molecules. Good correlations between experimental and theoretical spectra and ground-state dipoles were obtained.[284] This same model has been used to calculate the hyperpolarizability tensor β_{ijk} in a form related to SHG. Reasonable correlation between theory and experiment has been obtained.[284] The convergence of the excited state perturbation expansion used in the calculation of β_{ijk} tensor has also been extensively investigated and the results compared with previous calculations[285-288] on similar molecules.

The INDO-method Hamiltonian[294] has been used in an FPT-INDO approach to calculate molecular polarizabilities.[295-298] The various MINDO Hamiltonians—MINDO/1,[299, 300] MINDO/2,[301, 302] and MINDO/3[303]—have been used to calculate the polarizabilities and first hyperpolarizabilities of various molecules[304-307] in an FPT-MINDO treatment. The MNDO-method Hamiltonian developed by Dewar et al.[308] has been also used in an FPT-MNDO framework to calculate α and β of a large series of molecules.[307, 309-311] Consistently lower polarizabilities than experimental values were obtained[297] by the FPT-INDO method; however, polarizabilities and polarizability anisotropy have been predicted[295-298] better than with the CNDO methods. A reasonable agreement between FPT-MINDO/3-calculated β and experimental values has been obtained for molecules such as CH_3F, CH_2F_2, and CHF_3 (Ref. 306).

The finite-field perturbation theory within the framework of the AM1-method[312] and the PM3-method[313] Hamiltonians has been used to calculate polarizabilities and hyperpolarizabilities of mono-, di-, and trisubstituted benzenes[311a] and for the polyenes.[311b] In reporting the α values from the semi-empirical work, atomic correction factors developed by Dewar and Stewart[310] were included.[311b] These correction factors were developed only for the MNDO methods, but the same factors were used in the MNDO, AM1, and PM3 methods.

The AM1 and PM3 results were expected to be improved if atomic correction factors were developed for each method separately.[311b] The AM1 γ values were consistently higher than MNDO or PM3 methods. This was ascribed to differences in the geometries of the polyenes predicted by these different methods. The AM1 method does seem to give a slightly larger value when compared at the same geometry, but the differences were small. For long polyenes the long axis component of γ was found to dominate,[311b] and any errors in the other smaller components were not significant.

Hush and Williams[252] suggested that the CNDO-underestimated α values could be corrected by adding on a contribution $\delta\alpha$ for each atom in the molecule. This idea was also extended by Dewar et al.[307] who used anisotropic $\delta\alpha$'s with different values for atoms in different chemical environments. A similar scheme based on the INDO polarizabilities of benzene and biphenyl molecules has been performed.[297] Molecular polarizabilities calculated by the INDO, CNDO, and MNDO methods are rather different. Although CNDO/S, MINDO/3, or MNDO methods may give better values of the individual components of α and $\bar{\alpha}$ than the INDO method, the latter has been found to give better polarizability anisotropy.[297]

It is clear from the above literature survey that there exist a host of semiempirically calculated data of α, β, and γ for wide variety of molecules. A review of some of these data is given in Tables I, II, and III. The smallness of the basis set used and the inconsistency of the different approximations inherent in these ZDO Hamiltonians may be the reason for such a scatter of calculated values around the experimentally determined values, even though these experimental values are quite tenuous. Although the agreement between calculated and experimental values is reasonably good in some cases, one has to be careful when handling these calculated values. In semiempirical quantum treatments of electronic spectra (see, e.g., Refs. 276–278) the parametrizations of the method have been performed to fit a small number of low-lying excited states. On the other hand, in the calculations of polarizabilities, it is certainly not permissible to restrict the perturbation sums, to run over only those excited states that have been used to fix the parameters. The only valid procedure is to add terms until the series converges to whatever accuracy is required.

TABLE I
Calculated Polarizabilities (10^{-24} cm) Using Semiempirical Model
Hamiltonian Methods for Some Systems

Molecule	α_{xx}	α_{yy}	α_{zz}	$\bar{\alpha}$	Reference	Method
CO	1.14	3.70	1.14	1.99	247	PT-CNDO/2
	0.71	1.07	0.71	0.83	251, 252	FPT-CNDO/2
	1.31	2.44	1.31	1.68	281	PT-CNDO/S-CI
	0.95	2.85	0.95	1.58	304	FPT-MINDO/1
	0.90	2.26	0.90	1.35	304	FPT-MINDO/2
				1.95	315	Experimental
CO_2	1.34	6.45	1.34	3.04	247	PT-CNDO/2
	0.32	1.99	0.32	0.88	251, 252	FPT-CNDO/2
	0.51	6.38	0.51	2.46	281	PT-CNDO/S-CI
	0.73	4.44	0.73	1.97	304	FPT-MINDO/1
	0.50	3.93	0.50	1.64	304	FPT-MINDO/2
				2.63	316	Experimental
CH_4	3.10	3.10	3.10	3.10	247	PT-CNDO/2
	0.65	0.65	0.65	0.65	251, 252	FPT-CNDO/2
	1.13	1.13	1.13	1.13	281	PT-CNDO/S-CI
	1.64	1.64	1.64	1.64	304	FPT-MINDO/1
	1.05	1.05	1.05	1.05	304	FPT-MINDO/2
				2.56[a]	256	CHF-PT-EB-CNDO
				2.50[a]	256	CHF-PT-EB-CNDO
				2.70[a]	256	CHF-PT-EB-CNDO
				3.07	279	PT-EB-CNDO-CI
				2.30	254	FPT-EB-CNDO
				2.62	315	Experimental
CH_3F				2.65	265	CHF-PT-EB-CNDO
				2.82	120	Experimental
CH_2F_2				2.58	265	CHF-PT-EB-CNDO
				3.22	120	Experimental
CHF_3				2.47	265	CHF-PT-EB-CNDO
				3.51	120	Experimental
CH_3Cl				4.27	272	CHF-PT-EB-CNDO
				4.52	317	Experimental
CH_2Cl_2				5.87	272	CHF-PT-EB-CNDO
				6.64	317	Experimental
C_6H_5Cl				12.22	272	CHF-PT-EB-CNDO
				11.84	318, 60	Experimental
C_2H_6	5.68	5.84	5.84	5.79	247	PT-CNDO/2
	1.32	1.21	1.21	1.24	247	PT-CNDO/2
	1.97	2.34	2.34	2.21	281	PT-CNDO/S-CI
				5.32	280	PT-EB-CNDO-CI
				3.77	255	CHF-PT-EB-CNDO
				4.75	254	FPT-EB-CNDO
				4.47[b]	256	CHF-PT-EB-CNDO
				4.93[b]	256	CHF-PT-EB-CNDO
				4.83[b]	256	CHF-PT-EB-CNDO
				4.95[b]	256	CHF-PT-EB-CNDO
				4.48	315	Experimental

TABLE I   *(Continued)*

H_2CO	2.29	4.28	1.71	2.76	247	PT-CNDO/2
	0.66	1.48	0.66	0.93	247	PT-CNDO/2
	1.19	3.47	0.66	1.77	281	PT-CNDO/S-CI
				2.45	315	Experimental
HCOOH	2.11	3.79	0.75	2.22	281	PT-CNDO/S-CI
CH_3CHO	4.71	6.76	3.94	5.14	247	PT-CNDO/2
	1.30	2.14	0.99	1.48	247	PT-CNDO/2
	3.12	4.30	1.82	3.09	281	PT-CNDO/S-CI
				4.59	315	Experimental
$(CH_3)_2CO$	7.08	9.95	6.34	7.79	247	PT-CNDO/2
	1.96	3.17	1.56	0.93	247	PT-CNDO/2
	3.11	5.07	2.50	3.56	281	PT-CNDO/S-CI
				6.39	315	Experimental
$(CH_3)_2O$	5.45	6.13	5.28	5.62	247	PT-CDO/2
	1.28	1.96	1.18	1.47	247	PT-CNDO/2
	2.50	2.33	2.18	2.33	281	PT-CNDO/S-CI
				5.24	315	Experimental
C_2H_2	0.33	2.12	0.33	0.71	251, 252	FPT-CNDO/2
	0.70	6.44	0.70	2.62	281	PT-CNDO/S-CI
	0.73	5.09	0.73	2.18	304	FPT-MINDO/1
	0.54	3.86	0.54	1.65	304	FPT-MINDO/2
				3.48	265[c]	CHF-PT-EB-CNDO
				3.49	316	Experimental
C_2H_4	0.00	0.00	3.23	1.08	240	VPT-PPP
	1.09	2.46	0.49	1.35	304	FPT-CNDO/2
	2.18	7.29	1.05	3.51	281	PT-CNDO/S-CI
	3.72	5.45	1.02	3.40	304	FPT-MINDO/1
	2.00	4.29	0.79	2.36	304	FPT-MINDO/2
				4.40	257	CHF-PT-EB-CNDO
				4.22	316	Experimental
tert-Butadiene				3.03	229	VPT-HMO
				2.51	229	VPT-PPP
	12.63	2.06	0.0	4.89	240	VPT-PPP
				8.16	311b	FPT-MNDO
				8.16	311b	FPT-AM1
				7.66	311b	FPT-PM3
				10.33	257	CHF-PT-EB-CNDO
Benzene				2.50	229	VPT-HMO
				2.33	229	VPT-PPP
	6.84	6.84	0.0	4.56	240	VPT-PPP
	10.70	10.70	1.68	7.69	293	FPT-CNDO/S-CI
	13.19	13.50	0.90	9.20	281	PT-CNDO/S-CI
	5.34	5.34	1.94	4.21	297	FPT-INDO
	10.07	10.07	1.52	7.22	307	FPT-MINDO/3
	10.23	10.23	1.21	7.22	307	FPT-MNDO
				9.17	258	CHF-PT-EB-CNDO
				9.87	316	Experimental
				9.90	319	Experimental
				10.14	320	Crystal

TABLE I (Continued)

Naphthalene				5.86	229	VPT-HMO
				6.92	229	VPT-PPP
	18.01	12.08	0.0	10.03	240	VPT-PPP
	23.63	14.37	0.0	12.00	231	VPT-PPP
	20.60	12.60	0.44	11.21	282	PT-CNDO/S-CI
	16.30	22.60	2.90	13.90	292	FPT-CNDO/S-CI
	20.14	33.07	0.68	17.96	281	PT-CNDO/S-CI
	22.70	16.20	2.88	13.93	293	FPT-CNDO/S-CI
	12.05	9.05	3.36	8.15	297	FPT-INDO
				17.11[d]	258	CHF-PT-EB-CNDO
				17.01[e]	258	CHF-PT-EB-CNDO
				16.60	321	Experimental
				16.50	322	Experimental
				19.10	320	Crystal
Anthracene				10.60	229	VPT-HMO
				11.44	229	VPT-PPP
	40.30	26.70	4.10	23.70	293	FPT-CNDO/S-CI
	65.54	34.64	0.12	33.43	281	PT-CNDO/S-CI
	20.91	13.84	4.81	13.19	297	FPT-INDO
				27.22	258	CHF-PT-EB-CNDO
				25.40	232	Experimental
				26.90	320	Crystal
Phenanthrene				9.40	229	VPT-HMO
				10.50	229	VPT-PPP
	29.42	17.43	0.0	15.62	240	VPT-PPP
	30.00	18.00	0.0	16.00	241, 242	VPT-PPP
	73.61	36.64	0.27	36.84	281	PT-CNDO/S-CI
	18.75	12.73	4.75	12.08	297	FPT-INDO
				25.67	258	CHF-PT-EB-CNDO
				23.60	319	Experimental
				27.03	320	Crystal
Biphenyl	13.75	10.22	3.93	9.30	297	FPT-INDO
				21.50[f]	258	CHF-PT-EB-CNDO
				21.70[g]	258	CHF-PT-EB-CNDO
				19.60	324	Experimental
				22.56	320	Crystal
p-terphenyl	24.13	15.13	5.91	15.06	297	FPT-INDO
				35.56	320	Crystal
Chrysene				13.36	229	VPT-HMO
				14.36	229	VPT-PPP
				14.30	325	Experimental
Pyrene				14.61	229	VPT-HMO
				14.40	229	VPT-PPP
				12.94	325	Experimental
1,2-Benz-anthracene				14.27	229	VPT-HMO
				14.84	229	VPT-PPP
				14.10	325	Experimental

TABLE I *(Continued)*

Allene				7.14	257	CHF-PT-EB-CNDO
Cyclopropene				5.31	257	CHF-PT-EB-CNDO
Cyclobutadiene				7.51	257	CHF-PT-EB-CNDO
NH_3				2.25	266	CHF-PT-EB-CNDO
				2.22	316	Experimental
CH_3NH_2				3.79	266	CHF-PT-EB-CNDO
$(CH_3)_2NH$				6.37	266	CHF-PT-EB-CNDO
$(CH_3)_3N$				9.59	266	CHF-PT-EB-CNDO
Aniline				11.51[h]	266	CHF-PT-EB-CNDO
				12.05	311a	FPT-MNDO
				11.53	327	Experimental
N,N-Dimethyl-aniline				20.00[i]	266	CHF-PT-EB-CNDO
N,N-Dimethyl PNA[j]	19.21	8.57	0.0	9.29	240	VPT-PPP
Azulene	25.14	12.98	0.0	12.7	240	VPT-PPP
	48.10	25.00	1.40	24.8	281	PT-CNDO/S-CI
Phenol	13.42	15.20	1.21	9.94	281	PT-CNDO/S-CI
Toluene	13.27	14.45	1.08	9.60	281	PT-CNDO/S-CI
				12.30	311a	FPT-MNDO
Pyridine	12.16	12.82	1.02	8.67	281	PT-CNDO/S-CI
				9.72	267	CHF-PT-EB-CNDO
				9.50	328	Experimental
Pyrimidine	11.86	12.16	1.58	8.53	281	PT-CNDO/S-CI
				9.13	267	CHF-PT-EB-CNDO
Pyridazine	13.32	12.73	1.76	9.27	281	PT-CNDO/S-CI
				8.98	267	CHF-PT-EB-CNDO
				8.79	328	Experimental
Nitrobenzene				10.62	311a	FPT-MNDO
p-Nitroaniline				15.34	311a	FPT-MNDO
m-Nitroaniline				15.26	311a	FPT-MNDO
o-Nitroaniline				15.09	311a	FPT-MNDO
2-Methyl-4-nitroaniline				17.38	311a	FPT-MNDO
3-Methyl-4-nitroaniline				17.35	311a	FPT-MNDO
Pyrazine	12.06	13.34	1.14	8.85	281	PT-CNDO/S-CI
				8.96	267	CHF-PT-EB-CNDO
				8.98	328	Experimental
Hexatriene[k]				13.45	311b	FPT-MNDO
				13.46	311b	FPT-AM1
				12.61	311b	FPT-PM3
				18.99	257	CHF-PT-EB-CNDO
Octatetraene[k]				19.54	311b	FPT-MNDO
				19.65	311b	FPT-AM1
				18.35	311b	FPT-PM3

TABLE I   *(Continued)*

Decapentaene[k]	26.24	311b	FPT-MNDO
	26.53	311b	FPT-AM1
	24.69	311b	FPT-PM3
Dodecahexaene[k]	33.39	311b	FPT-MNDO
	33.95	311b	FPT-AM1
	31.49	311b	FPT-PM3

Note. 1 a.u. $= 0.148176 \times 10^{-24}$ esu
$\approx 0.164867 \times 10^{-40}$ $C^2m^2J^{-1}$.

[a] Different basis sets.
[b] Different molecular conformations; see Ref. 256.
[c] Calculated values for the series $H_2(C_2)_n$ with $n = 1$–8, acetylene and diacetylene derivatives, and also some amino-, fluora-, and methyl-substituted molecules have been reported.[265]
[d] Experimental bond angles and lengths are those of benzene from *Tables of Interatomic Distances and Configurations in Molecules and Ions*, The Chemical Society, London, 1965.
[e] Experimental geometry from D. W. J. Cruickshank, *Acta Crystallogr.* **10**, 504 (1957).
[f] Planar geometry.
[g] Twisted 45°.
[h] The angle between the plane of the NH_2 group and the ring plane is 39.35°, the minimum energy configuration.[326]
[i] The carbons of the two methyl groups lie in the ring plane.[266]
[j] *para*-Nitroaniline.
[k] The *trans* configuration. Calculated values for polyenes $[H(C_2H_2)_nH, n = 2, 20]$ have been also reported.[311b]

TABLE II
Calculated First Hyperpolarizability (10^{-32} esu) for Some Systems Using
Semiempirical Model Hamiltonian Methods

Molecule	β	Reference	Method
CH_3F	−19.6	306[a]	FPT-MINDO/3
	−2.6	253	FPT-CNDO/2
	−24.4	329	Experimental
CH_2F_2	−22.5	306[a]	FPT-MINDO/3
	−3.4	253	FPT-CNDO/2
	−18.0	329	Experimental
CHF_3	−7.5	306[a]	FPT-MINDO/3
	+3.0	253	FPT-CNDO/2
	−10.8	329	Experimental
NH_3	−25.9	266	CHF-PT-EB-CNDO
	−41.8	330	Experimental
CH_3NH_2	−39.5	266	CHF-PT-EB-CNDO
$(CH_3)_2NH$	−116.6	266	CHF-PT-EB-CNDO
$(CH_3)_3N$	−760.3	266	CHF-PT-EB-CNDO

TABLE II   (*Continued*)

Molecule	β	Reference	Method
Aniline	107	266[b]	CHF-PT-EB-CNDO
	142	237[c]	VPT-PPP
	249	239[d]	VPT-PPP
	222	239[e]	VPT-PPP
	159	288	FPT-CNDO/S-CI
	244.6	284[f]	FPT-CNDO/S-CI
	366.2	284[g]	FPT-CNDO/S-CI
	161.0	311a	FPT-MNDO
	79	331[e]	Experimental
	148	332[d]	Experimental
Nitrobenzene	-147	237	VPT-PPP
	235	239[d]	VPT-PPP
	216	239[e]	VPT-PPP
	-230	288	FPT-CNDO/S-CI
	25	284[f]	FPT-CNDO/S-CI
	62	284[g]	FPT-CNDO/S-CI
	52	311a	FPT-MNDO
	197	33[e]	Experimental
	227	332[d]	Experimental
PNA[h]	155	237	VPT-PPP
	1 230	239[d]	VPT-PPP
	1 050	239[e]	VPT-PPP
	570	288[i]	FPT-CNDO/S-CI
	702	284[f]	FPT-CNDO/S-CI
	1 218	284[g]	FPT-CNDO/S-CI
	142	311a	FPT-MNDO
	3 450	333[d]	Experimental
	2 110	331[e]	Experimental
m-Nitroaniline	644	239[d]	VPT-PPP
	568	239[e]	VPT-PPP
	352	284[f]	FPT-CNDO/S-CI
	558	284[g]	FPT-CNDO/S-CI
	184	311a	FPT-MNDO
	600	333[d]	Experimental
	420	331[e]	Experimental
O-Nitroaniline	223	239[d]	VPT-PPP
	161	239[e]	VPT-PPP
	164	311a	VPT-MNDO
	1 020	333[d]	Experimental
	640	331[e]	Experimental
N,N-Dimethylaniline	-293.8	266[j]	CHF-PT-EB-CNDO
4-Cyanoaniline	640	284[f]	FPT-CNDO/S-CI
	721	284[g]	FPT-CNDO/S-CI
	1 334	334	Experimental
Fluorobenzene	66	284[f]	FPT-CNDO/S-CI
	84	284[g]	FPT-CNDO/S-CI
	106	295, 296	Experimental

TABLE II   *(Continued)*

Toluene	21	311a	FPT-MNDO
2-Methyl-4-nitroaniline	114	311a	FPT-MNDO
3-Methyl-4-nitroaniline	115	311a	FPT-MINDO
Me_2N — (ring, O) — NO_2	$1\,290^k$	238a	FPT-CNDO/S-CI
Me_2N — (ring, N, H) — NO_2	$1\,350^k$	283a	FPT-CNDO/S-CI
Me_2N — (ring, S) — NO_2	$1\,350^k$	283a	FPT-CNDO/S-CI

Note. 1 a.u. $= 0.863993 \times 10^{-32}$ esu
$\approx 0.320662 \times 10^{-52} \, C^3m^3J^{-2}$.

[a] β in Ref. 306 is defined as $\frac{3}{5}(\beta_{zxx} + \beta_{zyy} + \beta zzz)$ with the z axis coinciding with the dipole moment vector.

[b] The angle between the plane of the NH_2 group and the ring plane is 39.35°, the minimum energy configuration.[326]

[c] Calculated properties of various transoid linear polyenes, quinodimethane, stilbene, and diphenylbenzo-bisthiazole chromophores have been also reported,[237] using the VPT-PPP method.

[d] At frequencies of 1.17 eV and the scheme of parametrizing the PPP-method has been given.[239] Calculated and experimental values are at the same frequency.

[e] The same as in (d) but at frequencies of 0.94 eV.

[f] At zero frequency; calculated values for a wide variety of molecules have been also reported.[284]

[g] At a frequency corresponding to that of the experimental measurement.

[h] *p*-Nitroaniline.

[i] The calculated value corresponds to a frequency of $\hbar\omega = 0.65$ eV.

[j] The carbons of the two methyl groups lie in the ring plane.[266]

[k] Calculated values for polythiophenes, polyfuranes, polypyrroles, substituted phenylsilanes, and some selected dyes and pigments have been also reported.[283b]

TABLE III
Calculated Second Hyperpolarizability (10^{-39} esu) for Some Systems Using
Semiempirical Model Hamiltonian Methods

Molecule	γ	Reference	Method
CH_4	$1\,230^a$	256	CHF-PT-EB-CNDO
	$1\,232^a$	256	CHF-PT-EB-CNDO
	$1\,040^a$	256	CHF-PT-EB-CNDO
	$1\,453$	120	Experimental
CH_3F	$1\,093$	265	CHF-PT-EB-CNDO
	$1\,436$	329	Experimental
CH_2F_2	932	265	CHF-PT-EB-CNDO
	922	329	Experimental

TABLE III   *(Continued)*

Molecule	γ	Reference	Method
CHF_3	871	265	CHF-PT-EB-CNDO
	816	329	Experimental
CH_3Cl	3 133	272	CHF-PT-EB-CNDO
	3 576	317	Experimental
CH_2Cl_2	5 893	272	CHF-PT-EB-CNDO
	5 692	317	Experimental
C_6H_5Cl	18 386	272	CHF-PT-EB-CNDO
	18 386	332	Experimental
C_2H_6	2 546[b]	256	CHF-PT-EB-CNDO
	3 026[b]	256	CHF-PT-EB-CNDO
	2 936[b]	256	CHF-PT-EB-CNDO
	3 046[b]	256	CHF-PT-CNDO
	1 939	119	Experimental
C_2H_4	4 162	257	CHF-PT-EB-CNDO
	− 336	236	VPT-PPP-EHT
	4 548	335	Experimental
C_2H_2	10 074	265[c]	CHF-PT-EB-CNDO
Allene	7 707	257	CHF-PT-EB-CNDO
Cyclopropene	5 625	257	CHF-PT-EB-CNDO
tert-Butadiene	− 1 890	226	VPT-HMO
	8 070	236	VPT-PPP-EHT
	16 848	257	CHF-PT-EB-CNDO
	2 143	311b	FPT-MNDO
	1 872	311b	FPT-AM1
	1 954	311b	FPT-PM3
	13 800	335	Experimental
Cyclobutadiene	20 727	257	CHF-PT-EB-CNDO
Benzene	466	226	VPT-HMO
	6 800	236	VPT-PPP-EHT
	55	295	FPT-INDO
	36 901	248	VPT-CNDO/2
	12 433	258	CHF-PT-EB-CNDO
	12 360	335	Experimental
Naphthalene	56 252	230	VPT-HMO
	45 482	230	VPT-PPP
	30 381[d]	258	CHF-PT-EB-CNDO
	29 419	258	CHF-PT-EB-CNDO
	31 201	336	Experimental
Anthracene	28 141	230	VPT-HMO
	151 685	230	VPT-PPP
	56 026	258	CHF-PT-EB-CNDO
Phenanthrene	24 721	230	VPT-HMO
	111 424	230	VPT-PPP
	51 087	258	CHF-PT-EB-CNDO
Pyrene	67 700	258	CHF-PT-EB-CNDO
Biphenyl	14 101	230	VPT-HMO
	59 462	230	VPT-PPP
	36 688	258[e]	CHF-PT-EB-CNDO
	38 076	258[f]	CHF-PT-EB-CNDO

TABLE III *(Continued)*

NH_3	3 052	266	CHF-PT-EB-CNDO
	3 068	330	Experimental
CH_3NH_2	2 997	266	CHF-PT-EB-CNDO
$(CH_3)_2NH$	5 289	266	CHF-PT-EB-CNDO
$(CH_3)_3N$	9 420	266	CHF-PT-EB-CNDO
Aniline	19 242[g]	266	CHF-PT-EB-CNDO
	16 421	327	Experimental
N,N-Dimethylaniline	22 617[h]	266	CHF-PT-EB-CNDO
Pyridine	420	228	VPT-HMO
	9 923	267	CHF-PT-EB-CNDO
Pyridazine	420	228	VPT-HMO
	9 772	267	CHF-PT-EB-CNDO
Pyrimidine	8 412	267	CHF-PT-EB-CNDO
Pyrazine	420	228	VPT-HMO
	7 959	267	CHF-PT-EB-CNDO
S-Triazine	6 951	267	CHF-PT-EB-CNDO
S-Tetrazine	9 420	267	CHF-PT-EB-CNDO
P-Benzoquinone	14 809	270	CHF-PT-EB-CNDO
Anthraquinone	30 576	270	CHF-PT-EB-CNDO
trans-Hexatriene	−6 057	226	VPT-HMO
	59 522	236	VPT-PPP-EHT
	36 337	257	CHF-PT-EB-CNDO
	15 176	311b	FPT-MNDO
	13 354	311b	FPT-AM1
	12 785	311b	FPT-PM3
Octatetraene[i]	52 071	311b	FPT-MNDO
	48 007	311b	FPT-AM1
	44 127	311b	FPT-PM3
Decapentaene[i]	125 461	311b	FPT-MNDO
	121 402	311b	FPT-AM1
	108 283	311b	FPT-PM3
Decahexaene[i]	243 330	311b	FPT-MNDO
	246 338	311 b	FPT-AM1
	214 243	311b	FPT-PM3

Note. 1 a.u. $= 0.503\,717 \times 10^{-39}$ esu
$\approx 0.623\,597 \times 10^{-64}\ C^4 m^4 J^{-3}$.

[a] Different basis sets; see Ref. 256.

[b] Different molecular conformations; see Ref. 256.

[c] Calculated values for the series $H_2(C_2)_n$, with $n = 1$–8, acetylene and diacetylene derivatives, and also some amino-, fluoro-, and methyl-substituted molecules have been reported.[265]

[d] Calculated values for various aromatic molecules using an empirical formula relating γ to the effective molecular length have been also reported.[258]

[e] Planar geometry.

[f] Twisted 45°.

[g] The angle between the plane of the NH_2 groups and the ring plane is 39.35°, the minimum energy configuration.[326]

[h] The carbons of the two methyl groups lie in the ring plane.[266]

[i] The *trans* configuration. Calculated values for polyenes [$H(C_2H_2)_2H$, $n = 2, 20$] have been also reported.[311b]

These defects led some authors (see, e.g., Ref. 284) to reparameterize some semiempirical methods by correlating computed and measured dipole moments together with transition wavelengths. Others[256-258, 279, 280] have used extended basis sets to allow for an extra flexibility, which is useful in polarizability calculations since it permits the correct description of the polarized charge structure. As an example, the effect of double-zeta sets and their approximation by single zetas have been considered.[257]

Although the ultimate test has been considered the reliability of the results produced rather than a theoretical justification of the model,[314] it is well known that the validity of semiempirical methods has often been questioned perhaps more severely than their ab initio counterparts. The use of a semiempirical model Hamiltonian and of a simple SCF-level ground state clearly limits the accuracy of the calculated values.[237] More extensive studies have been performed of small molecules involving ab initio methods with extensive basis sets and correlations introduced into the ground-state wave functions by using MC-SCF or CI or coupled-cluster procedures and also into the excited states in various ways. To this end, the use of the more accurate ab initio methods to calculate atomic as well as molecular electrical properties is reviewed in the next section.

VII. AB INITIO COMPUTATIONS OF ATOMIC AND MOLECULAR POLARIZABILITIES AND HYPERPOLARIZABILITIES

From the time of very early simple pioneering calculations of atomic polarizabilities[337-342] up to the present when high-accuracy ab initio calculations of static and dynamic electrical properties have become feasible for simple and even large molecules,[8, 54, 216, 343-367] there has been a sharp increase in the understanding of linear and nonlinear optical properties of atoms and molecules. A microscopic understanding of the origin of hyperpolarizabilities and its relationship with the molecular electronic and geometrical structure obtainable now from large one-particle basis sets ab initio calculations with accurate inclusion of electron correlation is of vital importance in understanding macroscopic optical nonlinearity of molecular materials. This subject is a key for molecular engineering in the new technology of photonics and of particular importance for the field of nonlinear optics, especially in areas such as optical disk storage and optical switches.[353, 363, 366]

A. Static Properties

Two main approaches are used in the ab initio computations of atomic and molecular polarizabilities and hyperpolarizabilities. The first is the finite difference method mentioned briefly at the end of Section V.C.

Calculations have been done whereby the electric field terms for a fixed field strength are incorporated directly into the Hamiltonian and the Schrödinger equation is then solved. An alternative approach, called the fixed-charge method,[368] uses one or more fixed charges placed in the vicinity of the molecule. The perturbed total energies or other properties of the system can be written as an expansion in terms of moment and polarizability components. If different values of field strength or charge positions are used, a system of simultaneous equations can be written from the truncated series. This system of equations must be chosen sufficiently large to ensure that the truncation error is minimized, and unknown polarizabilities are obtained by solving these equations. Ab initio calculations using the finite-field method of Cohen and Roothaan[202] have been performed for a wide variety of atoms and molecules.[207-215, 346-349, 352, 355]

If only uniform fields are used in the finite-field methodology, only dipole polarizabilities and hyperpolarizabilities can be calculated. Field gradients and further higher-order field gradients must be applied in order to obtain higher multipole contributions.[54] The fixed-charge method amounts to adding an extra nucleus with a negative or a positive charge in regions where the molecular wave functions are negligible. The basis set used has to be adequate to describe any polarization of the molecule in the presence of the field. The fields applied should be small enough to let a truncated expansion accurately represent the moment or energy, but also strong enough that calculated effects will be numerically significant.[54, 348] One of the advantages of both the finite-field or the fixed-charge methods is that either one can be easily carried out at all electronic structure levels, SCF, CI, or MBPT. The fixed-charge method does not seem to be used as widely as the finite-field method and few calculations can be found in the literature.[348, 369-371]

Another approach is the derivative Hartree-Fock (DHF) method.[54, 372-378] In this method, many important molecular properties are directly defined as a derivative of a suitable order of an energy. Formal differentiation of the Schrödinger equation has been easily accomplished and, as in perturbation theory, there is a $2n + 1$ rule.[372, 374, 378] Using the nth order wave function, energy derivatives up to $2n + 1$ may be evaluated directly and the derivative wave functions between the n and $2n + 1$ orders are not required explicitly. This $2n + 1$ rule can be used without any complication if the basis set used is not dependent on the differentiation parameters. King and Komornicki[372] have given a very general and well-organized development of the relationships leading to the $2n + 1$ rule and at the same time have derived expressions for errors in derivatives from using imperfectly optimized wave functions. Sadlej[379-382] has recommended the use of basis sets that are dependent on the perturbation because this may afford using a smaller basis set. For electric field properties, specific field dependence has been incorporated into Gaussian basis functions.[382] Hudis and Ditchfield[383] have also suggested a different way of selecting field-dependent bases, and both schemes seem quite promising.

Unique to the DHF approach[54, 373] is its open-endedness. Dipole hyperpolarizability of any high order could be computed. In the application of DHF to multipole polarizabilities, the parameters of differentiation are the elements of the expansion of the electrical potential. The general equations together with an effective logic procedure developed for DHF have been given in detail,[54] showing its attractive organization for the calculation of properties. It has been concluded[54] that open-shell wave functions can also be employed, but the Fock operator needs to be more generally defined. A one-Fock-operator method, which employs projection operators, is formally quite workable for DHF treatment of open-shell states. An alternative formulation of the general problem is given by Schaefer and coworkers,[384–386] where derivatives of SCF wave functions are presented. They elect to give expressions directly in terms of one- and two-electron integrals. Although it is tediously solved order by order, it has been successfully worked out to low order entirely for closed- as well as open-shell wave functions. Various tests of the basis set used in high-level ab initio calculations of electrical properties have been carried out.[387–396] The standard triple-zeta core-valence sets of Dunning[397] and Huzinaga[398] supplemented with diffuse valence and polarization functions were found to be of reasonably good quality and as economical as possible for electrical properties calculations.

DHF calculated polarizabilities and hyperpolarizabilities for various molecular systems have been reported.[54] The isotropic dipole polarizability of AH_n molecules has shown a regular decline with the atomic number of the atom A.[396] For most of the molecules studied[54] the dipole polarizability was found to be significantly determined by the heavy atom and not the hydrogens. Multiple bonds were also found to be less polarizable. Accuracy of the DHF calculated polarizabilities has been found to be limited by the neglect of correlation effects as well as by basis set quality.[54] Including electron correlation for covalent species resulted in 10–15% refinement in dipole polarizabilities.[208, 387, 399] As the covalency decreases, the correlation effects become larger and refinement may be $\sim$ 20–30% in a molecule such as LiH.[210, 212, 400] Correlation effects could be greater for higher derivative properties[54] and therefore high-quality basis sets are needed. Diffuse and higher l-type functions are essential for the description of intra-atomic polarization, just as they would be in describing polarization of an isolated atom. Interatomic valence polarizations do not need this same flexibility in basis set.

The Hartree-Fock method provides the basis for satisfactory evaluations of one-electron properties such as dipole polarizability. Within the general framework of this formalism, several alternative procedures have been used. An approximate method is the uncoupled Hartree-Fock (UCHF) method proposed by Dalgarno.[401] It has been used by Yoshimine and Hurst[402] to calculate dipole polarizabilities of various atomic systems. It has the disadvantage, however, that the perturbed functions are not required to be solutions of the perturbed Hartree-Fock equations.

The coupled perturbed Hartree-Fock method (CPHF) developed in its early formulation[401, 403, 404] has also been used to calculate dipole polarizabilities of many atomic systems.[405, 405] The fully coupled perturbed Hartree-Fock method,[193-195] which is equivalent to the finite-field approach, has been widely applied to calculate the static polarizabilities and hyperpolarizabilities[343-345, 407-411] of different molecules. Zyss and Berthier[298a] developed a mutually consistent field approach coupled with a finite-field dipolar perturbation within the framework of an ab initio SCF model to account for the influence of the crystalline environment on molecular properties. The basis set used by these authors does not include the diffuse functions necessary for a semiquantitative description of polarizabilities as demonstrated in more recent studies. The effect of varying the basis set on the second hyperpolarizability has been analyzed.[343]

Some model basis sets involving polarization (p/d, pp/dd) and diffuse functions have been used and good agreement with experimental values has been obtained. The calculated value of γ was found to be very sensitive to exponent change of both the polarization and diffuse functions. The effect of adding polarization and diffuse functions has also been investigated with the Dunning[397] and Huninaga[398] basis, while for the hydrogen atom the 4-31G basis[412, 413] has been used. The 4-31G basis was found to give considerably higher values of γ.[343] A method has been outlined for selecting the diffuse basis functions required for describing the response of the electron cloud in the presence of an electric field.[354] It has been shown that the CPHF based on ab initio wavefunctions is a suitable approach for calculating first and second hyperpolarizability tensors of molecules with nonlinear optical characteristics,[216, 408] and large molecularsystems can be handled by taking advantage of the extended computational resources of today's supercomputers.

It is now generally accepted that energy-optimized basis sets which yield satisfactory geometrical parameters in molecular structure determination need be augmented with diffuse functions for a more flexible description of the electron density away from the nuclei. For small molecules the basis set dependence of the calculated properties is already well documented.[389-396, 408] For short polyenes extra diffuse functions and diffuse polarization functions are crucial for describing the second hyper-polarizability. However, for the longer chain the importance of diffuse functions is notably reduced.[408]

Starting from the double-zeta quality basis set used for the geometry determination for nitrobenzene, three diffuse functions on the second-row atoms of p and/or d type have been added[216] to calculate polarizability and first and second hyperpolarizabilities. No diffuse functions were appended to the double-zeta basis set for the hydrogen atoms because effects associated with the out-of-plane π electrons were mainly concerned and α and β should not be affected much by the lack of diffuse functions on the hydrogen atoms. In contrast, at least two diffuse d-type functions appear to be necessary to obtain satisfactory values of γ. However, the trends extracted from calculated α, β, and γ are in accord with experimental observations. The CPHF method ab initio calculations of α, β, and γ for the haloform series CHX_3 have been performed[345, 410] using several basis sets. It was also found that α can be accurately calculated for these and similar molecules, provided that the basis set includes diffuse functions. For β and γ the agreement between experiment and theory was somewhat less satisfactory.

Fourth-order MBPT has been widely applied to calculate static α, β, and γ for atoms and molecules[274, 347, 348, 352, 355, 399, 414-417] where electron correlation effects have been considered. Other methods that include electron correlation have also been used to study static polarizabilities, such as the coupled electron pair approximation (CEPA) used by Werner and Meyer[418] to compute the dipole moments and polarizabilities of few atoms and small molecules, the modified coupled pair functional (MCPF) method used in conjunction with the finite-field method to compute (hyper)polarizabilities of Ne and HF,[346, 419] and the coupled-cluster single-and double-excitation (CCSD) wave functions with a perturbational estimate of connected triple excitations CCSD(T) used for the calculations of static (hyper)polarizabilities of atoms and simple molecules.[353, 420, 421]

An excellent agreement between the MCPF and CCSD results[346] indicated that MCPF is a convenient method for the calculation of such electrical properties, particularly considering that application to open-shell states is straightforward. Atomic natural orbital (ANO) basis sets[422] for the first row atoms, which are based on van Duijneveldt ($13s$ $8p$) primitive set[423] and an even-tempered ($6d$ $4f$) polarization set have been used.[346, 353] Up to three diffuse f functions have been added[353] and this large, flexible, one-particle basis set in conjunction with high-level treatments of electron correlation has been used to determine static α and γ for the noble gases.[353] SCF-second order Moller-Plesset (MP2) perturbation theory and CCSD methods were used[353] and the effect of connected triple excitations was explored using the CCSD(T) method[421] which is expected to give valence correlation energies very close to those from a full CI wave function.

The computed α and γ values are estimated to be accurate to within a few percent. Agreement with experimental data for the static γ is good for Ne to Xe, but for Ar and Kr the differences are larger than the combined theoretical and experimental uncertainties. Ne has shown the largest correlation contribution to γ (40%) compared to Ar (23%), Kr (20%), and Xe (18%). Relativistic effects on the polarizabilities of the heavier atoms have been examined,[353] but the perturbation theory estimates have indicated that these effects are negligible.

An extensive basis set study and a discussion of the importance of electron correlation for hyperpolarizability calculations has been given by Maroulis et al.[347, 348, 352, 355, 424, 427] SCF and complete fourth-order MBPT values of the independent components of α, β, and γ for various simple molecules have been calculated. For the ethyne molecule, the Dunning and Hay[428] split valence basis set of ($9s\ 5p/4s$) Gaussian-type functions (GTFs) contracted to [$3s\ 2p/2s$] and augmented by different s-, p-, and d-GTF on hydrogen atoms and s-, p-, d-, and f-GTF on carbon atoms giving nine different basis sets were used[347] to calculate α and γ for ethyne in its ground state. It has been found that basis sets that do not contain d-GTF on hydrogens do not predict accurate values for the transverse component of the α tensor. This may be the reason for the underestimation of the longitudinal and transverse components of the γ tensor of ethyne calculated by Jameson and Fowler.[429] Electron correlation was found to lower the isotropic averages $\bar{\alpha}$ and $\bar{\gamma}$ by only 3.4 and 1.8%, respectively.[347]

For CO_2, various basis sets with different numbers of contracted GFT have been used.[348] The calculated $\bar{\alpha}$ values were found in the range of 15.79 to 15.986 a.u., showing better than 1.5% agreement. These values also agree well with the SCF values of 15.76 a.u.[208] and 15.83 a.u.[430, 431] Spackman's[344] value of 15.42 a.u. is too small. The different basis sets used[348] have given calculated $\Delta\alpha$ values in the range 11.727 to 11.793 a.u., while the calculated SCF values 13.19 a.u.,[344] 12.09 a.u.,[208] and 12.07 a.u.[430, 431] are too anisotropic due to the lack of f-GTF in the basis sets used. Electron correlation was found[348] to alter $\bar{\alpha}$ and $\bar{\gamma}$ by approximately 11 and 42%, respectively, and, as in case of the dinitrogen,[424] correlation changes the longitudinal components more than the transverse ones. This indicates that electron correlation affects the anisotropies more than the isotropic averages.

Five different basis sets were used[352] to calculate α, β, and γ for H_2O in its ground state at both the SCF and fourth-order MBPT levels. A carefully optimized basis set consisting of 84 contracted GTF was employed in the fourth-order MBPT. Two larger basis sets were employed in a single-, double- and quadrupole (SDQ) fourth-order MBPT (SDQ-MBPT(4)) to check the stability of the second- and third-order corrections to the molecular properties. The best SCF calculated values were obtained with the large basis set $(13s\ 10p\ 6d\ 2f/9s\ 6p\ 2d)$ $[9s\ 7p\ 6d\ 2f/6s\ 5p\ 2d]$ comprising 136 contracted GTF. This basis set has given a value of $\bar{\alpha} = 8.531$ a.u.,[352] in good agreement with previous SCF calculations.[274, 387, 411, 432] The value of β was found to be -10.86 a.u.,[352] not showing the good agreement with other SCF calculated values -9.2 a.u.,[274] -9.1 a.u.[407] and -9.6 a.u.[373] as in case of $\bar{\alpha}$. The increase in basis set size was found to bring about small variations in the magnitude of the individual components of γ, but the mean $\bar{\gamma}$ displayed remarkable stability[352] and was calculated to be 979 a.u., which is higher than the previously calculated 526 a.u.[433] and 888.5 a.u.[373] values. Electron correlation was found to increase all components of α, the magnitude of β remarkably (73.2%), and $\bar{\gamma}$ value by an important 94.3%.

The CHF study of β reported by Lazzeretti and Zanasi[407] is one of the first systematic explorations of basis set effects. Their basis sets were $[9s\ 6p\ 3d/6s\ 2p\ 1d]$; $[9s\ 6p\ 3d\ 1f/6s\ 2p\ 1d]$, and a third set that represents a "tight" version of the second. Their results have shown that addition of an f-GTF on oxygen leads to a slight decrease in magnitude for all components of β, a trend found also by Maroulis.[352] The correlation corrections[352] follow closely the pattern predicted by the SDQ-MBPT calculations by Purvis and Bartlett.[274] Electron correlation effects and basis set dependence of the calculated electrical properties of H_2O_2 (Ref. 427) and linear polyynes[355] have been also investigated. Electron correlation was found to increase α and has a strong effect on both β and γ. The SCF value and has a strong effect on both β and γ. The SCF value of $\bar{\gamma}$ changes by a remarkable 67% from 959 to 1061 a.u.[427] On the other hand, calculated γ values for linear polyynes were found to be more sensitive to geometry changes than to electron correlation.[355]

A study of the effect of basis set augmentation on the calculated nonlinear optical properties of some organic molecules has been performed by Daniel and Dupuis[216] via ab initio CPHF method. Various basis sets have been used. Starting with a $(9s, 5p)/[3s, 2p]$ double-zeta quality basis set for the second-row atoms and a $(4s)/[2s]$ double-zeta basis set for H, four basis sets having no or at most one diffuse d set and five other basis sets with two diffuse d functions have been tested. For α and β these different basis sets yield similar results.[216] With the z axis coinciding with the molecular dipole moment vector, the diffuse functions most affect α_{yy} and β_{yyz}. The α_{xx} and α_{zz} components and $\bar{\alpha}$ changed by at most 15%. This illustrates how basis functions on neighboring atoms help describe the polarization of the valence electrons. The components β_{xxz} and β_{zzz} were found to be relatively insensitive to the basis set. The quantity β_{vec}, however, was found to be sensitive to the basis set owing to the negative sign of β_{zzz} which enters its definition.[216] In contrast, at least two diffuse d-type functions appear to be necessary to obtain satisfactory values of γ.

There have been several attempts to derive optimum exponents of different basis sets for polarizability calculations.[344, 387] Optimization of d-function exponents for Ne and the heavy atoms in HF, H_2O, NH_3, and CH_4 has been performed,[387] while d-function exponents on first- and second-row atoms and p-function exponents on hydrogen have been optimized[344] with respect to maximum mean polarizability for the first- and second-row AH_n hydrides. A relatively small 6-31G($+sd + sp$) basis set has been developed[344] and used in conjunction with SCF and second-order Moller-Plesset perturbation theory to calculate static dipole polarizabilities for 24 molecules ranging in size from HF- to cyclopropane. SCF results for $\bar{\alpha}$ were found generally between 5 and 15% below experimental static values. This 6-31G ($+sd + sp$) basis set[344] is probably the smallest that could be applied to a large class of polyatomic molecules. The effects due to basis set size and inclusion of diffuse and polarization functions of d and f type were examined for the haloform series CHX_3, where X = F, Cl, Br, and I.[345]

The effective core potential (ECP) approach[434-438] has been used with the nonlinearities calculated by means of the CPHF formalism. A relativistic ECP approach has been also used when dealing with Br- and I-containing molecules. Results similar in quality to the all-electron calculations have been obtained,[345] and calculated values of α, β, and γ appear to converge upon the addition of diffuse functions to the double-zeta or the triple-zeta quality valence basis. Contributions of f functions have been found to be negligible. The agreement between computed and experimental γ values was not good for $CHCl_3$, $CHBr_3$, and CHI_3, while for CHF_3 the calculated value of the γ tensor was two times smaller than the reported experimental value.[345] The calculations by Sekino and Bartlett[360] suggested that for the nonresonant values of the γ tensor in CHF_3 the dispersion effects are small. Due to their smaller basis set, their calculated γ value is about two-thirds smaller than the value reported by Dupuis and coworkers.[345] Better agreement between calculated and experimental γ values has been obtained for molecules such as benzene and styrene.[349]

The excited state electrical properties of molecules are usually quite different from those in their ground states. Although the theoretical evaluation of higher-order electrical properties in electronically excited states is nowadays feasible, few such calculations[356, 440] are found in the literature. Only a few semiempirical model Hamiltonian calculations (see, for example, Refs. 229 and 281) of excited state polarizabilities.

B. Dynamic Properties

Several factors[353] must be considered in the ab initio calculations of (hyper)polarizabilities and the subsequent comparison of theoretical and experimental results. First, calculations have to be done using large one-particle basis sets with an accurate treatment of electron correlation. This may allow for an accurate description of the response of the electron density to an applied electric field. Second, the frequency dependence has to be computed explicitly, since comparison is mostly made with frequency-dependent electric field experimental values. A third factor concerns molecular vibrations. For molecules a vibrational dependence may be large[54, 441] and should be evaluated. Another influential factor is the solvent effects on the measured properties, a parameter that is difficult to account for in computer simulations at present. Finally, for molecular crystals there are additionally the local field effects, which must be accounted for in the transformation of macroscopic susceptibilities to microscopic (hyper)polarizabilities. These factors may explain the fact that, although the experimental values are subject to potential uncertainties,[422] the agreement between computed and experimental values is less than the sophistication of the method might suggest.[211, 360, 367].

For the computations of frequency-dependent properties, the finite field method is not applicable and one is forced to use an alternative procedure. Two equivalent methods have gained an increasing interest during the past decade[358]: the time-dependent CHF (TDCHF) method[196, 442] and the random phase approximation (RPA).[443, 444] An extension of the TDHF approximation was obtained by using a multiconfiguration (MC) SCF state function to describe the system[361] where electron correlation is being considered. The resulting MCTDHF formalism has been developed by Yeager, Jorgensen, and coworkers[444] and shown to be equivalent to the MCCHF method in the time-independent case.

Various nonlinear optical polarizabilities were derived and evaluated by TDHF theory.[360] The frequency-dependent polarizabilities $\alpha(-\omega; \omega)$, $\beta(-2\omega, \omega, \omega)$, and $\gamma(-3\omega; \omega, \omega, \omega)$ have been evaluated for HF, CH_4, CH_3F, CH_2F_2, CHF_3, and CF_4 molecules. The (hyper)polarizabilities of the HF molecule have been calculated using DZ: DZ plus polarization functions (DZP) and $[6s\ 5p\ 4d\ 2f/5s\ 3p]$ contracted Gaussian basis set. Although the results by the small basis sets were unsatisfactory,[360] the predicted percentage deviation from the static case was almost constant regardless of the basis set used. Calculations of the frequency-dependent nonlinear optical properties of CH_4, CH_3F, CH_2F_2, CHF_3, and CF_4 have been performed[360] using the DZP basis set.

The calculated values of β and γ were found to be smaller than experiment. The small DZP basis not augmented by any diffuse functions describes the highly polarized space very poorly. Also, neglect of correlation effects is another reason for smaller calculated values than experimental values. A small change in the basis set by adding diffuse functions just on the carbon atom for these molecules drastically changes the calculated values of β and γ. Only one s (exponent 0.017) and a set of diffuse p functions (exponent 0.01) were placed on the carbon atoms, which resulted in a two- to seven-times increase in the $\gamma_\parallel$ value.

The importance of correlation contributions to the dynamic polarizability tensor and its derivatives with respect to the internuclear separation has been investigated [357] for N_2, CO, HCl, and Cl_2 molecules. The second-order polarization propagator approximation (SOPPA)[445] has been used in an ab initio calculation. The mean polarizability is affected, but little by inclusion of correlation effects beyond CHF. The ω dependence is expected to be most reliable in the second-order calculations, since the excitation energies of the molecules are considerably closer to experiments in a SOPPA than in a TDHF calculation.[445, 446] A basis set of 50 STOs has been used in the calculations whose size led to a significant error in $\alpha(\omega)$ and $\gamma(\omega)$ for most molecules, especially Cl_2 molecule. A basis set that is especially suited for calculations of induced electric properties, developed by Langhoff and Chong,[447, 448] has been used by Stroyer-Hansen and Svendsen[359] in an ab initio calculation of the dynamic polarizability of N_2. It consists of the contracted Gaussians [$5s$, $3p$, $2d$] where addition of an extra f function had no effect on the calculated polarizability values.[447] The study was performed within the SOPPA scheme. This method has been also applied to N_2 both to first-order (RPA) and second-order.[449, 450] The frequency dependence of α and β was in good agreement with experiment.[359] A small basis set [$4s$ $2p$ $1d/2s$ $1p$] has been used with an MCTDHF approach to the singlet ground state of the LiH molecule.[361] The MCSCF calculated correlation energy was found to be 97% of the full CI correlation energy, and its effect on the calculated value of dipole polarizability was considerably large.

Methods for determining the frequency-dependent hyperpolarizabilities $\beta(-2\omega; \omega, \omega)$, $\beta(-\omega; \omega, 0)$, and $\beta(0; \omega, -\omega)$ at the SCF level of theory have been discussed and compared.[362] The basis set requirement for convergence has been also investigated. Basis sets including up to d functions on first-row atoms, and p functions on hydrogen incorporating diffuse functions appeared to be adequate. The calculated $\bar{\alpha}(-\omega; \omega)$, $|\bar{\beta}(-\omega; \omega, 0)|$, and $|\bar{\beta}(-2\omega; \omega, \omega)|$ for the CH_3F molecule with electron correlations were found to underestimate the experimental measurements by 3, 27, and 19%, respectively, and inclusion of vibrational contributions to the hyperpolarizability are expected to reduce this discrepancy.

Ab initio computation of the frequency-dependent (hyper)polarizabilities of p-nitroaniline in the framework of the time-dependent CPHF approach has been reported.[363] The computed values of α are in good agreement with experiment, while for β and γ the computed values are considerably smaller than the respective experimentally determined values. This discrepancy has been attributed to two main reasons: the neglect of electron correlation in the CPHF method and the unsaturation of the basis set used. A DZ basis set that includes semi-diffuse polarization functions has been used[363]; however, a larger basis set with additional diffuse functions may improve the calculated results.[366] Frequency-dependent γ of the N_2 molecule has been calculated[367] using generalized TDHF theory for several frequencies of applied field. The correlation effects have been estimated using a second-order MBPT. A quite satisfactory theoretical estimate for this molecule has been obtained.

C. Properties of Interacting Systems

Diatom polarizabilities of simple gases can be obtained from measurements of their collision-induced Raman spectra. The field of collision-induced light scattering and the diatom polarizabilities has been reviewed by Frommhold.[8] Experimental tools are now well developed so that geometrical parameters of various van der Waals complexes involving different molecular systems, (N_2, F_2, HF, H_2O, CO_2, NH_3, and many others) have been established. Theoretical calculations provide an understanding of these long-range interactions, especially using ab initio methods with sufficiently large basis sets. The main purpose of this section is to review the principal ab initio investigations of (hyper)polarizabilities for the van der Waals dimers and trimers consisting of noble gas atoms, simple diatomic molecules, and even larger systems.

The diatom is meant to describe a quasimolecular complex of two interacting atoms or monomers, which may be either free and in a collisional encounter or bound together as a van der Waals dimer.[8] The diatom polarizability, sometimes referred to as collision-induced polarizability, is simply the excess polarizability or, in other words, the sum of the polarizabilities of the two interacting atoms minus the sum of the polarizabilities of the unperturbed atoms. It is a tensor and its components vanish as the internuclear separation approaches infinity. For a pair of interacting atoms the tensor α has two unique components, one parallel and the other perpendicular to the internuclear axis. The mean incremental pair polarizability can be written as

$$\bar{\alpha}^{(2)}(R) = \tfrac{1}{3}\big[\alpha_{\|}(R) + 2\alpha_{\perp}(R)\big] - (\alpha_A + \alpha_B) \tag{55}$$

where α_A and α_B are the polarizabilities of the isolated atoms, and R is the internuclear separation. The physical quantities that are accessible to measurement are the invariants of that tensor, the trace, and the anisotropy.[8] The anisotropy, which is related to field-induced birefringence and causes depolarization, is

$$B^{(2)}(R) = \alpha_{\|}(R) - \alpha_{\perp}(R) \tag{56}$$

The trace, which is the spherically symmetric part related to the polarized scattering of light and refraction at moderate to high gas densities, is given as

$$A^{(2)}(R) = \tfrac{1}{3}\big(\alpha_{\|}(R) + 2\alpha_{\perp}(R)\big) \tag{57}$$

These quantities are of considerable interest and have been the subject of many ab initio calculations. They can be used as input to statistical mechanical models, which then yield quantities that can be measured experimentally, such as the second Kerr coefficient, the depolarized light-scattering spectrum, and second dielectric virial coefficients. A wide variety of interacting atomic and molecular systems have been studied using ab initio methods.[354, 358, 451–466]

For inert gas pairs there are a number of ab initio calculations at various levels of sophistication.[358, 451–455] Ab initio results have also been reported for the interaction polarizability of two nitrogen molecules,[358, 462] two fluorine molecules,[463] various combinations of alkali metal ions and halide ions,[456–459, 461, 464] and the $CO_2 \cdots NH_3$ system.[466] O'Brien et al.,[75] using a 30-GTF basis set for the helium diatom, obtained polarizabilities believed to be accurate to $\pm 5\%$ for nine different separations. However, because of the dependence of calculated properties on the basis set, the numerical values must be used with caution. The dispersion interaction coefficients for van der Waals dimers consisting of He, Ne, H_2, and N_2 have been computed in the TDCHF approximation.[358] Static multipole polarizabilities have also been calculated. It has been concluded that the basis sets have to be very large to keep the basis set errors in the computed properties smaller than the correlation errors.

The difference between coupled and uncoupled Hartree-Fock results considered as the apparent correlation has been found to be large. In the case of Ne, the true correlation effects are larger where correlation errors up to 25% exist. Similar findings have been reported for the Ne and Ar diatom.[452, 453] The polarized Raman spectrum of helium has been computed from wave mechanics[8] using a model of the trace and the interaction potential[467, 468] and the static ab initio results for the trace of the helium diatom. A satisfactory agreement has been obtained,[8] noting that the polarized spectrum of the trace amounts to only ~ 1% of the total intensity at low frequency shifts, but dominates the signals at high frequencies. As a consequence, the low frequency falloff of the experimental spectrum is somewhat uncertain and may not be significant at all.[8] The spectra computed on the basis of Certain and Fortune's work[451] and Kress and Kozak's work[453] were consistent with the measurement[469] that shows the slower high-frequency falloff characteristic of Dacre's data.[454] Most measurements are taken at frequencies of the visible, where the use of static diatom polarizabilities causes some errors in the computed spectrum. Therefore, dynamic data have to be used for some of the most commonly used laser frequencies. Basis set and correlation effects have to be taken into consideration, together with the three-body and even higher contributions possible at high gas densities.

The same study has been done for Ne and Ar with similar findings.[8] For Ne, the ab initio computations of the anisotropy[453] have resulted in a depolarized spectrum that deviates too much from the measurement. The computed intensities for Ar using ab initio calculated diatom polarizability[452] are only 30% or less of the observed ones. Such computations have to be done using very accurate ab initio calculations, available now even for dynamic properties.

Light scattering intensities, the dielectric constant, the Kerr effect, and refractivity of fluids all depend on the polarizability tensor of the interacting species. Number density expansions of these properties may be used to analyze the contributions of 1-, 2-, ..., n-body terms to the total values. A study of the dependence of the pair polarizability tensor $\alpha^{(2)}$ and the triplet polarizability tensor $\alpha^{(3)}$ on the internuclear separation is important in understanding the various optical and dielectric properties of simple fluids.

Ab initio finite-field SCF calculations of the pair polarizability tensor $\alpha^{(2)}$ and pair potentials at various configurations for different systems have been performed.[460–466] Values of the pair potential, dipole moment, and polarizability tensor $\alpha^{(2)}$ were reported for LiF,[459] LiCl,[457] and LiBr.[461] The Li^+ Gaussian basis set ($10s, 2p$ contracted to $4s, 2p$)[457] was used and for the Br^- ion the basis set of del Conde and Bagus[470] was enhanced with extra diffuse s, p, and d primitives, whose exponents were chosen so as to follow the approximate geometric progression of the primitives.[461]

Also, f functions were added, so that ($16s$-, $13p$-, $6d$-, and $1f$-type) primitives were contracted to ($9s$, $7p$, $3d$, and $1f$) basis functions on Br$^-$. A total of 16 and 101 primitives were thus contracted to 10 and 58 bias functions for Li$^+$ and Br$^-$, respectively. The calculated $\alpha^{(2)}$ and $B^{(2)}$ for LiF, LiCl, and LiBr have shown the expected trend that Br$^-$ is the much more polarizable anion. The calculated polarizabilities are close to estimates of the CHF limits. The correlation effects were expected to be significant only at very short distances, since at separations around the equilibrium bond distance the first excited state lies well above the ground state and therefore the correlation contributions to any ground state property will be small.

Ab initio computations of the interaction polarizability $\bar{\alpha}^{(2)}$ for H$_2$ $\cdots$ H$_2$ (Ref. 471), N$_2$ $\cdots$ N$_2$ (Ref. 462), and F$_2$ $\cdots$ F$_2$ (Ref. 463) have been performed. Four different dimer structures—rectangle, linear, T-shape, and cross-T—have been considered. It has been established[472-476] that such dimer configurations include the most favorable orientations. For the three dimers the variation in $\alpha^{(2)}$ with geometry was found to be approximately similar. The Dunning[397] ($11s$, $6p$, $2d/5s$, $4p$, $1d$) basis set has been used in the calculations, since it has given the most reasonable polarizability properties for the monomers.

The results of large basis set ab initio calculations of the interaction of CO$_2$ and NH$_3$ (Ref. 466) and also in urea dimer[354] have been reported. Static and frequency-dependent second-order properties have been calculated[466] and then used to calculate properties that describe the long-range interactions in CO$_2$ $\cdots$ CO$_2$, NH$_3$ $\cdots$ NH$_3$, and CO$_2$ $\cdots$ NH$_3$ systems. The T-shaped geometry was predicted using 3-21G and 6-31G* basis sets.[412, 413] The (hyper)polarizabilities of the urea monomer and dimer have been calculated using five different basis sets.[354] All are DZV basis sets added to different polarization and diffuse functions on suitable atoms to well describe the response of the electron cloud in the presence of an electric field.

In dense gases and liquids, two-, three-, and even higher particle number clusters are probably important. Experimental measurements of properties that depend directly on three-body collisions, such as optical SHG,[17, 18] are now available. Similar to pair polarizability, the incremental triplet contribution to the total polarizability of an assembly of molecules plays an important role in various bulk properties. Mathematical expressions for the triplet polarizability $\alpha^{(3)}$ and the total anisotropy $B^{(3)}$ have been given.[477, 478]

Using O'Brien's[75] ab initio $\alpha^{(2)}$ values, Heller and Gelbart[477] calculated the long-range dispersion energy coefficient C_6 and found it to be consistent with experimental value.[74] The $\alpha^{(2)}$ (LiCl)[457] was found to be approximately two orders of magnitude greater than $\alpha^{(2)}$ $(He_2)^{75, 454, 455}$ in the region of internuclear separations that determine ensemble average. The $\alpha^{(3)}$ (Li_2Cl^+) has been calculated using the ab initio method at 20 selected configurations[478] and were as large as the sum of incremental pair polarizabilities and generally have the opposite sign. For this strongly interacting system, estimates of the triplet polarizability obtained from Heller and Gelbart[477] superposition approximation have been found to agree poorly with the ab initio calculated value.[478] The largest errors occur in the linear configuration.

A recent ab initio calculation on urea dimer and trimer has been performed by Perez and Dupuis,[354] where optical properties of urea crystals made up of molecular structures associated through hydrogen bonds have been investigated. The steps involved in selecting the diffuse basis functions that well describe the response of the electron cloud in the presence of an electric field have been outlined. Some individual components of α and γ are enhanced significantly in the clusters, while other components have shown the effect of destructive interference from the neighboring units. In spite of the large neighboring effects, $\bar{\alpha}$ and $\bar{\gamma}$ follow the additivity property within a few percent. The study by Perez and Dupuis can be considered a first step in assessing the magnitude of the contributions of the hydrogen bonds in the crystal to the nonlinear response of the urea crystal. Since their results on the timer suggested that the effects of hydrogen bonds on the (hyper)polarizabilities of the urea crystal may not be as significant as previously thought, they concluded that calculations on urea pentamer would be more informative in evaluating the compound effects of linear and transverse hydrogen bonds.

A complete compilation of the host of ab initio computed electric properties for various atoms and molecules existing in the literature, such as dipole and multipole moments, (hyper)polarizabilities and their different components, dynamic values, and properties of polarizabilities for interacting systems, is a difficult and tedious effort. I have tried to collect only the α, β, and γ ab initio calculated values for most of the systems studied and these are reported in Table IV. The method used in the calculation has been given without giving the basis set used to minimize the complexity of the table. The reader should consult the cited references for details. As is clear from the above discussion, the method using the more flexible basis set and in which the electron correlation has been better accounted for will give the better agreement between calculated and experimental values (Table IV).

Large flexible basis sets, including polarization and diffuse functions of *spd* type and also higher angular momentum functions in conjunction with high-level treatments of electron correlation using methods such as MP2,

TABLE IV
Ab Initio Computed (Hyper)polarizabilities (a.u.) for Some Systems
Using Different Approaches and Basis Sets

System	α	β	γ	Method[a]	Reference
He	1.322		36.2	SCF	353
	1.359		40.6 ⎫	MP2	353
	1.361		40.8 ⎭		353
	1.383		43.5 ⎫	CCSD	353
	1.384[b]		43.6[b] ⎭		353
	1.322			CHF	358
Ne	2.34		71.9		420
	2.374		63.9		417
	2.34		72.2	SCF	420
	2.377		70.8		426
	2.38		71.2		420
	2.373		68.9		346
	2.71		110.8 ⎫	MP2	353[c]
	2.721		115.7 ⎭		426
	2.607		94.7	MP3	426
	2.676			CEPA	418
	2.337			CHF	358
	2.717		95.8	MBPT(2)	417
	2.605		80.5	MBPT(3)	417
	2.712		104.6	MBPT(4)	417
	2.61		107.3		420
	2.61		108.1 ⎫		420
	2.64		108.7 ⎬	CCSD	420
	2.64		108.1 ⎪		420
	2.63[d]		107.6[d] ⎭		420
	2.69		118.3	CCSD(T)	420
	2.669			Experiment	479
			115.8 ⎫	Experiment	480
			119.0 ⎭		
Ar	10.73		967 ⎫		353
	10.75		965 ⎬	SCF	353
	10.76		966 ⎭		353
	11.19[e]		1220[e] ⎫		353
	11.20		1214 ⎬	MP2	353
	11.16[f]		1209[f] ⎭		353
	11.10		1177 ⎫		353
	11.11		1180 ⎪		353
	11.11		1166 ⎬	CCSD	353
	11.12		1164 ⎪		353
	11.08[f]		1152[f] ⎭		353
	11.08			Experiment	479
			1167	Experiment	480
Kr	16.47		2260 ⎫	SCF	353
	16.39[g]		2280[g] ⎭		353

TABLE IV (*Continued*)

System	α	β	γ	Method[a]	Reference
	17.09		2740		353
	16.89[h]		2700[h]	MP2	353
	16.90[h]		2700[h]		353
	17.01		2680	CCSD	353
	17.03		2700		353
	17.14[i]		2810[i]	CCSD(T)	353
	17.16		2830		353
	16.79			Experiment	479
			2600	Experiment	480
Xe	27.10		5870	SCF	353
	26.49[g]		5860[g]		353
	27.85		6900		353
	27.86		6890	MP2	353
	27.40[h]		6750[h]		353
	27.82		68.30	CCSD	353
	27.84		6880		353
	27.99		7110	CCSD(T)	353
	27.99[j]		7180[j]		353
	27.16			Experiment	479
			6888	Experiment	480
H_2	5.462				358
	5.352				358
	5.456			CHF	481
	5.395				482
	5.116			TDHF	364
	5.091				364
	5.428			Experiment	364
N_2	11.09				210
	11.34				209
	11.43			CHF	208
	11.40				358
	10.98				358
	11.43				208
	11.36				357
	10.773			TDHF	364
	11.435				364
	11.29			SOPPA	357
	11.74			Experiment	364
LiH	24.039			SCF	483
	24.178			CHF	484
	24.170				361
	24.150[k]				361
	24.152[l]			TDHF	361
	22.422[k]				361
	22.000[l]				361
	27.256[k]			DQ-MCTDHF	361
	25.774[l]				361

TABLE IV *(Continued)*

System	α	β	γ	Method[a]	Reference
HF	2.75				344
	4.35				344
	4.43				344
			292	SCF	343
			403		343
			334	SCF	371
	2.82				344
	4.99			MP2	344
	5.16				344
	4.409			TDHF	364
	4.751				364
			465	SDQ-MBPT(4)	399
			834	Experiment	124
	5.6			Experiment	364
HCl	8.46				344
	15.30			SCF	344
	15.48				344
			4140		343
	8.39				344
	16.47			MP2	344
	16.76				344
	15.532			TDHF	364
			4130	Experiment	124
	17.39			Experiment	364
	17.41			Experiment	326
H_2O	5.06				344
	7.94				344
	8.05				344
	8.53				274, 432
	8.74			SCF	411
	8.50	-9.2			387
	8.531	-10.86	979	SCF	352
	8.52	-11.20	942		352
		-9.178			407
		-9.124			407
		-7.834			407
		-9.628	888.5		373
			526		433
	5.16			MP2	344
	9.19				344
	9.44				344
	9.94	-19.4	1830	MP2	352
	9.50	-16.8	1628	SDQ-MP4	352
	9.82			MBPT(4)	432
	9.54	-13.7		SDQ-MBPT(4)	274
	9.68			CEPA	387
	8.018			TDHF	364

TABLE IV   *(Continued)*

System	α	β	γ	Method[a]	Reference
	8.422				364
	9.81			Experiment	485
		−21.8		Experiment	330
			2311	Experiment	330
H_2S	14.64				344
	22.35			SCF	344
	22.55				344
	14.41				344
	23.81			MP2	344
	24.16				344
	22.54			TDHF	364
	24.71			Experiment	364
NH_3	8.22				344
	12.30			SCF	344
	12.50				344
	8.23				344
	13.96			MP2	344
	14.33				344
	12.513			TDHF	364
	12.836				364
	14.56			Experiment	364
PH_3	20.40				344
	28.62			SCF	344
	28.95				344
	20.11				344
	29.89			MP2	344
	30.32				344
	32.03			Experiment	344
CH_4	12.53				344
	15.61			SCF	344
	15.69				344
	12.32				344
	16.23			MP2	344
	16.37				344
	18.10			CHF	349
	15.77			TDHF	364
	15.90				364
	17.27			Experiment	387, 344
SiH_4	21.10				344
	29.61			SCF	344
	29.76				344
	21.39				344
	30.95			MP2	344
	31.13				344
	31.97			Experiment	344
HCN	16.33			SCF	344
	16.11			MP2	344
	16.74			Experiment	344

TABLE IV *(Continued)*

System	α	β	γ	Method[a]	Reference
CO	12.21			CHF	369
	12.45			SOPPA	357
	11.56		⎫		357
	11.36		⎬	TDHF	364
	12.23		⎭		364
	13.08			Experiment	364
CO_2	15.42		⎫	SCF	344
	15.845		844 ⎬	SCF	348
	17.75			MP2	344
	17.626		1197	SDQ-MBPT	348
	15.043		⎫	TDHF	364
	15.731		⎬	TDHF	364
	17.51			Experiment	344
N_2O	18.55			SCF	344
	19.51			MP2	344
	17.624			TDHF	364
	19.71			Experiment	344
OCS	30.96			SCF	344
	34.09			MP2	344
	30.635			TDHF	364
	34.45			Experiment	344
	33.72			Experiment	364
CS_2	50.79			SCF	344
	54.76			MP2	344
	50.46			TDHF	364
	54.89			Experiment	344
	55.28			Experiment	364
SO_2	29.22			SCF	344
	22.87			MP2	344
	22.78			TDHF	364
	25.49			Experiment	344
	25.61			Experiment	364
H_2O_2	14.30	-6.26	959	WCF	427
	14.93	-8.42	1659	MP2	427
	14.49	-7.08	1337	MP3	427
	14.57	-7.16	1418	DQ-MP4	427
	14.95	-8.56	1601	SDQ-MP4	427
	9.46			TDHF	486
	15.43			Experiment	486
H_2CO	16.05		⎫		344
	16.58	45.76			362
	16.61	43.61			362
	16.63	43.39			362
	16.64	43.56			362
	16.37	48.58	⎬	SCF	487
	16.50				488
	12.50	34.75			362
	16.41	46.82			362
	16.56	45.99			362
	16.58	43.71	⎭		362

TABLE IV (Continued)

System	α	β	γ	Method[a]	Reference
	16.93				344
	12.42	18.9			362
	17.45	45.7			362
	17.48	44.8		MP2	362
	17.52	41.0			362
	17.47	43.7			362
	17.52	40.3			362
	18.80			Experiment	344
CH_3F	15.16				344
	13.87	26.58			362
	15.76	36.03		SCF	362
	15.16	35.64			362
	15.77	36.23			362
	16.17			MP2	344
	16.83	40.3			362
	17.32			Experiment	344
CH_2F_2	15.15			SCF	344
	16.64			MP2	344
	18.20			Experiment	344
CHF_3	15.50			SCF	344
	17.45			MP2	344
	15.52	-8.33	134.6	ECP	345
	18.69			Experiment	344
		12.50		Experiment	345
			270	Experiment	329
$CHCl_3$	53.99	0.12	1 130	ECP	345
	57.36	23.15	3 573	Experiment	345
$CHBr_3$	76.26	7.52	2 434	ECP	345
	79.64	96.45	7 941	Experiment	345
CHI_3	116.75	1.04	5 758	ECP	345
	121.48	324.08	40 102	Experiment	345
CF_4	16.04			SCF	344
	18.20			MP2	344
	19.53			Experiment	344
C_2H_2	23.32		5 400	SCF	347
	22.15				344
	21.51			MP2	344
	30.53		5 480	MP3	347
	30.32		5 160	DQ-MP4	347
	23.38		5 601	SDQ-MP4	355
	22.52		5 310	CCD	347
	20.51			PE-MCSCF	489
	21.21			MCTDHF	490
	22.81			Experiment	490
			20,000 $\pm(4000)$	Experiment	119

TABLE IV *(Continued)*

System	α	β	γ	Method[a]	Reference
C_2H_4	27.25			SCF	344
	26.73			MP2	344
	27.24		⎫	TDHF	364
	27.84		⎭		364
	27.70			Experiment	364
C_2H_6	26.94			SCF	344
	28.10			MP2	344
	29.92			CHF	349
	27.00			TDHF	364
	27.12				364
	29.61			Experiment	364
C_3H_4	40.47			SCF	344
(allene)	39.28			MP2	344
	40.40			Experiment	344
C_3H_6	34.32			SCF	344
(cyclo-	35.83			MP2	344
propane)	37.18			Experiment	344
C_4H_2	47.85		11 450	SDQ-MP4	355
C_6H_2	47.50		20 400	SDQ-MP4	355
C_8H_2	107.23		36 000	SDQ-MP4	355
Butadiene[m]	42.82		1 098 ⎫	SCF	311b
	53.27		14 846 ⎭		311b
Hexa-	73.09		9 878 ⎫	SCF	311b
triene[m]	87.92		35 118 ⎭		311b
Octatet-	109.84		40 775 ⎫	SCF	311b
raene[m]	129.02		82 212 ⎭		311b
Decapenta-	152.15		114 624 ⎫	SCF	311b
ene[m]	175.69		178 443 ⎭		311b
Decahexa	198.89		253843 ⎫	SCF	311b
ene[m]	226.81		345721 ⎭		311b
n-C_3H_8	40.87			CHF	349
	38.07			TDHF	364
	42.09			Experiment	364
n-C_4H_{10}	53.08			CHF	349
	49.48			TDHF	364
	54.07			Experiment	364
n-C_5H_{12}	65.16			CHF	349
	60.87			TDHF	364
	66.07			Experiment	364
n-C_6H_{14}	77.55			CHF	349
	79.51			Experiment	491
n-C_7H_{16}	90.02			CHF	349
	91.89			Experiment	491

TABLE IV   (Continued)

System	α	β	γ	Method[a]	Reference
Urea	31.9	18.4	780 ⎫		354
	31.6	16.1	803 ⎪		354
	31.6	15.3	804 ⎬	SCF	354
	29.8	22.2	803 ⎪		354
	22.7	18.5	126 ⎭		354
Benzene	62.76	0.0	2 580.8	SCF	216
	67.49		7 544	Experiment	216
Aniline	76.26	127.3	2 580.8	SCF	216
	84.36	162.0	11 316	Experiment	216
NB[n]	83.0	115.74	1 985	SCF	216
	83.0	231.5	10 720	Experiment	216
PNA[o]	96.5	509.3	3 970.5	SCF	216
		1111.12		Experiment	216
MNA[p]	108.65	555.56	4 367.5	SCF	216
		1099.5		Experiment	216
MNB[q]	109.33	694.45	4 963.1	SCF	216
		2268.5		Experiment	216
DMNB[r]	122.82	902.79	6 352.8	SCF	216
		2257		Experiment	216

Note. For α, β, and γ 1 a.u. = $1.481\,76 \times 10^{-25}$ esu, $8.639\,93 \times 10^{-33}$ esu, and $5.037]7 \times 10^{-40}$ esu, respectively, and 1 a.u. = $1.648\,67 \times 10^{-41}$ $C^2 m^2 J^{-1}$, $3.206\,62 \times 10^{-53} C^3 m^3 J^{-2}$, and $6.235\,97 \times 10^{-65} C^4 m^4 J^{-3}$, respectively.

[a] The different values given for the same method are for different basis sets; see the reference cited for the basis set used.

[b] Best estimates of α and γ^{353} for He.

[c] Best estimates of α and γ for Ne were found to be 2.63(± 0.03) and 119(± 4) a.u., respectively.[353]

[d] 10 electrons correlated.

[e] Best estimates of α and γ^{353} for Ar.

[f] 16 electrons correlated.

[g] Relativistic effects have been considered.

[h] 26 electrons correlated.

[i] Best estimates of α and γ^{353} for Kr.

[j] Best estimates of α and γ for Xe were found to be 27.4 (± 0.5) and 7030 (± 200) a.u., respectively.[353]

[k] TDHF, length.

[l] TDHF, velocity.

[m] trans configuration.

[n] Nitrobenzene.

[o] p-Nitroaniline.

[p] o-Methyl-p-nitroaniline.

[q] p-N-Methylanine nitrobenzene.

[r] p-N, N-Dimethyl amino nitrobenzene.

SDQ-MP4, CCSD, and CCSD(T), are necessary for obtaining accurate ab initio calculated electrical properties at the SCF level of theory. The MCTDHF(DD) and MCTDHF(DQ) methods (see, e.g., Ref. 361) with a large basis set can yield reliable dynamic polarizability functions and correlation effects up to 97% of the total full CI correlation energy, are being taken into consideration. Although dispersion effects are systematically underestimated, the TDHF theory yields generally accurate frequency-dependent polarizabilities with a high-quality basis set.

Acknowledgments

I am very much indebted to M. W. Evans for sending me many preprints which helped me very much in preparing this chapter. I also want to thank him for revising this work. Thanks are due to those who helped me during the course of preparing this work. The facilities made possible for me at the Department of Chemistry, College of Sciences, King Saud University, Riyadh, are very much appreciated.

References

1. G. E. Chamberlain and J. C. Zorn, *Phys. Rev.* **129**, 677 (1963).

2. A. Salop, E. Pollack, and B. Bederson, *Phys. Rev.* **124**, 1431 (1961).

3. E. Pollack, E. J. Robinson, and B. Bederson, *Phys. Rev.* **134** A1210 (1964).

4. B. Bederson and E. J. Robinson, *Adv. Chem. Phys.* **10**, 1 (1961).

5. J. P. C. McTague and G. Birnbaum, *Phys. Rev. Lett.* **21**, 661 (1968).

6. J. P. C. McTague and G. Birnbaum, *Phys. Rev.* **A3**, 1376 (1971).

7. W. S. Gelbart, *Adv. Chem. Phys.* **26**, 1 (1974).

8. L. Frommhold, *Adv. Chem. Phys.* **46**, 1 (1981).

9. C. G. Le Févre and R. J. W. Le Févre, in A. Weissberger (Ed.), *Techniques of Chemistry*, Vol. 1, Part IIIC, Wiley, New York, 1972, p. 399.

10. L. L. Boyle, A. D. Buckingham, R. L. Disch, and D. A. Dunmur, *J. Chem. Phys.* **45**, 1318 (1966).

11. G. Hauchecorne, F. Kerhervé, and G. Mayer, *J. Phys.* (*Paris*) **32**, 47 (1971).

12. S. Kielich, *IEEE J. Quantum Electron.* **QE5**, 562 (1969).

13. R. S. Finn and J. F. Ward, *Phys. Rev. Lett.* **26**, 285 (1971).

14. R. S. Finn and J. F. Ward, *J. Chem. Phys.* **60**, 454 (1974).

15. I. J. Bigio and J. F. Ward, *Phys. Rev. A* **9**, 35 (1974).

16. I. J. Bigio, Thesis, University of Michigan, Ann Arbor, 1974.

17. W. M. Gelbart, *Chem. Phys. Lett.* **56**, 303 (1973).

18. R. Samson and R. A. Pasmanter, *Chem. Phys. Lett.* **25**, 405 (1974).

19. J. F. Ward and G. H. C. New, *Phys. Rev.* **185**, 57 (1969).

20. K. M. Leung, Thesis, University of Michigan, Ann Arbor, 1972.

21. K. M. Leung, J. F. Ward, and B. J. Orr, *Phys. Rev. A* **9**, 2440 (1974).

22. S. E. Harris and R. B. Miles, *Appl. Phys. Lett.* **19** 385 (1971).

23. R. B. Miles and S. E. Harris, *IEEE J. Quantum Electron.* **9**, 470 (1973).

24. P. W. Atkins, *Molecular Quantum Mechanics*, 2d ed., Oxford University Press, Oxford, UK. 1983.

25. C. J. F. Böttcher, *Theory of Electric Polarization*, Vol. 1, Elsevier Scientific, Amsterdam, 1973.

26. A. D. Buckingham, in G. Allen (Ed.), *MTP International Review of Science, Physical Chemistry Series One, Molecular Structure and Properties*, Vol. 2, Butterworths, London, 1972, p. 241.

27. J. A. Stratton, *Electromagnetic Theory*, McGraw-Hill, New York, 1941, Chapter 3.

28. J. A. Osborn, *Phys. Rev.* **67**, 351 (1945).

29. E. C. Stoner, *Philos. Mag.* **36**, 803 (1945).

30. S. Kielich and A. Piekara, *J. Chem. Phys.* **29**, 1297 (1958).

31. S. Kielich and A. Piekara, *Acta Phys. Pol.* **18**, 439 (1959).

32. A. Chelkowski, *J. Chem. Phys.* **28**, 1249 (1958).

33. A. Piekara and A. Chelkowski, *J. Chem. Phys.* **25**, 794 (1956).

34. A. Piekara, A. Chelkowski, and S. Kielich, *Z. Phys. Chem.* **206**, 375 (1957).

35. J. Malecki, *J. Chem. Phys.* **36**, 2144 (1962).

36. J. Malecki, *Acta. Phys. Pol.* **21**, 13 (1962).

37. J. A. Schellman, *J. Chem. Phys.* **26**, 1225 (1957).

38. J. Barriol and J. L. Greffe, *J. Chem. Phys.* **65**, 575 (1969).

39. A. D. Buckingham, *J. Chem. Phys.* **26**, 428 (1956).

40. A. D. Buckingham and J. A. Pople, *Proc. Phys. Soc.* **A68**, 905 (1955).

41. M. P. Bogaard and B. J. Orr, in A. D. Buckingham (Ed.), *MTP International Review of Science, Physical Chemistry Series Two, Molecular Structure and Properties*, Vol. 2, Butterworths, London, 1975, p. 149.

42. B. J. Orr and J. F. Ward, *Mol. Phys.* **20**, 513 (1971).

43. P. Langevin, *Ann. Chem. Phys.* **5**, 70 (1905).

44. P. Langevin, *Le Radium* **7**, 249 (1910).

45. M. Born and W. Heisenberg, *Z. Phys.* **23**, 388 (1924).

46. M. Born and P. Jordan, *Elementare Quantenmechanik*, Springer, Berlin, 1930.

47. A. Dalgarno, *Adv. Phys.* **11**, 281 (1962).

48. A. D. Buckingham and B. J. Orr, *Quart. Rev. Chem. Soc.* **21**, 195 (1967).

49. W. Byers Brown and D. M. Whisnant, *Chem. Phys. Lett.* **7**, 329 (1970).

50. R. L. Matcha and R. K. Nesbet, *Phys. Rev.* **160**, 72 (1967).

51. Z. J. Kiss and H. L. Welsh, *Phys. Rev. Lett.* **2**, 166 (1959).

52. D. R. Bosomworth and H. P. Gush, *Can. J. Phys.* **43**, 751 (1965).

53. W. N. Lipscomb, in W. Byers Brown (Ed.), *MTP International Review of Science, Physical Chemistry Series One, Theoretical Chemistry*, Vol. 1, Butterworth, London, 1972, p. 167.

54. C. E. Dykstra, S.-Y. Liu, and D. J. Malik, *Adv. Chem. Phys.* **75**, 37 (1989).

55. A. D. Buckingham, *Proc. Phys. Soc. (London) Sect. A* **68**, 910 (1955).

56. A. D. Buckingham and M. J. Stephen, *Trans. Faraday Soc.* **53**, 884 (1957).

57. A. D. Buckingham and D. A. Dunmar, *Trans. Faraday Soc.* **64**, 1776 (1968).

58. M. N. Grasso, K. T. Chung, and R. P. Hurst, *Phys. Rev.* **167**, 1 (1968).

59. C. G. Le Févre and R. J. W. Le Févre, *Rev. Pure Appl. Chem.* (*Australia*) **5**, 261 (1955).

60. R. J. W. Le Févre, in V. Gold (Ed.), *Advances in Physical Organic Chemistry*, Vol. 3, Academic, London, 1965, p. 1.

61. C. G. Le Févre and R. J. W. Le Févre, *J. Chem. Soc.* 4041 (1953).

62. R. S. Armstrong, J. J. Aroney, C. G. Le Févre, R. J. W. Le Févre, and M. R. Smith, *J. Chem. Soc.* 1471 (1958).

63. R. J. W. Le Févre and A. J. Williams, *J. Chem. Soc.* 1671 (1961).

64. R. S. Armstrong, M. J. Aroney, and R. J. W. Le Févre, *Australian J. Chem.* **15**, 703 (1962).

65. R. J. W. Le Févre and A. J. Williams, *J. Chem. Soc.* 562 (1964).

66. R. J. W. Le Févre and G. L. D. Ritchie, *J. Chem. Soc.* 4933 (1963).

67. J. G. Kirkwood, *J. Chem. Phys.* **4**, 592 (1936).

68. A. N. Kaufman and K. M. Watson, *Phys. Fluids* **4**, 931 (1961).

69. A. D. Buckingham, *Trans. Faraday Soc.* **52**, 747 (1956).

70. R. H. Cole, D. R. Johnston, and G. J. Oudemans, *J. Chem. Phys.* **33**, 1310 (1960).

71. A. D. Buckingham, *Trans. Faraday Soc.* **52**, 1035 (1956).

72. D. A. McQuarrie and H. B. Levine, *Physica* **31**, 749 (165).

73. R. H. Orcutt and R. H. Cole, *J. Chem. Phys.* **46**, 697 (1967).

74. D. Vidal and P. M. Lallemand, *J. Chem. Phys.* **64**, 4293 (1976).

75. E. F. O'Brien, V. P. Gutschick, V. McKoy, and J. P. McTague, *Phys. Rev.* **A8**, 690 (1973).

76. C. K. Ingold, *Structure and Mechanism in Organic Chemistry*, Cornell University Press, Ithaca, NY 1953.

77. J. Hine, *Physical Organic Chemistry*, McGraw-Hill, New York, 1956.

78. D. A. Long, *Proc. R. Soc. London Ser. A* **217**, 203 (1953).

79. L. A. Woodward, *Quart. Revs.* (*London*) **10**, 185 (1956).

80. M. J. Crawford, E. J. Stansbury, and H. L. Welsh, *Can. J. Phys.* **31**, 954 (1953).

81. J. H. Brewster, *J. Am. Chem. Soc.* **81**, 5475 (1959).

82. R. L. Williams, *Ann. Rep. Progr. Chem.* **58**, 34 (1961).

83. W. G. Hammerle and I. Tinoco, *J. Phys. Chem.* **60**, 1619 (1956).

84. I. Tinoco, *J. Am. Chem. Soc.* **79**, 4248, 4336 (1957).

85. I. Tinoco, *J. Am. Chem. Soc.* **81**, 1540 (1959).

86. I. Tinoco and R. W. Woody, *J. Am. Chem. Soc.* **32**, 461 (1960).

87. A. D. Buckingham and R. L. Disch, *Proc. R. Soc. London Ser. A* **273**, 275 (1963).

88. A. D. Buckingham and K. A. McLauchlan, *Proc. Chem. Soc.* 144 (1963).

89. A. D. Buckingham, *Proc. Chem. Soc.* 336 (1963).

90. R. F. Zürcher, *Disc. Faraday Soc.* **34**, 66 (1962).

91. R. F. Zürcher, *J. Chem. Phys.* **37**, 2421 (1962).

92. J. P. Vinti, *Phys. Rev.* **41**, 813 (1932).

93. R. Gans and B. Mrowka, *Schriften Königsberg. gelehrten Ges. Naturw. Kl.* **12**, 1 (1935).

94. R. de Malleman, *J. Phys. Radium* **7**, 295 (1926).

95. R. de Malleman, *Trans. Faraday Soc.* **26**, 281 (1930).

96. W. Liptay, in E. Lim (Ed.), *Excited States*, Vol. 1, Academic, New York, 1973, p. 129.

97. B. Koutek, *Coll. Czech. Chem. Commun.* **43**, 2368 (1978).

98. A. T. Amos and B. L. Burrows, *Adv. Quant. Chem.* **7**, 303 (1973).

99. N. Mataga and K. T. Kubota, *Molecular Interactions and Electronic Spectra*, Dekker, New York, 1970.

100. M. Nicol, *Appl. Spectrosc. Rev.* **8**, 183 (1974).

101. S. J. Cyvin, J. E. Rauch, and J. C. Decius, *J. Chem. Phys.* **43**, 4083 (1965).

102. W. Kaiser and C. G. B. Garrett, *Phys. Rev. Lett.* **7**, 229 (1961).

103. I. D. Abella, *Phys. Rev. Lett.* **9**, 453 (1962).

104. S. Singh and L. T. Bradley, *Phys. Rev. Lett.* **12**, 612 (1964).

105. W. L. Peticolas, *Annu. Rev. Phys. Chem.* **18**, 233 (1967).

106. R. W. Terhune and P. D. Maker, in A. K. Levine (Ed.), *Lasers: A Series of Advances*, Vol. 2, Dekker, New York, 1968, p. 295.

107. J. A. Giordmaine, *Phys. Today* **22**(9), 39 (1969).

108. S. Kielich, *Opto-electronics* **2**, 125 (1970).

109. T. P. Hughes, *Nature* **194**, 268 (1962).

110. G. Fiocco and E. Thompson, *Phys. Rev. Lett.* **10**, 89 (1963).

111. P. D. Maker and R. W. Terhune, *Phys. Rev.* **137**, A801 (1965).

112. S Kielich, in D. Davies (Ed.), *Dielectric and Related Molecular Processes*, Vol. 1, Chemical Society, London, 1972, p. 192.

113. A. D. Buckingham, in D. A. Ramsey (Ed.), *MTP International Review of Science, Physical Chemistry*, Series One, Spectroscopy, Vol. 3, Butterworth, London, 1972, p. 73.

114. N. E. Hill, W. E. Vaughan, A. H. Price, and M. Davies, *Dielectric Properties and Molecular Behaviour*, Van Nostrand-Reinhold, London, 1969.

115. R. R. Teachout and R. T. Pack, *Atomic Data* **3**, 195 (1971).

116. A. D. Buckingham, *Quart. Rev. (London)* **13**, 189 (1959).

117. A. D. Buckingham, *Adv. Chem. Phys.* **12**, 107 (1967).

118. T. M. Miller and B. Bederson; *Adv. At. Mol. Phys.* **13**, 1 (1977).

119. A. D. Buckingham, M. P. Bogaard, D. A. Dunmur, C. P. Hobbs, and B. J. Orr, *Trans. Faraday Soc.* **66**, 1548 (1970).

120. A. D. Buckingham and B. J. Orr, *Trans. Faraday Soc.* **65**, 673 (1969).

121. D. W. Schaefer, R. E. J. Sears, and J. S. Waugh, *J. Chem. Phys.* **53**, 2127 (1970).

122. A. D. Buckingham and B. J. Orr, *Proc. R. Soc. London Ser. A* **305** 259 (1968).

123. M. P. Bogaard, A. D. Buckingham, and G. L. D. Ritchie *Mol. Phys.* **18**, 575 (1970).

124. J. W. Dudley II and J. F. Ward, *J. Chem. Phys.* **82**, 4673 (1985).

125. W. G. Rado, *Appl. Phys. Lett.* **11**, 123 (1967).

126. H. Scheffers and J. Stark, *Phys. Z.* **35**, 625 (1934).

127. B. Bederson, J. Eisinger, K. Rubin, and A. Salop *Rev. Sci. Instrum.* **31**, 852 (1960).

128. S. M. Bass, R. L. DeLeon, and J. S. Muenter, *J. Chem. Phys.* **86**, 4305 (1987).

129. N. Ramsay, *Molecular Beams*, Oxford, London, 1956.

130. J. S. Muenter and T. R. Dyke, in A. D. Buckingham (Ed.), *MTP International Review of Science, Physical Chemistry Series Two, Molecular Structure and Properties*, Vol. 2, Butterworth, London, 1975.

131. S. E. Choi and R. B. Bernstein, *J. Chem. Phys.* **85**, 150 (1986).

132. L. Wharton, R. Berg, and W. Klemperer, *J. Chem. Phys.* **39**, 2023 (1963).
133. R. Berg, L. Wharton, and W. Klemperer, *J. Chem. Phys.* **43**, 2416 (1965).
134. E. W. Kaiser, *J. Chem. Phys.* **53**, 1686 (1970).
135. B. Fabricant, D. Krieger, and J. S. Muenter, *J. Chem. Phys.* **67**, 1576 (1977).
136. W. L. Ebenstein and J. S. Muenter, *J. Chem. Phys.* **80**, 3989 (1984).
137. B. H. Ruessink and C. MacLean, *J. Chem. Phys.* **85**, 93 (1986).
138. B. H. Ruessink and C. MacLean, *Mol. Phys.* **60**, 1059 (1987).
139. S. Stolte and J. Reuss, in R. B. Bernstein (Ed.), *Atom-Molecule Collision Theory: A Guide for the Experimentalists*, Plenum, New York, 1979.
140. G. C. Maitland, M. Rigby, E. B. Smith, and W. A. Wakeham *Intermolecular Forces: Their Origin and Determination*, Clarendon, Oxford, UK, 1981.
141. M. Keil and G. A. Parker, *J. Chem. Phys.* **82**, 1947 (1986).
142. P. Isnard, D. Robert, and L. Galatry, *Mol. Phys.* **31**, 1789 (1976).
143. U. Buck, J. Schleusener, D. J. Malik, and D. Secrest, *J. Chem. Phys.* **74**, 1707 (1981).
144. N. Bloembergen, *Nonlinear Optics*, Benjamin, New York, 1965.
145. L. D. Barron, *Molecular Light Scattering and Optical Activity*, Cambridge University Press, Cambridge, UK, 1982.
146. Y. R. Shen, *The Principles of Non-linear Optics*, Wiley, New York, 1984.
147. H. F. Hameka, *Theory of Interactions Between Molecules and Electromagnetic Fields*, Addison-Wesley, Reading, MA, 1965.
148. M. W. Evans, S. Woźniak, and G. Wagniére, *Physica B* **176**, 33 (1992).
149. S. Woźniak, M. W. Evans, and G. Wagniére, *Mol. Phys.* **75**, 81 (1992).
150. S. Woźniak, M. W. Evans, and G. Wagniére, *Mol. Phys.* **75** 99 (1992).
151. M. Faraday, *Philos. Mag.* **28**, 294 (1846).
152. S. Kielich, *Proc. Phys. Soc. London* **86**, 709 (1965).
153. A. D. Buckingham and P. J. Stephens, *Annu. Rev. Phys. Chem.* **17**, 38 (1966).
154. E. D. Palik and B. W. Menvis, *Appl. Opt.* **6**, 603 (1967).
155. P. S. Pershan, *Phys. Res.* **130**, 919 (1963).
156. J. P. van der Ziel, P. S. Pershan, and L. D. Malmstrom *Phys. Rev. Lett.* **15**, 190 (1965).
157. P. S. Pershan, J. P. van der Ziel, and L. D. Malmstrom, *Phys. Rev.* **143**, 574 (1966).
158. P. W. Atkins and M. H. Miller, *Mol. Phys.* **15**, 503 (1968).
159. Y. R. Shen, *The Principles of Non-linear Optics*, Wiley Interscience, New York, 1984.
160. M. W. Evans, *Phys. Rev. Lett.* **64**, 2909 (1990).
161. M. W. Evans, *J. Phys. Chem.* **95**, 2256 (1991).
162. M. W. Evans, *Opt. Lett.* **15** 863 (1990).
163. M. W. Evans, *Phys. Lett. A* **41**, 4601 (1990).
164. M. W. Evans and G. Wagniére, *Phys. Rev. A* **42**, 6537 (1991).
165. S. Woźniak and R. Zawodny, *Acta Phys. Pol.* **A61**, 175 (1982).
166. S. Woźniak and R. Zawodny, *Acta, Phys. Pol.* **68**, 675 (1985).
167. G. Wagniéré and A. Meier, *Chem. Phys. Lett.* **93**, 78 (1982).
168. O. Wagniére, *Z. Naturf.* **39a**, 254 (1984).
169. L. D. Barron and J. Vrbancich, *Mol. Phys.* **51**, 715 (1984).

170. S. Woźniak, *Mol. Phys.* **59**, 421 (1986).

171. S. Woźniak, *J. Chem. Phys.* **85**, 4217 (1986).

172. S. Woźniak, *Acta Phys. Pol.* **A72**, 779 (1987).

173. G. Gagniére, *Z. Phys. D* **8**, 229 (1988).

174. G. Wagniére, *Phys. Rev. A* **40**, 2437 (1989).

175. S. Woźniak, G. Wagniére, and R. Zawodny, *Phys. Lett. A* **154**, 259 (1991).

176. M. W. Evans, G. C. Lie, and E. Clementi, *J. Chem. Phys.* **87**, 6040 (1987).

177. M. W. Evans and G. Wagniére, *Phys. Rev. A* **42**, 6732, (1990).

178. M. W. Evans, *Phys. Rev. A* **41**, 6041 (1990).

179. M. W. Evans, S. Woźniak, and G. Wagniére, *Physica B* **173**, 357 (1991).

180. I. I. Sobel'man, *Introduction to the Theory of Atomic Spectra*, Pergamon, Oxford, UK, 1972 (English translation).

181. E. Merzbacher, *Quantum Mechanics*, New York, Wiley, 1961.

182. L. H. Scharpen, J. S. Muenter, and V. W. Laurie, *J. Chem. Phys.* **53**, 2513 (1970).

183. J. S. Muenter and V. W. Laurie, *J. Am. Chem. Soc.* **86**, 3901 (1964).

184. K. B. MacAdam and N. F. Ramsay, *Phys. Rev. A* **6**, 898 (1972).

185. J. S. Muenter, in A. D. Buckingham (Ed.), *MTP International Review of Science Physical Chemistry Series Two, Molecular Structure and Properties*, Chapter 2, London, Butterworth, 1974.

186. M. W. Evans, *Z. Phys. B Condensed Matter* **85**, 135 (1991).

187. D. C. Hanna, M. A. Yuratich, and D. Cotter (Eds.), *Nonlinear Optics of Free Atoms and Molecules*, Springer Series in Optical Sciences, vol. 17, Springer, Berlin, 1979.

188. S. Kielich, *Nonlinear Molecular Optics*, Nauka, Moscow, 1981.

189. C. Kalpouzos, D. McMorrow, W. T. Lotshaw, and G. A. Kenney-Wallace, *Chem. Phys. Lett.* **150** 138 (1988).

190. M. W. Evans, private communication.

191. A. S. Davydov, *Quantum Mechanics*, Pergamon, Oxford, UK, 1965, Chapter 9.

192. A. Dalgorno, in C. H. Wilox (Ed.), *Perturbation Theory and Its Applications to Quantum Mechanics*, Wiley, New York, 1966, p. 145.

193. R. McWeeny, *Phys. Rev.* **126**, 1028 (1962).

194. G. Diereksen and R. McWeeny, *J. Chem. Phys.* **44**, 3554 (1966).

195. J. L. Dodds, R. McWeeny, W. T. Raines, and J. P. Riley *Mol. Phys.* **33**, 611 (1977).

196. P. W. Langhoff, S. T. Epstein, and M. Karplus, *Rev. Mod. Phys.* **44**, 602 (1972).

197. A. Dalgarno, A. L. Ford, and J. C. Browne, *Phys. Rev. Lett.* **27**, 1033 (1971).

198. L. D. Barron and C. G. Gray, *J. Phys. A* **6**, 59 (1973).

199. J. A. Armstrong, N. Bloembergen, J. Ducuing, and P. S. Pershan, *Phys. Rev.* **127**, 1918 (1962).

200. J. F. Ward, *Rev. Mod. Phys.* **37**, 1 (1965).

201. A. Dalgarno, in D. R. Bates (Ed.), *Quantum Theory*, Vol. 1, *Elements*, Academic, New York, 1961.

202. H. D. Cohen and C. C. J. Roothaan, *J. Chem. Phys.* **43**, 34 (1965).

203. K. A. Brueckner, *Phys. Rev.* **97**, 1353 (1955).

204. J. Goldstone, *Proc. R. Soc. London Ser. A* **239**, 267 (1957).

205. H. P. Kelly, *Adv. Chem. Phys.* **14**, 129 (1969).

206. H. P. Kelly, *Phys. Rev. Lett.* **23**, 455 (1969).

207. W. Müller, J. Flesch, and W. Meyer, *J. Chem. Phys.* **80**, 3297 (1984).

208. M. A. Morrison and P. J. Hay, *J. Chem. Phys.* **70**, 3034 (1979).

209. R. D. Amos, *Mol. Phys.* **39**, 1 (1980); *J. Phys. B* **12**, 1315 (1979).

210. J. E. Gready, G. B. Bacskay, and N. S. Hush, *Chem. Phys.* **24**, 333 (1977); **31**, 467 (1978).

211. H. Sekino and R. J. Bartlett, *J. Chem. Phys.* **84**, 2726 (1986).

212. J. E. Gready, G. B. Bacskay, and N. S. Hush, *Chem. Phys.* **23**, 9 (1977).

213. R. D. Amos and J. H. Williams, *Chem. Phys. Lett.* **66**, 471 (1979).

214. I. G. John, G. B. Bacskay, and N. S. Hush, *Chem. Phys.* **38**, 319 (1979).

215. I. G. John, G. B. Bacskay, and N. S. Hush, *Chem. Phys.* **51**, 49 (1980).

216. C. Daniel and M. Dupuis, *Chem. Phys. Lett.* **171**, 209 (1990).

217. D. S. Chemla and J. Zyss (Eds.), *Nonlinear Optical Properties of Organic Molecules and Crystals*, Vols. 1 and 2, Academic, New York, 1987.

218. D. Pugh and J. O. Morley, in D. S. Chemla and J. Zyss (Eds.), *Nonlinear Optical Properties of Organic Molecules and Crystals*, Vol. 1, Academic, New York, 1987, p. 193.

219. J. F. Nicoud and R. J. Tweigh, in D. S. Chemla and J. Zyss (Eds.), *Nonlinear Optical Properties of Organic Molecules and Crystals*, Vol. 2, Academic, New York, 1987, p. 255.

220. D. J. Williams (Ed.), *Nonlinear Optical Properties of Organic and Polymer Materials*, ACS Symp. Ser. No. 233, American Chemical Society, Washington DC, 1983.

221. E. Hückel, *Z. Phys.* **70**, 204 (1931).

222. A. Streitweiser, Jr., *Molecular Orbital Theory for Organic Chemists*, Wiley, New York, 1961.

223. W. M. Flygare, *Molecular Structure and Dynamics*, Prentice-Hall, Englewood Cliffs, NJ, 1978, p. 372.

224. H. F. Hameka, *J. Chem. Phys.* **67**, 2935 (1977).

225. E. F. McIntyre and H. F. Hameka, *J. Chem. Phys.* **68**, 3481 (1978).

226. E. F. McIntyre and H. F. Hameka, *J. Chem. Phys.* **69**, 4814 (1978).

227. E. F. McIntyre and H. F. Hameka, *J. Chem. Phys.* **68**, 5534 (1978).

228. E. F. McIntyre and H. F. Hameka, *J. Chem. Phys.* **70**, 2215 (1979).

229. P. Matzke, O. Chacon, E. Sanhueza, and M. Trsic, *Int. J. Quantum Chem.* **6**, 407 (1972).

230. O. Zamani-Khamini and H. F. Hameka, *J. Chem. Phys.* **73**, 5693 (1980).

231. O. Zamani-Khamini and H. F. Hameka, *J. Chem. Phys.* **71**, 1607 (1979).

232. R. Pariser and R. G. Parr, *J. Chem. Phys.* **21**, 466 (1953).

233. R. Pariser and R. G. Parr, *J. Chem. Phys.* **21**, 767 (1953).

234. J. A. Pople, *Trans. Faraday Soc.* **49**, 1375 (1953).

235. O. Zamani-Khamini, E. F. McIntyre, and H. F. Hameka, *J. Chem. Phys.* **72**, 1280 (1980).

236. O. Zamani-Khamini, E. F. McIntyre, and H. F. Hameka, *J. Chem. Phys.* **72**, 5906 (1980).

237. D. Li, M. A. Ratner and T. J. Marks, *J. Am. Chem. Soc.* **110**, 1707 (1988).

238. D. Li, T. J. Marks, and M. A. Ratner, *Chem. Phys. Lett.* **131**, 370 (1986).

239. C. W. Dirk, R. J. Twieg, and G. Wagniére, *J. Am. Chem. Soc.* **108**, 5387 (1986).

240. A. Schweig, *Mol. Phys.* **14**, 533 (1968).

241. A. Schweig, *Chem. Phys. Lett.* **1**, 195 (1967).

242. A. Schweig, *Chem. Phys. Lett.* **1**, 163 (1967).

243. R. Hoffmann, *J. Chem. Phys.* **39**, 1397 (1963).

244. J. A. Pople, D. P. Santry, and G. A. Segal, *J. Chem. Phys.* **43**, S 129 (1965).

245. J. A. Pople and G. A. Segal, *J. Chem Phys.* **43**, S136 (1965).

246. J. A. Pople and G. A. Segal, *J. Chem. Phys.* **44**, 3289 (1966).

247. O. Rinaldi and J. Rivail, *Theoret. Chim. Acta* **32**, 243 (1974).

248. E. N. Svendsen, T. Stroyer-Hansen, and H. F. Hameka, *Chem. Phys. Lett.* **54**, 217 (1978).

249. E. N. Svendsen and T. Stroyer-Hansen, *Theoret. Chim. Acta* **45**, 53 (1977).

250. H. F. Hameka and E. N. Svendsen, *Int. J. Quantum Chem.* **10**, 249 (1976).

251. N. S. Hush and M. L. Williams, *Chem. Phys. Lett.* **5**, 507 (1970).

252. N. S. Hush and M. L. Williams, *Chem. Phys. Lett.* **6**, 163 (1970).

253. N. S. Hush and M. L. Williams, *Theoret. Chim. Acta* **25**, 346 (1972).

254. J. J. C. Teixeira-Dias and J. N. Murrell, *Mol. Phys.* **19**, 329 (1970).

255. D. W. Davies, *Mol. Phys.* **17**, 473 (1969).

256. C. A. Nicolaides, M. G. Papadopoulos, and J. Waite, *Theoret. Chim. Acta* **61**, 427 (1982).

257. M. G. Papadopoulos, J. Waite, and C. A. Nicolaides, *J. Chem. Phys.* **77**, 2527 (1982).

258. J. Waite, M. G. Papadopoulos, and C. A. Nicolaides, *J. Chem. Phys.* **77**, 2536 (1982).

259. J. Waite and M. G. Papadopoulos, *J. Comput. Chem.* **4**, 578 (1983).

260. J. Waite and M. G. Papadopoulos, *J. Chem. Phys.* **80**, 3503 (1984).

261. J. Waite and M. G. Papadopoulos, *J. Org. Chem.* **49**, 3837 (1984).

262. J. Waite and M. G. Papadopoulos, *J. Mol. Struct.* **125**, 155 (1984).

263. J. Waite and M. G. Papadopoulos, *J. Mol. Struct.* (*THEOCHEM*) **108**, 247 (1984).

264. J. Waite and M. G. Papadopoulos, *J. Chem. Phys.* **83**, 4047 (1985).

265. J. Waite and M. G. Papadopoulos, *J. Chem. Soc. Faraday Trans. 2* **81**, 433 (1985).

266. J. Waite and M. G. Papadopoulos, *J. Chem. Phys.* **82**, 1427 (1985).

267. M. G. Papadopoulos and J. Waite, *J. Chem. Phys* **82**, 1435 (1985).

268. J. Waite and M. G. Papadopoulos, *Z. Naturforsch. A* **40**, 142 (1985).

269. M. G. Papadopoulos and J. Waite, *Chem. Phys. Lett.* **135**, 361 (1986).

270. M. G. Papadopoulos and J. Waite, *J. Phys. Chem.* **90**, 5491 (1986).

271. J. Waite and M. G. Papadopoulos, *Can. J. Chem.* **66**, 1440 (1988).

272. J. Waite and M. G. Papadopoulos, *J. Phys. Chem.* **93**, 43 (1989).

273. P. A. Christiansen and E. A. McCullough, Jr., *Chem. Phys. Lett.* **63**, 570 (1979).

274. G. D. Purvis III and R. J. Bartlett, *Phys. Rev.* **A23**, 1594 (1981).

275. M. J. S. Dewar, *Faraday Discuss. Chem. Soc.* **62**, 345 (1977).

276. J. Del Bene and H. H. Jaffé, *J. Chem. Phys.* **48**, 1807 (1968).

277. P. Francois, P. Carles, and M. Rajzman, *J. Chim. Phys.* **74**, 606 (1977).

278. P. Francois, P. Carles, and M. Rajzmann, *J. Chim. Phys.* **76**, 328 (1979).

279. H. Shinoda and T. Akutagawa, *Bull. Chem. Soc. Japan* **48**, 3431 (1975).

280. E. N. Svendsen and T. Stroyer-Hansen, *Int. J. Quantum Chem.* **13**, 235 (1978).

281. F. T. Marchese and H. H. Jaffé, *Theoret. Chim. Acta* **45**, 241 (1977).

282. F. T. Marchese and H. H. Jaffé, *J. Mol. Struct.* **86**, 97 (1981).

283. (a) J. O. Morley, *J. Chem. Soc. Faraday Trans.* **87**, 3009, 3015 (1991); (b) J. O. Morley and D. Pugh, *J. Chem. Soc. Faraday Trans.* **87**, 3021 (1991); (c) J. O. Morley, P. Pavlides, and D. Pugh, *J. Chem. Soc. Faraday Trans.* 2 **85**, 1789 (1989).

284. V. J. Docherty, D. Pugh, and J. O. Morley, *J. Chem. Soc. Faraday Trans.* 2 **81**, 1179 (1985).

285. C. C. Teng and A. F. Gorito, *Phys. Rev.* B **28**, 6766 (1983).

286. C. C. Teng and A. F. Garito, *Phys. Rev. Lett.* **50**, 350 (1983).

287. S. J. Lalama, K. D. Singer, A. F. Garito, and K. N. Desai, *Appl. Phys. Lett.* **39**, 940 (1981).

288. S. J. Lalama and A. F. Garito, *Phys. Rev.* A **20**, 1179 (1979).

289. J. A. Morrell and A. C. Albrecht, *Chem. Phys. Lett.* **64**, 46 (1979).

290. J. A. Morell, A. C. Albrecht, K. H. Levin, and C. L. Tang, *J. Chem. Phys.* **71**, 5063 (1979).

291. J. J. C. Teixeira-Dias and P. J. Sarre, *J. Chem. Soc. Faraday Trans.* 2 **71**, 906 (1975).

292. H. Meyer, K. W. Schulte, and A. Schweig, *Chem. Phys. Lett.* **31**, 187 (1975).

293. R. Mathies and A. C. Albrecht, *J. Chem. Phys.* **60**, 2500 (1974).

294. J. A. Pople, D. L. Beveridge, and P. A. Dobosh, *J. Chem. Phys.* **47**, 2026 (1967).

295. J. Zyss, *J. Chem. Phys.* **70**, 3333, 3341 (1979).

296. J. Zyss, *J. Chem. Phys.* **71**, 909 (1979).

297. P. J. Bounds, *Chem. Phys. Lett.* **70**, 143 (1980).

298. (a) J. Zyss and G. Berthier, *J. Chem. Phys.* **77**, 3635 (1982); (b) W. A. Parkinson and M. C. Zerner, *J. Chem. Phys.* **94**, 478 (1991).

299. N. C. Baird and M. J. S. Dewar, *J. Chem. Phys.* **50**, 1262 (1969).

300. N. C. Baird, M. J. S. Dewar and R. Sustmann, *J. Chem. Phys.* **50**, 1275 (1969).

301. M. J. S. Dewar and E. Haselbach, *J. Am. Chem. Soc.* **92**, 590 (1970).

302. N. Bodor, M. J. S. Dewar, A. Harget, and E. Haselbach, *J. Am. Chem. Soc.* **92**, 3854 (1970).

303. R. C. Bingham, M. J. S. Dewar, and D. H. Lo, *J. Am. Chem. Soc.* **97**, 1285, 1294, 1302, 1307 (1975).

304. H. Meyer and A. Schweig. *Theoret. Chim. Acta* **29**, 375 (1973).

305. M. J. S. Dewar, R. C. Haddon, and S. H. Suck, *J. Chem. Soc. Chem. Commun.* 612 (1974).

306. M. J. S. Dewar, S. H. Suck, P. K. Weiner, and J. B. Bergmann, Jr., *Chem. Phys. Lett.* **38**, 226 (1976).

307. M. J. S. Dewar, Y. Yamaguchi, and S. H. Suck, *Chem. Phys. Lett* **59**, 541 (1978).

308. M. J. S. Dewar and W. Thiel, *J. Am. Chem. Soc.* **99**, 4899 (1977).

309. H. A. Kurtz, J. J. P. Stewart, and K. Dieter, *J. Comput. Chem.* **11**, 82 (1990).

310. M. J. S. Dewar and J. J. P. Stewart, *Chem. Phys. Lett.* **111**, 416 (1984).

311. (a) B. H. Cardelino, C. E. Moore, and R. E. Stickel, *J. Phys. Chem.* **95**, 8645 (1991); (b) H. A. Kurtz, *Int. J. Quantum Chem. Quant. Chem. Symp.* **24**, 791 (1990).

312. M. J. S. Dewar, E. G. Zoebisch, E. F. Healy, and J. J. P. Stewart, *J. Am. Chem. Soc.* **107**, 3902 (1985).

313. J. J. P. Stewart, *J. Comput. Chem.* **10**, 209 (1989).

314. J. N. Murrell, *Struct. Bonding* (*Berlin*) **32**, 93 (1977).

315. J. Applequist, J. R. Carl, and K. K. Fong, *J. Am. Chem. Soc.* **94**, 2952 (1972).

316. N. J. Bridge and A. D. Buchingham; *Proc. R. Soc.* (*London*) *Ser. A* **295**, 334 (1966).

317. M. P. Bogaard, B. J. Orr, A. D. Buckingham, and G. L. D. Ritchie, *J. Chem. Soc. Faraday Trans.* 2 **74**, 1573 (1978).

318. R. J. W. Le Févre and B. P. Rao, *J. Chem. Soc.* 1465 (1958).

319. R. J. W. Le Févre and K. M. S. Sundaram, *J. Chem. Soc.* 4442 (1963).

320. P. J. Bounds and R. W. Munn, *Chem. Phys.* **24**, 343 (1977).

321. R. J. W. Le Févre and L. Radom, *J. Chem. Soc. B* 1295 (1967).

322. C. G. Le Févre and R. J. W. Le Févre, *J. Chem. Soc.* 1641 (1955).

323. R. J. W. Le Févre, L. Radom, and G. L. D. Ritchie, *J. Chem. Soc. B* 775 (1968).

324. R. J. W. Le Févre and D. S. N. Murthy, *Aust. J. Chem.* **21**, 1903 (1968).

325. J. Schuyer, L. Blom, and D. W. van Krevelen, *Trans. Faraday Soc.* **49**, 1391 (1953).

326. C. C. Strametz and H. H. Schmidtke, *Theoret. Chim. Acta* **42**, 13 (1976).

327. M. J. Aroney and R. J. W. Le Févre, *J. Chem. Soc.* 2161 (1960).

328. F. Mulder, G. Van Dijk, and C. Huiszoon, *Mol. Phys.* **38**, 577 (1979).

329. J. F. Ward and I. J. Bigio, *Phys. Rev. A* **11**, 60 (1975).

330. J. F. Ward and C. K. Miller, *Phys. Rev. A* **19**, 826 (1979).

331. B. F. Levine, *Chem. Phys. Lett* **37**, 516 (1976).

332. J. L. Oudar, D. S. Chemla, and E. Batifol, *J. Chem. Phys.* **67**, 1626 (1977).

333. J. L. Oudar and D. S. Chemla, *J. Chem. Phys.* **66**, 2664 (1977).

334. A. Dulcic and C. Sauteret. *J. Chem. Phys.* **69**, 3454 (1978).

335. J. F. Ward and D. S. Elliot, *J. Chem. Phys.* **69**, 5438 (1978).

336. C. G. Bethea, *J. Chem. Phys.* **69**, 1312 (1978).

337. E. Schrödinger, *Ann. Phys.* **80**, 437 (1926).

338. G. Wentzel, *Z. Phys.* **38**, 635 (1926).

339. P. S. Epstein, *Phys. Rev.* **28**, 695 (1926).

340. S. Doi, *Proc. Phys. Math. Soc.* (*Japan*) **10**, 223 (1928).

341. C. Schwartz, *Phys. Rev.* **123**, 1700 (1961).

342. H. P. Kelly, *Phys. Rev.* **136**, B896 (1964).

343. J. Waite and M. G. Papadopoulos, *J. Chem. Phys.* **85**, 2831 (1986).

344. M. A. Spackman, *J. Phys. Chem.* **93**, 7594 (1989).

345. S. P. Karna, M. Dupuis, E. Perrin, and P. N. Prasad, *J. Chem. Phys.* **92**, 7418 (1990).

346. D. P. Chong and S. R. Langhoff, *J. Chem. Phys.* **93**, 570 (1990).

347. G. Maroulis and A. J. Thakkar, *J. Chem. Phys.* **93**, 652 (1990).

348. G. Maroulis and A. J. Thakkar, *J. Chem. Phys.* **93**, 4164 (1990).

349. D. R. Beck and D. H. Gay, *J. Chem. Phys.* **93**, 7264 (1990).

350. K. E. Laidig and R. F. W. Bader, *J. Chem. Phys.* **93**, 7213 (1990).

351. P. L. Polavarapu, *Chem. Phys. Lett.* **174**, 511 (1990).

352. G. Maroulis, *J. Chem. Phys.* **94**, 1182 (1991).

353. J. E. Rice, P. R. Taylor, T. J. Lee, and J. Almlöf, *J. Chem. Phys.* **94**, 4972 (1991).

354. J. Perez and M. Dupuis, *J. Phys. Chem.* **95**, 6525 (1991).

355. G. Maroulis and A. J. Thakkar, *J. Chem. Phys.* **95**, 9060 (1991).

356. M. Urban and A. J. Sadlej, *Theoret. Chim. Acta* **78**, 189 (1990).

357. M. Urban, G. H. F. Diercksen, A. J. Sadlej, and J. Noga, *Theoret. Chim. Acta* **77**, 29 (1990).

358. F. Visser, P. E. S. Wormer, and P. Stam, *J. Chem. Phys.* **79**, 4973 (1983).

359. T. Stroyer-Hansen and E. N. Svendsen, *J. Chem. Phys.* **84**, 1950 (1986).

360. H. Sekino and R. J. Bartlett, *J. Chem. Phys.* **85**, 976 (1986).

361. K. Sasagane, K. Mori, A. Ichihara, and R. Itoh, *J. Chem. Phys.* **92**, 3619 (1990).

362. J. E. Rice, R. D. Amos, S. M. Colwell, N. C. Handy, and J. Sanz, *J. Chem. Phys.* **93**, 8828 (1990).

363. S. P. Karna, P. N. Prasad, and M. Dupuis, *J. Chem. Phys.* **94**, 1171 (1991).

364. M. A. Spackman, *J. Chem. Phys.* **94**, 1288 (1991).

365. M. A. Spackman, *J. Chem. Phys.* **94**, 1295 (1991).

366. S. P. Karna, E. Perrin, P. N. Prasad, and M. Dupuis, *J. Phys. Chem.* **95**, 4329 (1991).

367. H. Sekino and R. J. Bartlett, *J. Chem. Phys.* **94**, 3665 (1991).

368. A. D. McLean and M. Yoshimine, *J. Chem. Phys.* **46**, 3682 (1967).

369. R. D. Amos, *Mol. Phys.* **38**, 33 (1979).

370. D. M. Bishop and C. Pouchan, *J. Chem. Phys.* **80**, 789 (1984).

371. D. M. Bishop and G. Maroulis, *J. Chem. Phys.* **82**, 2380 (1985).

372. H. F. King and A. Komornicki, *J. Chem. Phys.* **84**, 5645 (1986).

373. C. E. Dykstra and P. G. Jasien, *Chem. Phys. Lett.* **109**, 388 (1984).

374. N. C. Handy and H. F. Schaefer III, *J. Chem. Phys.* **81**, 5031 (1984).

375. T. Takada, M. Dupuis, and H. F. King, *J. Comput. Chem.* **4**, 234 (1983).

376. M. Dupuis, *J. Chem. Phys.* **74**, 5758 (1981).

377. J. A. Pople, R. Krishnan, H. B. Schlegel, and J. S. Binkley, *Int. J. Quantum Chem.* **S13**, 225 (1979).

378. T. S. Nee, R. G. Parr, and R. J. Bartlett, *J. Chem. Phys.* **64**, 2216 (1976).

379. A. J. Sadlej, *Chem. Phys. Lett.* **47**, 50 (1977).

380. A. J. Sadlej, *Mol. Phys.* **34**, 731 (1977).

381. A. J. Sadlej, *Theoret. Chim. Acta* **47**, 205 (1978).

382. K. Szalewicz, L. Adamowicz, and A. J. Sadlej, *Chem. Phys. Lett.* **61**, 548 (1979).

383. J. A. Hudis and R. Ditchfield, *Chem. Phys.* **86**, 455 (1984).

384. Y. Osamura, Y. Yamaguchi, and H. F. Schaefer, *J. Chem. Phys.* **75**, 2919 (1981).

385. Y. Yamaguchi, M. Frisch, J. Gaw, H. F. Schaefer, and J. S. Binkley, *J. Chem. Phys.* **84**, 2262 (1986).

386. Y. Osamura, Y. Yamaguchi, and H. F. Schaefer, *Chem. Phys.* **103**, 227 (1986).

387. H.-J. Werner and W. Meyer, *Mol. Phys.* **31**, 855 (1976).

388. P. A. Christiansen and E. A. McCullough, *Chem. Phys. Lett.* **55**, 439 (1978).

389. D. J. Malik and C. E. Dykstra, *J. Chem. Phys.* **83**, 6307 (1985).

390. C. E. Dykstra, *J. Chem. Phys.* **82**, 4120 (1985).

391. S.-Y. Liu and C. E. Dykstra, *Chem. Phys. Lett.* **119**, 407 (1985).

392. C. E. Dykstra, S.-Y. Liu, and D. J. Malik, *J. Mol. Struct.* (*THEOCHEM*) **135**, 357 (1986).

393. S.-Y. Liu, C. E. Dykstra, K. Kolenbrander, and J. M. Lisy, *J. Chem. Phys.* **85**, 2077 (1986).

394. D. E. Bernholdt, S.-Y. Liu, and C. E. Dykstra, *J. Chem. Phys.* **85**, 5120 (1986).

395. S.-Y. Liu, C. E. Dykstra, and D. J. Malik, *Chem. Phys. Lett.* **130**, 403 (1986).

396. S.-Y. Liu and C. E. Dykstra, *J. Phys. Chem.* **91**, 1749 (1987).

397. T. H. Dunning, *J. Chem. Phys.* **55**, 716, 3958 (1971).

398. S. Huzinaga, *J. Chem. Phys.* **42**, 1293 (1965).

399. R. J. Bartlett and G. D. Rurvis III, *Phys. Rev. A* **20**, 1313 (1979).

400. G. Karlström, B. O. Roos, and A. J. Sadlej, *Chem. Phys. Lett.* **86**, 374 (1982).

401. A. Dalgarno, *Proc. R. Soc. London Ser. A* **251**, 282 (1959).

402. M. Yoshimine and R. P. Hurst, *Phys. Rev. A* **135**, 612 (1964).

403. S. Kaneko, *J. Phys. Soc. Japan* **14**, 1600 (1959).

404. L. C. Allen, *Phys. Rev.* **118**, 167 (1960).

405. A. Dalgarno and J. McNamee, *Proc. Phys. Soc.* (*London*) **77**, 673 (1961).

406. A. L. Stewart, *Proc. Phys. Soc.* (*London*) **77**, 447 (1961).

407. P. Lazzeretti and R. Zanasi, *J. Chem. Phys.* **74**, 5216 (1981).

408. J. B. Hurst, M. Dupuis, and E. Clementi, *J. Chem. Phys.* **89**, 385 (1988).

409. L. Adamowicz and R. J. Bartlett, *J. Chem. Phys.* **83**, 4988 (1986); *Phys. Rev. A* **37**, 1 (1988).

410. S. Karna and M. Dupuis, *Chem. Phys. Lett.* **171**, 201 (1990).

411. J. Almlof and P. R. Taylor, *J. Chem. Phys.* **92**, 551 (1990).

412. R. Ditchfield, W. J. Hehre, and J. A. Pople, *J. Chem. Phys.* **54**, 724 (1971).

413. W. J. Hehre, R. Ditchfield, and J. A. Pople, *J. Chem. Phys.* **56**, 2257 (1972).

414. G. H. F. Diercksen, V. Kellö, B. O. Roos, and A. J. Sadlej, *Chem. Phys.* **77**, 93 (1983).

415. G. H. F. Diercksen, V. Kellö, and A. J. Sadlej, *Chem. Phys. Lett.* **95**, 226 (1983).

416. S. A. Kucharski, Y. S. Lee, G. D. Purvis III, and R. J. Bartlett, *Phys. Rev. A* **29**, 1619 (1984).

417. I. Cernusak, G. H. F. Diercksen, and A. J. Sadlej, *Phys. Rev. A* **33**, 814 (1986).

418. H.-J. Werner and W. Meyer, *Phys. Rev. A* **13**, 13 (1976).

419. (a) D. P. Chong and S. R. Langhoff, *J. Chem. Phys.* **84**, 5606 (1986); (b) R. Ahlrichs, P. Scharf, and C. Ehrhardt, *J. Chem. Phys.* **82**, 890 (1985).

420. P. R. Taylor, T. J. Lee, J. E. Rice, and J. Almlöf *Chem. Phys. Lett.* **163**, 359 (1989).

421. K. Raghavachari, G. W. Trucks, J. A. Pople, and M. Head-Gordon, *Chem. Phys. Lett.* **157**, 479 (1989).

422. J. Almlöf and P. R. Taylor, *J. Chem. Phys.* **86**, 4070 (1987).

423. F. B. van Duijneveldt, IBM Research Report No. RJ 945 (1971).

424. G. Maroulis and A. J. Thakkar, *J. Chem. Phys.* **88**, 7623 (1988); **89**, 6558 (1988).

425. G. Maroulis and A. J. Thakkar, *J. Chem. Phys.* **90**, 366 (1989); **92**, 812 (1990).

426. G. Maroulis and A. J. Thakkar, *Chem. Phys. Lett.* **156**, 87 (1989).

427. G. Maroulis, *J. Chem. Phys.* **96**, 6048 (1992).

428. T. H. Dunning, Jr. and P. J. Hay, in H. F. Schaefer III (Ed.), *Methods of Electronic Structure Theory*, Plenum, New York, 1977.

429. C. J. Jameson and P. W. Fowler, *J. Chem. Phys.* **85**, 3432 (1986).

430. F. Visser and P. E. S. Wormer, *Chem. Phys.* **92**, 129 (1985).

431. A. J. Sadlej and B. O. Roos, *Theoret. Chim. Acta* **76**, 173 (1989).

432. G. H. F. Diercksen, V. Kellö, and A. J. Sadlej, *J. Chem. Phys.* **79**, 2918 (1983).

433. P. D. Dacre, *J. Chem. Phys.* **80**, 5677 (1984).

434. L. R. Kahn, P. Baybutt, and D. G. Trulhar, *J. Chem. Phys.* **65**, 3826 (1976).

435. W. J. Stevens, H. Basch, and M. Krauss, *J. Chem. Phys.* **81**, 6026 (1984).

436. D. D. Konowalow, M. E. Rosenkrantz, W. J. Stevens, and M. Krauss, *Chem. Phys. Lett.* **64**, 317 (1979).

437. M. E. Rosenkrantz, W. J. Stevens, M. Krauss, and D. D. Konowalow, *J. Chem. Phys.* **72**, 2525 (1980).

438. W. J. Stevens and M. Krauss, *J. Phys. B* **16**, 2921 (1983).

439. E. Perrin, P. N. Prasad, P. Mougenot, and M. Dupuis, *J. Chem. Phys.* **91**, 4728 (1989).

440. B. O. Roos, *Adv. Chem. Phys.* **69**, 399 (1987).

441. D. M. Bishop, *Rev. Mod. Phys.* **62**, 343 (1990).

442. P. Jorgensen, *Annu. Rev. Phys. Chem.* **26**, 359 (1975).

443. C. W. McCurdy, T. N. Rescigno, D. L. Yeager, and V. McKoy, in H. F. Schaefer III (Ed.), *Methods of Electronic Structure*, Plenum, New York, 1977.

444. J. Oddershede, P. Jorgensen, and D. L. Yeager, *Comp. Phys. Rep.* **2**, 33 (1984).

445. J. Oddershede, *Adv. Quantum Chem.* **11**, 275 (1978).

446. E. S. Nielsen, P. Jorgensen, and J. Oddershede, *J. Chem. Phys.* **73**, 6238 (1980).

447. S. R. Langhoff, C. W. Bauschlicher, and D. P. Chong, *J. Chem. Phys.* **78**, 5287 (1983).

448. G. D. Zeiss, W. R. Scott, N. Suzuki, D. P. Chong, and S. R. Langhoff, *Mol. Phys.* **37**, 1543 (1979).

449. E. N. Svendsen and J. Oddershede, *J. Chem. Phys.* **71**, 3005 (1979).

450. J. Oddershede and E. N. Svendsen, *Chem. Phys.* **64**, 359 (1982).

451. P. J. Fortune and P. R. Certain, *J. Chem. Phys.* **61**, 2620 (1974).

452. P. Lallemand, D. J. David, and B. Bigot, *Mol. Phys.*, **27**, 1029 (1974).

453. J. W. Kress and J. J. Kozak, *J. Chem. Phys.* **66**, 4516 (1977).

454. P. D. Dacre, *Mol. Phys.* **36**, 541 (1978).

455. D. W. Oxtoby and W. M. Gelbert, *Mol. Phys.* **29**, 1569, **30**, 535 (1975).

456. D. G. Bounds, A. Hinchliffe, and J. H. R. Clarke, *Chem. Phys. Lett.* **45**, 367 (1977).

457. D. G. Bounds and A. Hinchliffe, *Chem. Phys. Lett.* **54**, 289 (1977).

458. D. G. Bounds and A. Hinchliffe, *Chem. Phys. Lett.* **56**, 303 (1978).

459. D. G. Bounds, *Chem. Phys.* **42**, 405 (1979).

460. D. G. Bounds and A. Hinchliffe, *J. Chem. Phys.* **72**, 298 (1980).

461. A. Hinchliffe, *Chem. Phys. Lett.* **70**, 610 (1980).

462. D. G. Bounds, A. Hinchliffe, and C. J. Spicer, *Mol. Phys.* **42**, 73 (1981).

463. A. Hinchliffe, *Adv. Mol. Rel. Int. Processes* **20**, 63 (1981).

464. A. Hinchliffe, *Adv. Mol. Rev. Int. Processes* **22**, 251 (1982).

465. R. W. Munn, C. J. Spicer, and A. Hinchliffe, *Chem. Phys. Lett.* **90**, 417 (1982).

466. R. D. Amos, N. C. Handy, P. J. Knowles, J. E. Rice, and A. J. Stone, *J. Chem. Phys.* **89**, 2186 (1985).

467. L. Frommhold, K. H. Hong, and M. H. Proffitt, *Mol. Phys.* **35**, 665 (1978).

468. R. A. Aziz, V. P. S. Nain, J. S. Carley, W. L. Taylor, and G. T. McConville, *J. Chem. Phys.* **70**, 4330 (1979).

469. M. H. Proffitt and L. Frommhold, *Phys. Rev. Lett.* **42**, 1473 (1979); *J. Chem. Phys.* **72**, 1377 (1980).

470. G. del Conde and P. S. Bagus, *Theoret. Chim. Acta* **45**, 121 (1977).

471. D. G. Bounds, *Mol. Phys.* **38**, 2099 (1979).

472. A. A. Hasanein, M. Ferrario, and M. W. Evans, *Adv. Mol. Rel. Int. Processes* **20**, 47 (1981).

473. A. A. Hasanein, M. Ferrario, and M. W. Evans, *Adv. Mol. Rel. Int. Processes* **20**, 215 (1981).

474. A. A. Hasanein, *J. Comput. Chem.* **5**, 528 (1984).

475. A. A. Hasanein and M. W. Evans, *J. Mol. Liquids* **29**, 45 (1984).

476. J. C. Raich and N. S. Gillis, *J. Chem. Phys.* **66**, 846 (1977).

477. D. F. Heller and W. M. Gelbart, *Chem. Phys. Lett.* **27**, 359 (1974).

478. D. G. Bounds and A. Hinchliffe, *Mol. Phys.* **38**, 717 (1979).

479. A. Kumar and W. J. Meath, *Can J. Chem.* **63**, 1616 (1985).

480. D. P. Shelton, *Phys. Rev. Lett.* **62**, 2660 (1989); *Phys. Rev. A* **42**, 2578 (1990).

481. R. M. Berns and P. E. S. Wormer, *Mol. Phys.* **44**, 1215 (1981).

482. D. M. Bishop and L. M. Cheung, *J. Chem. Phys.* **72**, 5125 (1980).

483. G. Maroulis and D. M. Bishop, *Theoret. Chim. Acta* **69**, 161 (1986).

484. P. Lazzeretti, E. Rossi, and R. Zanassi, *J. Phys. B* **15**, 521 (1982).

485. E.-A. Reinsch, *J. Chem. Phys.* **83**, 5784 (1985).

486. E. Barbagli and M. Maestro, *Chem. Phys. Lett.* **24**, 567 (1974).

487. J. G. C. M. van Duijneveldt-van de Rijdt and F. B. van Duijneveldt *J. Mol. Struct.* **89**, 185 (1982).

488. P. W. Fowler, *Mol. Phys.* **47**, 355 (1982).

489. M. Duran, Y. Yamaguchi, R. B. Remington, and H. F. Schaefer III, *CHem. Phys.* **122**, 201 (1988).

490. M. Jaszunski, A. Rizzo, and D. Yeager, *Chem. Phys. Lett.* **149**, 79 (1988).

491. J. E. H.-Haverkort, F. Baas, and J. J. M. Beenakker, *Chem. Phys.* **79**, 105 (1983).

SELECTION RULES FOR COHERENT AND INCOHERENT NONLINEAR OPTICAL PROCESSES

G. E. STEDMAN

*Department of Physics, University of Canterbury,
Christchurch 1, New Zealand*

CONTENTS

Modern Nonlinear Optics, Part 2, Edited by Myron Evans and Stanisław Kielich. Advances in Chemical Physics Series, Vol. LXXXV.
ISBN 0-471-57546-1 © 1993 John Wiley & Sons, Inc.

I. INTRODUCTION

This chapter ranges widely with a light touch. It is an intimate mixture of three ingredients: undergraduate-level material, which hopefully enhances readability and which I have found hard to come by and useful for my students; a graduate-level review and critique of related theoretical analyses of field-dependent and nonlinear optical properties (never exhaustive, but at least defining my perspective); and some totally novel material particularly in regard to coherent optical processes, Berry phase in optics, and ring laser experiments. The last of these are far from being completely worked out, but rather open new doors to further selection rules and indicate the need for a far more thorough analysis than is attempted here. My main concern is to forge links between some of the very different theoretical analyses that are now available and to illustrate their wide ramifications. A recurrent theme is that of gyrotropic effects, and the extent to which the circumstances of their occurrence are dictated by time reversal considerations. We commence with that topic.

Gyrotropic optical effects, as defined by Stedman,[1] are those whose contributions to a quantum process are reversed on exchanging left and right circularly polarized light beams for all participating photons. They have previously been discussed under a wide variety of names, including chiral or chiroptical effects or asymmetries; see, for example, Stedman,[2] Blum,[3] and Silverman.[4] While it is an exaggeration to say that these effects give a unique window on parity violation, their widespread interest certainly reflects the signature they give of experimental conditions when the conventional symmetries associated with parity appear to be rescinded. Gyrotropy is certainly neither complete nor unique, nor even my preferred basis for classification; other indexes that usefully contribute to a complete classification are discussed later.

In the case of the natural circular birefringence induced by the weak interaction into atomic physics, this parity violation is real; the fundamental coupling Hamiltonian connecting the nuclear neutrons and the electronic electrons via the Z^0 boson is itself pseudoscalar. In the theory of anyonic systems, which is of possible relevance to the quantum Hall effect and even high-temperature superconductors, the parity violation whose detection via circular birefringence has been mooted is reflected in a pseudoscalar Lagrangian term in the $2 + 1$ dimensional field theory appropriate to the planar system, although the fundamental physics in the more standard $3 + 1$ quantum field theory underlying such experiments conserves parity.

However, in the vast majority of applications of everyday interest there is no question about the conservation of parity in the fundamental theory; yet the experimental conditions conspire to give the effect a gyrotropic character. There are three main reasons for this. First, the material system under study may have an intrinsic handedness, being an enantiomer of a molecule that is not equivalent to its mirror reflection. Second, an external electric field that is reversed under parity may induce gyrotropy. Third, the choice of interaction between matter and radiation may itself suffice to induce gyrotropy even in the absence of external fields and of enantiomeric forms. In this chapter I review and extend a wide-ranging literature on such matters as gyrotropic optical effects and the recognition of their occurrence and geometrical character using established time reversal, parity, and symmetry arguments in the notation of Stedman.[1, 2, 5]

As mentioned above, some topics discussed in this chapter are novel as far as I know. First, a selection rule based on time reversal considerations is given for all coherent optical effects; it gives, for example, the result that any coherent process involving an odd number of photons is forbidden within fluids for all multipoles. Second, selection rules are given for the conditions under which a given nonlinear optical medium will generate an optical effect leading to a nonreciprocal path length and so a signal in a ring laser gyro, and the extent to which this overlaps the concept of gyrotropy is explained. Third, we consider the reasons for the circumstances under which a beam traversing a medium, being reflected, and then returned will suffer a cumulative or canceling optical rotation, depending on the choice of mirror (a standard mirror versus an optical phase conjugation mirror) and of the source of chiral asymmetry (Faraday effect versus natural optical activity versus Berry phase). In such matters the answers, while elegant, are not simple.

In the works referenced I have developed and applied a general approach to the analysis of selection rules for optical effects that may be exhibited by material systems. The aim is to incorporate all the restrictions associated with the following:

- The point group symmetry of the material, including whether or not it is centric (that is, centrosymmetric), although little discussion will be given for other than fluid systems
- The choice of polarization and orientation of the light beams relative to the material
- The type of optical process under consideration, including whether it is gyrotropic
- What multipoles of coupling are involved, in particular whether interference between multipoles is required

The basic method of proof is standard: Take the various matrix elements contributing to the rate of the quantum process according to the extended Golden Rule (including for example a Kramers-Heisenberg term), and summarize the effects of point group symmetry, time reversal, and parity conjugation.

Previously I considered incoherent optical processes, in which the amplitudes for each scatterer in a fluid or other ensemble do not interfere. This applies when there is a net energy exchange between the electromagnetic field and the material system, and hence a net change of state of the system. When there is no energy interchange between these subsystems, the coherent terms in the sum over intermediate states will dominate. However an incoherent background will accompany them in practice. For example, hyper-Rayleigh scattering will accompany second-harmonic generation (see Andrews and Sherborne[65]; see also Andrews' contribution to this volume); in this review, hyper-Rayleigh scattering is regarded as an incoherent process.

In the case of coherent systems, two principal rules emerge, associated with time reversal and parity considerations, respectively, the latter being applicable only for a centric system (one that is equivalent to its mirror reflection, to within a rotation). In addition, the point group symmetry of individual systems, or the rotational symmetry of their net contributions in a fluid, lead to further selection rules as well as to precise formulae for the geometrical dependence of the process (upon orientation, choice of polarization vectors and the like). This has been developed previously for incoherent processes using a diagram formalism, which is a natural counterpart to and outcome of a perturbative scheme in nonrelativistic quantum field theory, as well as a convenient and general method of applying group theoretical constraints.

II. PERTURBATIVE APPROACH TO COHERENT AND INCOHERENT PROCESSES

A. Introduction

We first review some very general results in perturbative quantum mechanics. The aim here is nevertheless a novel one: to motivate and underpin the fundamental origins of certain connections between coherent and incoherent optical processes, which will be of interest when we compare the selection rules for these cases. It helps to do so in the context of simple examples. The results have far wider validity than these simple examples, and a good grasp of the underlying structure is a prerequisite to an intelligent use of our selection rules, and indeed the equivalent rules of other authors.

We choose two examples for this purpose: refraction and natural optical activity. In each case, an absorptive and dispersive physical property are included. Refraction is inevitably accompanied by absorption in the region of anomalous dispersion, and in quantum theory refraction may (as we describe later) be regarded as the consequence of virtual absorption processes. Optical activity covers both optical rotation (of the plane of polarization of the beam; also known as circular birefringence) and circular dichroism (preferential absorption of one handedness of polarization). Optical activity has over the last century been analyzed from a wide variety of viewpoints, as is evident from the compendium of techniques and discussions given in Lakhtakia.[6]

Any such refractive or dispersive effect inevitably has an absorptive counterpart. In each case the magnitude of the former involves such terms as

$$\sum_i \frac{|\langle \text{initial}|V|i\rangle|^2}{E_{\text{initial}} - E_i - \hbar\omega} \tag{1}$$

where in this example V is the matter–radiation interaction, $-q\mathbf{A}(\mathbf{r}) \cdot \mathbf{p}/m$, and ω is the frequency of the incoming photon, which in this example is destroyed in a virtual intermediate state, consequent on the nonresonant electronic absorption process $|\text{initial}\rangle \rightarrow |i\rangle$. In the refractive process, this virtual absorption is later time reversed to conserve energy and the photon proceeds on its merry way. But these events may be thought of as delaying it on its journey across the material, changing the net speed of light and thus the refractive index in proportion to this expression. Interesting physical explanations of virtual photons can be found in Feynman[7] and Lawson.[8] This effect requires forward scattering so that the photon has its original direction as well as its original energy, though not necessarily its original polarization state.

At resonance, when the denominator in Eq. (1) vanishes, real absorption can occur at the rate given by the Golden Rule:

$$\frac{2\pi}{\hbar} \sum_i |\langle \text{initial}|V|i\rangle|^2 \delta(E_{\text{initial}} - E_i - \hbar\omega). \tag{2}$$

B. Fluctuation–Dissipation Theorem

The generality of the link between such fluctuation and dissipation effects is enshrined in the fluctuation–dissipation theorem. This is well illustrated in the superb summary of Callen,[9] who lists as examples of fluctuation–dissipation partners Stokes' law retardation of a sphere in a viscous fluid and Brownian motion, and spontaneous emission of photons and vacuum excitations through virtual processes in quantum electrodynamics. Mathematically, both Eqs. (1) and (2) can be obtained from the complex susceptibility

$$\sum_i \frac{|\langle \text{initial}|V|i\rangle|^2}{E_{\text{initial}} - E_i - \hbar\omega - i\varepsilon} \tag{3}$$

with the use of the relation

$$\lim_{\varepsilon \to 0+} \frac{1}{\omega - i\varepsilon} = \mathscr{P}\frac{1}{\omega} + i\pi\delta(\omega) \tag{4}$$

Although Eq. (4) might be dismissed as a mere mathematical identity for addicts of complex variable theory, this equation is of basic importance in quantum field theory, for example, in distinguishing the self-energy or mass shift from its imaginary part, the linewidth, or decay constant. Such expressions as that for the complex susceptibility arise naturally as self-energy terms in a field theoretic approach to the Golden Rule (see Stedman[2, 10]).

We now explain the ubiquity of the fluctuation–dissipation theorem from the viewpoint of perturbation theory. Any dissipative process defined by a set of initial and final states will occur at a Golden Rule rate given in a thermally equilibrated system by a sum over final states and a thermal average over initial states:

$$2\pi/\hbar \sum_{\text{initial, final}} \exp(-\beta E_{\text{initial}})|\langle \text{initial}|V|\text{final}\rangle|^2 \delta(E_{\text{initial}} - E_{\text{final}}) \quad (5)$$

If we Fourier transform the delta function, with time as the variable conjugate to energy, $\delta(E) = (1/2\pi)\int \exp[iEt/\hbar]\,dt$, various exponential factors arise. The full system energies in these may be replaced by the full Hamiltonian, provided the exponential factors are appropriately ordered:

$$\sum_{\text{initial, final}} \exp(iE_{\text{initial}}t/\hbar)\exp(-\beta E_{\text{initial}})\langle \text{initial}|V|\text{final}\rangle$$

$$\times \exp(-iE_{\text{final}}t/\hbar)\langle \text{final}|V|\text{initial}\rangle$$

$$= \sum_{\text{initial, final}} \langle \text{initial}|\exp(-\beta H + iHt/\hbar)V$$

$$\times \exp(-iHt/\hbar)|\text{final}\rangle\langle \text{final}|V|\text{initial}\rangle \quad (6)$$

The sum over final states then disappears through completeness. This gives an expression proportional to $\langle\langle V(t)V(0)\rangle\rangle$ where $\langle\langle \cdots \rangle\rangle$ denotes the thermal average $\langle \exp(-\beta H) \cdots \rangle/\langle \exp(-\beta H)\rangle$, and $V(t)$ denotes the Heisenberg operator $\exp(itH/\hbar)V\exp(-itH/\hbar)$. This is a correlation function basic to the discussion of fluctuation effects.

C. Kramers-Kronig Relations

The Kramers-Kronig dispersion relations appear naturally within the field theoretic approach, and originate from the requirement of causality. They provide one way of quantifying the fluctuation–dissipation relation by imposing connections on the real and imaginary parts of any response function, such as the susceptibility of Eq. (3).

The Kramers-Kronig relations prohibit one seeing in the dark by wearing dark glasses. An ultrashort (say femtosecond) flash of a torch when Fourier analyzed in time contains all optical frequencies. Before the flash, these Fourier components conspire to cancel. If by wearing suitably colored dark glasses, one could remove certain Fourier components without affecting the remainder, the latter would not cancel at earlier times than the flash, and the flash would have visible precursors. If we accept causality, we must deny this possibility. Therefore, any passive system responding to a stimulus by attenuating some frequency components must also affect the relative phases of the Fourier components that do not get absorbed. The Kramers-Kronig relations

$$\text{Im } H(\omega) = -\frac{1}{\pi}\mathscr{P}\int d\omega' \frac{\text{Re } H(\omega')}{\omega' - \omega} \tag{7}$$

$$\text{Re } H(\omega) = \frac{1}{\pi}\mathscr{P}\int d\omega' \frac{\text{Im } H(\omega')}{\omega' - \omega} \tag{8}$$

between the dispersion or frequency dependence of the real and imaginary parts of a response function $H(t)$ express this essential link, showing that the specification of either the real or the imaginary part determines the other. It is perhaps amusing to note that at least in electronics the relationship imposed by causality between the real and imaginary parts of the response function of a passive circuit is not necessarily one-to-one; a discrete set of imaginary part functions may each be consistent with a given real part if a circuit is not a "minimum phase" circuit—that is, one whose response function $H(\omega)$ has zeros z_i and poles z_j in the complex plane satisfying Re $z_i \le 0$, Re $z_j \le 0$—and compatibility with causality is maintained in spite of the apparent constraints of the Kramers-Kronig relations by the existence of boundary terms in the integration. We ignore this complication below.

The simple proof of the Kramers-Kronig relations given by Yu-Kuang Hu[11] is worthy of a brief review. If a system provides a response $P(\omega)$ for a stimulus $E(\omega)$ via a passive linear causal response function $H(\omega)$, so that $P(\omega) = H(\omega)E(\omega)$, then the corresponding Fourier-transformed functions are related by convolution: $p(t) = h(t) * e(t)$ (we define the Fourier transform by $p(t) = \int_{-\infty}^{\infty} d\omega\, P(\omega)e^{i\omega t}$). From causality, the constraint $e(t) = 0$ for $t < 0$ imposes the same constraint on $p(t)$. Hence $h(t)$ always vanishes for negative times in a passive system, and we may write $h(t) = h_+(t)\theta(t)$, where $\theta(t)$ is the unit step function (unity for $t \ge 0$, zero for $t < 0$) and $h_+(t) = h(t)$ for $t > 0$. At this point $h_+(t)$ is arbitrary for negative times. The Fourier transformed step function is

$$\Theta(\omega) = \frac{1}{2\pi}\lim_{\varepsilon \to 0+}\int_0^\infty \exp[-i(\omega - i\varepsilon)t]\, dt = -\frac{-i}{2\pi(\omega - i\varepsilon)} \tag{9}$$

(the ε to retain convergence). This allows the representation

$$H(\omega) = \Theta(\omega) * H_+(\omega) = -i\int_{-\infty}^{\infty}\frac{1}{\omega - \omega' - i\varepsilon}H_+(\omega')\, d\omega' \tag{10}$$

where

$$H_+(\omega) = \int_0^\infty h(t)\exp(-i\omega t)\,dt \pm \int_{-\infty}^0 h(-t)\exp(-i\omega t)\,dt$$

$$= \int_0^\infty h(t)[\exp(-i\omega t) \pm \exp(i\omega t)]\,dt$$

(11)

This shows that $H_+(H_-)$ is real (imaginary). Since $h(t)$ is the average of $h_+(t)$ and $h_-(t)$, the Fourier transforms are related in the same way and $H_+(\omega), H_-(\omega)$ are respectively twice the real and imaginary parts of $H(\omega)$. We may equate the imaginary (real) parts of the two sides of Eq. (10) using Eq. (4) to derive the Kramers-Kronig relations (Eqs. (7) and (8)).

A full treatment of the various dispersion relations and sum rules for optical activity is given by Thomaz and Nussenweig,[12] and a general account of sum rules in linear and nonlinear optics is given by Bassani and Scandolo.[13]

D.   Optical Theorem

Another relevant topic is the optical theorem, in which a general relation can be found between forward scattering (as for contributions to the refractive index) and the integrated imaginary part of the collision cross section (absorption); see Newton.[14] A simple discussion of this is given from a wave viewpoint in Feynman's lectures.[15] This also serves to emphasize the connection between a coherent fluctuation process and an incoherent dissipative one.

E.   Coherent and Incoherent Processes: The Consequences

These general results indicate in particular the existence of a very deep-seated link between any coherent process and those incoherent processes whose rates may be derived from that for the coherent process in the limit that the denominator associated with an intermediate state tends to zero. This suggests links between the selection rules for the coherent process and those for certain associated incoherent processes. For each possible intermediate state i, any coherent process (such as refraction, in Eq. (1)) has a sequence of matrix elements in its amplitude

$$X_i = \langle \text{initial}|A|i\rangle\langle i|B|\text{initial}\rangle/(E_{\text{initial}} - E_i).$$

A and B are effective operators, which possibly include further intermediate states and appropriate denominators. The rotational average of X_i has precisely the same form as that of the corresponding term (an interference term, if the operators A, B are different) in the incoherent process $|\text{initial}\rangle \to |i\rangle$, since X_i appears in each expression. A coherent optical process must therefore survive the selection rules appropriate for the corresponding incoherent process for at least one such possible intermediate state i.

For example, virtual absorption to a single intermediate state as in refraction is uniquely related to the corresponding real absorption. From Eq. (1) $A = B = V$; if we specialize to a gyrotropic process, we will identify A and B more specifically as different multipole coupling terms within V (see Section III), and X_i will be an interference term. The one virtual intermediate state i for a given contribution to a refraction effect will be the final state of the corresponding absorptive process, the photon then matching the energy difference with the initial electronic state. A chiral asymmetry in one (say circular dichroism) will inevitably appear in the other (as optical rotation). Similarly, we may expect a connection between gyrotropy in transmission and reflection, if only from energy and angular momentum conservation considerations, which suffice to give constraints on amplitudes as well as intensities (see Section VI.B for a recent related discussion in anyon theory).

A more complicated example is that of second-harmonic generation. In this coherent process two identical photons are absorbed and one with double the energy created. The corresponding rate expression in perturbation theory will involve at least two intermediate states, whose nature depends on the temporal ordering of the three photon–matter interactions. In one of these, for example, the two incoming photons are annihilated before the creation of the outgoing photon. The virtual (two-photon annihilation) process $|\text{initial}\rangle \to |i\rangle$ (itself a two-stage process with its own intermediate state; we suppress this point) can be viewed as the fluctuation-type counterpart of a dissipative process, in this case two-photon absorption. The remaining virtual process contributing to second-harmonic generation, $|i\rangle \to |\text{final}\rangle$, is also the fluctuation-type counterpart of a dissipative process, in this case single-photon absorption. Since the sums over matrix elements involved are common to each, the selection rules for the coherent process of second-harmonic generation are arguably related to those for the incoherent processes of both one- and two-photon absorption. An extension of our earlier work on selection rules for incoherent processes may benefit from a consideration of such links. This will be assessed in Section III; for the moment I mention only that it is possible to find selection rules for coherent processes that have no counterpart for the allied incoherent processes.

III. BASIC SELECTION RULES FOR OPTICAL PROCESSES

A. Multipole Expansion

A standard time-dependent perturbation theoretic approach[2, 10, 16] gives the Fermi Golden Rule of Eq. (5), extended for higher order contributions, in the form

$$P = \frac{2\pi}{\hbar} \mathscr{A} v_{\text{initial}} \sum_{\text{final}} \left| \sum_\alpha \langle \text{initial} | O_\alpha^{\text{eff}} | \text{final} \rangle \right|^2 \delta(E_{\text{final}} - E_{\text{initial}}) \quad (12)$$

where

$$O_\alpha^{\text{eff}} = O_\alpha + \sum_{\beta_i} O_\alpha |i\rangle\langle i| O_\beta / [E_{\text{initial}} - E_i] + \cdots \quad (13)$$

$$O_\alpha = -q\mathbf{A}(\mathbf{r}_\alpha, t) \cdot \mathbf{p}/m \quad (14)$$

$$\mathbf{A}(\mathbf{r}_\alpha, t) = \sum_{\mathbf{k}j} \sqrt{(\hbar/2N\omega_{\mathbf{k}j})} \left[a_{\mathbf{k}j} \mathbf{e}_{\mathbf{k}j} \exp i(\mathbf{k} \cdot \mathbf{r}_\alpha - \omega_{\mathbf{k}j} t) + \text{h.c.})\right] \quad (15)$$

The form of O_α we have chosen is that of the velocity gauge. We note here the possibility of other interaction terms. The two-photon term proportional to $\mathbf{A}^2$ involves no new point of principle in the following development; its effects are readily integrated (see Section III.D) and it can effectively be treated on the same footing with second-order terms in O_α. When spin is considered, an extra term of the form $q\mathbf{s} \cdot \nabla \times \mathbf{A}/m$ (Ref. 17) should be added. Other spin-dependent terms appear even in the velocity gauge in a fully relativistic theory; under transformation to the length gauge, when spin is considered there arises even in the dipole approximation a term $(q/2m^2c^2)\mathbf{s} \cdot \dot{\mathbf{A}} \times \mathbf{p}$ which also should be included in the full radiation–matter interaction.[18] The second term in O_α^{eff} arises from the next order of perturbation theory and so is of Kramers-Heisenberg form.

When the interaction Hamiltonian is expanded in powers of the wave vector $\mathbf{k}$ and the various terms are coupled to irreducible tensors under the rotation group, they may then be categorized by multipole. Thus, the zeroth-order term includes the electric dipole (E1) operator $\mathbf{e} \cdot \mathbf{p}$, and the second-order term includes $(\mathbf{e} \cdot \mathbf{p})(\mathbf{k} \cdot \mathbf{r})$, which can be rewritten as $\sum_{j=0,1,2}[\mathbf{e}\ \mathbf{k}]^j \cdot [\mathbf{r}\ \mathbf{p}]^j$ where our notation $[\mathbf{a}\ \mathbf{b}]^j$ denotes a coupled tensor proportional to $\mathbf{a} \cdot \mathbf{b}$, $\mathbf{a} \times \mathbf{b}$, and the symmetric rank -2 tensor $[\mathbf{a}\ \mathbf{b}]^2$ respectively. These give for the case $\mathbf{a} = \mathbf{e}$, $\mathbf{b} = \mathbf{k}$, zero for $j = 0$, $\mathbf{m} \cdot \mathbf{L}$,

500   G. E. STEDMAN

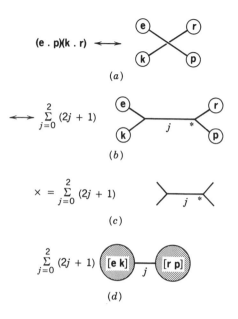

Figure 1. Angular momentum coupling diagrams for the analysis of the multipole coupling terms in the matter-radiation interaction. (a) Diagram representation, following the appendix, of the two scalar products. (b–d). Use of the Clebsch-Gordan series (unitarity of Clebsch-Gordan coefficients (c)) to expand the scalar products in terms of contractions of a matter and a radiation irreducible tensor operator $[\mathbf{e}\ \mathbf{k}]^j$ and $[\mathbf{r}\ \mathbf{p}]^j$, respectively.

where the magnetization vector $\mathbf{m} \equiv \mathbf{e} \times \mathbf{k}$, for $j = 1$ (magnetic dipole or M1), and the $j = 2$ case electric quadrupole (E2).

The recouplings intrinsic in this tensorial expansion are readily visualized and indeed executed using the diagram technique described, for example, in the appendix. The problem is to expand $(\mathbf{e} \cdot \mathbf{p})(\mathbf{k} \cdot \mathbf{r})^n$ in terms of irreducible electronic tensors $[\mathbf{p}\ \mathbf{r} \ldots]^j$. For the first-order term where $n = 1$, we can rephrase this by requiring that $(\mathbf{e} \cdot \mathbf{p})(\mathbf{k} \cdot \mathbf{r})$ should be expanded in terms of angular momentum coupling diagrams (Fig. 1a), and then the electronic operators recoupled $[\mathbf{p}\ \mathbf{r}]^j$ (Fig. 1b). This is achieved from the use of the diagram identity (see appendix) of Fig. 1c, and yields Fig. 1d.

B. Neumann's Principle

Several considerations show that the nature of an optical interaction effect, in particular, its geometrical dependence (on beam directions and polarizations, etc.) cannot be determined solely from a study of the uncoupled systems of matter and radiation. This observation is at odds with a trend toward appeals to Neumann's principle, and needs justification.

For this task, and also for the later task of deriving selection rules, we need to identify a number of parameters whose values can be used to categorize the effect. These are the parameter γ, whose value is $+1$ for a nongyrotropic and -1 for a gyrotropic process; the parameter K, whose value is $+1$ when no multipole interference terms are needed and -1 when interference is required; the parameter τ, which gives the net time reversal symmetry ($+1$ if time-even, -1 if time-odd) of all interactions in both amplitudes for the process; the parameters ε, β (equal to $1, 2, \ldots$) for the orders of coupling to an external electric or magnetic field respectively, a parameter Γ for the number of photon interactions in the combined amplitudes; Π for the net parity of the external field couplings (including the effect of ε or β being greater than 1); spin–parity labels J^π defining the irrep of $O(3)$ which reduces to the identity irrep of the symmetry group material system. Barron's[19] development emphasizes a concept of chirality whose distinctives may be enshrined by specifying these parameters and others that are in principle derivable from them. Any subset of these various parameters may be used to help define the optical process in question. There will be a minimal number of mutually independent parameters, of the order of three or four, whose specification will give an unambiguous definition of any effect. The choice of these independent classificatory parameters reflects the perspective and emphasis of the author in question. No intrinsic and canonical hierarchy for ranking these parameters exists; selection rules merely relate them.

Our initial statement is best justified by noting that of these parameters only J and π depend on the nature of the uncoupled systems and not on the form of their interaction as well, and that these parameters do not suffice to evaluate the remaining parameters, even given all the following selection rules. Nor would a specification of the orders $\Gamma, \varepsilon, \beta$ as well yet enable one to specify the process uniquely; the quantities τ and γ still depend on K. This vital information can be determined only from a knowledge of the details of the coupling. The Faraday effect occurs in E1 coupling, while natural optical activity requires interference between E1 and a higher multipole of odd order in $\mathbf{k}$. A knowledge of the subsystem symmetries alone does not suffice to indicate these characteristics. One might also mention in passing the problem that the separation between system Hamiltonian and interaction Hamiltonian is itself gauge-dependent; for a recent comment on and extension of this literature, see Wang and Stedman.[18]

We contrast this with Neumann's principle, a hoary epigram of physics folklore along the following lines: The symmetry of optical effects displayed by a system depends on and reflects the symmetry of that material system. Recent appeals to Neumann's principle in our context include Barron[19] and Evans.[20, 21] In quoting such a formulation of this principle, Voigt[22] adds his own cautions, which history has confirmed. Neumann's principle is arguably sterile and misleading, and much confusion in the literature can be traced in part to its uncritical acceptance. Opechowski[23] has already pointed out one circular argument implicit in Neumann's principle, namely that of identifying the appropriate supergroup, and has categorized the principle as useless. Opechowski gives an example of a more satisfactory approach, which incidentally also incorporates time reversal considerations. While all agree that the symmetries of a material system limit the kinds of optical effects it may exhibit, the symmetries alone do not dictate the optical effects, especially in such matters as chirality and its definition. As mentioned in the introduction, there are several reasons why a given process may be gyrotropic, and not all are defined by the symmetries of the material system.

For example, consider a chiral system, that is, one that is symmetric under time reversal but not parity, the parity-transformed system not being rotationally equivalent to the original. A chiral system lacks any S_n axis, or any symmetry operation including a reflection. Its antonym, achiral system, is symmetric under parity as well as time reversal (I note here that a printer's error in Stedman[1] unwittingly turned this into a converse statement!). Each of a chiral and achiral system may exhibit both gyrotropic and nongyrotropic effects. For example, in the Faraday effect, an achiral system may exhibit a gyrotropic effect; in natural optical activity, which is a gyrotropic effect, a chiral system is essential: Highlighting the symmetry of the material system in one's analysis, while important, is not definitive for the kind of optical effect exhibited.

Not all of these parameters need to be specified explicitly to identify an effect unambiguously. Four is usually sufficient, and a measure of conventionality is used to compress these definitions. Natural optical activity implies $\varepsilon = \beta = 0$ (since "natural" implies no field assistance) and so $\Pi = 1$; optical activity implies gyrotropy ($\gamma = -1$) and by default we may assume $\Gamma = 2$, as for two amplitudes contributing to the intensity or probability of a one-photon process. All other parameters will follow from the selection rules. Whether some of these parameters have an especially important role in defining the process is a matter of taste. Barron, for example, is concerned to emphasize the unique role of "true chirality" and the interplay between chirality and magnetic effects, and this leads to a specialized emphasis in his nomenclature. In the introduction I took a different emphasis, mentioning the inequivalent classification of gyrotropy. In general, I do not wish to be limited to any particular emphasis or choice of application, and the most that I shall aim for is an unambiguous definition of the effect.

C. Parity Versus Time Reversal

The basic ingredients to the selection rules we discuss involve the time reversal, parity, and point group operators. In many cases, interactions and quantum states may be classified as having definite transformation properties under these operators and the selection rules follow. Sometimes a given rule may be derived by more than one argument, and it then becomes unclear what precisely is tested by confirmation of that rule. There is a long-standing tendency to confuse the relative priority of these arguments, which is nicely pointed by an anecdote of Telegdi.[24] On the announcement of parity nonconversation in 1957, Dirac, unlike Pauli, was not caught with his guard down. He said, "If you look carefully, you will see that the concept [of parity] is not once used in my book." To Dirac, parity was always an unnecessary additional concept. In view of the relevance of parity nonconservation in the weak interaction for atomic physics, it seems wise to regard any selection rule that can be proved from either time reversal or parity considerations as being first and foremost a touchstone of time reversal invariance.

D. Definitions

First, we shall assume that all material systems under discussion are time-even, that is, that they have no internal magnetic or other time-odd fields to render time-reversed electronic eigenstates nondegenerate. There will be one brief exception to this: anyonic systems (Section VI.B); so long as we are not concerned to distinguish linear from quadratic field effects, we can cover this by thinking of the time reversal violation as associated with an external magnetic field.

Second, we need to distinguish systems that are symmetric with respect to parity from those that are not. As mentioned in Section III.B, one may use the names achiral and chiral, respectively, and this is a well-established practice. It suffers, however, from the defect that the term parity is sometimes used not so much in connection with the inversion operator P: $\mathbf{r} \to -\mathbf{r}$, to which it properly belongs, as to reflection, $P' = \sigma_m$ in a suitable plane m; and these lead to inequivalent definitions. A centric system is one which is symmetric under P. This usage accords with the standard identification of the pure rotation group as the chiral point groups. However the common definition of a chiral system as one that cannot be superposed on its mirror image, even after a rotation, is based on the use of P'. A molecule with a rotation–reflection symmetry (point groups $T_d, C_{\infty v}, C_{6v}, D_{3h}, C_{5v}, D_{2d}, C_{4v}, C_{3v}, C_{3h}, S_4, C_{2v}, C_s$; for all of which a spin-parity state $J^\pi = 0^-$ is not an invariant, but there does exist an invariant J^- for some $J > 0$), such as H_2O. This system is acentric, but is nevertheless superposable on its mirror image, given an appropriate rotation; if the mirror plane is suitably chosen, no rotation is necessary. In this intermediate case, then, certain gyrotropic effects characteristic of an acentric system are allowed, while in a fluid (where the spin-parity label $J^\pi = 0^-$ is an invariant) the results are more typical of centric systems (where $J^{\pi} = 0^+$ is an invariant). For this reason I shall now avoid as well the use of the terms chiral and achiral system, and use the terms (a)centric and rotation–reflection (a)symmetric, respectively, to distinguish these two usages of parity.

In Section III.B we briefly introduced a variety of parameters, each of which may be used to help classify any optical effect. We expand the definition of some of these.

The *orders* of the electric and magnetic *field coupling* (ε, β, respectively) give the number of matrix elements in the intensity contribution that contain the appropriate coupling. The inverse processes involving optical induction of magnetization, such as the inverse Faraday effect, also require a single matrix element coupling matter to low-frequency electromagnetic fields; we shall include such matrix elements in our count of ε, β, respectively.

The *photon interaction number* Γ is equal to the number of matrix elements in the intensity contribution that contain a one-photon coupling, and is increased by two for a matrix element involving the two-photon coupling $q^2 A^2 / 2m$. We distinguish this from the number n_γ of participating photons in any process. With the possibility of lone matrix elements involving coupling to low-frequency electromagnetic fields removed as above, Γ is restricted to optical photons and will be even, with $n_\gamma = \frac{1}{2}\Gamma$, since a photon absorption process then involves an appropriate matrix element for the amplitude and for its complex conjugate. The separation between ε and β on the one hand and Γ on the other is somewhat artificial from a QED viewpoint, and in practice is motivated by the convenience of classifying the low-frequency photons counted under ε and β as part of the nonoptical subsystems, whether field or matter.

The *wave-vector phase* K is defined for the full rate contribution (Eq. (12)) by $K \equiv (-1)^{n_k}$, where n_k is the total number of occurrences of all photon wave vectors $\mathbf{k}$ in all matrix elements as a result of multipole expansion. For example, if all interactions are electric dipole (E1), no $\mathbf{k}$ vectors occur in the interaction Hamiltonian, $n_k = 0$ and $K = 1$; whereas if the expression corresponds to an interference term between E1 and say magnetic dipole (M1), then $n_k = 1$ and $K = -1$.

We also define a *net time reversal phase* τ as the joint time reversal symmetry of all interactions, including each of the ε or β terms in the count. The contributions of the Γ interactions with optical photons in principle should be included, but in practice can be ignored, since regardless of the multipole choice each interaction is time odd, and since there is an even number of such terms. Therefore, τ is equal to -1 if the effect in question is linear or cubic (not quadratic) in an external time-odd field, such as a magnetic field, and $+1$ otherwise. In more extended applications, the effect of an electric field when mediated by the resulting current may be interpreted as time-odd because of the accompanying dissipation; see Kriplovich and Pospelev.[25]

A *gyrotropic* (or nongyrotropic) process is one for which $\gamma = -1$ ($+1$, respectively), where γ is the phase of any contribution $I(\{e, k\})$ to the Golden Rule rate of an optical effect under the joint complex conjugation of all polarization vectors e: $I(\{e^*, k\}) = \gamma I(\{e, k\})$. Some background is given throughout earlier sections.

Π is the *net parity* of all external field interactions, again summed over all $(\varepsilon + \beta)$ relevant matrix elements. It is equal to -1 if the effect in question is linear or cubic (not quadratic) in an external odd-parity field such as an electric field, and $+1$ otherwise.

We may also profitably include the spin-parity labels J^π defining the irrep of O(3) which reduces to the identity irrep of the symmetry group material system.

E. Rules

Several basic selection rules follow:

Time Reversal Rule 1

$K\tau = \gamma$.

Parity Rule 1

For a centric system, $K\Pi = +1$.

Symmetry Rule 1

For incoherent optical processes involving randomly oriented systems, Parity rule 1 holds for rotation–reflection symmetric systems as well as centric systems.

Such rules have been thought less than simple.[26] They could hardly be stated more briefly. As the examples will show, they are easy to apply. Better, they are both general and comprehensive, and suffice to exhaust the constraints on the vast majority of incoherent processes. The difficulty with simple pictorial arguments (which might otherwise have been expected to carry the day with me!) is that they are not readily generalized to higher-order processes or to systems of lower symmetry than SO(3). Many of the rules used and conclusions developed in Evans[20, 26] hold only for fluid systems.

We outline the proofs of these rules; the analysis of other symmetry, such as the effects of point group symmetry of a molecule or crystal, has been given for many incoherent processes by Stedman[2, 5] and will be adapted to certain coherent processes in Section V.

F. Proofs

Consider any term X contributing to the Golden Rule rate or probability. In general it will have the form of a product of any of the terms A, such as

$\langle i|\mathbf{e}\cdot\mathbf{p}|f\rangle$ say, in the amplitude O_α^{eff} (Eq. (13)) and the complex conjugate B^* of any other term B (such as $\langle i|\mathbf{im}\cdot\mathbf{l}|f\rangle$). A might be a product of various E1 matrix elements connecting the initial and final states; in an interference term, B will differ from A in having one M1 or E2 interaction matrix element. Since X is real, the complex conjugate expression $A*B$ must also appear in X.

Time Reversal Rule 1 (Refs. 2 and 5) follows from applying the constraint

$$\langle i|V|j\rangle = \langle \bar{i}|\bar{V}|\bar{j}\rangle^* \qquad (16)$$

itself resulting from the antilinearity of the time reversal operator T in quantum mechanics (see Messiah[26a]) to all matrix elements in both amplitudes A and B^*. The overbar represents the action of T: $|\bar{i}\rangle \equiv T|i\rangle$, $\bar{V} = TVT^{-1}$. Each interaction operator V is assumed to be Hermitian. It is then time-even ($\tau_V = +1$) or time-odd ($\tau_V = -1$) as $\bar{V} = \pm V$. (I am explicit about this assumption since the components of irreducible tensor operators O_l^λ in a complex basis $\lambda\ell$ are not necessarily individually Hermitian, and their time reversal signature is best defined by the relation $\bar{O}_l^\lambda\dagger = O_l^\lambda$; see Wang and Stedman[27] for an application where such distinctions are important.) Once interference terms of complex conjugate character are combined, the exchange of i for j is inconsequential. In a time-even system (barring the action of external fields, whose effects may be included otherwise via the relevant perturbation matrix elements) the summation degenerate initial and over final states allows time reversal partners $|i\rangle$, $|\bar{i}\rangle$ to be treated on an equal footing.

A contribution to the rate P, which is necessarily real, will arise only from the real part of each term in X. In fact, each term is either real or imaginary, when we use an expansion such as the multipole expansion in which each term has a definite symmetry under complex conjugation. Time Reversal Rule 1 identifies those terms that are pure imaginary (and so cannot contribute) from those that are real.

Each term X contains some mix of all photon polarization vectors $\mathbf{e},\mathbf{e}',\dots$, which itself can be made symmetric or antisymmetric under complex conjugation $\mathbf{e}\to\mathbf{e}^*$, etc. For any polarization vector $\mathbf{e}$ in A, its conjugate $\mathbf{e}^*$ necessarily appears in B^*, so that the initial and final radiation states are common to A and B. The same kind of recoupling of electronic operators that led to the multipole operators En, Mn ($n = 1, 2, \dots$) can be used to symmetrize or antisymmetrize their product. One simply couples $\mathbf{e}$ and $\mathbf{e}^*$ to a rank J tensor $[\mathbf{e}\,\mathbf{e}^*]^J$ (Fig. 2a) and we identify each term $J = 0, 1, 2$ as gyrotropic or nongyrotropic according to the symmetry of the coupling: $\gamma = (-1)^J$.

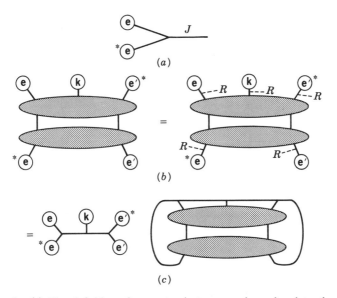

Figure 2. (a) The definition of a gyrotropic term can be reduced to the coupling symmetry J of each polarization vector and its conjugate. (b) For an incoherent process in a fluid, the orientational averaging can be executed by simultaneous rotations of all optical vectors and then pinching the legs as in (c) according to the appropriate diagram reduction theorem.[2] A particular sequence of couplings is used to achieve this particular geometric form, and it is this sequence that enables one to distinguish gyrotropic from nongyrotropic processes (through the coupling symmetries of the photon polarization vectors in conjugate pairs). This simplified presentation omits a number of technicalities, such as the weighted sum over intermediate ranks.

In addition, the action of time reversal will also complex conjugate the imaginary unit i which accompanies each wavevector **k** in the multipole expansion, and will also change the overall sign according to the number of these vectors present, that is, by a factor K. Time Reversal Rule 1 simply requires that all these sign changes consequent on complex conjugating a term AB^* must cancel.

Parity Rule 1 is a straightforward consequence of tallying parities of all interactions in all amplitudes, without regard to any restrictions on the parity of intermediate or of the final states. It assumes only that each state has a definite parity.

Fluids introduce an especially important symmetry. If each molecule has a random orientation and contributes additively to the rate P, the effect on P is as if all photon polarization vectors and wave vectors are jointly rotated by an arbitrary SO(3) operation (Fig. 2b). On averaging over this rotation, we obtain a strong requirement that these vectors by themselves should combine to form a rotational invariant. Formally, this follows from such theorems as Schur's lemma, and the Wigner-Eckart theorem, which may be represented in the diagram notation reviewed in the appendix as pinching theorems (Fig. 2c). In effect such symmetry permits the problem of finding P to be factorized into two problems: that of finding the reduced matrix elements, namely those factors that depend on energies, oscillator strengths, etc., but not the symmetry aspects or choice of beam geometry, and that of finding the geometrical factor that depends not on the detailed dynamics but only on the symmetry of the systems. The group-theoretic couplings of operators that are involved are necessarily executed in mirror symmetric fashion in these two factors, and in particular the resulting final coupling for both the geometrical factor and the dynamical constant will be identical; in addition, for a symmetry group, it will be the identity irrep of the group in question.

In the case of fluids, this says that the effective operator O_α^{eff} is a rotational invariant, and also that the various field- and photon-related vectors, such as the polarization vector and the wave vector (when multipoles are included), must be coupled using the quantum theory of angular momentum to a rotational invariant. It then becomes obvious that for rotationally averaged scatterers only those effects associated with spin-parity labels J^π for which $J = 0$ will survive. This means that Parity Rule 1 may be extended in such cases to rotation–reflection symmetric as well as centric systems, thus deriving Symmetry Rule 1.

G. Examples

These rules suffice to show that, for example, natural optical activity requires both interference between multipoles in the expansion of the modulus of Eq. (12) and an acentric system, but is allowed for a fluid of such systems. We repeat this proof as a simple but testing example of the application of these rules, and consider optical activity in more detail.

If the refractive index of left and right circularly polarized light is different in a medium displaying this effect, so is the absorption (Section II.E). Hence, the generalized Golden Rule rate P of absorption of light (Eq. (12)) is different for a polarization choice $\mathbf{e}$ and its complex conjugate. Term by term, as we have indicated in our discussion of Time Reversal Rule 1, the various contributions to P are either symmetric or antisymmetric under the replacement $\mathbf{e} \rightarrow \mathbf{e}^*$. Hence, only the antisymmetric or *gyrotropic* terms contribute to optical activity.

This then requires from Time Reversal Rule 1 that $K\tau = -1$. For natural optical activity no time-odd fields are available, and $\tau = +1$, so that $K = -1$; i.e., the net order of the matrix elements in both amplitudes must be odd, and we must consider interference between multipoles. This still permits at this stage either E1/M1 or E1/E2 interference contributions. Parity Rule 1 then disqualifies the effect in centric systems, since $\Pi = +1$ in the absence of external fields.

For a fluid, Symmetry Rule 1 requires that the various photon vectors form a rotational invariant. This means that the polarization vector $\mathbf{e}$ from the E1 amplitude must combine with the vectors $[\mathbf{e}\ \mathbf{k}]^j$ appropriate for E2 ($j = 2$) or M1 ($j = 1$) (the latter are strictly complex conjugated, since they arise from the other term in the square modulus). This can be achieved only for M1, since two vectors can form a scalar product. Its form, $\mathbf{e} \cdot [\mathbf{e}^* \times \mathbf{k}]$, can be transmuted to the clearly gyrotropic form $[\mathbf{e} \times \mathbf{e}^*] \cdot \mathbf{k}$, and this finally gives confirmation of the exhibition of natural optical activity in a sugar solution. Exactly parallel arguments to the above can be extended to all orders of nonlinear processes.

It is curious that no such systematic application of time reversal considerations seems to have been made in nonlinear optics. The applications by Chiu[28, 29] are an exception; while correct as far as they go, they do not exhaust the content of time reversal in the manner achieved by our time reversal selection rule.

Gyrotropic effects are defined with regard to the use of circularly polarized light beams. However the latter are not essential for the identification of gyrotropic effects if sufficient additional information is available. In time-even fluids for example, from Time Reversal Rule 1 gyrotropic terms may be defined as those in which the sum over the coupled angular momenta j (describing the coupling of each polarization vector with its complex conjugate) is odd. When at a given order of multipole interaction such terms are unambiguously associated with certain geometric factors, their relevance may be adequately tested in linear polarization. This is illustrated by the analysis of Hecht and Nafie[29a] for Raman optical activity using linear intensity differentials.

H. Relativistic Analogues

The kind of arguments used to arrive at our selection rules typically involve the application of mirror reflection or parity inversion or time reversal to just part of a system, usually to either the radiation (photon) or material system. This contrasts with the kinds of argument used in relativistic physics in which parities are multiplied over all contributing excitations.

Eimerl[30] has noted a link between crossing symmetries of the relativistic S matrix and permutation symmetries of the dielectric susceptibility tensors of nonlinear optics. Clearly there are links between his formalism and ours. In addition, he discusses reciprocity theorems (his Eq. (11)) in which amplitudes for time-reversed histories are related if the photons are also time reversed. The exact form of the reciprocity theorems of nonlinear optics have been the subject of some debate, and results related to and derived from our selection rules have clarified the situation.[2] It turns out that intensity contributions are always the same for experiments in which incoming and outgoing states are exchanged without further modification, while the intensity contributions from two experiments in which incoming and outgoing states are individually time reversed (and exchanged as well if desired, since this is of no consequence) are equal or opposite as $\tau = +1, -1$ respectively. It is strictly incorrect, therefore, to regard the "reciprocity theorem of Raman scattering" or "time reversal symmetry" or "detailed balance" to apply regardless of sign when component states are time reversed.[5]

This distinction is not made in much previous nonrelativistic work on Rayleigh and Raman scattering. Nor is it apparent in the work of Eimerl,[30] who certainly time reverses his photons (by complex conjugating the polarization vectors and reversing the momenta). Indeed, there is an apparent conflict with the crossing symmetry announced in Eimerl's Eq. (11): According to this equation, amplitudes are preserved under crossing any photon from in to out (or vice versa) if its polarization vector is conjugated and its wave vector reversed. Incidentally, this is further confused by Eimerl's identification of this photon mode transformation as being a reversal of the sense of circular polarization, which it is not! Clearly the relation between crossing symmetry and the nonrelativistic forms of the reciprocity theorems needs a fresh analysis, and it is not certain that the symmetries Eimerl discusses will survive such an analysis.

Part of the difficulty in reconciling these approaches is the lack of distinction within a thorough-going QED formalism such as Eimerl's between electric ($\tau = 1$) and magnetic ($\tau = -1$) interactions: Both are (virtual-) photon-induced, and only the overall time reversal characteristics of the interaction, rather than those of a subsystem, are explored.

IV. SOME INCOHERENT FIELD-INDUCED EFFECTS

Some discussion has arisen recently in connection with effects that depend on a static or optically induced magnetic field, and this is an appropriate point to comment on these matters, incidentally illustrating the significance of the above material for incoherent processes. We must clearly delineate the range of effects under discussion. The selection rules previously given, in particular Time Reversal Rule 1, are invaluable both in disentangling various effects, some of which have been lumped together in the literature, as well as in identifying effects thought to be distinct.

A. Magnetic Linear Dichroism

If we include interference terms such as E1/M1, we may also have a nongyrotropic effect that is linear in an applied magnetic field, now necessarily in a chiral system (since $K = -1$ in both Time Reversal Rule 1 and Parity Rule 1). We refer to this effect as linear magnetic linear dichroism (linear MLD, or magnetic linear birefringence): That part of the absorption (or refraction, respectively) of a light beam that is linear in a magnetic field depends on the direction of linear polarization of the light beam, provided the system is chiral and provided we allow for E1-M1/E2 interference terms.

Its complicated history is well illustrated by the galaxy of names that have been employed for its description. MLD is an obvious contender, since the effect as an optical effect stands in the same relation to the Faraday effect or MCD as does linear to circular dichroism. As early as 1976 Judd and Runciman[35] discussed such effects under the title of transverse Zeeman effect; see also van Siclen.[36] The term MLD has been used for the corresponding quadratic effect previously; see, for example, Vala et al.,[31, 32] Moreau and Boccara,[33] and Briat.[34] My use of the term[2, 5] for the linear counterpart must be distinguished from the quadratic effect by adding that adjective; the fundamental selection rules show that linear and quadratic MLD make different demands on the system symmetry. The quadratic effect does not require either a chiral system or multipole interference. Evans[26] uses the term forward–backward birefringence for the effect discussed by Wagnière and Meier.[37] This nomenclature highlights a geometrical behavior that is appropriate for fluids only, and that also embraces (as did Wagnière and Meier) other effects than (the linear and nongyrotropic) MLD. Barron and Vrbancich[38] chose magnetochiral dichroism, Wozniak and Zawodny[39] magnetospatial dispersion, and Andrews and Bittner[40] gyrotropic dressing; I note also the work of Andrews[41, 42] on magnetic effects in two-photon spectra. Incidentally, a quick study of this reference list is sufficient to dispose of all the various recent claims to originality in the literature; the diversity of nomenclature and the differing precisions of the definitions obviously contribute to this confusion. Briat[34] traces the reports of MLD in the literature to 1969. Linear MLD may be distinguished from the Cotton-Mouton effect as now understood by the latter being quadratic in magnetic field and gyrotropic, although the early history of this topic did not have the same clarity of distinction.

Consider first a one-photon incoherent absorptive optical effect that is linear in a magnetic field. Since for any such effect that is linear in a magnetic field $\tau = -1$, and $\Pi = +1$, we conclude from Time Reversal

Rule 1 and Parity Rule 1 that for a gyrotropic process K can be $+1$, thus allowing MCD (magnetic circular dichroism, or the Faraday effect) in E1 coupling in all systems including centric systems.

B. Electric Linear and Circular Dichroism (ELD, ECD)

We now work through a couple of simple examples, so that the reader can gain familiarity with our method. First let us show that ECD cannot occur in a fluid (random array) of molecules, even if each lacks a center of symmetry. ECD is an incoherent process, so that the full intensity must be rotationally averaged. This intensity is linear in each of the following external vectors, fixed in the lab frame: $\mathbf{e}$ (the polarization vector), $\mathbf{e}^*$ (from the conjugate amplitude), the field $\mathbf{E}$, and any $\mathbf{k}$ vector from higher multipole coupling. We search for rotational invariants $I(\mathbf{e})$ that correspond to rate contributions, which satisfy $I(\mathbf{e}) = -I(\mathbf{e}^*)$.

Regarding the induction of linear birefringence (a dependence of $I(\mathbf{e})$ on the direction of linear polarization) by an electric field, we must ask, What level of multipole expansion, if any, is needed? What restrictions, if any, are there on the inversion symmetry of the absorbing molecule? Could this process occur in a fluid?

We answer these questions as follows. ECD needs an intensity term that is rotationally invariant. Since $\tau = +1$ (the electric field is time-even), and CD is gyrotropic ($\gamma = -1$), from Time Reversal Rule 1 K must be -1, and so E1/M1 or E2 interference is necessary. Let us take M1 multipole coupling, to be precise. For a fluid, the external vectors $\mathbf{e}, \mathbf{m}^*$ (where $\mathbf{m} \equiv \mathbf{e} \times \mathbf{k}$ and is parallel to the magnetic field of the electromagnetic wave) and $\mathbf{E}$ must couple to a rotational invariant. On recoupling, this means that we must form $[[\mathbf{e}\ \mathbf{e}^*]^j[\mathbf{E}\ \mathbf{k}]^j]^0$, and j must be 1 rather than 0 or 2 for gyrotropy; two vectors ($J = 1$) such as $\mathbf{e}$ and $\mathbf{e}^*$ can couple to angular momentum $j = 0, 1,$ or 2. The interchange symmetry of the coupling is given by the sum of the j values, so that $\gamma = (-1)^{2j+J}$. Hence, the only possible rotational invariant has the form $[\mathbf{e} \times \mathbf{e}^*] \cdot [\mathbf{E} \times \mathbf{k}]$. This necessarily contains $\mathbf{e} \cdot \mathbf{k}$, which is zero from transversality.

For ELD, with $\tau = +1$, $\gamma = +1$, it follows that $K = +1$ and E1 coupling suffices for ELD. ΠK must be $+1$ in a centric system, but for ELD $K = +1$ and $\Pi = -1$ ($\mathbf{E}$ is a polar vector), so ELD is forbidden in a centric system. In a fluid, $\mathbf{E}$, $\mathbf{e}$, and $\mathbf{e}^*$ must combine to a rotational invariant. This can only be the triple scalar product which involves $\mathbf{e} \times \mathbf{e}^*$, which is of gyrotropic form contrary to hypothesis. So ELD also is impossible in a fluid.

C. Electromagnetic Optical Activity (EMOA)

Now we consider EMOA, optical activity of a system that is linear in each of an external electric and a magnetic field (see, for example, Kielich and

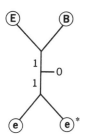

Figure 3.   A coupling diagram for determining the polarization dependence of electromagnetic optical activity.

Zawodny[43] and Ross et al.[44] What level of multipole expansion, if any, is needed? What restrictions, if any, are there on the inversion symmetry of the absorbing molecule? Can this process occur in a fluid, and if so what is the best choice of geometry (directions of the fields compared to the polarization and wave vector of the light beam) that would make this process easy to distinguish from ECD and MCD?

In answer, EMOA has $\tau = -1$ (from **B**) and $\tau K = \gamma = -1$ (since CD is gyrotropic); hence, $K = +1$, and E1 is adequate. Since $\Pi = -1$ (from the **E**; **B** being an axial vector does not contribute), while $K = +1$, ELD is forbidden in a centric system. In a fluid, $\mathbf{E}, \mathbf{B}, \mathbf{e}, \mathbf{e}^*$ must couple to a rotational invariant. In the coupling $[[\mathbf{E}\ \mathbf{B}]^j[\mathbf{e}\ \mathbf{e}^*]^j]^0$, j must be odd, i.e., 1, for the interchange of $\mathbf{e}$ and $\mathbf{e}^*$ to give a negative sign (OA) (Fig. 3). Hence, the only possible geometrical dependence is $(\mathbf{E} \times \mathbf{B}) \cdot (\mathbf{e} \times \mathbf{e}^*) = (\mathbf{E} \times \mathbf{B}) \cdot \mathbf{k}$, so that with **E**, **B**, and **k** mutually perpendicular we get maximal effect. As such any optical activity is distinguishable from say MCD, where **B** must have a component in the direction **k**.

D.   Pump Effects

The inverse Faraday effect[45-48] and kindred phenomena such as inverse magnetochiral birefringence[49] involve the induction of a magnetic field from the interaction of photons with matter. These involve a novel feature that is not accommodated by my previous presentation of time reversal selection rules,[2, 5] but that has been anticipated in the definitions of Section III.D: Even in the absence of a static field it is possible to have an odd time reversal symmetry for the net interactions, since Γ is odd. One matrix element must involve an unpaired interaction of the matter with a low-frequency (static or nearly static) electromagnetic field which is time-odd at the electronic level. This is achieved by coupling to orbital or spin angular momentum. The papers of Pershan et al.[45] and Wozniak et al.,[49] for example, exhibit the relevant interactions (of the form $\mathbf{E} \times \mathbf{E}^* \cdot \mathbf{s}$ or $\mathbf{M} \cdot \mathbf{B}$) and matrix elements. In the formalism proposed here, the time reversal phase of such a matrix element is categorized as giving a unit contribution to β. The restrictions on gyrotropy, etc., are exactly the same as for the analogous (inverse) effects such as the Faraday effect, the counterpart matrix element being that involving the interaction with a static field. Evans[20, 26, 50–53] considers several possibilities for such effects associated with circularly polarized laser beams.

In these connections I note first the existence of several spin-dependent interactions between radiation and matter that might give such terms, depending on the choice of gauge (Section III.A), some of which have been ignored in other contexts.[18] Second, many of the restrictions discussed in these papers pertain only to fluids, such as the case discussed by Evans[51]; a much wider range of possibilities occurs in systems of lower symmetry, as shown by Table 6 of Ref. 5. The combination $\mathbf{e} \times \mathbf{e}^*$ arises in the geometrical factor for a number of well-documented effects, such as optical activity of all kinds. In my view it has nothing to do with an intrinsic field of the photon, the latter being fully described by Maxwell's well-known theory. Other geometrical factors are appropriate in other circumstances, examples being latent in the above discussion; fuller details are given in Stedman.[2, 5] This is consistent with the manner in which Evans and others in effect note the relevance of $\mathbf{e} \times \mathbf{m}^*$ for other radiation-induced effects. However, I would not deny the possibility of identifying such components of a geometric factor with an appropriate identification of a "photon" within a medium. Some works referenced above distinguish the potential of a static and a time-varying electric field in inducing circular birefringence, for example. A time variation per se in a Stark matrix element does nothing to conquer PT-based selection rules. At optical frequencies the Faraday dynamo effect will give a magnetic field as the result of a time variation in an electric field, which may then be assessed under MCD, MLD, etc. It is proved in Section IV (see also Ref. 5) that the static field effects of ELD and ECD both vanish for fluids, but not for other systems. Even in those cases, circular birefringence holds for all ensembles, and it is the linear birefringence that requires a chiral ensemble. A static field may be more effective than a time-varying field in polarizing the atoms that can then orient in the field, removing the rotational symmetry. Finally, I mention that various claims for novelty in some of these papers, particularly with regard to effects related to MLD (whether linear or quadratic), arise only because the variety of names for allied processes has made tracing of the original ideas a major task in itself. The selection rules of this script are presented as a cataloguing tool for elucidating and disentangling these various contributions.

E. Optical-Detection of Parity Violation in Atomic Physics

There is by now a wide variety of methods of detection of parity noncon-servation in atoms (Bouchiat and Pottier,[56] Sandars[57]. These indeed depend on looking for an optical effect such as MLD which requires for its existence a chiral system (and for a gas or beam of disoriented atoms, an acentric system). What exactly does this permit and even require generally for any detection scheme?

From Parity Rule 1, we see that the detection of parity nonconservation in atoms requires an effect for which $K\Pi = -1$, and thus from Time Reversal Rule 1 an effect that is gyrotropic or nongyrotropic, since the PT signature of any external field is $+1$, -1 respectively.

There are now two cases. If there is an external field acting linearly, whether it be electric or magnetic, its PT parity is -1, so that the detection of parity nonconservation in atomic physics then requires the search for a nongyrotropic effect. Evans[26] mentions a process such as MLD as a possible method of detection of parity nonconservation in atoms (incidentally the opening statement of Section 7 of that paper confuses P and T violation). While MLD is one possibility, it is apparently not the best. Field-dependent measurements of nongyrotropic effects have indeed been made in cesium in some beautiful work by the Paris group, but these have generally used ELD, the nongyrotropic effect linear in an electric field (see Bouchiat et al.,[58] or an effect that was also linear in an electric field, but also depended on a magnetic field (see Bouchiat and Pottier[56]). In general, use of external field immediately reduces the magnitude of the observable, but may compensate for this by permitting a reduction in noise through combining a modulation with phase-sensitive detection loops.

The strategy of the Oxford and other groups was to avoid an external field. In this case PT $= +1$, and the detection of parity nonconservation in atomic physics then requires the search for a gyrotropic effect. The obvious choice is that of (natural, i.e., field-independent) circular dichroism. Such experiments have been done at Oxford for bismuth,[59] and more recently for thallium vapor, e.g., Wolfenden et al.[60]

V. COHERENT PROCESSES

A. General Conditions

In principle, then, the summation over participating ion $\alpha, \beta, \ldots$ should be taken coherently. However, if the interaction with radiation leaves the material system changed in at least one ionic state, the incoherence of the sum over final states reduces the sum over α, etc., to a single term.

The subject of coherence is a complex and fascinating one; see, for example, Andrews and Sherborne,[62] Sherborne,[1] Freund,[63-66] Bogaturov,[67] and Scully.[68] Many linear and nonlinear optical processes, for example, those preserving the momentum of the radiation field, are coherent. The contributions of individual scatterers add with a definite phase, since the associated wave factor $\exp(i\Sigma \mathbf{k} \cdot \mathbf{r}_\alpha)$ is unity. This leads to stronger selection rules.

A *general condition* for coherence is that the net transfer of energy between radiation and matter vanishes, the final and initial states then being degenerate. Otherwise, the fact that the sum over final states is outside the square modulus expression will decohere the atomic contributions.

A further condition is that the algebraic sum of the factors $\exp i(\mathbf{k} \cdot \mathbf{r}_\alpha)$ for all photons involved should equal unity. Otherwise, the random relation between the phases of successive terms will destroy coherence effects when summing over ions. Three ways of achieving this result are worthy of mention:

1. The algebraic sum of the photon wave vectors (with opposite signs for creation and annihilation) cancels. This, together with the requirement for an elastic process, indicates the following cases as being of special interest. At second order forward Rayleigh scattering is relevant; thus at third or higher order, harmonic generation such as second-harmonic generation (SHG) and its time inverse of parametric down-conversion (PDC), and at fourth order coherent anti-Stokes Raman scattering (CARS) or four-wave mixing (4WM).

2. An alternative is that the ions form a regular lattice, when the coincidence of $\Sigma \mathbf{k}$ with a reciprocal lattice vector allows the coherence to be maintained. This is the case of Bragg scattering.

3. In the case of specular reflection at an interface between dielectric media, the algebraic sum of wave vectors is perpendicular to the interface. In that case layers of atoms at a given depth into the medium contribute coherently, since their vector separation is parallel to the interface.

B. Anderson Localization, Disordered Systems, and Invisible Paint

Time reversal arguments have found a significant application in transport theory for electrons in a disordered system. Quantum interference can occur between two different paths C, D by which an electron propagates, through a series of random collisions, from point A back to itself in a metal, and in particular the time-reversed paths $D = \bar{C}$ interfere destructively. The proof of this involves a straightforward application of Eq. (16) to each collision matrix element of the form $\langle \bar{\psi} | T | \phi \rangle$, together with the fact that under double time reversal a one-electron state, such as a Kramers system, reverses sign. Such effects lead to a variety of physical phenomena for microscopic and thin film systems, for example, a halved periodicity in the flux quantum appropriate to the Aharonov-Bohm effect, and various effects in resistivity, thermopower, etc., in such systems (see Al'tshuler and Lee[69] and Kearney et al.[70]). One effect is that of weak localization, in that backscattering becomes twice as probable as diffuse scattering. This tends to localize the electrons even in a thoroughly disordered system, and lies at the root of several important physical effects. The naive descriptions paint this as a coherent process over randomly positioned scatterers. However, the full proof of the extent to which the various ensemble averages combine to give this measure of coherence is not simple, as the analysis of Freund[63] indicates.

Freund[63] raises the possibility, as a question to the students who detected the flaws in the usual assertions of coherence in this context, that one could construct a system in which the backscattering contributions interfered destructively—a system whose paint rendered it invisible.

C. Second-Harmonic Generation and Parametric Downconversion

Let us start the discussion using a particular example, that of second-harmonic generation (SHG), in which two photons of frequency ω, wave vector $\mathbf{k}$, and polarization $\mathbf{e}$ are absorbed, and one photon of frequency 2ω, wave vector $2\mathbf{k}$, and polarization $\mathbf{e}'$ is emitted. All wave vectors are parallel since then both energy and momentum can be conserved. (The neglect of dispersion can be overcome.) Hence, both $\mathbf{e}$ and $\mathbf{e}'$ are perpendicular to $\mathbf{k}$.

We also bear in mind its time-inverse, parametric down-conversion (PDC): A photon of frequency 2ω (polarization $\mathbf{e}$) splits into two photons of frequency ω (polarization $\mathbf{e}'$). Parametric downconversion is an optical example of a parametric oscillation phenomenon in physics, in which the variation of some parameter of a resonant system at twice the resonant frequency will build up energy of oscillation. Adler and Breazeale[71] give some interesting examples. For example, a child can (but probably doesn't!) pump a swing by alternately standing (at midpoint) and sitting (at end points), and so injecting a mechanical frequency twice the natural frequency of the swing. Faraday noted that a cylinder floating on water will, if oscillated at 2ω, generate water waves of frequency ω. A heteroparametric generator can be made from a capacitor in parallel with two sets of coils. The sets have a gap between, in which a toothed metal wheel rotates, parametrically varying the inductance. In 1935 Mandelstam got 12 kV and 4 kW from such a device!

These processes conserve electronic and radiation subsystem energies separately, returning the atomic system to its original state. Hence, the amplitudes over different atoms are indistinguishable and cohere.

D.   Second Parity Rule

Parity Rule 1 was originally devised for, and particularly useful for, incoherent processes. In the case of refraction or one-photon absorption, it imposes no restriction on E1 matrix elements, and merely forbids E1/M1 interference. By itself, it allows SHG/PDC in centric systems, provided multipole interference is chosen accordingly, Since there is no external field, $\Pi = +1$, and so Parity Rule 1 requires $K = +1$; no interference terms between multipoles of different order in wave vector are allowed for centric systems. Hence, from Time Reversal Rule 1, the absence of static fields ($\tau = 1$), and $K = +1$, SHG/PDC must then be nongyrotropic, and linear polarization is adequate. In acentric systems, Parity Rule 1 does not apply, and hence Time Reversal Rule 1 permits gyrotropic SHG effects.

However, there is a new feature. Parity Rule 1 made no assumptions about any connection between the parities of the initial and final (or indeed intermediate) states, only that each was a state of definite parity; any further constraints lead to further rules. Since for coherent processes the initial and final atomic states are the same, there is a new rule, Parity Rule 2, gained by considering parity for each amplitude. It has the form

Parity Rule 2

$$N_A K_A \Pi_A = +1 \tag{17}$$

for a centric system, where N_A is the number of photon interactions (since $\mathbf{p}$ is of odd parity, and the various $\mathbf{A} \cdot \mathbf{p}$ interactions no longer appear quadratically) in the set of matrix elements comprising amplitude A, the wave-vector phase K_A is $(-1)^{n_k}$ with n_k being the order of the amplitude A in all photon wave vectors, and Π_A is the product of all external field parities in the amplitude A.

For SHG/PDC each amplitude involves two photon absorptions and one emission, giving $N_A = 3$. In the absence of external fields, $\Pi_A = +1$. Hence, K_A must be -1 in a centric system. This is, of course, consistent with Parity Rule 1 for the total SHG rate, since $K = K_A K_{A*} = 1$. However, it goes beyond that rule, in demonstrating that multipole interference is an absolute requirement for SHG/PDC in centric systems, and that in both amplitudes. Clearly SHG and PDC will be very much easier to obtain in chiral media, and this is reflected in the choice of materials.

It is now of interest to compare these novel consequences of Parity Rule 2 with the applications of Parity Rule 1 to the incoherent processes of one- and two-photon absorption, which by the argument of Section II are related to SHG/PDC. The answer is perhaps surprising: There is no connection! Neither one- nor two-photon absorption are forbidden by Parity Rule 1 within centric systems, even in E1 coupling. (We have not taken into account in any of these parity rules the possibility that the intermediate states of the coherent process, corresponding to the final states of the incoherent processes, might be restricted to the same configuration and so have the same parity as the initial state; that restriction itself spawns new parity rules in all cases with their own set of interrelationships.) The reason for this is that the subdivision of matrix elements is different; in effect, SHG/PDC is an interference between the two incoherent processes. A one-photon absorption rate depends on the square modulus of a matrix element, and Parity Rule 1 deliberately looks at the parity restrictions only on the final squared form. However, one amplitude will contribute to K_A, and the other to K_{A^*}. Therefore, Parity Rule 2, based as it is on the equality of initial and final states in a coherent process, is in general a significant contribution to the restrictions arising from parity considerations.

E.   Second Time Reversal Rule

We search now for a possible extension of Time Reversal Rule 1 to individual amplitudes for a coherent effect. In the incoherent case, an amplitude A (a generalization of that in Section III.F to a product of matrix elements) consists of a set of matrix elements

$$A = \langle \text{initial} | V_1 | i \rangle \langle i | V_2 | j \rangle \cdots \langle n | V_n | \text{final} \rangle \qquad (18)$$

(we suppress the denominators for simplicity). In proving Time Reversal Rule 1, we demanded that the product AB^* for any two such terms must be real if this interference term were to contribute, and noted that this required the counting of phases arising from the imaginary units accompanying each wave vector, the time reversal character of the interactions, and the effect of complex conjugating photon polarization vectors. It was necessary to combine two such amplitudes since it is only the products AB^* that need to be real, and at first sight there would appear to be no special requirement on any amplitude A.

However, an important new feature in a coherent process is that the final and initial electronic states are the same, or at least degenerate within a time-even basis set (that is, a set of states that includes the time-reversed state of every one of its members). In boson or even-electron (non-Kramers) systems, there may be no need to distinguish a state and its time reverse, the distinction that arises in a complex basis (such as the $SO(3) - SO(2)|JM\rangle$ basis) being avoidable by converting to a real basis (e.g., $|JM\rangle \pm |J - M\rangle$), and T^2 (double time reversal) is an invariance operation. However, in an odd-electron or Kramers system, this distinction is unavoidable; every such state has a distinct time reversal partner, since from the Dirac relativistic field theory of the electron, each has an eigenvalue $\tau_\Lambda = -1$ under T^2 (Refs. 2 and 72).

This situation now permits the study of a new type of time reversal selection rule whose counterparts, while not widely known, have been most powerful in a variety of systems: the theory of the Jahn-Teller effect, Raman scattering, virtual phonon exchange, Berry phase, and many other topics. A review is given in Moore and Stedman.[73] Central requirements for its use are a time-even and degenerate basis set from which both initial and final states are drawn, and an effective operator O_α^{eff} with definite time-reversal signature: $\overline{O_\alpha^{\text{eff}}}\dagger = \tau_0 O_\alpha^{\text{eff}}$. One need not worry about asymmetries in such an effective operator arising through the very different subamplitudes for one-photon creation and two-photon absorption invalidating this result, since the Hermitian conjugated terms must also appear in the same operator. If we take the initial state (without loss of generality) as the time reverse $|\overline{\lambda \ell}\rangle$ of one member of the basis set inquestion, and the final state as some other member $|\lambda \ell'\rangle$, the amplitude A may be written as proportional to

$$A = M_{ll'} = \langle \overline{\lambda \ell}|O_\alpha^{\text{eff}}|\lambda \ell'\rangle \tag{19}$$

From this, one may perform a standard manipulation[72]; the application now of both Hermitian and T conjugation shows that $M_{ll'} = \tau_0 \tau_\Lambda M_{l'l}$. This in turn restricts the $SO(3)$ tensorial rank (or for a general symmetry group, the irreducible representation or irrep μ) of the operator O_α^{eff} to the symmetric or antisymmetric square of the angular momentum (in general, the irrep) λ of the basis set, as $\tau_0 \tau_\Lambda = +1$ or -1, respectively. This gives the result that coherent optical processes require the use of coupling operators satisfying the constraint, which we dub Time Reversal Rule 2:

Time Reversal Rule 2 $$\mu \in [\lambda \times \lambda]_{\tau_0 \tau_\Lambda} \tag{20}$$

where the square bracket denotes the symmetrized Kronecker product.

As with Parity Rule 2, this does not duplicate the information of Time Reversal Rule 1 for the incoherent processes. Its closest parallel is with the rule for nonresonant phonon Raman scattering,[74] where, however, μ

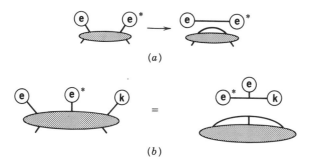

Figure 4. Rotational averaging for a coherent process in a fluid: (a) refraction in E1 approximation; (b) gyrotropic refraction.

includes the symmetry of the participating phonon mode as well as that of the two photons involved in the transition.

F. Fluids

Symmetry Rule 1 for rotational symmetry may in a coherent process be applied to each amplitude; since the sum over randomly oriented systems is coherent within each amplitude, the latter must each form rotational invariants separately. In all such applications of symmetry, the effect is to couple both electronic operators to an O_α^{eff} that is invariant under the point group, and all purely photon- and field-related vectors in a mirror image fashion, also to a rotational invariant.

In the case of refraction within E1 coupling, this rotational average (Fig. 4a) gives the geometrical factor invariant $\mathbf{e} \cdot \mathbf{e}^*$, which is unity through normalization. This is consistent with the selection rules for incoherent absorption. Extending this to a term in the amplitude involving one E1 and one M1 coupling, we would get on fluid averaging $\mathbf{e} \cdot \mathbf{e}^* \times \mathbf{k}$ (Fig. 4b), which is exactly the form found for the incoherent process of circular dichroism.

G. Fluids and Time Reversal Rule 3

Consider now the case of coherent optical processes in fluids with relation to time reversal symmetry. For time-even effective operators, whether the system is Kramers or non-Kramers, this rule implies that the coupling symmetry must be even; $[j \times j]_+$ contains even and odd ranks only since j is integral and half-integral, respectively, and conversely for $[j \times j]_-$; note that $\tau_j = (-1)^{2j}$. This is in any case guaranteed by, and not additional to, the requirement that the coherent averaging over scatterers gives geometrical factors that are rotational scalars and so couple to zero angular momentum; the same coupling that assures this result also (by the mirror symmetry of the pinching operation) guarantees that the associated electronic multipole coupling operators also couple to zero angular momentum. And zero is an even number. Hence, for example, fluids with time-even constituents and environment happily refract light! Also, four-photon processes such as coherent anti-Stokes Raman scattering and four-wave mixing are in principle possible in fluids of time-even molecules, as far as this consideration is concerned.

If, however (still for the case of coherent optical processes in fluids), the effective operator is time-odd, the symmetry constraints on μ are now incompatible, and the same arguments when suitably modified now combine to outlaw such an effect. The Time Reversal Rule 2 requires the coupling to be odd, but zero is still an even number. The characteristic of being time-odd is dictated entirely by the number N_A of photon interactions in the amplitude A, regardless of multipole expansion terms, since all the c numbers and operators associated with the expansion terms are HT-symmetric. Hence, we can conclude the following:

Time Reversal Rule 3

All coherent optical processes in a fluid must involve an even number of photons. This is an unexpected and powerful result. It gives, for example, a new and elegant proof of the result of Andrews and Blake[75, 76] that second-harmonic generation is forbidden in a fluid of randomly oriented scatterers. Like the more conventional proofs, it holds for all orders of multipole coupling, and indeed as for the proof of Andrews[77] to even harmonics of arbitrary order, and to the inclusion of the weak diamagnetic interaction as well.

It may be salutary to compare the above one-line proof with the proof by more conventional techniques but in our diagram notation. Consider first SHG within E1 coupling. The photon vectors $\mathbf{e}$, $\mathbf{e}$, and $\mathbf{e}'^*$ must form a rotational invariant. This allows only the triple scalar product $[\mathbf{e}\mathbf{e}\mathbf{e}'^*]$, which vanishes on account of its antisymmetry in each pair of polarization vectors. Now go to E1/M1 interference. The vectors $\mathbf{e}$, $\mathbf{e}$, $\mathbf{e}'^*$ and $\mathbf{k}$ (or $\mathbf{k}'$ for the emergent photon, which is parallel to $\mathbf{k}$) must be coupled to a rotational invariant. An even-rank rotational invariant can involve only scalar products of vectors, so that such a scalar is certain to contain $\mathbf{e}^{()} \cdot \mathbf{k}^{()}$, which vanishes from transversality. The same is true at all higher orders.

The above proof may be compared with the more laborious proof, using standard techniques, of the result of Andrews and Blake[75] and Andrews[77]: Second-harmonic generation is forbidden to all orders in fluids. The

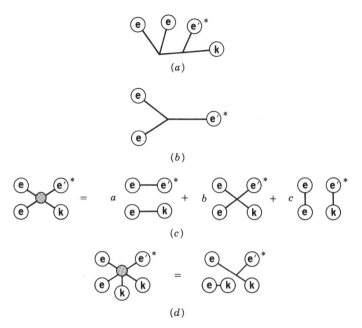

Figure 5. Coupling diagrams for the polarization dependence of second-harmonic generation (SHG): (a) a schematic for general multipole coupling; (b) SHG in E1 coupling only; (c) SHG at first order in the wave vector, that is, in E1/M1-E2 interference: the only possible terms are those products of scalar products listed on the right; (d) SHG at second order in wave vector.

following amounts to a reinterpretation of the proof of Andrews and Blake using diagram methods.

For a coherent process, in each amplitude within the expression for the intensity of a second-harmonic generation process, the polarization vectors and photon wave vectors can be jointly averaged over all orientations; this is equivalent to combining coherently the contributions of equivalent and randomly oriented scatterers. Hence, the photon wave vectors and polarization vectors must couple to a rotational invariant. This is depicted in Fig. 5a.

In E1 coupling, no wave vectors are present; the photon vectors $\mathbf{e}$ and $\mathbf{e}$ (for the two annihilated photons) and $\mathbf{e}'^*$ (for the created photon) cannot form an invariant, since it would have to be the triple scalar product, which vanishes when two vectors are equal (Fig. 5b).

At the next order of multipole, the above three vectors and a fourth, the wave vector $\mathbf{k}$ (for either an annihilated or the created photon), might be expected to produce an invariant. However, by energy-momentum conservation all photon wave vectors are parallel to **k**, and so by transversality all polarization vectors are perpendicular to it. An invariant composed of four vectors can consist only of scalar products, and one of these (that involving **k**) must vanish (Fig. 5c). Hence, the process is forbidden in E2/M1 interference.

At the next order, two wave vectors and the three polarization vectors, two of them equal, must form an invariant if there is to be a contribution. The invariant must involve one triple scalar product, which therefore can involve only the three nonparallel vectors $\mathbf{e}, \mathbf{e'}^*, \mathbf{k}$ if it is not to vanish, but the remaining scalar product $\mathbf{e} \cdot \mathbf{k}$ itself then vanishes (Fig. 5d). Iterating this, we see that at every order no invariant is possible and the process is generally forbidden, despite numerous claims to the contrary (for a fuller review see the chapter by Andrews in this volume).

H. Coherent Processes in Systems with Lower Symmetry

Consider now a coherent process in a system of scatterers with lower symmetry than a fluid, characterized by a point group G. If the various scatterers are oriented in the same direction, they will act like a single unit. The amplitude will scale according to the number of sites, and the theory for a single scatterer will apply when analyzing symmetry restrictions. Within O(3), Time Reversal Rule 2 tells us that the coupling symmetry J of the photon- and field-related operators must still be even, but not necessarily zero, and the two such coupling symmetries J_A, J_B in the two amplitudes A and B must in turn couple and branch to the invariant or identity irrep of G. The last, but not the first, part of this program is precisely as for the standard analysis of incoherent processes.[5]

If, on the other hand, the various molecules randomly adopt one of a set of possible orientations that span the operations of the symmetry group, as for a point defect in a variety of possible substitutional sites, then an analogous result holds as for fluids. The coherent sum in any amplitude amounts to a group average over all group elements, and we obtain another time reversal rule within the point group: The coupling symmetry μ of Time Reversal Rule 2 must correspond to the identity irrep 0(G) of G.

Now the product $[\lambda \times \lambda]_{\tau_\lambda}$ always contains 0(G), whereas $[\lambda \times \lambda]_{-\tau_\lambda}$ never contains 0(G) for any real (orthogonal or symplectic) irrep λ of any point group G. The proof follows from the Frobenius-Schur theorem (e.g., Stedman[2]). If $\chi^\lambda(g)$ is the character of λ for operation $g \in G$, the character orthogonality theorem states that $\Sigma_g \chi^\lambda(g)\chi^\mu(g)^* / |G| = \delta_{\lambda\mu}$, and the number of occurrences of 0 in $[\lambda \times \lambda]_\pm$ is therefore $\Sigma_g \chi^{[\lambda \times \lambda]\pm}(g)/ |G| = \frac{1}{2}\Sigma_g[\chi^\lambda(g)^2 \pm \chi^\lambda(g^2)]/ |G|$. The Frobenius-Schur theorem states that $\Sigma_g \chi^\lambda(g^2)/ |G|$ is the $2j$ phase $\{\lambda\}$, which in turn is $+1$ (-1) for an orthogonal (symplectic) irrep λ, which in all but the lowest symmetries will correspond to non-Kramers and Kramers systems, respectively. Again, for a real irrep $\Sigma_g \chi^\lambda(g)^2 = 1$ from the character orthogonality theorem. It is assumed that λ, the group label for the ground level of each scatterer, is an irreducible representation of the same group G that represents the orientational diversity. Hence, Time Reversal Rule 3 also holds in this case: Only in the case of an even photon process can a coherent process survive.

I. Symmetric Coherent Processes

Let us consider the case of a coherent process in which both A and B have the same form, and in which each term A, B is Hermitian-conjugation symmetric, the virtual excitation and deexcitation being mutually time inverse. The simplest example is that of field-independent refraction, or forward Rayleigh scattering, in the E1 coupling approximation. Both parity rules are innocuous; all component signs N_A, Π_A, K_A are $+1$. Time Reversal Rule 1 merely tells us that $\gamma = 1$, and the process is nongyrotropic. Time Reversal Rule 2 then comes into play by requiring that the coupling symmetry μ of the two one-photon interactions $\mathbf{e} \cdot \mathbf{p}$ that correspond to excitation and deexcitation of the virtual intermediate state must appear in the symmetric or antisymmetric square of the state irrep for non-Kramers or Kramers systems, respectively (the two one-photon interactions combine to give a time-even effective operator O_α^{eff}). The electronic operators are identical in E1 coupling, each containing terms with transformation character under G according to the various irrep labels ν which appear in the branching in the O(3)-G chain of the spin-parity label $J^\pi = 1^-$ appropriate for the momentum operator $\mathbf{p}$ which appears in the E1 interaction; we label their reducible sum by the O(3) parent label 1^-. Using irrep labels of G (including 0 for its identity irrep, as in Butler[78]), and suppressing denominators, the operator O_α^{eff}, which is time-even, must have the form $[[\mathbf{e}\ \mathbf{e}^*]^\mu[\mathbf{p}\ \mathbf{p}]^{\mu^*}]^0$ with the irrep μ belonging both to the symmetric coupling $[1^- \times 1^-]_+$, i.e., to the reps transforming as 0^+ and 2^+, and to the symmetrized product $[\lambda \times \lambda]_{\tau_\lambda}$. The compatibility of these restrictions is guaranteed since the latter product will always include the identity irrep transforming as 0^+. The latitude of freedom that they give for choices of electronic operator $[\mathbf{p}\ \mathbf{p}]^\mu$ will be reflected in the possible variety of geometric factors $[\mathbf{e}\ \mathbf{e}^*]^\mu$. The procedure for determining these is exactly parallel to those used in discussing geometrical factors for the corresponding incoherent absorption processes in Stedman,[5] and the conclusions are unchanged.

The fact that the coupling symmetry of the polarization vectors is necessarily the same as that of the electronic operators means that this conclusion neatly accords with the argument from Time Reversal Rule 1 that the process must be nongyrotropic. It is not so obvious whether the conclusions from all these selection rules are so closely interrelated in subgroups of O(3).

J. Example of TRR2: Gyrotropic Refraction

Now consider the situation where terms in which E1 and M1 matrix elements are mixed in the amplitude A for the coherent process of refraction. Let us concentrate on one such term A_{E1M1}, summed coherently over all ions α, and its interference I_{AB}^1, I_{AB}^2 with an E1 term B_{E1} or a similar mixed term B_{E1M1}, respectively, in the conjugate amplitude. We ignore external fields. We assume that the resulting contribution X to any rate is symmetrized by complex conjugation, etc., in the appropriate way.

Time Reversal Rule 1 tells us that I_{AB}^1, I_{AB}^2 are respectively gyrotropic and nongyrotropic. Parity Rule 1 tells us that I_{AB}^1 is forbidden for centric systems. Parity Rule 2 tells us that A and B_{E1M1}, but not B_{E1}, are forbidden for centric systems, and hence shows that I_{AB}^2 is also forbidden for centric systems in such a coherent process. Time Reversal Rule 2 tells us that when the two interactions in either A or B are coupled to rank μ, this irrep label must be restricted to the appropriately symmetrized part of the Kronecker square of the ground state irrep.

What geometrical factors does this leave us? Let us use the O(3) rotation group for coupling the operators, and in the case of fluids for averaging A_{E1M1}. O_α^{eff} has the form $\mathbf{e} \cdot \mathbf{p}|i\rangle\langle i|\mathbf{m}^* \cdot \mathbf{L}/(E_{\text{initial}} - E_i)$ (which can then be symmetrized by exchanging the E1 and M1 interactions).

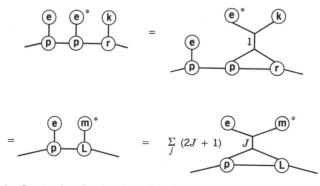

Figure 6. Gyrotropic refraction through higher order terms in the multipole expansion.

Time Reversal Rule 2 requires the coupling angular momentum $\mu \rightarrow J$ to be even in either Kramers or non-Kramers systems. The coherent averaging over scatterers with Time Reversal Rule 2 requires it to be invariant. Hence, the terms of the form $O_\alpha^{\text{eff}} \propto \Sigma_J c_J [\text{e m}^*]^J$: $[\text{p L}]^J$ as represented and shown in Fig. 6 are constrained to $J = 0$ for fluids, and to $J = 0, 2$ for other systems. For $J = 0$, we obtain a feasible and suitably gyrotropic factor $[\text{e m}^*]^0 = [\text{e e}^* \text{ k}] \sim \text{k} \cdot \text{k}$. For $J = 2$, we get a similar expression, which can then be particularized to the point group in question by the methods of Stedman.[5]

K.   Example of TRR2: Second-Harmonic Generation in Solids

In the case of SHG/PDC, fluids are uninteresting for the reasons given in Section V.G. μ itself is the coupled rank from three photon interactions. It must correspond at the O(3) level to an odd angular momentum J to satisfy Time Reversal Rule 3. Since three vectors are coupled in E1 multipole coupling, J must be 1 or 3. Since the two incoming polarization vectors must be combined symmetrically and can couple only to an even rank, we may write the possible geometric factors as a linear combination of the form $[[\text{e e}]^K \text{e}'^*]^{J \downarrow 0(G)}$ for $K = 0, 2$, and $J = 1, 3$. For example, with $K = 0$, $J = 1$ we have $(\text{e} \cdot \text{e})\text{e}'^*$ for that part of the geometrical factor from one amplitude, while $K = 2$ permits a $J = 3$ term as well. In this way the work performed in this chapter for coherent processes in fluids may be extended to coherent processes in oriented molecules or crystals, in the same manner as incoherent processes in oriented molecules or crystals were treated in Stedman.[5] We shall not go any further into this problem in this work.

L.   Coherent Anti-Stokes Raman Scattering

As another example of a coherent process, we consider coherent anti-Stokes Raman scattering (CARS), in which two laser photons of frequency ω are absorbed, and a photon in each of a Stokes and anti-Stokes Raman sideband appropriate to the material are emitted at frequencies $\omega - \Omega$, $\omega + \Omega$, respectively, Ω being the frequency of, say, a vibrational mode, which like the electronic system suffers no net change. Momentum conservation is guaranteed in a suitable geometry. Clearly this allows a coherent process, since the associated exponential factors $\exp(i\Sigma\text{k} \cdot \text{r})$ then cancel.

This now permits the answering of the obvious questions: Can CARS be exhibited by a fluid? Can the individual molecules be centric? What multipoles of coupling are required? Under what conditions is the process gyrotropic?

Let us assume first that E1 coupling is adequate, and $K = 1$. Then, since there is no external field and an even number of photon interactions,

Parity Rule 2 tells us that centric systems are allowed, Time Reversal Rule 1 that the process is nongyrotropic, and Time Reversal Rule 3 that the process is permitted in fluids. A possible geometrical factor for a CARS amplitude would be $[e \; e]^J$: $[e^+ \; e^-]^J$, for $J = 0, 2$, where $e^\pm$ refer to the frequency-shifted photons, and the colon to a full contraction of the tensors to an invariant.

M. Four-Wave Mixing

The same broad conclusions hold for other four-wave mixing phenomena, in which, for example, two pump photons with opposite wave vectors $\mathbf{k}$, $-\mathbf{k}$ but identical polarization $\mathbf{e}$ interact with a probe photon of polarization $\mathbf{e}'$ and wave vector $\mathbf{k}'$ to produce a photon with wave vector $-\mathbf{k}'$. Again E1 coupling in a fluid system suffices, and a typical geometrical factor will be $[e \; e]^J$: $[e' \; e']^J$ for $J = 0, 2$.

N. Reflection

The coherent process of refraction by forward scattering and in particular its polarization dependence has already been presented as the counterpart of the incoherent process of absorption. Insofar as the polarization dependence of refraction also is relevant to that of transmission (see Sections II.E and VI.B), we may expect it to be paralleled also by that of reflection, if only for energy conservation reasons (see Ou and Mandel,[79] also Mascarenhas[80]). Indeed, various authors have studied the polarization dependence induced in reflection at an interface by the parity and time reversal symmetries of the materials.

Reflection is also a coherent process, as discussed in Section V.A. Many general and important quantum results are assembled in Lekner.[81] Recently Buckingham (unpublished) has considered the observability of the reflection analogue of the ECD refraction signal obtained by Buckingham and Shatwell[54] in oriented molecules. It is valuable and also closer to the work of Buckingham and Shatwell to choose a geometry with the external field normal to the plane of incidence. Silverman[4, 83-85] discussed gyrotropic effects from chiral media, calculating Fresnel coefficients from the Maxwell equations for various choices of constitutive relations, and showed that those analyses based on constitutive relations that ignored light-induced anisotropy in the magnetization were experimentally distinguishable and defective. Related analyses of reflection from chiral media are given by Miteva[82] and Cory and Rosenhouse.[88] Silverman et al.[87] discuss experimental schemes to observe gyrotropic effects (called by these authors chiral asymmetries) in chiral media (called by these authors gyrotropic media). The basic conclusion, that chiral media can induce gyrotropic effects, is entirely in accord with Time Reversal Rule 1 and Parity Rule 1 as applied to the related effect of circular birefringence. We mention also Silverman's intriguing work[84, 85] on generation of optical activity by rotation, which nevertheless appears difficult to test, at least when induced by the earth rotation.[86]

O. Exercises

The student reader may like to consider the popular article by Collett[89] on the experiment of Grangier et al.[90] in the light of our earlier sections. While the present article does not cover anything like the necessary background to understand the full details of Grangier's experiment, it certainly impinges on its geometry. What is the dissipative physical process involved? What is its dispersive/fluctuation counterpart? Which is of importance in this experiment? Is it a coherent process or not? Would a fluid suffice as medium? Why would one expect the intensity in one beam to affect the phase of the other?

As explained by Collett, the experiment involves two light beams. The signal whose intensity we wish to measure comes from a visible frequency dye laser; the probe comes from an infrared dye laser. The beams interact in a sodium atomic beam. Resonant two-photon processes gives a phase shift in each beam proportional to the intensity of the other. Measurement of the phase of the probe beam introduces noise in the intensity of the probe beam, and hence into the phase of the signal beam, since these conjugate actions are prescribed by Heisenberg uncertainty; however, this causes no feedback into the signal intensity, and it is then possible to measure the signal intensity without degrading its constancy. The phase measurement of the probe beam is made by enclosing that beam in a resonant cavity, which has the effect of exchanging phase and intensity fluctuations and permitting measurement of the phase by intensity measurements outside the cavity.

The basic process is two-photon absorption, with one photon from each laser beam. Two photon absorption (TPA) as such is clearly dissipative; the atom changes state, and the answer depends on the intensity of each beam, as Collett[89] says. The corresponding fluctuation process is the contribution of virtual TPA transitions to refraction, which implies a phase change of an optical wave after passing through the refractive medium. In effect, a combination of these dissipative and fluctuating effects (the first being dependent on intensity of beam 1 and the second measuring the phase of beam 2) is what the experiment was designed to detect. Since the final atomic state is not the same, the process is not coherent. Whether a fluid suffices then depends on whether we can form a rotational invariant out of the two polarization vectors $\mathbf{e}, \mathbf{e}^*, \mathbf{e}', \mathbf{e}'^*$ of the two beams, and of any necessary $\mathbf{k}$ vectors. The answer is affirmative, and indeed the experiment uses an atomic beam of Na, i.e., randomly oriented scatterers. In E1 coupling $|\mathbf{e} \cdot \mathbf{e}'|^2$ would be an acceptable geometric factor. In E1 coupling, the fact that Na atoms are centric is compatible with Parity Rule 1: $\Pi = +1$ (no fields), and $K = +1$. So we have checked that Na beams can exhibit TPA.

As another example, consider an experiment by Jeff Kimble, reviewed by Barnett and Gilson,[91] in which parametric downconversion was used to generate squeezed light. A beam from a Nd:YAG laser at 1.06 μm was sent through a material to induce SHG and so a 0.53-μm beam. The 1.06-μm beam was rerouted by a mirror, while the 0.53-μm beam was sent into a cavity, which contained a material to stimulate parametric downconversion and so generation of a new 1.06-μm beam. This was mixed at a beamsplitter with the original beam, and homodyne detection was applied (the outputs from each exit port of the beamsplitter were subtracted). This eliminated quantum noise in the original beam and exposed the squeezing (sub-vacuum-level quantum noise) in the new 1.06-μm beam. What conclusions from the present article are relevant to the setup used by Kimble?

Both SHG and PDC demand a linear geometry ($\mathbf{k}'\|\mathbf{k}$) as in the Kimble experiment; linear polarization is adequate; they also demand a crystal, not a fluid, and the crystal must be acentric, as are $Ba_2NaNb_5O_{15}$ and Mg:LiNbO$_3$ as used by Kimble.

Time Reversal Rule 2 now enables a novel check: Are the electronic ground states in these crystals of appropriate symmetry? It is necessary that the effective operator correspond to a point group irrep which arises in the branching of an *odd* angular momentum $J = 1, 3$ (not 5 or above, since there are just three photon vectors $\mathbf{e}, \mathbf{e}, \mathbf{e}'^*$ and by mirror symmetry in the pinching theorems just three electron operator vectors that can so couple). For $J = 1$, we may have $[\mathbf{e} \ \mathbf{e}]^0 \mathbf{e}'^*$, while for $J = 3$, it will be possible to manufacture a more involved invariant $[[\mathbf{e} \ \mathbf{e}]^0 \mathbf{e}'^*]^3$, which will allow more latitude in geometry.

If the electronic ground state is nondegenerate, the Kronecker square when symmetrized appropriately for its time reversal signature will contain only the identity irrep. In that case, for $J = 1$, the polarization vector $\mathbf{e}'$ must be in the direction of any rotational axis of the crystal, so that its reduction is a point group invariant. The crystal must then be oriented with its symmetry axis appropriately placed. It is more likely that $J = 3$ will reduce to the point group identity irrep and without such geometric constraints. The odds are therefore good that virtually any acentric crystal system will satisfy Time Reversal Rule 2 for SHG.

VI. OTHER APPLICATIONS

A. Topological Effects and Phase Conjugation Mirrors

Tompkin et al.[92] show that Berry's phase is time even, and that the way par excellence to demonstrate this is to use an optical phase conjugating element for a mirror (which reverses wave vector and also complex conjugates the polarization vector, thus, indeed, time reversing the photon). On the phase-conjugated beam retracing the original path backward through the quarter and half wave plates, e.g., in a Michelson setup, the Berry phases cancel; nevertheless, for an ordinary mirror, which reverses wave vector while leaving polarization unaffected, the Berry phases accumulate.

In MCD, which is time-odd, with retracement of the path after reflection, the optical rotation accumulates, whereas in natural circular birefringence (NCB), which involves only time-even couplings, it cancels. This raises the question, Why is the Berry rotation—a time-even quantity—not of the latter character?

For both MCD and for NCB:

1. The optically active component of the absorption/refraction $I(\mathbf{e}, \mathbf{k})$ is gyrotropic.

2. For the normal mirror, the polarization $\mathbf{e}$ is the same before and after reflection.

3. The only distinction between the effects of the mirrors is that $\mathbf{e} \rightarrow \mathbf{e}^*$ for the phase conjugating mirror (PCM). Both types of mirror reverse the wave vector $\mathbf{k}$. Hence, an ordinary mirror, but not a PCM mirror, reverses handedness, and the PCM mirror precisely time reverses the photon.

Since MCD has no dependence on the direction of the wave vector, and since for a normal mirror the polarization $\mathbf{e}$ does not change on reflection, a given beam (of definite polarization $\mathbf{e}$) suffers the same attenuation on each pass, and the refractive effect accompanying this means that it also suffers the same speed change on each pass. Hence, the relative effects of MCD on two beams that have opposite handedness ($\mathbf{e}$ and $\mathbf{e}^*$) at the beginning accumulate on the second pass; one beam gets slowed both ways, the other one speeded. [A popular but rather more confusing way of saying this is that the Faraday or MCD optical rotation effect—the angle of rotation in a sense defined by the current direction of $\mathbf{k}$—depends on $\mathbf{B} \cdot \mathbf{k}$, and $\mathbf{k}$ reverses, so the angles are in opposite senses for the two passes, but since they are relative to opposite directions for $\mathbf{k}$, the final angle accumulates. This involves a few white lies; strictly, the $\mathbf{k}$ appearing in the geometrical factor $\mathbf{B} \cdot \mathbf{k}$ used for the Faraday effect does not contain the direction of the wave vector, since it originates in $\mathbf{e} \times \mathbf{e}^*$. The incipient error is canceled by a similar subtlety in defining a rotation angle relative to the current beam direction from $I(\mathbf{e}, \mathbf{k})$.]

With a PCM, MCD, or Faraday effect, optical rotation angles still accumulate, in spite of the conjugation of the polarization vector $\mathbf{e}$ at reflection. This was demonstrated by Tompkin et al. One simple way of making this plausible is that a time-odd (magnetic) field induces an effect that is not reversed by time-reversing the other subsystem. But how does one see this in detail? If a beam polarization $\mathbf{e}$ suffers a refraction $I(\mathbf{e}, \mathbf{k})$ outward differing from $I(\mathbf{e}^*, \mathbf{k})$, making it say faster than the beam $\mathbf{e}^*$, they arrive at the mirror with a phase difference $e^{i\phi}$. On reflection this accumulated wave phase factor is complex conjugated; in addition, the polarization $\mathbf{e}$ is conjugated, so that what was a faster beam is now a slower beam, giving a sign change to the reflected wave phase factor accumulated on the return journey. The net effect is again that these wave phase factors accumulate, as for an ordinary mirror.

Contrast this with the NCB: This effect depends on the direction of the wave vector $\mathbf{k}$. At an ordinary mirror, $\mathbf{e}$ is unchanged, the sense of optical rotation relative to the beam direction is unchanged, but either this or the linear dependence on $\mathbf{k}$ shows that this means that $I(\mathbf{e}, \mathbf{k}) = -I(\mathbf{e}, -\mathbf{k})$, so that the absorption/refraction (speed) differential on the first pass is canceled on the reverse pass, and the net effect is the same for either circular polarization.

At a PCM mirror, conjugating the polarization $\mathbf{e}$ both reverses the differential effect (exchanges fast for slow) for a gyrotropic process and reverses the differential already induced, so that for either a PCM or a normal mirror, the phase cancels, as might be expected for a "time-even effect." As Tompkin et al.[92] say, quoting the experiment of Boyd et al.,[93] "a uniform, time-reversible dynamical phase shift can be negated by a (PCM)".

But the topological Berry phase has the unique property of being different for the different mirrors, as demonstrated by Tompkins et al., so it cannot be assigned either of the above categories. The time-even character dictates that it must cancel with a PCM mirror, but the behavior for a normal mirror ($\mathbf{e}$ unchanged) is not simply predictable on this analysis. This is because we are looking at a topological effect, realized by the noncommuting behavior of different optical elements in a cascade, for which our essentially local analysis fails, based as it is on the assumption that $I(\{\mathbf{e}, \mathbf{k}\})$ of Section III is a constant throughout the material. Another factor is the difference in mirror properties; it is of interest to note from Boyd et al.[93] that Wiener's $\lambda/2$ distance from the metal mirror to the photosensitive antinode of E no longer holds in PCM.

B. Optical Detection of the P, T Violation in Anyonic Systems

In two dimensions, other statistics than pure boson or fermion are permitted for the quanta of relativistic field theory. The concept of anyons is reviewed in Wilczek,[94] Stone,[95] and Aitchison and Mavramatos,[96] for example.

Parity and time reversal symmetry are both broken in principle in anyonic systems. A summary of the consequences of this is given by Halperin et al.,[97] who explain that only a violation of both P and T symmetry permit a thermal Hall effect, or an asymmetry in the thermal conductivity tensor. They discuss the Wen-Zee suggestion of measuring optical rotation of the material. While they show that for a simple anyon gas, the optical rotation is zero in first order, they expect to see it from higher order corrections.

Kitazawa[98] discusses parity and time reversal violation and their optical detection for anyonic systems. The effective Lagrangian has the form $L = kA_\mu^2 + l\varepsilon^{\mu\nu\rho}\partial_\mu A_\nu A_\rho$, where A_μ is the vector potential in the $2 + 1$ dimensional field theory (x, y being the spatial dimensions of the system), and the second, Chern-Simons, term is the origin of the P and T violation in anyonic field theory. Kitazawa shows that such a system displays circular birefringence to a light beam propagating on the z axis. He gives an encouraging estimate for the observability of such an effect and deduces that its observation in either reflection or transmission would be evidence of P and T breaking.

From the viewpoint of Section IV, we might regard such a signal as evidence for T violation (as for MCD, which requires a time-odd field) or P violation (as for natural optical activity, which forbids use of a centric system), but not necessarily both; if such an effect is seen, it might be taken as confirmation of T violation, but is then not a separate test of P violation as well. This illustrates the discussion of Section III concerning the care that may be taken profitably over the assignment of optical experiments as tests of fundamental symmetries. It is notable that such ambiguities are allowed for in the first report of the observation of circular dichroism in high-T_c superconductors, by Lyons et al.[99] (This particular experimental result has been controversial; see Wilczek[94] and Sulewski et al.[100].)

It is no longer controversial that quantum Hall effect systems have P- and T-violating terms in the effective Hamiltonian, and that this

G. E. STEDMAN

breakdown permits characteristic "parity violating" effects in light scattering, such as gyrotropy in reflection, as in Ishikawa[101]; once again, the connection can be justified using the basic selection rules of Section III. Canright and Rojo[102] discuss PT symmetry in this connection, and show that gyrotropy in reflection must be linked to gyrotropy in transmission. This is an automatic consequence of the reciprocity relations of optics (see, for example, Ou and Mandel[79] and Stedman[2]), which, contrary to the statements of Canright and Rojo,[102] constrain amplitude transmission and reflection factors and not merely the corresponding intensities.

C. Ring Lasers

In a ring laser gyroscope, countercirculating beams are interfered at one mirror, and the beat frequency taken as the signal. The ring laser gyroscope reflects the absolute rotation of the laser as a whole, according to the Sagnac effect, and more generally is a detector of any nonreciprocal effect. The latter is defined by Statz et al.[104] as one that reverses with the sense of propagation of the beam.

To be more precise, consider a traveling electromagnetic wave in a ring laser, for which the electric field is proportional to the real part of $\mathbf{e}_{kj} \exp i(\omega t - \mathbf{k} \cdot \mathbf{x})$; as in Section III.A, $\mathbf{k}$ is the wave vector, j is the polarization, and ω is the frequency. Consider the circularly polarized beam with $\mathbf{k} \| + z$ and $\mathbf{e} = (1, i, 0)/\sqrt{2}$. For given t, the tip of $\mathbf{E}$ follows a right-handed spiral as a function of z; as t increases, the spiral advances bodily in the direction of increasing z. For given z, the tip forms a left-handed spiral as a function of t (in place of z). Looked at end on, from $z = +\infty$, as it advances toward the observer, the tip of $\mathbf{E}$ follows a clockwise path. We shall use the convention of calling this beam right circularly polarized. Now consider the beam obtained from this, for example, by normal reflection off a mirror: $\mathbf{e} = (1, i, 0)/\sqrt{2}, \mathbf{k} \| - z$. The wave factor is $\exp i(\omega t + kz)$. Both the t (fixed z) and z (fixed t) plots are right-handed spirals. As t increases, the xyz spiral travels in the $-z$ direction. Looked at end on from $z = -\infty$, the tip of $\mathbf{E}$ moves anticlockwise. Let us call this left circularly polarized.

A ring laser has four possible modes: clockwise (CW), and counterclockwise (CCW), and for each, two orthogonal polarizations. In a linear polarization basis we denote these as s (sagittal or senschrit) for the out-of-plane component of electric field) and p (parallel or in-plane component); note that the s mode usually is excited rather than p since according to the Fresnel reflection coefficients the p reflection is inevitably more lossy in typical situations. In a circularly polarized basis we denote these both as LCP and RCP.

A nonzero beat frequency between modes, thus giving a signal for the ring laser (as in the standard gyro application) $\omega_{CW} - \omega_{CCW}$, depends on having a nonreciprocal component of the optical path length: $n_{CCW} \neq n_{CW}$.

Consider first linearly polarized light; let us call this a search for *linear nonreciprocality*. Since $e \rightarrow e^*$ must then be an invariance operation, we can only have a nongyrotropic effect. Nonreciprocality then amounts to an absorptive and refractive effect $I(e, k)$, which reverses in sign with k: $I(e, k) = I(e^*, k) = -I(e, -k)$. Hence, the combined matrix elements must be of odd order in the optical wave vector and K must be -1.

An isolated time-even system, or one subjected to a time-even field such as an electric field, whatever its symmetry or centric character, cannot display linear nonreciprocity. For the requirements $K = -1, \tau = 1$ conflict with the nongyrotropic character of the effect.

We are forced to look at effects that are linear in an applied magnetic field to satisfy Rule I. Parity Rule 1 then tells us that this can occur only in acentric media. The lowest order such effect is MLD (Section IV.A).

Because of the parity rule, neon will not suffice as a medium, at least in this order of approximation. Hence, a helium-neon ring laser does not spontaneously produce strong signals from stray magnetic fields, such as the earth field or those in the plasma. The story is different if we allow the interaction with the neon, particularly in the gain tube, to change the cavity eigenmode to elliptical polarization.

Even linear nonreciprocity is allowed for fluids. The angular momentum coupling diagram (Fig. 7) gives on calculation the geometrical factor $\mathbf{B} \cdot \mathbf{k}$, since of the various possible scalar products only this (times $e \cdot e^*$, which is unity) survives; $e \cdot k = 0$. While this is the same geometrical factor that occurs in the Faraday effect, this is a totally different process (involving E1-M1/E2 interference, and also needing a chiral system), and is unobserved as yet.

Now consider circularly polarized light, and let us call any nonreciprocity induced by some medium in the effective refractive index for a

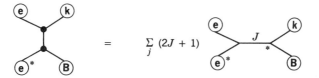

Figure 7. The nonreciprocity of magnetic linear dichroism. The invariants in the expansion on the right side reduce to $\mathbf{B} \cdot \mathbf{k}$ through transversality ($e \cdot k = 0$), as for the Faraday effect.

circularly polarized beam *circular nonreciprocality*. Note that a circularly polarized beam reverses its handedness upon reflection. Hence, the legs of a ring for any mode have alternating polarizations.

How in practice does one induce circular, or at least elliptical, polarization, in a ring laser? There are several methods, which produce different results and so make a ring laser sensitive to different optical effects.

First, we might use the *interaction with the medium*; applying a magnetic field to the plasma gives a "Zeeman laser" (see Statz et al.[104]). This throws the ring into circular polarization, since, in principle, the slightest optical rotation invalidates linear polarization as a candidate for an eigenmode, but is perfectly compatible with circular polarization for an eigenmode. Since the appropriate geometrical factor has the form $\mathbf{m} \cdot \mathbf{B}$ where $\mathbf{m} \equiv \mathbf{e} \times \mathbf{k}$ (Section III.A), the refractive index change depends on the polarization vector $\mathbf{e}$ on which the definition of $\mathbf{m}$ is based, and not on the sense of the wave vector $\mathbf{k}$. This means that within any leg it is the modes whose polarizations differ according to their polarization vectors $\mathbf{e}, \mathbf{e}^*$, rather than differing by being left or right circularly polarized, that are distinguished in frequency; the difference in effective optical path length implies a differential equivalent refractive index $n(\mathbf{e}) \neq n(\mathbf{e}^*)$. In general, one may use a *nonreciprocal polarization rotator*, that is, a medium for which $n(\mathbf{e}) \neq n(\mathbf{e}^*)$, whatever the sense of $\mathbf{k}$. In such cases LCP-CW and RCP-CCW modes are degenerate, but are split from RCP-CW and LCP-CCW.

Second, we might use a *nonplanar ring*; Statz et al.[104] discuss image rotation effects, which then make the eigenmodes circularly polarized. Bilger et al.[105] have discussed the transition from planarity to nonplanarity in some detail. The ring must have at least four mirrors for the beam to be nonplanar and to have circularly polarized eigenstates. At reflection from a mirror, in an *s-p* basis, the Jones matrix is $\begin{pmatrix} 1 & 0 \\ 0 & -1 \end{pmatrix}$; image rotation in the next leg of a nonplanar ring adds $\begin{pmatrix} 1 & \theta \\ -\theta & 1 \end{pmatrix}$. A second reflection is followed by an image rotation by the *opposite* angle. Squaring the resulting product of four matrices gives a total Jones matrix for the ring, $\begin{pmatrix} 1 & 4\theta \\ -4\theta & 1 \end{pmatrix}$. It has eigenvalues $\lambda = 1 \pm 4i\theta$, with eigenvectors $\mathbf{e} \equiv (1, i)$ or $\mathbf{e}^* = (1, -i)$. Again the effective refractive index depends on the polarization vector and not the wave vector, and the effect of this on the mode degeneracies is exactly the same as for the Zeeman laser: LCP-CW and RCP-CCW modes are degenerate, but split from RCP-CW and LCP-CCW.

Third, one may use *quarter wave plates* (see Bilger and Stedman[106]), transforming (*s*) linear into circular at the start of a "research station" in one leg, and vice versa at the other end; the axes of these plates are at 45° to the s axis, and at 90° to each other. Then a polarization vector $\mathbf{e} \equiv (a, b)$ for the incident beam is transformed progressively into $(a, \mathrm{i}b)$ and $(a, -\mathrm{i}(\mathrm{i}b)) = \mathbf{e}$. A circularly nonreciprocal medium in the research station (where the polarization is $(a, \mathrm{i}b)$) will then induce a beat frequency associated with an effectively linear nonreciprocality in the ring as a whole.

Fourth, we may use a *reciprocal polarization rotator* element such as a system displaying natural optical activity (NOA) or natural circular bire-fringence (NCB).[104] Anything that rotates linear polarization ruins its candidature for an eigenmode.

In the last two cases, a nonvanishing signal requires the introduction of a medium for which $n(\mathrm{RCP}) \neq n(\mathrm{LCP})$, regardless of direction. Note that this is not the same criterion as for the first two cases. In the last two cases, LCP-CW and LCP-CCW are degenerate, but split from RCP-CW and RCP-CCW. Would an electric field give such an effect? This requires ECD, which is disallowed for a fluid such as neon gas.

A review of the progress of and plans for a major ring laser experiment, in which the geometrical results derived above will be of vital importance, is given in Stedman et al.[103]

APPENDIX: DIAGRAM NOTATION

We give here a minimal introduction to the diagram notation of our earlier work.[2] The discussion is heavily restricted, essentially to diagram-matic vector algebra, which is understandable by a first-year undergradu-ate, in a real (say Cartesian) coordinate system. The applications are therefore restricted to fluid systems. Such SO(3) (rotation group) vector and tensor coupling diagrams and associated graphical methods, originat-ing with Jucys and collaborators, have proved outstandingly useful in atomic physics, and indeed reflect the topology of field theoretic or Feynman-type diagrams. They form an appropriate technique to extract symmetry restrictions in the field theoretic approach.

A vector $\mathbf{a}$ is represented by a node that is appropriately labeled, and that has a leg carrying component labels (Fig. 8a). The scalar product of two vectors involves a sum over matched component labels, which is simply depicted by uniting the legs, as in Fig. 8b. Finally, the vector product of two vectors may be used to define an appropriate node, whose numerical value is that of the antisymmetric symbol $\varepsilon_{\alpha\beta\gamma}$ on the compo-nent labels α, β, γ (Fig. 8c), so that the vector identity $\mathbf{a} \times (\mathbf{b} \times \mathbf{c}) = \mathbf{b}(\mathbf{c} \cdot \mathbf{a}) - \mathbf{c}(\mathbf{a} \cdot \mathbf{b})$ has the diagram representation of Fig. 8d. In diagrams, we have Fig. 8e, which in transliteration reads $\varepsilon_{\alpha\beta\gamma}\varepsilon_{\lambda\mu\gamma} = \delta_{\alpha\lambda}\delta_{\beta\mu} - \delta_{\alpha\mu}\delta_{\beta\lambda}$.

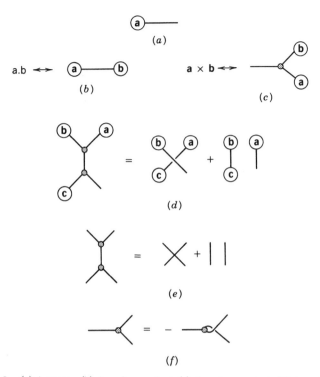

Figure 8. (a) A vector. (b) A scalar product. (c) A vector product. (d) A vector identity. (e) A diagram representation of (d). (f) Antisymmetry of the node representing $\varepsilon_{\alpha\beta\gamma}$.

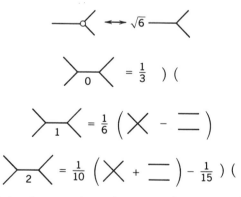

Figure 9. Relation between the node representing $\varepsilon_{\alpha\beta\gamma}$ and the $3jm$ symbol $\begin{pmatrix} 1 & 1 & 1 \\ \alpha & \beta & \gamma \end{pmatrix}$, and the expansions of the angular momentum coupling trees of order 4 for vector terminals.

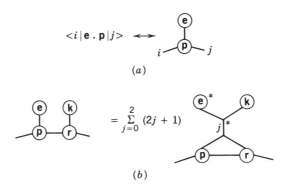

$$\langle i\,|\,\mathbf{e}\cdot\mathbf{p}\,|\,j\rangle \longleftrightarrow$$

(a)

$$= \sum_{j=0}^{2} (2j+1)$$

(b)

Figure 10. Application of the formalism to the multipole expansion.

Exchange of two component labels changes the sign of the antisymmetric symbol, as in Fig. 8f. A loop carries an implied summation over component label, and so has the numeric value 3.

The extension of this from vector algebra $(j = 1)$ to any angular momentum j is done by using the key results of Fig. 9. This apparatus is quite sufficient to reformulate much of the foregoing. Consider for example the tensors arising in multipole expansion. The E1 term is proportional to $\mathbf{e}\cdot\mathbf{r}$, and its matrix elements $\langle i\,|\,\mathbf{e}\cdot\mathbf{r}\,|\,j\rangle$ may be depicted by Fig. 10a. The next term in the expansion has the form $\langle i\,|\,(\mathbf{e}\cdot\mathbf{p})(\mathbf{k}\cdot\mathbf{r})\,|\,j\rangle$, but is not as it stands an irreducible tensor operator of the rotation group. We can combine the electronic operators $\mathbf{p},\mathbf{r}$ into an irreducible operator $[\mathbf{r}\,\mathbf{p}]^{j}$ for $j = 0, 1, 2$, and at the same time similarly combine $\mathbf{e}$ and $\mathbf{k}$, by use of Fig. 9, and the result of Fig. 10b is obtained. Using Fig. 9, these various terms can be rewritten. In the term $j = 0$, the expression $(\mathbf{e}\cdot\mathbf{k})(\mathbf{r}\cdot\mathbf{p})$ is obtained. This vanishes since the polarization $\mathbf{e}$ and wave vector $\mathbf{k}$ are necessarily orthogonal for an electromagnetic wave. The term $j = 1$ can be rewritten, and becomes the M1 term (Fig. 6), while the $j = 2$ term is the E2 coupling. Similarly, at higher orders n we may take n factors of $\mathbf{k}\cdot\mathbf{r}$ with $\mathbf{e}\cdot\mathbf{p}$ and construct tensors out of the $\mathbf{p}$ and $\mathbf{r}$ vectors, in a tree structure. All such angular momentum coupling trees can be reduced to sums over products of scalar products, since these are the only independent even-rank rotational invariants associated with vectors. Had there been an odd number of vectors, one triple product would be permitted. In general, all terms obeying these restrictions appear, each with some numerical coefficient.

The remaining major operation, used on occasion in the text, is the action of a rotation by some group on all of n lines, followed by a summation over that rotation. This has the consequence in diagram terms of pinching together the n lines involved. A full account with many simple examples are given in Stedman.[5]

Acknowledgments

I am grateful to Laurence Barron, Myron Evans, and Wes Sandle and particularly David Andrews for useful comments on related matters.

References

1. G. E. Stedman, *Phys. Lett. A* **152**, 19 (1991).
2. G. E. Stedman, *Diagram Techniques in Group Theory*, Cambridge University Press, Cambridge, UK, 1990.
3. K. Blum, *J. Phys. B* **23**, L253, (1990).
4. M. P. Silverman, *J. Opt. Soc. Am. A* **3** 830 (1986).
5. G. E. Stedman, *Adv. Phys.* **34**, 513 (1987).
6. A. Lakhtakia (Ed.), *Selected Papers on Natural Optical Activity*, MSIS, 1990.
7. R. P. Feynman, *QED: The Strange Theory of Light and Matter*. Princeton University Press, Princeton, NJ, 1985.
8. J. D. Lawson, *Contemp. Phys.* **11**, 575 (1970).
9. H. B. Callen, in D. ter Haar (Ed.), in *Fluctuations, Relaxation and Resonance in Magnetic Systems* Oliver & Boyd, Edinburgh, 1961.
10. G. E. Stedman, *Am. J. Phys.* **51**, 750 (1983).
11. Yu-Kuang Hu, *Am. J. Phys.* **57**, 821 (1989).
12. M. T. Thomaz and H. M. Nussensweig, *Ann. Phys.* **139**, 14 (1982).
13. F. Bassani and S. Scandolo, *Phys. Rev. B* **44**, 8446 (1991).
14. R. G. Newton, *Am. J. Phys.* **44**, 629 (1976).
15. R. P. Feynman, R. B. Leighton, and M. Sands, *The Feynman Lectures on Physics*, Addison-Wesley, Reading, MA, 1990.
16. J. Yu and W. P. Su, *Phys. Rev. B* **44**, 13315 (1991).
17. D. M. Brink and G. M. Satchler, *Angular Momentum*, 2d ed., Oxford University Press, London, 1968, p. 88.
18. Q. Wang and G. E. Stedman, *J. Phys. B.* **26** to be published.
19. L. D. Barron, in P. G. Mezey (Ed.), *New Developments in Molecular Chirality*, Kluwer, Dordrecht, 1991.
20. M. W. Evans, *J. Mod. Opt.* **37**, 1655 (1990).
21. M. W. Evans, *Mod. Phys.* **71**, 193 (1990).
22. W. Voigt, *Lehrbuch der Kristallphysik*, Teubner, Leipzig, 1910.
23. W. Opechowski, *Int. J. Magn.* **5**, 317 (1974).
24. V. L. Telegdi, in J. Mehra (Ed.), *The Physicist's Conception of Nature*, Reidel, Dordrecht, 1973, p. 459.

25. I. B. Kriplovich and M. E. Pospelev, *Z. Phys. D* **17**, 81 (1990).

26. M. W. Evans, *Int. J. Mod. Phys. B* **5**, 1963 (1991).

26a. A. Messiah, *Quantum Mechanics* vol. 2, North-Holland, Amsterdam, 1962.

27. Q. Wang and G. E. Stedman, *J. Phys. B* **25**, L157, L167 (1992).

28. Y-N Chiu, *Phys. Rev. A* **32**, 2257 (1985).

29. Y-N Chiu, *J. Chem. Phys.* **86**, 1686 (1987).

29a. L. Hecht and L. A. Nafie, *Chem. Phys. Lett.*, **174**, 575 (1990).

30. D. Eimerl, *Phys. Rev. A* **19**, 816 (1979).

31. M. Vala, J. C. Rivoal, and J. Badoz, *Mol. Phys.* **30**, 1325 (1975).

32. M. Vala, J. C. Rivoal, C. Grizolia, and J. Pyka, *J. Chem. Phys.* **82**, 4376 (1985).

33. N. Moreau and A. C. Boccara, *J. Chem. Phys.* **64** 961 (1981).

34. B. Briat, *Mol. Phys.* **42**, 347 (1981).

35. B. R. Judd and W. A. Runciman, *Proc. R. Soc. London Ser. A* **352**, 91 (1976).

36. C. deW van Siclen, *J. Phys. Chem. Solids* **48**, 497 (1987).

37. G. Wagnière and A. Meier, *Chem. Phys. Lett.* **93**, 78 (1982).

38. L. D. Barron and J. Vrbancich, *Mol. Phys.* **51**, 715 (1984).

39. S. Wozniak and R. Zawodny, *Acta Phys. Pol. A* **61**, 175 (1982).

40. D. L. Andrews and A. M. Bittner, *J. Chem. Soc. Farady. Trans.* **87**, 513 (1991).

41. D. L. Andrews, *J. Phys. B* **19**, L613 (1986).

42. D. L. Andrews, *Chem. Phys.* **112**, 61 (1987).

43. S. Kielich and R. Zawodny, *Acta Pol. Phys.* **42**, 337 (1972).

44. H. J. Ross, B. S. Sherborne, and G. E. Stedman, *J. Phys. B* **22**, 459 (1989).

45. P. S. Pershan, J. P. van der Ziel, and L. D. Malmstrom, *Phys. Rev.* **143**, 574 (1966).

46. N. L. Manakov, V. D. Ovsiannikov, and S. Kielich, *Acta Phys. Pol. A* **53**, 581 (1978).

47. N. L. Manakov, V. D. Ovsiannikov, and S. Kielich, *Acta Phys. Pol. A* **53**, 595 (1978).

48. S. Kielich, N. L. Manakov, and V. D. Ovsiannikov, *Acta Phys. Pol. A* **53**, 737 (1978).

49. S. Wozniak, M. W. Evans, and G. Wagniere, *Mol. Phys.* **75**, 81 (1992).

50. M. W. Evans, *J. Mol. Spectrosc.* **146**, 351 (1991).

51. M. W. Evans, *Mod. Phys. Lett. B* **5**, 1065 (1991).

52. M. W. Evans, *Phys. Rev. Lett.* **64**, 2909 (1990).

53. M. W. Evans, *Physica B* **168**, 9 (1991).

54. A. D. Buckingham and R. A. Shatwell, *Phys. Rev. Lett.* **45**, 21 (1980).

55. M. W. Evans, *Physica B* **168**, 9 (1991).

56. M. A. Bouchiat and L. Pottier, *Science* **234**, 1203 (1986).

57. P. G. H. Sandars, *Phys. Scripta* **36**, 904 (1987).

58. M. A. Bouchiat, Ph. Jacquier, M. Lintz, and L. Pottier, *Opt. Commun.* **56**, 100 (1985).

59. D. N. Stacey, *Phys. Scripta* **T40**, 15 (1992).

60. T. D. Wolfenden, P.E.G. Baird, and P.G.H. Sandars, *Eur. Lett.* **15**, 731 (1991).

61. B. S. Sherborne, *J. Phys. C. M.* **1**, 4825 (1990).

62. D. L. Andrews and B. S. Sherborne, *J. Chem. Phys.* **86**, 4011 (1987).

63. I. Freund, *Phys. Today* **42**(8), 88 (1989).
64. I. Freund, *Phys. Rev. A* **37**, 1007 (1988).
65. I. Freund, *J. Opt. Soc. Am. A* **9**, 456 (1992).
66. I. Freund, *Opt. Commun.* **87**, 5 (1992).
67. A. N. Bogaturov, *Opt. Commun.* **87**, 1 (1991).
68. M. O. Scully, *Phys. Rev. Lett.* **67**, 1855 (1991).
69. B. L. Al'tshuler and P. A. Lee, *Phys. Today* **41**(12), 36 (1988).
70. M. T. Kearney, R. T. Syme, and M. Pepper, *Phys. Rev. Lett.* **66**, 1622 (1991).
71. L. Adler and M. A. Breazeale, *Am. J. Phys.* **39**, 1522 (1971).
72. A. A. Abragam and B. Bleaney, *Electron Paramagnetic Resonance of Transition Metal Ions*, Clarendon, Oxford, 1970.
73. D. J. Moore and G. E. Stedman, *J. Phys. Cond. Matter* **2**, 2559 (1990).
74. C. D. Churcher and G. E. Stedman, *J. Phys. C* **14**, 2237 (1981).
75. D. L. Andrews and N. Blake, *Phys. Rev. A* **38**, 3113 (1988).
76. D. L. Andrews and N. Blake, *Phys. Rev. A* **41**, 4550 (1990).
77. D. L. Andrews, *J. Phys. B* **13**, 4091 (1980).
78. P. H. Butler, *Point Group Applications, Methods and Tables*, Plenum, New York, 1981.
79. Z. Y. Ou and L. Mandel, *Am. J. Phys.* **57**, 66 (1989).
80. K. S. Mascarenhas, *Am. J. Phys.* **59**, 1150 (1991).
81. J. Lekner, *Theory of Reflection*, Nijhoff/Kluwer, Dordrecht, 1987.
82. A. I. Miteva, *J. Phys. C* **3**, 529 (1991).
83. M. P. Silverman, *Phys. Lett. A* **126**, 171 (1987).
84. M. P. Silverman, *Am. J. Phys.* **58**, 310 (1990).
85. M. P. Silverman, *Phys. Lett. A* **146**, 175 (1991).
86. G. E. Stedman, *Phys. World* **3**, Nov. 23 (1990).
87. M. P. Silverman, N. Ritchie, G. M. Cushman, and B. Fisher, *J. Opt. Soc. Am. A* **5**, 1852 (1988).
88. H. Cory and I. Rosenhouse, *J. Mod. Opt.* **38**, 1229 (1991).
89. M. Collett, *Phys. World* **4**(6), 28 (1991).
90. P. Grangier, J-F Roch, and G. Roger, *Phys. Rev. Lett.* **66**, 1418 (1991).
91. S. M. Barnett and C. R. Gilson, *Eur. J. Phys.* **9**, 257 (1988).
92. W. R. Tompkin, M. S. Malcuit, R. W. Boyd, and R. Y. Chiao, *J. Opt. Soc. Am. B* **7**, 230 (1990).
93. R. W. Boyd, T. M. Habashy, A. A. Jacobs, L. Mandel, M. Nieto-Vesperinas, W. R. Tompkin, and E. Wolf, *Opt. Lett.* **12**, 42 (1987).
94. F. Wilczek, *Phys. World* **4**(1), 40 (1991).
95. M. Stone, *Int. J. Mod. Phys. B* **4**, 1465 (1990).
96. I. J. R. Aitchison and N. E. Mavromatos, *Contemp. Phys.* **32**, 219 (1991).
97. B. I. Halperin, J. March-Russell, and F. Wilczek, *Phys. Rev. B* **40**, 8726 (1989).
98. Y. Kitazawa, *Phys. Rev. Lett.* **65**, 1275 (1990).
99. K. B. Lyons, J. Kwo, J. F. Dillon, Jr., G. P. Espinosa, M. McGlashan-Powell, A. P. Ramirez, and L. F. Schneemeyer, *Phys. Rev. Lett.* **64**, 2949 (1990).

100. P. E. Sulewski, P. A. Fleury, K. B. Lyons, and S-W Cheng, *Phys. Rev. Lett.* **67**, 3864 (1991).

101. K. Ishikawa, *Phys. Rev. D* **31**, 1432 (1985).

102. G. S. Canright and A. G. Rojo, *Phys. Rev. Lett.* **68**, 1601 (1992).

103. G. E. Stedman, H. R. Bilger, Li Ziyuan, M. P. Poulton, C. H. Rowe, and P. V. Wells, *Aust. J. Phys.* (to be published).

104. H. Statz, T. A. Dorschner, M. Holtz, and I. W. Smith, in M. L. Stitch and M. Bass (Eds.), *Laser Handbook*, Vol. 4, Elsevier, Amsterdam, 1985.

105. H. R. Bilger, G. E. Stedman, and P. V. Wells, *Opt. Commun.* **80**, 133-7 (1990).

106. H. R. Bilger and G. E. Stedman, *Phys. Lett. A* **122**, 289 (1987).

MOLECULAR THEORY OF HARMONIC GENERATION

DAVID L. ANDREWS

*School of Chemical Sciences, University of East Anglia,
Norwich, England*

CONTENTS

I. INTRODUCTION

The generation of optical harmonics accompanies the passage of light through various kinds of medium. It originates in processes whereby the

Modern Nonlinear Optics, Part 2, Edited by Myron Evans and Stanisław Kielich. Advances in Chemical Physics Series, Vol. LXXXV.
ISBN 0-471-57546-1 © 1993 John Wiley & Sons, Inc.

energy of two or more photons of incident light emerges in the creation of single photons of output, the original photons being annihilated in the process. It is easily calculated that the probability of any such process requiring the local coincidence of two or more photons is significant only at the levels of intensity associated with laser light. This is reflected in the fact that observations of harmonic generation were first made only once laser sources became available.[1]

The nature of the medium in which harmonic conversion occurs strongly influences the efficiency of the process and the character of the harmonic emission. In particular, laser-induced second-harmonic generation (SHG) produces tightly collimated emission in noncentrosymmetric crystals and is widely used as a means of laser frequency conversion. Ionic crystalline materials, such as potassium dihydrogen phosphate (KDP), are commonly used to obtain 532-nm radiation from the 1064-nm output of powerful Nd:YAG lasers, with a conversion efficiency that is often around 25%. The combination of multiple-stage harmonic generation with tunable dye laser downconversion thus opens enormous windows on the spectrum of wavelengths that can be derived from fixed-wavelength laser sources.

Harmonic conversion processes are frequently either forbidden or occur only weakly in molecular gases and liquids. The principle of harmonic generation in molecules has nonetheless become an area of increasing interest for a number of reasons. It transpires that many organic crystals and polymeric solids not only have usefully large nonlinear optical susceptibilities, but are also surprisingly robust and thermally stable. Consequently, the fabrication of organic materials for laser frequency conversion has become a growth area, and developments in the associated theory are continuing apace.[2, 3] Although molecular gases and liquids cannot usually compete in terms of efficiency, their vibronic structure often leads to intense and essentially discrete absorption features, and can be associated with a wavelength dependence of refractive index, which can expedite harmonic conversion.[4, 5] Moreover, such media are themselves of considerable scientific interest through the information that can be derived from studying the harmonic processes permitted in them.

Any process in which each output photon is created at the expense of two or more photons of incident light can be fundamentally categorized according to two basic criteria. One consideration is whether the process is elastic or inelastic, i.e., whether the emergent photons contain precisely the summed energy of the annihilated photons, or if there is some either uptake or loss of energy by the material in which the process occurs. The term parametric is often used to describe elastic processes, for which the susceptibility properties of the material parametrically determine the rate of conversion. Inelastic harmonic processes in which an exchange of

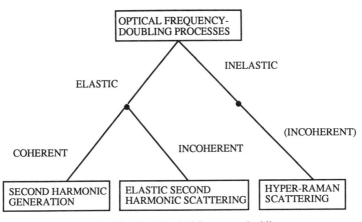

Figure 1. Classification of optical frequency-doubling processes.

energy does take place are accompanied by quantum transitions in the nonlinear medium, and are thus termed nonparametric.

The other major factor in determining the characteristics of a harmonic process is whether the response is coherent or incoherent in its nature, that is to say, whether the light emergent from individual points within the sample interferes constructively or in a random fashion. Only elastic conversion processes can be coherent. Coherent harmonic generation processes are associated with laser-like emission, providing intensity levels that far surpass those associated with incoherent phenomena. It is much to be regretted that the very term *coherent* is immensely overworked in this area of physics, being variously applied in different contexts to at least two other quite distinct aspects of light–matter interaction. It is worth emphasising, therefore, that coherence in the sense meant here has nothing to do with the special statistical properties of the laser radiation, nor has it to do with the concerted nature of multiphoton excitation, although both of these of course play a part in the theory of nonlinear optical phenomena. The single factor that determines whether a nonlinear scattering process is coherent or not is wave-vector matching.

Figure 1 illustrates the classification of processes entailing the destruction of two laser photons. Broadly these may be regarded as optical frequency-doubling phenomena. For the inelastic case, the term frequency-doubling is of course only an approximate description. This nonparametric process, which is accompanied by quantum transitions in the nonlinear material, is normally referred to as hyper-Raman scattering,[6] by extension of the nomenclature applied to inelastic scattering close to the fundamental optical frequency. The counterpart process of elastic frequency doubling can be further classified as either coherent second-harmonic generation or incoherent elastic second-harmonic light scattering, the latter also being known as hyper-Rayleigh scattering. This review is limited to consideration of the elastic processes of harmonic emission.

In much of the existing literature, the theory of harmonic production is based on the conventional formulation for solids, originating in the elegant and pioneering work of Bloembergen and others.[7] Such theories are generally cast in terms of a bulk nonlinear susceptibility, related to microscopic properties through incorporation of the appropriate Lorentz field factors. Since the majority of theoretical applications have concerned crystalline media, such an approach is entirely appropriate. In considering the much weaker harmonic processes that occur in gases and liquids, it has become common to adopt the conventional theoretical formulation, taking account of the fluid isotropy only by considering the implications for the bulk susceptibility of the macroscopic symmetry. However, such considerations do not adequately account for the effects of local fluid structure and molecular tumbling. These exert a powerful influence on the selection rules and polarization dependence of nonlinear optical processes (see, for example, Stedman[8].

Kielich and his coworkers have played a major role in the development of theories that properly take these factors into account. As will be demonstrated, only theories properly cast in terms of molecular nonlinear optical response can accommodate both parametric and nonparametric processes and clarify the distinction between coherent and incoherent phenomena. A molecule-based theoretical description also facilitates the resolution of recent controversies over the issue of surface SHG. In this connection, several unfounded conjectures that appeared in the literature have served to highlight the shortcomings of the traditional approach in application to molecular fluids, principally through neglect of the distinctively molecular mechanism for nonlinear optical interactions in fluids. Despite proper quantum-mechanical development of the appropriate nonlinear susceptibility tensors, major problems can arise through the inappropriate application of theoretical methods designed for application to crystals, where random phase properties are not necessarily encountered.

Where the detailed nonlinear optical response of molecules is required, recourse to molecular quantum electrodynamics brings both rigor and conceptual facility. Its relative unfamiliarity to many involved in the field invites an appraisal in which its distinctive elements can be highlighted and compared to the classical approach, and this is done in Section II. In Section III it is shown how the collective response of an ensemble of molecules is formally derived from the molecular probability amplitudes for harmonic emission. Section IV deals with the coherent process of harmonic generation, with both second- and higher-order harmonics being considered. Particular attention is paid to the production of SHG at surfaces or where an electric field is applied. Section V deals with the much weaker incoherent phenomenon of elastic second-harmonic light scattering.

II. MOLECULAR FORMULATION OF THEORY

There are two distinct approaches to the molecular description of optical processes. Clearly the electronic behavior of any medium at the atomic or molecular level should normally require quantum-mechanical treatment. However, there remain the possibilities of either retaining a classical description of the radiation, or of also treating it quantum mechanically. The former choice, representing what is usually known as a semiclassical formulation of the theory, has the strength of relating clearly to classical electrodynamics. The alternative procedure in which both matter and radiation are treated quantum mechanically is known as quantum electrodynamics. Before introducing the quantum electrodynamical basis for the theory of harmonic generation to be delineated subsequently, it is useful to begin with an outline of the corresponding classical description. This will provide an opportunity to define terms that are common to both theories, and to draw attention to features in which they essentially differ.

A. Classical Formulation

The starting point for the classical description of optical response is invariably the introduction of an electric polarization $\mathbf{P}$. This represents the relative displacement of positive and negative charges of a medium on application of an electric field, associated with the appearance of an induced electric dipole moment. For common electric field strengths, this is generally described by the constitutive equation

$$\mathbf{P} = \varepsilon_0 \mathbf{X} : \mathbf{E} \tag{1}$$

where $\mathbf{E}$ is the applied field, $\mathbf{X}$ is a second-rank tensor representing the electric susceptibility of the medium, and ε_0 is the vacuum permittivity; $\mathbf{P}$ then represents the induced dipole per unit volume. (In SI units, X is dimensionless and P has units of C m^{-2}). For the polarization component in a particular direction, we thus have

$$P_i = \varepsilon_0 \sum_j X_{ij} E_j \qquad (i, j = x, y, z) \tag{2}$$

or, adopting the convention of implied summation over repeated indices to be utilized throughout this review,

$$P_i = \varepsilon_0 X_{ij} E_j \tag{3}$$

Equation (3) represents the classical linear response of a medium to an applied electric field. When this field is time-dependent, as in the case of electromagnetic radiation, then for a given point $\mathbf{r}$ within the medium we can write

$$\mathbf{E}(\mathbf{r}, t) = \mathbf{E}_0 \cos(\mathbf{k} \cdot \mathbf{r} - \omega t) \tag{4}$$

Here $\mathbf{k}$ is the wave vector which points in the direction of propagation and has magnitude

$$k = |\mathbf{k}| = \frac{n_\omega \omega}{c} \tag{5}$$

ω is the circular frequency given by

$$\omega = 2\pi\nu = \frac{2\pi c}{\lambda} \tag{6}$$

and λ is the wavelength; n_ω is the refractive index of the medium at frequency ω. Clearly the electric polarization given by Eqs. (1) and (4) also oscillates with circular frequency ω. The classical picture of light scattering depicts the radiation of this fluctuating dipole as the source of emergent light with the same frequency as the incident light; the tensor character of Eq. (1) allows for light to emerge in directions other than the propagation direction of the incident beam, so producing the effect of elastic (Rayleigh) scattering.

A significant but often overlooked difference between the applications of Eq. (1) to static and radiative electric fields is that the equation is based on the concept of a dipolar description of charge distribution, which takes no account of any spatial variation in the field. Whereas static fields may well be at least approximately spatially homogeneous, the same is certainly not true for the electromagnetic fields described by Eq. (4). Only over regions of physical dimension much smaller than the optical wavelength is the spatial variation of the electric field negligible, and the dipole approximation is thus not always strictly appropriate for bulk media at optical frequencies, nor can higher-order multipolar correction terms readily be incorporated.[9] It will prove most convenient to introduce multipolar contributions within the context of the quantum electrodynamical treatment in Section II.B, though such considerations are of course equally important in the classical picture.

Having established the classical picture of conventional light scattering, we can now consider nonlinear optical response. At high field strengths the normal description fails, since it provides only for Rayleigh scattering at frequency ω (or, through coupling with molecular transitions, at Raman frequencies $\omega \pm \Delta\omega$). The origin of the harmonic is generally understood through an extension of the theory as follows. Even within the dipole approximation, Eq. (1) must be recognized as only an approximation based on the expectation of linear response to the applied electric field. In general, however, the response of a material may be more accurately represented in terms of a power series in $\mathbf{E}$, with the right side of (3) as the leading term. Thus, in general we may write

$$P_i = \varepsilon_0 \left(X_{ij}^{(1)} E_j + X_{ijk}^{(2)} E_j E_k + X_{ijkl}^{(3)} E_j E_k E_l + \cdots \right) \qquad (7)$$

with $\mathbf{X}$ now designated the first-order electric susceptibility $\mathbf{X}^{(1)}$. The second term in Eq. (7) represents a correction due to quadratic coupling with the electric field through the second-order susceptibility, the third term a cubic response, and so on; in SI units $\mathbf{X}^{(n)}$ has units of $(m/V)^{n-1}$. When the electric field is not too large, the correction terms are negligible and the response is accurately given by the leading linear term. However, at the high field strengths provided by intense lasers, the higher-order terms cannot be ignored.

With an electric field described by Eq. (4), the response given by Eq. (7) at $\mathbf{r} = 0$ may be expressed as follows:

$$P_i(t) = \varepsilon_0 \left[X_{ij}^{(1)} E_{0j} \cos \omega t + \tfrac{1}{2} X_{ijk}^{(2)} E_{0j} E_{0k} (\cos 2\omega t + 1) \right.$$
$$\left. + \tfrac{1}{4} X_{ijkl}^{(3)} E_{0j} E_{0k} E_{0l} (\cos 3\omega t + 3 \cos \omega t) + \ldots \right] \qquad (8)$$

From this it transpires that it is the quadratic term involving the second-order susceptibility which constitutes the source at frequency 2ω, higher-order harmonics being associated with the following terms. Hence, harmonic generation represents a type of light scattering that depends on a nonlinear optical response to the electric field of the laser radiation.

To complete the classical description, it is worth noting that there is a microscopic counterpart to Eq. (7) expressed in molecular terms, and represented by

$$\mu_i^{\text{ind}} = \left(\varepsilon_0^{-1} \alpha_{ij} d_j + \varepsilon_0^{-2} \beta_{ijk} d_j d_k + \varepsilon_0^{-3} \gamma_{ijkl} d_j d_k d_l + \ldots \right) \qquad (9)$$

where μ^{ind} is the induced molecular dipole moment, α is the molecular polarizability ($Jm^2\ V^{-2}$), β is the hyperpolarizability ($Jm^3\ V^{-3}$), and γ is the second hyperpolarizability ($Jm^4\ V^{-4}$), etc. The only significant differences to note at this point are first that non-centrosymmetric molecules may additionally possess an intrinsic dipole moment in the absence of any applied field, so that the total molecular dipole should be represented as

$$\mu = \mu_0 + \mu^{ind} \qquad (10)$$

and second, that in Eq. (9) the power series is expressed not in terms of E but d, the electric displacement field; this is the local electric field experienced by each molecule, modified by the polarization field due to neighboring molecules:

$$d = \varepsilon_0 e + p \qquad (11)$$

The use of lowercase symbols in Eqs. (9) and (11) signifies that both the field and material tensor parameters refer to the optical response at the local microscopic level; the electric field e is the counterpart of the macroscopic field E, while the local polarization p in any isotropic or cubic medium is related to the macroscopic polarization P by $p = \frac{1}{3}P$ (Ref. 10).

It is now appropriate to consider the strikingly different concepts involved in the quantum electrodynamical description of harmonic production. One illustration of the essential differences between the classical and quantum electrodynamical pictures of the processes to be described below can be found in the nomenclature itself. We have seen how the descriptor *nonlinear* originates directly from the classical picture; the quantum treatment shows that an equally suitable term is *multiphoton*.

B. Quantum Electrodynamics

Quantum electrodynamics correctly describes not only all the processes amenable to semiclassical methods, but also others, such as spontaneous emission, which are not. Its principal characteristic is that light and matter together constitute a closed dynamical system that is treated with full quantum-mechanical rigor. In this framework, second-harmonic generation is envisaged as the three-photon process wherein two frequency-ω photons of incident light are annihilated and a frequency-2ω photon carrying the sum of their energies is created. There is overall energy conservation by both the matter involved in the process and the radiation field; the interaction may thus be viewed as a transition from one state of the radiation field to another, supported by the medium. The definitive molecular formulation of quantum electrodynamics provided by Craig and

Thirunamachandran[11] forms the basis for much of the theory developed below.
The framework of molecular quantum electrodynamics provides for the direct calculation of the tensor parameters involved in linear and nonlinear optical interactions, which are essentially the same polarizability, hyperpolarizability, etc., as appear in Eq. (9). The approach to nonlinear optical calculations has a very different basis from Eq. (9), however, and generally involves determining quantum probability amplitudes for transitions between states of the radiation field. The full quantum electrodynamical Hamiltonian for a system of indistinguishable atoms or molecules labeled ξ may be represented as follows:

$$H = H_{\text{rad}} + \sum_{\xi} H_{\text{mol}}^{(\xi)} + \sum_{\xi} H_{\text{int}}^{(\xi)} \tag{12}$$

Here H_{rad} is the Hamiltonian for the radiation field in vacuo, $H_{\text{mol}}^{(\xi)}$ is the field-free Hamiltonian for molecule ξ, and $H_{\text{int}}^{(\xi)}$ is a term representing the complete multipolar interaction of molecule ξ with the radiation. One of the key features of this equation should be stressed before proceeding further. Neither the eigenstates of H_{rad} nor those of $H_{\text{mol}}^{(\xi)}$ are stationary states for the system described by Eq. (12). Thus, the presence of the radiation field modifies the form of the molecular wave functions, and the presence of matter modifies the form of the radiation wave functions. It is perhaps worth remarking that irrespective of the state, the Hamiltonian remains the same, and thus even when no light is present the coupling still produces a modification of molecular wave functions. This is, for example, manifest in the occurrence of spontaneous emission (luminescence) from isolated molecules in excited states, and in the lifting of degeneracy between the 2^2S and 2^2P states of atomic hydrogen (the Lamb shift). The correct predictions and interpretations of these features based on zero-point fluctuations of the radiation field represent two of the most significant successes of quantum electrodynamics.

Returning to Eq. (12), we can now consider the detailed nature of the three terms in the Hamiltonian. The simplest to deal with is the middle term, which denotes a sum of the normal nonrelativistic Schrödinger operators $H_{\text{mol}}^{(\xi)}$ for each molecule, and requires no further elaboration. The radiation field term H_{rad} is the operator equivalent of the classical expression for electromagnetic energy, and is given by

$$H_{\text{rad}} = \frac{1}{2}\varepsilon_0 \int \left\{ \varepsilon_0^{-2} \, \mathbf{d}^{\perp 2}(\mathbf{r}) + c^2 \mathbf{b}^2(\mathbf{r}) \right\} d^3\mathbf{r} \tag{13}$$

where $\mathbf{d}^{\perp}$ and $\mathbf{b}$ are the transverse electric displacement field and magnetic field operators, respectively. In the second-quantized formalism of quantum electrodynamics, each of these fields is expressible as a summation over radiation modes. The explicit expressions are as follows:

$$\mathbf{d}^{\perp}(\mathbf{r}) = i \sum_{\mathbf{k},\lambda} \left(\frac{\hbar\omega_{\mathbf{k}}\varepsilon_0}{2V} \right)^{1/2} \left\{ \mathbf{e}^{(\lambda)}(\mathbf{k})a^{(\lambda)}(\mathbf{k})e^{i\mathbf{k}\cdot\mathbf{r}} - \bar{\mathbf{e}}^{(\lambda)}(\mathbf{k})a^{\dagger(\lambda)}(\mathbf{k})e^{-i\mathbf{k}\cdot\mathbf{r}} \right\}$$

$$(14)$$

$$\mathbf{b}(\mathbf{r}) = i \sum_{\mathbf{k},\lambda} \left(\frac{\hbar\omega_{\mathbf{k}}}{2c\varepsilon_0 V} \right)^{1/2} \left\{ \mathbf{b}^{(\lambda)}(\mathbf{k})a^{(\lambda)}(\mathbf{k})e^{i\mathbf{k}\cdot\mathbf{r}} - \bar{\mathbf{b}}^{(\lambda)}(\mathbf{k})a^{\dagger(\lambda)}(\mathbf{k})e^{-i\mathbf{k}\cdot\mathbf{r}} \right\}$$

$$(15)$$

Here V denotes the quantization volume, and $\mathbf{e}^{(\lambda)}(\mathbf{k})$ is the unit polarization vector for a mode characterized by wave vector $\mathbf{k}$, polarization λ, and circular frequency $\omega_{\mathbf{k}} = c'|\mathbf{k}|$ (where $c' = c/n_\omega$ is the velocity of light in a medium of refractive index n_ω). The polarization vector is considered a complex quantity so as to admit the possibility of circular polarizations, and the complex unit vector $\mathbf{b}^{(\lambda)}(\mathbf{k})$ is defined by

$$\mathbf{b}^{(\lambda)}(\mathbf{k}) = \hat{\mathbf{k}} \times \mathbf{e}^{(\lambda)}(\mathbf{k}) \qquad (16)$$

Associated with each mode $(\mathbf{k}, \lambda)$ are photon creation and annihilation operators, $a^{\dagger(\lambda)}(\mathbf{k})$ and $a^{(\lambda)}(\mathbf{k})$, respectively. These operate upon eigenstates of H_{rad} with $n(\mathbf{k}, \lambda)$ photons as follows:

$$a^{\dagger(\lambda)}(\mathbf{k})|n(\mathbf{k},\lambda)\rangle = (n+1)^{1/2}|(n+1)(\mathbf{k},\lambda)\rangle \qquad (17)$$

$$a^{(\lambda)}(\mathbf{k})|n(\mathbf{k},\lambda)\rangle = n^{1/2}|(n-1)(\mathbf{k},\lambda)\rangle \qquad (18)$$

increasing the number of $(\mathbf{k}, \lambda)$ photons by one in the first case and reducing it by one in the second. These operators satisfy the commutation relations

$$\left[a^{(\lambda)}(\mathbf{k}), a^{(\lambda')}(\mathbf{k}') \right] = 0 \qquad (19)$$

$$\left[a^{\dagger(\lambda)}(\mathbf{k}), a^{\dagger(\lambda')}(\mathbf{k}') \right] = 0 \qquad (20)$$

$$\left[a^{(\lambda)}(\mathbf{k}), a^{\dagger(\lambda')}(\mathbf{k}') \right] = \delta_{\mathbf{k},\mathbf{k}'}\delta_{\lambda,\lambda'} \qquad (21)$$

Using these results the radiation Hamiltonian of Eq. (13) may alternatively be expressed as

$$H_{\text{rad}} = \sum_{\mathbf{k},\lambda} \left\{ a^{\dagger(\lambda)}(\mathbf{k})a^{(\lambda)}(\mathbf{k}) + \tfrac{1}{2} \right\} \hbar\omega_{\mathbf{k}} \qquad (22)$$

The H_{rad} eigenstates $|n(\mathbf{k}, \lambda)\rangle$ are number states, for which the number of photons is a quantum number with a sharp value. Such states have completely undefined phase, and so are not ideal for all physical applications, but they are useful for development of the theory. Other states that more closely model the coherence properties of laser light will be introduced later.

The explicit form of the interaction Hamiltonian $H_{int}^{(\xi)}$ consists of an infinite multipolar series of terms, the leading contributions to which are

$$
\begin{aligned}
H_{int}^{(\xi)} = {} & -\varepsilon_0^{-1}\mu(\xi) \cdot \mathbf{d}^{\perp}(\mathbf{R}_\xi) - \varepsilon_0^{-1}Q_{ij}(\xi)\nabla_i^{\perp}\mathbf{d}_j(\mathbf{R}_\xi) \\
& - \mathbf{m}(\xi) \cdot \mathbf{b}(\mathbf{R}_\xi) - \cdots
\end{aligned}
\tag{23}
$$

Here $\mu(\xi)$ is the electric dipole (E1) operator for molecule ξ located at position $\mathbf{R}_\xi$, $Q_{ij}(\xi)$ is the corresponding electric quadrupole (E2) operator, and $\mathbf{m}(\xi)$ is the magnetic dipole (M1) operator. The electric dipole term, which represents the strongest coupling with the radiation, is sufficient for the majority of cases in which the electronic excitations of molecules are restricted to regions much smaller than a typical wavelength for optical-frequency radiation. The electric quadrupole and magnetic dipole terms together are then smaller by a factor typically of the order of the fine structure constant $\alpha \approx \frac{1}{137}$. The E1 approximation suffices for the majority of topics in nonlinear light scattering, principally because most of the phenomena concerned are sufficiently weak that small corrections would normally be unobservable. However, consideration of the higher-order terms does become important in processes where the leading electric dipole term vanishes for symmetry reasons, as will be seen later.

C. Time-Dependent Perturbation Theory

With the full Hamiltonian given by Eq. (12), the time evolution of the system wave function Ψ is determined by the time-dependent Schrödinger equation:

$$
\frac{i\hbar\partial\Psi(t)}{\partial t} = H\Psi(t)
\tag{24}
$$

The exact solution of this equation is, not surprisingly, impossible in closed form. Nonetheless, very good approximate solutions can be derived, based on the assumption that the coupling between matter and radiation can be treated as a perturbation on the product eigenstates of H_0, where

$$
H_0 = H_{rad} + \sum_{\xi} H_{mol}^{(\xi)}
\tag{25}
$$

Assuming that the system is in an eigenstate of H_0 at time 0, the wave function at a later time t may first be expressed by the relation

$$|\Psi(t)\rangle = \exp\left(-\frac{iH_0 t}{\hbar}\right) U(t,0)|\Psi(0)\rangle \tag{26}$$

which serves as the defining equation for a time evolution operator $U(t,0)$ describing how the system evolves in the time interval $(0, t)$. Substitution of Eq. (26) into Eq. (24) leads to a series expansion for $U(t,0)$:

$$U(t,0) = 1 + \sum_{n=1}^{\infty} (i\hbar)^{-n} \int_{t_0}^{t} \int_{t_0}^{t_1} \cdots \int_{t_0}^{t_{n-1}} \tilde{H}_{\text{int}}(t_1)\tilde{H}_{\text{int}}(t_2)$$

$$\times \cdots \tilde{H}_{\text{int}}(t_n)\, dt_1\, dt_2 \cdots dt_n \tag{27}$$

where $\tilde{H}_{\text{int}}(t)$ is the (interaction picture) representation of the operator responsible for the coupling between light and matter, given by the expression

$$\tilde{H}_{\text{int}}(t) = \exp\left(\frac{iH_0 t}{\hbar}\right) H_{\text{int}} \exp\left(-\frac{iH_0 t}{\hbar}\right) \tag{28}$$

In the electric dipole approximation this results in the expression

$$\tilde{H}_{\text{int}}(t) = -\varepsilon_0^{-1}\boldsymbol{\mu} \cdot \tilde{\mathbf{d}}(\mathbf{r}, t) \tag{29}$$

where the interaction representation of the microscopic electric displacement vector, $\tilde{\mathbf{d}}(\mathbf{r}, t)$, may be expressed as a sum of two parts:

$$\tilde{\mathbf{d}}(\mathbf{r}, t) = \tilde{\mathbf{d}}^{(+)}(\mathbf{r}, t) + \tilde{\mathbf{d}}^{(-)}(\mathbf{r}, t) \tag{30}$$

$$\tilde{\mathbf{d}}^{(+)}(\mathbf{r}, t) = i \sum_{\mathbf{k},\lambda} \left\{\frac{\hbar\omega_{\mathbf{k}}\varepsilon_0}{2V}\right\}^{1/2} \mathbf{e}^{(\lambda)}(\mathbf{k}) a^{(\lambda)}(\mathbf{k}) e^{i(\mathbf{k}\cdot\mathbf{r}-\omega_{\mathbf{k}}t)} \tag{31}$$

$$\tilde{\mathbf{d}}^{(-)}(\mathbf{r}, t) = -i \sum_{\mathbf{k},\lambda} \left\{\frac{\hbar\omega_{\mathbf{k}}\varepsilon_0}{2V}\right\}^{1/2} \bar{\mathbf{e}}^{(\lambda)}(\mathbf{k}) a^{\dagger(\lambda)}(\mathbf{k}) e^{-i(\mathbf{k}\cdot\mathbf{r}-\omega_{\mathbf{k}}t)} \tag{32}$$

For a process associated with an initial state $|i\rangle$ and a final state $|f\rangle$, the probability amplitude can now be calculated through the equation

$$c_{fi} = \langle f|U(t,0)|i\rangle \tag{33}$$

The substitution of Eq. (27) into Eq. (33) produces a series of terms in increasing powers of the electric displacement field **d**, as in the classical picture. However, for a specific pair of initial and final states, only certain terms in the series can be nonzero because of simple quantum-mechanical selection rules. In particular, an n-photon interaction is described by the term involving the appropriate combination of n photon operators, and thus has its leading contribution from the term involving the nth power of $\tilde{H}_{\text{int}}$. This highlights the fact that the classical or semiclassical equation (9) has no proper foundation in the full quantum description of nonlinear optics, since it would amount to the adding together of probability amplitudes connected with different final states. This remark is not simply an abstract one, but is one that is in principle experimentally verifiable. The rate of any nonlinear process is related to the square modulus of the induced dipole represented by (9); however, cross-terms between contributions corresponding to different processes are not observed. Where radiation is treated classically, the disappearance of cross-terms has to be attributed to destructive interference and the significance of the point is lost.

One possible way forward involves treating the molecules as possessing a small, discrete number of energy states, perhaps no more than two. This N-level approximation method offers the advantage of facilitating closed-form analytic solutions in many cases, and it is a technique widely used in the study of coherent interactions of light with atoms. However, molecules generally possess a very large number of energy levels, which often crowd close enough together to form quasi-continua at comparatively low energies; similar remarks apply *a fortiori* to solids. The alternative, and much more widely used method, involves time-dependent perturbation theory. Here all the energy states of the molecules are included in the calculation, producing a result cast in terms of an infinite series. However, at all but the highest laser intensities the series represented by Eq. (27) converges rapidly, and for any given process it is normally sufficient to consider only the leading nonzero term. The resulting expression for the rate of a given optical process is expressed by the Golden Rule:

$$\Gamma = \frac{2\pi}{\hbar} \left| \sum_{\xi} M_{fi}^{(\xi)} \right|^2 \rho_f \tag{34}$$

where ρ_f is a density of final states for the process, and the transition matrix element (probability amplitude) M_{fi} connecting the initial state $|i\rangle$

with final state $|f\rangle$ for any given molecule is given by

$$
\begin{aligned}
M_{fi} = \langle f|H_{\text{int}}|i\rangle &+ \sum_r \frac{\langle f|H_{\text{int}}|r\rangle\langle r|H_{\text{int}}|i\rangle}{E_0 - E_r} \\
&+ \sum_{s,r} \frac{\langle f|H_{\text{int}}|s\rangle\langle s|H_{\text{int}}|r\rangle\langle r|H_{\text{int}}|i\rangle}{(E_0 - E_s)(E_0 - E_r)} \\
&+ \sum_{t,s,r} \frac{\langle f|H_{\text{int}}|t\rangle\langle t|H_{\text{int}}|s\rangle\langle s|H_{\text{int}}|r\rangle\langle r|H_{\text{int}}|i\rangle}{(E_0 - E_t)(E_0 - E_s)(E_0 - E_r)} + \cdots
\end{aligned} \tag{35}
$$

Here all states and energies are eigenstates of H_0 and thus relate to the *total system* comprising both radiation and matter, and the summations over the virtual intermediate states r, s, t, etc., are taken over all such states excluding i or f. Since the total system is closed, the initial and final energies must always be equal (to within a quantum uncertainty limit) and hence

$$
E_i = E_f \equiv E_0 \tag{36}
$$

When the electric dipole approximation for H_{int} is employed, it follows from Eq. (14) that each Dirac bracket must be associated with the creation or annihilation of one photon. Hence the mth term in Eq. (35), corresponding to mth order perturbation theory, is the dominant term for a nonlinear m-photon process. This statement remains true if, for example, E2 and M1 couplings are included in the calculation.

It is worth noting that the absence of longitudinal interactions between molecules in the Coulomb gauge means that there is no term in Eq. (23) representing interactions with any applied static electric or magnetic field. While such interactions could in principle be modeled by summing the couplings of each sample molecule with each constituent particle of the field source, it is simpler in practice to "dress" the molecular states with a suitable time-independent perturbation. In the case of a static electric displacement field **D**, these dressed states are given by the following equation:

$$
|r'\rangle = |r\rangle - \varepsilon_0^{-1} \sum_{s \neq r} (\boldsymbol{\mu}^{sr} \cdot \mathbf{D}) E_{rs}^{-1}|s\rangle + \cdots \tag{37}
$$

where $|r\rangle$ and $|s\rangle$ represent eigenstates of the conventional Hamiltonian operator H_{mol} in the absence of the applied electric field, and E_{rs} is the difference between the corresponding zeroth-order energies $E_r - E_s$; $\boldsymbol{\mu}^{sr}$

is the transition electric dipole moment for the $|s\rangle \leftarrow |r\rangle$ transition. A harmonic emission process that takes place in the presence of a static electric field may then be modeled using Eqs. (34) and (35) with the dressed states as basis.

Returning to the general matrix element result (35), we can now consider in more detail the leading contribution for an m-photon process. For the present it is convenient to retain a generalized nonlinear optical formulation which embraces not only harmonic emission but other parametric processes such as four-wave mixing. For n-harmonic generation, clearly we have $m = n + 1$. Writing the system states explicitly as products of molecular and radiation states, and restricting consideration to the case where the final state of the molecule is identical to its initial state as in all elastic harmonic emission processes, we have

$$
\begin{aligned}
M_{fi}^{(m)} = \sum_{r_{\text{mol}}^{(1)}} \cdots \sum_{r_{\text{mol}}^{(m-1)}} \sum_{r_{\text{rad}}^{(1)}} \cdots \sum_{r_{\text{rad}}^{(m-1)}} &\langle f_{\text{rad}}; i_{\text{mol}} | H_{\text{int}} | r_{\text{mol}}^{(m-1)}; r_{\text{rad}}^{(m-1)} \rangle \\
&\times \langle r_{\text{rad}}^{(m-1)}; r_{\text{mol}}^{(m-1)} | H_{\text{int}} | r_{\text{mol}}^{(m-2)}; r_{\text{rad}}^{(m-2)} \rangle \\
&\cdots \times \langle r_{\text{rad}}^{(1)} r_{\text{mol}}^{(1)} | H_{\text{int}} | i_{\text{mol}}; i_{\text{rad}} \rangle \\
&\times \left[\left(E_{i_{\text{mol}}} - E_{r_{\text{mol}}^{(m-1)}} \right) + \left(E_{i_{\text{rad}}} - E_{r_{\text{rad}}^{(m-1)}} \right) \right]^{-1} \\
&\cdots \times \left[\left(E_{i_{\text{mol}}} - E_{r_{\text{mol}}^{(1)}} \right) + \left(E_{i_{\text{rad}}} - E_{r_{\text{rad}}^{(1)}} \right) \right]^{-1}
\end{aligned}
\tag{38}
$$

In this equation, it should be noted that in each of the $(m - 1)$ summations over the intermediate radiation states $|r_{\text{rad}}^{(j)}\rangle$, only a limited number of possibilities can make nonvanishing contributions. These are entirely determined by the sequencing of the creation and annihilation events for the photons emitted and absorbed during the overall interaction. Each of these possible sequences is conveniently represented using a method based on the Feynman time-ordered diagrams of elementary particle physics. There are a number of variations on the precise details of this method, all of which necessarily lead to the same final results; the formalism employed here follows the tradition established by Wallace[12] and Hanna et al.[13] Other authors whose theory is based on density matrix formalism employ a modified version of this diagrammatic method, a good description of which has recently been given by Boyd.[14]

A time-ordered diagram can be regarded as a space–time representation of the possible sequences of photon interactions involved in a given physical process. With the conventions adopted here, time progresses upward, and a straight vertical line represents the world-line of an atom or molecule; photons are represented by diagonal wavy lines, and static fields by horizontal lines. No importance is attached to the precise displacements on either the vertical or horizontal axis: it is only the ordering of radiation–molecule interactions that is significant. Each diagram is thus a purely illustrative device, representing one of the ultimately indeterminable sequences of photon creation and annihilation events involved in a particular process. Calculations based on this method necessitate the construction of all topologically different diagrams connecting the same initial and final states: the summations over the intermediate states $|r_{\mathrm{rad}}^{(1)}\rangle$ to $|r_{\mathrm{rad}}^{(m-1)}\rangle$ in Eq. (38) are then equivalent to summations over the complete set of time orderings.

A pertinent case to illustrate the method is frequency doubling. Since the matrix element for the process is constructed to represent the optical response of individual molecules, questions of coherence are at this stage irrelevant; the same molecular matrix element and the same molecular nonlinear susceptibility tensor is involved in both coherent second-harmonic generation and the corresponding incoherent hyper-Rayleigh scattering process. The initial and final states for the interaction may be written as

$$|i_{\mathrm{mol}}; i_{\mathrm{rad}}\rangle = |0; n(\mathbf{k}, \lambda), n'(\mathbf{k}', \lambda')\rangle \qquad (39)$$

$$|i_{\mathrm{mol}}; f_{\mathrm{rad}}\rangle = |0; (n-2)(\mathbf{k}, \lambda), (n'+1)(\mathbf{k}', \lambda')\rangle \qquad (40)$$

assuming that the molecule is initially in its ground state, and that there are initially n photons of the pump mode $(\mathbf{k}, \lambda)$ and n' photons of the harmonic mode $(\mathbf{k}', \lambda')$. Three possible sequences of photon annihilation and creation can provide a route from the initial to the final state, as represented by the three time-ordered diagrams of Fig. 2.

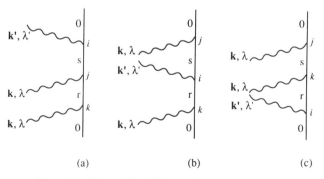

(a) (b) (c)

Figure 2. Time-ordered diagrams for frequency doubling.

In Fig. 2a, the sequence of interactions is as follows. First, a photon of the pump mode is absorbed by a molecule in its ground state $|0\rangle$, which thereby undergoes a transition to a state $|r\rangle$. Next, a second pump photon is absorbed, and the molecule proceeds to a state $|s\rangle$. Finally, a harmonic frequency photon is emitted, and the molecule returns to its ground state. The two intermediate states for this sequence are thus represented by

$$|r_{\text{mol}}^{(1)}; r_{\text{rad}}^{(1)}\rangle = |r;(n-1)(\mathbf{k},\lambda), n'(\mathbf{k}',\lambda')\rangle \qquad (41)$$

$$|r_{\text{mol}}^{(2)}; r_{\text{rad}}^{(2)}\rangle = |s;(n-2)(\mathbf{k},\lambda), n'(\mathbf{k}',\lambda')\rangle \qquad (42)$$

Figures 2b and c represent the other possible sequences in which emission of the harmonic photon precedes one or both pump photon interactions.

It is important to emphasize that no single time-ordered diagram represents a physically distinguishable process: the diagrams are ultimately only calculational aids based on the approximations of perturbation theory. In fact, the combination of photon creation and annihilation events that take place at each molecule appear simultaneous, as far as real experimental measurements with finite time resolution are concerned. However, the time–energy uncertainty relation does permit short-lived intermediate states that are not properly energy-conserving. This helps explain why it is necessary to include diagrams corresponding to apparently acausal time sequences such as that of Fig. 2c. It nonetheless always transpires that such apparently physically unreasonable cases are associated with the smallest (though seldom entirely negligible) contributions.

D. Nonlinear Susceptibility Tensors

Time-ordered diagrams of the type introduced in the last section facilitate exact evaluation of the susceptibility tensors involved in any given optical process. The technique is a powerful one that can readily provide the detailed structure of tensor response including high-order multipolar contributions. For simplicity, however, the electric dipole approximation is retained for the present. To proceed further, it is necessary to specify in more detail the nature of the m-photon processes of interest. Harmonic emission involves one radiation mode $(\mathbf{k}, \lambda)$ for the annihilated photons and another $(\mathbf{k}', \lambda')$ for the harmonic, containing before the interaction n and n' photons, respectively. Equation (38) may thus in general be re-expressed as follows:

$$M_{fi}\{(\mathbf{k}',\lambda'),(\mathbf{k},\lambda)\} = -\varepsilon_0^{-m}\chi(-\omega',\omega,\ldots,\omega)\odot^m\varphi\{(\mathbf{k}',\lambda'),(\mathbf{k},\lambda)\}$$

$$(43)$$

which represents the m-fold tensor contraction of $\chi(-\omega',\omega,\ldots,\omega)$, the appropriate rank-$m$ microscopic nonlinear susceptibility tensor containing all the molecular variables, with $\varphi\{(\mathbf{k}', \lambda'), (\mathbf{k}, \lambda)\}$, a tensor constructed from radiation parameters. It should be emphasized that here and henceforth, the use of lower-case Greek symbols is consistently associated with molecular and local field variables.

Explicit expressions for the tensors appearing in Eq. (43) are as follows. For the radiation tensor we have

$$\varphi\{(\mathbf{k}', \lambda'), (\mathbf{k}, \lambda)\} = \langle (n' + 1)(\mathbf{k}', \lambda') / \mathbf{d}^{\perp} | n'(\mathbf{k}', \lambda') \rangle$$
$$\times \cdots \langle (n - 1)(\mathbf{k}, \lambda) / \mathbf{d}^{\perp} | n(\mathbf{k}, \lambda) \rangle \quad (44)$$

and the susceptibility tensor is given by

$$\chi(-\omega', \omega, \ldots, \omega) = \sum_{r_{\text{mol}}^{(1)}} \cdots \sum_{r_{\text{mol}}^{(m-1)}} \sum_{r_{\text{rad}}^{(1)}} \cdots \sum_{r_{\text{rad}}^{(m-1)}} \langle i_{\text{mol}} | \boldsymbol{\mu} | r_{\text{mol}}^{(m-1)} \rangle$$
$$\times \langle r_{\text{mol}}^{(m-1)} | \boldsymbol{\mu} | r_{\text{mol}}^{(m-2)} \rangle \cdots \langle r_{\text{mol}}^{(1)} | \boldsymbol{\mu} | i_{\text{mol}} \rangle$$
$$\times \left[\left(\tilde{E}_{r_{\text{mol}}^{(m-1)}} - \tilde{E}_{i_{\text{mol}}} \right) + \left(E_{r_{\text{rad}}^{(m-1)}} - E_{i_{\text{rad}}} \right) \right]^{-1}$$
$$\cdots \times \left[\left(\tilde{E}_{r_{\text{mol}}^{(1)}} - \tilde{E}_{i_{\text{mol}}} \right) + \left(E_{r_{\text{rad}}^{(1)}} - E_{i_{\text{rad}}} \right) \right]^{-1} \quad (45)$$

where the summation over intermediate radiation states is represented by the time-ordered diagrams. A number of further remarks need to be made before proceeding to evaluate Eq. (45) using these diagrams, however.

First, it will be evident that a modification of the energy denominators has been made, as indicated by the tilde designation of the molecular state energies. This signifies the time-honored method of including damping factors to model the combined effect of the various decay processes responsible for the limited lifetime and finite linewidth of each molecular excited state: a full quantum electrodynamical treatment of these matters has recently been given by Blake.[15] Thus, each molecular state energy becomes a complex quantity, i.e., $\tilde{E}_{\mathrm{r}} = (E_r \pm \frac{1}{2} i \hbar \gamma_r)$, where γ_r is the corresponding homogeneous linewidth, and the choice of sign is considered in the next section. In energy terms $\hbar \gamma_r$ is typically several orders of magnitude smaller than E_r, with the precise value determined by the nature of the molecule and the bulk phase (gas, liquid, solid). The effect of the damping correction is thus in many circumstances insignificant, and the susceptibility can be regarded as a real quantity. Exceptions arise in the case of large organic structures whose ultrafast excited state decay mechanisms produce damping factors on the terahertz scale.

Considerations of damping are more generally important near to resonance, which occurs when there exists a molecular state differing in energy from the initial state by an amount equal to the energy of one or more of the photons involved. Under such circumstances, amplification of the rate is certainly observed, but not to the infinite degree that theory would predict if such damping were ignored. Clearly the inclusion of damping in the theory is crucial in such cases, and the susceptibility must be treated as a complex quantity. Detailed consideration of lineshape is also necessary to properly accommodate in the theory the dispersion behavior that features in the realization of wave-vector matching (see Section III.C).

Nonlinear susceptibility tensors are normally characterized by listing the frequencies of the photons involved in the processes they represent. This is necessary because the effects of dispersion in any real medium result in differences between susceptibilities of the same rank but associated with different processes. In the special case where all fields involved in a process are static, then there is of course a unique susceptibility for each rank, equivalent to the parameters in the classical expansion of Eq. (7). The convention adopted here for differentiating nonlinear susceptibilities is to write after the susceptibility symbol χ each photon frequency within brackets, first the frequency of the emitted photon with a negative sign, then the absorbed photon frequencies with implied positive signs. The ordering of subscripts on any susceptibility or radiation tensor is assumed to relate identically to the ordering of the frequency arguments. Since the tensors are seldom completely index-symmetric (see below), a correlation between indices and photon frequencies is a very necessary consideration.

This brings us back to the time-ordered diagrams, and the rules for their application to the derivation of nonlinear susceptibilities. Each vertex first needs to be labeled with the appropriate subscript, using the same subscript for the corresponding photons in each diagram. Thus, the subscript ordering varies from diagram to diagram, as in Fig. 2. The tensor is then given by the summation over all molecular intermediate states of a sum of terms, one from each time-ordered diagram, written down according to the following prescription. Each term has as numerator a product of transition dipole moment components, read off sequentially down the diagram. For example, from Fig. 2a we obtain the numerator $\mu_i^{0s}\mu_j^{sr}\mu_k^{r0}$. The corresponding denominator is then obtained by taking a product of factors, one for each intermediate state, in each of which the energy of the initial state is subtracted from the (complex) intermediate state energy. Again, in the case of Fig. 2a, we find that for the intermediate state $|s\rangle$, the difference in molecular energies is $\tilde{E}_{s0}$ $(= \tilde{E}_s - \tilde{E}_0)$, and the difference in photon energies $-2\hbar\omega$ (compare Eqs. (42) and (39)), thus giving a factor of $(\tilde{E}_{s0} - 2\hbar\omega)$: for the intermediate state $|r\rangle$, the difference in molecular energies is $\tilde{E}_{r0}$ and the difference in photon energies is $-\hbar\omega$ (Eqs. (41) and (39)), giving a factor of $(\tilde{E}_{r0} - \hbar\omega)$. Proceeding in a similar way from Fig. 2b and c, we thus obtain the following complete expression for the frequency-doubling susceptibility tensor (hyperpolarizability):

$$
\begin{aligned}
\chi_{ijk}(-2\omega, &\omega, \omega) \\
= \sum_r \sum_s \Bigg[& \frac{\mu_i^{0s}\mu_j^{sr}\mu_k^{r0}}{\left(\tilde{E}_{s0} - 2\hbar\omega\right)\left(\tilde{E}_{r0} - \hbar\omega\right)} + \frac{\mu_j^{0s}\mu_i^{sr}\mu_k^{r0}}{\left(\tilde{E}_{s0} + \hbar\omega\right)\left(\tilde{E}_{r0} - \hbar\omega\right)} \\
& + \frac{\mu_j^{0s}\mu_k^{sr}\mu_i^{r0}}{\left(\tilde{E}_{s0} + \hbar\omega\right)\left(\tilde{E}_{r0} + 2\hbar\omega\right)} \Bigg]
\end{aligned}
$$

E. Resonance Conditions

(46)

The inclusion of damping factors in the energy denominators of nonlinear susceptibility tensors produces the resonant Lorentzian lineshape characteristic of homogeneous line-broadening processes. The parameter γ may be considered a sum of the inverse lifetimes associated with each line-broadening mechanism, and represents the FWHM linewidth of the nonlinear response near resonance. The sign of each damping factor $\pm \frac{1}{2}i\hbar\gamma_r$ included in each complex energy $\tilde{E}_r$ is ultimately dictated by the causality condition that all singularities of the tensor must lie in the lower half of the complex frequency plane.[16] For multifrequency processes, this rule is not one that is amenable for application by inspection. A simpler prescription is to give a negative sign to the damping in denominator terms that can exhibit resonance, and a positive sign in terms that can exhibit only anti-resonance (equivalent to resonance at a negative frequency).

For all parametric processes involving the emission of a single photon, and also for most nonparametric types of scattering, the above rules are neatly encapsulated in a prescription given by Hanna et al.[13] If the resonant denominator is associated with a line segment of the corresponding time-ordered diagram lying below the vertex for the emitted photon, the damping factor appears with a negative sign; if the denominator corresponds to a segment above the emission vertex, the factor takes a positive sign. Hence, the explicitly damped expression for the molecular hyperpolarizability tensor is

$$
\begin{aligned}
\chi_{ijk}(-2\omega, \omega, \omega) = \sum_r \sum_s \Bigg[& \frac{\mu_i^{0s}\mu_j^{sr}\mu_k^{r0}}{\left(E_{s0} - 2\hbar\omega - \frac{1}{2}i\hbar\gamma_s\right)\left(E_{r0} - \hbar\omega - \frac{1}{2}i\hbar\gamma_r\right)} \\
& + \frac{\mu_j^{0s}\mu_i^{sr}\mu_k^{r0}}{\left(E_{s0} + \hbar\omega + \frac{1}{2}i\hbar\gamma_s\right)\left(E_{r0} - \hbar\omega - \frac{1}{2}i\hbar\gamma_r\right)} \\
& + \frac{\mu_j^{0s}\mu_k^{sr}\mu_i^{r0}}{\left(E_{s0} + \hbar\omega + \frac{1}{2}i\hbar\gamma_s\right)\left(E_{r0} + 2\hbar\omega + \frac{1}{2}i\hbar\gamma_r\right)} \Bigg]
\end{aligned}
$$

(47)

For a given nonlinear scattering process under near-resonance conditions, there will generally be a particular excited state $|a\rangle$ of the material which contributes strongly to the sums over intermediate molecular states in one or more of the terms in Eq. (45). Consider, for example, the case of frequency doubling in a medium possessing an excited state $|a\rangle$ of energy close to that of the emitted harmonic $2\hbar\omega$. Clearly in the summation over intermediate states $|s\rangle$ in Eq. (46), the energy denominator of the first term will become very small in the contribution corresponding to $|s\rangle = |a\rangle$, and the tensor may be conveniently represented as the sum of a resonant and a nonresonant part:

$$\chi_{ijk}(-2\omega, \omega, \omega) = \chi_{ijk}^{\text{res}}(-2\omega, \omega, \omega) + \chi_{ijk}^{\text{non-res}}(-2\omega, \omega, \omega) \quad (48)$$

where

$$\chi_{ijk}^{\text{res}}(-2\omega, \omega, \omega) = \frac{\mu_i^{0a}}{\tilde{E}_{a0} - 2\hbar\omega} \sum_r \left(\frac{\mu_j^{ar}\mu_k^{r0}}{\tilde{E}_{r0} - \hbar\omega} \right) \quad (49)$$

$$\chi_{ijk}^{\text{non-res}}(-2\omega, \omega, \omega) = \sum_r \sum_{s \neq a} \left[\frac{\mu_i^{0s}\mu_j^{sr}\mu_k^{r0}}{(\tilde{E}_{s0} - 2\hbar\omega)(\tilde{E}_{r0} - \hbar\omega)} \right]$$

$$+ \sum_r \sum_s \left[\frac{\mu_j^{0s}\mu_i^{sr}\mu_k^{r0}}{(\tilde{E}_{s0} + \hbar\omega)(\tilde{E}_{r0} - \hbar\omega)} \right.$$

$$\left. + \frac{\mu_j^{0s}\mu_k^{sr}\mu_i^{r0}}{(\tilde{E}_{s0} + \hbar\omega)(\tilde{E}_{r0} + 2\hbar\omega)} \right] \quad (50)$$

The resonant denominator factor $(\tilde{E}_{a0} - 2\hbar\omega)$ may alternatively be expressed as

$$\left(\tilde{E}_{a0} - 2\hbar\omega\right) = \hbar\left(\Delta\omega - \tfrac{1}{2}i\gamma_a\right) \quad (51)$$

with $\Delta\omega$ representing the detuning from resonance. When $\Delta\omega$ is small, the resonant part of the susceptibility strongly dominates over the nonresonant background contribution, and the resonant term factorizes into the product of a one-photon term (associated with the $|0\rangle \leftarrow |a\rangle$ transition) and a two-photon term (associated with the $|a\rangle \leftarrow |r\rangle \leftarrow |0\rangle$ transition).

The above example reflects the general feature that under resonance conditions, optical processes acquire the characteristics associated with coupled sequential processes. In the particular instance cited, the limiting behavior is that of two-photon absorption followed by emission of the harmonic. The difference between this process and normal frequency doubling lies in the fact that an appreciable, and in principle measurable, time-lag may exist between the absorption and the emission. This is because the approximate energy conservation in the two-photon absorption stage allows the intermediate state $|a\rangle$ to exist for a time interval that is no longer severely constrained by the time–energy uncertainty principle.

F. Index Symmetry Considerations

The last major consideration in arriving at a suitable representation of nonlinear light scattering at the molecular level is that of index symmetry in the nonlinear susceptibility and radiation tensors in Eq. (43). There are two aspects to this: One is the rigorous index symmetry that can ensue when two or more photons involved in a given process belong to the same radiation mode; the other is the approximate index symmetry often assumed for calculational simplicity, but which is seldom well justified.

First consider the case of rigorous index symmetry. This is most simply dealt with by referring to the radiation tensor $\varphi\{(\mathbf{k}', \lambda'), (\mathbf{k}, \lambda)\}$ defined by Eq. (44). If evaluation of two or more of the composite Dirac brackets results in factors with the same vector character, then the tensor must possess a symmetry with respect to permutations of the corresponding indices. Once again the case of frequency doubling serves to illustrate the point. Here, we have

$$\varphi_{ijk} = \langle (n' + 1)(\mathbf{k}', \lambda')/d_i^{\perp}|n'(\mathbf{k}', \lambda')\rangle$$

$$\times \langle (n - 2)(\mathbf{k}, \lambda)/d_j^{\perp}|(n - 1)(\mathbf{k}, \lambda)\rangle$$

$$\times \langle (n - 1)(\mathbf{k}, \lambda)/d_k^{\perp}|n(\mathbf{k}, \lambda)\rangle \qquad (52)$$

Evaluation of the Dirac brackets using Eqs. (14), (17), and (18) leads to the result

$$\varphi_{ijk} = \left\{ -i\left(\frac{\hbar\omega'\varepsilon_0}{2V}\right)^{1/2} (n' + 1)^{1/2} \bar{e}'_i \, e^{-i\mathbf{k}'\cdot\mathbf{r}} \right\}$$

$$\times \left\{ i\left(\frac{\hbar\omega\varepsilon_0}{2V}\right)^{1/2} (n - 1)^{1/2} e_j e^{i\mathbf{k}\cdot\mathbf{r}} \right\} \left\{ i\left(\frac{\hbar\omega\varepsilon_0}{2V}\right)^{1/2} n^{1/2} e_k e^{i\mathbf{k}\cdot\mathbf{r}} \right\}$$

$$= \frac{1}{2} i\left(\frac{\hbar\omega\varepsilon_0}{V}\right)^{3/2} \{(n' + 1)(n - 1)n\}^{1/2} \bar{e}'_i e_j e_k e^{i(2\mathbf{k}-\mathbf{k}')\cdot\mathbf{r}} \qquad (53)$$

From examination of the right side of Eq. (53), it is clear that φ_{ijk} is symmetric with respect to exchange of the indices j and k, i.e., $\varphi_{ijk} \equiv \varphi_{ikj}$. This feature of the tensor is conveniently represented by placing brackets around the indices i and j, i.e., $\varphi_{i(jk)}$.

We can now consider the effects of this index symmetry in the radiation tensor on the matrix element given in the general case by Eq. (43), and in the specific case of frequency doubling by

$$M_{fi} = -\varepsilon_0^{-3}\chi_{ijk}(-2\omega, \omega, \omega)\varphi_{i(jk)} \qquad (54)$$

Because of the summation over the indices i, j, and k involved in the contraction of the tensors χ and φ, it now becomes evident that only the j,k-index symmetric part of the susceptibility tensor can contribute to the process (note that as given by Eq. (46), $\chi(-2\omega, \omega, \omega)$ is not index-symmetric *per se*). This is readily shown by first writing the susceptibility tensor as a sum of two parts, one of which, $\chi_{i(jk)}(-2\omega, \omega, \omega)$, is j, k-symmetric, and the other of which, $\chi_{i[jk]}(-2\omega, \omega, \omega)$, is j, k-antisymmetric, i.e., changes sign on exchange of the indices j and k:

$$\chi_{ijk}(-2\omega, \omega, \omega) = \chi_{i(jk)}(-2\omega, \omega, \omega) + \chi_{i[jk]}(-2\omega, \omega, \omega) \qquad (55)$$

where

$$\chi_{i(jk)}(-2\omega, \omega, \omega) = \tfrac{1}{2}\{\chi_{ijk}(-2\omega, \omega, \omega) + \chi_{ikj}(-2\omega, \omega, \omega)\} \qquad (56)$$

$$\chi_{i[jk]}(-2\omega, \omega, \omega) = \tfrac{1}{2}\{\chi_{ijk}(-2\omega, \omega, \omega) - \chi_{ikj}(-2\omega, \omega, \omega)\} \qquad (57)$$

From Eq. (54) we thus have

$$M_{fi} = -\varepsilon_0^{-3}\chi_{i(jk)}(-2\omega, \omega, \omega)\varphi_{i(jk)} - \varepsilon_0^{-3}\chi_{i[jk]}(-2\omega, \omega, \omega)\varphi_{i(jk)} \qquad (58)$$

and the second term must be zero because it is equal to its own negative. (Since in Eq. (54) the indices j and k are repeated and therefore summed, they are dummy indices and can be interchanged throughout, the result is that the first term remains the same but the second reverses its sign.) Consequently it is correct to rewrite Eq. (54) as

$$M_{fi} = -\varepsilon_0^{-3}\chi_{i(jk)}(-2\omega, \omega, \omega)\varphi_{i(jk)} \qquad (59)$$

It is worth noting that the index-symmetric hyperpolarizability tensor defined by Eqs. (56) and (46) could be represented by six time-ordered diagrams, each corresponding to one of the permutations of the indices i, j, and k. Having regard to the explicit form of the damping as expressed by Eq. (47), it becomes evident that $\chi_{i(jk)}(-2\omega, \omega, \omega)$ satisfies the formal relation

$$\bar{\chi}_{i(jk)}(-2\omega, \omega, \omega) = \chi_{i(jk)}(+2\omega, -\omega, -\omega) \qquad (60)$$

leading to the sum rule[17]

$$\int_0^\infty \mathrm{Re}\, \chi_{i(jk)}(-2\omega, \omega, \omega)\, d\omega = 0 \qquad (61)$$

The construction of the appropriately index-symmetric susceptibility for a process where three or more identical photons are absorbed follows along similar lines, and the shortcut to the result is obvious. For example, for frequency tripling, we have six permutations of the three indices relating to the annihilated photons, and hence write

$$M_{fi} = -\varepsilon_0^{-4} \chi_{i(jkl)}(-3\omega, \omega, \omega, \omega) \varphi_{i(jkl)} \qquad (62)$$

where

$$\chi_{i(jkl)} = \tfrac{1}{6}\{\chi_{ijkl} + \chi_{ijlk} + \chi_{ikjl} + \chi_{iklj} + \chi_{iljk} + \chi_{ilkj}\} \qquad (63)$$

The question of approximate index symmetry in the susceptibility tensors can now be addressed. As noted earlier, any nonlinear process involving only static (zero-frequency) fields is correctly described in terms of a classical susceptibility tensor that is always fully index-symmetric. When optical frequencies are involved, the differences between the energy denominators of the various terms in the tensor remove this symmetry. If, however, the photon frequencies all fall well below any electronic transition frequencies of the material, all energy denominators become approximately equal, ($= \tilde{E}_{s0}\tilde{E}_{r0}$ in the case of frequency doubling), and the tensor in effect becomes fully index-symmetric. Index symmetry based on the assumption that such an approximation is valid, normally referred to as Kleinman symmetry,[18] is unjustified in most applications.[19]

G. Relation to Molecular Structure

Much work has been done on the relationship between molecular structure and nonlinear optical properties (see, for example, Chemla and Zyss[2] and Prasad and Williams[3]). Early work concentrated on simple models that would allow useful predictions to be made, the one-dimensional box model of conjugated polyenes being the *exemple par excellence*. From such studies it was established that hyperpolarizabilities should have a cubic dependence on conjugated chain length, for example. Although much more accurate molecular wave-function packages are now commonly employed to derive results for these and other kinds of molecules, much of the folklore of the subject that is ultimately attributable to the early models persists, and some words of caution are appropriate.

The one-dimensional model of molecular nonlinear optical response necessarily requires all transition dipoles to be associated with axial shifts in electron distribution, and in this sense it seems logical to assume that only z components of the nonlinear susceptibility tensors are significant. Indeed, this is the assumption in much recent work on surface SHG. Work on conjugated polymer crystals certainly shows that the magnitude of tensor components associated with transverse (x or y) directions is often far smaller than for the axial components (see, for example, Flytzanis[20]). However, in such crystals the intrinsic molecular orientation is fixed. Both static (environmentally and intramolecularly determined) and dynamic bending are significant features in free molecules, and neither linearity nor even rigidity can necessarily be assumed.

Secondly, the argument that long rod-like molecules should have a nonlinear susceptibility dominated by z components is valid only if the electronic structure is dominated by axially delocalized states. Nevertheless, even a colorless compound may have an active localized multiphoton resonance in the UV, and it is often the case that nonlinear optical response is strongly associated with chromophore sites c within the molecule. These contributions may be regarded as additive through an equation of the following form:

$$\chi = \sum_c \chi(c)\exp(i\Delta k \cdot R_c) \tag{64}$$

where R_c is the vector displacement of each chromophore within the molecule and Δk is the mismatch between wave vectors of the absorbed and emitted photons,

$$\Delta k = \sum_r^m g_r k_r \tag{65}$$

with g_r taking the value of $+1$ for each absorbed photon and -1 for the emitted photon. For coherent harmonic emission where $\Delta\mathbf{k}$ is small (see Section III.C) the phase factor in Eq. (64) can be omitted. The validity of the resultant group or bond additivity model has been extensively tested for a wide range of substances; see, for example, Chemla et al.[21] and Levine and Bethea.[22] The crucial point is that the additivity applies not only for the z components. For example, the value of χ_{zxx} is a sum of the $\chi_{(c)zxx}$ components for the contributing chromophores.

Finally, it is worth noting that calculational simplicity is occasionally achieved at an expense of physical realism which amounts to gross caricature of the nonlinear medium. In this context it is hard to find a model less credible for representing the nonlinear optical response of fluids than the free electron gas model (see, for example, Heinz and DiVincenzo[23]), appropriate as it is for describing other kinds of medium with extensively delocalized electronic structure. Not only does the application to fluids sacrifice proper modeling of the electronic integrity of their constituent molecules, it also rules out any consideration of local symmetry constraints and thus leads to entirely incorrect conclusions (see Section IV.B,C).

III. COLLECTIVE RESPONSE OF A MOLECULAR ENSEMBLE

In constructing a theory of harmonic emission by real materials, the collective response of an ensemble of molecules or scattering centers obviously needs careful consideration. There are three main aspects to this. One is the fact that scattering processes at individual centers are generally modified by the electrodynamic influences of the surrounding material; another is that in media where there is a distribution of molecular orientations, the molecular response may differ from molecule to molecule. Finally, scattering at different centers can result in synergistic effects and result in a coherent response. The consideration of these features occupies this section, which represents the link between the molecular and macroscopic theories of harmonic emission. It is perhaps worth re-emphasizing that the theoretical framework described here is nonetheless one that is specifically cast in terms of molecular properties, in contrast to most classical approaches which take account of macroscopic symmetry but not local molecular structure.

To begin a consideration of optical nonlinearity in a molecular ensemble, we first return to the general Fermi rule rate equation of Eq. (34):

$$\Gamma = \frac{2\pi}{\hbar} \left| \sum_{\xi}^{N} M_{fi}^{(\xi)} \right|^2 \rho_f \qquad (66)$$

where N is the number of scattering centers in the appropriate ground state $|i\rangle$ within the interaction volume. Writing the individual matrix element for a specific m-photon process at molecule ξ as the tensor contraction of a molecular nonlinear susceptibility and a radiation tensor according to Eq. (43), we have

$$\Gamma = \frac{2\pi\rho_f}{\hbar}\left|\sum_{\xi}^{N}\varepsilon_0^{-m}\chi(\xi)(-\omega',\omega,\ldots,\omega)\odot^m\varphi(\xi)\{(\mathbf{k}',\lambda'),(\mathbf{k},\lambda)\}\right|^2 \quad (67)$$

The molecular susceptibility tensor carries the label ξ because the values of its components in a space-fixed frame will generally depend on molecular orientation, which may differ from molecule to molecule: the fact that the radiation tensor is also labeled ξ may at first sight be surprising. It results, however, from the fact that this tensor has a spatial dependence associated with the phase factors in the mode expansions of Eqs. (14) and (15); for example, the radiation tensor for frequency doubling carries the exponential $e^{i(2\mathbf{k}-\mathbf{k}')\cdot\mathbf{r}}$, as seen in Eq. (53). This feature gives rise to the all-important issue of coherence, as will be seen below.

Without necessary regard to multipolar approximation, the tensor representing the radiation field for the interaction at molecule ξ may be written in general form as follows:

$$\varphi(\xi) = \sigma\exp(i\Delta\mathbf{k}\cdot\mathbf{R}_\xi) \quad (68)$$

where $\mathbf{R}_\xi$ is the position vector of the molecule relative to an arbitrary fixed origin, and $\Delta\mathbf{k}$ is the wave-vector mismatch as defined by (65). Equation (67) may thus be written as

$$\Gamma = \frac{2\pi\rho_f}{\hbar}\left|\sum_{\xi}^{N}\varepsilon_0^{-m}\chi(\xi)\odot^m\sigma\exp(i\Delta\mathbf{k}\cdot\mathbf{R}_\xi)\right|^2 \quad (69)$$

where the explicit parametric dependences of χ and σ given in Eq. (67) have been dropped simply to avoid cluttering this and subsequent equations. Equation (69) is the key equation for determining ensemble response. For application to specific processes in specific media, however, it needs further theoretical development, and at this point it is appropriate to draw a distinction between the response from molecular solids and fluids.

A. Molecular Crystals

Take first the case of a molecular crystal. It is traditional to represent the nonlinear optical response of such a system in terms of the summed responses of its constituent unit crystallographic cells.[24] The shortcomings of this model in its application to certain types of crystal, such as its failure to incorporate the effects of bulk excitations (excitons) are obvious, and have been spelled out in detail by Meredith.[25] The way to properly account for polariton and retarded intermolecular interactions within the framework of quantum electrodynamics has been spelled out by Knoester and Mukamel.[26] However, the traditional approach does clarify the correlation between local microscopic and bulk macroscopic response and, provided the approximate nature of the theory is recognized, it is an approach that has many advantages.

Consider a crystal with n molecules per unit cell. Since each cell is identically oriented, the sum over molecules in Eq. (69) can be rewritten as a double sum, first over the molecules labeled s in each cell, then over the $M = (N/n)$ cells labeled u. For this purpose each molecular position vector $\mathbf{R}_\xi$ can be written as a sum of $\mathbf{R}_u$, a position vector for the unit cell to which it belongs, and $\mathbf{R}_s$, a vector denoting the position of the molecule within the unit cell. Hence, we obtain the following result:

$$\Gamma = \frac{2\pi\rho_f}{\hbar\varepsilon_0^{2m}} \left| \sum_u^M \exp(i\Delta\mathbf{k} \cdot \mathbf{R}_u) \sum_s^n \boldsymbol{\chi}(s)\odot^m\boldsymbol{\sigma}\exp(i\Delta\mathbf{k} \cdot \mathbf{R}_s) \right|^2 \quad (70)$$

It is convenient to introduce a cell susceptibility, which entails a structure factor for nonlinear optical response, through the defining equation

$$\hat{\boldsymbol{\chi}} = \sum_s^n \boldsymbol{\chi}(s)\exp(i\Delta\mathbf{k} \cdot \mathbf{R}_s) \quad (71)$$

We thus have

$$\Gamma = \frac{2\pi\rho_f}{\hbar\varepsilon_0^{2m}} \left| \sum_u^M \hat{\boldsymbol{\chi}}\odot^m\boldsymbol{\sigma}\exp(i\Delta\mathbf{k} \cdot \mathbf{R}_u) \right|^2 \quad (72)$$

However, the tensors $\hat{\boldsymbol{\chi}}$ and $\boldsymbol{\sigma}$ have the same value for each unit cell, so that the result simplifies to

$$\Gamma = \frac{2\pi\rho_f}{\hbar\varepsilon_0^{2m}} |\hat{\boldsymbol{\chi}}\odot^m\boldsymbol{\sigma}|^2 \left| \sum_u^M \exp(i\Delta\mathbf{k}\cdot\mathbf{R}_u) \right|^2 \tag{73}$$

It is helpful to rewrite the result of Eq. (73) as a sum of diagonal and off-diagonal terms as follows:

$$\Gamma = \frac{2\pi\rho_f}{\hbar\varepsilon_0^{2m}} |\hat{\boldsymbol{\chi}}\odot^m\boldsymbol{\sigma}|^2 \left[M + \sum_{u\neq}^M \sum_{u'}^{M-1} \exp(i\Delta\mathbf{k}\cdot\mathbf{R}_{uu'}) \right] \tag{74}$$

where $\mathbf{R}_{uu'} = \mathbf{R}_u - \mathbf{R}_{u'}$. The first term in this equation represents an incoherent contribution to the rate, which is clearly independent of any correlation in phase between processes occurring in different unit cells:

$$\Gamma_{\text{inc}} = M\left(\frac{2\pi\rho_f}{\hbar\varepsilon_0^{2m}}\right) |\hat{\boldsymbol{\chi}}\odot^m\boldsymbol{\sigma}|^2 \tag{75}$$

The second term in Eq. (74) is an interference term which depends through the phase factor $\exp(i\Delta\mathbf{k}\cdot\mathbf{R}_{uu'})$ on the relative displacement of each pair of unit cells with respect to the wave-vector mismatch; this term therefore represents a coherent contribution, and can be written as

$$\Gamma_{\text{coh}} = \frac{2\pi\rho_f}{\hbar\varepsilon_0^{2m}} |\hat{\boldsymbol{\chi}}\odot^m\boldsymbol{\sigma}|^2 (\eta_M - M) \tag{76}$$

where

$$\eta_M = \left| \sum_u^M \exp(i\Delta\mathbf{k}\cdot\mathbf{R}_u) \right|^2 \tag{77}$$

The detailed form of the important parameter η_M, which is a sensitive function of wave-vector mismatch, is discussed in Section III.C.

Finally, evaluation of the rate of nonlinear scattering using Eqs. (75) and (76) necessitates further consideration of the tensor $\hat{\boldsymbol{\chi}}$. The components of this microscopic cell tensor are expressed in terms of molecular susceptibility components through the relation

$$\hat{\chi}_{i_1\cdots i_m} = \sum_s^n \chi_{(s)\lambda_1\cdots\lambda_m} l_{i_1\lambda_1}^{(s)} \cdots l_{i_m\lambda_m}^{(s)} \exp(i\Delta\mathbf{k}\cdot\mathbf{R}_s) \tag{78}$$

once again adopting the convention of implied summation over repeated indices; here l_{i,λ_r} denotes the direction cosine between the crystal axis i_r and the molecular axis λ_r.

B. Gases and Liquids

In molecular fluids the individual molecular centers are responsible for harmonic emission, and it is crucial to take account of the powerful symmetry constraints that are associated with random molecular orientation. At first sight, for media with random time-averaged molecular positions, it is tempting to directly replace the sum over ξ in Eq. (69) with a volume integral, an approximation that would lead to a Dirac delta function in $\Delta\mathbf{k}$, and so imply a nonvanishing rate only in the case $\Delta\mathbf{k} = 0$. Such a result would be fallacious, however, since it would be based on the assumption of a spherically symmetric distribution of molecules with infinite extent. In practice the interaction volume is restricted by the focal volume of the laser beam, and is in general markedly axial.

To take into account the tumbling motions experienced by each molecule in a fluid, we again begin by rewriting Eq. (69) as a sum of diagonal and off-diagonal terms as follows:

$$
\Gamma = \frac{2\pi\rho_f}{\hbar\varepsilon_0^{2m}}\left[\sum_\xi^N |\chi(\xi)\odot^m\boldsymbol{\sigma}|^2 \right.
$$
$$
\left. + \sum_{\xi\ne}^N \sum_{\xi'}^{N-1} (\chi(\xi)\odot^m\boldsymbol{\sigma})(\bar{\chi}(\xi')\odot^m\bar{\boldsymbol{\sigma}})\exp(i\Delta\mathbf{k}\cdot\mathbf{R}_{\xi\xi'})\right] \quad (79)
$$

where $\mathbf{R}_{\xi\xi'} = \mathbf{R}_\xi - \mathbf{R}_{\xi'}$. The first term once again represents the incoherent contribution, and by representing the appropriate average over the orientations of molecule ξ by angular brackets $\langle \cdots \rangle_\xi$, we thus have

$$
\Gamma_{\text{inc}} = \frac{2\pi\rho_f}{\hbar\varepsilon_0^{2m}}\sum_\xi^N \left\langle |\chi(\xi)\odot^m\boldsymbol{\sigma}|^2 \right\rangle_\xi \quad (80)
$$

Although at any given time the response from each molecule depends on its orientation with respect to the incident light, the averaged response from all N molecules must be identical, and Eq. (80) may therefore be more simply expressed as

$$
\Gamma_{\text{inc}} = N\left(\frac{2\pi\rho_f}{\hbar\varepsilon_0^{2m}}\right)\left\langle |\chi\odot^m\boldsymbol{\sigma}|^2 \right\rangle \quad (81)
$$

Two important features can immediately be identified in this result. One is the obvious but significant fact that incoherent scattering depends linearly on the number of scatterers N, and therefore delineates the process as colligative. Secondly, the rotational average in Eq. (81) is taken over the modulus square of the matrix element. In both respects, these features mark important differences from the coherent term to be examined next.

The second term in Eq. (79) represents the coherent contribution to the scattering rate. If there is no time-averaged orientational correlation between different molecules, then independent rotational averaging can be performed for molecules ξ and ξ', giving

$$\Gamma_{coh} = \frac{2\pi\rho_f}{\hbar\varepsilon_0^{2m}} \sum_{\xi}^{N} \langle (\boldsymbol{\chi}(\xi) \odot^m \boldsymbol{\sigma}) \rangle_\xi \sum_{\xi' \neq \xi}^{N-1} \langle (\bar{\boldsymbol{\chi}}(\xi') \odot^m \bar{\boldsymbol{\sigma}}) \rangle_{\xi'} \exp(i\Delta\mathbf{k} \cdot \mathbf{R}_{\xi\xi'})$$

(82)

(The case of correlated molecules requires a different treatment, which has been described in detail by Kielich.[27]) Again, the rotational average of each matrix element is molecule-independent, so that (82) can be rewritten as

$$\Gamma_{coh} = \frac{2\pi\rho_f}{\hbar\varepsilon_0^{2m}} |\langle \boldsymbol{\chi} \odot^m \boldsymbol{\sigma} \rangle|^2 \sum_{\xi}^{N} \sum_{\xi' \neq \xi}^{N-1} \exp(i\Delta\mathbf{k} \cdot \mathbf{R}_{\xi\xi'})$$

(83)

or

$$\Gamma_{coh} = \frac{2\pi\rho_f}{\hbar\varepsilon_0^{2m}} |\langle \boldsymbol{\chi} \odot^m \boldsymbol{\sigma} \rangle|^2 (\eta_N - N)$$

(84)

where η_N is as defined by Eq. (77), but with the summation taken over molecules rather than crystal units. In contrast to the case of incoherent scattering, the rotational average in Eq. (84) is taken over the matrix element *before* it is squared. This apparently minor difference results in markedly different behavior, as will be seen.

Calculation of the tensor orientational averages in Eqs. (81) and (84) necessitates more detailed analysis of the tensor inner product $\boldsymbol{\chi} \odot^m \boldsymbol{\sigma}$, which can be written as

$$\boldsymbol{\chi} \odot^m \boldsymbol{\sigma} = \chi_{i_1 \cdots i_m} \sigma_{i_1 \cdots i_m}$$

(85)

with components referred to a reference frame in which the radiation tensor components are fixed, and the molecular nonlinear susceptibility components thus necessarily vary with molecular orientation. It is generally more convenient to refer the molecular tensor to a molecule-fixed frame, denoted below by indices λ_r, in terms of which its own components are rotation-invariant. A suitable choice of molecular frame moreover facilitates consideration of the implications of molecular symmetry for relationships between the tensor components. We then have

$$\chi \odot^m \sigma = \chi_{\lambda_1 \cdots \lambda_m \sigma_{i_1} \cdots i_m} l_{i_1 \lambda_1} \cdots l_{i_m \lambda_m} \tag{86}$$

where only the direction cosines $l_{i_r \lambda_r}$ vary with molecular rotation. The orientational averages in (81) and (84) are thus obtained by averaging over the direction cosine products:

$$\left\langle |\chi \odot^m \sigma| \right\rangle^2 = \chi_{\lambda_1 \cdots \lambda_m} \bar{\chi}_{\lambda_{m+1} \cdots \lambda_{2m} \sigma_{i_1} \cdots i_m} \bar{\sigma}_{i_{m+1} \cdots i_{2m}}$$
$$\times \left\langle l_{i_1 \lambda_1} \cdots l_{i_{2m} \lambda_{2m}} \right\rangle \tag{87}$$

$$\left| \left\langle \chi \odot^m \sigma \right\rangle \right|^2 = \left| \chi_{\lambda_1 \cdots \lambda_m \sigma_{i_1} \cdots i_m} \left\langle l_{i_1 \lambda_1} \cdots l_{i_m \lambda_m} \right\rangle \right|^2 \tag{88}$$

and hence the former, incoherent case involves a rank-$2m$ average, but the latter, coherent case a rank-m average. Explicit results for these averages for isotropic fluids are given by Andrews and Thirunamachandran,[28a] Andrews and Ghoul,[29] and Andrews and Blake[30]; results for anisotropic fluids are given by Andrews and Harlow.[31]

C. Coherence and Wave-Vector Matching

At this juncture it is worth exploring the relationship between coherence and wave-vector matching, an issue that has a very significant bearing on both the directionality and intensity of nonlinear scattering. As seen above, while incoherent scattering is independent of any relationship between the wave vectors of absorbed and emitted photons, coherent scattering both in solids, Eq. (76), and fluids, Eq. (84), involves a factor of the form

$$\eta_N = \left| \sum_{\xi}^{N} \exp(i\Delta k \cdot R_{\xi}) \right|^2 \tag{89}$$

(with ξ, N replaced by u, M for the case of a molecular crystal). Assuming a completely random distribution for the molecular displacement vectors

R_ξ leads to the simple result[32]

$$\eta_N \approx N \qquad (\Delta k \neq 0) \qquad (90)$$

$$\eta_N = N^2 \qquad (\Delta k = 0) \qquad (91)$$

The former case, which applies if the value of Δk is sufficiently large for $\exp(i\Delta k \cdot R_\xi)$ to be a rapidly oscillating function within the boundaries of the interaction volume, thus results in the linear dependence on the number of scatterers that corresponds to incoherent response; the latter case produces a totally different quadratic dependence on the number of scatterers, which characterizes coherent response. As a result of this difference, processes that can satisfy the wave-vector matching condition $\Delta k = 0$ invariably produce much stronger signals than those that do not. Clearly, then, the question of whether the wave-vector mismatch can be zero or not is one of immense significance.

In passing, it may be pointed out that for any nonlinear process accompanied by an exchange of energy between the radiation field and the matter, as, for example, in hyper-Raman scattering where the final molecular state $|f\rangle$ differs from the initial state $|i\rangle$, it is not generally possible to fulfill the wave-vector matching condition. In fact, even if such a condition could be fulfilled, consideration of the uncorrelated quantum-mechanical phase factors associated with the final state in different molecules would reveal that Eq. (90) still effectively applies. From Eqs. (76) and (84) it is therefore apparent that the coherent part of the response always vanishes, and as a consequence such inelastic nonlinear scattering processes are invariably incoherent.

By contrast, parametric or elastic scattering processes in which the material medium returns to its original state can always, in principle, satisfy wave-vector matching. At the simplest level, it is clear that the appropriate conditions could be achieved in a dispersion-free medium with all wave vectors involved in the process arranged collinearly. If the refractive index for all frequencies is unity, each wave vector is related to its corresponding frequency by the relation $k_r = (\omega_r/c)\hat{k}$, and the condition $\Delta k = 0$ is immediately satisfied by virtue of energy conservation. Although no truly dispersion-free materials exist, it is still often possible to satisfy the wave-vector condition if the frequencies involved are such that the appropriate refractive indices match.

The effects of dispersion, especially in condensed media, are often such that exact wave-vector matching cannot be achieved, but conditions can be produced where the value of Δk is small. Under such circumstances where $\exp(i\Delta k \cdot R_\xi)$ is a slowly varying function of R_ξ, the approximation of

Eq. (90) is inappropriate, particularly since the interaction volume is not spherically symmetric. It is then best to proceed by defining a mismatch direction z, such that $\Delta \mathbf{k} \cdot \mathbf{R}_\xi = |\Delta \mathbf{k}| z_\xi$, where z_ξ denotes the displacement of molecule ξ along the z axis, subsequently converting the summation over molecules in (89) to an integral between limits 0 and L, corresponding to the boundaries of the nonlinear material. Thus, we have

$$\eta_N = \left| \sum_\xi^N \exp\left(i|\Delta \mathbf{k}|z_\xi\right) \right|^2 \tag{92}$$

$\therefore$

$$\eta_N \approx \left| \frac{N}{L} \int_0^L \exp(i|\Delta \mathbf{k}|z)\, dz \right|^2 \tag{93}$$

Evaluation of the integral and use of the identity $e^{ix} - 1 = 2ie^{ix/2} \sin(x/2)$ then leads to the result

$$\eta_N = \frac{N^2 \sin^2\left(\frac{1}{2}|\Delta \mathbf{k}|L\right)}{\left(\frac{1}{2}|\Delta \mathbf{k}|L\right)^2} \tag{94}$$

giving the familiar sinc^2 dependence. Clearly this function attains its largest value in the limit $\Delta \mathbf{k} \to 0$, where the result reduces to that given as Eq. (91).

For efficient harmonic conversion, it is important to maximize the value of η_N, which essentially means minimizing the wave-vector mismatch. Generally it is regarded as necessary to ensure that the value of $\frac{1}{2}|\Delta \mathbf{k}|L$ is limited to values in the interval $(-\frac{1}{2}\pi, \frac{1}{2}\pi)$, within which η_N remains above ~ 0.4 of its peak value. Maximum harmonic emission clearly occurs where the pump and harmonic waves propagate collinearly, and we then have

$$|\Delta \mathbf{k}| = |n\mathbf{k} - \mathbf{k}'| = \frac{n\omega}{c}(n_\omega - n_{\omega'}) \tag{95}$$

where n_ω and $n_{\omega'}$ are the refractive indices for the pump and harmonic frequencies, respectively. One of the most important criteria for efficient harmonic production is therefore the degree of match between these two refractive indices. For example, in the earliest experiments on second-harmonic generation,[1] the mismatch of pump and harmonic refractive indices was such that the conversion efficiency was $\sim 10^{-5}\%$; however, figures of up to $\sim 25\%$ or more are now feasible when precise index matching is accomplished.

In fluids or optically isotropic solids where refraction is independent of propagation direction, it is impossible to guarantee wave-vector matching unless the refractive indices n_ω and $n_{\omega'}$ are equal. The implications of this are clearly less serious for gases than for condensed matter, since refractive indices vary much less with frequency in the former. Although the effects of dispersion seldom enable precise index matching to be satisfied, it can be possible in special cases where either the pump or harmonic lies close to an absorption band and the refractive index rapidly varies with frequency; this generally means working in a region of anomalous dispersion.[5] However, this situation carries the disadvantage of introducing possible intensity loss through absorption itself. If index matching is not possible, the condition $L \leq \pi / |\Delta \mathbf{k}|$ places a constraint on the distance within which harmonic conversion is an efficient process. The harmonic intensity ceases to grow beyond this limit, known as the coherence length, which typically lies between 1 and 100 mm in the case of gases, but only a few micrometers in a condensed phase.[33]

In optically anisotropic solids, however, where refractive index is dependent on direction of propagation and polarization (the phenomenon known as birefringence), it is often possible to obtain index matching by judicious choice of crystal orientation. In a uniaxial crystal, waves polarized with their electric vector perpendicular to the optic axis (ordinary-, or o-waves) have a refractive index that is independent of the direction of propagation *per se*. However, waves polarized with a finite electric vector component along the optic axis (extraordinary-, or e-waves) have a refractive index that depends on the angle between the direction of propagation and the optic axis.[34] This effect can be utilized in two types of phase-matching conditions. In type I phase matching the pump beam propagates as an ordinary wave and the harmonic as an extraordinary wave; in type II phase matching the fundamental is split into o-wave and e-wave components, and the harmonic propagates as an e-wave with stochastic polarization. Obtaining the index-matching condition that maximizes harmonic conversion thus becomes a matter of choosing a suitable configuration for the crystal, depending on the precise wavelengths of light involved. Such considerations are of course also important in the design of nonlinear optical materials. The optimization of second-harmonic generation in molecular crystals of monoclinic symmetry, for example, has been discussed in detail by White and Eckhardt.[35]

D. Molecular and Macroscopic Susceptibilities

Establishing a link with a macroscopic description of nonlinear response is principally a matter of considering local field effects. From the outset, these have been fully incorporated into the theory by the quantization of the electric displacement field, rather than the applied electric field, in Eq. (14). Further attention to local fields at this stage is therefore necessary only in so far as it enables microscopic susceptibilities to be related to bulk susceptibilities, a consideration that is primarily relevant to the solid state.

To begin, it is worth emphasizing that a description of harmonic emission in terms of bulk susceptibilities is valid only for coherent processes, where the factor η introduces a quadratic dependence of the scattering rate on the number of scatterers, M or N. Since this corresponds to a linear dependence in the matrix element, we can therefore write

$$N\chi(-\omega', \omega, \ldots, \omega)\odot^m\sigma\{(\mathbf{k}', \lambda'), (\mathbf{k}, \lambda)\}$$
$$= V\rho\chi(-\omega', \omega, \ldots, \omega)\odot^m\sigma\{(\mathbf{k}', \lambda'), (\mathbf{k}, \lambda)\} \qquad (96)$$

where ρ is the number density and V the interaction volume. It is convenient to introduce a new radiation tensor, σ_0, defined by

$$\boldsymbol{\varphi}_0(\xi) = \sigma_0 \exp(i\Delta\mathbf{k} \cdot \mathbf{R}_\xi) \qquad (97)$$

with $\boldsymbol{\varphi}_0(\xi)$ exactly as given by (44), save for the replacement of the electric displacement field operator $\mathbf{d}^\perp$ by the electric field operator $\mathbf{e}^\perp$. To a first approximation, recasting Eq. (96) in terms of this new tensor results in the following expression:

$$N\chi(-\omega', \omega, \ldots, \omega)\odot^m\sigma\{(\mathbf{k}', \lambda'), (\mathbf{k}, \lambda)\}$$
$$= V\mathbf{X}(-\omega', \omega, \ldots, \omega)\odot^m\sigma\{(\mathbf{k}', \lambda'), (\mathbf{k}, \lambda)\} \qquad (98)$$

where

$$\mathbf{X}(-\omega', \omega, \ldots, \omega) \equiv \rho\mathbf{L}_m(\omega', \omega)\odot^m\chi(-\omega', \omega, \ldots, \omega) \qquad (99)$$

and $\mathbf{L}_m(\omega', \omega)$ is a Lorentz factor tensor of rank $2m$, given by the outer tensor product of m rank-2 tensors as follows:

$$\mathbf{L}_m(\omega', \omega) = \left(\tfrac{1}{3}\right)^m (\boldsymbol{\kappa}_{\omega'} + 2\boldsymbol{\delta})(\boldsymbol{\kappa}_\omega + 2\boldsymbol{\delta})^{m-1} \qquad (100)$$

Here $\boldsymbol{\chi}_{\omega_r}$ is the dielectric tensor for frequency ω_r, and $\boldsymbol{\delta}$ is the unit

(Kronecker delta) tensor. In dielectrically isotropic media, Eq. (99) simply amounts to multiplying the microscopic tensor by a local field factor $\frac{1}{3}(\kappa_{\omega_r} + 2)$ for each photon involved in the process. For example, in the case of frequency doubling we obtain

$$X(-2\omega, \omega, \omega) = \tfrac{1}{27}\rho(\kappa_{2\omega} + 2)(\kappa_\omega + 2)^2 \chi(-2\omega, \omega, \omega) \quad (101)$$

The result for $\mathbf{X}(-\omega', \omega, \ldots, \omega)$ as given by Eq. (99) can be identified with the macroscopic nonlinear susceptibility. However, certain assumptions underlie the result for the Lorentz field tensor given by (100), which can be summarized by saying that the electric displacement field at each frequency is assumed to be linearly related to the true electric field for the same frequency. Thus, no account is taken of nonlinear corrections, or of higher-order multipolar corrections. A useful discussion of the shortcomings of the molecular model of macroscopic polarization has been given by Meredith et al.,[36] and Knoester and Mukamel[37] have shown how to incorporate local field effects from a more fundamental viewpoint.

Finally, it should be noted that for disordered or polymeric substances the link between molecular and macroscopic susceptibilities is rather more intricate. In particular, the formulae from which molecular susceptibilities are normally derived can be calculationally awkward for many-electron systems. A neat solution to this problem has recently been offered by Yu and Su,[38] who have demonstrated the utility of their method by reference to third-harmonic generation. The nonlinear susceptibilities of molecular aggregates also require careful treatment. Spano and Mukamel[39] have shown how the third-order susceptibility has resonant terms that depend quadratically on the number of molecules assembled. Off-resonance, the result scales linearly with the number of monomer units, and so relates to the form of result given by Eq. (64).

IV. COHERENT HARMONIC EMISSION

The formalism developed in previous sections can now be applied to a number of processes involving coherent harmonic emission in molecular systems. The harmonic generation process may be defined as a coherent elastic nonlinear scattering in which n photons of laser light with frequency ω (wave vector $\mathbf{k}$ and polarization vector $\mathbf{e}^{(\lambda)}$) are converted into a single photon of frequency $n\omega$ (wave vector $\mathbf{k}'$ and polarization vector $\mathbf{e}'^{(\lambda')}$). The specific case of second-harmonic generation has to some extent already been described by way of example in the theoretical development of Sections II and III, and the general treatment follows along very similar lines. The starting point is the rate equation, which for a molecular crystal is given by (76) and for a fluid by (84). In the case of fluids, the explicit result is as follows:

$$\Gamma_{NHG} = \frac{2\pi\rho_f}{\hbar\varepsilon_0^{2n+2}}\left|\left\langle \chi(-n\omega,\omega,\dots,\omega)\odot^{n+1}\sigma\{(\mathbf{k}',\lambda'),(\mathbf{k},\lambda)\}\right\rangle\right|^2(\eta_N - N)$$

(102)

where the coherence factor η_N is given by (94), σ is obtained by factorizing out the wave-vector matching phase factor from the radiation tensor φ according to Eqs. (44) and (68), and χ is the appropriate molecular nonlinear susceptibility tensor determined by Eq. (45). The corresponding result for a crystalline medium is obtained by replacing N, the number of molecules, by M, the number of unit cells, in Eq. (102); also χ, the molecular nonlinear susceptibility, is replaced by $\hat{\chi}$, the unit cell susceptibility, and the angular brackets which denote molecular orientational averaging are removed.

The explicit expression for the density of final radiation states in Eq. (102) is given by[40]

$$\rho_f = (8\pi^3\hbar c'')^{-1}k'^2 V\,d\Omega$$

(103)

where c'' is the velocity of propagation of the harmonic, and $d\Omega$ is an element of solid angle around the wave-vector matching direction of emission. Assuming that the number of scattering centers is large so that $\eta_N \gg N$, then recasting the result in terms of a radiant intensity of harmonic emission, $I^{(n)}$, we obtain

$$I^{(n)} = (4\pi^2\hbar\varepsilon_0^{2n+2})^{-1}k'^3 V\eta_N\left|\left\langle \chi(-n\omega,\omega,\dots,\omega)\odot^{n+1}\sigma\{(\mathbf{k}',\lambda'),(\mathbf{k},\lambda)\}\right\rangle\right|^2$$

(104)

The structure of the nonlinear susceptibility tensor χ, as given by Eq. (45) for the generation of harmonics of arbitrary order, is most easily discussed with reference to the corresponding time-ordered diagrams. The typical diagram shown in Fig. 3 represents the successive absorption of q photons, followed by emission of the harmonic photon, followed by a further $(n - q)$ absorptions before the molecule returns to its ground state. There are $(n + 1)$ topologically distinct diagrams of this type to consider, each of which can be labeled with an index q in the range $0 \le q \le n$. Interpreting these by the rules of Section II.D, we obtain a general result which can

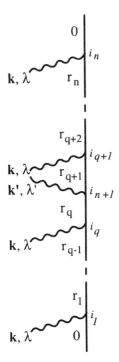

Figure 3. One of the $(n + 1)$ topologically distinct time-ordered diagrams for n-harmonic generation.

most compactly be expressed as follows[41]:

$$\chi_{i_1 \cdots i_{n+1}}(-n\omega, \omega, \ldots, \omega)$$

$$= \sum_{q=0}^{n} \sum_{r_1} \cdots \sum_{r_n} \frac{\left(\prod_{j=q+1}^{n} \mu_{i_j}^{r_{j+1}r_j}\right) \mu_{i_{n+1}}^{r_{q+1}r_q} \left(\prod_{k=1}^{q} \mu_{i_k}^{r_k r_{k-1}}\right)}{\prod_{s=q+1}^{n} \left[\tilde{E}_{r_s r_0} + (n - s + 1)\hbar\omega\right] \prod_{t=1}^{q} \left[\tilde{E}_{r_t r_0} - t\hbar\omega\right]}$$

$$(105)$$

In writing the result in this form, both $|r_0\rangle$ and $|r_{n+1}\rangle$ are identified with the ground state $|0\rangle$, and use is made of the convention

$$\prod_{l=p}^{p-1} f_l(l) = 1 \qquad (106)$$

Consider next the radiation tensor σ in Eq. (104), which is given by the following expression:

$$\sigma_{i_1(i_2 \cdots i_{n+1})}\{(\mathbf{k}', \lambda'), (\mathbf{k}, \lambda)\}$$

$$= -i^{n+1}\left(\frac{\hbar\omega\varepsilon_0}{2V}\right)^{(n+1)/2}\left[\frac{n_\omega!n}{(n_\omega - n)!}\right]^{1/2}\bar{e}'_{i_1}e_{i_2} \cdots e_{i_{n+1}} \qquad (107)$$

where n_ω is the number of pump photons in the quantization volume V, and the brackets around the indices $i_2 \cdots i_{n+1}$ signify permutational symmetry. Introducing a polarization tensor, s, defined by

$$s_{i_1(i_2 \cdots i_{n+1})}\{(\mathbf{k}', \lambda'), (\mathbf{k}, \lambda)\} = \bar{e}'_{i_1}e_{i_2} \cdots e_{i_{n+1}} \qquad (108)$$

we then obtain

$$l^{(n)} = (4\pi^2\hbar)^{-1}k'^3V\left(\frac{\hbar\omega}{2\varepsilon_0 V}\right)^{n+1}\left[\frac{n_\omega!n}{(n_\omega - n)!}\right]\eta_N$$

$$\times \left|\left\langle \chi(-n\omega, \omega, \ldots, \omega)\odot^{n+1}s\{(\mathbf{k}', \lambda'), (\mathbf{k}, \lambda)\}\right\rangle\right|^2 \qquad (109)$$

A. Quantum Optical Considerations

As the applications of Eq. (109) unfold, one of the first considerations is how to apply the result to the conditions relating to a given laser source. For this purpose it is necessary to recast the result in terms of physically meaningful pump radiation parameters, in place of the artificial quantization volume V and photon number n_ω which appear in (109). The procedure for this reformulation allows consideration of pump radiation states characterized by various forms of photon statistics.

The number states $|n(\mathbf{k}, \lambda)\rangle$ employed in Section II, which are the most usual basis for quantum electrodynamical calculations based on time-dependent perturbation theory, are eigenstates of the unperturbed radiation Hamiltonian. As such they represent states of the radiation field for which there is a precise nonfluctuating value for the number of photons. Such states have completely undefined phase, however, and are clearly a poor representation of coherent laser light. An alternative set of states which is certainly much better suited to modeling laser radiation is the overcomplete set composed of the coherent states $|\alpha(\mathbf{k}, \lambda)\rangle$, which minimize the uncertainty in phase and occupation number of a given radiation mode.[33, 42] These states are eigenstates of the corresponding annihilation operators, satisfying the result

$$a^{(\lambda)}(\mathbf{k})|\alpha(\mathbf{k}, \lambda)\rangle = \alpha(\mathbf{k}, \lambda)|\alpha(\mathbf{k}, \lambda)\rangle \tag{110}$$

where $\alpha(\mathbf{k}, \lambda)$ is a complex number whose modulus relates to the mean photon number m through

$$m = \langle \alpha(\mathbf{k}, \lambda)|a^{+(\lambda)}(\mathbf{k})a^{(\lambda)}(\mathbf{k})|\alpha(\mathbf{k}, \lambda)\rangle = |\alpha(\mathbf{k}, \lambda)|^2 \tag{111}$$

Thus, while a process such as n-harmonic generation involving the absorption of n photons from a single beam contributes a factor of $\{n_\omega(n_\omega - 1) \cdots (n_\omega - n)\}^{1/2}$ to the matrix element if a number state with n_ω photons of pump radiation is employed for the calculation, according to Eq. (18), a factor of α^n arises where a coherent state is employed. The corresponding rate factors are then $\zeta_n(n_\omega) = n_\omega!/(n_\omega - n)!$ and $\zeta_n(n_\omega) = |\alpha|^{2n} = n_\omega^n$, respectively.

A different perspective is obtained by considering the photon number to be subject to fluctuations that satisfy particular types of statistical distribution. By suitably weighting rate equations calculated on the basis of number states, various kinds of radiation can then be modeled. Thus, if P_q is defined as the time-averaged probability of finding q photons in the quantization volume, we have

$$\frac{n_\omega!}{(n_\omega - n)!} \rightarrow \zeta_n(n_\omega) = \sum_{q=0}^{\infty} P_q(n_\omega)\frac{q!}{(q - n)!} \tag{112}$$

For number states $P_q(n_\omega) = \delta_{n_\omega, q}$ and the result $n_\omega!/(n_\omega - n)!$ is recovered. For coherent radiation, however, the appropriate form of photon distribution function is a Poisson distribution, expressible in terms of its mean n_ω by the relation

$$P_q(n_\omega) = \exp(-n_\omega)\frac{n_\omega^q}{q!} \tag{113}$$

Substitution of this result in Eq. (112), followed by a little algebraic manipulation, readily reproduces the result $\zeta_n(n_\omega) = n_\omega^n$. A convenient generalization is

$$\zeta_n(n_\omega) = g^{(n)}n_\omega^n \tag{114}$$

where $g^{(n)}$ is the degree of nth-order coherence. For coherent light this parameter takes the value of unity for all q; other values typify different kinds of photon distribution. The fact that only for $n = 1$ is $g^{(n)}$ unity for all types of radiation serves as a reminder that conventional optical processes, which involve the absorption or scattering of photons singly, are unique in being insensitive to photon statistics, depending only on the mean number of photons present.

The above treatment of the phase properties of the pump does not of course address the issue of the corresponding properties of the emitted harmonic. While space does not permit their discussion here, they are of considerable interest and reflect a number of distinctive quantum optical features, such as photon antibunching and squeezing.[43, 44] In connection with second-harmonic generation, squeezing effects arise in crystals or media whose isotropy is removed by application of a static electric field[45, 46]; in the case of third harmonic emission it occurs even in genuinely isotropic media.[47] The way in which such properties evolve in time is itself the subject of much ongoing research, and Gantsog et al.[48] have recently shown how it is possible for the harmonic to evolve, initially into a superposition of two states characterized by different phases, leading eventually to a complete randomization of phase.

Returning to the development of the equations for the harmonic intensity, it is next pertinent to consider expressing the results in terms of the pump irradiance. Rate equations expressed in terms of mean photon number and quantization volume are not directly applicable to experiment. Moreover, since the quantization volume is merely a theoretical artifact, it must invariably cancel out in the final rate equation. However, the photon density, which is the ratio of these two quantities, is directly related to the mean irradiance (power per unit beam cross-sectional area) through the relation

$$I_\omega = \frac{n_\omega \hbar c' \omega}{V} \tag{115}$$

From (115) and (112) we thus obtain the following algorithm for replacement of any quantum electrodynamical rate factor based on number states by a more general parameter cast in terms of mean irradiance and degree of coherence:

$$\left[\frac{n_\omega!}{(n_\omega - n)!} \right] \rightarrow \left(\frac{I_\omega V}{\hbar c' \omega} \right)^n g_\omega^{(n)} \tag{116}$$

Substitution into Eq. (109) thus produces the following general result:

$$l^{(n)} = \left(\frac{c''k'^4}{8\pi^2\varepsilon_0}\right)\left(\frac{I_\omega}{2c'\varepsilon_0}\right)^n$$

$$\times g_\omega^{(n)}\eta_N\left|\left\langle\chi(-n\omega,\omega,\ldots,\omega)\odot^{n+1}\mathbf{s}\{(\mathbf{k}',\lambda'),(\mathbf{k},\lambda)\}\right\rangle\right|^2 \quad (117)$$

B. Symmetry Criteria

One of the most crucial features of harmonic generation is its sensitivity to symmetry criteria both on the local molecular and also the macroscopic level. Sound parity arguments prove that the electric dipole generation of even harmonics is forbidden in any centrosymmetric crystal, and the same arguments can be applied to individual atoms or centrosymmetric molecules (see below). However, such processes are known to remain unobservable even in gases and liquids composed of noncentrosymmetric molecules. By extension of the classical electrodynamical principles normally reserved for solids, it is commonly concluded that the generation of even harmonics is forbidden in fluids because these too possess macroscopic inversion symmetry. The argument nonetheless obscures the fact that the coherent process is in any case forbidden by the random rotational motions of the molecules.

Before considering these issues in detail, it is worth establishing the conditions for the existence of the molecular or unit cell nonlinear susceptibility $\overset{(\cdot)}{\chi}(-n\omega,\omega,\ldots,\omega)$. As befits a parametric process, the initial and final molecular states are identical and normally carry the full ground-state symmetry of the molecule or unit cell. Thus, since each term in the explicit expression of Eq. (105) contains a product of $(n + 1)$ transition electric dipole moments, the tensor can be nonvanishing only if the totally symmetric representation of the appropriate point group or space group is spanned by the product of $(n + 1)$ translations. Where the molecule or crystal possesses a center of symmetry, this condition can be met only in the generation of odd harmonics, where $(n + 1)$ is even and the product of translations is thus of gerade (even) symmetry.

Where local symmetry permits harmonic generation based exclusively on electric dipole coupling, the inclusion of higher-order multipolar contributions in the coupling equation (23) produces additional terms which are normally negligible. However, if electric dipole harmonic generation is forbidden, as is the case for even harmonics in a centrosymmetric species, these higher-order terms can become significant and may be manifest in weak harmonic emission (though not in fluids, for other reasons outlined below). For example, if one of the three ungerade electric dipole interactions involved in second-harmonic generation is replaced by a gerade magnetic dipole or electric quadrupole interaction, the operator product spans the totally symmetric representation even in centrosymmetric materials, and the corresponding response tensor is nonzero.

In general, if the multipolar expansions of the interactions involving the harmonic and pump waves are taken to order s and t respectively, then with $p = t - n$ and $q = s - 1$, we obtain the following generalization of Eq. (117)[27, 41]:

$$
I^{(n)} = \left(\frac{c'' k'^4}{8\pi^2 \varepsilon_0} \right) \left(\frac{I_\omega}{2c' \varepsilon_0} \right)^n g_\omega^{(n)} \eta_N
$$

$$
\times \left| \left\langle \sum_p \sum_q k^p k'^q \Lambda^{p,q}(-n\omega, \omega, \ldots, \omega) \odot^{n+p+q+1} \prod_i^{n+1} \mathbf{w}^{(i)} \prod_j^{p+q} \hat{\mathbf{k}}^{(j)} \right\rangle \right|^2
$$

$$(118)$$

where $\mathbf{w}^{(i)}$ are unit vectors perpendicularly disposed to the propagation direction and $\Lambda(-n\omega, \omega, \ldots, \omega)$ is a generalized form of response tensor which accommodates any combination of multipolar couplings. For example, second-harmonic generation with electric or magnetic dipole coupling gives $s = 1$ and $t = 2$, so that $p = 0$ and $q = 0$; however, if one electric quadrupole interaction is included in the interaction with the pump, we have $s = 1$ and $t = 3$, giving $p = 1$ and $q = 0$. For application to crystalline media, where the local response is associated with the unit cell and there is no rotational averaging to perform, analysis of the above result reveals that there must always be certain multipolar nonlinear susceptibility tensors that are nonzero, so that generation of every harmonic is at least weakly allowed.

When fluids are considered, the rotational averaging denoted by angular brackets imposes further constraints, however. This can first be illustrated by taking the simplest case, that of second-harmonic generation in the electric dipole approximation. The result requires evaluation of the rotational average in Eq. (117), which for the particular case under consideration gives the following expression:

$$
\langle A \rangle \equiv \left\langle \chi(-n\omega, \omega, \ldots, \omega) \odot^{n+1} s\{(\mathbf{k}', \lambda'), (\mathbf{k}, \lambda)\} \right\rangle
$$

$$
= \left\langle \chi(-2\omega, \omega, \omega) \odot^3 \bar{\mathbf{e}}' \mathbf{e} \mathbf{e} \right\rangle
$$

$$
= \left\langle \chi_{i(jk)}(-2\omega, \omega, \omega) \bar{e}'_i e_j e_k \right\rangle .
$$

$$(119)$$

Evaluation of the requisite third rank tensor average, using a method described in detail elsewhere,[28a] yields the result

$$\langle A \rangle = \tfrac{1}{6}\chi_{\lambda(\mu\nu)}(-2\omega, \omega, \omega)\bar{e}'_i e_j e_k \varepsilon_{ijk}\varepsilon_{\lambda\mu\nu} \qquad (120)$$

where Greek indices denote components referred to molecular axes. The index contraction between the fully index-antisymmetric Levi-Civita tensor ε_{ijk} and the j, k-symmetric product of polarization components $e_j e_k$ invokes the cross product of e with itself and is clearly zero. Thus, the macroscopic isotropy forbids the observation of a coherent second-harmonic signal even in the forward direction. Exceptions to this rule can arise only where there is an induced anisotropy, as, for example, may be conferred on any polar fluid by application of a static electric field (see Section IV.D). A quadratic interaction with the sizeable electric fields of intense pump laser radiation, or charge separation through multiphoton ionization, can also induce a degree of local anisotropy; this appears to be the explanation of the even harmonics observed in atomic gases under extreme conditions.[49, 50, 90-92]

Similar conclusions have to be drawn when higher-order multipolar contributions such as the magnetic dipole and electric quadrupole coupling terms in Eq. (23) are considered. Take for example the possibility of a contribution to coherent SHG associated with one electric quadrupolar interaction in the pump photon annihilation process. This requires evaluation of a fourth rank orientational average

$$\begin{aligned}
\langle A' \rangle &\equiv \left\langle \tilde{\chi}_{i(jk)l}(-2\omega, \omega, \omega)\bar{e}'_i e_j e_k k_l \right\rangle \\
&= \tfrac{1}{15}\Big[\big(3\tilde{\chi}_{\lambda(\lambda\mu)\mu} - \tilde{\chi}_{\lambda(\mu\mu)\lambda}\big)(\bar{e}' \cdot e)(k \cdot e) \\
&\quad + \big(2\tilde{\chi}_{\lambda(\mu\mu)\lambda} - \tilde{\chi}_{\lambda(\lambda\mu)\mu}\big)(k \cdot \bar{e}')(e \cdot e) \Big]
\end{aligned} \qquad (121)$$

and the result is again transparently zero since for emission in the forward direction, which is necessary for phase-matched coherent response, k is orthogonal to both e and e'. Analogous remarks apply when the contribution involving electric quadrupole emission of the harmonic photon is considered.

In fact rigorous treatment of the problem shows that to all orders of multipolar approximation, the generation of all even harmonics is forbidden to fluids.[41, 51] The same conclusion can be drawn more elegantly on the basis of angular momentum considerations.[52] The result has interesting practical implications, since it facilitates and vindicates the use of SHG as a surface-specific probe (see Section IV.C). This whole issue has become obscured by some treatments which, by dealing with nonlinear polarization rather than the observable signal intensity, have failed to differentiate coherent and incoherent response.[23, 53] At this stage, it should simply be noted that within the bulk of an isotropic medium, any longitudinal polarization component (i.e., one that lies along the propagation vector of the pump radiation) can produce only incoherent harmonic emission (see Section V).[93]

Isotropic averaging also places constraints on polarization dependence of coherent emission. When the pump radiation is plane polarized, for example, odd harmonics are necessarily produced with a retention of polarization. It furthermore transpires that the generation of any odd harmonic is forbidden if the pump beam is circularly polarized.[54, 55] This is because odd harmonics are associated with even-rank rotational averages, which lead to scalar products ($\mathbf{e} \cdot \mathbf{e}$) in the polarization tensor. For circular polarization, this scalar product is zero and so the harmonic intensity necessarily vanishes. Consideration of angular momentum conservation provides an alternative viewpoint which leads to the same conclusions. It should be noted, however, that these arguments are applicable only to coherent harmonic production; incoherent harmonic scattering remains finite both for even harmonics and for circularly polarized radiation, whether for forward or nonforward emission (see Section V).

C. Second-Harmonic Generation at Surfaces and Interfaces

As seen above, the generation of even harmonics is forbidden in a bulk medium possessing isotropic or cubic symmetry. However, at the boundary of any such medium, as, for example, at a fluid interface, local symmetry is lowered and the exclusion of even harmonics no longer applies. For this reason extensive use has been made of the surface selectivity of second-harmonic generation, and it has become an important tool for studying the chemistry and physics of interfaces,[56] both with regard to surfaces and surface-adsorbed layers. Although the harmonic is often studied by reflection, a host of experimental geometries is possible and these have been systematically discussed by Mizrahi and Sipe.[57]

At any two-dimensional boundary, the general expression for the intensity of n-harmonic emission associated with the surface can be obtained from (102), in which the coherence factor $\eta_N - N$ defined by

$$\eta_N - N = \sum_{\xi}^{N} \sum_{\xi' \neq \xi}^{N-1} \exp(i\Delta\mathbf{k} \cdot \mathbf{R}_{\xi\xi'}) \tag{122}$$

should be interpreted as being summing over surface molecules. Conse-

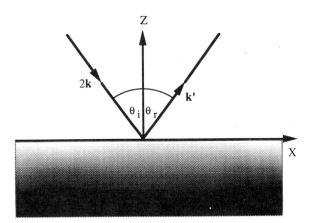

Figure 4. Geometry for surface harmonic generation. The X component of $(2\mathbf{k} - \mathbf{k}')$ must vanish for phase-matched emission.

quently, each $\mathbf{R}_{\xi\xi'}$ vector lies in the (X, Y) plane, where Z specifies the surface normal and where X is defined through identifying (X, Z) with the $(\mathbf{k}, \mathbf{k}')$ scattering plane, as in Fig. 4. For any emission for which $\Delta\mathbf{k} \equiv n\mathbf{k} - \mathbf{k}'$ is disposed in the Z direction, the argument of the exponential in (122) becomes zero and the factor $\eta_N - N$ attains the value $N(N - 1)$ associated with coherent emission. If dispersion is neglected, this amounts to the usual rule for reflection at an angle equal to the angle of incidence. For SHG however, as can be understood from Fig. 4, it is generally necessary to fulfil the condition

$$2k \sin \theta_i = k' \sin \theta_r \qquad (123)$$

where θ_i and θ_r are the angles of incidence and reflection as normally defined. Consequently, we obtain

$$\frac{\sin \theta_i}{\sin \theta_r} = \frac{n_{2\omega}}{n_\omega} \qquad (124)$$

which is a special case of a general relationship for frequency addition first obtained by Bloembergen and Pershan.[58] The phase-matching requirement that leads to the reflection law applies to layers of molecules down to within a depth of approximately k^{-1} below the surface.[59] At optical frequencies, any coherent signal from layers further beneath the surface would invariably be destroyed through interference. In this connection some, but not all, of the confusion that has arisen over the issue of surface harmonic emission can be traced to the different usage of the term "bulk" by various authors. Principally, the ambiguity arises where surface SHG is used in the study of adsorbed molecules, and the term "bulk" is used to differentiate the substrate material. For such a purpose the term "substrate" is much to be preferred, however, since it does not imply the major part of the solid material. Such a distinction is crucial where the substrate has a bulk constitution that is isotropic (see below).

Consider first the case of second-harmonic reflection from the surface of a solid, or from molecules adsorbed on a solid. The strength of the harmonic intensity is powerfully influenced by orientational factors, as determined by $A = \chi(-2\omega, \omega, \omega) \odot^3 \sigma\{(\mathbf{k}', \lambda'), (\mathbf{k}, \lambda)\} = \chi \odot^3 \bar{\mathbf{e}}' \mathbf{e}\mathbf{e}$ in Eq. (102). Plane polarized components of the pump and harmonic waves are conventionally defined as s- or p-polarized by the relations

$$\mathbf{e}_s = \hat{\mathbf{J}} \tag{125}$$

$$\mathbf{e}_p = \cos\theta_i \hat{\mathbf{I}} + \sin\theta_i \hat{\mathbf{K}} \tag{126}$$

$$\mathbf{e}'_s = \hat{\mathbf{J}} \tag{127}$$

$$\mathbf{e}'_p = -\cos\theta_r \hat{\mathbf{I}} + \sin\theta_r \hat{\mathbf{K}} \tag{128}$$

where $(\hat{\mathbf{I}}, \hat{\mathbf{J}}, \hat{\mathbf{K}})$ are unit vectors in the X, Y, Z directions, respectively. Hopf and Stegeman[60] use similar relations but expressed in terms of glancing angles. Substitution of these results in the expression for A is straightforward. For example, with an s-polarized pump, and resolving the harmonic also for its s component, $A = \chi_{YYY}$. Clearly experiments with different polarization conditions (including circular polarizations not dealt with here) facilitate evaluation of the various components of the susceptibility tensor, as referred to the (X, Y, Z) frame. These in turn are related to the components in the molecular frame through the relation

$$\chi_{ijk} = \chi_{\lambda\mu\nu} \langle R_{i\lambda} R_{j\mu} R_{k\nu} \rangle \tag{129}$$

where $R_{i\lambda}$ is the (i, λ) element of the Euler angle matrix relating the molecule-based to the laboratory-fixed frame. On this occasion the angular brackets denote a distributional, rather than an isotropic, average over molecular orientations. This reflects the fact that, at least for surface-adsorbed molecules, there will generally be a preferred but not a fixed orientation with respect to the surface.

Frequently, calculations are based on the premise that certain kinds of molecules have susceptibilities dominated by axial components, and as such the caveat of Section II.G should be borne in mind. A great many experimental studies have utilized exactly these principles in the determination of surface molecular orientation.[56] Extensions of the theory to surface dimers have also been derived.[61]

Orientational factors are also manifest in other ways. For example, the SHG signal reflected from the (111) face of a cubic crystal depends on the angle the $(\mathbf{k}, \mathbf{k}')$ plane makes with the C_3 axis, and a plot of the harmonic intensity against this azimuthal angle nicely displays the threefold symmetry. Moreover, it provides a sensitive measure of adsorbed species.[62] As with surface-enhanced Raman scattering, there is often a significant enhancement of the SHG signal associated with molecules adsorbed on suitable metal surfaces, thought to be associated with coupling to surface plasmons and also influenced by surface roughness. The detailed theory is described by Chen et al.,[63] Aktsipetrov et al.,[64] and Li and Seshadri.[65]

When the symmetry implications of reflection from the surface of an isotropic fluid are considered, it once again transpires that the molecular rotational average annihilates the signal associated with pure electric dipole coupling (if the molecules at the surface are isotropically oriented). This is because Eq. (120), which entails the factor $(\mathbf{e} \times \mathbf{e})$, still applies. However, when the possible involvement of electric quadrupole coupling is entertained, the second term of (121) associated with the polarization factor $(\mathbf{k} \cdot \bar{\mathbf{e}}')(\mathbf{e} \cdot \mathbf{e})$ will generally not vanish, since $\mathbf{k}$ is not parallel to $\mathbf{k}'$. From a classical viewpoint, this signal is inferred as resulting from a longitudinal nonlinear polarization; however, such signals arise only at or close to the surface. For similar reasons there is a term arising from electric quadrupole emission that will also contribute, associated in this case with a factor $(\mathbf{k}' \cdot \mathbf{e})(\bar{\mathbf{e}}' \cdot \mathbf{e})$.

Summarizing the above results, it is clear that when SHG is studied in a transmission mode there can be no coherent (phase-matched) signal from the bulk of any isotropic medium. There is, of course, an incoherent (nondirected) signal associated with pure electric dipole (E1^3) coupling (see Section V). However, when SHG is studied by reflection from the surface of an isotropic medium, there is a coherent signal associated with an interaction involving one electric quadrupole (E1^{2}E2). Both the coherent (E1^{2}E2) and the incoherent (E1^3) signals are classically interpreted as arising from an induced longitudinal polarization. If the bulk medium from whose surface the harmonic is studied is structurally isotropic throughout its bulk, the SHG signal has to originate from molecules within a depth of $\sim k^{-1}$ of the surface. However, the nature of many fluid surfaces is such that a degree of orientational order may be present both at and just below the surface. In such circumstances a mechanism similar to that associated with electric field-induced second-harmonic generation (EFISHG) may produce a signal from within the bulk. This may account for some experimental results that appear to show a harmonic signal associated with the bulk of an isotropic medium. The "skin depth" associated with this effect is in urgent need of further experimental study; nonetheless, the crucial point is that the signal in such cases derives from a region that is not truly isotropic.

D. Electric Field-Induced Second-Harmonic Generation

Although SHG is normally forbidden in isotropic media, the application of a static electric field can remove the forbidden character of the process by two distinct mechanisms. In one mechanism the static field induces an electro-optical contribution to the harmonic signal at the molecular level. This contribution is formally associated with fourth-order perturbation theory involving the static coupling term in Eq. (37), and is represented by time-ordered diagrams of the kind shown in Fig. 5. By this mechanism SHG becomes allowed both in centrosymmetric crystals and in fluids.

The second mechanism applies to fluids consisting of polar (and thus necessarily noncentrosymmetric) molecules. Here, the electric displacement field D exerts a torque on each permanent molecular dipole μ determined by the vector product $\mu \times D$. This removes bulk isotropy by inducing a degree of molecular alignment,[66] governed under equilibrium conditions by the Boltzmann distribution. Hence, in fluid media a weighted molecular rotational average has to be effected with regard to a temperature-dependent distribution of molecular orientations.

The basic rate equation for SHG in the presence of an applied static electric field D can be obtained from Eq. (117), where $n = 2$ for SHG, by

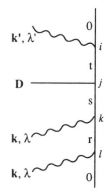

Figure 5. One of the 12 time-ordered diagrams for EFISHG. The horizontal line denotes interaction with the static electric field.

incorporating the term from fourth-order perturbation theory alluded to above, and also incorporating the Boltzmann weighting term $\exp(\boldsymbol{\mu} \cdot \mathbf{D}/\varepsilon_0 kT)$. The result is a radiant second-harmonic intensity given by

$$
I_{\text{EFISHG}} = \left(\frac{c'' k'^4}{8\pi^2 \varepsilon_0}\right)\left(\frac{I_\omega}{2c'\varepsilon_0}\right)^2 g_\omega^{(2)} \eta_N
$$

$$
\times \left|\left\langle \left\{ \chi(-2\omega, \omega, \omega)\odot^3 \bar{\mathbf{e}}'\mathbf{ee} + \varepsilon_0^{-1}\check{\chi}(-2\omega, 0, \omega, \omega) \right.\right.\right.
$$

$$
\left.\left.\left. \odot^4 \bar{\mathbf{e}}'\mathbf{Dee} \right\} \exp\left(\frac{\boldsymbol{\mu} \cdot \mathbf{D}}{\varepsilon_0 kT}\right) \right\rangle \times \left\langle \exp\left(\frac{\boldsymbol{\mu} \cdot \mathbf{D}}{\varepsilon_0 kT}\right) \right\rangle^{-1} \right|^2 \quad (130)
$$

where the correction term involving the fourth-order tensor $\check{\chi} \equiv \chi(-2\omega, 0, \omega, \omega)$ represents the local electro-optical interaction.[67-69] The full structure of the result including the form of the molecular tensors is given elsewhere.[31, 70, 71] Here it is appropriate to draw out a few salient features of the dependence of the SHG intensity on experimental variables. These are principally the laser polarization, electric field strength, and temperature. The last two of these come into play through the Boltzmann factor γ given by

$$
\gamma = \frac{\mu D}{\varepsilon_0 kT} \quad (131)
$$

Under normal low-field conditions where $\gamma \ll 1$, the harmonic intensity may be expressed by the following leading terms in a Taylor series:

$$
I_{\text{EFISHG}} = \left(\frac{I_\omega^2 D^2 g_\omega^{(2)} c'' k'^4}{450\pi^2 \varepsilon_0^3 c'^2}\right)\eta_N
$$

$$
\times \left|(3p_1 - p_2)\left(\check{\chi}_{\lambda\lambda(\mu\mu)} + \frac{\chi_{\lambda(\lambda\mu)}\mu_\mu}{kT}\right)\right.
$$

$$
\left. + (2p_2 - p_1)\left(\check{\chi}_{\lambda\mu(\mu\lambda)} + \frac{\chi_{\lambda(\mu\mu)}\mu_\lambda}{kT}\right)\right|^2 \quad (132)
$$

Here p_1 and p_2 are polarization parameters given by

$$
p_1 = (\hat{\mathbf{D}} \cdot \mathbf{e})(\bar{\mathbf{e}}' \cdot \mathbf{e}) \quad (133)
$$

$$
p_2 = (\hat{\mathbf{D}} \cdot \bar{\mathbf{e}}')(\mathbf{e} \cdot \mathbf{e}) \quad (134)
$$

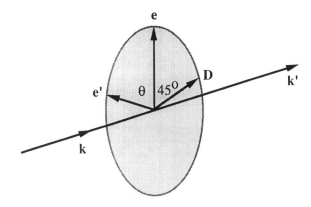

Figure 6. One possible geometry for the characterization of EFISHG. The vectors **e**, **e′**, and **D** lie in the plane perpendicular to the pump and harmonic propagation vectors.

Clearly the harmonic signal disappears entirely when the static field is applied parallel to the pump beam, since the pump and harmonic polarization vectors **e** and **e′** are necessarily perpendicular to this direction. This conclusion remains true even when all higher orders in the Taylor series are considered. It is therefore normal practice to apply the field perpendicularly to maximize the conversion efficiency. Moreover, the effect can be heightened if the field is spatially periodic.[72] One striking polarization characteristic of electric field-induced harmonic generation is the possibility of producing a harmonic signal from a circularly polarized pump. Indeed, if the harmonic is resolved for its y-polarized component when the pump beam and the static field are applied in the z and x directions, ($p_1 = p_2 = \frac{1}{2}i$), it transpires that harmonic generation necessitates use of a chiral beam.

A combined study of the polarization- and temperature- or field-dependence of the harmonic intensity can provide important information on molecular nonlinear optical and electro-optical parameters.[73] An optimum geometry is as shown in Fig. 6, where the electric field is applied at right angles to the beam propagation direction and at 45° to the pump polarization vector. The harmonic signal is then analyzed for two orthogonal polarization components. If θ is the angle between the pump and harmonic polarization vectors, two special cases arise. With $\theta = -63.4°$, then $(3p_1 - p_2) = 0$ and we obtain leading terms

$$I_{\text{EFISHG}}^{-63.4} = \left(\frac{I_\omega^2 D^2 g_\omega^{(2)} c'' k'^4}{180 \pi^2 \varepsilon_0^3 c'^2} \right) \eta_N \left(\check{\chi}_{\lambda\mu(\lambda\mu)} + \frac{2\chi_{\lambda(\mu\mu)}\mu_\lambda}{kT} \right) \check{\chi}_{\nu_0(\nu_0)} \quad (135)$$

However, with $\theta = 26.6°$, we have $(2p_2 - p_1) = 0$ and thus

$$l_{\text{EFISHG}}^{26.6} = \left(\frac{I_\omega^2 D^2 g_\omega^{(2)} c'' k'^4}{180\pi^2 \varepsilon_0^3 c'^2}\right) \eta_N \left(\check{\chi}_{\lambda\lambda(\mu\mu)} + \frac{2\chi_{\lambda(\lambda\mu)}\mu_\mu}{kT}\right)\check{\chi}_{\nu\nu_{(00)}} \quad (136)$$

Selective temperature experiments in these two geometries thus make possible the direct evaluation of the parameters $\check{\chi}_{\lambda\lambda(\mu\mu)}$, $\check{\chi}_{\lambda\mu(\lambda\mu)}$, $\chi_{\lambda(\lambda\mu)}\mu_\mu$, and $\chi_{\lambda(\mu\mu)}\mu_\lambda$. Each of these is readily expressible in terms of components. If the permanent moment lies in the z direction then $\chi_{\lambda(\mu\mu)}\mu_\lambda = \chi_{zxx} + \chi_{zyy} + \chi_{zzz}$, for example, and for a molecule with a threefold or higher principal axis of symmetry, this reduces to $2\chi_{zxx} + \chi_{zzz}$. Individual components are not directly amenable to evaluation, however, in contrast to the case of a rigidly oriented system.

For centrosymmetric molecules, where the third rank susceptibility (hyperpolarizability) vanishes, the theory simplifies considerably and in an obvious way. Here, measurement of both the real and imaginary parts of the tensor $\chi(-2\omega, 0, \omega, \omega)$ are possible.[74] All such parameters provide crucial information on molecular electronic properties, and in particular the extent of delocalization in extended conjugated species.[2]

E. Third-Harmonic Generation

In contrast to SHG, third harmonic generation is a process that, like Rayleigh scattering, is on the basis of molecular or bulk phase symmetry universally allowed. Firstly, all molecules have a nonzero susceptibility tensor $\chi(-3\omega, \omega, \omega, \omega)$, since it is of even parity. Secondly, the rotational average that it is appropriate to perform on the expression for the harmonic intensity for any fluid sample gives a nonvanishing result. Specifically, from (117) we obtain

$$l_{\text{THG}} = T|\langle \chi(-3\omega, \omega, \omega, \omega)\odot^4 \mathbf{s}\{(\mathbf{k}', \lambda'), (\mathbf{k}, \lambda)\}\rangle|^2 \quad (137)$$

where T is a parameter whose value is determined by the laser beam characteristics and given by

$$T = \left(\frac{c'' k'^4}{8\pi^2 \varepsilon_0}\right)\left(\frac{I_\omega}{2c'\varepsilon_0}\right)^3 g^{(3)}\eta \quad (138)$$

$\chi(-3\omega, \omega, \omega, \omega)$ is a fourth-rank nonlinear susceptibility constructed according to the principles of Section II, and the polarization tensor $\mathbf{s} = \bar{\mathbf{e}}'\mathbf{eee}$. For fluids, evaluation by the usual methods of the rotational average denoted by the angular brackets in (137) gives the following result:

$$l_{\text{THG}} = \left(\frac{T}{5}\right)\chi_{\lambda(\lambda\mu\mu)}(\bar{\mathbf{e}}' \cdot \mathbf{e})(\mathbf{e} \cdot \mathbf{e}) \tag{139}$$

The factor $(\mathbf{e} \cdot \mathbf{e})$ attests to the fact that in the singular case of a circularly polarized pump the intensity of third-harmonic emission is zero, as established in Section IV.B. Indeed, the realization of this unusual condition provides the only means of totally suppressing the harmonic. The result for systems with partial orientational order is somewhat more complex, and has been discussed by Mazely and Hetherington.[75]

The cubic dependence on laser intensity displayed in (138) indicates that THG should be a weaker process than SHG $(\sim I_\omega^2)$, though, of course, only when both are allowed. Since SHG is forbidden in gases, THG is here usefully employed as a means of frequency upconversion, though its phase-matching requirements often necessitate generation in regions of anomalous dispersion close to resonance. In this respect two-photon resonances are more helpful than one-photon resonances, since the latter can result in strong absorption of the pump radiation. The detailed structure of the two-photon resonant THG tensor, incorporating the effect of rotational structure for diatomic species, has been given by Aguillon et al.[76] For a full discussion of THG the reader is referred to the excellent monograph by Reintjes.[77]

V. ELASTIC SECOND-HARMONIC LIGHT SCATTERING

Having discussed the principal features of the coherent harmonic signal, it is now appropriate to return to the general results of Sections II and III to develop the equations describing incoherent response. For this much weaker kind of phenomenon consideration is restricted to the second-harmonic process known as either elastic second-harmonic light scattering (ESHLS) or hyper-Rayleigh scattering, and once again this is an area in which Kielich played a pioneering role in developing the theory.[78] In this process, two photons of laser light with frequency ω (wave vector $\mathbf{k}$ and polarization vector $\mathbf{e}^{(\lambda)}$) are converted into a single photon of frequency 2ω (wave vector $\mathbf{k}'$ and polarization vector $\mathbf{e}'^{(\lambda')}$), but without imposing the wave-vector matching condition discussed in Section III.C. As will be shown below this leads to one of the key features of the incoherent process, which is harmonic emission over 4π steradians, though not with an isotropic intensity distribution. This contrasts markedly with the laser-like emission associated with coherent response.

The basic rate equation for incoherent nonlinear response in a molecular fluid is given by Eq. (81). From this result, the explicit equation for the rate of second-harmonic scattering in a fluid is

$$\Gamma_{inc} = N\left(\frac{2\pi\rho_f}{\hbar\varepsilon_0^6}\right)\left\langle|\chi(-2\omega,\omega,\omega)\odot^3\sigma\{(\mathbf{k}',\lambda'),(\mathbf{k},\lambda)\}|^2\right\rangle \quad (140)$$

and as discussed previously there is no coherent signal. Here $\chi(-2\omega,\omega,\omega)$ is the same hyperpolarizability tensor as appears in the theory of SHG, with components as defined by Eq. (46) and (56), which is nonzero only for noncentrosymmetric molecules. The tensor is symmetric in two indices and, under the conditions in which ESHLS is usually studied, can be treated as a real quantity (see Section II.D). It should not in general be regarded as fully index-symmetric (see Section II.F).

Once again the result corresponding to Eq. (140) for a molecular crystal, which follows from (75), differs only in that the number of unit cells M appears in place of N, the unit cell susceptibility $\hat{\chi}$ instead of χ, and the angular brackets denoting orientational averaging disappear. In this case it can be expected that emission in the forward direction will be dominated by the much stronger coherent signal, if the crystal symmetry permits. Under such circumstances the incoherent signal will nonetheless be detectable in other directions.

Using the methods of Section IV.A, the above result for the rate of scattering can be transformed into an expression for the second-harmonic intensity:

$$l_{ESHLS} = D\left\langle|\chi(-2\omega,\omega,\omega)\odot^3\mathbf{s}\{(\mathbf{k}',\lambda'),(\mathbf{k},\lambda)\}|^2\right\rangle \quad (141)$$

where the beam parameter D is defined by

$$D = N\left(\frac{c''k'^4}{8\pi^2\varepsilon_0}\right)\left(\frac{I_\omega}{2c'\varepsilon_0}\right)^2 g_\omega^{(2)} \quad (142)$$

and the polarization tensor $\mathbf{s} = \bar{\mathbf{e}}'\mathbf{ee}$. Equation (141) differs crucially from the corresponding result for coherent response (see Eq. (117) for the general case) in the positioning of the angular brackets denoting the rotational average required for application to fluid media. For incoherent processes such as the one now under consideration, the rotational average is performed on the modulus square of the inner product of the nonlinear susceptibility and polarization tensors; the converse holds for coherent processes.

The rotational average required in (141) is thus

$$
\begin{aligned}
\langle A \rangle &\equiv \left\langle \chi(-2\omega,\omega,\omega)\bar{\chi}(-2\omega,\omega,\omega)\odot^6\bar{\mathbf{e}}'\mathbf{eee}'\overline{\mathbf{ee}} \right\rangle \\
&= \left\langle \chi_{i(jk)}(-2\omega,\omega,\omega)\bar{e}'_i e_j e_k \bar{\chi}_{l(mn)}(-2\omega,\omega,\omega)e'_l \bar{e}_m \bar{e}_n \right\rangle
\end{aligned}
\tag{143}
$$

The sixth-rank rotational average in Eq. (143) comprises products of Kronecker delta triplets,[28a, 79] each of which is referred to either the laboratory or the molecular frame. The laboratory frame deltas in particular contract with the polarization vector components to give scalar products among $\mathbf{e}$, $\bar{\mathbf{e}}$, $\mathbf{e}'$, and $\bar{\mathbf{e}}'$. The full result for the ESHLS intensity is as follows:

$$
\begin{aligned}
I_{\mathrm{ESHLS}} = \left(\frac{D}{105}\right) &\bigl\{ [30(\bar{\mathbf{e}}'\cdot\mathbf{e})(\mathbf{e}\cdot\mathbf{e}')(\bar{\mathbf{e}}\cdot\bar{\mathbf{e}}) - 12(\bar{\mathbf{e}}'\cdot\mathbf{e})(\mathbf{e}'\cdot\bar{\mathbf{e}}) \\
&\quad - 12(\mathbf{e}\cdot\mathbf{e}')(\bar{\mathbf{e}}'\cdot\bar{\mathbf{e}}) - 10(\mathbf{e}\cdot\mathbf{e})(\bar{\mathbf{e}}\cdot\bar{\mathbf{e}}) + 8]\chi_{\lambda(\lambda\mu)}\chi_{\mu(\nu\nu)} \\
&\quad + [-12(\bar{\mathbf{e}}'\cdot\mathbf{e})(\mathbf{e}\cdot\mathbf{e}')(\bar{\mathbf{e}}\cdot\bar{\mathbf{e}}) + 16(\bar{\mathbf{e}}'\cdot\mathbf{e})(\mathbf{e}'\cdot\bar{\mathbf{e}}) \\
&\quad + 2(\mathbf{e}\cdot\mathbf{e}')(\bar{\mathbf{e}}'\cdot\bar{\mathbf{e}}) + 4(\mathbf{e}\cdot\mathbf{e})(\bar{\mathbf{e}}\cdot\bar{\mathbf{e}}) - 6]\chi_{\lambda(\lambda\mu)}\chi_{\nu(\mu\nu)} \\
&\quad + [-10(\bar{\mathbf{e}}'\cdot\mathbf{e})(\mathbf{e}\cdot\mathbf{e}')(\bar{\mathbf{e}}\cdot\bar{\mathbf{e}}) + 4(\bar{\mathbf{e}}'\cdot\mathbf{e})(\mathbf{e}'\cdot\bar{\mathbf{e}}) \\
&\quad + 4(\mathbf{e}\cdot\mathbf{e}')(\bar{\mathbf{e}}'\cdot\bar{\mathbf{e}}) + 8(\mathbf{e}\cdot\mathbf{e})(\bar{\mathbf{e}}\cdot\bar{\mathbf{e}}) - 5]\chi_{\lambda(\mu\mu)}\chi_{\lambda(\nu\nu)} \\
&\quad + [8(\bar{\mathbf{e}}'\cdot\mathbf{e})(\mathbf{e}\cdot\mathbf{e}')(\bar{\mathbf{e}}\cdot\bar{\mathbf{e}}) - 6(\bar{\mathbf{e}}'\cdot\mathbf{e})(\mathbf{e}'\cdot\bar{\mathbf{e}}) \\
&\quad - 6(\mathbf{e}\cdot\mathbf{e}')(\bar{\mathbf{e}}'\cdot\bar{\mathbf{e}}) - 5(\mathbf{e}\cdot\mathbf{e})(\bar{\mathbf{e}}\cdot\bar{\mathbf{e}}) + 11]\chi_{\lambda(\mu\nu)}\chi_{\lambda(\mu\nu)} \\
&\quad + [-12(\bar{\mathbf{e}}'\cdot\mathbf{e})(\mathbf{e}\cdot\mathbf{e}')(\bar{\mathbf{e}}\cdot\bar{\mathbf{e}}) + 2(\bar{\mathbf{e}}'\cdot\mathbf{e})(\mathbf{e}'\cdot\bar{\mathbf{e}}) \\
&\quad + 16(\mathbf{e}\cdot\mathbf{e}')(\bar{\mathbf{e}}'\cdot\bar{\mathbf{e}}) + 4(\mathbf{e}\cdot\mathbf{e})(\bar{\mathbf{e}}\cdot\bar{\mathbf{e}}) - 6]\chi_{\lambda(\mu\nu)}\chi_{\mu(\lambda\nu)} \bigr\}
\end{aligned}
\tag{144}
$$

where use has been made of the symmetry in the last two indices of the nonlinear susceptibility tensor, whose components are for simplicity assumed to be real. The five independent quadratic parameters of the form $\chi \cdots \chi \cdots$ appearing in Eq. (144) were first identified and discussed by Bersohn et al.,[80] abolishing the assumption of Kleinman symmetry inherent in other early treatments.

The polarization vector scalar products in (144) are generally nonzero, except that $(\mathbf{e}\cdot\mathbf{e})(\bar{\mathbf{e}}\cdot\bar{\mathbf{e}})$ is zero when the pump is circularly polarized, and the result faithfully represents the detailed polarization- and angle-dependence of incoherent second-harmonic emission.[81-83] Since emission can occur in a nonforward direction, it is associated with what would classically be regarded as a longitudinal polarization of the medium at the harmonic frequency.[93] As noted earlier, arguments about the possible role of higher multipoles in harmonic emission from fluids have tended to obscure the fact that such longitudinal polarizations arise even in the electric dipole approximation, and are invariably associated with incoherent emission.

When plane polarized radiation is employed, one of the most salient features of the harmonic scattering is its degree of polarization, expressed in terms of a depolarization ratio conventionally defined by

$$\rho_\perp = l(\perp \to \|)/l(\perp \to \perp) \tag{145}$$

where $\|$ and $\perp$ denote polarizations lying in, and perpendicular to, the scattering plane which contains the vectors $\mathbf{k}, \mathbf{k}'$. The result is

$$\rho_\perp = \frac{3\chi_{\lambda(\mu\mu)}\chi_{\lambda(\nu\nu)} + 6\chi_{\lambda(\mu\nu)}\chi_{\lambda(\mu\nu)} - 2\chi_{\lambda(\mu\nu)}\chi_{\mu(\lambda\nu)} - 2\chi_{\lambda(\lambda\mu)}\chi_{\mu(\nu\nu)} - 2\chi_{\lambda(\lambda\mu)}\chi_{\nu(\mu\nu)}}{\chi_{\lambda(\mu\mu)}\chi_{\lambda(\nu\nu)} + 2\chi_{\lambda(\mu\nu)}\chi_{\lambda(\mu\nu)} + 4\chi_{\lambda(\mu\nu)}\chi_{\mu(\lambda\nu)} + 4\chi_{\lambda(\lambda\mu)}\chi_{\mu(\nu\nu)} + 4\chi_{\lambda(\lambda\mu)}\chi_{\nu(\mu\nu)}}$$

$$\tag{146}$$

or, in special cases where the Kleinman approximation holds and full index symmetry obtains in the molecular tensor,

$$\rho_\perp = \frac{4\chi_{(\lambda\mu\nu)}\chi_{(\lambda\mu\nu)} - \chi_{(\lambda\lambda\mu)}\chi_{(\nu\nu\mu)}}{6\chi_{(\lambda\mu\nu)}\chi_{(\lambda\mu\nu)} + 9\chi_{(\lambda\lambda\mu)}\chi_{(\nu\nu\mu)}} \tag{147}$$

Relations equivalent to (146) and (147), but cast in terms of irreducible tensor invariants, have been variously obtained by Maker,[84] Andrews and Thirunamachandran,[83] and Kielich and Bancewicz.[85] Such expressions are directly amenable to the incorporation of rotational line structure (see also Bancewicz et al.[86]). Moreover they greatly facilitate the derivation of the selection rules for inelastic second harmonic (hyper-Raman) scattering.

One curious feature of ESHLS concerns the polarization of the harmonic when the pump radiation is unpolarized, u (or radially polarized). As with Rayleigh scattering, the process itself produces a degree of polarization in the scattered light. It may be shown from Eq. (144) that the harmonic intensity is an unusual linear combination of the intensities produced by the two orthogonal pump polarizations $\|$, $\perp$, and by circular pump polarizations of opposite helicity L, R. For the harmonic component of polarization μ,

$$l(u \to \mu) = \tfrac{1}{8}[2l(\| \to \mu) + 2l(\perp \to \mu) + l(L \to \mu) + l(R \to \mu)]$$

$$(\omega' = 2\omega) \tag{148}$$

which holds for any scattering angle.[28b] The result should be compared to the following expressions which are well known to hold for Rayleigh scattering:

$$l(u \to \mu) = \tfrac{1}{2}[l(\| \to \mu) + l(\perp \to \mu)] = \tfrac{1}{2}[l(L \to \mu) + l(R \to \mu)]$$
$$(\omega' = \omega) \quad (149)$$

The interesting fact is that the relations (149) do not hold for ESHLS. The reason is that here the possibility of the two incident photons having different polarizations has to be entertained. Consequently, the relationship

$$\rho_u\left(\frac{\pi}{2}\right) = \frac{2\rho_\perp}{1 + \rho_\perp} \quad (\omega' = \omega) \quad (150)$$

for right-angled Rayleigh scattering also has no counterpart for ESHLS. In fact, the result for ρ_u is as follows:

$$\rho_u = \frac{\chi_{\lambda(\mu\mu)}\chi_{\lambda(\nu\nu)} + 23\chi_{\lambda(\mu\nu)}\chi_{\lambda(\mu\nu)} - 10\chi_{\lambda(\mu\nu)}\chi_{\mu(\lambda\nu)} + 4\chi_{\lambda(\lambda\mu)}\chi_{\mu(\nu\nu)} - 10\chi_{\lambda(\lambda\mu)}\chi_{\nu(\mu\nu)}}{3\chi_{\lambda(\mu\mu)}\chi_{\lambda(\nu\nu)} + 13\chi_{\lambda(\mu\nu)}\chi_{\lambda(\mu\nu)} + 5\chi_{\lambda(\mu\nu)}\chi_{\mu(\lambda\nu)} - 2\chi_{\lambda(\lambda\mu)}\chi_{\mu(\nu\nu)} + 5\chi_{\lambda(\lambda\mu)}\chi_{\nu(\mu\nu)}}$$
$$(151)$$

or, again where Kleinman index symmetry holds in the hyperpolarizability tensor,

$$\rho_u = \frac{13\chi_{(\lambda\mu\nu)}\chi_{(\lambda\mu\nu)} - 5\chi_{(\lambda\lambda\mu)}\chi_{(\nu\nu\mu)}}{18\chi_{(\lambda\mu\nu)}\chi_{(\lambda\mu\nu)} + 6\chi_{(\lambda\lambda\mu)}\chi_{(\nu\nu\mu)}} \quad (152)$$

A more extensive treatment in terms of Stokes parameters has been given by Kozierowski and Kielich.[87] This incorporates not only the case of unpolarized but also circularly polarized light, and gives results for the depolarization ratio, reversal ratio, and degree of circularity of the harmonic.

It should be noted that ESHLS can occur even in fluid media comprising centrosymmetric molecules, through the involvement of fourth-rank molecular response tensors. One mechanism involves the application of a static electric field and entails the EFISHG tensor $\check{\chi} \equiv \chi(-2\omega, 0, \omega, \omega)$ introduced in Section IV.D. Even in the absence of any such applied field, the involvement of electric quadrupolar interactions as in the tensor $\tilde{\chi}(-2\omega, \omega, \omega)$ of Section IV.B can produce a signal. The calculation in this case calls for the evaluation of an eighth-rank tensor average; details are given by Kielich et al.[88] The generation of local anisotropy through molecular orientational correlation can also play a role.[82, 89] Although the theory underlying the operation of this mechanism has largely been devised from a molecular statistical point of view, the effect can also be understood on the basis of interactions between neighboring molecules effectively reducing their inversion symmetry, leading to finite values for their hyperpolarizabilities.

VI. CONCLUSION

The molecular theory of harmonic generation embraces an enormously wide range of disciplines, from chemistry to optics, and as such it has had a significant input from researchers with a diversity of backgrounds. This review has attempted a brief systematization of this body of knowledge, and a feature that will have struck the reader is the frequency with which necessary reference has been made to the work of the Poznań group. It is hoped that this will contribute to a wider appreciation of their incisive contributions to the subject as a personal tribute to Stanisław Kielich.

References

1. P. A. Franken, A. E. Hill, C. W. Peters, and G. Weinreich, *Phys. Rev. Lett.* **7**, 118 (1961).

2. D. S. Chemla and J. Zyss, *Nonlinear Optical Properties of Organic Molecules and Crystals*, Vols. 1 and 2, Academic, Orlando, FL, 1987.

3. P. N. Prasad and D. J. Williams, *Introduction to Nonlinear Optical Effects in Molecules and Polymers*, Wiley, New York, 1991.

4. M. N. R. Ashfold and J. D. Prince, *Contemp. Phys.* **29**, 125 (1988).

5. M. N. R. Ashfold and J. D. Prince, *Mol. Phys.* **73**, 297 (1991).

6. S. J. Cyvin, J. E. Rauch, and J. C. Decius, *J. Chem. Phys.* **43**, 4085 (1965).

7. N. Bloembergen, *Nonlinear Optics*, Benjamin, New York, 1965.

8. G. E. Stedman, *Adv. Phys.* **34**, 513 (1985).

9. L. D. Landau and E. M. Lifshitz, *Electrodynamics in Continuous Media*, Pergamon, New York, 1960, p. 252.

10. J. D. Jackson, *Classical Electrodynamics*, Wiley, New York, 1975, p. 117.

11. D. P. Craig and T. Thirunamachandran, *Molecular Quantum Electrodynamics*, Academic, London, 1984.

12. R. Wallace, *Mol. Phys.* **11**, 457 (1966).

13. D. C. Hanna, M. A. Yuratich, and D. Cotter, *Nonlinear Optics of Free Atoms and Molecules*, Springer, Berlin, 1979.

14. R. W. Boyd, *Nonlinear Optics*, Academic, Boston, 1992.

15. N. P. Blake, *J. Chem. Phys.* **93**, 6165 (1990).

16. P. N. Butcher, R. Loudon, and T. P. McLean, *Proc. Phys. Soc.* **85**, 565 (1965).
17. K.-E. Peiponen, E. M. Vartiainen, and T. Asakura, *J. Phys.: Cond. Matt.* **4**, 299 (1992).
18. D. A. Kleinman, *Phys. Rev.* **126**, 1977 (1962).
19. G. Wagnière, *Appl. Phys. B* **41**, 169 (1986).
20. C. Flytzanis, in L. Neel (Ed.) *Nonlinear Behaviour of Molecules, Atoms and Ions in Electric, Magnetic or Electromagnetic Fields*, Elsevier, Amsterdam, 1979, pp. 185–206.
21. D. S. Chemla, J. L. Oudar, and J. Jerphagnon, *Phys. Rev.* **B12**, 4534 (1975).
22. B. F. Levine and C. G. Bethea, *J. Chem. Phys.* **63**, 2666 (1975).
23. T. F. Heinz and D. P. DiVincenzo, *Phys. Rev.* **A42**, 6249 (1990).
24. J. Zyss and J. L. Oudar, *Phys. Rev.* **A26**, 2028 (1982).
25. G. R. Meredith, *Proc. SPIE* **824**, 126 (1987).
26. J. Knoester and S. Mukamel, *J. Chem. Phys.* **91**, 989 (1989).
27. S. Kielich, in E. Wolf (Ed.), *Progress in Optics*, Vol. 20, North-Holland, New York, 1983, pp. 157–261.
28. D. L. Andrews and T. Thirunamachandran, (a) *J. Chem. Phys.* **67**, 5026 (1977); (b) *Opt. Commun.* **22**, 312 (1977).
29. D. L. Andrews and W. A. Ghoul, *J. Phys. A* **14**, 1281 (1981).
30. D. L. Andrews and N. P. Blake, *J. Phys. A* **22**, 49 (1989).
31. D. L. Andrews and M. J. Harlow, *Phys. Rev.* **A29**, 2796 (1984).
32. D. Marcuse, *Principles of Quantum Electronics*, Academic, New York, 1980.
33. R. Loudon, *The Quantum Theory of Light*, 2d ed., Clarendon, Oxford, 1983.
34. J. Wilson and J. F. B. Hawkes, *Optoelectronics: An Introduction*, Prentice-Hall, London, 1983, pp. 87–91.
35. K. M. White and C. J. Eckhardt, *Phys. Rev.* **A36**, 3885 (1987).
36. G. R. Meredith, B. Buchalter, and C. Hanzlik, *J. Chem. Phys.* **78**, 1533 (1983).
37. J. Knoester and S. Mukamel, *Phys. Rev.* **A39**, 1899 (1989).
38. J. Yu and W. P. Su, *Phys. Rev.* **B44**, 13315 (1991).
39. F. C. Spano and S. Mukamel, *Phys. Rev.* **A40**, 5783 (1989).
40. E. A. Power, *Introductory Quantum Electrodynamics*, Longmans, London, 1964, p. 15.
41. D. L. Andrews, *J. Phys. B* **13**, 4091 (1980).
42. W. H. Louisell, *Quantum Statistical Properties of Radiation*, Wiley, New York, 1973, pp. 104–109.
43. M. Kozierowski and R. Tanaś, *Opt. Commun.* **21**, 229 (1977).
44. L. Mandel, *Opt. Commun.* **42**, 437 (1982).
45. S. Kielich, R. Tanaś, and R. Zawodny, *J. Mod. Opt.* **34**, 979 (1987).
46. S. Kielich, R. Tanaś, and R. Zawodny, *Appl. Phys. B* **45**, 249 (1988).
47 S. Kielich, R. Tanaś, and R. Zawodny, *J. Opt. Soc. Am. B* **4**, 1627 (1987).
48. Ts. Gantsog, R. Tanaś, and R. Zawodny, *Phys. Lett.* **A155**, 1 (1991).
49. T. Mossberg, A. Flusberg, and S. R. Hartmann, *Opt. Commun.* **25**, 121 (1978).
50. K. Miyazaki, T. Sato, and H. Kashiwagi, *Phys. Rev.* **A23**, 1358 (1981).
51. D. L. Andrews, and N. P. Blake, *Phys. Rev.* **A38**, 3113 (1988).
52. G. E. Stedman, *Adv. Chem. Phys.*, in press.

53. T. F. Heinz, H. W. K. Tom, and Y. R. Shen, *Phys. Rev.* **A28**, 1883 (1983).

54. N. Bloembergen, in Société Française de Physique (Ed.), *Polarisation Matière et Rayonnement*, Universitaires de France, Paris, 1969, p. 109.

55. C. L. Tang and H. Rabin *Phys. Rev.* **B3**, 4025 (1971).

56. Y. R. Shen, *Nature* **337**, 519 (1989).

57. V. Mizrahi and J. E. Sipe, *J. Opt. Soc. Am. B* **5**, 660 (1988).

58. N. Bloembergen and P. S. Pershan, *Phys. Rev.* **128**, 606 (1962).

59. Y. R. Shen, *The Principles of Nonlinear Optics*, Wiley, New York, 1984, p. 76.

60. F. A. Hopf, and G. I. Stegeman, *Applied Classical Electrodynamics*, Vol. 1, Wiley, New York, 1985, pp. 172–174.

61. E. S. Peterson and C. B. Harris, *J. Chem. Phys.* **91**, 2683 (1989).

62. R. M. Corn, *Anal. Chem.* **63**, 285A (1991).

63. C. K. Chen, T. F. Heinz, D. Ricard, and Y. R. Shen, *Phys. Rev.* **B27**, 1965 (1983).

64. O. A. Aktsipetrov, A. A. Nikulin, V. I. Panov, S. I. Vasil'ev, and A. V. Petukhov, *Solid State Commun.* **76**, 55 (1990).

65. G. Li and S. R. Seshadri, *Phys. Rev.* **B44**, 1240 (1991).

66. G. Mayer, *C. R. Acad. Sci.* **B267**, 54 (1968).

67. S. Kielich, *Acta Phys. Pol.* **36**, 621 (1969).

68. S. Kielich, in C. T. O'Konski (Ed.), *Molecular Electro-Optics*, Dekker, New York, 1976, pp. 391–444.

69. S. Kielich, in L. Neel (Ed.), *Nonlinear Behaviour of Molecules, Atoms and Ions in Electric, Magnetic or Electromagnetic Fields*, Elsevier, Amsterdam, 1979, pp. 111–124.

70. Y. T. Lam and T. Thirunamachandran, *J. Chem. Phys.* **77**, 3810 (1982).

71. D. L. Andrews and B. S. Sherborne, *J. Phys. B* **19**, 4265 (1986).

72. D. P. Shelton and A. D. Buckingham, *Phys. Rev.* **A26**, 2787 (1982).

73. D. L. Andrews, *Proc. Tech. Conf. Electro-Optics/Laser Int. '84 UK*, Cahners, Twickenham, 1984, pp. 435–439.

74. F. Kajzar, I. Ledoux, and J. Zyss, *Phys. Rev.* **A36**, 2210 (1987).

75. T. L. Mazely and W. M. Hetherington III, (a) *J. Chem. Phys.* **86**, 3640 (1987); (b) *J. Chem. Phys.* **87**, 1962 (1987).

76. F. Aguillon, A. Lebéhot, J. Rousseau, and R. Campargue, *J. Chem. Phys.* **86**, 5246 (1987).

77. J. F. Reintjes, *Nonlinear Optical Parametric Processes in Liquids and Gases*, Academic, Orlando, FL, 1984.

78. S. Kielich, *Acta Phys. Pol.* **33**, 141 (1968).

79. S. Kielich, *Acta. Phys. Pol.* **20**, 433 (1961).

80. R. Bersohn, Y.-H. Pao, and H. L. Frisch, *J. Chem. Phys.* **45**, 3184 (1966).

81. V. L. Strizhevskii and V. M. Klimenko, *Sov. Phys. JETP* **26**, 163 (1968).

82. S. Kielich and M. Kozierowski, *Acta Phys. Pol.* **A45**, 231 (1974).

83. D. L. Andrews and T. Thirunamachandran, *J. Chem. Phys.* **68**, 2941 (1978).

84. P. D. Maker, *Phys. Rev.* **A1**, 923 (1970).

85. S. Kielich and T. Bancewicz, *J. Raman Spectrosc.* **21**, 791 (1990).

606　　　　　DAVID L. ANDREWS

86. T. Bancewicz, Z. Ożgo, and S. Kielich, *Phys. Lett.* **44A**, 407 (1973).
87. M. Kozierowski and S. Kielich, *Acta Phys. Pol.* **A66**, 753 (1984).
88. S. Kielich, M. Kozierowski, Z. Ożgo, and R. Zawodny, *Acta Phys. Pol.* **A45**, 9 (1974).
89. S. Kielich, J. R. Lalanne, and F. B. Martin, *Phys. Rev. Lett.* **26**, 1295 (1971).
90. M. S. Malcuit, R. W. Boyd, W. V. Davis and K. Rzązewski, *Phys. Rev.* **A41**, 3822 (1990).
91. L. Marmet, K. Hakuta and B. P. Stoicheff, *Opt. Lett.* **16**, 261 (1991).
92. Y. Liang, J. M. Watson and S. L. Chin, *J. Phys. B* **25**, 2725 (1992).
93. D. L. Andrews, *J. Mod. Opt.*, in press.

THE INTERACTION OF SQUEEZED LIGHT WITH ATOMS

A. S. PARKINS

Joint Institute for Laboratory Astrophysics, and Department of Physics, University of Colorado, Boulder, Colorado

CONTENTS

Modern Nonlinear Optics, Part 2, Edited by Myron Evans and Stanisław Kielich. Advances in Chemical Physics Series, Vol. LXXXV.
ISBN 0-471-57546-1   © 1993 John Wiley & Sons, Inc.

I. INTRODUCTION

The electric field for a given mode of the electromagnetic field can be decomposed into two quadrature components with time dependence $\cos(\omega t)$ and $\sin(\omega t)$, respectively. The amplitudes of the two quadratures of the field are noncommuting Hermitian operators and thus obey an uncertainty principle. When the field is in the vacuum state or a coherent state, the uncertainty product for the variances of the quadrature amplitudes is a minimum, with the uncertainties in the two quadratures being equal. When the field is in a squeezed state, the variance of one quadrature is less than that of the vacuum, while the variance of the other quadrature is increased in accordance with the uncertainty principle.

Following a considerable amount of theoretical research, squeezed states of the electromagnetic field were first realized experimentally in 1985 via four-wave mixing in an atomic beam.[1] In the ensuing years, schemes for the generation of squeezed light have been refined, with a noteworthy recent development being the availability of a tunable source of squeezed light exhibiting a noise reduction 5.0 dB ($\sim 70\%$) below the vacuum level.[2]

Such advances in squeezed light generation have encouraged theoretical research into possible applications of squeezed light in optical systems. The field of atomic spectroscopy arises as an obvious application, since the radiative properties of atoms are intimately related to fluctuations in the electromagnetic field. That squeezed light can indeed alter the fundamental radiative properties of an atom was first pointed out by Gardiner,[3] whose paper on spontaneous emission of a two-level atom in a squeezed vacuum can be regarded as the seminal work in this field. The dramatic prediction put forward in this work was that squeezed light can, in principle, induce a reduction in the linewidth of the spectrum of spontaneous emission. Following on from this work, a variety of classic and standard problems in quantum optics have been re-examined with squeezed light included in the formulation. In this way, a significant body of work on the interaction of squeezed light with atoms has accumulated. The purpose of the present article is to review the major results of this work.

Some description of the computational tools required for producing these results is appropriate, and so we begin in Section II with an overview of the formalism used for describing the interaction of squeezed light with atoms. The vast majority of analyses have made use of a master equation formulation, which requires that the squeezed light be broadband. In practice, this may not necessarily be so, and thus, in addition to presenting the master equation, we also describe a more general approach based on the so-called adjoint equation, which has been employed to study the interaction of atoms with finite-bandwidth squeezed light.

In Section III, we begin the review of squeezed-light-induced phenomena with a consideration of the most fundamental of all quantum optics problems. In particular, we consider the effect of squeezed light on a single atom in the context of spontaneous emission, resonance fluorescence, and probe field absorption. These are also the problems for which the effects of finite-bandwidth squeezing have been studied, and so this aspect is also reviewed in this section, with some mention given to the methods employed for the finite-bandwidth calculations, since these methods are, we believe, of considerable interest in their own right. A brief section on laser cooling in a squeezed vacuum is also included in Section III. Section IV describes some novel phenomena that arise with three-level atoms and with collections of atoms, while Section V is dedicated to modifications of laser behavior caused by squeezed light.

It is important to emphasize that this field of research is currently only theoretical, and experiments have yet to be attempted. In the recent experiment of Polzik et al.,[2] squeezed light was used to enhance the sensitivity of spectroscopic measurements of atomic cesium, but the squeezed light was in the form of a plane wave and atomic radiative processes were not fundamentally altered. The major caveat to the majority of the predictions made regarding the alteration of radiative properties is the requirement that the atoms couple exclusively to squeezed modes of the radiation field (hence, a solitary plane wave has a negligible effect on radiative processes); this is perhaps the primary obstacle facing experimentalists at present. However, this obstacle may in principle be avoided by making use of optical cavities, and, indeed, it is most likely that the first experiments will be carried out using such configurations. Hence, we conclude in Section VI with a review of the possibilities offered in the field of cavity quantum electrodynamics.

II. FORMALISM FOR TREATING SQUEEZED LIGHT INPUTS

A. Squeezed-Light: Degenerate Parametric Amplification

The degenerate parametric amplifier is the prototype for all squeezing devices and has, to date, been the most successful source of squeezed light.[2, 4] It is the most likely source of squeezed light for experiments with atoms. Theoretical analyses of this device are numerous, so a detailed exposition of its properties is not necessary. However, in view of its fundamental importance to this particular review, a brief description is appropriate.

The systematic Hamiltonian for the process of degenerate parametric amplification, with the pump field treated classically, can be written as

$$H_{\text{sys}} = \hbar\omega_0 a^\dagger a + \tfrac{1}{2} i\hbar \left[\varepsilon_D \, e^{-i\omega_p t} (a^\dagger)^2 - \varepsilon_D^* \, e^{i\omega_p t} a^2 \right] \qquad (1)$$

where a is the annihilation operator for the cavity mode of frequency ω_0, ω_p is the frequency of the pump beam, and ε_D is the pump field driving strength. It is assumed that the cavity and pump are tuned so that $\omega_p = 2\omega_0$. Downconversion from frequency $2\omega_0$ to frequency ω_0 occurs through the action of a nonlinear optical medium placed inside the cavity and produces highly correlated pairs of photons. The cavity itself is considered to be a one-sided cavity; that is, it possesses one near-perfect mirror and one output mirror of finite transmissivity (at the frequency ω_0). This configuration yields the optimum amount of squeezing in the output field (i.e., in the field emerging from the cavity), the properties of which have been computed by Collett and Gardiner[5] using an input–output formulation (see also Ref. 6). The correlation functions of the output field operators take the form (in a frame rotating at frequency ω_0)

$$\left\langle a_{\text{out}}^\dagger(t + \tau) a_{\text{out}}(t) \right\rangle = \frac{b_+^2 - b_-^2}{4} \left(\frac{e^{-b_-|\tau|}}{2b_-} - \frac{e^{-b_+|\tau|}}{2b_+} \right) \qquad (2)$$

$$\left\langle a_{\text{out}}(t + \tau) a_{\text{out}}(t) \right\rangle = \frac{b_+^2 - b_-^2}{4} \left(\frac{e^{-b_-|\tau|}}{2b_-} + \frac{e^{-b_+|\tau|}}{2b_+} \right) \qquad (3)$$

where

$$b_+ = \tfrac{1}{2}\kappa_D + |\varepsilon_D| \qquad b_- = \tfrac{1}{2}\kappa_D - |\varepsilon_D| \qquad (b_\pm > 0) \qquad (4)$$

with κ_D the loss rate through the output mirror. The effect of squeezing is best seen through the quadrature phase operators $E_1(t) = -i[a_{out}(t) - a_{out}^\dagger(t)]$ and $E_2(t) = a_{out}(t) + a_{out}^\dagger(t)$, which exhibit the correlation functions

$$\langle E_1(t+\tau)E_1(t) \rangle = -\frac{b_+^2 - b_-^2}{2b_+}\, e^{-b_+|\tau|} + \delta(\tau) \qquad (5)$$

$$\langle E_2(t+\tau)E_2(t) \rangle = \frac{b_+^2 - b_-^2}{2b_-}\, e^{-b_-|\tau|} + \delta(\tau) \qquad (6)$$

The exponential terms in these expressions give the effect of squeezing, while the δ-correlated terms represent vacuum fluctuations. Strong squeezing (or noise reduction below the vacuum level) occurs in the quadrature E_1, over a bandwidth b_+, as one approaches the threshold for parametric oscillation, i.e., as $|\varepsilon_D| \to \tfrac{1}{2}\kappa$. Fluctuations in the unsqueezed quadrature E_2 naturally become very large in the same limit. It is worth noting that the general forms of these correlation functions have been confirmed in the experiments of Wu et al.[4]

Two distinct timescales characterize the correlation functions, corresponding to the two decay rates b_+ and b_-. In the case that both of these decay rates are very large (i.e., the bandwidth of squeezing is large) the exponentials may be approximated by δ functions to give the "squeezed white noise limit":

$$\langle E_1(t+\tau)E_1(t) \rangle = (2N - 2M + 1)\delta(\tau) \qquad (7)$$

$$\langle E_2(t+\tau)E_2(t) \rangle = (2N + 2M + 1)\delta(\tau) \qquad (8)$$

where we have introduced the squeezing parameters N and M defined by

$$N = \frac{b_+^2 - b_-^2}{4}\left(\frac{1}{b_-^2} - \frac{1}{b_+^2}\right) \qquad M = \frac{b_+^2 - b_-^2}{4}\left(\frac{1}{b_-^2} + \frac{1}{b_+^2}\right) \qquad (9)$$

The parameters N and M have become standard in work on the interaction of squeezed light with atoms, and will be used frequently in this article. With these particular definitions, N and M satisfy $M^2 = N(N + 1)$, corresponding to a minimum uncertainty state, or ideal squeezing. Strong squeezing corresponds to the high intensity limit $N \to \infty$, in which case

$$2N - 2M + 1 \to \frac{1}{4N} \tag{10}$$

and fluctuations in the quadrature phase $E_1(t)$ can become arbitrarily small. Once again, we note that the minimum uncertainty state description of squeezing produced in degenerate parametric amplification is supported by the experiments of Wu et al.

Finally, it is important to note that while the squeezed white noise limit corresponds to a flat spectrum, there is still an important frequency dependence implicit in the model. Squeezing results from correlations between modes that are detuned equally above and below a reference, or carrier, frequency (i.e., ω_0). Thus, the detuning of broadband squeezed light from an atomic resonance can still play an important role.

B. Master Equation for a System Driven by Broadband Squeezed Light

The master equation occupies a central role in theoretical quantum optics. It is an equation describing the evolution of the reduced density operator $\rho(t)$ of a "small" system, such as an atom, that is interacting with a much larger system referred to as the reservoir or bath. In quantum optics this reservoir is most commonly the electromagnetic field. Given a Hamiltonian of the form

$$H = H_{sys} + H_{res} + H_{int} \tag{11}$$

where the three terms represent the system, reservoir, and interaction, respectively, the density operator ρ_{tot} for the system plus the reservoir obeys the Schrödinger equation

$$\dot{\rho}_{tot} = \frac{1}{i\hbar}[H, \rho_{tot}] \tag{12}$$

The reduced density operator is defined by $\rho = \mathrm{Tr}_R(\rho_{tot})$; i.e., it is the density operator obtained by tracing over the bath variables. The derivation of the quantum optical master equation and the approximations involved are well known, and details can be found, for example, in the standard text of Louisell[7] (see also Ref. 6, which includes a consideration of squeezing). As an example, we consider a single two-level atom (at rest) coupled to a one-dimensional squeezed electromagnetic field, for which the system, reservoir (or field), and interaction Hamiltonians can be written

$$H_{sys} = \tfrac{1}{2}\hbar\omega_a\sigma_z \qquad H_{res} = \hbar\int_0^\infty d\omega\ \omega a^\dagger(\omega)a(\omega) \tag{13}$$

$$H_{int} = i\int_0^\infty d\omega\kappa(\omega)\left(\frac{\hbar\omega}{2}\right)^{1/2}\left[a(\omega) - a^\dagger(\omega)\right](\sigma^+ + \sigma^-) \tag{14}$$

where σ^+, σ^-, and σ_z are the atomic pseudo-spin operators, $a(\omega)$ are boson annihilation operators for the field, and $\kappa(\omega)$ is the dipole coupling coefficient. Following the standard approaches and approximations, and given that the squeezed white noise limit applies, the master equation can be derived in the form

$$\begin{aligned}
\dot{\rho} = &\frac{1}{i\hbar}\left[H_{sys}, \rho\right] \\
&+ \frac{\gamma_a}{2}(N + 1)(2\sigma^-\rho\sigma^+ - \sigma^+\sigma^-\rho - \rho\sigma^+\sigma^-) \\
&+ \frac{\gamma_a}{2}N(2\sigma^+\rho\sigma^- - \sigma^-\sigma^+\rho - \rho\sigma^-\sigma^+) \\
&- \gamma_a M\, e^{-2i\omega_0 t}\sigma^+\rho\sigma^+ - \gamma_a M^*\, e^{2i\omega_0 t}\sigma^-\rho\sigma^-
\end{aligned} \tag{15}$$

where γ_a is the atomic transition linewidth. An equation of this form, describing a system driven by broadband squeezed light, was first presented by Gardiner and Collett.[8] The final two terms in this equation distinguish it from more traditional master equations. They arise from correlation functions of field operators of the form (3), which characterize a squeezed vacuum, and do not appear, for instance, when the system is driven by a thermal field.

A major assumption required in the derivation of this master equation is that the correlation time of the electromagnetic field is much shorter than the timescale of radiative processes of the atom (Markov approximation). That is, the white-noise description of the squeezed field that was introduced in the previous section is fundamental to the master equation formulation.

A further important assumption, implicit in this model, is that the atom interacts exclusively with squeezed modes of the electromagnetic field. If this one-dimensional model is to be applicable to the actual situation of an atom in free (three-dimensional) space, then this assumption corresponds to an incident squeezed electric dipole wave; i.e., squeezed light must be incident on the atom from a 4π solid angle. This obviously represents a formidable practical problem, but schemes can be devised that circumvent this problem and thus offer a much more optimistic picture. Such schemes are discussed in a later section.

C. A More General Approach: The Adjoint Equation

The quantum optical master equation is a very powerful calculational tool for situations in which it is valid, more particularly, when the incident radiation field is broadband. An alternative but more general approach, which has been developed largely in connection with problems involving squeezed light, is that based on the so-called adjoint equation.[6, 9, 10] This equation arises from a treatment based on quantum Langevin equations, in which straightforward connections can be made between input field, output field, and system variables. In this way, the adjoint equation incorporates an important body of theory that has been developed to describe in a precise manner inputs and outputs in damped quantum systems.[8, 9, 11] This body of theory has been an essential element in the theory of squeezed light generation, in which intracavity fields have to be related to the externally measured fields.[5, 12]

Our review of this approach will concentrate on the example of a single two-level atom coupled to a one-sided, one-dimensional electromagnetic field modeled by a bath of harmonic oscillators. The Hamiltonian for the interacting atom and field can be written

$$H = H_{\text{sys}} + \frac{1}{2} \int_0^\infty d\omega \left\{ \left[p(\omega) - \kappa(\omega)\sigma_x \right]^2 + \omega^2 q(\omega)^2 \right\} \quad (16)$$

where $H_{\text{sys}} = \hbar \omega_a \sigma_z / 2$, $\sigma_x = \sigma^+ + \sigma^-$, and $p(\omega)$ and $q(\omega)$ are related to the field annihilation operator by $a(\omega) = (2\hbar\omega)^{-1/2}[\omega q(\omega) + i p(\omega)]$. The Heisenberg equation of motion for an arbitrary system operator $Y(t)$ is given by

$$\dot{Y} = \frac{i}{\hbar}\left[H_{\text{sys}}, Y \right] + \frac{i}{2\hbar} \int_0^\infty d\omega \left[\left(p(\omega, t) - \kappa(\omega)\sigma_x \right)^2, Y \right] \quad (17)$$

$$= \frac{i}{\hbar}\left[H_{\text{sys}}, Y \right]$$
$$+ \frac{i}{2\hbar} \int_0^\infty d\omega \left[p(\omega, t) - \kappa(\omega)\sigma_x, \left[p(\omega, t) - \kappa(\omega)\sigma_x, Y \right] \right]_+ \quad (18)$$

which can be simplified somewhat by noting that the atom and bath represent different degrees of freedom, and hence the equal-times commutators of atomic operators with bath operators must be zero; i.e., $[p(\omega, t), Y(t)] = 0$.

To derive the quantum Langevin equation one solves formally for the field operators in terms of initial conditions at time t_0 in the remote past (i.e., t_0 is the time at which the interaction is assumed to begin) and substitutes these solutions into the equation of motion for the atomic operator Y. Assuming that $\kappa(\omega)$ is constant over the bandwidth of interest (electric dipole approximation), the quantum Langevin equation can be written

$$\dot{Y} = \frac{i}{\hbar}\left[H_{\text{sys}}, Y\right] + \frac{i}{2}\left[\frac{\gamma_a}{2\omega_a}\dot{\sigma}_x + \sqrt{\frac{\gamma_a}{\hbar\omega_a}}\,E_{\text{in}}(t), [\sigma_x, Y]\right]_+ \quad (19)$$

where

$$E_{\text{in}}(t) = i\int_0^\infty d\omega\sqrt{\frac{\hbar\omega}{2\pi}}\,\{a(\omega, t_0)e^{-i\omega(t-t_0)} - a^\dagger(\omega, t_0)e^{i\omega(t-t_0)}\} \quad (20)$$

The term proportional to $\dot{\sigma}_x$ represents radiation damping. The term $E_{\text{in}}(t)$ depends on the initial quantum state of the bath through the operators $a(\omega, t_0)$ and $a^\dagger(\omega, t_0)$ ($t_0 < t$) and can be specified quite arbitrarily. In particular, there is no restriction on the correlation time of the input field that $E_{\text{in}}(t)$ represents.

For the purpose of this review it is not necessary to present in any detail the theory of inputs and outputs alluded to earlier, but some mention at this stage is appropriate, having introduced the input field $E_{\text{in}}(t)$ above. The primary result of the input–output formalism takes the form of a boundary condition relating input and output fields to each other via the internal dynamics of the system. The output field $E_{\text{out}}(t)$ has the form of Eq. (20) but with the mode operators evaluated at some time in the distant future (i.e., $t_1 > t$). For our particular example, the boundary condition is simply

$$E_{\text{out}}(t) = E_{\text{in}}(t) + \sqrt{\frac{\hbar\gamma_a}{\omega_a}}\,\dot{\sigma}_x(t) \quad (21)$$

That is, the output field consists of a reflection of the input field plus a radiated field, which is determined by the behavior of the atomic operator

$\sigma_x(t)$. Detailed descriptions of the input–output formalism can be found in Refs. 6, 8, and 9.

The quantum Langevin equation (19) is evidently a nonlinear operator equation, which makes it difficult to deal with directly. The adjoint equation, which is a direct consequence of the quantum Langevin equation, is far more amenable to calculation. This equation describes the evolution of a quantum stochastic density operator $\mu(t)$, which is a function of the impressed quantum noise $E_{in}(t)$. The density operator $\mu(t)$ is introduced in the following way. It is assumed that initially the system and bath are independent, so that the density operator factorizes in the Heisenberg picture. Let $Y(t)$ be an arbitrary system operator in the Heisenberg picture, and Y the corresponding Schrödinger picture operator. Then the operator $\mu(t)$ is defined by

$$\text{Tr}_{\text{sys}}\{Y(t)\rho_{\text{sys}}\} = \text{Tr}_{\text{sys}}\{Y\mu(t)\} \tag{22}$$

given that the equality is true for all system operators Y and $Y(t)$. An explicit definition of $\mu(t)$ is given in Ref. 6, together with a derivation of the equation of motion for $\mu(t)$, which can be written in the compact form

$$\dot{\mu}(t) = [A_0 + \beta(t)A_1]\mu(t) \tag{23}$$

where A_0 and A_1 are linear operators in the system space given by

$$A_0\mu(t) = -\frac{i}{\hbar}[H_{\text{sys}}, \mu(t)] + \frac{i}{2}\left[\left[\frac{\gamma_a}{2\omega_a}\dot{\sigma}_x, \mu(t)\right]_+, \sigma_x\right] \tag{24}$$

$$A_1\mu(t) = i[\sigma_x, \mu(t)] \tag{25}$$

with $\dot{\sigma}_x = (i/\hbar)[H_{\text{sys}}, \sigma_x]$, and where we have introduced the operator $\beta(t)$ defined by

$$\beta(t)\mu(t) = -\frac{1}{2}\sqrt{\frac{\gamma_a}{\hbar\omega_a}}[E_{in}(t), \mu(t)]_+ \tag{26}$$

The introduction of the operator $\beta(t)$ marks a crucial step in this approach, since, noting that the commutator $[E_{in}(t), E_{in}(t')]$ is a complex number, it can be shown that

$$\beta(t)\beta(t') = \beta(t')\beta(t) \tag{27}$$

for all t and t'. Hence, we have a commuting form of quantum noise, and

$\beta(t)$ can be treated as a c-number random function. The adjoint equation can in turn be treated as a classical stochastic differential equation, amenable to solution using any of the methods of classical stochastic theory. A similar reduction of the quantum problem to an essentially classical one has been presented by Ritsch and Zoller,[13] based on the existence of a positive P representation[6] for the incident light field. With no conditions having been made on the nature of the input fields, it follows that the adjoint equation approach is not restricted to white-noise approximations to the input fields, and that it provides a starting point for investigations of the effect of finite-bandwidth squeezed noise inputs. Such investigations are the subject of Section III.D.

1.   The Master Equation

In the case that a white-noise approximation to the input field is appropriate, then the familiar master equation can be derived from the adjoint equation using the cumulant expansion technique of van Kampen.[14] This method yields an equation of motion for the density operator $\rho(t) = \mathrm{Tr}_{\mathrm{res}}\{\mu(t)\rho_{\mathrm{res}}\}$ in the form of a perturbation expansion in a smallness parameter proportional to the correlation time of the noise $\beta(t)$. The equation, to second order in this parameter, is

$$\dot{\rho}(t) = A_0\rho(t) + \int_0^\infty d\tau \langle \beta(t)\beta(t-\tau)\rangle A_1 e^{A_0\tau} A_1 e^{-A_0\tau}\rho(t) \quad (28)$$

With various standard approximations (rotating-wave approximation, weak atom–field coupling, neglect of frequency shift terms), this equation can be shown to lead to the master equation (15) when the input is squeezed white noise.

2.   Reflections of the Input Squeezed Field

Once the master equation has been formulated, it is a straightforward matter to compute one-time and two-time averages (using the quantum regression theorem) of the atomic variables, and thus to describe the behavior of the atom. However, a further consideration in the computation of output spectra from systems that are subjected to squeezed light is the effect of reflections of the input squeezed light. One must allow not only for the nonzero power spectrum of the squeezed vacuum, but also for correlations (or interferences) that are established between the reflected squeezed light and the light radiated from the atom. These correlations can substantially affect the total fluorescence spectrum.

This aspect of squeezed light spectroscopy has been dealt with in detail by Gardiner et al.[9] Computations of spectra incorporating reflections of the incident squeezed light have also been presented by Kennedy and Walls,[15] in the context of intracavity atom–field interactions with squeezed light. The input–output formulation specifies that the output electric field $E_{out}(t)$ is related to the input and atomic fields by Eq. (21). Commutation relations between atomic and input field operators can also be derived, and using these, and the boundary condition (21), the correlation function for the output field can be expressed in the form[9]

$$
\begin{aligned}
\langle E_{out}(t+\tau)E_{out}(t)\rangle &= \langle E_{in}(t+\tau)E_{in}(t)\rangle \\
&+ \frac{1}{2}\frac{\hbar\gamma_a}{\omega_a}\langle[\dot{\sigma}_x(t+\tau),\dot{\sigma}_x(t)]_+\rangle \\
&+ \frac{1}{2}\sqrt{\frac{\hbar\gamma_a}{\omega_a}}\ \{\langle[\dot{\sigma}_x(t+\tau),E_{in}(t)]_+\rangle \\
&\quad+\langle[\dot{\sigma}_x(t),E_{in}(t+\tau)]_+\rangle\}
\end{aligned}
\tag{29}
$$

The last two correlation functions can be computed using the general formula (for an arbitrary system operator $Y(t)$)

$$
\begin{aligned}
&\langle[Y(t+\tau),E_{in}(t)]_+\rangle \\
&= i\sqrt{\frac{\gamma_a}{\hbar\omega_a}}\int_{-\infty}^{t+\tau}ds\langle[E_{in}(t),E_{in}(s)]_+\rangle\langle[\sigma_x(s),Y(t+\tau)]\rangle
\end{aligned}
\tag{30}
$$

which Gardiner et al.[9] have shown to be valid to the same order as that to which the master equation is valid.

The added complication of reflections could, in principle, be avoided with the introduction of a small "window" of unsqueezed vacuum modes through which to observe the fluorescence (small because one assumes that the atom sees predominantly squeezed field modes), and, indeed, this has been the configuration most commonly assumed in the variety of applications considered since Gardiner's original work on spontaneous emission. In practice this may well be a reasonable assumption, since the first experiments with squeezed light are likely to employ cavity configurations in which the squeezed light is incident on the atoms through the cavity modes. The reflections could then be avoided by simply observing the fluorescent light out the side of the cavity, or by observing the light transmitted through one side of the cavity upon which squeezed light is not incident.

III. SQUEEZED LIGHT AND A SINGLE TWO-LEVEL ATOM

A. Spontaneous Emission: Inhibition of Atomic Phase Decay

The master equation (15) describes spontaneous emission of a two-level atom in a squeezed vacuum, given that the atom interacts only with squeezed modes.[3] We shall assume a resonant interaction ($\omega_0 = \omega_a$), and make a choice of phase such that M is real and positive. Defining atomic polarization quadratures σ_x and σ_y by

$$\sigma^{\pm} = \frac{1}{2}(\sigma_x \pm i\sigma_y) \tag{31}$$

and working in a frame rotating at frequency ω_0, equations of motion for the averaged atomic variables can be derived in the form

$$\langle \dot{\sigma}_x \rangle = -\gamma_a\left(N + M + \tfrac{1}{2}\right)\langle \sigma_x \rangle \equiv -\gamma_x\langle \sigma_x \rangle$$

$$\langle \dot{\sigma}_y \rangle = -\gamma_a\left(N - M + \tfrac{1}{2}\right)\langle \sigma_y \rangle \equiv -\gamma_y\langle \sigma_y \rangle \tag{32}$$

$$\langle \dot{\sigma}_z \rangle = -\gamma_a - \gamma_a(2N + 1)\langle \sigma_z \rangle \equiv -\gamma_a - \gamma_z\langle \sigma_z \rangle \tag{33}$$

from which it is clear that the two polarization quadratures are damped at different rates, given by γ_x and γ_y. These rates are proportional, respectively, to the variances of the maximally unsqueezed and maximally squeezed quadrature phases of the incoming squeezed vacuum field. Hence, while the quadrature $\langle \sigma_x \rangle$ decays rapidly to its steady state value of zero, the decay of the $\langle \sigma_y \rangle$ quadrature is inhibited. In fact, the decay rate γ_y can, in principle, approach zero, as evidenced by the large squeezing limit

$$N - M + \frac{1}{2} \rightarrow \frac{1}{8N} \quad \text{as } N \rightarrow \infty \tag{34}$$

Of course, one must keep in mind that the limit $N \rightarrow \infty$ corresponds to the approach to threshold in the parametric amplifier providing the squeezed light. In this limit, the time constant $(\tfrac{1}{2}\kappa - |\varepsilon_D|)^{-1}$, which characterizes the unsqueezed quadrature phase, approaches infinity, thereby invalidating the squeezed white-noise formulation. It is possible to estimate more precisely the range of validity of the white-noise theory, and, as one might expect, it is found[9] that the approximation requires that the largest correlation time of the squeezed field, $(\tfrac{1}{2}\kappa - |\varepsilon_D|)^{-1}$, be much smaller than the smallest characteristic time constant of the atom, γ_x^{-1}.

With the above limitation understood, we now turn to the observable consequences of inhibited phase decay. In particular, we consider the spectrum of fluorescent light. The stationary correlation functions of the atomic variables can be computed with the aid of the quantum regression theorem, after which the correlation function of the total output field incorporating reflections can be found using (30) and (29). The fluorescence spectrum is given by the Fourier transform of the correlation function $\langle E_{\text{out}}^{(-)}(t + \tau)E_{\text{out}}^{(+)}(t)\rangle$, which is found to be

$$
\langle E_{\text{out}}^{(-)}(t + \tau)E_{\text{out}}^{(+)}(t)\rangle
$$
$$
= \hbar\omega_0\, e^{i\omega_0\tau}\left\{N\delta_s(\tau) + \frac{\gamma_a}{2}\frac{M}{2N + 1}\left[e^{-\gamma_y|\tau|} - e^{-\gamma_x|\tau|}\right]\right\} \quad (35)
$$

The spectrum thus consists of a flat background arising from the squeezed vacuum field (note that $\delta_s(\tau)$ denotes a δ function with respect to the "slow" atomic timescale as defined by $\gamma_{x,y}^{-1}$), plus two Lorentzians centered at the transition frequency. One of these is a negative peak of width γ_x, and consequently is very broad, whereas the other is a positive peak of width γ_y, possessing a subnatural linewidth. The extent of the narrowing of this linewidth is directly proportional to the degree of squeezing in the input light, and thus provides a direct measurement of squeezing.

B. Resonance Fluorescence

Resonance fluorescence from a two-level atom is a classic problem in quantum optics and has been well studied theoretically and experimentally (for references see Ref. 16). For strong driving fields, the spectrum of fluorescent light exhibits three peaks, which are often referred to as the Mollow triplet. A study of the effect of squeezed light on resonance fluorescence represents a natural progression from the work on spontaneous emission, and such a study was first carried out in detail by Carmichael et al.[17]

The relevant master equation for this situation is simply

$$
\dot{\rho} = -i\frac{\Omega_0}{2}\left[e^{-i\omega_0 t + i\phi_0}\sigma^+ + e^{i\omega_0 t - i\phi_0}\sigma^-, \rho\right] + (\dot{\rho})_{\text{spon}} \quad (36)
$$

where $(\dot{\rho})_{\text{spon}}$ is given by (15), and $\Omega_0 \propto \sqrt{\gamma_a}\,|\langle E_{\text{in}}\rangle|$ and ϕ_0 are the Rabi frequency and phase, respectively, of the coherent driving field. The frequency of the coherent field is assumed to be resonant with the central squeezing frequency. From an experimental point of view this is a reasonable assumption, since one would envisage using a single laser to provide the pump for both the parametric amplifier (after frequency doubling, as is actually done in experiments[2, 4]) and for the atoms. In this way, one also has control of the relative phase between the coherent and squeezed fields, which, as we shall see, plays an important role in the atomic dynamics.

From (36), one derives the Bloch equations for the atom. For simplicity, we ignore detuning between the atomic and driving field frequencies. In a frame rotating at frequency ω_0, and choosing polarization quadratures that are in-phase and out-of-phase with the coherent field,

$$\langle \sigma_x \rangle = \langle \sigma^+ \rangle e^{-i\omega_0 t + i\phi_0} + \langle \sigma^- \rangle e^{i\omega_0 t - i\phi_0} \tag{37}$$

$$\langle \sigma_y \rangle = -i(\langle \sigma^+ \rangle e^{-i\omega_0 t + i\phi_0} - \langle \sigma^- \rangle e^{i\omega_0 t - i\phi_0}) \tag{38}$$

we find

$$\langle \dot{\sigma}_x \rangle = -\gamma_a [N + M\cos(2\phi_0) + \tfrac{1}{2}]\langle \sigma_x \rangle - \gamma_a M \sin(2\phi_0)\langle \sigma_y \rangle \tag{39}$$

$$\langle \dot{\sigma}_y \rangle = -\gamma_a [N - M\cos(2\phi_0) + \tfrac{1}{2}]\langle \sigma_y \rangle - \gamma_a M \sin(2\phi_0)\langle \sigma_x \rangle - \Omega_0\langle \sigma_z \rangle \tag{40}$$

$$\langle \dot{\sigma}_z \rangle = -\gamma_a - \gamma_a(2N + 1)\langle \sigma_z \rangle + \Omega_0\langle \sigma_y \rangle \tag{41}$$

where, again, we have adopted a choice of phase for the squeezed vacuum such that M is real and positive. This does not lead to a loss of generality, since the dynamics are sensitive only to the relative phase between the coherent field and the squeezed field, which can obviously be varied through ϕ_0.

This dependence of the dynamics on phase represents a striking departure from traditional studies of resonance fluorescence, and leads, of course, to some interesting new phenomena. These phenomena are described in detail in Ref. 17. Perhaps the most significant of these is the phase dependence exhibited by the fluorescence spectrum under strong driving field conditions (such that Rabi splitting occurs). Two notable limiting cases arise for the choices of phase $\phi_0 = 0$ and $\phi_0 = \pi/2$, for which the equations (39–41) take the form

$$\langle \dot{\sigma}_x \rangle = -\gamma_a [N \pm M + \tfrac{1}{2}]\langle \sigma_x \rangle \equiv -\gamma_x\langle \sigma_x \rangle \tag{42}$$

$$\langle \dot{\sigma}_y \rangle = -\gamma_a [N \mp M + \tfrac{1}{2}]\langle \sigma_y \rangle - \Omega_0\langle \sigma_z \rangle \equiv -\gamma_y\langle \sigma_y \rangle - \Omega_0\langle \sigma_z \rangle \tag{43}$$

$$\langle \dot{\sigma}_z \rangle = -\gamma_a - \gamma_a(2N + 1)\langle \sigma_z \rangle + \Omega_0\langle \sigma_y \rangle \equiv -\gamma_a - \gamma_z\langle \sigma_z \rangle + \Omega_0\langle \sigma_y \rangle \tag{44}$$

The fluorescence spectrum that one then computes, after use of the quantum regression theorem to determine atomic correlation functions (for simplicity, we omit reflections, although these have been dealt with in the context of resonance fluorescence in a squeezed vacuum in Ref. 18), exhibits three peaks at the frequencies $\omega = \omega_0$ and $\omega = \omega_0 \pm \Omega_0$ ($\Omega_0 \gg \gamma_a$), with linewidths determined by the decay rates γ_x, γ_y, and γ_z. In particular, the halfwidth of the central peak is given by γ_x, while that of the two Rabi sidebands is given by the averaged value $(\gamma_y + \gamma_z)/2$. If we again consider the large squeezing limit, $N \gg 1$, we find, for $\phi_0 = 0$,

$$\gamma_x \simeq 2N\gamma_a \qquad \frac{\gamma_y + \gamma_z}{2} \simeq N\gamma_a \qquad (45)$$

while for $\phi_0 = \pi/2$,

$$\gamma_x \simeq \frac{\gamma_a}{8N} \qquad \frac{\gamma_y + \gamma_z}{2} \simeq 2N\gamma_a \qquad (46)$$

Hence, the Mollow triplet is now sensitively dependent on the phase of the coherent field, with the width of the central peak varying between subnatural and supernatural values. Noticeably, however, the sidebands only exhibit broadening compared to their normal (unsqueezed) vacuum profile. To illustrate these features, we plot the fluorescence spectrum in Fig. 1, for the two limiting choices of phase.

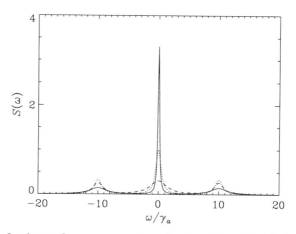

Figure 1. Incoherent fluorescence spectrum for $\Omega_0 = 10\gamma_a$, $N = 0.4$, $\phi_0 = 0$ (dashed line), and $\phi_0 = \pi/2$ (solid). The dotted curve is the ordinary fluorescence spectrum ($N = 0$).

With a nonzero detuning the decay rates of the Bloch vector components become more complicated and simple analytical results are no longer possible. However, following the approach of Courty and Reynaud,[19] one can uncover some interesting effects if one looks at the results from the point of view of the dressed state picture, in which one considers the energy levels of the combined atom-plus-laser system.[20]

In the limit of large generalized Rabi frequency $\Omega = (\Omega_0^2 + \Delta_a^2)^{1/2}$, one can make use of the secular approximation to obtain solutions for the Bloch vector components in which the following two decay rates appear

$$\lambda_1 = -\frac{\gamma_a}{2}\left[\left(1 + \frac{\Omega_0^2}{2\Omega^2}\right) + N\left(2 + \frac{\Omega_0^2}{\Omega^2}\right) - M\cos(2\phi_0)\left(1 - \frac{\Delta_a^2}{\Omega^2}\right)\right] \quad (47)$$

$$\lambda_2 = -\frac{\gamma_a}{2}\left[\left(2 - \frac{\Omega_0^2}{\Omega^2}\right) + 2N\left(1 + \frac{\Delta_a^2}{\Omega^2}\right) + 2M\cos(2\phi_0)\left(1 - \frac{\Delta_a^2}{\Omega^2}\right)\right]$$
$$(48)$$

with $\Delta_a = \omega_a - \omega_0$. In the steady state, the dressed-state inversion w_{dr} is related to $\langle\sigma_x(\infty)\rangle$ and $\langle\sigma_z(\infty)\rangle$ via

$$w_{dr} = \frac{\Delta_a}{\Omega}\langle\sigma_z(\infty)\rangle + \frac{\Omega_0}{\Omega}\langle\sigma_x(\infty)\rangle \quad (49)$$

which, in the secular approximation, can be worked out explicitly as

$$w_{dr} = \frac{\Delta_a}{\Omega}\frac{\gamma_a}{\lambda_2} \quad (50)$$

Clearly, for $\Delta_a = 0$ the population is evenly distributed between the dressed states. However, for $\Delta_a \neq 0$, and for λ_2 of the form (48), it is possible to have $|w_{dr}| = 1$, specifically, when $|\Delta_a/\Omega| = (2N + 2M + 1)^{-1}$, in which case the population is "trapped" in one of the dressed states. The condition just given is derived using $M^2 = N(N + 1)$, and is unique to squeezing; in particular, it is not possible to have $|w_{dr}| = 1$ for a purely thermal field ($N \neq 0$, $M = 0$).

A dramatic consequence of this population trapping is the disappearance of certain peaks from the fluorescence spectrum, as evidenced by the following expression for the spectrum

$$
\begin{aligned}
S(\omega) = &\frac{\pi}{2}\left(\frac{\Omega_0}{\Omega}w_{dr}\right)^2\delta(\omega) + \frac{\gamma_a}{(-\lambda_2)}\frac{\Omega_0^2}{2\Omega^2}\left[1 - (w_{dr})^2\right]\frac{\lambda_2^2}{\lambda_2^2 + \omega^2} \\
&+ \frac{\gamma_a}{(-\lambda_1)}\frac{1}{4}\left(1 + \frac{\Delta_a}{\Omega}\right)^2(1 + w_{dr})\frac{\lambda_1^2}{\lambda_1^2 + (\omega - \Omega)^2} \\
&+ \frac{\gamma_a}{(-\lambda_1)}\frac{1}{4}\left(1 - \frac{\Delta_a}{\Omega}\right)^2(1 - w_{dr})\frac{\lambda_1^2}{\lambda_1^2 + (\omega + \Omega)^2}
\end{aligned}
\tag{51}
$$

Hence, for $|w_{dr}| = 1$ two of the incoherent components of the spectrum vanish (of course, since we are observing the spectrum through a window of unsqueezed modes, the fluorescence lines will not completely disappear). As one might expect, this is accompanied by an increase in coherent scattering, as represented in the expression for $S(\omega)$ by the δ-function term.

In Fig. 2, we plot the variation in the incoherent fluorescence spectrum as a function of the ratio of the detuning Δ_a to the bare Rabi frequency Ω_0 (keeping the generalized Rabi frequency Ω fixed). The ordinary vacuum spectrum retains its symmetry as Δ_a is increased, and the sidebands decrease in size monotonically. With squeezing the spectrum becomes both phase dependent and asymmetric. In addition, we see, of course, the disappearance of two spectral peaks for a particular value of the detuning (and of the phase ϕ_0). This is a clear signature of squeezing, which is made somewhat appealing from an experimental point of view by the fact that large squeezing is not required to produce it.

We conclude this section on resonance fluorescence by re-emphasizing that the results presented thus far have been derived in the squeezed white noise limit. In fact, the conditions for the validity of the white noise theory are more stringent for strongly driven resonance fluorescence than for spontaneous emission, since the squeezed noise must appear δ-correlated even on the timescale of the Rabi oscillations. In frequency space, this corresponds to a squeezing bandwidth that is much broader than the Rabi splitting appearing in the fluorescence triplet.

C. Probe Absorption Spectra

Rather than observing the fluorescence emitted by the atom, it is also possible to consider measurements of the absorption of energy from a weak probe field that is incident on the atom in addition to a stronger "pump" field.[21-23] In the master equation, the addition of a probe field

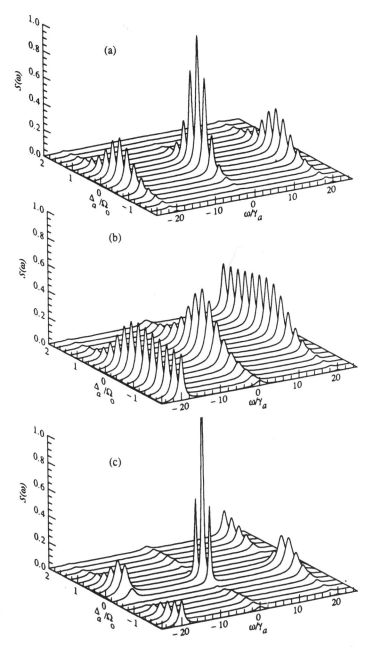

Figure 2. Incoherent fluorescence spectrum as a function of the detuning Δ_a (in units of Ω_0, keeping $\Omega = 20\gamma_a$), for (a) $N = 0$, (b) $N = 0.125$, $\phi_0 = 0$, and (c) $N = 0.125$, $\phi_0 = \pi/2$. In (c) the central peak for $\Delta_a = 0$ reaches the value 2.

corresponds to a modification of the Rabi frequency as follows:

$$\Omega_0\, e^{i\phi_0} \to \Omega_0\, e^{i\phi_0} + \Omega'\, e^{i\phi'}\, e^{-i(\Delta\nu)t} \tag{52}$$

where Ω' and ϕ' are the Rabi frequency and phase of the probe field respectively, and $(\Delta\nu) = \nu - \omega_0$ is the detuning between the probe and pump field frequencies. The effect of the weak probe field on the steady-state atomic density matrix is to induce small components oscillating at frequencies that differ by $\pm(\Delta\nu)$ from the frequencies of the unperturbed (i.e., $\Omega' = 0$) components. Small corrections are also induced in the unperturbed components themselves. The modified density matrix elements can thus be written (to first order in the perturbation) in the form[24]

$$\langle\sigma^-(t)\rangle = [\langle\sigma^-\rangle_{\Omega'=0} + (\delta\sigma^-)]e^{-i\omega_0 t} + (\delta\alpha_+)e^{-i\nu t} + (\delta\alpha_-)^* e^{-i(2\omega_0-\nu)t} \tag{53}$$

$$\langle\sigma_z(t)\rangle = [\langle\sigma_z\rangle_{\Omega'=0} + (\delta\sigma_z)] + (\delta\eta)e^{-i(\Delta\nu)t} + (\delta\eta)^* e^{i(\Delta\nu)t} \tag{54}$$

where $\langle\sigma^-\rangle_{\Omega'=0}$ and $\langle\sigma_z\rangle_{\Omega'=0}$ are steady-state values determined in the presence of the pump field alone, and $(\delta\sigma^-)$, $(\delta\alpha_+)$, $(\delta\sigma_z)$, and $(\delta\eta)$ are small constant parameters. Substituting (53) and (54) into their respective equations of motion (derived from the master equation), and retaining terms to lowest order in the probe field strength, these parameters are found to satisfy the equations

$$\left[-i(\Delta\nu) + i\Delta_a + \gamma_a\left(N + \frac{1}{2}\right)\right](\delta\alpha_+) + \gamma_a M(\delta\alpha_-) + \frac{i}{2}\Omega_0\, e^{i\phi_0}(\delta\eta)$$
$$= -\frac{i}{2}\Omega'\, e^{i\phi'}\langle\sigma_z\rangle_{\Omega'=0} \tag{55}$$

$$\left[-i(\Delta\nu) - i\Delta_a + \gamma_a\left(N + \frac{1}{2}\right)\right](\delta\alpha_-) + \gamma_a M(\delta\alpha_+)$$
$$- \frac{i}{2}\Omega_0\, e^{-i\phi_0}(\delta\eta) = 0 \tag{56}$$

$$[-i(\Delta\nu) + \gamma_a(2N + 1)](\delta\eta) + i\Omega_0\, e^{-i\phi_0}(\delta\alpha_+) - i\Omega_0\, e^{i\phi_0}(\delta\alpha_-)$$
$$= \Omega'\, e^{i\phi'}\langle\sigma^+\rangle_{\Omega'=0} \tag{57}$$

The sensitivity of the atomic polarization to the correlation between modes in a squeezed vacuum appears in these equations as a coupling between the "sidebands" at frequencies $\omega_0 \pm (\Delta\nu)$ (i.e., between $(\delta\alpha_+)$ and $(\delta\alpha_-)$).

The rate at which energy is absorbed from the probe field, or the probe absorption spectrum, is given by

$$W'(\nu) = \mathrm{Re}\left[-i\Omega'\,e^{i\phi'}(\delta\alpha_+)\right] \qquad (58)$$

Firstly, we note that the solution for $W'(\nu)$ is in fact independent of the probe phase ϕ'. In the absence of a coherent pump field ($\Omega_0 = 0$) the spectrum is

$$W'(\nu) = \frac{1}{4}(\Omega')^2 \frac{2}{2N+1}\left[\frac{\gamma_x}{(\Delta\nu)^2 + \gamma_x^2} + \frac{\gamma_y}{(\Delta\nu)^2 + \gamma_y^2}\right] \qquad (59)$$

with $\gamma_x = \gamma_a(N + M + \frac{1}{2})$ and $\gamma_y = \gamma_a(N - M + \frac{1}{2})$. Hence, the absorption spectrum exhibits the same subnatural-linewidth profile found in the spectrum of spontaneous emission.

When a coherent pump field is present the most interesting results are again found for resonant excitation ($\Delta_a = 0$), with the choices of coherent field phase $\phi_0 = 0$ or $\phi_0 = \pi/2$. For these choices, the absorption spectrum is

$$W'(\nu) = -\frac{1}{2}(\Omega')^2\langle\sigma_z\rangle_{\Omega'=0} \qquad (60)$$

$$\times \mathrm{Re}\left\{\frac{-i(\Delta\nu) + \gamma_z - \Omega_0^2/\gamma_k}{[-i(\Delta\nu) + \gamma_z][-i(\Delta\nu) + \gamma_k] + \Omega_0^2} + \frac{1}{-i(\Delta\nu) + \gamma_j}\right\}$$

where $\gamma_z = \gamma_a(2N + 1)$, and $(j, k) = (x, y)$ for $\phi_0 = 0$ and (y, x) for $\phi_0 = \pi/2$. For $\phi_0 = \pi/2$, it follows that the spectrum exhibits a line of subnatural width γ_y at $(\Delta\nu) = 0$, while for $\phi_0 = 0$ the spectral features are broadened in comparison to the ordinary vacuum spectrum. Examples of these features are shown in Fig. 3. It is interesting to note that for choices of phase inbetween 0 and $\pi/2$ the absorption spectrum is asymmetric (even though $\Delta_a = 0$). To conclude, the same dramatic phase dependence shown by the spectrum of resonance fluorescence also appears in the probe absorption spectrum.

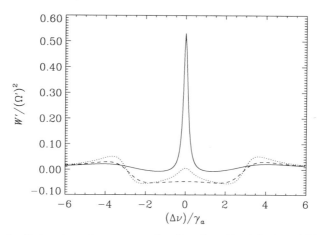

Figure 3. Probe absorption spectrum for $\Omega_0 = 3\gamma_a$, $N = 0.5$, $\phi_0 = 0$ (dashed line), $\phi_0 = \pi/2$ (solid line). The dotted curve is the ordinary vacuum absorption spectrum ($N = 0$).

D. Effects of Finite-Bandwidth Squeezed Light

The squeezed white-noise formulation, from which various predictions have been made in the previous sections, is, of course, an idealization. Indeed, present squeezed light sources, most notably the degenerate parametric amplifier, exhibit bandwidths only of the order of typical atomic transition linewidths. The purpose of this section is to review approaches taken to, and results obtained from, investigations of the effects of finite-bandwidth squeezing on the problems just discussed.

1. Formulation

Studies of the effects of finite-bandwidth squeezed light were first presented by Parkins and Gardiner[25] and Ritsch and Zoller.[13] Both studies were based on the set of stochastic Bloch equations that can be derived from the adjoint equation, which, as described earlier, gives the evolution of the density matrix $\mu(t)$ which is a 2×2 matrix functional of the incoming electric field operator $E_{in}(t)$ (Ritsch and Zoller actually obtain the equations from a stochastic density matrix equation whose derivation is based on the assumption of a positive P representation for the incident squeezed field). To set up these equations, we express $E_{in}(t)$ (which we shall assume represents only the incoherent portion of the field) in terms of its quadrature phase operators as

$$E_{in}(t) = \sqrt{\hbar\omega_0}\left[E_1(t)\cos(\omega_0 t) + E_2(t)\sin(\omega_0 t)\right] \qquad (61)$$

and work in a frame rotating at the carrier squeezing frequency ω_0, once again defining atomic polarization quadratures that are in phase and out of phase with the coherent driving field. Defining $\bar{\sigma}_i(t) = \text{Tr}\{\sigma_i \mu(t)\}$ as the atomic average of the spin operators, the stochastic Bloch equations can be derived in the form

$$\dot{\bar{\sigma}}_x = -\Delta_a \bar{\sigma}_y - \beta_X(t)\bar{\sigma}_z \tag{62}$$

$$\dot{\bar{\sigma}}_y = \Delta_a \bar{\sigma}_x - \Omega_0 \bar{\sigma}_z - \beta_Y(t)\bar{\sigma}_z \tag{63}$$

$$\dot{\bar{\sigma}}_z = -\gamma_a + \Omega_0 \bar{\sigma}_y + \beta_X(t)\bar{\sigma}_x + \beta_Y(t)\bar{\sigma}_y \tag{64}$$

where the rotating-wave approximation has been made, and $\beta_X(t)$ and $\beta_Y(t)$ are defined by

$$\beta_X(t)\rho = \tfrac{1}{2}\sqrt{\gamma_a}\left\{-\sin(\phi_0)[E_1(t),\rho]_+ + \cos(\phi_0)[E_2(t),\rho]_+\right\}$$
$$\equiv \left\{-\sin(\phi_0)\beta_1(t) + \cos(\phi_0)\beta_2(t)\right\}\rho \tag{65}$$

$$\beta_Y(t)\rho = \tfrac{1}{2}\sqrt{\gamma_a}\left\{\cos(\phi_0)[E_1(t),\rho]_+ + \sin(\phi_0)[E_2(t),\rho]_+\right\}$$
$$\equiv \left\{\cos(\phi_0)\beta_1(t) + \sin(\phi_0)\beta_2(t)\right\}\rho \tag{66}$$

Because of the properties of commutativity shown by $\beta_X(t)$ and $\beta_Y(t)$, these equations can be regarded as classical stochastic differential equations with the statistics of $\beta_1(t)$ and $\beta_2(t)$ specified by

$$\langle \beta_1(t)\rangle = \langle \beta_2(t)\rangle = 0 \tag{67}$$

$$\langle \beta_1(t+\tau)\beta_1(t)\rangle = \gamma_a\left\{\frac{b_-^2 - b_+^2}{2b_+}\,e^{-b_+|\tau|} + \delta(\tau)\right\} \tag{68}$$

$$\langle \beta_2(t+\tau)\beta_2(t)\rangle = \gamma_a\left\{\frac{b_+^2 - b_-^2}{2b_-}\,e^{-b_-|\tau|} + \delta(\tau)\right\} \tag{69}$$

$$\langle \beta_1(t+\tau)\beta_2(t)\rangle = 0 \tag{70}$$

The classical nature of the equations enables the use of well-established stochastic methods (or variations thereof) to obtain solutions.

In Ref. 25 the most versatile, and indeed the most direct approach taken to solving the stochastic Bloch equations was numerical simulation. For convenience, the white-noise part of $\beta_X(t)$ and $\beta_Y(t)$ can be averaged over analytically, leaving the following equations to be simulated:

$$\dot{\bar{\sigma}}_x = -\frac{\gamma_a}{2}\bar{\sigma}_x - \Delta_a\bar{\sigma}_y - \beta_X^c(t)\bar{\sigma}_z \tag{71}$$

$$\dot{\bar{\sigma}}_y = -\frac{\gamma_a}{2}\bar{\sigma}_y + \Delta_a\bar{\sigma}_x - \Omega_0\bar{\sigma}_z - \beta_Y^c(t)\bar{\sigma}_z \tag{72}$$

$$\dot{\bar{\sigma}}_z = -\gamma_a - \gamma_a\bar{\sigma}_z + \Omega_0\bar{\sigma}_y + \beta_X^c(t)\bar{\sigma}_x + \beta_Y^c(t)\bar{\sigma}_y \tag{73}$$

where the superscript c denotes the colored noise portion of the input field, so, in particular,

$$\langle \beta_1^c(t+\tau)\beta_1^c(t) \rangle = \gamma_a \frac{b_-^2 - b_+^2}{2b_+} e^{-b_+|\tau|} \tag{74}$$

$$\langle \beta_2^c(t+\tau)\beta_2^c(t) \rangle = \gamma_a \frac{b_+^2 - b_-^2}{2b_-} e^{-b_-|\tau|} \tag{75}$$

Noise sources with the correct statistics can be constructed using summations of suitably weighted Gaussian-distributed random numbers. The negative correlations that characterize squeezing require that $\beta_1(t)$ be pure imaginary, which enables $\bar{\sigma}_x(t)$, $\bar{\sigma}_y(t)$, and $\bar{\sigma}_z(t)$ to develop imaginary parts. However, in practice one takes only the real parts of the computed averages, since the imaginary parts average to zero after a sufficient number of trajectories, provided the initial conditions are real and the integration routine is stable. In Ref. 25, methods were also presented, based on the adjoint equation, with which correlation functions (and thus spectra) can be computed from simulations of the stochastic Bloch equations.

In their study of finite bandwidth effects on absorption spectra, Ritsch and Zoller[13] adopted a generating function technique to derive, using Ito calculus, an infinite hierarchy of differential equations for atom–field averages. In the stationary limit, these equations can be solved numerically with truncation at a suitably high level. To outline this approach, one notes firstly that $\beta_1^c(t)$ and $\beta_2^c(t)$ are simply independent Ornstein-Uhlenbeck processes, so that the problem can in fact be formulated as a

Markovian system with the following set of stochastic differential equations:

$$d\beta_1^c(t) = -b_1\beta_1^c(t)\,dt + \left[2b_1\langle(\beta_1^c)^2\rangle\right]^{1/2}dW_1(t) \qquad (76)$$

$$d\beta_2^c(t) = -b_2\beta_2^c(t)\,dt + \left[2b_2\langle(\beta_2^c)^2\rangle\right]^{1/2}dW_2(t) \qquad (77)$$

$$d\mu(t) = \left[A + B_1\beta_1^c(t) + B_2\beta_2^c(t)\right]\mu(t)\,dt \qquad (78)$$

where $b_1 = b_+$ and $b_2 = b_-$, $\langle(\beta_{1,2}^c)^2\rangle$ are the stationary variances, and $dW_1(t), dW_2(t)$ are independent white-noise increments. The operators A, B_1, and B_2 are obtained from (23) after transforming to the rotating frame and making the rotating wave approximation.

One then defines a marginal characteristic function by

$$u(\boldsymbol{\lambda}, t) = \left\langle \exp\left[i \sum_{j=1,2} \lambda_j\beta_j^c(t)\right]\mu(t)\right\rangle \qquad (79)$$

with $\boldsymbol{\lambda} = (\lambda_1, \lambda_2)$. Using Ito calculus,[26] one can derive an equation of motion for $u(\boldsymbol{\lambda}, t)$, solutions of which are found with an ansatz of the form

$$u(\boldsymbol{\lambda}, t) = \left\langle \exp\left[i \sum_{j=1,2} \lambda_j\beta_j^c(t)\right]\right\rangle g(\boldsymbol{\lambda}, t)$$

$$= \exp\left[-\frac{1}{2}\sum_{j=1,2}\lambda_j^2\langle(\beta_j^c)^2\rangle\right]g(\boldsymbol{\lambda}, t) \qquad (80)$$

Expanding $g(\boldsymbol{\lambda}, t)$ in a power series, and using this in the equation of motion for $u(\boldsymbol{\lambda}, t)$, one derives the infinite system of coupled differential equations

$$\dot{u}_{n_1, n_2} = (A - n_1b_1 - n_2b_2)u_{n_1, n_2}$$

$$+ B_1\left[\frac{\langle(\beta_1^c)^2\rangle}{2}\right]^{1/2}u_{n_1+1, n_2} + B_1n_1\left[2\langle(\beta_1^c)^2\rangle\right]^{1/2}u_{n_1-1, n_2} \qquad (81)$$

$$+ B_2\left[\frac{\langle(\beta_2^c)^2\rangle}{2}\right]^{1/2}u_{n_1, n_2+1} + B_2n_2\left[2\langle(\beta_2^c)^2\rangle\right]^{1/2}u_{n_1, n_2-1}$$

for the system–field averages

$$
u_{n_1, n_2}(t) = \left\langle H_{n_1}\!\left[\frac{\beta_1^c(t)}{\left(2\langle(\beta_1^c)^2\rangle\right)^{1/2}}\right] H_{n_2}\!\left[\frac{\beta_2^c(t)}{\left(2\langle(\beta_2^c)^2\rangle\right)^{1/2}}\right]\mu(t)\right\rangle \quad (82)
$$

where H_n are Hermite polynomials. The average $\langle\mu(t)\rangle$ is thus obtained by solving for $u_{00}(t)$ with the initial condition $u_{n_1, n_2}(0) = \delta_{n_1 0}\delta_{n_2 0}u(0)$.

By truncating the infinite set of equations at a sufficiently high level, Ritsch and Zoller solved for $\langle\mu(t)\rangle$ in the steady state, and, with a straightforward modification to this approach, were also able to compute probe absorption spectra. We note that the white-noise limit (i.e., the master equation), corresponding to $b_1, b_2 \to \infty$, is recovered from (81) by keeping only the lowest-order couplings. As b_1, b_2 are decreased, the values of n_1, n_2 at which a suitable truncation can be made increase.

2. Spontaneous Emission

Using the simulation method, Parkins and Gardiner[25] examined spontaneous emission of a two-level atom in a finite-bandwidth squeezed vacuum, comparing their results with the white-noise predictions of Section III.A. Their primary finding was that, while the effect of inhibition of phase decay is degraded somewhat, the basic effect persists for bandwidths of squeezing only a few times the atomic linewidth. A comparison of the behavior of the Bloch vector components for a finite-bandwidth squeezed

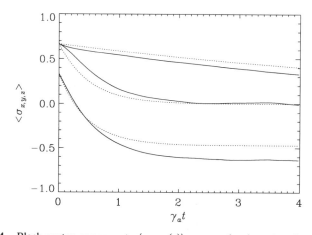

Figure 4. Bloch vector components $\langle\sigma_{x, y, z}(t)\rangle$ vs. $\gamma_a t$, for $b_+ = 6\gamma_a$, $b_- = 3\gamma_a$, computed via simulation (solid line) and in the squeezed white noise approximation (dotted line).

light input, as computed by simulation, and for a white-noise input is shown in Fig. 4. The agreement between approaches is not tremendous, but the finding mentioned above is emphasized, in that the inhibition of phase decay is still significant. Parkins and Gardiner also developed some improved approximate analytical results which give much better agreement with the simulated fast decays of $\langle \sigma_x \rangle$ and $\langle \sigma_z \rangle$. Modified (deterministic) Bloch equations giving similarly improved agreement with simulation have also been presented by Cirac and Sánchez-Soto.[27]

The degradation of the inhibition effect with decreasing squeezing bandwidth suggests that in the perfect squeezing limit, $b_- \to 0$, the effect ultimately ceases to occur. Using both analytical and simulation approaches, Parkins and Gardiner were also able to show that this is indeed the case, and that the decay rate of $\langle \sigma_y \rangle$ returns to its ordinary vacuum value, $\gamma_a/2$, as $b_- \to 0$.

3. Probe Absorption Spectra

The findings of Ritsch and Zoller regarding probe absorption spectra in the presence of finite-bandwidth squeezed light are basically in accord with the findings of Parkins and Gardiner reviewed above. For sufficiently small b_-, the lineshape of the absorption spectrum only broadens as the squeezing is increased; so, for a particular value of the DPA cavity linewidth, there is a particular optimum linewidth reduction that can be achieved rather than the complete linewidth narrowing predicted by the white-noise theory in the perfect squeezing limit. An example demonstrating the initial narrowing and eventual broadening of the spectrum as b_- decreases is shown in Fig. 5. Ritsch and Zoller also computed absorption spectra for the case in which a coherent driving field is present. The fast timescale associated with a strong Rabi frequency puts a more stringent condition on the bandwidth of squeezing if the white noise picture is to be valid. In particular, Ritsch and Zoller showed that the subnatural linewidths seen in earlier white-noise analyses of absorption spectra are broadened, and ultimately lost once the Rabi frequency becomes comparable to or greater than the bandwidth of squeezing.

4. Resonance Fluorescence Spectra

The main aim of the work just reviewed was to identify the detrimental effects that a finite bandwidth of squeezing has on the effects predicted using the white-noise formulation. Similar degradation of the phenomena seen in resonance fluorescence spectra can be expected, but the presence of a reservoir spectral structure on the scale of the Rabi frequency also allows for some very interesting dynamical effects to come into play. Parkins[28] has investigated such effects, concentrating on the particular

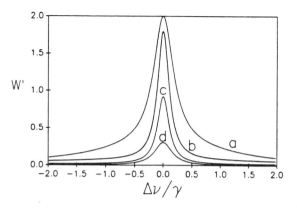

Figure 5. Probe absorption spectrum for (a) $b_+ = 6.2\gamma_a$, $b_- = 3.8\gamma_a$, (b) $b_+ = 7.5\gamma_a$, $b_- = 2.5\gamma_a$, (c) $b_+ = 8.8\gamma_a$, $b_- = 1.2\gamma_a$, (d) $b_+ = 9.6\gamma_a$, $b_- = 0.4\gamma_a$. (Reproduced with permission from Ritsch and Zoller.[13])

scenarios depicted in Fig. 6, where the squeezed vacuum power spectrum is superimposed on the spectrum of fluorescence emitted by the atom. The first scenario represents a situation in which the bandwidth of squeezing, though possibly broad compared to the natural atomic linewidth, falls well within the Rabi splitting of the fluorescence triplet. In the second scenario, the squeezed vacuum is assumed to exhibit two spectral peaks that we shall take to be centered on the two sidebands of the fluorescence triplet. This kind of squeezing is what one expects from (frequency) nondegenerate parametric amplification, and we shall refer to it as two-

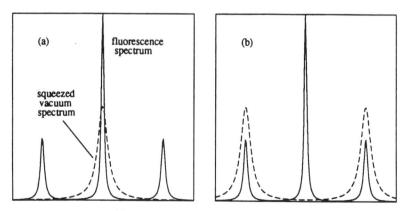

Figure 6. (a) Single-mode and (b) two-mode squeezing scenarios.

mode squeezing, in the sense that two cavity modes are excited in the parametric amplification process. Correspondingly, we shall refer to the squeezing in the first scenario as single-mode squeezing. It should be added that two-mode squeezing is not merely a construction of convenience, since squeezing spectra with this double-peaked form are found in the output of a number of practical squeezing devices.

A straightforward interpretation of the effects found by Parkins can be gained from a qualitative inspection of the stochastic Bloch equations ((71)–(73)). In lowest order (neglecting noise terms), and for zero detuning, $\Delta_a = 0$, $\bar{\sigma}_x(t)$ displays a simple exponential decay, while $\bar{\sigma}_y(t)$ and $\bar{\sigma}_z(t)$ undergo Rabi oscillations at frequency Ω_0 and also decay exponentially. The contribution to the time development of $\bar{\sigma}_x(t)$ from the additional noise terms is proportional to the time average of $\beta_X^c(t)\bar{\sigma}_z(t)$. Since $\bar{\sigma}_z(t)$ undergoes Rabi oscillations, it follows that this contribution will be significant only if $\beta_X^c(t)$ contains Fourier components at the frequencies $\pm\Omega_0$. A similar argument can be applied to $\bar{\sigma}_y(t)$ and $\bar{\sigma}_z(t)$. The time development of both $\bar{\sigma}_y(t)$ and $\bar{\sigma}_z(t)$ will be sensitive to the spectral components of $\beta_Y^c(t)$ around zero frequency, while in the equation of motion for $\bar{\sigma}_z(t)$, the term $\beta_X^c(t)\bar{\sigma}_x(t)$ will be significant only if, once again, $\beta_X^c(t)$ has nonnegligible frequency components at $\pm\Omega_0$. In summary, the decay rates of the Bloch vector components are sensitive to the spectral components of $\beta_X^c(t)$ at the frequencies $\pm\Omega_0$, and to the spectral components of $\beta_Y^c(t)$ at zero frequency.

Consider first the single-mode squeezing scenario, with $\beta_X^c(t)$ and $\beta_Y^c(t)$ possessing significant spectral components only at frequencies much smaller than Ω_0. Then, from the arguments given above, it is clear that it should be possible to effectively decouple the noise source $\beta_X^c(t)$ from the dynamics of the atom. This possibility takes on special significance in the case of a squeezed vacuum input since, through an appropriate choice of phase ϕ_0, the unsqueezed (noisy) quadrature can be made to correspond to $\beta_X^c(t)$. In the case that the squeezing bandwidth is still large compared to the radiative linewidth, i.e., $\Omega_0 \gg b_\pm \gg \gamma_a$, one can apply a decorrelation approximation to the equations (one might also refer to this as a "local" white-noise approximation) to obtain

$$\langle \dot{\sigma}_x \rangle \simeq -\gamma_x \langle \sigma_x \rangle \tag{83}$$

$$\langle \dot{\sigma}_y \rangle \simeq -\gamma_y \langle \sigma_y \rangle - \Omega_0 \langle \sigma_z \rangle \tag{84}$$

$$\langle \dot{\sigma}_z \rangle \simeq -\gamma_a - \gamma_z \langle \sigma_z \rangle + \Omega_0 \langle \sigma_y \rangle \tag{85}$$

where, for the two limiting choices of phase, $\phi_0 = 0$ and $\phi_0 = \pi/2$, the decay rates are

$$\gamma_x = \frac{\gamma_a}{2} \qquad \gamma_y = \frac{\gamma_a}{2}[1 + 2(N \mp M)] \qquad \gamma_z = \gamma_a[1 + (N \mp M)] \qquad (86)$$

Hence, the component $\langle \sigma_x \rangle$ is damped at its normal vacuum rate independent of the choice of phase, and, in the (incoherent) fluorescence spectrum, the central peak of the triplet is unchanged from its normal vacuum profile. Meanwhile, the components $\langle \sigma_y \rangle$ and $\langle \sigma_z \rangle$ are damped at the rate

$$\tfrac{1}{2}(\gamma_y + \gamma_z) = \gamma_a(N \mp M + \tfrac{3}{4}) \qquad (87)$$

which can be greater or less than in a normal vacuum, depending on the choice of phase. In fact, for strong squeezing ($N \gg 1$), and the choice $\phi_0 = 0$, this rate can approach one-third of the normal vacuum decay rate. Hence, the dramatic phase sensitivity shown in the spectrum transfers from the central fluorescence peak to the Rabi sidebands as one moves from the broadband squeezing limit of Section III.B to the narrowband (on the scale of the Rabi frequency) limit considered here.

For the two-mode squeezing scenario, $\beta_X^c(t)$ and $\beta_Y^c(t)$ have significant spectral density in two well-separated peaks that are assumed to be centered on the frequencies $\pm \Omega_0$. Hence, one now expects that $\beta_Y^c(t)$ will be decoupled from the dynamics, and that $\beta_X^c(t)$ will be the dominant noise term. Using the decorrelation approximation again, one can derive equations as above where now

$$\gamma_x = \frac{\gamma_a}{2}[2(N \pm M) + 1] \qquad \gamma_y = \frac{\gamma_a}{2} \qquad \gamma_z = \gamma_a[(N \pm M) + 1] \qquad (88)$$

The decay rate γ_x is identical to that found in the (single-mode) white-noise limit of Section III.B, and so the linewidth of the central fluorescence peak again varies between supernatural and subnatural values as the phase ϕ_0 is varied from 0 to $\pi/2$. The new feature added by the two-mode squeezing scenario is that now the Rabi sidebands simultaneously exhibit the same variation between supernatural and subnatural linewidths (although with a smaller degree of broadening and narrowing). Hence, in principle it is possible for all three peaks in the fluorescence triplet to exhibit subnatural linewidths for a particular choice of phase.

To confirm these predictions, Parkins carried out stochastic simulations of the equations for both scenarios and for a range of parameters in each case. The result of one such simulation is shown in Fig. 7 (for the single-mode squeezing scenario). It was also shown that the predictions of the simple decorrelation approximation are quite robust in the face

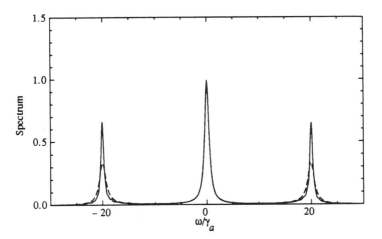

Figure 7. Incoherent fluorescence spectrum for the single-mode squeezing scenario with $\Omega_0 = 20\gamma_a$, $b_+ = 2.25\gamma_a$, $b_- = 0.75\gamma_a$, and $\phi_0 = 0$ (solid line), compared with the normal vacuum result (dashed line).

of nonideal squeezing bandwidths, i.e., with bandwidths only of the order of the natural atomic linewidth.

E.   Laser Cooling in a Squeezed Vacuum

Laser cooling of atoms is presently of considerable experimental and theoretical interest,[29] and recently several authors have modified analyses of standard cooling configurations to incorporate squeezed light. Shevy[30] computed the force on a two-level atom moving in a laser standing wave and damped by a squeezed vacuum. He demonstrated that the unique features exhibited by an atom in a squeezed vacuum can lead to greatly enhanced cooling forces (in particular, forces in a standing wave due to stimulated processes can be related to the lineshape of the probe absorption spectrum, which is dramatically altered in a squeezed vacuum). Graham et al.[31] extended this work to the computation of steady-state temperatures and found that under near-resonant conditions atoms can be cooled to temperatures below the Doppler limit, $T_D = \hbar\gamma_a/2k_B$. However, both analyses were highly idealized, in that a position-independent squeezed vacuum was assumed (allowing the use of relatively simple Bloch equations similar to those introduced earlier in this review). The practical implementation of such a scheme would be extremely difficult.

Alternatively, one can consider laser cooling of trapped ions, where the vibrational amplitude of the ion in the trap is much less than the wavelength of the cooling laser (Lambe-Dicke limit). The assumption of position-independent squeezing is therefore more appropriate. Cirac and Zoller[32] have examined such a situation in which the ion is subject to a squeezed vacuum and to a laser standing wave. Under suitable conditions, and with the ion located at the node of the standing wave, they find that sub-Doppler-limit temperatures are achievable with the ion essentially in the trap ground state.

IV.   MULTILEVEL AND MULTIATOM EFFECTS IN SQUEEZED LIGHT

A.   Three-Level Atom in a Broadband Squeezed Vacuum

The squeezed vacuum is characterized by correlated pairs of photons. Therefore, it is quite natural to expect that the behavior of two-photon transitions in atomic systems should be influenced in a significant way by the presence of squeezed light. Such a situation does indeed arise in the interaction of a three-level "ladder" system with squeezed light when the double carrier frequency of the squeezing, $2\omega_0$, is tuned close to the two-photon transition frequency (i.e., to the frequency between the lower and upper atomic states) of the ladder system.[33, 34]

Consider the system depicted in Fig. 8, where ω_{ij} are the transition frequencies and γ_{ij} are the spontaneous emission rates between states $|i\rangle$ and $|j\rangle$. Under the assumption of broadband squeezing (ideally coupled to the atom), one can follow a master equation treatment to solve for the atomic dynamics. Arguably the most interesting features of this model are the steady-state atomic level populations, which give a clear signature of the effect of two-photon correlations in the squeezed vacuum. The populations of states $|2\rangle$ and $|3\rangle$ are[33, 34]

$$\rho_{22} = \frac{N(N+1) - |M|^2}{3N^2 + 3N + 1 - 3|M|^2} \qquad \rho_{33} = \frac{N^2 w - |M|^2(w - 1 - \alpha)}{3N^2 + 3N + 1 - 3|M|^2}$$

$$(89)$$

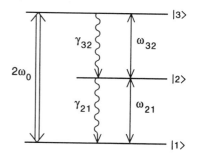

Figure 8.   Three-level "ladder" system.

where $\alpha = \gamma_{32}/\gamma_{21}$, $w = \alpha + N(1 + \alpha)$, and $|M|^2 \le N(N + 1)$. For a thermal field, one sets $M = 0$, and the level populations can be shown to obey a Boltzmann distribution with $\rho_{11} > \rho_{22} > \rho_{33}$. However, with a minimum uncertainty squeezed vacuum field, for which $|M|^2 = N(N + 1)$, one has

$$\rho_{22} = 0 \qquad \rho_{33} = \frac{N}{\alpha + N(\alpha + 1)} \qquad (90)$$

and so clearly the level populations no longer obey a Boltzmann distribution. This is in distinct contrast to the steady-state populations of a two-level atom in a squeezed vacuum, which are indistinguishable from those in a thermal field. Furthermore, if $\alpha < N/(N + 1)$ then $\rho_{33} > \frac{1}{2}$, corresponding to a population inversion between levels $|3\rangle$ and $|1\rangle$.

These features are manifestations of the two-photon correlations in a squeezed vacuum field, which facilitate an enhancement of two-photon transitions in the ladder system (i.e., one photon of a correlated pair excites the atom to the intermediate state, while its twin rapidly completes the two-photon transition), transferring population from the ground state $|1\rangle$ to the second excited state $|3\rangle$ without population of the state $|2\rangle$, and thus producing a higher population in state $|3\rangle$.

Another distinctive consequence of two-photon correlations in the incident field arises from a consideration of the time evolution of the population of state $|3\rangle$ in the low intensity limit, $N \ll 1$. In a thermal field ($M = 0$), one can write[33]

$$\rho_{33}(\tau) \simeq \frac{N^2}{1 - \alpha}\left[(1 - \alpha) - e^{-\alpha\tau} + \alpha e^{-\tau}\right] \qquad (91)$$

which shows that the two-photon transition rate exhibits a quadratic dependence on intensity (τ is a dimensionless time). This follows from the fact that in a thermal field the transition $|1\rangle \to |3\rangle$ corresponds to a two-step process, $|1\rangle \to |2\rangle \to |3\rangle$. At low intensity each step is proportional to the intensity N, resulting in a quadratic dependence on N of the two-photon transition rate.

However, in a squeezed vacuum field, two-photon correlations effectively enable the transition $|1\rangle \to |3\rangle$ to occur in a single step proportional to N, and in fact one can derive

$$\rho_{33}(\tau) \simeq \frac{N}{\alpha}(1 - 2e^{-\alpha\tau/2} + e^{-\alpha\tau}) \qquad (92)$$

Hence, at low intensity, the two-photon transition rate exhibits a linear dependence on the intensity N of the squeezed vacuum.[33, 35, 36]

Finally, it has been pointed out that in an ideally squeezed vacuum ($|M|^2 = N(N + 1)$), the three-level atom can relax into a highly correlated pure state.[34] In particular, for $\alpha = 1$, the stationary level populations are

$$\rho_{11} = \frac{N + 1}{2N + 1} \qquad \rho_{22} = 0 \qquad \rho_{33} = \frac{N}{2N + 1} \qquad (93)$$

and there is a nonzero stationary correlation between the states $|1\rangle$ and $|3\rangle$ given by

$$\rho_{13} = \frac{M}{2N + 1} \qquad (94)$$

These results correspond to the stationary (pure) state

$$|\Psi\rangle_{st} = \left[\frac{N + 1}{2N + 1}\right]^{1/2} |1\rangle + \left[\frac{N}{2N + 1}\right]^{1/2} |3\rangle \qquad (95)$$

Hence, the correlations and phase information contained in the squeezed vacuum field can be transferred to the atomic system.

B. The Two-Atom Dicke Model in a Broadband Squeezed Vacuum

The two-atom Dicke model has close parallels with the three-level ladder system. In this model, two (two-level) atoms are taken to be separated by a distance much smaller than their resonant wavelength (so that spatial effects can be ignored), but sufficiently large that any exchange interaction between the two atoms due to overlap of atomic wave functions is negligible. Under these conditions, the two-atom system can be represented by a single three-level system, with a ground state $|1\rangle = |1\rangle_1|1\rangle_2$, an upper excited state $|2\rangle = |2\rangle_1|2\rangle_2$, and an intermediate state $|+\rangle = (|2\rangle_1|1\rangle_2 + |1\rangle_1|2\rangle_2)/\sqrt{2}$, where $|1\rangle_i$ and $|2\rangle_i$ are the ground and excited states of the ith atom, respectively. The additional intermediate state $|-\rangle = (|2\rangle_1|1\rangle_2 - |1\rangle_1|2\rangle_2)/\sqrt{2}$ is completely decoupled from the other triplet of states (which follows from the imposed symmetry under the exchange $1 \leftrightarrow 2$).

As one would therefore expect, the two-atom Dicke model exhibits the same features displayed by the three-level ladder system under the influence of squeezing. The population decay constants and evolution show dependence on the magnitude of the squeezing parameter $|M|$ (Ref. 37) (but not on the phase of squeezing), and for a minimum uncertainty squeezed state, $|M|^2 = N(N + 1)$, the final equilibrium atomic state is a pure state of the form

$$|\Psi\rangle_{eq} = \left[\frac{N + 1}{2N + 1} \right]^{1/2} |1\rangle_1 |1\rangle_2 + \left[\frac{N}{2N + 1} \right]^{1/2} |2\rangle_1 |2\rangle_2 \qquad (96)$$

again reflecting the transfer of the correlations from the field to the atomic system. In fact, this creation of a highly correlated pure steady-state is a general property of (even-numbered) collections of atoms interacting with a broadband ideal squeezed vacuum field.[38]

The Dicke model assumes interatomic spacings much less than the resonant wavelength. With interatomic spacings comparable to the wavelength (such that spatial effects are significant), Ficek[39] has examined a variety of phenomena related to the effects of squeezed light. For example, where the interatomic spacing cannot be neglected, he finds that the simple three-state picture of the atomic pair cannot be applied (i.e., the state $|-\rangle$ is now populated), and the steady state is no longer the correlated pure state of the Dicke model.

C. Photon Echoes with Squeezed Light

In the simplest case, a photon echo is produced by a sequence of two coherent pulses incident upon a system of inhomogeneously broadened atoms.[40] The first pulse serves to induce a macroscopic polarization in the sample. Inhomogeneous broadening then causes dephasing of the individual dipole moments throughout the sample in a process that manifests itself as a rapid dissipation of the macroscopic dipole moment. Hence, the collection of atoms as a whole may cease radiating in only a fraction of the natural lifetime of the atomic transition involved. If a second coherent pulse (of suitable intensity) is applied at a time τ after the first pulse, the dephasing process may be reversed, resulting in a momentary reformation of the macroscopic dipole moment at a time 2τ after the first excitation pulse. The pulse of light that is subsequently emitted is known as a photon echo.

Of course, throughout this sequence of events each individual dipole moment is incoherently damped by spontaneous emission at the rate $\gamma_a/2$. By varying the pulse delay time, it is therefore possible to change the amplitude of the echo, and so in fact obtain a measure of γ_a. In previous sections, we have seen how squeezed light can significantly alter decay processes in atoms, and hence it seems logical to expect that squeezed light should have a substantial effect on the photon echo phenomenon. In fact, this is the case, but, in addition, it can be shown that the difference in transverse decay rates induced by squeezed light can in principle lead to a new form of photon echo that does not rely on a second coherent pulse to initiate the rephasing process.[41]

The mechanism for producing such an echo is quite different from that of a normal echo, and is illustrated (for the idealized case of a spatially degenerate two-level system) in the sequence of diagrams shown in Fig. 9. These diagrams show the time development of the $\langle \sigma_x \rangle$ and $\langle \sigma_y \rangle$ components corresponding to 100 random detunings Δ_a chosen from a Gaussian distribution with a standard deviation $10\gamma_a$. An initial $\pi/2$ coherent pulse prepares the atoms in a state with $\langle \sigma_y \rangle \simeq -1$, producing a macroscopic polarization. The individual Bloch vectors then precess freely up to the

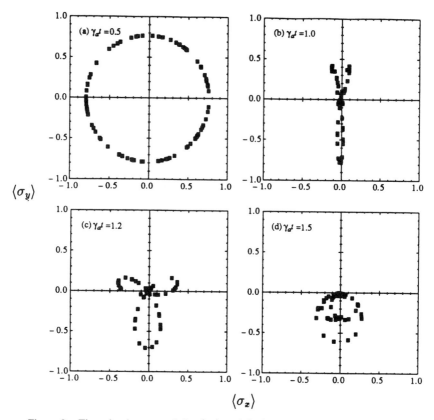

Figure 9. Time development of the $\langle \sigma_x \rangle$ and $\langle \sigma_y \rangle$ components corresponding to 100 (Gaussian) random detunings, for a single $\pi/2$ coherent pulse ($t = 0$) followed by a squeezed pulse. See text for explanation.

time $\gamma_a t = 0.5$, at which stage they are distributed uniformly about the origin (giving rise to zero net polarization), as shown in Fig. 9a. When squeezing is applied, the two-level equations of motion possess eigenvalues of the form

$$\lambda_{1,2} = -\gamma_a \left[N \mp \sqrt{M^2 - (\Delta_a/\gamma_a)^2} + \tfrac{1}{2} \right] \qquad (97)$$

from which it is apparent that squeezing offers the possibility of both inhibiting the phase decay and preventing the precession of the Bloch vector about the z axis. This leads to a rapid alignment of the vectors along the line $\langle \sigma_x \rangle = 0$, with vectors corresponding to larger detunings quickly attenuated to zero (Fig. 9b). Once the squeezing is removed at time $\gamma_a t = 1.0$, a separation of vectors with different rates of precession is revealed (Fig. 9c), and a partial realignment occurs, giving rise to a partial reformation (echo) of the initial macroscopic polarization (Fig. 9d). From a practical point of view, the simplified model used to obtain the above results suffers from the same problem as the models used for laser cooling, in that it assumes a position-independent squeezed vacuum.

V. SQUEEZED-RESERVOIR LASERS

In view of the interesting phenomena exhibited by single atoms driven by squeezed light, it was a natural progression for theorists to consider more complicated systems in which spontaneous emission and vacuum fluctuations still play an important role. The laser is such a system, and in this section we review the work that has been done on so-called "squeezed-reservoir" lasers.

A. Squeezed-Pump Lasers

Following a standard approach to the quantum theory of the laser, Marte and Walls[42] have studied the characteristics of a laser pumped with broadband squeezed light, i.e., a laser with reduced pump fluctuations. The laser is modeled as an ensemble of $\mathcal{N}$ two-level atoms coupled to a single cavity mode resonant with the atomic transition frequency. Such an interaction is represented by the Hamiltonian

$$H_0 = \hbar\omega_0 a^\dagger a + \tfrac{1}{2}\hbar\omega_0 S_z + i g\hbar \left(a^\dagger S_- - a S_+ \right) \qquad (98)$$

where a is the cavity mode annihilation operator, S_z, S_+, S_- are collective atomic operators for $\mathcal{N}$ atoms, and g is the dipole coupling strength. The spontaneous decay of the atoms, and the incoherent pumping mechanism

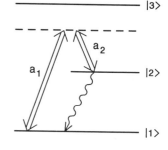

Figure 10. Pumping scheme for a three-level atom with two squeezed cavity modes a_1 and a_2.

that produces the required atomic inversion, is modeled by coupling the atoms to an "inverted" harmonic oscillator heat bath. This is an established means of representing pumping in a laser,[43] but the assumption of a squeezed pump leads to some modification, and in the laser master equation the relevant atomic pumping term takes the form

$$\left[\frac{\partial \rho}{\partial t}\right]_A = \sum_\nu \left\{ \frac{1}{2}\gamma_a N_p (2\sigma_\nu^- \rho \sigma_\nu^+ - \rho \sigma_\nu^+ \sigma_\nu^- - \sigma_\nu^+ \sigma_\nu^- \rho) \right.$$

$$\left. + \frac{1}{2}\gamma_a (N_p + 1)(2\sigma_\nu^+ \rho \sigma_\nu^- - \rho \sigma_\nu^- \sigma_\nu^+ - \sigma_\nu^- \sigma_\nu^+ \rho) \right\} \quad (99)$$

$$- \sum_\nu \left\{ \gamma_a M_p \sigma_\nu^+ \rho \sigma_\nu^+ + \gamma_a M_p^* \sigma_\nu^- \rho \sigma_\nu^- \right\}$$

where σ_ν^+ and σ_ν^- are atomic operators for the νth atom.

A scheme for the practical realization of such a pumping mechanism has been put forward by Haake et al.[44] This scheme consists of three-level atoms interacting with two cavity modes whose difference frequency is resonant with the active laser transition, as depicted in Fig. 10. If the inputs to the two cavity modes (a_1 and a_2) are taken to be independent squeezed vacuum fields, and if the cavity mode decay rates are much larger than the dipole coupling constant, then a master equation of the form (99) can be derived which produces atomic inversion on the lasing transition and squeezes the pump fluctuations.

Finally, taking into account losses through the cavity mirrors of the laser via the term

$$\left[\frac{\partial \rho}{\partial t}\right]_F = \kappa(2a\rho a^\dagger - a^\dagger a \rho - \rho a^\dagger a) \quad (100)$$

with κ the cavity mode decay rate, the full laser master equation can thus be written as

$$\frac{\partial \rho}{\partial t} = \frac{i}{\hbar}[\rho, H_0] + \left[\frac{\partial \rho}{\partial t}\right]_A + \left[\frac{\partial \rho}{\partial t}\right]_F \tag{101}$$

The quantum statistics of the laser can be computed by transforming the master equation into a c-number Fokker-Planck equation using the methods developed by Haken.[45] This Fokker-Planck equation is equivalent to a set of coupled stochastic differential equations for the cavity field amplitude, the atomic polarization, and the atomic inversion. With various approximations, these stochastic equations can be manipulated or solved to give information on, for instance, the laser intensity and linewidth. A standard approximation in laser theory that we shall make is that the atomic variables relax to their steady-state values much faster than the cavity field variables. Thus, one may adiabatically eliminate the atomic variables and derive equations for the field variables alone.

A further simplification corresponds to taking the semiclassical limit, in which noise terms in the stochastic differential equations are dropped on the assumption that quantum fluctuations are small. In this limit, the equations of motion for the real and imaginary parts of the cavity field amplitude reveal some of the most interesting features of the squeezed-pump laser model. These equations can be written in terms of the gradient of a potential as

$$\dot{\alpha}_X = -\frac{\partial \Phi(\alpha_X, \alpha_Y)}{\partial \alpha_X} \qquad \dot{\alpha}_Y = -\frac{\partial \Phi(\alpha_X, \alpha_Y)}{\partial \alpha_Y} \tag{102}$$

with

$$\Phi(\alpha_X, \alpha_Y) = \tfrac{1}{2}\kappa\left[\alpha_X^2 + \alpha_Y^2 - Cn_0 \ln R(\alpha_X, \alpha_Y)\right] \tag{103}$$

Here $C = 2g^2 \mathcal{N}/\kappa(2N_p + 1)^2$ is the cooperativity parameter, $n_0 = \gamma_a^2(2N_p + 1)^2/8g^2$ is the saturation photon number, and

$$R(\alpha_X, \alpha_Y) = (1 - m^2) + \frac{\alpha_X^2}{n_0}(1 - m) + \frac{\alpha_Y^2}{n_0}(1 + m) \tag{104}$$

with $m = M_p/(N_p + 1/2)$. Stable steady-state solutions of the equations are found to be $\alpha = 0$ for $C \leq 1 - m$, and $\alpha = \pm i\sqrt{n_0[C - (1 - m)]}$ for $C > 1 - m$. In the usual incoherently pumped laser (i.e., without squeezing), the solution above threshold (where $\alpha \neq 0$) is independent of

an experimental point of view, this would certainly be more amenable to realization than the squeezed-pump lasers of the previous section. In the master equation, this corresponds to setting $M_p = 0$ and modifying the cavity mode decay contribution to

$$
\begin{aligned}
\left[\frac{\partial \rho}{\partial t}\right]_F &= \kappa(N_c + 1)(2a\rho a^\dagger - a^\dagger a\rho - \rho a^\dagger a) \\
&+ \kappa N_c(2a^\dagger \rho a - aa^\dagger \rho - \rho aa^\dagger) \\
&- \kappa M_c(2a^\dagger \rho a^\dagger - a^\dagger a^\dagger \rho - \rho a^\dagger a^\dagger) \\
&- \kappa M_c^*(2a\rho a - aa\rho - \rho aa)
\end{aligned}
\tag{106}
$$

The approach to analyzing this system is similar to that described in the previous section. As before, one finds that the phase symmetry of the laser is broken with the addition of squeezing, and stable steady-state solutions are again in phase with the low-noise quadrature of the squeezed field. Detailed studies of the properties of the squeezed-cavity laser have been presented by Ginzel et al.[47] and Marte and Wall.[42] Substantial linewidth narrowing (up to 50% in principle) is found to occur only about an unstable steady-state and is thus a transient effect. For stable steady states, the linewidth is generally increased, and any decrease that may occur for suitable parameters is very small and of little practical interest.

VI. CAVITY QED WITH SQUEEZED LIGHT INPUTS

The inhibition of atomic phase decay by squeezed light, and related effects such as those described in previous sections, have yet to be confirmed experimentally. The major stumbling block to most of this work is the requirement of an ideal squeezed-vacuum-atom coupling so that the atom interacts only (or at least primarily) with squeezed modes of the radiation field. This is a significant practical problem which Gardiner[3] pointed out in his original paper, stating the need for either an incoming squeezed electric dipole wave, or an appropriate one-dimensional situation.

The first suggestion presents somewhat formidable practical problems in holding atoms still at the focus, and in minimizing Doppler effects. The second approach is considerably more realistic and corresponds to the use of optical cavities, which modify the properties of the vacuum modes in a particular region of space and allow one to contemplate experiments with essentially one-dimensional beams of squeezed light. In addition, there are different types of cavities that one might wish to consider. The division that we shall make in this section is between two particular cavity configurations, both of which are of current experimental interest (and have in fact been realized), and both of which offer the possibility of seeing effects of squeezed light in atomic spectroscopy.

The first is the microcavity, or half-wavelength Fabry-Pérot cavity, in which two parallel plane mirrors are separated by only one half of a wavelength. In this regime, the dynamics of the atom are essentially the same as in free space, only now a strong spatial selection of radiation modes coupling to the atom can be achieved, so that only a relatively small solid angle of incoming squeezed light is necessary.

In the other scenario, we consider another (larger) kind of cavity that is being used at present in experiments testing the predictions of cavity QED in the strong-coupling regime.[48] In this regime, the dipole coupling strength between an atom and the field mode of an optical cavity is comparable to or larger than the dissipative rates associated with spontaneous emission and cavity losses. The atom-plus-cavity should therefore be regarded as a single multilevel system, with its own characteristic properties and responses to squeezed light.

A. The Microscopic Fabry-Pérot Cavity

Experimental support for this approach is provided by the work of De Martini and collaborators[49] on enhanced and inhibited spontaneous emission in microscopic cavities consisting of plane mirrors separated by a distance L of the order of the transition wavelength λ. The principle behind the approach is as follows. For $L = \lambda/2$, atoms whose dipole moments are parallel to the mirrors couple strongly and exclusively (for a cavity of sufficiently high finesse) to the modes whose propagation vectors lie within a small solid angle about a line perpendicular to the mirror surfaces. Hence, one need only squeeze these modes to achieve an effective squeezed-vacuum-atom coupling.[50] Such a scheme should also suit the likely source of squeezed light, a degenerate parametric oscillator, which in present experiments produces a near-plane-wave output.

The cavity configuration that we have in mind is depicted in Fig. 13, where R denotes the (real) reflectivity of the mirror. To illustrate the scheme quantitatively, we begin with the mode function for the field inside the cavity, which can be represented in terms of a sum of incident and reflected waves as

$$\mathbf{f}_{\mathbf{k}s}(\mathbf{r}) = V^{-1/2} g(k_z) \left(\boldsymbol{\varepsilon}_{\mathbf{k}s} \, e^{i\mathbf{k}\cdot\mathbf{r}} + \boldsymbol{\varepsilon}'_{\mathbf{k}s} \, e^{i\mathbf{k}'\cdot\mathbf{r}} \right) \tag{107}$$

where V is the quantization volume, $\boldsymbol{\varepsilon}_{\mathbf{k}s}$ and $\mathbf{k}$ are polarization and wave

A. S. PARKINS

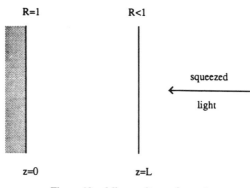

Figure 13. Microcavity configuration.

vectors, respectively (the prime denotes the reflected wave), and

$$|g(k_z)|^2 = \frac{1 - R^2}{(1 - R)^2 + 4R\sin^2(k_z L)} \tag{108}$$

The effect of the cavity is most clearly exhibited in the form of $|g(k_z)|^2$, which is identified as the Airy function of the cavity. If R is close to one, then this function displays a series of sharp peaks for angles of incidence such that $\sin(k_z L) = \sin(kL \cos\theta_k) = 0$. If, therefore, $L = \lambda/2$, the function $|g(k_z)|^2$ will exhibit a sharp peak centered at $\cos\theta_k = 1$ (the peak at $\cos\theta_k = 0$ can be ignored, since other factors entering the calculations at a later stage are zero at this angle); that is, a strong coupling is effected only with those modes in a small solid angle about the z axis.

To describe the incoming squeezed field, the one-dimensional models of earlier sections must be generalized to three dimensions, and this is achieved with the definitions

$$[a(\mathbf{k}, s), a^\dagger(\mathbf{k}', s')] = \delta^3(\mathbf{k} - \mathbf{k}')\delta_{ss'} \tag{109}$$

$$\langle a^\dagger(\mathbf{k}, s)a(\mathbf{k}', s')\rangle = N(k)U_s^*(\mathbf{k})U_{s'}(\mathbf{k}')\delta(k - k')/k^2 \tag{110}$$

$$\langle a(\mathbf{k}, s)a(\mathbf{k}', s')\rangle = M(k)U_s(\mathbf{k})U_{s'}(\mathbf{k}')\delta(2k_0 - k - k')/kk' \tag{111}$$

where $k_0 = \omega_0/c$, $|M(k)|^2 \le N(k)[N(k) + 1]$, and $U_s(\mathbf{k})$ is the square-normalized mode function of the input squeezed field. This mode function is assumed to be zero outside a cone centered on the z axis and bounded by the angle θ_{sq}. In this way, a squeezed field of limited angular extent can be modeled.

The choice of $U_s(\mathbf{k})$ within the cone of squeezing is critical. Put simply, the incident squeezed field needs to be as closely mode-matched to the cavity as possible in order for the effects of the squeezing to be maximized. In particular, optimal matching is achieved with the choice

$$U_s(\mathbf{k}) = [\mathcal{N}(k)]^{-1/2}\mathbf{d}^* \cdot \mathbf{f}^*_{\mathbf{k}s}(\mathbf{h}_0) \qquad \text{for } \theta_k \leq \theta_{\mathrm{sq}} \qquad (112)$$

where $\mathbf{d}$ is the transition dipole moment and $\mathcal{N}(k)$ is a normalization constant. Here $\mathbf{h}_0 = (0, 0, h_z)$ is the position of the point at which the squeezed light is considered to be focused.

Because of the small mirror spacing, the microcavity will in general be a very low Q cavity, and thus the process of spontaneous emission inside the cavity can be regarded as irreversible, and we can employ the familiar Born and Markov approximations to derive equations of motion for the atomic variables. With the additional assumption that the squeezed field is broadband, one can derive equations of the form (32), with, in particular,

$$\langle \dot{\sigma}_y \rangle = -\frac{\gamma_a}{2} B(\mathbf{h}) \langle \sigma_y \rangle \qquad (113)$$

where $\mathbf{h}$ is the position of the atom, and

$$B(\mathbf{h}) = 3(N - M)\frac{1 + R}{1 - R}\frac{1}{\mathcal{N}'}$$

$$\times \left\{ \left[\int_{\cos\theta_{\mathrm{sq}}}^1 du(1 + u^2)\frac{\sin^2(k_0 h_z u)}{1 + F\sin^2(k_0 Lu)}J_0\!\left(k_0 h_x\sqrt{1 - u^2}\right) \right]^2 \right\}$$

$$+ \frac{3}{2}\frac{1 + R}{1 - R}\int_0^1 du(1 + u^2)\frac{\sin^2(k_0 h_z u)}{1 + F\sin^2(k_0 Lu)}J_0\!\left(k_0 h_x\sqrt{1 - u^2}\right)$$

$$(114)$$

with

$$\mathcal{N}' = \int_{\cos\theta_{\mathrm{sq}}}^1 du(1 + u^2)\frac{\sin^2(k_0 h_z u)}{1 + F\sin^2(k_0 Lu)}J_0\!\left(k_0 h_x\sqrt{1 - u^2}\right) \quad (115)$$

and $F = 4R/(1 - R)^2$. J_0 is the zeroth-order Bessel function. Free space conditions can be modeled by the limits $R = 0$ and $h_z \to \infty$, and without squeezing ($N = M = 0$) in these limits one finds $B(\mathbf{h}) = 1$.

Considering the case where $k_0 L = \pi$ ($L = \lambda/2$), the optimum squeezing effect occurs for $h_x = 0$ and $h_z = L/2$ (i.e., the atom is at the center of the cavity). An approximate expression for B can be derived in the form

$$B(h_x = 0, h_z = L/2) \simeq \frac{3}{\pi} \frac{1+R}{2\sqrt{R}} \Big\{ \tan^{-1}(\pi\sqrt{F})$$

$$+ 2(N - M)\tan^{-1}\big[\pi\sqrt{F}(1 - \cos\theta_{sq})\big]\Big\} \quad (116)$$

Remembering that $N - M \to -\frac{1}{2}$ in the strong squeezing limit, the term in brackets may become very small if one chooses $\theta_{sq} = \pi/2$. However, if F is large, this requirement on the angular extent of squeezing can be relaxed; for instance, if $R = 0.99$ and $\theta_{sq} = 0.2$ radians, then $\tan^{-1}(\pi\sqrt{F}) = 1.57$, while $\tan^{-1}[\pi\sqrt{F}(1 - \cos\theta_{sq})] = 1.49$. Hence, a substantial reduction in the decay rate is still possible with a squeezed light input of quite modest angular dimensions.

The exact solution for $B(\mathbf{h})$, as a function of h_x and h_z, is shown in Fig. 14. It is also clear from this figure that the squeezing effect persists in the x direction for several times the width L of the cavity.

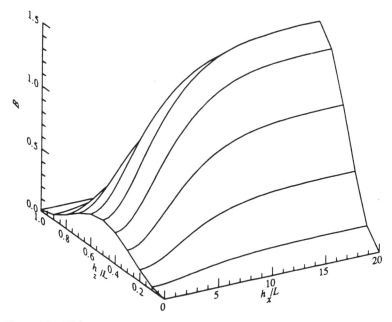

Figure 14. $B(\mathbf{h})$ as a function of h_x and h_z for $L = \lambda/2$, $R = 0.99$, $N = 1$, and $\theta_{sq} = 0.3$.

A more detailed study of the effects of varying parameters such as R, L, and θ_{sq} is presented by Parkins and Gardiner.[50] These authors also consider the effect of nonideal mode-matching of the squeezed field to the cavity, demonstrating the importance of good matching to achieving a sizeable reduction in the decay rate. Ficek[39] carried out a similar analysis in the context of two-atom systems in squeezed vacuum fields.

B. Cavity QED in the Strong-Coupling Limit

1. The Jaynes-Cummings Model: Collapses and Revivals

The resonant interaction of a two-level atom with a single mode of the radiation field is a perennial problem in quantum optics. The most well-known feature arising from this problem is the collapse and revival of the Rabi oscillations of the atomic inversion. This results from the quantum statistical description of the single-mode field in terms of a superposition of number states $|n\rangle$ weighted according to some distribution $P(n)$. The behavior of the atomic inversion is found from a summation of its responses to each of the number states, and, within this summation, dephasing of the individual responses leads to interferences which result in collapses and revivals of the atomic inversion.

The Hamiltonian for this system is, in the rotating-wave approximation,

$$H_0 = \tfrac{1}{2}\hbar\omega_a\sigma_z + \hbar\omega_a a^\dagger a + i\hbar g(a^\dagger\sigma^- - a\sigma^+) \tag{117}$$

where a is the annihilation operator for the field mode, and g is the dipole coupling strength. For an atom initially in the ground state, a solution for the inversion can be derived in the form

$$\langle\sigma_z(t)\rangle = -\frac{1}{2}\sum_{n=0}^{\infty} P(n)\cos(2gn^{1/2}t) \tag{118}$$

Usually, no analytic expression for this sum exists, but the general behavior of $\langle\sigma_z(t)\rangle$ can be described by noting that (118) can be written as the product of a rapidly oscillating term of the form $\cos(2g\bar{n}^{1/2}t)$ and a slowly varying envelope function $E(t)$, i.e.,

$$\langle\sigma_z(t)\rangle = -\tfrac{1}{2}\operatorname{Re}\left[\exp(2ig\bar{n}^{1/2}t)E(t)\right] \tag{119}$$

where

$$E(t) = \sum_{n=0}^{\infty} P(n)\exp\left[2ig(n^{1/2} - \bar{n}^{1/2})t\right] \tag{120}$$

and $\bar{n}$ is the mean photon number of the field. Interferences between the different components (with different oscillation frequencies) in the summation cause $E(t)$ to first approach zero and then increase again, hence the collapse and revival. The time taken for the most complete revival is determined by the rate of oscillation of the slowest components of $E(t)$, i.e., by the components with $n = \bar{n} \pm 1$. If $\bar{n} \gg 1$, this revival time is given by

$$t_R = \frac{2\pi\bar{n}^{1/2}}{g} \tag{121}$$

To estimate the collapse time, we assume that $P(n)$ has a characteristic width (Δn), and that the characteristic time taken for the envelope function to collapse to zero is given by the time taken for those components with $n = \bar{n} \pm (\Delta n)$ to rotate through an angle of π. Given $\bar{n} \gg (\Delta n)$, the collapse time is thus

$$t_C \simeq \frac{\pi\bar{n}^{1/2}}{g(\Delta n)} \tag{122}$$

which can evidently depend on the photon number variance $(\Delta n)^2$.

Taking note of this result, Milburn[51] considered the situation in which the initial state of the field is a squeezed state $|\alpha, r\rangle$, where $|\alpha, r\rangle$ is generated from the vacuum via the transformation

$$|\alpha, r\rangle = D(\alpha)S(r)|0\rangle \tag{123}$$

where $S(r) = \exp\{\frac{1}{2}r[a^2 - (a^\dagger)^2]\}$ $(r \geq 0)$ is the "squeeze" operator, and $D(\alpha)$ is the coherent displacement operator (α is the coherent amplitude). The mean photon number for this state is

$$\bar{n} = \langle a^\dagger a\rangle = |\alpha|^2 + \sinh^2(r) \tag{124}$$

where the two contributions to $\bar{n}$ are from the coherent excitation and the squeezing, respectively. The variance of the photon number distribution for the state $|\alpha, r\rangle$ can be shown to be

$$(\Delta n)^2 = \tfrac{1}{2}\sinh^2(2r) + |\alpha|^2[\cosh(2r) - \sinh(2r)\cos(2\theta)] \tag{125}$$

where θ is the phase of α. If the coherent contribution to $\bar{n}$ is dominant, then for the two limiting choices of phase $\theta = \pi/2$ and $\theta = 0$, one can write

$$(\Delta n)^2 \simeq \bar{n}e^{\pm 2r} \qquad (126)$$

Hence, the photon number distribution can be broader ($\theta = \pi/2$) or narrower ($\theta = 0$) than for a coherent state alone ($r = 0$), corresponding to superpoissonian or subpoissonian statistics, respectively.

It follows from (122) and (126) that the collapse time for Rabi oscillations in the Jaynes-Cummings model with the field initially in a squeezed state as described above is approximated by $t_C \simeq \pi e^{\pm r}/g$, which may be a shorter or longer time than for an initial coherent state. In a similar fashion, one may also show that the time taken for the onset of the first revival is also proportional to $1/(\Delta n)$, so, for instance, an increase in the collapse time is accompanied by a delay in the onset of the revival. Numerical evaluation of the exact expression for the atomic inversion (118), with appropriately chosen parameters, confirms these features.[51]

An interesting feature of the photon number distribution for a squeezed coherent state is that it is oscillatory for sufficiently strong squeezing (i.e., for sufficiently large r). In such a regime, the coherent contribution to the photon number variance $(\Delta n)^2$ may no longer be assumed to dominate the squeezed contribution. It follows that the predictions of Milburn are no longer valid, and that new phenomena may arise. This is indeed the case, as shown by Satyanarayana et al.,[52] who demonstrated that the oscillatory photon number distribution produces an oscillatory envelope, or echoes, during the collapse that follows each revival. These echoes are essentially the result of the different peaks of the photon number distribution, which define different rephasing times for the cosines $\cos(2gn^{1/2}t)$ in the expression for $\langle \sigma_z(t) \rangle$ (we note that some subtle interferences between the contributions from different peaks are also important to the formation of echoes in the revivals).

2. Fluorescence Spectrum of an Atom Strongly Coupled to a Cavity Driven by Broadband Squeezed Light

The models of the previous section are idealized, in that they neglect dissipation arising, for example, from spontaneous emission into other modes of the radiation field. Indeed, for the observation of emitted spectra, dissipative channels of some sort must be introduced. A more realistic and experimentally relevant model is thus one that incorporates couplings of the field mode and atom to reservoirs, enabling cavity losses and spontaneous emission to be taken into account, and providing the input channel for a multimode squeezed vacuum (as will likely be used in any experiment). Savage[53] has presented such a model in a study of the fluorescence spectrum of an atom strongly coupled to a cavity driven by broadband squeezed light. As in work described earlier in this paper, the starting point for calculations is the master equation, which for this particular configuration takes the form (in the interaction picture)

$$\frac{\partial \rho}{\partial t} = \frac{1}{i\hbar}[H_0, \rho] + \left(\frac{\partial \rho}{\partial t}\right)_A + \left(\frac{\partial \rho}{\partial t}\right)_F \qquad (127)$$

where

$$H_0 = i\hbar g(a^\dagger \sigma^- - a\sigma^+) + i\hbar E(a - a^\dagger) \qquad (128)$$

and

$$\left(\frac{\partial \rho}{\partial t}\right)_A = \frac{\gamma_a}{2}(2\sigma^- \rho \sigma^+ - \sigma^+ \sigma^- \rho - \rho \sigma^+ \sigma^-) \qquad (129)$$

$$\begin{aligned}
\left(\frac{\partial \rho}{\partial t}\right)_F = \ & \kappa(N+1)(2a\rho a^\dagger - a^\dagger a\rho - \rho a^\dagger a) \\
& + \kappa N(2a^\dagger \rho a - aa^\dagger \rho - \rho aa^\dagger) \\
& - \kappa M(2a^\dagger \rho a^\dagger - a^\dagger a^\dagger \rho - \rho a^\dagger a^\dagger) \\
& - \kappa M^*(2a\rho a - aa\rho - \rho aa)
\end{aligned} \qquad (130)$$

That is, a broadband squeezed vacuum (whose bandwidth is much larger than the cavity mode decay rate κ) is injected into the cavity through one of its mirrors (the other mirror is assumed to be perfect). The first term in H_0 is the Jaynes-Cummings dipole interaction term, while the second term describes a resonant coherent field of real amplitude E driving the cavity mode. The term (129) describes spontaneous emission of the atom into modes other than the cavity mode (i.e., out the sides of the cavity).

To solve the master equation and compute spectra via the quantum regression theorem, Savage employed numerical integration and a truncated basis set. The focus of his attention was on the Rabi sidebands of the fluorescence triplet emitted by the atom out the sides of the cavity under conditions of strong dipole coupling and strong coherent driving field strength, $E \gg g \gg \kappa, \gamma_a$. For different choices of the relative phase between the coherent and squeezed fields, the Rabi sidebands can be broader or narrower than those found in the absence of squeezing, as shown in Fig. 15 (in fact, the parameters for this figure do not fully satisfy the condition above, but the effects are still very clear). This effect can be directly related to the work of Milburn[51] described in the previous section.

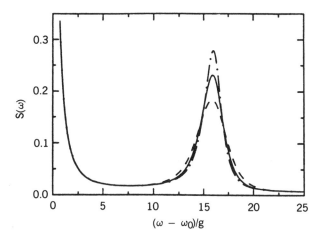

Figure 15. Incoherent fluorescence spectrum $S(\omega)$ for $\kappa/g = \gamma_a/g = 1$, $E/g = 8$, with no squeezing (solid line), with in-phase squeezing, $N = 0.125$ (dot-dashed line), and with in-quadrature squeezing, $N = 0.125$ (dashed line). (Reproduced with permission from Savage.[53])

Under the conditions of strong coupling and strong coherent driving field considered by Savage, the widths of the sidebands are in fact determined by the collapse time for Rabi oscillations in the Jaynes-Cummings model, which, as pointed out by Milburn, varies depending on whether the input field has superpoissonian, poissonian, or subpoissonian statistics. It should also be pointed out that, under such conditions of strong coupling and driving, the widths of the sidebands are no longer determined by linewidths associated with dissipation (in fact, the linewidths of the sidebands are much larger than κ and γ_a).

The above model has been extended to the many-atom case by Kennedy and Walls,[15] but linewidth narrowing, which one might consider to be the most dramatic signature of squeezing, is no longer a significant feature. Kennedy and Walls compute the optical spectra of the light transmitted through, and reflected from, the cavity. In the transmitted spectra, they find that it is possible to partially suppress the vacuum Rabi peaks in the strong coupling regime, while in the reflected spectra interference between the reflected squeezed light and the light exiting the cavity is found to produce some interesting (phase-dependent) spectral structures.

3.   Squeezed Light Excitation of a "Vacuum" Rabi Resonance

Another scheme for observing effects of squeezed light has been proposed, which, although still operating in the strong-coupling regime, recovers the essential features of the original two-level-atom formulations of Section III.[54] Of course, this scheme retains the advantage that one need only drive a single cavity mode with squeezed light. The proposal is based on the observation by Tian and Carmichael[55] that for strong dipole coupling the atom–cavity system behaves as a two-state system when excited by a coherent field near one of the "vacuum" Rabi resonances. This behavior is a consequence of the unequal spacing of the energy levels of the coupled atom and cavity mode. It follows that if we replace the coherent field by a squeezed field then we should realize a two-state system coupled to a squeezed field, provided that other energy levels are not appreciably populated. Population of other levels is avoided by employing finite-bandwidth squeezed light which, because of the unequal spacing of the energy levels, facilitates a selective excitation of the "vacuum" Rabi transition.

In particular, one can take as the source of squeezed light the output from a degenerate parametric amplifier, in which case the input to the atom–cavity system is described by the correlation functions

$$\langle a_{\text{in}}^{\dagger}(t)a_{\text{in}}(0)\rangle = \frac{b_+^2 - b_-^2}{4}\left[\frac{e^{-b_-|t|}}{2b_-} - \frac{e^{-b_+|t|}}{2b_+}\right] \tag{131}$$

$$\langle a_{\text{in}}(t)a_{\text{in}}(0)\rangle = \frac{b_+^2 - b_-^2}{4}\left[\frac{e^{-b_-|t|}}{2b_-} + \frac{e^{-b_+|t|}}{2b_+}\right] \tag{132}$$

where b_+, b_- are the bandwidths of the two quadratures.

The squeezed light is incident upon one side of a (two-sided) cavity containing a single two-level atom that is resonant with the cavity mode at frequency ω_0. The interaction between the atom and cavity mode is treated in the rotating-wave and dipole approximations, and is again described by the Jaynes-Cummings Hamiltonian (in a frame rotating at frequency ω_0):

$$H_0 = \hbar\delta\left(a^{\dagger}a + \tfrac{1}{2}\sigma_z\right) + i\hbar g(a^{\dagger}\sigma^- - a\sigma^+) \qquad \delta = \omega_a - \omega_0 \tag{133}$$

The ground state of the Hamiltonian H_0 is $|g\rangle = |0\rangle|-\rangle$, with energy $E_g = -\hbar\delta/2$, and the excited states are the "dressed-states" $(n = 1, 2, \ldots)$

$$|n, u\rangle = \frac{1}{\sqrt{2}}(|n-1\rangle|+\rangle + i|n\rangle|-\rangle) \tag{134}$$

$$|n, l\rangle = \frac{1}{\sqrt{2}}(|n-1\rangle|+\rangle - i|n\rangle|-\rangle) \tag{135}$$

with energies $E_{n,u} = \hbar[\delta(n - \tfrac{1}{2}) + g\sqrt{n}]$ and $E_{n,l} = \hbar[\delta(n - \tfrac{1}{2}) - g\sqrt{n}]$. Here $|+\rangle$ and $|-\rangle$ are the upper and lower states of the atom, and $|n\rangle$,

$n = 0, 1, 2, \ldots$, are the Fock states for the cavity field mode. Single photon transitions between the energy eigenstates of H_0 occur at the frequencies $\delta \pm [\sqrt{n} \pm \sqrt{n-1}]g$. Transitions between the ground and first excited states $(|1, u\rangle, |1, l\rangle)$ generate the "vacuum" Rabi peaks at frequencies $(E_{1,u} - E_g)/\hbar = \delta + g$ and $(E_{1,l} - E_g)/\hbar = \delta - g$. These peaks have been observed in recent experiments.[48]

The cavity is assumed to consist of two mirrors with loss rates κ_1 and κ_2, respectively. Squeezed light enters the cavity through mirror 1, while the input to mirror 2 is simply vacuum fluctuations. The spontaneous emission rate of the atom into modes other than the cavity mode is given by γ_a. With the assumption that b_- and b_+ are much larger than the characteristic decay rates of the system (Markov approximation), and using standard methods (see, for example, Ref. 6), one can derive the following master equation describing the coupled atom and cavity mode with a squeezed input of the sort described above:

$$
\begin{aligned}
\dot{\rho}(t) = &-\frac{i}{\hbar}[H_0, \rho(t)] + \frac{\gamma_a}{2}(2\sigma^-\rho\sigma^+ - \sigma^+\sigma^-\rho - \rho\sigma^+\sigma^-) \\
&+ (\kappa_1 + \kappa_2)(2a\rho a^\dagger - a^\dagger a\rho - \rho a^\dagger a) \\
&+ \frac{\kappa_1}{2}(N - M)\int_0^\infty d\tau \, b_+ e^{-b_+\tau} \\
&\quad \times \big[[\rho(t), a(-\tau) - a^\dagger(-\tau)], a - a^\dagger\big] \\
&- \frac{\kappa_1}{2}(N + M)\int_0^\infty d\tau \, b_- e^{-b_-\tau} \\
&\quad \times \big[[\rho(t), a(-\tau) + a^\dagger(-\tau)], a + a^\dagger\big]
\end{aligned}
\tag{136}
$$

where $N - M = -(b_+^2 - b_-^2)/2b_+^2$, $N + M = (b_+^2 - b_-^2)/2b_-^2$, and

$$
\begin{aligned}
a(-\tau) &= e^{-iH_0\tau/\hbar}a e^{iH_0\tau/\hbar} \\
&= \frac{1}{2}\sum_{n=1}^\infty (\sqrt{n} + \sqrt{n+1})|n, u\rangle\langle n+1, u|e^{i(E_{n+1,u}-E_{n,u})\tau/\hbar} \\
&\quad + \frac{1}{2}\sum_{n=1}^\infty (\sqrt{n} + \sqrt{n+1})|n, l\rangle\langle n+1, l|e^{i(E_{n+1,l}-E_{n,l})\tau/\hbar} \\
&\quad + \frac{1}{2}\sum_{n=1}^\infty (\sqrt{n} - \sqrt{n+1})|n, u\rangle\langle n+1, l|e^{i(E_{n+1,l}-E_{n,u})\tau/\hbar} \\
&\quad + \frac{1}{2}\sum_{n=1}^\infty (\sqrt{n} - \sqrt{n+1})|n, l\rangle\langle n+1, u|e^{i(E_{n+1,u}-E_{n,l})\tau/\hbar} \\
&\quad + \frac{i}{\sqrt{2}}|g\rangle\langle 1, u|e^{i(E_{1,u}-E_g)\tau/\hbar} - \frac{i}{\sqrt{21}}|g\rangle\langle 1, l|e^{i(E_{1,l}-E_g)\tau/\hbar}
\end{aligned}
\tag{137}
$$

It should be noted that while it has been assumed that the squeezed-input bandwidths are larger than the cavity and atomic dissipative rates, no assumption has been made about the relative sizes of b_-, b_+ and δ, g, hence the form of $a(-\tau)$. In the expression for $a(-\tau)$ the characteristic dressed-state transition frequencies appear, and each transition contributes an integral of the form $\int_0^\infty d\tau \exp[-b_{\pm}\tau + i(E_k - E_j)\tau/\hbar]$ to the master equation, where the subscripts k, j denote particular eigenstates. Hence, each integral is related to the spectral density of the input squeezed field at a particular frequency.

Following the approach of Tian and Carmichael,[55] the carrier frequency of the squeezed light is tuned to the transition $|g\rangle \rightarrow |1, l\rangle$, as depicted in Fig. 16. To avoid population of higher states one requires that $g \gg b_X, b_Y$. Because of the unequal spacing of the energy levels in the coupled atom–cavity system, this means that the squeezed light input field is not resonant with any other transitions in the dressed-state ladder, and hence populations in the higher dressed states should be negligible. To be a little more precise, the next closest transition frequency corresponds to the transition $|1, l\rangle \rightarrow |2, l\rangle$, with $(E_{2,l} - E_{1,l})/\hbar = \delta - (\sqrt{2} - 1)g \simeq 0.6g$ for $\delta = g$; so, in particular, if $0.6g \gg b_X, b_Y$, then the state $|2, l\rangle$ should not be appreciably populated and the two-state approximation holds. In

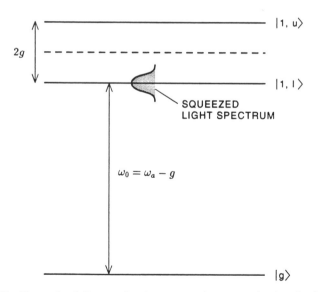

Figure 16. Energy level diagram for the atom–cavity system showing the first pair of dressed states $\{|1, l\rangle, |1, u\rangle\}$. The carrier squeezing frequency is tuned to the lower "vacuum" Rabi resonance.

this way it is possible to effect a two-state system coupled to a squeezed vacuum exhibiting all of the features expounded in the original work of Gardiner.[3]

The two-state approximation follows from the master equation by neglecting all of the integrals except for the one corresponding to the resonant interaction, $(E_{1,l} - E_g)/\hbar = \delta - g = 0$. Neglecting population in all but the two lowest levels, and defining $u = \langle g|\rho|1,l \rangle$ + c.c., $v = -i\langle g|\rho|1,l \rangle$ + c.c., and $w = \langle 1,l|\rho|1,l \rangle - \langle g|\rho|g \rangle$, one finds

$$\dot{u} \simeq -\frac{1}{2}\left[\kappa_1(2N + 2M + 1) + \kappa_2 + \frac{\gamma_a}{2}\right]u \equiv -\gamma_x u \qquad (138)$$

$$\dot{v} \simeq -\frac{1}{2}\left[\kappa_1(2N - 2M + 1) + \kappa_2 + \frac{\gamma_a}{2}\right]v \equiv -\gamma_y v \qquad (139)$$

$$\dot{w} \simeq -\left(\kappa_1 + \kappa_2 + \frac{\gamma_a}{2}\right) - \left[\kappa_1(2N + 1) + \kappa_2 + \frac{\gamma_a}{2}\right]w \qquad (140)$$

These equations are of the form derived in Section III.A, except for the additional decay rates κ_2 and γ_a, which correspond to the decay channels that are not subject to squeezed light. For strong squeezing $N - M \to -\frac{1}{2}$ and the decay of v is inhibited. For an appreciable effect to be visible one obviously requires that κ_1 be at least of the order of $(\kappa_2 + \gamma_a/2)$.

Remaining in the two-state approximation, and using the quantum regression theorem, the following two-time correlation functions can be evaluated as

$$\langle \sigma^+(\tau)\sigma^-(0) \rangle \simeq \langle a^\dagger(\tau)a(0) \rangle$$
$$\simeq \frac{\kappa_1 N/4}{\kappa_1(2N + 1) + \kappa_2 + \gamma_a/2}[e^{-\gamma_x|\tau|} + e^{-\gamma_y|\tau|}] \qquad (141)$$

Hence, in this approximation the spectrum of light radiated by the atom out the sides of the cavity, and the spectrum of light transmitted through mirror 2 of the cavity (given by the Fourier transforms of the above correlation functions), will exhibit a peak of subnatural linewidth $2\gamma_y$ (i.e., $2\gamma_y$ is less than the natural linewidth, $(\kappa_1 + \kappa_2 + \gamma_a/2)$, of the cavity-plus-atom system), superimposed on a broad peak of supernatural linewidth $2\gamma_x$.

The two-state model is of course an idealization, and for more realistic or practical situations account must be taken of higher energy levels. To this end one can compute spectra from steady-state (numerical) solutions to the master equation (136) using a truncated basis set. Results for the

where

$$\gamma_x = \frac{\gamma_a}{2} + \frac{\gamma_c}{2}\frac{\kappa^2}{\kappa^2 + \Omega_0^2}(2N - 2\varepsilon M + 1) \tag{149}$$

$$\gamma_y = \frac{\gamma_a}{2} + \frac{\gamma_c}{2}(2N + 2\varepsilon M + 1) \qquad \gamma_z = \gamma_x + \gamma_y \tag{150}$$

$$\Omega_1 = \Omega_0\frac{\kappa\gamma_c/2}{\kappa^2 + \Omega_0^2} \qquad \Omega_2 = \Omega_0\left[1 + \frac{\kappa\gamma_c/2}{\kappa^2 + \Omega_0^2}(2N - 2\varepsilon M + 1)\right] \tag{151}$$

$$\gamma_1 = \gamma_a + \gamma_c\frac{\kappa^2 + \Omega_0^2/2}{\kappa^2 + \Omega_0^2} \tag{152}$$

and $\varepsilon = \pm 1$, corresponding to the coherent driving field being in phase or out of phase, respectively, with the maximally squeezed quadrature of the squeezed input field.

Equations of this form, but without squeezing ($N = M = 0$), have been used by Lewenstein et al.[60] to describe the dynamical suppression of spontaneous emission in a cavity, whereby for $\Omega_0 \gg \kappa$ decay into the cavity modes is inhibited. This is because, for $\Omega_0 \gg \kappa$, the cavity does not support the fluctuations at frequencies $\omega_0 \pm \Omega_0$ which normally induce the decay of the component $\langle\sigma_x\rangle$ (i.e., the second term in the expression for γ_x is negligible). Fluctuations at the cavity resonance frequency ω_0 may still induce the decay of $\langle\sigma_y\rangle$ and $\langle\sigma_z\rangle$, but with the addition of squeezing as above these fluctuations may be reduced (for an appropriate choice of driving field phase, i.e., for $\varepsilon = -1$), and consequently, with respect to the cavity modes, the decay of all three components of the Bloch vector may be inhibited. In terms of the fluorescence spectrum, one would therefore expect all three components of the Mollow triplet to exhibit narrowing, provided, of course, that γ_c is of the order of γ_a.

Acknowledgments

Over the period of time covered by the work presented in this review, I have benefited immensely from collaborations and discussions with H. J. Carmichael, J. I. Cirac, M. J. Collett, C. W. Gardiner, T. A. B. Kennedy, A. S. Lane, M. A. M. Marte, M. D. Reid, H. Ritsch, B. C. Sanders, C. M. Savage, D. F. Walls, and P. Zoller. The work at JILA is supported in part by the NSF.

References

1. For references and examples, see the following special issues on squeezed states: *J. Opt. Soc. Am. B* **4**(10) (1987); and *J. Mod. Opt.* **34**(6/7) (1987).

2. E. S. Polzik, J. Carri, and H. J. Kimble, *Phys. Rev. Lett.* **68**, 3020 (1992).

3. C. W. Gardiner, *Phys. Rev. Lett.* **56**, 1917 (1986).

4. L.-A. Wu, H. J. Kimble, J. L. Hall, and H. Wu, *Phys. Rev. Lett.* **57**, 2520 (1986); L.-A. Wu, M. Xiao, and H. J. Kimble, *J. Opt. Soc. Am. B* **4**, 1465 (1987).

5. M. J. Collett and C. W. Gardiner, *Phys. Rev. A* **30**, 1386 (1984).

6. C. W. Gardiner, *Quantum Noise*, Springer, Berlin, 1991.

7. W. Louisell, *Quantum Statistical Properties of Radiation*, Wiley, New York, 1974.

8. C. W. Gardiner and M. J. Collett, *Phys. Rev. A* **31**, 3761 (1985).

9. C. W. Gardiner, A. S. Parkins, and M. J. Collett, *J. Opt. Soc. Am. B* **4**, 1683 (1987).

10. C. W. Gardiner, *IBM J. Res. Dev.* **32**, 127 (1988).

11. M. J. Collett, *Input, Output, and Dissipation in the Theory of Quantum and Classical Amplifiers*, Thesis, University of Waikato, 1983.

12. B. Yurke, *Phys. Rev. A* **29**, 408 (1984).

13. H. Ritsch and P. Zoller, *Phys. Rev. Lett.* **61**, 1097 (1988); H. Ritsch and P. Zoller, *Phys. Rev. A* **38**, 4657 (1988).

14. N. G. van Kampen, *Stochastic Processes in Physics and Chemistry*, North-Holland, Amsterdam, 1981.

15. T. A. B. Kennedy and D. F. Walls, *Phys. Rev. A* **42**, 3051 (1990).

16. J. D. Cresser, J. Häger, G. Leuchs, M. Rateike, and H. Walther, in R. Bonifacio (Ed.), *Dissipative Systems in Quantum Optics*, Springer, Berlin, 1982.

17. H. J. Carmichael, A. S. Lane, and D. F. Walls, *Phys. Rev. Lett.* **58**, 2539 (1987); *J. Mod. Opt.* **34**, 821 (1987).

18. A. S. Lane, *Generation and Applications of Squeezed Coloured Noise*, Thesis, University of Waikato, 1988.

19. J.-M. Courty and S. Reynaud, *Europhys. Lett.* **10**, 237 (1989).

20. C. Cohen-Tannoudji and S. Reynaud, *J. Phys. B* **10**, 345 (1977).

21. H. Ritsch and P. Zoller, *Opt. Commun.* **64**, 523 (1987); **66**, 333(E) (1988).

22. S. An, M. Sargent, and D. F. Walls, *Opt. Commun.* **67**, 373 (1988); S. An and M. Sargent, *Phys. Rev. A* **39**, 3998 (1989).

23. W. Zhang and W. Tan, *Opt. Commun.* **69**, 135 (1988).

24. B. R. Mollow, *Phys. Rev. A* **5**, 2217 (1972).

25. A. S. Parkins and C. W. Gardiner, *Phys. Rev. A* **37**, 3867 (1988).

26. C. W. Gardiner, *Handbook of Stochastic Methods*, Springer, Berlin, 1983.

27. J. I. Cirac and L. L. Sánchez-Soto, *Phys. Rev. A* **44**, 1948 (1991).

28. A. S. Parkins, *Phys. Rev. A* **42**, 4352 (1990); **42**, 6873 (1990).

29. See *J. Opt. Soc. Am. B* **6**, No. 11 (1989), special issue on laser cooling and trapping of atoms.

30. Y. Shevy, *Phys. Rev. Lett.* **64**, 2905 (1990).

31. R. Graham, D. F. Walls, and W. Zhang, *Phys. Rev. A* **44**, 7777 (1991).

32. J. I. Cirac and P. Zoller, *Phys. Rev. A* **47**, 2191 (1993).

33. Z. Ficek and P. D. Drummond, *Phys. Rev. A* **43**, 6247 (1991); **43**, 6258 (1991).

34. V. Bŭzek, P. L. Knight, and I. K. Kudryavtsev, *Phys. Rev. A* **44**, 1931 (1991).

35. J. Gea-Banacloche, *Phys. Rev. Lett.* **62**, 1603 (1989).

36. J. Javanainen and P. L. Gould, *Phys. Rev. A* **41**, 5088 (1990).

37. G. M. Palma and P. L. Knight, *Phys. Rev. A* **39**, 1962 (1989); Z. Ficek and P. D. Drummond, *Phys. Rev. A* **42**, 1826 (1990).

38. G. S. Agarwal and R. R. Puri, *Opt. Commun.* **69**, 267 (1989); *Phys. Rev. A* **41**, 3782 (1990).

39. Z. Ficek, *Phys. Rev. A* **42**, 611 (1990); **44**, 7759 (1991); *Opt. Commun.* **82**, 130 (1991); Z. Ficek and B. C. Sanders, *Quantum Opt.* **2**, 269 (1990).

40. L. Allen and J. H. Eberly, *Optical Resonance and Two-Level Atoms*, Wiley, New York, 1975, Chap. 9 and references therein.

41. A. S. Parkins and C. W. Gardiner, *Phys. Rev. A* **40**, 2534 (1989).

42. M. A. M. Marte and D. F. Walls, *Phys. Rev. A* **37**, 1235 (1988); M. A. M. Marte, H. Ritsch, and D. F. Walls, *Phys. Rev. Lett.* **61**, 1093 (1988); *Phys. Rev. A* **38**, 3577 (1988).

43. R. J. Glauber, in E. R. Pike and S. Sarkar (Eds.), *Frontiers in Quantum Optics*, Adam Hilger, Boston, 1986.

44. F. Haake, D. F. Walls, and M. J. Collett, *Phys. Rev. A* **39**, 3211 (1989).

45. H. Haken, *Laser Theory*, Vol. 2, Springer, Berlin, 1984.

46. J. Gea-Banacloche, *Phys. Rev. Lett.* **59**, 543 (1987).

47. C. Ginzel, J. Gea-Banacloche, and A. Schenzle, *Acta Phys. Pol. A* **78**, 123 (1990); *Phys. Rev. A* **42**, 4164 (1990); C. Ginzel, R. Schack, and A. Schenzle, *J. Opt. Soc. Am. B* **8**, 1704 (1991).

48. M. G. Raizen, R. J. Thompson, R. J. Brecha, H. J. Kimble, and H. J. Carmichael, *Phys. Rev. Lett.* **63**, 240 (1989); R. J. Thompson, G. Rempe, and H. J. Kimble, *Phys. Rev. Lett.* **68**, 1132 (1992).

49. F. De Martini, G. Innocenti, G. R. Jacobovitz, and P. Mataloni, *Phys. Rev. Lett.* **59**, 2955 (1987); see also W. Jhe, A. Anderson, E. A. Hinds, D. Meschede, L. Moi, and S. Haroche, *Phys. Rev. Lett.* **58**, 666 (1987).

50. A. S. Parkins and C. W. Gardiner, *Phys. Rev. A* **40**, 3796 (1989); **42**, 5765(E) (1990).

51. G. J. Milburn, *Opt. Acta* **31**, 671 (1984).

52. M. V. Satyanarayana, P. Rice, R. Vyas, and H. J. Carmichael, *J. Opt. Soc. Am. B* **6**, 228 (1989).

53. C. M. Savage, *Quantum Opt.* **2**, 89 (1990).

54. A. S. Parkins, P. Zoller, and H. J. Carmichael, to appear in *Phys. Rev. A*.

55. L. Tian and H. J. Carmichael, *Phys. Rev. A* **46**, R6801 (1992).

56. P. R. Rice and L. M. Pedrotti, *J. Opt. Soc. Am. B* **9**, 2008 (1992).

57. H. J. Carmichael, Interaction of nonclassical states of light and atoms, lecture notes presented at the Australian Summer School on Quantum Optics, Canberra, January 1992, unpublished.

58. J. I. Cirac, *Phys. Rev. A* **46**, 4354 (1992).

59. P. R. Rice and H. J. Carmichael, *IEEE J. Quantum Electron.* **24**, 1351 (1988).

60. M. Lewenstein, T. W. Mossberg, and R. J. Glauber, *Phys. Rev. Lett.* **59**, 775 (1987); M. Lewenstein and T. W. Mossberg, *Phys. Rev. A* **37**, 2048 (1988).

THE EFFECTIVE EIGENVALUE METHOD AND ITS APPLICATION TO STOCHASTIC PROBLEMS IN CONJUNCTION WITH THE NONLINEAR LANGEVIN EQUATION

W. T. COFFEY

School of Engineering
Department of Microelectronics & Electrical Engineering
Trinity College Dublin, Ireland

YU. P. KALMYKOV

Institute of Radio Engineering and Electronics
Russian Academy of Sciences
Fryazino, Moscow Region, Russia

E. S. MASSAWE

School of Engineering
Department of Microelectronics & Electrical Engineering
Trinity College Dublin, Ireland

CONTENTS

Modern Nonlinear Optics, Part 2, Edited by Myron Evans and Stanisław Kielich. Advances in Chemical Physics Series, Vol. LXXXV.
ISBN 0-471-57546-1 © 1993 John Wiley & Sons, Inc.

I. INTRODUCTION

The concept of the effective eigenvalue appears to have been originally introduced into the study of relaxation problems in statistical physics by Leontovich.[1] It was later developed and applied (sometimes implicitly) to a variety of stochastic problems in laser physics,[2-4] magnetic domains,[5, 6] polar fluids,[7] polymers,[8] nematic liquid crystals,[9, 10] etc. In the present context, namely the theory of the Brownian motion, the method constitutes a truncation procedure which allows one using simple assumptions to obtain closed-form approximations to the solution of certain infinite hierarchies of differential-difference equations in the time variables. These equations govern the time behavior of the statistical averages characterizing the relaxation of nonlinear stochastic systems. Thus, their solution is needed to calculate observable quantities such as the relaxation times and dynamic susceptibilities of the system.

The infinite hierarchies of differential-difference equations may arise in each of three entirely equivalent ways from the nonlinear Langevin equation (it should be understood before proceeding any further that in the present review we are concerned only with the response in the long-time or noninertial limit):

1. By averaging the noninertial Langevin equation regarded as a stochastic differential-difference equation of the Stratonovich type[11]

2. By averaging the inertial Langevin equation and proceeding to the noninertial limit[11-13]

3. By constructing from the noninertial Langevin equation the Fokker-Planck equation for the transition probability $W(\{x\}, t)$ in configuration space $\{x\} = \{x_1, x_2, \ldots, x_n\}$ and separating the variables by assuming a solution in the form of a sum of known functions of the configuration and unknown functions of the time.[3]

The Fokker-Planck equation approach is a very powerful method for the analysis of nonlinear stochastic systems.[3] However, this approach is extremely lengthy to use in practice, especially in the case of multidimensional systems, since it involves many mathematical manipulations. Hence, an alternative approach is desirable. One of the main purposes of this review is to show how the relaxation behavior may be obtained in a simple manner directly from the nonlinear Langevin equation of the process. The work of the previous authors is also expanded upon and treated in some detail to provide an introduction to the subject.

In general, the differential-difference equations take the form of three-(or higher-order) term recurrence relations[3] between the set of statistical averages describing the dynamical behavior. Thus, the behavior of any selected average is coupled to that of all the others so forming a hierarchy of averages. The time behavior of the first-order average, for example, involves that of the second-order average, which in turn involves the third-order average, and so on.

If the recurrence relation between the averages is a three-term one, the Laplace transform of the solution for a step stimulus may be expressed as an infinite continued fraction. The characteristic equation or secular determinant of the system (the continued fraction having been expressed as a rational function of the complex frequency) will in general possess an infinite number of roots. Thus, the system will have a denumerable set of relaxation times and corresponding relaxation modes. In an actual numerical solution of the problem successive convergents of the continued fraction are calculated until convergence is reached when a solution is deemed to have been obtained.[3] If the underlying recurrence relation is not a three-term one, numerical solutions may be obtained (albeit not in as convenient a representation as the continued fraction) by writing the set of the differential-difference equations as a first-order matrix differential equation.[14] The size of the system matrix is then successively increased until convergence is attained.

The numerical approach to the problem has the disadvantage that closed form solutions are not available. Furthermore, the qualitative behavior is not at all obvious. Several investigators (see, e.g., Refs. 6, 15, and 16) have attempted to overcome this problem by recasting the solution of the Fokker-Planck equation as a Sturm-Liouville problem and evaluating the first few eigenvalues of that equation by various methods. It is then supposed that the longest relaxation time (corresponding to the reciprocal of the lowest eigenvalue) is the only one of significance, so that the system may be described by a single relaxation mode. This always involves the assumption that the contribution of the higher-order eigenfunctions to the relaxation process is negligible.

To pose the solution as a Sturm-Liouville problem the solution of the noninertial Fokker-Planck equation which may be written down symbolically as[3]

$$\frac{\partial}{\partial t}W(\{\mathbf{x}\},t) = L_{\mathrm{FP}}(\{\mathbf{x}\})W(\{\mathbf{x}\},t) \tag{1}$$

is expressed as an infinite sum of products of unknown functions $\{y_k\}$ in the configuration space and exponential time decays $\{e^{-\lambda_k t}\}$, viz.,

$$W(\{\mathbf{x}\},t) = \sum_k c_k(t)y_k(\{\mathbf{x}\})$$

where the functions $c_k(t)$ and $y_k(\{x\})$ obey the equations

$$\frac{\mathrm{d}}{\mathrm{d}t}c_k(t) + \lambda_k c_k(t) = 0 \tag{2}$$

and

$$L_{\mathrm{FP}}(\{\mathbf{x}\})y_k(\{\mathbf{x}\}) = \lambda_k y_k(\{\mathbf{x}\}) \tag{3}$$

The λ_k are the eigenvalues of the Sturm-Liouville equation for which solutions $\{\lambda_k\}$ exist.

Thus, the time behavior of the average $\langle A \rangle$ of any quantity $A(\{x\})$ of interest can be evaluated by means of the formula

$$\langle A \rangle = \int A(\{x\}) W(\{x\}, t) \, d\{x\} \tag{4}$$

Because of Eq. (1), the average $\langle A \rangle$ from Eq. (4) has the form

$$\langle A \rangle = \sum_k a_k e^{-\lambda_k t} \tag{5}$$

where the a_k are the weight coefficients (amplitudes) corresponding to the eigenvalues λ_k. The reformulation of the task of calculating the relaxation behavior as a Sturm-Liouville problem allows us to give the formal definition of the effective eigenvalue.

To achieve this let us now suppose that the time behavior of $\langle A \rangle$ may be approximated by

$$\langle A(t) \rangle = \langle A(0) \rangle e^{-\lambda_{\text{eff}} t}$$

whence according to Eq. (5)

$$\lambda_{\text{eff}} = -\frac{\langle \dot{A}(0) \rangle}{\langle A(0) \rangle} = \frac{\sum_k a_k \lambda_k}{\sum_k a_k} \tag{6}$$

and τ_{eff}, the effective relaxation time, is then

$$\tau_{\text{eff}} = \lambda_{\text{eff}}^{-1} \tag{7}$$

It is difficult to evaluate λ_{eff} from this formula using the Sturm-Liouville equation because a knowledge of the law of formation of the eigenvalues λ_k and their corresponding amplitudes a_k is rarely available. The approach taken in this review does not attempt to calculate λ_{eff} by explicitly calculating the eigenvalue spectrum as required by Eq. (5); rather it gives λ_{eff} in terms of equilibrium averages of the distribution function.

The concept of the effective eigenvalue so far does not particularly rely on the linearity of the response. In many cases the linear response may effectively be described by a single relaxation time. This is not true for processes such as the Kerr effect relaxation of an assembly of dipolar molecules under the influence of a step stimulus. Here two effective eigenvalues are needed[14] to describe the response because it is intrinsically nonlinear. Another example where two effective eigenvalues are needed is ferromagnetic resonance[6, 11] of assemblies of single-domain ferromagnetic particles where "entanglement" of the dipole and quadrupole moments is required to produce the resonance even in the linear response approximation. If the response is linear it is possible to considerably simplify the expression for effective relaxation times using linear response theory, as we now demonstrate.

According to linear response theory[3, 10] (where we suppose for the purpose of illustration that $\mathbf{m}(t)$ represents the instantaneous electric dipole moment of a system) the autocorrelation function of the dipole moment is given by[10, 17]

$$C_m(t) = \frac{\langle \mathbf{m}(0)\mathbf{m}(t)\rangle_0 - \langle \mathbf{m}(0)\rangle_0^2}{\langle \mathbf{m}^2(0)\rangle_0 - \langle \mathbf{m}(0)\rangle_0^2} \tag{8}$$

where the subscript zero on the angular braces denotes that the ensemble averages are to be evaluated in the absence of the external stimulus which yields the relaxation behavior.

The autocorrelation function $C_m(t)$ has a clear physical meaning. It describes relaxation of the average dipole moment $\langle \mathbf{m} \rangle$ (having switched off at $t = 0$ a small constant external electric field $\mathbf{E}$, which has been applied in the infinite past). We have

$$\langle \mathbf{m} - \mathbf{m}_0 \rangle = \mathbf{E}\big(\langle \mathbf{m}^2(0)\rangle_0 - \langle \mathbf{m}(0)\rangle_0^2\big)C_m(t)$$

where $\mathbf{m}_0$ is the equilibrium value of $\langle \mathbf{m} \rangle$.

The effective eigenvalue method implies that $C_m(t)$ may be represented as a pure exponential decay. We then have with the aid of Eqs. (7) and (8)

$$\tau_{\text{eff}} = -\frac{\langle \mathbf{m}^2(0)\rangle_0 - \langle \mathbf{m}(0)\rangle_0^2}{\langle \mathbf{m}(0)\dot{\mathbf{m}}(0)\rangle} \tag{9}$$

The significance of the foregoing is that we can evaluate λ_{eff} from the underlying Langevin equation for $\mathbf{m}(t)$ without solving it since we require only the value of $\dot{\mathbf{m}}(t)$ at time $t = 0$. Thus, all the desired properties of the system in the linear response approximation can be evaluated. For example, the complex susceptibility $\chi(\omega) = \chi'(\omega) - i\chi''(\omega)$ obeys the relation[10]

$$\chi(\omega) = \chi'(0)\left[1 - i\omega\int_0^\infty C(t)e^{-i\omega t}\right] \qquad (10)$$

with $C(t)$ defined as in Eq. (8) and $\chi'(0) = \langle \mathbf{m}^2(0)\rangle_0 - \langle \mathbf{m}(0)\rangle_0^2$. If the process can be described by a single effective relaxation time then the effective eigenvalue method immediately leads to the Debye equation

$$\chi(\omega) = \frac{\chi'(0)}{1 + i\omega\tau_{\text{eff}}} \qquad (11)$$

We note (for convenience we refer to Eqs. (5) and (6)) that a global characterization of the decay of the electric polarization is given by the relaxation time T defined as

$$T = \int_0^\infty C(t)\,dt = \frac{\sum_k a_k \lambda_k^{-1}}{\sum_k a_k} \qquad (12)$$

namely the area under the autocorrelation function curve. This correlation time also includes contributions from all the eigenvalues but it gives no information on the different time regions of relaxation that are possible; the behavior of T and τ_{eff} is sometimes similar. If different time scales are involved, however, τ_{eff} differs from T and in view of the definition

$$\tau_{\text{eff}} = \frac{\sum_k a_k}{\sum_k a_k \lambda_k} \qquad (13)$$

accurately represents the initial slope of the decay of the polarization, thereby giving precise information on the initial decay of the polarization in the noninertial limit.

We have also mentioned that it is often assumed that the longest relaxation time of the distribution function $W(\{\mathbf{x}\}, t)$ accurately represents that of the polarization decay. It is apparent that this is true only if a single eigenvalue determines the decay process (see Section II.J).

We began by supposing that the decay process in the time domain was a pure exponential, thus allowing us to derive Eq. (9) for the effective eigenvalue. Another entirely equivalent way of describing the concept was introduced by Morita[7] in the context of the hierarchy of differential-difference equations. We suppose that in the continued fraction (or failing

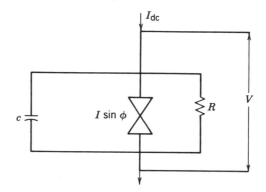

Figure 1. Equivalent circuit of Josephson junction.

ψ_L pair wave functions for the right and left superconductor, respectively. The phase difference $\phi = \phi_L - \phi_R$ between these wave functions (of the Cooper pairs) is given by the Josephson equation[22, 23]

$$\dot{\phi}(t) = \frac{2eV(t)}{\hbar} \tag{17}$$

where $V(t)$ is the potential difference across the junction, e is the charge on the electron, and $\hbar = h/2\pi$, where h is is the Planck constant. The junction is now modeled (Fig. 1)[3, 23] by a resistance R in parallel with a capacitance C, across which is connected a dc current generator I_{dc} (representing the bias current applied to the junction). At the other end of the junction (across the resistance R) is connected a phase-dependent current generator, $I \sin \phi$, representing the Josephson supercurrent due to the Cooper pairs tunneling through the junction. Since the junction operates at a temperature above absolute zero, there will be a noise current $i(t)$ superimposed on the bias current. The circuit thus behaves according to the Langevin equation[3]

$$I_{dc} + i(t) = C \frac{dV(t)}{dt} + \frac{1}{R} V(t) + I \sin \phi(t) \tag{18}$$

Using the Josephson equation (17), Eq. (18) may be cast into the form[3, 22]

$$\left(\frac{\hbar}{2e} \right)^2 C \ddot{\phi}(t) + \left(\frac{\hbar}{2e} \right)^2 \frac{1}{R} \dot{\phi}(t) + \frac{\hbar}{2e} I \sin \phi(t) = \frac{\hbar}{2e} [I_{dc} + i(t)] \tag{19}$$

Equation (19) has the same form as that for a Brownian particle moving under the influence of both a cosine potential and a linear potential. The behavior of the system will be determined by the relative amplitudes of the constant and periodic terms. A more complete discussion of Eq. (19) is given in Refs. 3 and 23, and in the papers quoted therein. The quantities of physical interest are the mean value of the voltage $\langle V \rangle = (\hbar/2e)\langle \dot{\phi} \rangle$ and/or junction impedance to external high-frequency current. We present exact and approximate calculations of the junction impedance to an external small-signal current using Eq. (19) in the noninertial limit when we can neglect the capacitative term $C\ddot{\phi}$.

5. The use of the theory for the description of quantum noise in ring laser gyros arises because the phase difference $\phi(t)$ between the clockwise and counterclockwise running waves in such a gyro operating in the steady state is governed by the Langevin equation[19, 20]

$$\dot{\phi}(t) = a + b \sin \phi(t) + F(t) \tag{20}$$

where

$$a = -\frac{8\pi}{\lambda L}(\mathbf{A} \cdot \mathbf{\Omega})$$

$|\mathbf{A}|$ is the area covered by the optical path of length L along the ring cavity, Ω is the rotation rate of the gyro, λ is the wavelength of the laser, b is the backscattering coefficient, $F(t)$ is a white-noise source term. The quantity of interest in this application is the beat signal $\cos \phi(t)$, in particular, its spectrum $\alpha(\omega)$, defined as[19, 20]

$$\alpha(\omega) = \int_{-\infty}^{\infty} e^{i\omega\tau}\langle \cos \phi(0)\cos \phi(\tau) \rangle \, d\tau \tag{21}$$

The beat signal spectrum $\alpha(\omega)$ for the ring laser obtained using the numerical exact solution of Eq. (20) given in Ref. 20 is also compared with the result yielded by the effective eigenvalue method.

We note that when the tilt is equal to zero (the parameter a of Eq. (17)) the model may be applied (with appropriate changes of notation) to the theory of dielectric relaxation of an assembly of noninteracting two-dimensional dipoles, discussed briefly in Section III.

Problems 4 and 5 are of the upmost importance in the context of this review, since they can be solved exactly. A comparison of the exact solutions and those yielded by the effective eigenvalue method allows us to comment on the accuracy of the latter. Using the foregoing problems as an example, we shall demonstrate that the effective eigenvalue method in certain circumstances yields a simple and extremely concise analytical description of the solution of complex stochastic problems. Our method of obtaining the exact solution of these problems has the merit of being considerably simpler than the previously available algorithms described in Refs. 3, 20, and 53.

Thus, the purpose of the review is twofold:

1. To demonstrate (using as examples various well-known models) how the quantities characterizing the dynamical behavior of stochastic systems in the time and/or frequency domains may be calculated by direct averaging of nonlinear Langevin equations. The problem of constructing and solving the underlying Fokker-Planck equations is thus bypassed entirely.

2. To show that the effective eigenvalue method is a valuable and powerful tool for the purpose of obtaining simple analytic solutions to nonlinear stochastic problems in certain ranges of the parameter values and for certain potentials.

II.   THREE-DIMENSIONAL ROTATIONAL BROWNIAN MOTION IN AXIALLY SYMMETRIC POTENTIALS

A.   The Noninertial Langevin Equation for a Nematic Liquid Crystal in the Meier-Saupe Mean Field Approximation

As our first example of Brownian motion in axially symmetric potentials we reconsider the theory of dielectric relaxation of nematic liquid crystals due to Martin et al.[16] This proceeds from the Fokker-Planck equation without explicit reference to the underlying Langevin equation. The aim is to extend the Debye theory of dielectric relaxation of assemblies of noninteracting polar molecules subjected to a weak alternating (ac) field to include the effects of a strong intermolecular potential giving rise to the nematic state.

The essence of the diffusion equation method[3] is to write down the Fokker-Planck equation (Smoluchowski equation), for the transition probability of orientations of dipoles in configuration space. This is solved[16] by the method of separation of the variables. The separation procedure yields a Sturm-Liouville equation in the space variables. The reciprocal of the lowest eigenvalue of this equation yields the longest relaxation time of the probability distribution of orientations. Furthermore, on expanding the dipole moment as a series of eigenfunctions of the Sturm-Liouville equation and averaging over the distribution function, the orientational polarization may be expressed as an infinite set of discrete Debye type relaxation mechanisms. The relaxation times and amplitudes are then determined by the eigenvalues of the Sturm-Liouville equation. Approximate analytic solutions for the lowest eigenvalue of the distribution function may be found for high and low nematic potential barriers.[16] We term the Sturm-Liouville method method I. This is commonly used in the study of nematic liquid crystals.[16]

An alternative approach to the problem is to expand the transition probability as a series of spherical harmonics. This yields the time behavior of the transition probability as an infinite hierarchy of differential-difference equations. The lowest-order member of the hierarchy governs the time dependence of the polarization but is coupled to all the higher members by the differential-difference scheme, thus giving an infinite number of relaxation modes. In the frequency domain the hierarchy may often be written as an infinite continued fraction which facilitates its solution. The most general method of solution is effected by converting the hierarchy into the set of ordinary differential equations (see Section II.J)

$$\dot{\mathbf{X}} = \mathbf{A}\mathbf{X} + \mathbf{B}U$$

and successively increasing the size of the system matrix $\mathbf{A}$ (Ref. 3) by means of the recurrence relations of the hierarchy until convergence is attained. The reciprocal of the lowest eigenvalue of this set yields the longest relaxation time of the system. We term this method II.

Method I, based on the Sturm-Liouville equation, has also been used in the study of the analogous relaxation process in ferromagnetic domains in conjunction with various asymptotic methods[5, 6, 15] to obtain analytic solutions for the lowest eigenvalue of the distribution function for high and low anisotropy. The contribution of higher-order eigenvalues to the polarization is neglected so that the polarization can be described by a single eigenvalue. It is then assumed that the reciprocal of the lowest eigenvalue may be identified with the relaxation time characterizing a flip of the magnetization.[6] Both methods I and II have been used in Ref. 11 in conjunction with linear response theory[10] in the study of the dispersion of the magnetic susceptibility of fine ferromagnetic particles.

The disadvantage of the diffusion equation method is that in all applications we first have to derive that equation from the Chapman-Kolmogorov equation in curvilinear coordinates and the underlying Langevin equation. Next we must either use elaborate mathematical formulae involving spherical harmonics to deduce the set of differential-difference equations or we must study the properties of Sturm-Liouville equations, the solution of which cannot generally be given in terms of known functions. In neither method is it easy to generalize the results to an arbitrary nematic potential.

It is the purpose of this section to show how the differential-difference equations (Method II) for a nematic liquid crystal for the Meier-Saupe potential arise naturally from the noninertial Langevin equation written in the vector form and defined as a Stratonovich equation, thus bypassing the diffusion equation entirely. Having derived the differential-difference equations by averaging the Langevin equation we then show how closed form expressions of the Debye type may be obtained for the complex susceptibility. The results are given for an ac field applied parallel and perpendicular to the axis of symmetry. We then show how the results may be generalized to an arbitrary uniaxial nematic potential. The availability of these expressions rests on the assumption that the contribution to the dynamical behavior of all processes that occur on a timescale $< \frac{1}{3}\tau_D$, where τ_D is the Debye relaxation time, may be adequately approximated by their equilibrium values. This assumption may be stated more precisely, as that in the Laplace transform of the differential-difference equation describing the behavior of the mean dipole moment (here the first spherical harmonic average) the ratio of the Laplace transforms of the averages of the first- and higher-order spherical harmonics may be replaced by their final equilibrium values. This reduces the nth-order characteristic equation of the system to one of the first order. The growth and decay of the mean dipole moment is thus characterized by a single exponential with an relaxation time (the reciprocal of the effective eigenvalue λ_{eff}) which is a function of the nematic potential. A numerical analysis of our results and a comparison with the exact solution is given in Section II.J, where it is shown that the present approach yields an excellent description of the transverse relaxation and accurately describes the longitudinal relaxation for low barrier heights only.

We commence our study by considering the rotational Brownian movement of a linear molecule subject to the mean nematic field $\mathbf{E}_0$ and an external electric field $\mathbf{E}_1(t)$. The total electric field $\mathbf{E}(t)$ acting on the molecule is

$$\mathbf{E}(t) = \mathbf{E}_0 + \mathbf{E}_1(t) \qquad (22)$$

The molecule contains a rigid electric dipole $\boldsymbol{\mu}$ along the axis of symmetry. The angular velocity $\boldsymbol{\omega}(t)$ of the molecule satisfies the kinematic relation[24, 32]

$$\frac{d\boldsymbol{\mu}(t)}{dt} = \boldsymbol{\omega}(t) \times \boldsymbol{\mu}(t) \qquad (23)$$

We specialize Eq. (23) to the rotational Brownian motion of a molecule by supposing that ω obeys the Euler-Langevin equation[17]:

$$I \frac{d\omega(t)}{dt} + \zeta\omega(t) = \lambda(t) + \mu(t) \times E(t) \qquad (24)$$

where I is the moment of inertia of the molecule about any line through the origin perpendicular to the line of symmetry, $\zeta\omega$ is the damping torque, and $\lambda(t)$ is the white-noise driving torque, both due to the Brownian movement. $\lambda(t)$ satisfies

$$\overline{\lambda_i(t)} = 0 \qquad (25)$$

$$\overline{\lambda_i(t)\lambda_j(t')} = 2kT\zeta\delta_{ij}\delta(t - t') \qquad (26)$$

where δ_{ij} is Kronecker's delta, $i, j = 1, 2, 3$, which correspond to Cartesian axes x, y, z fixed in the molecule. This is the assumption that the random torques about different axes are statistically independent. $\delta(t)$ is the Dirac delta function. The term $\mu \times E(t)$ in Eq. (24), is the torque due to the nematic mean field E_0 and external field $E_1(t)$. The overbar indicates a statistical average.

Equation (24) includes the inertia of the molecule. The noninertial response occurs when I tends to zero or when ζ, the friction coefficient, becomes very large. In this limit the angular velocity vector is

$$\omega(t) = \frac{\lambda(t)}{\zeta} + \frac{\mu \times E(t)}{\zeta} \qquad (27)$$

We combine this with the kinematic relation (23), yielding

$$\frac{d\mu(t)}{dt} = \frac{\lambda(t)}{\zeta} \times \mu(t) + \frac{\{\mu(t) \times E(t)\} \times \mu(t)}{\zeta} \qquad (28)$$

or

$$\frac{d\mu(t)}{dt} = \frac{\lambda(t)}{\zeta} \times \mu(t) + \frac{\mu^2 E(t)}{\zeta} - \frac{\mu(t)\{\mu(t) \cdot E(t)\}}{\zeta} \qquad (29)$$

which is the Langevin equation for the rotational motion of the dipole moment μ of the molecule in the noninertial limit.

B. Equation of Motion for a Step Change in Field Applied Along the Polar Axis

To achieve the simplest possible presentation of the problem let us first suppose that the molecule is subject to a mean nematic field $\mathbf{E}_0 = -\operatorname{grad} V$, where $V(\vartheta) = -V_0 \cos^2 \vartheta$ in spherical polar coordinates, $V_0 = 3AS/2V_m^2$, where S is the order parameter, V_m is the molar volume, and A is a constant of interactions.[16] Hence,

$$\mathbf{E}_0 = \mathbf{k}\left(-\frac{\partial V}{\mu \partial \cos \vartheta}\right) = \mathbf{k}\frac{2V_0}{\mu}\cos \vartheta(t) = \mathbf{k}E_0 \cos \vartheta(t) \qquad (30)$$

where $E_0 = 2V_0/\mu$.

Let us suppose that at $t = 0$ a small field $\mathbf{E}_1 U(t)$, where $U(t)$ is the unit step function, is applied along the z axis, so that

$$\mathbf{E}(t) = \mathbf{E}_0 + \mathbf{E}_1 U(t)\mathbf{k} \qquad (31)$$

$$= \frac{E_0}{\mu}\mu_z(t)\mathbf{k} + E_1 U(t)\mathbf{k} \qquad (32)$$

Equation (29) then becomes, with the aid of Eq. (30),

$$\dot{\mu}_x(t) = \frac{1}{\zeta}\left(\lambda_y(t)\mu_z(t) - \lambda_z(t)\mu_y(t)\right)$$

$$- \frac{\mu_x(t)\mu_z(t)}{\zeta}\left[\frac{E_0\mu_z(t)}{\mu} + E_1 U(t)\right] \qquad (33)$$

$$\dot{\mu}_y(t) = \frac{1}{\zeta}\left(\lambda_z(t)\mu_x(t) - \lambda_x(t)\mu_z(t)\right)$$

$$- \frac{\mu_y(t)\mu_z(t)}{\zeta}\left[\frac{E_0\mu_z(t)}{\mu} + E_1 U(t)\right] \qquad (34)$$

$$\dot{\mu}_z(t) = \frac{1}{\zeta}\left(\lambda_x(t)\mu_y(t) - \lambda_y(t)\mu_x(t)\right)$$

$$+ \frac{\left(\mu^2 - \mu_z^2(t)\right)}{\zeta}\left[\frac{E_0\mu_z(t)}{\mu} + E_1 U(t)\right] \qquad (35)$$

The desired quantity is the average behavior of the component of the dipole moment in the field direction, that is, the longitudinal component.

Equation (35) contains multiplicative noise terms: $\lambda_x(t)\mu_y(t)$ and $\lambda_y(t)\mu_x(t)$. Taking the Langevin equation for N stochastic variables $\{\xi(t)\}$ $= \{\xi_1(t), \xi_2(t), \xi_3(t), \ldots, \xi_N(t)\}$ as

$$\frac{d}{dt}\xi_i(t) = h_i(\{\xi(t)\}, t) + g_{ij}(\{\xi(t)\}, t)\Gamma_j(t) \qquad (36)$$

with

$$\overline{\Gamma_j(t)} = 0 \qquad (37)$$

$$\overline{\Gamma_i(t)\Gamma_j(t')} = 2\delta_{ij}\delta(t - t') \qquad (38)$$

and interpreting it as a Stratonovich[3] equation, Risken[3] has shown that the drift coefficient is

$$D_i(\{x\}, t) = \dot{x}_i = \lim_{\tau \to 0} \frac{1}{\tau}\overline{(\xi_i(t + \tau) - x_i)}\Big|_{\xi_k(t)=x_k}$$

$$= h_i(\{x\}, t) + g_{kj}(\{x\}, t)\frac{\partial}{\partial x_k}g_{ij}(\{x\}, t) \qquad (39)$$

$$k = 1, 2, \ldots, N$$

The last term in Eq. (39) is called the noise-induced or spurious drift.[11] $\xi_i(t + \tau)$, $\tau > 0$, is a solution of Eq. (36), which has the sharp value $\xi_k(t) = x_k$ for $k = 1, 2, \ldots, N$. The quantities x_k in Eq. (39) are themselves random variables with the probability density function $W(\{x\}, t)$ defined such that $W\, dx_k$ is the probability of finding x_k in the range x_k to $x_k + dx_k$. Instead of using different symbols we have distinguished the starting values at time t from the stochastic variables by deleting the time argument as in Ref. 11.

We now use this theorem to evaluate the average of the multiplicative noise terms in Eq. (35). We have nine tensor components:

$$g_{11} = 0 \qquad g_{12} = \frac{\mu_z}{\zeta} \qquad g_{13} = \frac{-\mu_y}{\zeta}$$

$$g_{21} = \frac{-\mu_z}{\zeta} \qquad g_{22} = 0 \qquad g_{23} = \frac{\mu_x}{\zeta} \qquad (40)$$

$$g_{31} = \frac{\mu_y}{\zeta} \qquad g_{32} = \frac{-\mu_x}{\zeta} \qquad g_{33} = 0$$

whence with the aid of Eq. (39)

$$
kT\zeta g_{xj}(\{\mu\},t)\frac{\partial}{\partial x_x}g_{3j}(\{\mu\},t)
$$

$$
= kT\zeta\left[g_{11}\frac{\partial}{\partial x_1}g_{31} + g_{12}\frac{\partial}{\partial x_1}g_{32} + g_{13}\frac{\partial}{\partial x_1}g_{33}\right.
$$

$$
+ g_{21}\frac{\partial}{\partial x_2}g_{31} + g_{22}\frac{\partial}{\partial x_2}g_{32} + g_{23}\frac{\partial}{\partial x_2}g_{33}
$$

$$
\left. + g_{31}\frac{\partial}{\partial x_3}g_{31} + g_{32}\frac{\partial}{\partial x_3}g_{32} + g_{33}\frac{\partial}{\partial x_3}g_{33}\right] \tag{41}
$$

$$
= kT\zeta\left[\frac{\mu_z}{\zeta}\frac{\partial}{\partial\mu_x}\left(\frac{-\mu_x}{\zeta}\right) + \left(\frac{-\mu_z}{\zeta}\right)\frac{\partial}{\partial\mu_y}\left(\frac{\mu_y}{\zeta}\right)\right.
$$

$$
\left. + \frac{\mu_z}{\zeta}\frac{\partial}{\partial\mu_z}\left(\frac{\mu_y}{\zeta}\right) + \left(\frac{-\mu_x}{\zeta}\right)\frac{\partial}{\partial\mu_z}\left(\frac{-\mu_x}{\zeta}\right)\right]
$$

$$
= \frac{-2kT}{\zeta}\mu_z \tag{42}
$$

which is the noise-induced drift. We must now further average Eq. (42) over the density distribution function $W(\{\mu\},t)$ of dipole orientations in configuration space at time t. Equation (42) then becomes $(-2kT/\zeta)$ $\langle\mu_z\rangle$,where $\langle f\rangle$ denotes averaging a function f over the density function $W(\{\mu\},t)$, namely $\langle f(\{\mu\})\rangle = \int f(\{\mu\})W(\{\mu\},t)\,d\mu$. The averaged deterministic drift, from Eq. (35) is

$$
\langle h_3\{\mu\}\rangle = \frac{\mu^2 - \langle\mu_z^2\rangle}{\zeta}E_1 U(t) + \frac{\langle\mu^2\mu_z - \mu_z^3\rangle E_0}{\zeta\mu} \tag{43}
$$

Thus, the averaged equation of motion of the dipole is

$$
\frac{d}{dt}\langle\mu_z\rangle + \frac{2kT}{\zeta}\langle\mu_z\rangle = \frac{E_1 U(t)}{\zeta}\langle\mu^2 - \mu_z^2\rangle + \frac{E_0}{\zeta}\langle\mu^2\mu_z - \mu_z^3\rangle \tag{44}
$$

Now

$$
\mu_z = \mu\cos\vartheta = \mu P_1(\cos\vartheta) \tag{45}
$$

and so

$$\frac{d}{dt}\langle P_1(\cos\vartheta)\rangle + \frac{2kT}{\zeta}\langle P_1(\cos\vartheta)\rangle$$

$$= \frac{2\mu[E_1U(t)]}{3\zeta}\left[1 - \langle P_2(\cos\vartheta)\rangle\right] \tag{46}$$

$$+ \frac{2}{5}\frac{\mu E_0}{\zeta}\left[\langle P_1(\cos\vartheta)\rangle - \langle P_3(\cos\vartheta)\rangle\right]$$

$$P_2(u_z) = \tfrac{1}{2}(3u_z^2 - 1) \tag{47}$$

$$P_3(u_z) = \tfrac{1}{2}(5u_z^3 - 3u_z) \tag{48}$$

are the Legendre polynomials of order 1, 2, and 3, $u_z = \cos\vartheta$.

C. Equation of Motion of the Transverse or x Component of the Dipole Moment

In the transverse case the step change in the field is applied parallel to the x axis so that we need to determine the behavior of $\langle\mu_x\rangle$. We find as before that the x component of the dipole moment satisfies

$$\frac{d}{dt}\langle\mu_x\rangle + \frac{2kT}{\zeta}\langle\mu_x\rangle = \frac{\mu^2}{\zeta}E_1U(t)$$

$$- \frac{\langle\mu_x^2\rangle}{\zeta}E_1U(t) - \frac{\langle\mu_z^2\mu_x\rangle}{\mu\zeta}E_0 \tag{49}$$

Now

$$\mu_x = \mu\cos\vartheta\cos\phi = \mu X_{11}(\vartheta,\phi) \tag{50}$$

where X_{11} is the spherical harmonic of order 1 (Ref. 25). Thus, Eq. (49) becomes

$$\frac{d}{dt}\langle X_{11}\rangle + \frac{2kT}{\zeta}\langle X_{11}\rangle = -\frac{\mu E_0}{\zeta}\left(\frac{2}{15}\langle X_{31}\rangle + \frac{1}{5}\langle X_{11}\rangle\right)$$

$$+ \frac{\mu E_1U(t)}{3\zeta}\left[2 - \frac{1}{2}\langle X_{22}\rangle + \langle P_2\rangle\right] \tag{51}$$

$$X_{31} = \tfrac{3}{2}(5\cos^2\vartheta - 1)\sin\vartheta\cos\phi \tag{52}$$

$$X_{22} = 3\sin^2\vartheta\cos 2\phi \tag{53}$$

Recalling that

$$\langle P_1(\cos \vartheta(0)) \rangle_1 = 0 \qquad (63)$$

the Laplace transform of Eq. (62) is

$$\left(s + 2\frac{kT}{\zeta} \right) \mathscr{L}\langle P_1 \rangle_1 = \frac{2\mu E_0}{5\zeta} \left[\mathscr{L}\langle P_1 \rangle_1 - \mathscr{L}\langle P_3 \rangle_1 \right] \qquad (64)$$

yielding

$$\left(s + \frac{1}{\tau_D} - \frac{2\sigma}{5\tau_D} \right) \mathscr{L}\langle P_1 \rangle_1 = -\frac{2\sigma}{5\tau_D} \mathscr{L}\langle P_3 \rangle_1 \qquad (65)$$

where $\tau_D = \zeta/2kT$ is the Debye relaxation time. The characteristic equation of the system is

$$s + \frac{1}{\tau_D} - \frac{2\sigma}{5\tau_D} \left[1 - \frac{\mathscr{L}\langle P_3 \rangle_1}{\mathscr{L}\langle P_1 \rangle_1} \right] = 0 \qquad (66)$$

We have stated that this will lead to a polynomial equation of high order in s yielding a discrete set of relaxation times and that one cannot in general obtain a closed form result for the lowest root. If we suppose, however, following the approach outlined in Refs. 7 and 24, that the term

$$\frac{\mathscr{L}\langle P_3 \rangle_1}{\mathscr{L}\langle P_1 \rangle_1}$$

may be replaced by its final (equilibrium) value (i.e., its value as t tends to infinity)

$$\frac{\lim_{t \to \infty} \langle P_3 \rangle_1}{\lim_{t \to \infty} \langle P_1 \rangle_1} = \frac{\lim_{s \to 0} s\mathscr{L}\langle P_3 \rangle_1}{\lim_{s \to 0} s\mathscr{L}\langle P_1 \rangle_1}$$

then Eq. (66) may be evaluated in terms of a single relaxation time that depends on the strength of the nematic potential. This is accomplished as follows. At equilibrium, which occurs when $t \gg \tau_D$,

$$\lim_{t \to \infty} \frac{\langle P_3 \rangle_1}{\langle P_1 \rangle_1} = \frac{\int_{-1}^{+1} \frac{1}{2}(5x^3 - 3x)e^{\sigma x^2 + \xi_1 x}\,dx}{\int_{-1}^{+1} x e^{\sigma x^2 + \xi_1 x}\,dx} \qquad (67)$$

On expanding Eq. (67) as a Taylor series in ξ_1 we have (we retain only terms linear in ξ_1 since we are concerned with the linear response)

$$\lim_{t \to \infty} \frac{\langle P_3 \rangle_1}{\langle P_1 \rangle_1} = \frac{1}{2} \frac{\int_{-1}^{+1}(5x^4 - 3x^2)e^{\sigma x^2}\,dx}{\int_{-1}^{+1}x^2 e^{\sigma x^2}\,dx} \tag{68}$$

Furthermore, on defining the function F by[6]

$$F(\sigma) = \int_0^1 e^{\sigma x^2}\,dx \tag{69}$$

we have

$$\int_0^1 x^2 e^{\sigma x^2}\,dx = F'(\sigma) = \frac{1}{2\sigma}[1 - F(\sigma)] \tag{70}$$

$$\int_0^1 x^4 e^{\sigma x^2}\,dx = F''(\sigma) = \frac{1}{2\sigma}\left\{1 - \frac{3}{2\sigma}[1 - F(\sigma)]\right\} \tag{71}$$

The function $F(\sigma)$ is related to Dawson's integral $D(\sigma)$,[26]

$$D(\sigma) = \exp(-\sigma^2)\int_0^\sigma \exp(t^2)\,dt$$

by

$$D(\sqrt{\sigma}) = \sqrt{\sigma}\,e^{-\sigma}F(\sigma) \tag{72}$$

Equation (68) may now be written

$$\lim_{t \to \infty} \frac{\langle P_3 \rangle_1}{\langle P_1 \rangle_1} = -\frac{3}{2} + \frac{5}{2}\frac{F''(\sigma)}{F'(\sigma)} \tag{73}$$

Thus, the effective eigenvalue is

$$\lambda_\parallel = \frac{1}{\tau_D} - \frac{2\sigma}{5\tau_D}\left[1 - \lim_{t \to \infty} \frac{\langle P_3 \rangle_1}{\langle P_1 \rangle_1}\right] \tag{74}$$

or

$$\lambda_\parallel = \frac{1}{\tau_D}\left(1 - \sigma + \frac{\sigma F''(\sigma)}{F'(\sigma)}\right) = \frac{1}{2\tau_D}\left(\frac{F(\sigma)}{F'(\sigma)} - 1\right) \tag{75}$$

in agreement with the result of Ref. 6. There the Fokker-Planck equation combined with Leontovich's method[1] was used for the calculation of the magnetization relaxation times of single-domain ferromagnetic particles (see Section II.I). Thus,

$$\tau_{\parallel} = \frac{1}{\lambda_{\parallel}} = \frac{\tau_D}{1 - \sigma + \sigma[F''(\sigma)/F'(\sigma)]} = 2\tau_D \frac{F'(\sigma)}{F(\sigma) - F'(\sigma)} \quad (76)$$

or

$$\tau_{\parallel} = 2\tau_D \frac{\sqrt{\sigma} - D(\sqrt{\sigma})}{D(\sqrt{\sigma})(1 + 2\sigma) - \sqrt{\sigma}} \quad (77)$$

Since[6]

$$\frac{\sigma F''(\sigma)}{F'(\sigma)} = \sigma \frac{\frac{1}{5} + \frac{1}{7}\sigma \cdots}{\frac{1}{3} + \frac{3}{5}\sigma} \approx \frac{3}{5}\sigma \quad (78)$$

for small σ, we have

$$\tau_{\parallel} = \tau_D \quad (79)$$

If terms $O(\sigma^2)$ are retained

$$\tau_{\parallel} = \frac{\tau_D}{1 - \frac{2}{5}\sigma} = \tau_D\left(1 + \frac{2}{5}\sigma\right) \quad (80)$$

On the other hand, for very large σ, $\tau_{\parallel}$ becomes

$$\tau_{\parallel} = 2\sigma\tau_D \quad (81)$$

in sharp contrast to the result for the large dc bias field where $\tau_{\parallel}$ vanishes as the reciprocal of the field strength[24] and in qualitative agreement with Refs. 6 and 29, where the longitudinal relaxation time increases as the potential strength is increased.

E. Calculation of the Effective Relaxation Time for the Transverse Component of the Dipole Moment

We assume as before that

$$\langle X_{nm} \rangle = \langle X_{nm} \rangle_0 + \langle X_{nm} \rangle_1 \quad (82)$$

which leads us to the linearized equation for $\langle X_{nm} \rangle$:

$$\langle \dot{X}_{11} \rangle_1 + \frac{2kT}{\zeta} \langle X_{11} \rangle_1 = -\frac{\mu E_0}{\zeta} \left(\frac{2}{15} \langle X_{31} \rangle_1 + \frac{1}{5} \langle X_{11} \rangle_1 \right)$$
$$+ \frac{\mu E_1 U(t)}{3\zeta} \left[2 - \frac{1}{2} \langle X_{22} \rangle_0 + \langle P_2 \rangle_0 \right] \tag{83}$$

and as in Eq. (59)

$$\langle X_{11} \rangle_0 = 0 \tag{84}$$

Thus, the eigenvalue equation is

$$s + \frac{1}{\tau_D} + \frac{\sigma}{5\tau_D} + \frac{2\sigma}{15\tau_D} \frac{\mathscr{L} \langle X_{31} \rangle_1}{\mathscr{L} \langle X_{11} \rangle_1} = 0 \tag{85}$$

Now

$$\lim_{t \to \infty} \langle X_{11} \rangle = \langle \sin \vartheta \cos \phi \rangle$$
$$= \frac{\int_0^{2\pi} \int_0^{\pi} \sin \vartheta \cos \phi \, e^{\sigma \cos^2 \vartheta + \xi_1 \sin \vartheta \cos \phi} \sin \vartheta \, d\vartheta \, d\phi}{\int_0^{2\pi} \int_0^{\pi} e^{\sigma \cos^2 \vartheta + \xi_1 \sin \vartheta \cos \phi} \sin \vartheta \, d\vartheta \, d\phi} \tag{86}$$

which in the linear approximation of ξ_1 is

$$\langle X_{11} \rangle_1 = \frac{\xi_1 \int_{-1}^{+1} (1 - x^2) e^{\sigma x^2} \, dx}{\int_{-1}^{+1} e^{\sigma x^2} \, dx}$$
$$= \xi_1 \frac{F(\sigma) - F'(\sigma)}{F(\sigma)} \tag{87}$$

Likewise,

$$\langle X_{31} \rangle_1 = \frac{3}{2} \frac{\xi_1 \int_{-1}^{+1} (5x^2 - 1)(1 - x^2) e^{\sigma x^2} \, dx}{\int_{-1}^{+1} e^{\sigma x^2} \, dx}$$
$$= \frac{3}{2} \xi_1 \frac{\int_{-1}^{+1} (-5x^4 + 6x^2 - 1) e^{\sigma x^2} \, dx}{\int_{-1}^{+1} e^{\sigma x^2} \, dx} \tag{88}$$
$$= \frac{3}{2} \xi_1 \frac{(-5F'' + 6F' - F)}{F}$$

Thus,

$$\lim_{t \to \infty} \frac{\langle X_{31} \rangle_1}{\langle X_{11} \rangle_1} = \frac{3}{2} \frac{6F' - 5F'' - F}{F - F'} \tag{89}$$

whence

$$-s = \lambda_\perp = \frac{1}{\tau_D} \left[1 + \frac{\sigma}{5} + \frac{2\sigma}{15} \frac{3}{2} \frac{6F' - 5F'' - F}{F - F'} \right] \tag{90}$$

$$= \frac{1}{\tau_D} \left[1 + \sigma \frac{F' - F''}{F - F'} \right] \tag{91}$$

Now

$$\lambda_\perp = \tau_\perp^{-1} \tag{92}$$

whence

$$\tau_\perp(\sigma) = \frac{\tau_D}{1 + \sigma \dfrac{F' - F''}{F - F'}} = 2\tau_D \frac{F(\sigma) - F'(\sigma)}{F(\sigma) + F'(\sigma)} \tag{93}$$

or

$$\tau_\perp = 2\tau_D \frac{D(\sqrt{\sigma})(2\sigma + 1) - \sqrt{\sigma}}{D(\sqrt{\sigma})(2\sigma - 1) + \sqrt{\sigma}} \tag{93a}$$

Let us suppose that σ is so small that only terms $O(\sigma^2)$ are retained. Then

$$\tau_\perp(\sigma) = \tau_D \left(1 - \frac{\sigma}{5} \right) \tag{94}$$

while for very large σ the behavior is similar to that of a dipole in a dc bias field, viz.,

$$\tau_\perp(\sigma) = \frac{\tau_D}{\sigma} \tag{95}$$

in contrast to the result for $\tau_\parallel$ and again in agreement with the result of Refs. 6 and 29.

F.   Calculation of the Frequency Dependence of the Susceptibility

Having determined the effective relaxation times for transverse and longitudinal fields, we may calculate the frequency dependence of the polarization as follows. We recall that Eq. (61) is

$$\langle \dot{P}_1 \rangle_1 + \frac{2kT}{\zeta} \langle P_1 \rangle_1 = \frac{2\mu E_0}{5\zeta} \left[\langle P_1 \rangle_1 - \langle P_3 \rangle_1 \right]$$
$$+ \frac{2\mu E_1 U(t)}{3\zeta} \left[\langle P_0 \rangle_0 - \langle P_2 \rangle_0 \right] \tag{96}$$

and that the effective eigenvalue procedure reduces this equation to

$$\langle \dot{P}_1 \rangle_1 + \frac{1}{\tau_\parallel(\sigma)} \langle P_1 \rangle_1 = \frac{2\mu E_1 U(t)}{3\zeta} \left[\langle P_0 \rangle_0 - \langle P_2 \rangle_0 \right]$$
$$= \frac{2\mu E_1 U(t)}{\zeta} \left[1 - \frac{F'(\sigma)}{F(\sigma)} \right] \tag{97}$$

where we have used Eqs. (69) and (70) to simplify the right side of Eq. (96). Equation (97) may be further simplified to

$$\frac{d}{dt} \langle P_1 \rangle_1 + \frac{1}{\tau_\parallel(\sigma)} \langle P_1 \rangle_1 = \frac{\xi_1 U(t)}{2\tau_D} \left[1 - \frac{F'(\sigma)}{F(\sigma)} \right] \tag{98}$$

by using Eq. (75). This is a first-order ordinary differential equation in $\langle P_1 \rangle_1$ with solution

$$\langle P_1 \rangle_1 = \frac{\xi_1 F'(\sigma)}{F(\sigma)} U(t) \left\{ 1 - \exp\left[-t/\tau_\parallel(\sigma) \right] \right\} \tag{99}$$

The rise transient of the polarization arising from $\mathbf{E}_1$, assuming N noninteracting dipoles, is then

$$N\mu \langle P_1 \rangle_1 = \frac{N\mu^2 E_1 F'(\sigma)}{kT F(\sigma)} U(t) \left\{ 1 - \exp\left[-t/\tau_\parallel(\sigma) \right] \right\} \tag{100}$$

or

$$P_r(t) = \chi'_\parallel(0) E_1 U(t) \left\{ 1 - \exp\left[-t/\tau_\parallel(\sigma) \right] \right\} \tag{101}$$

where

$$\chi'_{\parallel}(0) = \frac{N\mu^2}{kT}\frac{F'(\sigma)}{F(\sigma)}$$

is the static susceptibility. N is the number of molecules per unit volume. This is the rise transient of the polarization. The decay transient following the removal of E_1 is

$$P_d(t) = \chi'_{\parallel}(0)E_1 U(t)e^{-t/\tau_{\parallel}(\sigma)} \tag{102}$$

The after-effect function $f(t)$ is thus

$$f(t) = \chi'_{\parallel}(0)^{-t/\tau_{\parallel}(\sigma)} \tag{103}$$

The frequency-dependent susceptibility arising from the imposition of an ac field $E_1 e^{i\omega t}$ may then be written (since we have limited the solution to terms linear in ξ_1) from linear response theory[10] as

$$\chi_{\parallel}(\omega) = \chi'_{\parallel}(\omega) - i\chi''_{\parallel}(\omega) = f(0) - i\omega\int_0^\infty f(t)e^{-i\omega t}\,dt$$

$$= \chi'_{\parallel}(0)\left[1 - i\omega\int_0^\infty e^{-t/\tau_{\parallel}(\sigma)}\,e^{-i\omega t}\,dt\right] \tag{104}$$

Thus,

$$\chi_{\parallel}(\omega) = \frac{\chi'_{\parallel}(0)}{1 + i\omega\tau_{\parallel}} \tag{105}$$

This is the result for a longitudinal field and is in agreement with an earlier result of Raĭkher and Shliomis[6] for magnetic relaxation of ferromagnetic domains. From Eq. (83), the result for a transverse field is

$$\frac{d}{dt}\langle X_{11}\rangle_1 + \frac{2kT}{\zeta}\langle X_{11}\rangle_1 = -\frac{\mu E_0}{\zeta}\left[\frac{2}{15}\langle X_{31}\rangle_1 + \frac{1}{5}\langle X_{11}\rangle_1\right]$$

$$+ \frac{\mu E_1 U(t)}{3\zeta}\left[2 - \frac{1}{2}\langle X_{22}\rangle_0 + \langle P_2\rangle_0\right] \tag{106}$$

which with our closure procedure becomes

$$
\frac{d}{dt}\langle X_{11}\rangle_1 + \lambda_\perp \langle X_{11}\rangle_1
$$
$$
= \frac{\mu E_1 U(t)}{3\zeta}\left[2 - \frac{1}{2}\langle X_{22}\rangle_0 + \langle P_2\rangle_0\right]
\tag{107}
$$

where

$$
\langle X_{22}\rangle_0 = 0
\tag{108}
$$

since

$$
\int_0^{2\pi} \cos 2\phi \, d\phi = 0
\tag{109}
$$

This is a consequence of the axial symmetry of the nematic potential. Equation (107) then simplifies with the aid of Eq. (69) to

$$
\frac{d}{dt}\langle X_{11}\rangle_1 + \frac{1}{\tau_\perp}\langle X_{11}\rangle_1 = \frac{\xi_1 U(t)}{2\tau_D}\left[1 + \frac{F'(\sigma)}{F(\sigma)}\right]
\tag{110}
$$

We then find, just as for the parallel case, that the complex susceptibility for a small transverse field $E_1 e^{i\omega t}$ is

$$
\chi_\perp(\omega) = \frac{\chi'_\perp(0)}{1 + i\omega\tau_\perp}
\tag{111}
$$

where

$$
\chi'_\perp(0) = \frac{\mu^2 N}{2kT}\left[1 - \frac{F'(\sigma)}{F(\sigma)}\right]
$$

and $\tau_\perp$ is given by Eq. (93).

G. The Relaxation Times for An Arbitrary Uniaxial Nematic Potential

The analysis so far has been restricted to the simplest nematic potential, namely

$$
V(\vartheta) = -V_0 \cos^2 \vartheta
\tag{112}
$$

We now show how the analysis may be extended to yield the relaxation times for an arbitrary nematic potential. The nematic field E_0 is the gradient of a potential $V(\{\mu\})$ so that

$$E_0(\{\mu\}) = -\frac{\partial V}{\partial \mu} \tag{113}$$

where

$$\frac{\partial}{\partial \mu} = i\frac{\partial}{\partial \mu_x} + j\frac{\partial}{\partial \mu_y} + k\frac{\partial}{\partial \mu_z} \tag{114}$$

Because $|\mu|$ = constant = μ, V is indeterminate by an arbitrary function of μ^2, and E_0 by an arbitrary vector along μ, which contribute nothing to $\mu \times E_0$. We confine ourselves to an uniaxial potential, where $E_0(\{\mu\})$ has only a k component so that

$$E_0(\{\mu\}) = -k\frac{\partial}{\partial \mu_z}V(\{\mu\}) \tag{115}$$

The total electric field acting on the molecule is

$$E_T(t) = E_0(\{\mu(t)\}) + E_1(t) \tag{116}$$

This consists of the axially symmetric field E_0 and the small externally applied probe field $E_1(t)$, where $\mu E_1(t) \ll kT$.

By using Eq. (116) we can rewrite the Langevin equation (24) as

$$\frac{d\mu(t)}{dt} = -\zeta^{-1}\big(\mu(t) \times \big(\mu(t) \times E_0(\{\mu(t)\})\big)\big)$$
$$+ F(\{\mu(t)\})E_1(t) + L(t) \tag{117}$$

where

$$F(\{\mu(t)\}) = -\zeta^{-1}\big(\mu(t) \times (\mu(t) \times e)\big) \qquad e = \frac{E_1(t)}{E_1(t)} \tag{118}$$

$$L(t) = \zeta^{-1}\big(\lambda(t) \times \mu(t)\big) \tag{119}$$

On averaging (c.f., Eqs. (44) and (49)),

$$\frac{d}{dt}\langle \mu_x \rangle + \frac{1}{\tau_D}\langle \mu_x \rangle = -\zeta^{-1}\langle E_0 \mu_x \mu_z \rangle + \langle F_x \rangle E_1(t) \qquad (120)$$

$$\frac{d}{dt}\langle \mu_y \rangle + \frac{1}{\tau_D}\langle \mu_y \rangle = -\zeta^{-1}\langle E_0 \mu_y \mu_z \rangle + \langle F_y \rangle E_1(t) \qquad (121)$$

$$\frac{d}{dt}\langle \mu_z \rangle + \frac{1}{\tau_D}\langle \mu_z \rangle = \zeta^{-1}\langle E_0(\mu^2 - \mu_z^2) \rangle + \langle F_z \rangle E_1(t) \qquad (122)$$

We first consider the equation of motion $\langle \mu_z \rangle$. We suppose that a small field $E_1(t) = E_1(t)\mathbf{k}$ is applied along the z axis at time $t = 0$. We require as before the linear response to $\mathbf{E}_1(t)$. We therefore assume that in Eq. (122) $\langle \mu_z \rangle$ and $\langle E_0(\mu^2 - \mu_z^2) \rangle$ can be represented as

$$\langle \mu_z \rangle = \langle \mu_z \rangle_0 + \langle \mu_z \rangle_1 \qquad (123)$$

$$\langle E_0(\mu^2 - \mu_z^2) \rangle = \langle E_0(\mu^2 - \mu_z^2) \rangle_0 + \langle E_0(\mu^2 - \mu_z^2) \rangle_1 \qquad (124)$$

As before, the subscript 0 denotes the equilibrium ensemble average in the absence of the field $\mathbf{E}_1(t)$ and the subscript 1 denotes the portion of the ensemble average that is linear in $\mathbf{E}_1(t)$. Thus, we have from Eqs. (122)–(124)

$$\frac{d}{dt}\left(\langle \mu_z \rangle_0 + \langle \mu_z \rangle_1\right) + \frac{1}{\tau_D}\left(\langle \mu_z \rangle_0 + \langle \mu_z \rangle_1\right)$$
$$= \zeta^{-1}\left(\langle E_0(\mu^2 - \mu_z^2) \rangle_0 + \langle E_0(\mu^2 - \mu_z^2) \rangle_1\right) \qquad (125)$$
$$+ \left(\langle F_z \rangle_0 + \langle F_z \rangle_1\right)E_1(t)$$

Now $\langle \mu_z \rangle_0$ has reached equilibrium, so that

$$\langle \mu_z \rangle_0 = \int_0^{2\pi} \int_0^{\pi} \mu \cos \vartheta W_0(\vartheta)\sin \vartheta \, d\vartheta \, d\phi \qquad (126)$$

where

$$W_0(\vartheta) = C \exp\left(-\frac{V(\vartheta)}{kT}\right) \qquad (127)$$

is the equilibrium distribution function, ϑ and ϕ are the polar and azimuthal angles, respectively, and C is the normalizing constant. The ensemble average $\langle \mu_z \rangle_1$ satisfies

$$\frac{d}{dt}\langle \mu_z \rangle_1 + \frac{1}{\tau_D}\langle \mu_z \rangle_1 = \zeta^{-1}\langle E_0(\mu^2 - \mu_z^2)\rangle_1 + \zeta^{-1}\langle \mu^2 - \mu_z^2 \rangle_0 E_1(t) \tag{128}$$

To determine $\tau_{\parallel}$ we consider the unforced equation:

$$\frac{d}{dt}\langle \mu_z \rangle_1 + \frac{1}{\tau_D}\langle \mu_z \rangle_1 = \zeta^{-1}\langle E_0(\mu^2 - \mu_z^2)\rangle_1 \tag{129}$$

The characteristic equation of the system is then

$$s + \tau_D^{-1} - \zeta^{-1}\frac{\mathscr{L}\langle E_0(\mu^2 - \mu_z^2)\rangle_1}{\mathscr{L}\langle \mu_z \rangle_1} = 0 \tag{130}$$

where the symbol $\mathscr{L}$ is the Laplace transform.
If we suppose, again following Morita,[7] that

$$\frac{\mathscr{L}\langle E_0(\mu^2 - \mu_z^2)\rangle_1}{\mathscr{L}\langle \mu_z \rangle_1} = \frac{\mathscr{L}\langle E_0(\mu^2 - \mu_z^2)\rangle - \mathscr{L}\langle E_0(\mu^2 - \mu_z^2)\rangle_0}{\mathscr{L}\langle \mu_z \rangle - \mathscr{L}\langle \mu_z \rangle_0} \tag{131}$$

may be replaced by its final (equilibrium) value (i.e., its value as t tends to infinity),

$$\lim_{t \to \infty}\frac{\langle E_0(\mu^2 - \mu_z^2)\rangle_1}{\langle \mu_z \rangle_1} = \lim_{s \to 0}\frac{s\mathscr{L}\langle E_0(\mu^2 - \mu_z^2)\rangle_1}{s\mathscr{L}\langle \mu_z \rangle_1} \tag{132}$$

then Eq. (132) may be evaluated as follows. At equilibrium ($t \to \infty$)

$$\langle \mu_z \rangle = \mu \frac{\int_0^\pi \cos\vartheta \exp[-V/kT + (\mu E_1/kT)\cos\vartheta]\sin\vartheta\,d\vartheta}{\int_0^\pi \exp[-V/kT + (\mu E_1/kT)\cos\vartheta]\sin\vartheta\,d\vartheta}$$

which becomes in the linear approximation of E_1

$$\langle \mu_z \rangle \cong \mu \frac{\int_0^\pi \cos \vartheta [1 + (\mu E_1/kT) \cos \vartheta] \exp(-V/kT) \sin \vartheta \, d\vartheta}{\int_0^\pi (1 + (\mu E_1/kT) \cos \vartheta) \exp(-V/kT) \sin \vartheta \, d\vartheta}$$

$$= \mu \langle \cos \vartheta \rangle_0 + \frac{\mu^2 E_1}{kT} \left(\langle \cos^2 \vartheta \rangle_0 - \langle \cos \vartheta \rangle_0^2 \right) \tag{133}$$

$$= \langle \mu_z \rangle_0 + \frac{E_1}{kT} \left(\langle \mu_z^2 \rangle_0 - \langle \mu_z \rangle_0^2 \right)$$

Further,

$$\left\langle E_0 (\mu^2 - \mu_z^2) \right\rangle$$

$$= \mu^2 \frac{\int_0^\pi (1 - \cos^2 \vartheta) E_0 \exp[-V/kT + (\mu E_1/kT) \cos \vartheta] \sin \vartheta \, d\vartheta}{\int_0^\pi \exp[-V/kT + (\mu E_1/kT) \cos \vartheta] \sin \vartheta \, d\vartheta}$$

$$\cong \mu^2 \frac{\int_0^\pi (1 - \cos^2 \vartheta) E_0 [1 + (\mu E_1/kT) \cos \vartheta] \exp(-V/kT) \sin \vartheta \, d\vartheta}{\int_0^\pi \exp(-V/kT) [1 + (\mu E_1/kT) \cos \vartheta] \sin \vartheta \, d\vartheta}$$

$$\tag{134}$$

$$= \left\langle E_0 (\mu^2 - \mu_z^2) \right\rangle_0 + \frac{\mu^3 E_1}{kT} \left\langle E_0 (\cos \vartheta - \langle \cos \vartheta \rangle_0)(1 - \cos^2 \vartheta) \right\rangle_0$$

The second term on the right side of Eq. (134) can be evaluated by noting that

$$E_0 (\cos \vartheta - \langle \cos \vartheta \rangle_0)(1 - \cos^2 \vartheta) \rangle_0$$

$$= \frac{\int_0^\pi - [\partial V/(\mu \partial \cos \vartheta)][\cos \vartheta - \langle \cos \vartheta \rangle_0][1 - \cos^2 \vartheta] \exp(-V/kT) \sin \vartheta \, d\vartheta}{\int_0^\pi \exp(-V/kT) \sin \vartheta \, d\vartheta}$$

$$= \frac{kT}{\mu} \frac{-[\cos \vartheta - \langle \cos \vartheta \rangle_0][1 - \cos^2 \vartheta] e^{-V/kT} \big|_0^\pi}{- \int_0^\pi (1 - 3 \cos^2 \vartheta + 2 \cos \vartheta \langle \cos \vartheta \rangle_0) e^{-V/kT} \sin \vartheta \, d\vartheta}{\int_0^\pi e^{-V/kT} \sin \vartheta \, d\vartheta}$$

$$= -\frac{kT}{\mu} \left[1 - 3 \langle \cos^2 \vartheta \rangle_0 + 2 \langle \cos \vartheta \rangle_0^2 \right] \tag{135}$$

In the nematic phase $\langle \cos \vartheta \rangle_0$ is equal to zero. Substituting Eq. (135) into (134) we obtain

$$\left\langle E_0\left(\mu^2 - \mu_z^2\right)\right\rangle = \left\langle E_0\left(\mu^2 - \mu_z^2\right)\right\rangle_0 - E_1\left(\mu^2 - 3\langle \mu_z^2 \rangle_0\right) \qquad (136)$$

The effective eigenvalue $\lambda_{\parallel}$ with this procedure is then

$$\lambda_{\parallel} = \tau_D^{-1} - \zeta^{-1} \frac{\left\langle E_0\left(\mu^2 - \mu_z^2\right)\right\rangle - \left\langle E_0\left(\mu^2 - \mu_z^2\right)\right\rangle_0}{\langle \mu_z \rangle - \langle \mu_z \rangle_0}$$

$$= \tau_D^{-1} - \zeta^{-1} \frac{\mu^2 E_1\left(1 - 3\langle \cos^2 \vartheta \rangle_0\right)}{\mu^2 E_1/kT\left(\langle \cos^2 \vartheta \rangle_0 - \langle \cos \vartheta \rangle_0^2\right)} \qquad (137)$$

$$= \left(2\tau_D\right)^{-1} \frac{1 - \langle \cos^2 \vartheta \rangle_0}{\langle \cos^2 \vartheta \rangle_0}$$

We have used Eqs. (133) and (136) here. Thus, the longitudinal relaxation time $\tau_{\parallel} = \lambda_{\parallel}^{-1}$ may be expressed in terms of the equilibrium averages as

$$\tau_{\parallel} = 2\tau_D \frac{\langle \cos^2 \vartheta \rangle_0}{1 - \langle \cos^2 \vartheta \rangle_0}$$

$$= \tau_D \frac{2\langle P_2 \rangle_0 + 1}{1 - \langle P_2 \rangle_0} = \tau_D \frac{2S + 1}{1 - S} \qquad (138)$$

where P_2 is the Legendre polynomial of order 2 and $S = \langle P_2 \rangle_0$ is the order parameter. Equation (138) was first obtained by Meier[9, 10] from the Fokker-Planck equation.

We now calculate the transverse relaxation time $\tau_{\perp}$. We consider the same problem as above but this time the step change in the field $\mathbf{E}_1(t) = iU(t)E_1(t)$ is applied parallel to the x axis so that we need to determine the behavior of $\langle \mu_x \rangle$ from Eq. (120). We find, just as before, that the eigenvalue equation is

$$s + \tau_D^{-1} - \zeta^{-1} \frac{\mathscr{L}\langle E_0 \mu_x \mu_z \rangle_1}{\mathscr{L}\langle \mu_x \rangle_1} = 0 \qquad (139)$$

or

$$s + \tau_D^{-1} - \zeta^{-1} \frac{\langle E_0 \mu_x \mu_z \rangle_1}{\langle \mu_x \rangle_1} = 0 \qquad (t \to \infty) \qquad (140)$$

Now

$$\langle \mu_x \rangle = \mu \frac{\int_0^\pi \int_0^{2\pi} \sin \vartheta \cos \phi \exp[-V/kT + (\mu E_1/kT)\sin \vartheta \cos \phi]\sin \vartheta \, d\vartheta \, d\phi}{\int_0^\pi \int_0^{2\pi} \exp[-V/kT + (\mu E_1/kT)\sin \vartheta \cos \phi]\sin \vartheta \, d\vartheta \, d\phi} \qquad (141)$$

which in the linear approximation of E_1 is

$$\langle \mu_x \rangle \cong \frac{\mu^2 E_1}{kT} \langle \sin^2 \vartheta \cos^2 \phi \rangle_0 = \frac{\mu^2 E_1}{2kT} (1 - \langle \cos^2 \vartheta \rangle_0) \qquad (142)$$

In the same way

$$\langle E_0 \mu_x \mu_z \rangle \cong \frac{\mu^3 E_1}{kT} \langle E_0 \cos \vartheta \sin^2 \vartheta \cos^2 \phi \rangle_0$$

$$= -\mu^2 E_1 (1 - 3\langle \cos^2 \vartheta \rangle_0)/2 \qquad (143)$$

$$\langle E_0 \mu_y \rangle = 0 \qquad (144)$$

We have used the same procedure as in Eq. (135) to obtain Eq. (143). Thus, the effective eigenvalue $\lambda_\perp$ is given by

$$\lambda_\perp = \tau_D^{-1} - \frac{\langle E_0 \mu_x \mu_z \rangle}{\langle \mu_x \rangle} = \tau_D^{-1} - \zeta^{-1} \frac{\mu^2 E_1 (1 - 3\langle \cos^2 \vartheta \rangle_0)}{(\mu^2 E_1/kT)(1 - \langle \cos^2 \vartheta \rangle_0)}$$

$$= (2\tau_D)^{-1} \frac{1 + \langle \cos^2 \vartheta \rangle_0}{1 - \langle \cos^2 \vartheta \rangle_0} \qquad (145)$$

whence the transverse relaxation time $\tau_\perp = \lambda_\perp^{-1}$ may be expressed in terms of the equilibrium averages as

$$\tau_\perp = 2\tau_D \frac{1 - \langle \cos^2 \vartheta \rangle_0}{1 + \langle \cos^2 \vartheta \rangle_0}$$

$$= 2\tau_D \frac{1 - \langle P_2 \rangle_0}{2 + \langle P_2 \rangle_0} = 2\tau_D \frac{1 - S}{2 + S} \qquad (146)$$

Equations (138) and (146) for the relaxation times $\tau_\|$ and $\tau_\perp$ are valid for any axially symmetric nematic potential.

The retardation factors $g_\|$ and $g_\perp$ defined by

$$\tau_\| = g_\| \tau_D \qquad \tau_\perp = g_\perp \tau_D \qquad (147)$$

may now be expressed entirely in terms of the order parameter S as

$$g_\| = \frac{2S + 1}{1 - S} \qquad g_\perp = \frac{2 - 2S}{2 + S} \qquad (148)$$

On eliminating the order parameter S in Eq. (148) one may deduce that

$$g_\| = \frac{2 - g_\perp}{g_\perp} \qquad (149)$$

$$g_\perp = \frac{2}{g_\| + 1} \qquad (150)$$

In accordance with the results of Section II.F, the longitudinal and transverse components of the complex susceptibility are given by the Debye equations

$$\chi_\|(\omega) = \frac{\chi_\|'(0)}{1 + i\omega\tau_\|} \qquad \chi_\perp(\omega) = \frac{\chi_\perp'(0)}{1 + i\omega\tau_\perp} \qquad (151)$$

with relaxation times given by Eqs. (138) and (146); also

$$\chi_\|'(0) = \frac{\mu^2 N}{3kT}(2S + 1) \qquad \chi_\perp'(0) = \frac{\mu^2 N}{3kT}(1 - S) \qquad (152)$$

We have shown how general formulae for $\tau_\|$ and $\tau_\perp$ (valid for an arbitrary uniaxial potential of the crystalline anisotropy) may be calculated directly in terms of the equilibrium order parameter S from the Langevin equation where that equation is regarded as a stochastic nonlinear equation of the Stratonovich type. This eliminates the complicated mathematical analysis arising from the Fokker-Planck equation. Our approach is based on the effective eigenvalue method which enables us in this instance to close the hierarchy of differential-difference equations by reducing the nth-order characteristic equation of the system to one of the first order. The relaxation of the polarization components is thus characterized by a single effective eigenvalue or weighted decay rate. The effective eigenvalues $\lambda_{\parallel}$ and $\lambda_{\perp}$ give precise information on the initial decay of the polarization components. Equation (148) is in qualitative agreement with available experimental data (e.g., Ref. 28) and with previous theoretical estimates,[16, 29, 30] from which it follows that the relaxation time $\tau_{\parallel}$ increases and $\tau_{\perp}$ decreases in nematic liquid crystals as compared to the Debye relaxation time in the isotropic phase.

A comparison of the results predicted by Eq. (34) with experimental data for p, p'-heptylcyanobiphenyl[31] is given in Fig. 2. Experimental points for $g_{\perp}$ and $g_{\parallel}$ have been evaluated from Eqs. (147), where $\tau_{\perp}$ and $\tau_{\parallel}$ are regarded as experimental values and the values of the Debye relaxation times τ_D were extrapolated from those of the isotropic liquid. One may see from Fig. 2 that the theory deviates from the experimental observations. This is due to the fact that the effective eigenvalue for the longitudinal component of the polarization in a nematic liquid crystal yields a smaller relaxation time than that given by the lowest eigenvalue of the distribution function[9, 10, 16] as obtained by solution of the Sturm-Liouville equation. This is corroborated by numerical solution of the Fokker-Planck equation (transformed into a Schrödinger equation) used in laser physics for the calculation of the intensity correlation functions of the light field.

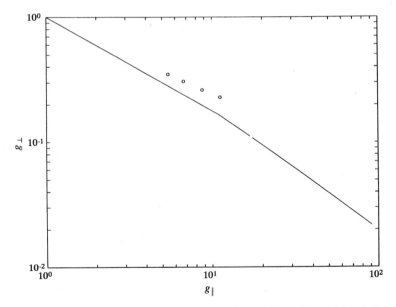

Figure 2. The retardation factor $g_{\perp}$ versus $g_{\parallel}$ for p, p'-heptylcyanobiphenyl. Experimental data (circles) from ref. [31].

Here λ_{eff} may be up to 25% greater than the lowest eigenvalue. The deviation has also been experimentally verified by Arecchi et al.[52] In Section II.J we compare the relaxation behavior yielded by the effective eigenvalue approach with the exact and asymptotic solutions of the problem.

H. Analogy with Magnetic Relaxation of Single-Domain Ferromagnetic Particles

The treatment of the relaxation behavior of nematic liquid crystals just given bears a close resemblance to the theory of magnetic relaxation of ferromagnetic domains. The theory has been reviewed by us in Volume 83 of this series. This study was initiated by W. F. Brown, Jr., who based his approach on the Fokker-Planck equation. The form of the Fokker-Planck equation obtained by Brown[15, 33] closely resembles that obtained by Debye[34] in his study of the dielectric relaxation of polar fluids. For convenience, the Fokker-Planck equation for the probability density of orientations of the magnetization $\mathbf{M}(t)$ of a single-domain ferromagnetic particle is called Brown's equation. Having obtained this equation, Brown applied the method of separation of the variables to convert the solution to a Sturm-Liouville problem.

The Sturm-Liouville method involves calculations of considerable mathematical complexity.[6, 15, 33] In view of this difficulty we have reconsidered in Ref. [11] the relaxation problem in the light of the new approach described here. This procedure, based on averaging the stochastic Gilbert equation for the magnetization $\mathbf{M}(t)$ allowed us to derive equations for the time evolution of the averages of magnetization components directly from that equation. The problem of constructing and solving the Fokker-Planck equation was thus bypassed entirely.

To discuss the Langevin equation for a single-domain ferromagnetic particle we first consider Gilbert's equation for the dynamic behavior of the magnetization vector of the particle $\mathbf{M}(t)$ in the presence of thermal agitation[11, 15]:

$$\frac{d}{dt}\mathbf{M}(t) = \gamma\mathbf{M}(t) \times \left(\mathbf{H}_T(t) - \eta\frac{d}{dt}\mathbf{M}(t)\right) \qquad (153)$$

where γ is the "gyromagnetic" ratio and η is a phenomenological damping parameter,

$$\mathbf{H}_T = \mathbf{H}_{\text{ef}} + \mathbf{H}(t) + \mathbf{h}(t) \qquad (154)$$

is the total magnetic field, which consists of an axially symmetrical field

H_{ef} and a small external applied field $\mathbf{H}(t)$ ($\nu(\mathbf{M} \cdot \mathbf{H}) \ll kT$, k is the Boltzman constant, T is the temperature, ν is the volume of the particle) and $\mathbf{h}(t)$ is a random field term as used by Brown.[15] The random field $\mathbf{h}(t)$ has the properties

$$\overline{h_i(t)} = 0 \qquad \overline{h_j(t)h_i(t')} = \frac{2kT\eta}{\nu}\delta(t-t')\delta_{ij} \qquad (i,j = x,y,z) \quad (155)$$

where M_s is the (constant) magnitude of the magnetization, only its direction varies with time, the overbar indicates a statistical average,[3] $\delta(t)$ is the Dirac-delta function, and δ_{ij} is Kronecker's delta, so that we have assumed isotropy of the statistical properties of $h_i(t)$.

The effective field $\mathbf{H}_{ef}$ is due to the effect of the crystalline anisotropy.[33] If $V(\mathbf{M})$ is the free energy per unit volume expressed as a function of $\mathbf{M}$, then[33]

$$\mathbf{H}_{ef}(\mathbf{M}) = -\frac{\partial V}{\partial \mathbf{M}}$$

where

$$\frac{\partial}{\partial \mathbf{M}} = \mathbf{i}\frac{\partial}{\partial M_x} + \mathbf{j}\frac{\partial}{\partial M_y} + \mathbf{k}\frac{\partial}{\partial M_z}$$

If we confine ourselves to an uniaxial potential, so that $\mathbf{H}_{ef}(\mathbf{M})$ has only a $\mathbf{k}$ component, then

$$\mathbf{H}_{ef}(\mathbf{M}) = -\mathbf{k}\frac{\partial}{\partial M_z}V(\mathbf{M}) \qquad (156)$$

In the context of relaxation a single-domain ferromagnetic particle may reverse the direction of its magnetization due to thermal fluctuations inside the particle (spontaneous remagnetization). This is the solid-state mechanism known as Néel relaxation.[11] The Néel relaxation can be considered as a rotational diffusion of the magnetization $\mathbf{M}(t)$ with respect to the body of the particle in the crystalline field, so that it is an analog of the rotational Brownian motion of molecules in nematic liquid crystals.

Gilbert's Eq. (153) may be rearranged explicitly as shown in Ref. 35 to yield that equation in the Landau-Lifshitz form[36]:

$$\frac{d}{dt}\mathbf{M}(t) = g'M_s\mathbf{M}(t) \times \mathbf{H}_T(t) + h'(\mathbf{M}(t) \times \mathbf{H}_T(t)) \times \mathbf{M}(t) \quad (157)$$

where $M_s = |\mathbf{M}(t)|$, and the constants g' and h' are

$$g' = \frac{\gamma}{\left(1 + \eta^2\gamma^2 M_s^2\right)M_s} \qquad h' = \frac{\eta\gamma^2}{1 + \eta^2\gamma^2 M_s^2} \qquad (158)$$

Using Eq. (154) we can rewrite Eq. (157) as

$$\frac{d\mathbf{M}(t)}{dt} = g'M_s\mathbf{M}(t) \times \mathbf{H}_{ef}(t) - h'\mathbf{M}(t) \times \left(\mathbf{M}(t) \times \mathbf{H}_{ef}(t)\right) \\ + \mathbf{F}(t)H(t) + \mathbf{L}(t) \qquad (159)$$

where

$$\mathbf{F}(t) = g'M_s\mathbf{M}(t) \times \mathbf{e} - h'\mathbf{M}(t) \times \left(\mathbf{M}(t) \times \mathbf{e}\right) \qquad \mathbf{e} = \frac{\mathbf{H}(t)}{H(t)} \qquad (160)$$

$$\mathbf{L}(t) = g'M_s\mathbf{M}(t) \times \mathbf{h}(t) - h'\mathbf{M}(t) \times \left(\mathbf{M}(t) \times \mathbf{h}(t)\right)$$

We may further rewrite Eqs. (159) and (160) as

$$\frac{d}{dt}M_x(t) = g'H_{ef}(t)M_sM_y(t) - h'H_{ef}(t)M_x(t)M_z(t) + F_x(t)H(t) \\ + h'\left(M_s^2 - M_x^2(t)\right)h_x(t) \\ - \left(g'M_sM_z(t) + h'M_x(t)M_y(t)\right)h_y(t) \\ + \left(g'M_y(t)M_s - h'M_z(t)M_x(t)\right)h_z(t) \qquad (161)$$

$$\frac{d}{dt}M_y(t) = -g'H_{ef}(t)M_sM_x(t) - h'H_{ef}(t)M_z(t)M_y(t) + F_y(t)H(t) \\ + h'\left(M_s^2 - M_y^2(t)\right)h_y(t) \\ - \left(g'M_sM_x(t) + h'M_z(t)M_y(t)\right)h_z(t) \\ + \left(g'M_sM_z(t) - h'M_x(t)M_y(t)\right)h_x(t) \qquad (162)$$

$$\frac{d}{dt}M_z(t) = h'H_{ef}(t)\left(M_s^2 - M_z^2(t)\right) + F_z(t)H(t) \\ + h'\left(M_s^2 - M_z^2(t)\right)h_z(t) \\ - \left(g'M_sM_y(t) + h'M_z(t)M_x(t)\right)h_x(t) \\ + \left(g'M_sM_x(t) - h'M_z(t)M_y(t)\right)h_y(t) \qquad (163)$$

where $F_i(t)$ and $L_i(t)$ are the projections of the vectors $\mathbf{F}(t)$ and $\mathbf{L}(t)$ onto the x, y, and z axes. These are the Langevin equations for the problem under consideration.

It is evident that Eqs. (161)–(163) have the form of the set of nonlinear stochastic equations:

$$\frac{d}{dt'}M_i(t') = \Phi_i(\{\mathbf{M}(t')\},t') + g_{ik}(\{\mathbf{M}(t')\},t')h_k(t') \quad (i = x, y, z)$$

$$(164)$$

$$\overline{h_k(t)} = 0 \qquad \overline{h_k(t)h_m(t')} = \delta_{km}\mu\delta(t - t') \qquad (165)$$

where summation over k is understood and $\mu = 2kT\eta/\nu$.

Equation (164) contains multiplicative noise terms $g_{ik}(\{\mathbf{M}(t')\},t')h_k(t')$. This poses an interpretation problem as discussed in Section II.B.

We can use Eq. (39) to evaluate the statistical average of the multiplicative noise terms in Eqs. (161)–(165). In our case the stochastic variables are $\xi_1(t') = M_x(t')$, $\xi_2(t') = M_y(t')$, $\xi_3(t') = M_z(t')$. Thus, on averaging Eqs. (161)–(163) we may obtain the equations of motion for the starting values of the components $M_i = M_i(t)$, $(i = x, y, z)$ of the magnetization, namely[11]

$$\frac{d}{dt}M_x = g'H_{\text{ef}}M_sM_y - h'H_{\text{ef}}M_xM_z + F_xH(t) + \frac{\mu}{2}\left(g_{jk}\frac{\partial}{\partial M_j}g_{1k}\right) \quad (166)$$

$$\frac{d}{dt}M_y = -g'H_{\text{ef}}M_sM_x - h'H_{\text{ef}}M_yM_z + F_yH(t) + \frac{\mu}{2}\left(g_{jk}\frac{\partial}{\partial M_j}g_{2k}\right)$$

$$(167)$$

$$\frac{d}{dt}M_z = h'H_{\text{ef}}\left(M_s^2 - M_z^2\right) + F_zH(t) + \frac{\mu}{2}\left(g_{jk}\frac{\partial}{\partial M_j}g_{3k}\right) \quad (168)$$

The evaluation of the noise induced drift $(\mu/2)[g_{jk}(\partial/\partial M_j)g_{ik}]$ leads to[11]

$$\frac{\mu}{2}g_{jk}\frac{\partial}{\partial M_j}g_{ik} = -\mu\left(g'^2 + h'^2\right)M_s^2M_i = -\tau_N^{-1}M_i \qquad (169)$$

where $\tau_N = \nu/(2kTh')$ is the relaxation time for $H_{\text{ef}} = 0$, and then to the equations for the averages $\langle M_i \rangle$:

$$\frac{d}{dt}\langle M_x \rangle + \frac{1}{\tau_N}\langle M_x \rangle = g'M_s\langle H_{ef}M_y \rangle - h'\langle H_{ef}M_xM_z \rangle + \langle F_x \rangle H(t)$$

(170)

$$\frac{d}{dt}\langle M_y \rangle + \frac{1}{\tau_N}\langle M_y \rangle = -g'M_s\langle H_{ef}M_x \rangle - h'\langle H_{ef}M_yM_z \rangle + \langle F_y \rangle H(t)$$

(171)

$$\frac{d}{dt}\langle M_z \rangle + \frac{1}{\tau_N}\langle M_z \rangle = h'\langle H_{ef}(M_s^2 - M_z^2) \rangle + \langle F_z \rangle H(t)$$ (172)

These equations govern the Néel relaxation of a single-domain ferromagnetic particle. They resemble Eqs. (120)–(122) for the Debye relaxation of a nematic liquid crystal. They differ from the nematic equations, however, in so far as they contain precessional terms $g'M_s\langle H_{ef}M_x \rangle$ and $g'M_s\langle H_{ef}M_y \rangle$, which are absent from the nematic equations. The precessional terms are responsible for ferromagnetic resonance which is due to precessional motion of the magnetization $\mathbf{M}(t)$ in the anisotropy field $\mathbf{H}_{ef}$.

Equations (170)–(172) (v. Section VII of Ref. 11) are the first terms in an infinite hierarchy of differential equations. It is obvious that to solve these equations, we must also obtain equations for $\langle H_{ef}M_x \rangle$, $\langle H_{ef}M_yM_z \rangle$, $\langle H_{ef}M_y \rangle$, $\langle H_{ef}M_xM_z \rangle$, $\langle H_{ef} \rangle$, $\langle H_{ef}M_z^2 \rangle$ and so on. This difficulty may also be circumvented by means of the effective eigenvalue method.

On using this method we have shown in Ref. 11 that the longitudinal $(\tau_\parallel)$ and transverse $(\tau_\perp)$ magnetization relaxation times may be expressed in terms of the equilibrium averages $\langle \cos \vartheta \rangle_0$ and $\langle \cos^2 \vartheta \rangle_0$. We have

$$\tau_\parallel = 2\tau_N \frac{\langle \cos^2 \vartheta \rangle_0 - \langle \cos \vartheta \rangle_0^2}{1 - \langle \cos^2 \vartheta \rangle_0}$$

(173)

$$\tau_\perp = 2\tau_N \frac{1 - \langle \cos^2 \vartheta \rangle_0}{1 + \langle \cos^2 \vartheta \rangle_0}$$

(174)

Equations (173) and (174) for the relaxation times $\tau_\parallel$ and $\tau_\perp$ are valid for any axially symmetric potential of the crystalline anisotropy. We also note that the term $\langle \cos \vartheta \rangle_0$ in Eq. (176) appears due to the fact that the effective field H_{eff} may also include an applied constant magnetic field $\mathbf{H}_0$.

In the simplest uniaxial case, the crystalline anisotropy energy density is $K \sin^2 \vartheta$ $(K > 0)$, where ϑ is the angle between $\mathbf{M}$ and the positive z

axis.[33] In the absence of the external field $\mathbf{H}_0 = 0$, so that for relaxation in the purely crystalline potential,

$$V(\vartheta) = -K \cos^2 \vartheta + K \qquad (175)$$

which is entirely analogous to the Meier-Saupe nematic potential, we have

$$\langle \cos \vartheta \rangle_0 = 0 \qquad (176)$$

and

$$\langle \cos^2 \vartheta \rangle_0 = \frac{1}{2\sqrt{\sigma} \, D(\sqrt{\sigma})} - \frac{1}{2\sigma}$$

or

$$\langle \cos^2 \vartheta \rangle_0 = \frac{F'(\sigma)}{F(\sigma)} \qquad (177)$$

where $\sigma = \nu K / kT$, $D(x)$ is the Dawson integral Eq. (72), and $F(\sigma)$ is defined by Eq. (69). The symbol $\langle f \rangle_0$ now means

$$\langle f \rangle_0 = \frac{\int_0^\pi f(\vartheta) e^{-\nu V(\vartheta)/kT} \sin \vartheta \, d\vartheta}{\int_0^\pi e^{-\nu V(\vartheta)/kT} \sin \vartheta \, d\vartheta} \qquad (178)$$

According to Eqs. (173) and (174) we then have[11]

$$\tau_{\parallel} = 2\tau_{\mathrm{N}} \frac{\sqrt{\sigma} - D(\sqrt{\sigma})}{D(\sqrt{\sigma})(1 + 2\sigma) - \sqrt{\sigma}} = 2\tau_{\mathrm{N}} \frac{F'(\sigma)}{F(\sigma) - F'(\sigma)} \qquad (179)$$

$$\tau_{\perp} = 2\tau_{\mathrm{N}} \frac{D(\sqrt{\sigma})(2\sigma + 1) - \sqrt{\sigma}}{D(\sqrt{\sigma})(2\sigma - 1) + \sqrt{\sigma}} = 2\tau_{\mathrm{N}} \frac{F(\sigma) - F'(\sigma)}{F(\sigma) + F'(\sigma)} \qquad (180)$$

Equations (179) and (180), that is, the magnetization relaxation times $\tau_{\parallel}$ and $\tau_{\perp}$ coincide precisely with the dielectric relaxation times of a nematic obtained in Sections II.D and II.E (if we put $\tau_{\mathrm{N}} = \tau_{\mathrm{D}}$). They also coincide with results of Refs. 6 and 11.

Just as in Section II.D, Eq. (179) allows us to calculate from Eq. (172) the longitudinal susceptibility $\chi_{\parallel}(\omega)$ of the particle, which is given by[6]:

$$\chi_{\parallel}(\omega) = \frac{\chi'_{\parallel}(0)}{1 + i\omega\tau_{\parallel}} \qquad (181)$$

where[11]

$$\chi_{\parallel}(0) = \frac{\nu}{kT}\langle M_z^2 \rangle_0 = \frac{\nu M_s^2}{kT}\frac{F'(\sigma)}{F(\sigma)}$$

is the static longitudinal susceptibility. Calculations of the transverse susceptibility $\chi_{\perp}(\omega)$ (when the probe field $\mathbf{H}(t)$ is applied along the x axis) have been given in the context of the Fokker-Planck equation approach in Ref. 6. We give an equivalent derivation based on the Langevin equation. As shown in Appendix B the average equations of motion when expressed in terms of spherical harmonics are for the dipole component $\langle X_{11} \rangle$

$$\frac{d}{dt}\langle X_{11} \rangle + \frac{1}{\tau_N}\left(1 + \frac{\sigma}{5}\right)\langle X_{11} \rangle + \frac{2\sigma}{15\tau_N}\langle X_{31} \rangle - \frac{\sigma}{3\alpha\tau_N}\langle X_{2,-1} \rangle$$

$$= \frac{\xi(t)}{6\tau_N}[2 - \tfrac{1}{2}\langle X_{22} \rangle + \langle P_2 \rangle]$$

(182)

and for the quadrupole component $\langle X_{2,-1} \rangle$

$$\frac{d}{dt}\langle X_{2,-1} \rangle + \frac{1}{\tau_N}\left(3 - \frac{\sigma}{7}\right)\langle X_{2,-1} \rangle + \frac{2\sigma}{5\alpha\tau_N}\langle X_{31} \rangle$$

$$+ \frac{3\sigma}{5\alpha\tau_N}\langle X_{11} \rangle + \frac{12\sigma}{35\tau_N}\langle X_{4,-1} \rangle$$

$$= \frac{\xi(t)}{\tau_N}\left[\frac{6}{5}\langle P_1 \rangle + \frac{3}{10}\langle P_3 \rangle - \frac{1}{20}(\langle X_{32} \rangle + \langle X_{3,-2} \rangle)\right.$$

$$\left. - \frac{1}{2\alpha}\left(1 - \langle P_2 \rangle - \frac{1}{2}\langle X_{22} \rangle\right)\right]$$

(183)

where $\alpha = h'/g' = \gamma M_s \eta$ is the dimensionless damping parameter (for all ferromagnets $\alpha \ll 1$ [6]),

$$\xi(t) = \frac{\nu M_s H(t)}{kT}$$

$$\langle X_{nm} \rangle = \langle P_n^{|m|} \cos m\phi \rangle, \langle X_{n,-m} \rangle = \langle P_n^{|m|} \sin m\phi \rangle \qquad m \leq n$$

and $P_n^{|m|}$ are the associated Legendre functions of order n. These are the exact average equations of motion.

To determine the frequency dependence of the transverse susceptibility $\chi_\perp(\omega)$ let us now suppose that a small constant magnetic field $\mathbf{H} = \mathbf{i}H$ (where $\mathbf{i}$ denotes the transverse x direction) has been applied to the system for a long time and that it is suddenly switched off at time $t = 0$. Our task is to calculate the one-sided Fourier transform of the magnetization decay function $f(t)$ defined as

$$f_\perp(t) = \frac{M_s \langle X_{11} \rangle}{H} \tag{184}$$

which according to linear response theory[49] allows us to calculate the transverse susceptibility $\chi_\perp(\omega)$ as follows:

$$\chi_\perp(\omega) = f_\perp(0) - i\omega \int_0^\infty f_\perp(t) e^{-i\omega t}\, dt \tag{185}$$

The effective eigenvalue procedure now reduces Eqs. (182) and (183) for $t > 0$ to

$$\frac{d}{dt}\langle X_{11} \rangle + \frac{1}{\tau_\perp}\langle X_{11} \rangle - \frac{\sigma}{3\alpha\tau_N}\langle X_{2,-1} \rangle = 0 \tag{186}$$

$$\frac{d}{dt}\langle X_{2,-1} \rangle + \frac{1}{\tau_2}\langle X_{2,-1} \rangle + \frac{3}{\alpha}\left(\frac{1}{\tau_\perp} - \frac{1}{\tau_N}\right)\langle X_{11} \rangle = 0 \tag{187}$$

with initial conditions

$$\langle X_{11} \rangle \cong \frac{\nu M_s H}{kT}\langle \sin^2 \vartheta \cos^2 \phi \rangle_0$$

$$= \frac{\nu M_s H}{2kT}\left(\frac{F(\sigma) - F'(\sigma)}{F(\sigma)}\right) = \frac{\chi_\perp'(0)H}{M_s} \tag{188}$$

$$\langle X_{2,-1} \rangle = 0 \tag{189}$$

where $\chi_\perp'(0)$ is the static transverse susceptibility.

At the derivation of Eq. (187) from Eq. (183) we have taken into account that in accordance with Eqs. (182) and (186)

$$\langle X_{31} \rangle = \frac{15}{2\sigma}\left[\frac{\tau_N}{\tau_\perp} - 1 - \frac{\sigma}{5}\right]\langle X_{11} \rangle$$

The effective relaxation time $\tau_\perp$ has been calculated before[6, 11] and is given by Eq. (180). The second effective relaxation time τ_2 is evaluated as follows. In the absence of the probe field $\mathbf{H}(t)$, $\langle X_{2, -1} \rangle$ is proportional in the time domain to the equilibrium correlation function $C_{2, -1}(t)$, defined as

$$C_{2, -1}(t) = \langle X_{2, -1}(0) X_{2, -1}(t) \rangle_0$$

Therefore, we have

$$\tau_2 = -\frac{\langle X_{2, -1} \rangle}{(d/dt)\langle X_{2, -1} \rangle} = -\frac{C_{2, -1}(0)}{\dot{C}_{2, -1}(0)} \tag{190}$$

The equation for $\dot{C}_{2, -1}(0)$ can be obtained from Eq. (B.13) for the starting value of $M_y H_{ef} = (2K/3M_s)X_{2, -1}$ at $H(t) = 0$ on multiplying this equation by $X_{2, -1}(0)$ and averaging over the equilibrium distribution function at $t = 0$. The result is

$$\dot{C}_{2, -1}(0) + \frac{1}{\tau_N}\left(3 - \frac{\sigma}{7}\right)C_{2, -1}(0) = -\frac{12\sigma}{35\tau_N}\langle X_{2, -1}(0) X_{4, -1}(0) \rangle_0 \tag{191}$$

Thus, we have from Eqs. (190) and (191)

$$\begin{aligned}
\tau_2 &= \tau_N \frac{\langle X_{2, -1}^2 \rangle_0}{(3 - \sigma/7)\langle X_{2, -1}^2 \rangle_0 + (12/35)\sigma\langle X_{4, -1} X_{2, -1} \rangle_0} \\
&= 2\tau_N\left(3 - \sigma + 2\sigma\frac{F''(\sigma) - F'''(\sigma)}{F'(\sigma) - F''(\sigma)}\right) \\
&= 2\tau_N\left(\sigma - 2 - \frac{2\sqrt{\sigma} - 2D(\sqrt{\sigma})}{3\sqrt{\sigma} - (3 + 2\sigma)D(\sqrt{\sigma})}\right)
\end{aligned} \tag{192}$$

Equation (192) coincides precisely with Eq. (34) of Ref. 6.

On applying the one-sided Fourier transform to Eqs. (186) and (187) at $t > 0$ and taking into account Eqs. (188) and (189), we obtain

$$\left(i\omega + \frac{1}{\tau_\perp}\right)\langle \tilde{X}_{11} \rangle - \frac{\sigma}{3\alpha\tau_N}\langle \tilde{X}_{2, -1} \rangle = \frac{\chi'_\perp(0)H}{M_s} \tag{193}$$

$$\frac{3}{\alpha}\left(\frac{1}{\tau_\perp} - \frac{1}{\tau_N}\right)\langle \tilde{X}_{11} \rangle + \left(i\omega + \frac{1}{\tau_2}\right)\langle \tilde{X}_{2, -1} \rangle = 0 \tag{194}$$

where the one-sided Fourier transforms $\langle \tilde{X}_{nm} \rangle$ are defined as

$$\langle \tilde{X}_{nm} \rangle = \int_0^\infty e^{-i\omega t} \langle X_{nm} \rangle \, dt$$

On solving these equations and using Eqs. (184) and (185) we obtain

$$\tilde{f}(\omega) = \frac{M_s \langle \tilde{X}_{11} \rangle}{H} = \frac{\chi_\perp(0)(i\omega + 1/\tau_2)}{(i\omega + 1/\tau_2)(i\omega + 1/\tau_\perp) + \dfrac{\sigma}{\tau_N \alpha^2}\left(\dfrac{1}{\tau_\perp} - \dfrac{1}{\tau_N}\right)} \tag{195}$$

$$f(0) = \chi'_\perp(0)$$

Thus, on using Eq. (185) we obtain the transverse susceptibility

$$\frac{\chi_\perp(\omega)}{\chi'_\perp(0)} = \frac{(1 + i\omega\tau_2) + \Delta}{(1 + i\omega\tau_2)(1 + i\omega\tau_\perp) + \Delta} \tag{196}$$

where

$$\Delta = \frac{\sigma\tau_2(\tau_N - \tau_\perp)}{\alpha^2 \tau_N^2}$$

Equation (196) is in full agreement with Eq. (37) of Ref. 6, as may be seen by substituting for τ_2, τ_N, and Δ. For small σ Eq. (196) reduces to Eq. (7.67) of Ref. 11. In the limit of high damping when the dimensionless damping constant $\alpha = \gamma\eta M_s$ becomes very large, Eq. (196) reduces to the purely relaxation equation

$$\chi_\perp(\omega) = \frac{\chi'_\perp(0)}{1 + i\omega\tau_\perp} \tag{197}$$

In the other important limit, $\sigma \to \infty$ (high potential barrier) when $\tau_\perp, \tau_2 \cong \tau_N/\sigma$, Eq. (196) yields the Landau-Lifshitz formula for ferromagnetic resonance[36]:

$$\chi_\perp(\omega) = \gamma M_s \frac{(1 + \alpha^2)\omega_0 + i\alpha\omega}{(1 + \alpha^2)\omega_0^2 - \omega^2 + 2i\alpha\omega\omega_0} \tag{198}$$

have[24] (denoting in this instance final equilibrium values by $\langle P_n \rangle_\xi$)

$$\langle P_1 \rangle_\xi = L(\xi) \tag{209}$$

$$\langle P_2 \rangle_\xi = 1 - \frac{3L(\xi)}{\xi} \tag{210}$$

so that

$$\mathscr{L}\langle P_1 \rangle = \frac{\xi}{3s(s\tau_D + \xi/3L(\xi))} \tag{211}$$

so that the rise transient has the effective time constant

$$\tau_{\text{eff}} = \frac{3L(\xi)\tau_D}{\xi} \tag{212}$$

in agreement with the result of Morita.[7] Since we have supposed that in the infinite continued fraction (Eq. (207)) that the second-order Legendre polynomial (describing the dynamic Kerr effect average) has reached its equilibrium value, it is necessary to proceed further down the continued fraction in order to calculate the rise transient of the dynamic Kerr effect. We have from Eq. (207)

$$\frac{\mathscr{L}\langle P_2 \rangle}{\mathscr{L}\langle P_1 \rangle} = \frac{3\xi/5}{s\tau_D + 3(1 + (\xi/5)(\mathscr{L}\langle P_3 \rangle/\mathscr{L}\langle P_2 \rangle))} \tag{213}$$

We now again truncate by supposing that the ratio $\langle P_3 \rangle/\langle P_2 \rangle$ has reached its equilibrium value. The simplest way to calculate this ratio is to recall that at equilibrium

$$\langle P_2 \rangle_\xi = \frac{\xi}{5}\left[\langle P_1 \rangle_\xi - \langle P_3 \rangle_\xi\right] \tag{214}$$

so that

$$\frac{\langle P_1 \rangle_\xi}{\langle P_2 \rangle_\xi} - \frac{5}{\xi} = \frac{\langle P_3 \rangle_\xi}{\langle P_2 \rangle_\xi} \tag{215}$$

And recalling that

$$\frac{\langle P_1 \rangle_\xi}{\langle P_2 \rangle_\xi} = \frac{L(\xi)}{1 - 3L(\xi)/\xi} \tag{216}$$

we have

$$\frac{\langle P_3 \rangle_\xi}{\langle P_2 \rangle_\xi} = \frac{\xi^2 L(\xi) - 5[\xi - 3L(\xi)]}{\xi[\xi - 3L(\xi)]} \tag{217}$$

so that at this level of truncation

$$\frac{\mathscr{L}\langle P_2 \rangle}{\mathscr{L}\langle P_1 \rangle} = \frac{3\xi/5}{s\tau_D + \frac{3}{5}\{[\xi^2 L(\xi) - 5(\xi - 3L(\xi))]/(\xi - 3L(\xi))\}}$$

$$= \frac{3\xi/5}{s\tau_D + 3\xi^2 L(\xi)/(\xi - 3L(\xi))} \tag{218}$$

Thus,

$$\mathscr{L}\langle P_2 \rangle = \frac{3\xi}{5\tau_D} \frac{\mathscr{L}\langle P_1 \rangle}{s + (3/\tau_D)[\xi^2 L(\xi)/(\xi - 3L(\xi))]} \tag{219}$$

and on using Eq. (195)

$$\mathscr{L}\langle P_2 \rangle = \frac{\frac{1}{5}(\xi^2/\tau_D^2)}{s(s + \xi/L(\xi)\tau_D)\{s + (3/\tau_D)[\xi^2 L(\xi)/(\xi - 3L(\xi))]\}} \tag{220}$$

Equation (220) shows clearly that two effective eigenvalues λ_1 and λ_2, namely

$$\lambda_1 = \frac{\xi}{L(\xi)\tau_D} \qquad \lambda_2 = \frac{3}{\tau_D}\left(\frac{\xi^2 L(\xi)}{\xi - 3L(\xi)}\right) \tag{221}$$

are needed to describe the transient Kerr effect response due to the application of a strong dc field. This point was first emphasized by Morita and Watanabe[14] who derived Eq. (20) in their review of the dynamic Kerr effect. Calculations performed by them suggest that Eq. (220) provides a satisfactory approximation to the response computed by taking successive convergents of the continued fraction—Eq. (207). The calculation described here is valuable as yet another simple example of a situation where two effective eigenvalues are required to describe the response. The procedure may be continued by iteration to obtain the cubic term in the dielectric response. We note that we are calculating the response to a strong dc field; thus, there is no connection between any of the above formulae and the ac response, unlike the nematic liquid crystal problem.

J.   Comparison of Effective and Longest Relaxation Times

It is important to compare the relaxation times yielded by Eqs. (138) and (146) with the numerical and asymptotic results for the largest relaxation time. We consider the simplest potential of uniaxial anisotropy, namely

$$\frac{V}{kT} = \sigma(1 - \cos^2 \vartheta)$$

For a longitudinal applied field Brown[15] has shown that the longest relaxation time in the limit of high potential barriers, $\sigma \gg 1$, is well approximated by

$$\tau_\| \simeq \tau_D \frac{\sqrt{\pi}}{2} \sigma^{-3/2} \exp \sigma \tag{222}$$

For small potential barriers, $\sigma \ll 1$, on the other hand, one may show by perturbation theory[15] that

$$\tau_\| \simeq \frac{\tau_D}{1 - \frac{2}{5}\sigma + 48\sigma^2/875} \tag{223}$$

The exact solution for all the relaxation times and relaxation modes may be found by numerical solution of the set of differential-difference equations:

$$\frac{2\tau_D}{n(n+1)}\langle \dot{P}_n \rangle + \left[1 - \frac{2\sigma}{(2n-1)(2n+3)} \right]\langle P_n \rangle$$

$$= \frac{2\sigma}{2n+1}\left[\frac{n-1}{2n-1}\langle P_{n-2} \rangle - \frac{n+2}{2n+3}\langle P_{n+2} \rangle \right] \tag{224}$$

$$\langle P_n \rangle = \langle P_n(\cos \vartheta(t)) \rangle \quad t > 0$$

These govern the linear response of the system following the removal of a steady dc field at time $t = 0$. The set of Eqs. (224) may be written either by successively averaging the Langevin equation as described here and in Appendix A or by separating the variables in the Fokker-Planck equation, as described in Section III of Ref. 11. The Fokker-Planck equation is [11]

$$2\tau_D \sin \vartheta \frac{\partial W}{\partial t} = \frac{\partial}{\partial \vartheta}\left[\sin \vartheta \left(\frac{v}{kT}\frac{\partial V}{\partial \vartheta}W + \frac{\partial W}{\partial \vartheta} \right) \right] \tag{225}$$

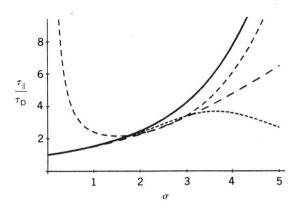

Figure 3. Longest relaxation time (solid line) from Eq. (224), Brown's asymptotic expression Eq. (222) (--- line), and perturbation solution Eq. (223) (------ line) compared with relaxation time rendered by the effective eigenvalue (——— line) Eq. (138) as a function of the barrier height parameter σ.

The differential-difference equations (224) are solved by writing them as

$$\dot{\mathbf{X}} = \mathbf{A}\mathbf{X} \qquad (226)$$

where (it will only be necessary to consider the vector of odd averages)

$$\mathbf{X} = \left(\langle P_1 \rangle, \langle P_3 \rangle, \ldots, \langle P_{2n+1} \rangle\right)^T \qquad (227)$$

and forming the characteristic equation

$$\det(s\mathbf{I} - \mathbf{A}) = 0 \qquad (228)$$

The longest relaxation time is proportional to the reciprocal of the lowest root of this equation, the size of the matrix being adjusted until convergence is obtained. Satisfactory results are usually obtained by taking $\mathbf{A}$ as 20×20. In Figs. 3 and 4 we show the results for the longest relaxation time yielded by the three Eqs. (222), (223), and (224) compared with the effective relaxation time rendered by Eq. (138). It is evident that $\tau_\parallel$ as rendered by Eq. (138) provides a poor approximation to the longest relaxation time in the limit of high potential barriers. This is a direct consequence of the exponential character of the longest relaxation time (causing it to dominate all the others in the high σ limit). The effective eigenvalue provides a poor approximation in the high σ limit because it depends on the equilibrium averages, which have *no exponential terms*.

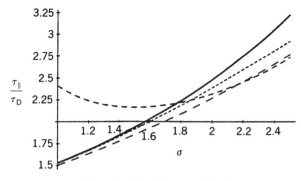

Figure 4. Magnification of Fig. 3.

The relaxation modes of $\langle P_1 \rangle$ may be found from Eq. (224) by assuming that $\mathbf{A}$ has a linearly independent set of n eigenvectors ($\mathbf{p}_1 \cdots \mathbf{p}_n$), so that

$$\mathbf{P}^{-1}\mathbf{A}\mathbf{P} = \mathbf{\Lambda} \qquad (229)$$

where $\mathbf{P}$ is the matrix consisting entirely of the n eigenvectors of $\mathbf{A}$ and

$$\mathbf{\Lambda} = \begin{pmatrix} \lambda_1 & & & \\ & \lambda_2 & & 0 \\ & & \lambda_3 & \\ & 0 & & \ddots \end{pmatrix} \qquad (230)$$

whence

$$\mathbf{X}(t) = \mathbf{P} \exp \mathbf{\Lambda} t \mathbf{P}^{-1} \mathbf{X}_0 \qquad (231)$$

where

$$\exp \mathbf{\Lambda} t = \begin{pmatrix} e^{\lambda_1 t} & & & \\ & e^{\lambda_2 t} & & 0 \\ 0 & & \ddots & \\ & & & e^{\lambda_n t} \end{pmatrix}$$

and $\mathbf{X}_0$ is the matrix of the initial values of the $\langle P_n \rangle$. The solution, Eq. (231), may then be exhibited in the form

$$\mathbf{X}(t) = b_1 e^{\lambda_1 t} \mathbf{p}_1 + b_2 e^{\lambda_2 t} \mathbf{p}_2 + \cdots + b_n e^{\lambda_n t} \mathbf{p}_n \qquad (232)$$

where the b_n are to be determined from the initial conditions. The initial value vector $\mathbf{X}_0$ is determined as follows. We have at time $t = 0$ when the steady field $\mathbf{E}_1(t)$ is switched off

$$\langle P_1(\cos \vartheta) \rangle = \frac{\int_{-1}^{+1} x \, e^{\sigma x^2 + \xi_1 x} \, dx}{\int_{-1}^{+1} e^{\sigma x^2 + \xi_1 x} \, dx} \tag{233}$$

which in the linear approximation is

$$\langle P_1(x) \rangle = \frac{\int_{-1}^{+1} x \, e^{\sigma x^2}[1 + \xi_1 x] \, dx}{\int_{-1}^{+1} e^{\sigma x^2}[1 + \xi_1 x] \, dx} = \xi_1 \frac{\int_0^1 x^2 \, e^{\sigma x^2} \, dx}{\int_0^1 e^{\sigma x^2} \, dx}$$

$$= \xi_1 \langle x^2 \rangle_0 = \frac{\xi_1}{3} [1 + 2\langle P_2 \rangle_0] \tag{234}$$

In general, we will have

$$\langle P_n(x) \rangle = \frac{\int_{-1}^{+1} P_n(x) e^{\sigma x^2} \, dx}{\int_{-1}^{+1} e^{\sigma x^2} \, dx} + \xi_1 \frac{\int_{-1}^{+1} x P_n(x) e^{\sigma x^2} \, dx}{\int_{-1}^{+1} e^{\sigma x^2} \, dx} \tag{235}$$

Now

$$x P_n(x) = \frac{1}{2n + 1} [(n + 1) P_{n+1}(x) + n P_{n-1}(x)] \tag{236}$$

so that for even $n = 2k$, say,

$$\langle P_{2k} \rangle = \langle P_{2k} \rangle_0$$

Thus, the even averages will not be represented in the initial value vector $\mathbf{X}_0$. For the averages of odd functions

$$\langle P_{2k+1} \rangle = \xi_1 \frac{\int_{-1}^{+1} [1/(4k + 3)][(2k + 2) P_{2k+2} + (2k + 1) P_{2k}] e^{\sigma x^2} \, dx}{\int_{-1}^{+1} e^{\sigma x^2} \, dx} \tag{237}$$

so that

$$\langle P_{2k+1} \rangle = \frac{\xi_1}{4k + 3} [(2k + 2)\langle P_{2k+2} \rangle_0 + (2k + 1)\langle P_{2k} \rangle_0] \tag{238}$$

Since

$$\langle P_2 \rangle_0 = \frac{1}{2} [3\langle \cos^2 \vartheta \rangle_0 - 1]$$

$$= \frac{1}{2} \left(\frac{3}{2\sqrt{\sigma} D(\sqrt{\sigma})} - \frac{3}{2\sigma} - 1 \right) \tag{239}$$

We may calculate $\langle P_{2k} \rangle_0$ in terms of Dawson's integral from the recurrence relation (Eq. (224) with $\langle \dot{P}_n \rangle = 0$)

$$\left[1 - \frac{2\sigma}{(2n - 1)(2n + 3)} \right] \langle P_n \rangle_0$$

$$= \frac{2\sigma}{2n + 1} \left[\frac{n - 1}{2n - 1} \langle P_{n-2} \rangle_0 - \frac{n + 2}{2n + 3} \langle P_{n+2} \rangle_0 \right] \tag{240}$$

and Eq. (239). Hence we have the elements of the vector $\mathbf{X}_0$. This procedure with the aid of Eq. (232) now yields $\langle P_1 \rangle$ in the form

$$\mu \langle P_1 \rangle = \frac{\mu^2 E_1}{kT} \sum_{k=1,3,5} a_k e^{-\lambda_k t} \tag{241}$$

The amplitudes a_k are given in Table I and are in close agreement with those given by Martin et al.,[16] who used the Sturm-Liouville method to

TABLE I

σ	k	a_k
0.1	1	0.342
	3	0.93×10^{-5}
	5	0.843×10^{-10}
2	1	0.5275
	3	0.36×10^{-2}
	5	0.146×10^{-4}
5	1	0.754
	3	0.9×10^{-2}
	5	0.31×10^{-3}
8	1	0.856
	3	0.52×10^{-2}
	5	0.7×10^{-3}

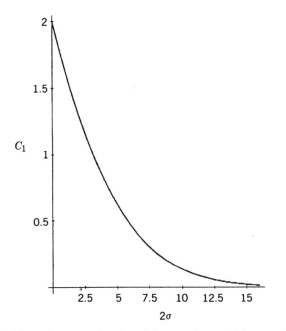

Figure 5. Eigenvalue c_1 as a function of the nematic potential parameter $A = 2\sigma$.

calculate the a_k. The effective eigenvalue

$$\lambda_{\text{eff}} = \frac{\Sigma_k a_k \lambda_k}{\Sigma_k a_k} \qquad (242)$$

may now be evaluated from the numerical results presented here. We write in order to utilize the presentation of Martin et al.[16]

$$2\tau_D \lambda_{\text{eff}} = \frac{\Sigma_{k=1,3,5} c_k a_k}{\Sigma_{k=1,3,5} a_k} \qquad (243)$$

Here we have supposed that $\lambda_k = c_k (2\tau_D)^{-1}$. For $\sigma = 5$ we have from Fig. 2 of Martin et al.[16] (see also Figs. 5–7)

$$c_1 \simeq 0.1 \qquad c_3 \simeq 13 \qquad c_5 = 31$$

The Fokker-Planck equation following the removal of a transverse field is [11]

$$
2\tau_D \frac{\partial W}{\partial t} = \frac{1}{\sin \vartheta} \frac{\partial}{\partial \vartheta} \left(\sin \vartheta \frac{\partial W}{\partial \vartheta} \right) + \frac{1}{\sin^2 \vartheta} \frac{\partial^2 W}{\partial \phi^2}
$$
$$
+ 2\sigma \sin \vartheta \cos \vartheta \frac{\partial W}{\partial \vartheta} + 2\sigma (3\cos^2 \vartheta - 1) W \tag{249}
$$

which on assuming that

$$
W(\vartheta, \phi, t) = \sum_{n=0}^{\infty} \sum_{m=-n}^{n} a_{nm}(t) P_n^{|m|}(\cos \vartheta) e^{im\phi} \qquad |m| \le n \tag{250}
$$

results in the set of differential-difference equations [11]:

$$
2\tau_D \frac{da_{nm}(t)}{dt} = \left[-n(n+1) + 2\sigma \frac{n(n+1) - 3m^2}{(2n+3)(2n-1)} \right] a_{nm}(t)
$$
$$
+ 2\sigma \left[\frac{(n - |m|)(n+1)(n - |m| - 1)}{(2n-1)(2n-3)} a_{n-2,m} \tag{251} \right.
$$
$$
\left. - \frac{n(n + |m| + 1)(n + |m| + 2)}{(2n+3)(2n+5)} a_{n+2,m} \right]
$$

with

$$
\langle P_n^{|m|}(x) e^{-im\phi} \rangle = \frac{1}{2n+1} \frac{(n + |m|)!}{(n - |m|)!} \frac{a_{n,-m}}{a_{00}} \tag{252}
$$

$$
\langle P_n^{|m|}(x) e^{im\phi} \rangle = \frac{1}{2n+1} \frac{(n + |m|)!}{(n - |m|)!} \frac{a_{nm}}{a_{00}} \tag{253}
$$

$$
\langle P_n^{|m|}(x) \cos m\phi \rangle = \frac{1}{2n+1} \frac{(n + |m|)!}{(n - |m|)!} \frac{a_{nm} + a_{n,-m}}{2a_{00}} \tag{254}
$$

$$
\langle P_n^{|m|}(x) \sin m\phi \rangle = \frac{1}{2n+1} \frac{(n + |m|)!}{(n - |m|)!} \frac{a_{nm} - a_{n,-m}}{2ia_{00}} \tag{255}
$$

and

$$
a_{n,-m} = a_{nm}^* \tag{256}
$$

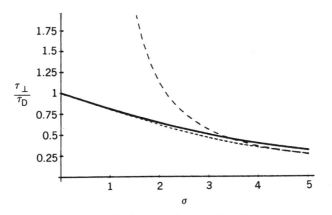

Figure 8. Numerical solution for longest relaxation time for a transverse applied field from Eq. (251) (solid line) compared with Storonkin's asymptotic solution Eq. (248) (dashed line) and compared with the relaxation time rendered by the effective eigenvalue Eq. (146) as a function of σ. Note the close agreement of the effective eigenvalue with the numerical solution for all σ values.

The longest relaxation time of Eq. (251) (shown in Fig. 8) may be found just as for $m = 0$ by forming the characteristic equation

$$\det(s\mathbf{I} - \mathbf{A}) = 0$$

The effective relaxation time is compared with Eq. (248) and that yielded by the exact numerical solution for the longest relaxation time in Fig. 8, showing that the effective relaxation time yields a very close approximation to the longest relaxation time for all values of σ in this case. We would like to thank John Waldron for the numerical solutions presented in this section.

III. TWO-DIMENSIONAL ROTATIONAL BROWNIAN MOTION IN N-FOLD COSINE POTENTIALS

A comprehensive numerical study of the rotational Brownian motion of the two-dimensional rotator in N-fold cosine potentials in both time and frequency domains has been given by Reid,[37] with particular reference to the behavior of the dielectric dispersion and absorption spectra. He compares the spectra computed from the model with those obtained from experimental observations of rotator-phase furan and CH_2Cl_2. There is reasonable agreement with experiment. Further, he finds that (unlike the free rotator or the harmonic potential version) this model can reproduce both relaxation and resonant (when inertial effects are included) behavior. This is due to the use of a periodic rather than a parabolic potential. Such a potential allows the flipping of rotators to neighboring wells, thus permitting both relaxation and oscillatory behavior[37] in the same model. We apply the effective eigenvalue method to this model in the noninertial limit. Such a model has also been studied (again in the noninertial limit) by Lauritzen and Zwanzig[38] in connection with site models of dielectric relaxation in molecular crystals. They converted the solution to a Sturm-Liouville problem. The governing Sturm-Liouville equation is the particular form of the Hill equation[39] known as Ince's equation,[40] which they integrated using Runge-Kutta methods. They then determined the eigenvalues by the trial and error method. They also obtained using asymptotic methods a closed-form expression for the lowest eigenvalue for high potential barriers, which is the analog of that given by Brown[15] for rotation in three dimensions.

A. The Langevin Equation for Rotation in Two Dimensions: Application to the Dielectric Relaxation of an Assembly of Two-Dimensional Rotators

In this section we deal with the rotational motion of a two-dimensional rotator with dipole moment μ in the N-fold cosine potential

$$V(\theta) = -V_0 \cos N\theta \qquad (257)$$

where θ is the angle between the dipole vector μ and the z axis. The quantities of interest are the longitudinal and transverse components of the polarization $P_{\parallel}(t)$ and $P_{\perp}(t)$, respectively, where we apply a small additional time-dependent field $\mathbf{F}(t)$ along either the z axis or the x axis. These polarization components are determined by

$$P_{\parallel}(t) = \mu N \big[\langle \cos \theta(t) \rangle - \langle \cos \theta \rangle_{eq} \big] \qquad (258)$$

$$P_{\perp}(t) = \mu N \langle \sin \theta(t) \rangle \qquad (259)$$

where $\langle \cos \theta \rangle_{eq}$ is the equilibrium average of $\cos \theta$ in the absence of the field $F(t)$, and N is the number of dipoles per unit volume. The one-sided Fourier transforms of $P_{\parallel}(t)$ and $P_{\perp}(t)$ are proportional to the components of the complex susceptibility $\chi_{\parallel}(\omega)$ and $\chi_{\perp}(\omega)$, respectively.

The Langevin equation for a dipole μ to rotate about an axis normal to the xz plane is[37]

$$I\ddot{\theta}(t) + \zeta\dot{\theta}(t) + NV_0 \sin N\theta(t) - \frac{\partial}{\partial \theta}(\mu(t) \cdot \mathbf{F}(t)) = \lambda(t) \qquad (260)$$

In Eq. (260), I is the moment of inertia of the rotator about the axis of rotation, and $\zeta\dot\theta$ and $\lambda(t)$ are the frictional and white noise torques due to the Brownian motion of the surroundings. The random torque $\lambda(t)$ has the property

$$\overline{\lambda(t)} = 0 \qquad \overline{\lambda(t)\lambda(t')} = 2\zeta kT\delta(t - t') \qquad (261)$$

To specialize Eq. (260) to the step-on field we write

$$\mathbf{F}(t) = \mathbf{F}_0 U(t) \qquad (262)$$

where $U(t)$ is the unit step function, and F_0 is the amplitude we require to calculate, for this model, the statistical averages in Eqs. (258) and (259) when the inertial effects are ignored.

The problem that presents itself when treating the model using the Langevin equation in the form of Eq. (260) is that it is not apparent how that equation may be linearized to yield the solution for small $\mu F_0/kT$. This difficulty may be circumvented by rewriting Eq. (260) as an equation of motion for

$$p = \cos\theta \qquad (263)$$

so that

$$\dot\theta = -\dot p(1 - p^2)^{-1/2} = -\dot p(\sin\theta)^{-1} \qquad (264)$$

$$\ddot\theta = -\ddot p(1 - p^2)^{-1/2} - \dot p^2 p(1 - p^2)^{-3/2} \qquad (265)$$

$$= -\ddot p(\sin\theta)^{-1} - \dot\theta^2 p(\sin\theta)^{-1} \qquad (266)$$

The Langevin equation (260) with the field $\mathbf{F}$ applied along the z axis and with this change of variable becomes

$$I\frac{d^2}{dt^2}\cos\theta(t) + \zeta\frac{d}{dt}\cos\theta(t) + I\dot\theta^2(t)\cos\theta(t)$$
$$+ \frac{NV_0}{2}[\cos(N + 1)\theta(t) - \cos(N - 1)\theta(t)] \qquad (267)$$
$$= \mu F_0(1 - \cos^2\theta(t)) - \sin\theta(t)\lambda(t)$$

which is the Langevin equation for the motion of the instantaneous dipole moment. To solve it we first form its statistical average over a large number of rotators

$$I\frac{d^2}{dt^2}\langle\cos\theta\rangle + \zeta\frac{d}{dt}\langle\cos\theta\rangle + \langle I\dot\theta^2(t)\cos\theta\rangle$$

$$+ \frac{NV_0}{2}[\langle\cos(N+1)\theta\rangle - \langle\cos(N-1)\theta\rangle]$$

$$= \mu F_0(1 - \langle\cos^2\theta\rangle) \qquad (268)$$

We have stipulated in Eq. (268) that

$$\langle\sin\theta(t)\lambda(t)\rangle = 0 \qquad (269)$$

We also note that Eq. (268) may be written as

$$I\frac{d^2}{dt^2}\langle\cos\theta\rangle + \zeta\frac{d}{dt}\langle\cos\theta\rangle + \langle I\dot\theta^2\cos\theta\rangle$$

$$+ \frac{NV_0}{2}[\langle\cos(N+1)\theta\rangle - \langle\cos(N-1)\theta\rangle] \qquad (270)$$

$$= \frac{1}{2}\mu F_0[1 - \langle\cos 2\theta\rangle]$$

We note further that when the field $\mathbf{F}$ is applied along the x axis the quantity of interest $q = \sin\theta(t)$ obeys the similar equation

$$I\frac{d^2}{dt^2}\langle\sin\theta\rangle + \zeta\frac{d}{dt}\langle\sin\theta\rangle + \langle I\dot\theta^2\sin\theta\rangle$$

$$\qquad (271)$$

$$+ \frac{NV_0}{2}[\langle\sin(N+1)\theta\rangle - \langle\sin(N-1)\theta\rangle] + \mu F_0[\langle\sin^2\theta\rangle - 1] = 0$$

The remaining terms in Eqs. (270) and (271) that cause difficulty are

$$\langle I\dot\theta^2\sin\theta\rangle \quad\text{and}\quad \langle I\dot\theta^2\cos\theta\rangle \qquad (272)$$

Since the noninertial response pertains to the situation where

$$t \gg \frac{I}{\zeta} \qquad (273)$$

which implicitly means[41] that a Maxwellian distribution of angular velocities has been achieved, we may now write

$$\langle I\dot{\theta}^2 \cos \theta \rangle = kT\langle \cos \theta \rangle \qquad (274)$$

and

$$\langle I\dot{\theta}^2 \sin \theta \rangle = kT\langle \sin \theta \rangle \qquad (275)$$

since the orientation and the angular velocity variables, when equilibrium
of the angular velocities has been reached, are decoupled from each other,
as far as the time behavior of the orientations is concerned.[42, 43] In the
noninertial limit

$$I\langle \ddot{p} \rangle = 0 \qquad (276)$$

in Eqs. (270) and (271), so that finally

$$\frac{d}{dt}\langle \cos \theta \rangle + \frac{kT}{\zeta}\langle \cos \theta \rangle = \frac{NV_0}{2\zeta}[\langle \cos(N-1)\theta \rangle - \langle \cos(N+1)\theta \rangle]$$
$$+ \frac{1}{2\zeta}\mu F(t)[1 - \langle \cos 2\theta \rangle] \qquad (277)$$

and

$$\frac{d}{dt}\langle \sin \theta \rangle + \frac{kT}{\zeta}\langle \sin \theta \rangle = \frac{NV_0}{2\zeta}[\langle \sin(N-1)\theta \rangle - \langle \sin(N+1)\theta \rangle]$$
$$+ \frac{\mu}{\zeta}F(t)[1 - \langle \sin^2 \theta \rangle] \qquad (278)$$

B. General Expressions for the Longitudinal and Transverse Effective Relaxation Times

Equations (277) and (278) are the first terms in the infinite hierarchy of
differential-difference equations, which describe the ensemble averages
$\langle \cos n\theta \rangle$ and $\langle \sin n\theta \rangle$. We have mentioned that the standard approach to
calculating the longitudinal and transverse relaxation times is accom-
plished by rewriting the infinite hierarchy as a set of ordinary differential
equations of the form

$$\dot{\mathbf{X}} = \mathbf{A}\mathbf{X} + \mathbf{B}U(t) \qquad (279)$$

where **A** is the transition matrix and **B** is the driving force matrix, and truncating at a given size of **A**. The longest relaxation time is then the reciprocal of the lowest root of the characteristic equation

$$\det\{s\mathbf{I} - \mathbf{A}\} = 0 \tag{280}$$

where s denotes the complex frequency.

The disadvantage of this method is that it is, in general, impossible to obtain a closed-form expression for the longest relaxation time. This difficulty may be circumvented in certain cases (see Section II.J) by means of the effective eigenvalue method.

We first consider the parallel equation of motion, Eq. (277), and recall that, just as in Section II, we are usually interested only in the response linear in F. We therefore assume that in Eq. (277)

$$\langle \cos n\theta \rangle = \langle \cos n\theta \rangle_{\text{eq}} + \langle \cos n\theta \rangle_1 \tag{281}$$

where the subscript eq denotes the equilibrium ensemble average in the absence of a perturbing constant field F and subscript 1 denotes the portion of the ensemble average that is linear in F. On substituting this equation into Eq. (277) we find that

$$\frac{d}{dt}\langle \cos \theta \rangle_1 + \frac{kT}{\zeta}\langle \cos \theta \rangle_1 = \frac{NV_0}{2\zeta}\left[\langle \cos(N-1)\theta \rangle_1 - \langle \cos(N+1)\theta \rangle_1\right]$$

$$+ \frac{1}{2\zeta}\mu F(t)[1 - \langle \cos 2\theta \rangle_0] \tag{282}$$

whence the effective eigenvalue method leads us to

$$\lambda^{\parallel}_{\text{eff}} = -\frac{(d/dt)\langle \cos \theta \rangle_1}{\langle \cos \theta \rangle_1}$$

$$= \frac{(kT/\zeta)\langle \cos \theta \rangle_1 + (NV_0/2\zeta)[\langle \cos(N+1)\theta \rangle_1 - \langle \cos(N-1)\theta \rangle_1]}{\langle \cos \theta \rangle_1} \tag{283}$$

We now have at $t = 0$

$$\langle \cos \theta \rangle_1 = \langle \cos \theta \rangle_0 - \langle \cos \theta \rangle_{\text{eq}} \tag{284}$$

At equilibrium

$$\langle \cos\theta\rangle_{eq} = \frac{\int_0^{2\pi} \cos\theta \; e^{(V_0\cos N\theta + \mu F_0\cos\theta)/kT}\,d\theta}{\int_0^{2\pi} e^{(V_0\cos N\theta + \mu F_0\cos\theta)/kT}\,d\theta}$$

$$\cong \langle\cos\theta\rangle_0 + \frac{\mu F_0}{kT}\Big[\langle\cos^2\theta\rangle_0 - \langle\cos\theta\rangle_0^2\Big] \tag{285}$$

$$= \langle\cos\theta\rangle_0 + \frac{\mu F_0}{2kT}\Big[1 + \langle\cos 2\theta\rangle_0 - 2\langle\cos\theta\rangle_0^2\Big]$$

Further, at $t = 0$

$$\frac{NV_0}{2}\Big[\langle\cos(N+1)\theta\rangle_1 - \langle\cos(N-1)\theta\rangle_1\Big]$$

$$= \frac{NV_0}{2}\Big[\langle\cos(N+1)\theta\rangle_0 - \langle\cos(N-1)\theta\rangle_0\Big] \tag{286}$$

$$- \frac{NV_0}{2}\Big[\langle\cos(N+1)\theta\rangle_{eq} - \langle\cos(N-1)\theta\rangle_{eq}\Big]$$

$$= -NV_0\Big[\langle\sin N\theta \sin\theta\rangle_0 - \langle\sin N\theta \sin\theta\rangle_{eq}\Big]$$

At equilibrium

$$NV_0\langle\sin N\theta \sin\theta\rangle_{eq} \cong NV_0\langle\sin N\theta \sin\theta\rangle_0$$

$$+ \frac{NV_0\mu F_0}{kT}\langle\sin N\theta \sin\theta(\cos\theta - \langle\cos\theta\rangle_0)\rangle_0 \tag{287}$$

The last term on the right side of Eq. (287) can be simplified as follows:

$$\frac{NV_0\mu F_0}{kT}\langle\sin N\theta \sin\theta(\cos\theta - \langle\cos\theta\rangle_0)\rangle_0$$

$$= \frac{\mu F_0}{2kT}\left\langle(\sin 2\theta - 2\sin\theta\langle\cos\theta\rangle_0)\frac{\partial}{\partial\theta}V(\theta)\right\rangle_0$$

$$= \mu F_0\Big[\langle\cos 2\theta\rangle_0 - \langle\cos\theta\rangle_0^2\Big] \tag{288}$$

Thus, taking account of Eqs. (285)–(288) we obtain

$$
\begin{aligned}
\lambda_{\text{eff}}^{\parallel} &= \frac{kT}{\zeta} - \frac{2kT}{\zeta} \frac{\langle \cos 2\theta \rangle_0 - \langle \cos \theta \rangle_0^2}{1 + \langle \cos 2\theta \rangle_0 - 2\langle \cos \theta \rangle_0^2} \\
&= \frac{kT}{\zeta} \frac{1 - \langle \cos 2\theta \rangle_0}{1 + \langle \cos 2\theta \rangle_0 - 2\langle \cos \theta \rangle_0^2}
\end{aligned}
\tag{289}
$$

so that the effective longitudinal relaxation time $\tau_{\parallel} = \lambda_{\text{eff}}^{\parallel}$ is given by

$$
\tau_{\parallel} = \tau_D \frac{1 + \langle \cos 2\theta \rangle_0 - 2\langle \cos \theta \rangle_0^2}{1 - \langle \cos 2\theta \rangle_0}
\tag{290}
$$

where

$$
\tau_D = \frac{\zeta}{kT}
\tag{291}
$$

is the Debye relaxation time for planar rotators.[12]

To calculate the transverse relaxation time $\tau_{\perp}$ we consider the same problem as above, but this time the step change in the field is applied parallel to the x axis so that we need to determine the behavior of $\langle \mu_x \rangle$. We assume as before that

$$
\langle \sin n\theta \rangle = \langle \sin n\theta \rangle_{\text{eq}} + \langle \sin n\theta \rangle_1
$$

which leads us to the linearized equation for $\langle \sin \theta \rangle_1$;

$$
\frac{d}{dt}\langle \sin \theta \rangle_1 + \frac{kT}{\zeta}\langle \sin \theta \rangle_1 = \frac{NV_0}{2\zeta}\left[\langle \sin(N-1)\theta \rangle_1 - \langle \sin(N+1)\theta \rangle_1 \right]
$$
$$
+ \frac{\mu}{\zeta}F(t)\left[1 - \langle \sin^2 \theta \rangle_0 \right]
\tag{292}
$$

Likewise, we find that the eigenvalue equation is

$$
\lambda_{\text{eff}}^{\perp} = \frac{(kT/\zeta)\langle \sin \theta \rangle_1 + (NV_0/2\zeta)\left[\langle \sin(N+1)\theta \rangle_1 - \langle \sin(N-1)\theta \rangle_1 \right]}{\langle \sin \theta \rangle_1}
\tag{293}
$$

Noting that

$$\langle \sin \theta \rangle_0 = 0 \qquad (294)$$

we obtain in the linear approximation of $\mu F_0 / kT$ that at $t = 0$

$$\langle \sin \theta \rangle_1 = -\langle \sin \theta \rangle_{eq} = -\frac{\mu F_0}{2kT}(1 - \langle \cos 2\theta \rangle_0) \qquad (295)$$

and similarly

$$\frac{NV_0}{2\zeta}[\langle \sin(N+1)\theta \rangle_1 - \langle \sin(N-1)\theta \rangle_1] = -\frac{\mu F_0}{\zeta}\langle \cos 2\theta \rangle_0 \qquad (296)$$

Thus,

$$\lambda_{eff}^{\perp} = \frac{kT}{\zeta}\frac{1 + \langle \cos 2\theta \rangle_0}{1 - \langle \cos 2\theta \rangle_0} \qquad (297)$$

so that the effective transverse relaxation time is given by

$$\tau_{\perp} = \tau_D \frac{1 - \langle \cos 2\theta \rangle_0}{1 + \langle \cos 2\theta \rangle_0} \qquad (298)$$

C. Effective Dielectric Relaxation Times in a Twofold Cosine Potential: Calculation of $\tau_{\parallel}$

Having determined general formulae for the effective relaxation time in an N-fold cosine potential, we now calculate the relaxation times for a dipole rotating in a $\cos 2\theta$ potential. This potential is of interest for the Brownian motion of a particle in an uniaxial crystalline potential created by its neighbors. The formulas for $\tau_{\parallel}$ and $\tau_{\perp}$ (Eqs. (290) and (298), respectively) are of course still valid.

Equation (290) will hold for the effective relaxation time for any potential of the form $\cos N\theta$, so that once again

$$\tau_{\parallel} = \tau_D \frac{1 + \langle \cos 2\theta \rangle_0 - 2\langle \cos \theta \rangle_0^2}{1 - \langle \cos 2\theta \rangle_0} \qquad (299)$$

In Eq. (299) the quantities $\langle \cos n\theta \rangle_0$ are given by

$$\langle \cos \theta \rangle_0 = \frac{\int_0^{2\pi} \cos \theta\, e^{\alpha \cos 2\theta}\, d\theta}{\int_0^{2\pi} e^{\alpha \cos 2\theta}\, d\theta} = 0 \tag{300}$$

$$\langle \cos 2\theta \rangle_0 = \frac{\int_0^{2\pi} \cos 2\theta\, e^{\alpha \cos 2\theta}\, d\theta}{\int_0^{2\pi} e^{\alpha \cos 2\theta}\, d\theta} \tag{301}$$

where

$$\alpha = \frac{V_0}{kT} \tag{302}$$

From the theory of Bessel functions[44] we know that

$$e^{\alpha \cos 2\theta} = \sum_{m=-\infty}^{\infty} I_m(\alpha) e^{im2\theta} \tag{303}$$

and

$$I_{-m}(\alpha) = I_m(\alpha) \tag{304}$$

where $I_m(\alpha)$ is the modified Bessel function of the first kind of integer order. Equation (301) thus becomes

$$\langle \cos 2\theta \rangle_0 = \frac{\sum_{m=-\infty}^{\infty} \int_0^{2\pi} I_m(\alpha) e^{im2\theta} \left[(e^{i2\theta} + e^{-i2\theta})/2 \right] d\theta}{\sum_{m=-\infty}^{\infty} \int_0^{2\pi} I_m(\alpha) e^{im2\theta}\, d\theta}$$
$$= \frac{I_{-1}(\alpha) + I_1(\alpha)}{2 I_0(\alpha)} \tag{305}$$

By using Eq. (304), Eq. (305) simplifies to

$$\langle \cos 2\theta \rangle_0 = \frac{I_1(\alpha)}{I_0(\alpha)} \tag{306}$$

Therefore,

$$\tau_\parallel = \tau_D \frac{I_0(\alpha) + I_1(\alpha)}{I_0(\alpha) - I_1(\alpha)} \tag{307}$$

For small values of α (low potential barrier) we have

$$\langle \cos 2\theta \rangle_0 \cong \frac{\int_0^{2\pi} \cos 2\theta [1 + \alpha \cos 2\theta]\, d\theta}{\int_0^{2\pi} [1 + \alpha \cos 2\theta]\, d\theta} = \frac{\alpha}{2} \tag{308}$$

and hence

$$\tau_\parallel \cong \tau_D \frac{1 + \alpha/2}{1 - \alpha/2} \cong \tau_D (1 + \alpha) \tag{309}$$

For large values of α we evaluate $\langle \cos 2\theta \rangle_0$ in the following way: The asymptotic expansion of $I_n(\alpha)$ for large values of α is[44]

$$I_n(\alpha) = \frac{e^\alpha}{\sqrt{2\pi\alpha}} \sum_{k=0}^{\infty} \frac{(-1)^k}{(2\alpha)^k} \frac{\Gamma(n + k + \frac{1}{2})}{k!\,\Gamma(n - k + \frac{1}{2})}$$

$$+ \frac{e^{-\alpha \pm (n + \frac{1}{2})i\pi}}{\sqrt{2\pi\alpha}} \sum_{k=0}^{\infty} \frac{(-1)^k}{(2\alpha)^k} \frac{\Gamma(n + k + \frac{1}{2})}{k!\,\Gamma(n - k + \frac{1}{2})} \tag{310}$$

For $\alpha \to \infty$ only the first term is of significance. On noting that[44]

$$\Gamma\left(n + \frac{1}{2}\right) = \frac{\sqrt{\pi}}{(2^n)}(2n - 1)!! \tag{311}$$

$$\Gamma\left(\frac{1}{2} - n\right) = (-1)^n \frac{2^n \sqrt{\pi}}{(2n - 1)!!} \tag{312}$$

we obtain

$$I_0(\alpha) = \frac{e^\alpha}{\sqrt{2\pi\alpha}}\left[1 + \frac{1}{8\alpha} + \cdots\right] \tag{313}$$

Similarly,

$$I_1(\alpha) = \frac{e^\alpha}{\sqrt{2\pi\alpha}} \sum_{k=0}^{\infty} \frac{(-1)^k}{(2\alpha)^k} \frac{\Gamma(\frac{3}{2} + k)}{k!\,\Gamma(\frac{3}{2} - k)}$$

$$= \frac{e^\alpha}{\sqrt{2\pi\alpha}}\left[\frac{\Gamma(\frac{3}{2})}{\Gamma(\frac{3}{2})} - \frac{\Gamma(\frac{5}{2})}{(2\alpha)\Gamma(\frac{1}{2})} + \frac{1}{2(2\alpha)^2}\frac{\Gamma(\frac{7}{2})}{\Gamma(-\frac{1}{2})} + \cdots\right] \tag{314}$$

$$= \frac{e^\alpha}{\sqrt{2\pi\alpha}}\left[1 - \frac{3}{8\alpha} + \cdots\right]$$

Using the expressions for $I_0(\alpha)$ and $I_1(\alpha)$ in Eq. (307) we get

$$
\begin{aligned}
\tau_{\parallel} &= \tau_D \frac{(1 + 1/8\alpha + \cdots) + (1 - 3/8\alpha - \cdots)}{(1 + 1/8\alpha + \cdots) - (1 - 3/8\alpha - \cdots)} \\
&= \tau_D \frac{2 - 1/4\alpha - \cdots}{1/2\alpha + \cdots} \cong 4\alpha\tau_D
\end{aligned}
\tag{315}
$$

The arguments of Section II.J again applies to this result since in the limit of high potential barriers[38] the longest relaxation time is:

$$
\tau_{\parallel} \simeq \tau_D \frac{\pi}{4\alpha} \exp \alpha
\tag{316}
$$

D.   Calculation of $\tau_{\perp}$

The relaxation time $\tau_{\perp}$ is given by Eq. (298):

$$
\tau_{\perp} = \tau_D \frac{1 - \langle \cos 2\theta \rangle_0}{1 + \langle \cos 2\theta \rangle_0}
$$

On taking account of Eq. (306) we have

$$
\tau_{\perp} = \tau_D \frac{I_0(\alpha) - I_1(\alpha)}{I_0(\alpha) + I_1(\alpha)}
\tag{317}
$$

On comparing with the results for $\tau_{\parallel}$, we see that

$$
\tau_{\parallel}\tau_{\perp} = (\tau_D)^2
$$

Therefore,

$$
\tau_{\perp} = \frac{(\tau_D)^2}{\tau_{\parallel}}
$$

Using the expression for $\tau_{\parallel}$ we then obtain for $\alpha \gg 1$

$$
\tau_{\perp} = \frac{\tau_D}{4\alpha}
\tag{318}
$$

Similarly, for small values of α we obtain

$$\tau_\perp = \tau_D \frac{1 - \alpha/2}{1 + \alpha/2} \cong \tau_D(1 - \alpha) \tag{319}$$

Thus, the results obtained are in full agreement with those for the three-dimensional counterpart of this model—the Meier-Saupe model.

At the time of going to Press, we, together with J. T. Waldron, have shown from the definition of the correlation time given in Eq. (12), that the exact solutions for the longitudinal and transverse correlation times are

$$T_\parallel = \frac{\tau}{2\alpha}(e^{2\alpha} - 1)\left[1 + \sum_{p=1}^{\infty} \frac{(-1)^p}{(2p + 1)} \frac{f_{2p+1}(0)}{f_1(0)} \frac{I_{p+1/2}(\alpha)}{I_{1/2}(\alpha)}\right] \tag{320}$$

and

$$T_\perp = \frac{\tau}{2\alpha}(1 - e^{-2\alpha})\left[1 + \sum_{p=1}^{\infty} \frac{(-1)^p}{(2p + 1)} \frac{g_{2p+1}}{g_1(0)} \frac{I_{p+1/2}(\alpha)}{I_{1/2}(\alpha)}\right] \tag{321}$$

where

$$\frac{f_{2p+1}(0)}{f_1(0)} = \frac{I_{p+1}(\alpha) + I_p(\alpha)}{I_1(\alpha) + I_0(\alpha)}$$

and

$$\frac{g_{2p+1}(0)}{g_1(0)} = \frac{I_{p+1}(\alpha) - I_p(\alpha)}{I_1(\alpha) - I_0(\alpha)}$$

Eq. (320) reduces to Eq. (316) in the limit of very high potential barriers while Eq. (321) reduces to Eq. (318) in the same limit as has been shown by Crothers using the method of steepest descents.

IV. THE BROWNIAN MOTION IN A TILTED COSINE POTENTIAL

We recall the relaxation problems to which this model may be applied before describing the solution procedure. Among these we mention the current–voltage characteristics of a Josephson tunneling junction,[23, 45] the mobility of superionic conductors,[3] cycle slips in second-order phase-locked loops,[46] the ring laser gyro,[19, 20, 47] etc. We consider the Josephson junction as a definite example. A related problem, although easier than the junction problem, is to calculate the relaxation times for a dipolar molecule in a $\cos N\theta$ potential. This has been considered in Section III. The Josephson junction represents the Brownian motion of a particle in a tilted cosine potential, while the dielectric problem represents the motion of a particle in a level cosine potential. We note that setting the tilt equal to zero in the Josephson problem and applying the effective eigenvalue procedure yields the longitudinal relaxation time for dielectric relaxation of an assembly of two-dimensional rotators in a cosine potential, in other words, the relaxation in a constant electric field $\mathbf{E}_0$ applied in the z direction. The transverse relaxation time is that associated with a small field applied perpendicular to the z direction, that is, the x direction.

A. Application to the Josephson Junction

The main aim of this section is to evaluate the effective eigenvalues for the Josephson junction and to demonstrate the accuracy of the effective eigenvalue method by comparing the exact and approximate calculations of the junction impedance for an externally applied small-signal alternating current in the presence of noise. We recall from the introduction that the Josephson tunneling junction is made up of two superconductors separated from each other by a thin layer of oxide. We label ψ_{R}, and ψ_{L} the wave functions for the right and left superconductors, respectively. The Josephson tunneling junction can be represented by the so called RCSJ (resistance–capacitance shunted junction) model, as shown in Fig. 1[23, 45, 50] where

$$\phi = \phi_{\mathrm{R}} - \phi_{\mathrm{L}} \qquad (322\text{-}332)$$

ϕ_{R} and ϕ_{L} are the phase angles associated with the wave functions ψ_{R} and ψ_{L} respectively, C and R are the capacitance and resistance of the junction, the term $I \sin \phi$ is the phase-dependent current generator representing the Josephson supercurrent due to the Cooper pairs tunneling through the junction, and I_{dc} is the bias current applied to the junction. Since the junction operates at temperatures above absolute zero there is a white noise current $L(t)$ superimposed on the bias current satisfying the conditions

$$\overline{L(t_1)L(t_2)} = \frac{2kT}{R}\delta(t_1 - t_2) \qquad (333)$$

$$\overline{L(t)} = 0 \qquad (334)$$

The overbar denotes the statistical average, $\delta(t)$ is the Dirac delta function, and t_1 and t_2 are distinct times.

The current balance equation for the junction is

$$I_{dc} + L(t) = C\frac{dV(t)}{dt} + \frac{V(t)}{R} + I\sin\phi(t) \qquad (335)$$

We note that $(\hbar/2e)\dot{\phi}(t)$ is the time-dependent voltage across the junction,[23] where $\hbar = h/2\pi$ (h is the Planck constant) and e is the charge of an electron. Thus,[23]

$$V(t) = \left(\frac{\hbar}{2e}\right)\dot{\phi}(t) \qquad (336)$$

Substitution of Eq. (336) in Eq. (335) yields

$$\left(\frac{\hbar}{2e}\right)^2 C\ddot{\phi}(t) + \frac{1}{R}\left(\frac{\hbar}{2e}\right)^2 \dot{\phi}(t) + \left(\frac{\hbar}{2e}\right)I\sin\phi(t) = \frac{\hbar}{2e}(I_{dc} + L(t)) \qquad (337)$$

Equation (337) has the same form as that for a Brownian particle of mass $(\hbar/2e)^2 C$ moving in the tilted cosine potential[22, 23]:

$$U(\phi) = -\frac{\hbar}{2e}(I_{dc}\phi + I\cos\phi) \qquad (338)$$

On further simplification Eq. (337) yields

$$C\ddot{\phi}(t) + \frac{1}{R}\dot{\phi}(t) + \frac{2eI}{\hbar}\sin\phi(t) = \frac{2e}{\hbar}(I_{dc} + L(t)) \qquad (339)$$

This is the Langevin equation of the Josephson junction.

B. Reduction of the Averaged Langevin Equation to a Set of Differential-Difference Equations

To proceed we change the variable in Eq. (339) by writing

$$r^n = e^{-in\phi} \qquad n = \ldots, -1, 0, 1, \ldots \qquad (340)$$

Consequently,

$$\frac{d}{dt} r^n = -in\dot{\phi}\, e^{-in\phi} \tag{341}$$

so that

$$\dot{\phi} = \frac{i\dot{r}^n}{nr^n} \tag{342}$$

Similarly,

$$\ddot{r}^n = -n^2\dot{\phi}^2 r^n - in\ddot{\phi} r^n \tag{343}$$

so that

$$\ddot{\phi} = \frac{i\left(\ddot{r}^n + n^2\dot{\phi}^2 r^n\right)}{nr^n} \tag{344}$$

Substituting expressions (342) and (344) into Eq. (339) and using the relation

$$\sin\phi = \frac{e^{i\phi} - e^{-i\phi}}{2i}$$

we obtain

$$\frac{Ci\left(\ddot{r}^n(t) + n^2\dot{\phi}^2(t)r^n(t)\right)}{nr^n(t)} + \frac{i\dot{r}^n(t)}{Rnr^n(t)} - \frac{2eI}{\hbar}\left(\frac{r(t) - r(t)^{-1}}{2i}\right)$$
$$= \frac{2e}{\hbar}\left(I_{dc} + L(t)\right) \tag{345}$$

or

$$C\frac{d^2 r^n(t)}{dt^2} + \frac{1}{R}\frac{dr^n(t)}{dt} + n^2 C\dot{\phi}^2(t)r^n(t)$$
$$= \frac{enI}{\hbar}\left(r^{n-1}(t) - r^{n+1}(t)\right) - \frac{i2en}{\hbar} r^n(t)\left(I_{dc} + L(t)\right) \tag{346}$$

On averaging and utilizing the properties of Eq. (334), Eq. (346) becomes

$$
\begin{aligned}
C\frac{d^2}{dt^2}\langle r^n \rangle &+ \frac{1}{R}\frac{d}{dt}\langle r^n \rangle + n^2 C\langle \dot{\phi}^2 r^n \rangle \\
&= \frac{enI}{\hbar}(\langle r^{n-1} \rangle - \langle r^{n+1} \rangle) - \frac{i2enI_{dc}}{\hbar}\langle r^n \rangle
\end{aligned} \tag{347}
$$

We shall consider Eq. (347) in the diffusion (noninertial or low-frequency) limit, where we can neglect the capacitive term $C\ddot{\phi}(t)$. The Maxwell-Boltzmann distribution[22]

$$
W_0(\dot{\phi}) = C_0 \exp\left[-\left(\frac{\hbar}{2e}\right)^2 \frac{C\dot{\phi}^2}{kT} \right] \tag{348}
$$

with C_0 the normalizing constant, has then set in so that the phase angle ϕ and its derivative $\dot{\phi}$ are decoupled from each other as far as the time behavior of $\phi(t)$ is concerned. Thus, on setting $C = 0$ in Eq. (347) and noting that

$$
\langle \dot{\phi}^2 r^n \rangle = \langle \dot{\phi}^2 \rangle_0 \langle r^n \rangle
$$

where

$$
\langle \dot{\phi}^2 \rangle_0 = \int_{-\infty}^{\infty} \dot{\phi}^2 C_0 \exp\left[-\left(\frac{\hbar}{2e}\right)^2 \frac{C\dot{\phi}^2}{kT} \right] d\dot{\phi} = \left(\frac{2e}{\hbar}\right)^2 \frac{kT}{C}
$$

we have

$$
\frac{d}{dt}\langle r^n \rangle + \frac{1}{\tau_0}\left(n^2 + \frac{inx\gamma}{2} \right)\langle r^n \rangle = \frac{n\gamma}{4\tau_0}(\langle r^{n-1} \rangle - \langle r^{n+1} \rangle) \tag{349}
$$

where

$$
x = \frac{I_{dc}}{I} \tag{350}
$$

which is the ratio of bias current amplitude to supercurrent amplitude (bias or tilt parameter),

$$
\gamma = \frac{\hbar I}{ekT} \tag{351}
$$

is the ratio of Josephson coupling energy to thermal energy (barrier height parameter), and

$$\tau_0 = \left(\frac{\hbar}{2e} \right)^2 \frac{1}{kTR} \tag{352}$$

is the Debye relaxation time. Note that the quantity $(\hbar/2e)^2(1/R)$ has the meaning of a damping coefficient.

Equation (349) is a well-known result,[3] which may be obtained from the underlying Fokker-Planck equation. The relevant Fokker-Planck equation in the noninertial limit is[3, 45]

$$\zeta \frac{\partial W}{\partial t} = \frac{\partial}{\partial \phi} \left(W \frac{\partial}{\partial \phi} U \right) + kT \frac{\partial^2 W}{\partial \phi^2} \tag{353}$$

where W is the transition probability of the phase,

$$\zeta = \left(\frac{\hbar}{2e} \right)^2 \frac{1}{R}$$

is the damping coefficient, and

$$U(\phi) = -\frac{\hbar}{2e} (I \cos \phi + I_{dc} \phi)$$

is the tilted cosine potential.

The distribution function W must be periodic in such a way that it can be expanded in a Fourier series as[3]

$$W(\phi, t) = \sum_{p=-\infty}^{\infty} a_p(t) e^{ip\phi} \tag{354}$$

Substituting Eq. (354) into Eq. (353) and using the orthogonality properties of the circular functions we find that the coefficients $a_p(t)$ satisfy

$$\frac{d}{dt} a_p(t) + \frac{kT}{\zeta} \left(p^2 + \frac{ip\hbar I_{dc}}{2ekT} \right) a_p(t) = \frac{p\hbar I}{4e\zeta} \left(a_{p-1}(t) - a_{p+1}(t) \right) \tag{355}$$

It can be easily shown that the $a_p(t)$, Eq. (354), are related to $\langle r^p \rangle$ by

$$a_p(t) = \frac{1}{2\pi}\langle e^{-ip\phi} \rangle = \frac{1}{2\pi}\langle r^p \rangle \tag{356}$$

Thus, Eq. (355) coincides precisely with Eq. (349).

We have found the average of Eq. (349) by noting that $\langle r^n(t)L(t) \rangle$ will vanish throughout, because in the inertial Langevin equation, $r^n(t)$ and $L(t)$ are statistically independent. However, this is not true for the noninertial Langevin equation[11]:

$$\frac{d}{dt}\phi(t) + \frac{2eR}{\hbar}I \sin\phi(t) = \frac{2eR}{\hbar}(I_{dc} + L(t)) \tag{357}$$

On transforming Eq. (357) as above we obtain

$$\frac{d}{dt}r^n(t) = \frac{enIR}{\hbar}(r^{n-1}(t) - r^{n+1}(t)) - \frac{i2enR}{\hbar}r^n(t)(I_{dc} + L(t)) \tag{358}$$

Just as in Section II, the multiplicative noise term $r^n(t)L(t)$ contributes a noise-induced drift term to the average. The multiplicative noise term poses an interpretation problem in averaging Eq. (358), as discussed in Section II.B. We recall that, taking the Langevin equation for a stochastic variable $\xi(t)$ as

$$\frac{d}{dt}\xi(t) = h(\xi(t),t) + g(\xi(t),t)L(t) \tag{359}$$

with

$$\overline{L_j(t)} = 0 \qquad \overline{L(t)L(t')} = 2\delta(t - t') \tag{360}$$

and interpreting it as a Stratonovich stochastic equation,[11] we have

$$\dot{x} = \lim_{\tau \to 0}\left\{\frac{1}{\tau}\overline{(\xi(t + \tau) - x)}\right\}\bigg|_{\xi(t)=x}$$
$$= h(x,t) + g(x,t)\frac{\partial}{\partial x}g(x,t) \tag{361}$$

where $\xi(t + \tau)$, $\tau > 0$ is a solution of Eq. (395), which at time t has the sharp value $\xi(t) = x$. It should be noted that the quantity x in Eq. (359) is itself a random variable with probability density function $W(x, t)$ defined such that $W(x, t)\,dx$ is the probability of finding x in the interval $(x, x + dx)$. Thus, on averaging Eq. (361) over $W(x, t)$ we obtain

$$\frac{d}{dt}\langle x \rangle = \langle h(x, t) \rangle + \left\langle g(x, t)\frac{\partial}{\partial x}g(x, t) \right\rangle \tag{362}$$

where the angular braces mean averaging over $f(x, t)$.

We may use the above results to evaluate the average of the multiplicative noise term in Eq. (358). We have

$$g(r^n) = -\frac{i2\,enRr^n}{\hbar}$$

$$g(r^n)\frac{\partial}{\partial r^n}g(r^n) = -\left(\frac{2neR}{\hbar}\right)^2 r^n \tag{363}$$

and

$$\frac{d}{dt}r^n = \frac{eInR}{\hbar}(r^{n-1} - r^{n+1}) - \frac{Rin2eI_{dc}r^n}{\hbar} - kTR^2\left(\frac{2en}{\hbar}\right)^2 r^n \tag{364}$$

Thus, the average equation of motion is

$$\frac{d}{dt}\langle r^n \rangle = \frac{ReIn}{\hbar}(\langle r^{n-1} \rangle - \langle r^{n+1} \rangle)$$

$$- \frac{Rin2eI_{dc}}{\hbar}\langle r^n \rangle - n^2 kTR^2\left(\frac{2e}{\hbar}\right)^2 \langle r^n \rangle \tag{365}$$

which is Eq. (349).

C. Calculation of the DC Current–Voltage Characteristics

Ambegaokar and Halperin[45] computed the dc current–voltage characteristic (when the capacitance is neglected) by solving the time independent Smoluchowski Eq. (353) associated with the Langevin equation (339). Having noted that the problem is formally the same as that of the translational Brownian motion of a particle in a tilted cosine potential, where the assumption of zero capacitance corresponds to ignoring inertial effects in the mechanical analogy, they found that (for details, see Ref. 3,

p. 289; for a further discussion of mechanical analogs see Ref. 50)

$$\nu = -\frac{4\pi}{\gamma}\left\{(e^{-\pi\gamma x} - 1)\int_0^{2\pi}\int_0^{2\pi}\frac{f(\phi)}{f(\phi')}\,d\phi\,d\phi' + \int_0^{2\pi}\int_0^{\phi'}\frac{f(\phi)}{f(\phi')}\,d\phi\,d\phi'\right\}^{-1}$$
(366)

where

$$\nu = \frac{V}{IR}$$
(367)

and

$$f(\phi) = \exp\left(\tfrac{1}{2}\gamma(x\phi + \cos\phi)\right)$$
(368)

They plotted these results using a numerical method to evaluate the integrals and also gave simple approximations for extreme cases:

$\gamma \to 0$ and $\nu = x$, i.e., $V/R = I_{dc}$ (369)
    This corresponds to Ohm's law.

$\gamma \to \infty$ and $x < 1$; then $\nu = 0$
    This is pure superconduction.

$\gamma \to \infty$ and $x > 1$; then $\nu = (x^2 - 1)^{1/2}$, i.e., $V = RI(x^2 - 1)^{1/2}$
    This is Ohm's law with a correction factor.

There are two major problems associated with the method of Ambegaokar and Halperin:

1. There is no simple form for the $I - V$ curve. The integrals in Eq. (366) have to be evaluated using a numerical method.

2. There is no obvious way of extending their results to the case where the input current contains a time-dependent part.

These difficulties may be overcome using a continued fraction method.[3] We may implement this method by recalling that for the stationary case we now have instead of Eq. (365)

$$\langle r^p \rangle_0\left(p^2 + \frac{i\gamma px}{2}\right) = \frac{p\gamma}{4}\left(\langle r^{p-1}\rangle_0 - \langle r^{p+1}\rangle_0\right)$$
(370)

where the symbol $\langle\ \rangle_0$ means averaging over the stationary distribution function $W_0(\phi)$, which is given by[3]

$$W_0(\phi) = C_0 e^{-U(\phi)/kT} \left[1 - \frac{(1 - e^{-\pi h I_{dc}/kT}) \int_0^\phi e^{U(\phi')/kT} d\phi'}{\int_0^{2\pi} e^{-U(\phi')/kT} d\phi'} \right] \quad (371)$$

where C_0 is the normalizing constant defined in such a way that

$$\int_0^{2\pi} W_0(\phi)\, d\phi = 1$$

We note also that the function $W_0(\phi)$ satisfies the condition[3]

$$W_0(\phi) = W_0(\phi + 2\pi)$$

Following the calculation of Risken,[3] we introduce the quantity

$$S_p = \frac{\langle r^p \rangle_0}{\langle r^{p-1} \rangle_0} \quad (372)$$

So, on dividing by $\langle r^{p-1} \rangle_0$, Eq. (370) becomes

$$S_p \left(p^2 + i\frac{x\gamma p}{2} \right) = \frac{p\gamma}{4}(1 - S_p S_{p+1})$$

or

$$S_p = \frac{0.5}{\dfrac{2p}{\gamma} + ix + 0.5 S_{p+1}} \quad (373)$$

For $p = 1$ we have

$$S_1 = \frac{0.5}{\dfrac{2}{\gamma} + ix + 0.5 S_2} \quad (374)$$

We can write S_2 from Eq. (373) as

$$S_2 = \frac{0.5}{\dfrac{4}{\gamma} + ix + 0.5 S_3} \quad (375)$$

and so on.

Substituting this into Eq. (374) and similarly for S_3, S_4, etc. we obtain the well-known result[3]

$$S_1 = \cfrac{0.5}{\cfrac{2}{\gamma} + \mathrm{i}x + \cfrac{0.25}{\cfrac{4}{\gamma} + \mathrm{i}x + \cfrac{0.25}{\cfrac{6}{\gamma} + \mathrm{i}x + \cfrac{0.25}{\cfrac{8}{\gamma} + \mathrm{i}x \cdots}}}} \qquad (376)$$

To calculate the current–voltage characteristics using Eq. (376) we first recall that the current balance equation is

$$I_{\mathrm{dc}} + L(t) = C\frac{\mathrm{d}V(t)}{\mathrm{d}t} + \frac{1}{R}V(t) + I\sin\phi(t) \qquad (377)$$

If we ignore capacitive effects and take expected values at the stationary state, we then obtain

$$I_{\mathrm{dc}} - I\langle\sin\phi\rangle_0 - \frac{\langle V\rangle_0}{R} = 0 \qquad (378)$$

or

$$\frac{I_{\mathrm{dc}}}{I} - \langle\sin\phi\rangle_0 - \frac{\langle V\rangle_0}{IR} = 0 \qquad (379)$$

which is

$$x - \langle\sin\phi\rangle_0 = \nu \qquad (380)$$

and is the equation that determines the dc current–voltage characteristics. We may find $-\langle\sin\phi\rangle_0$ merely by extracting the imaginary part of S_1 from Eq. (376), since

$$S_1 = \langle\mathrm{e}^{-\mathrm{i}\phi}\rangle_0 = \langle\cos\phi\rangle_0 - \mathrm{i}\langle\sin\phi\rangle_0 \qquad (381)$$

The plots yielded by this equation appear to be in agreement with those of Ambegaokar and Halperin.[45] The results are shown in Fig. 9. Equation (380) also allows us to write an expression for the resistance R_{J} of the

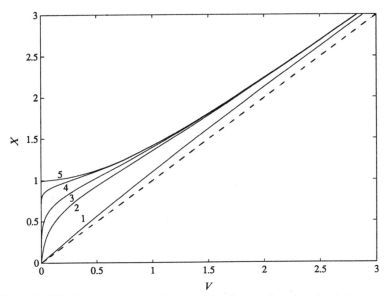

Figure 9. The dc current–voltage characteristics of a Josephson junction in the presence of noise. Curves 1–5 correspond to values of γ (ratio of Josephson coupling energy to thermal energy) = 1, 5, 10, 50, and 100, respectively. Note that for $\gamma = 0$ (dashed line) the behavior corresponds to Ohm's law.

junction arising from the superconductivity. We have from Eq. (378)

$$RI_{dc}\left(1 - \frac{I}{I_{dc}}\langle \sin \phi \rangle_0\right) = \langle V \rangle_0$$

so that

$$R_J = R(1 - x^{-1}\langle \sin \phi \rangle_0) \tag{382}$$

D.   Exact Calculation of the Linear Response to an Applied Alternating Current

If we suppose that the current is now $I_{dc} + I_m \exp(-i\omega t)$, we may no longer ignore the term $d\langle r^n \rangle/dt$ in Eq. (349), since

$$\frac{d\langle r^n \rangle}{dt} \neq 0$$

We also suppose that $\hbar I_m/2ekT \ll 1$ so that we can make the perturbation expansion (we may use the exponential form for the ac current since we seek only the linear response)

$$\langle r^n \rangle = \langle r^n \rangle_0 + \tilde{A}_n(\omega) I_m \exp(-i\omega t)/I + \cdots \tag{383}$$

On substituting Eq. (383) into Eq. (355) we obtain for the linear response

$$\left(-i\omega\tau_0 + n^2 + \frac{in\gamma}{2}x\right)\tilde{A}_n(\omega) = \frac{\gamma n}{4}\left(\tilde{A}_{n-1}(\omega) - \tilde{A}_{n+1}(\omega)\right) - \frac{in\gamma}{2}\langle r^n \rangle_0 \tag{384}$$

with $A_0(\omega) = 0$. Eq. (384) may be written as a three-term recurrence relation:

$$Q_n^-\tilde{A}_{n-1}(\omega) + \hat{Q}_n\tilde{A}_n(\omega) + Q_n^+\tilde{A}_{n+1}(\omega) = -B_n \tag{385}$$

where

$$Q_n^- = -\frac{1}{2} \quad Q_n^+ = \frac{1}{2} \quad \hat{Q}_n = \left[\frac{-2i\omega\tau_0}{\gamma n} + \frac{2n}{\gamma} + ix\right]$$

$$B_n = i\langle r^n \rangle_0$$

Equation (385) can readily be solved for the case $B_n = 0$. Denoting the solution of the homogeneous equation by $\tilde{r}_n(\omega)$ and introducing the quantity $\tilde{S}_n(\omega)$ defined as

$$\tilde{S}_n(\omega) = \frac{\tilde{r}_n(\omega)}{\tilde{r}_{n-1}(\omega)} \tag{386}$$

we have the continued fraction solution

$$\tilde{S}_n(\omega) = \cfrac{0.5}{-\cfrac{2i\omega\tau_0}{\gamma n} + \cfrac{2n}{\gamma} + ix + 0.5\tilde{S}_{n+1}(\omega)} \tag{387}$$

In particular for $n = 1$ we obtain

$$\tilde{S}_1(\omega) = \cfrac{0.5}{i\left(x - \cfrac{2\omega\tau_0}{\gamma}\right) + \cfrac{2}{\gamma} + \cfrac{0.25}{i\left(x - \cfrac{2\omega\tau_0}{2\gamma}\right) + \cfrac{4}{\gamma} + \cfrac{0.25}{i\left(x - \cfrac{2\omega\tau_0}{3\gamma}\right) + \cdots}}}$$

All other $\tilde{r}_n(\omega)$ can be calculated from Eq. (384) by iteration. On comparing Eqs. (373), (376) and (386), (387) it is obvious that

$$\tilde{S}_n(0) = S_n \equiv \langle r^n \rangle_0 \tag{388}$$

The inhomogeneous Eq. (385) can also be solved by the continued fraction method as described in Ref. 3. If we truncate the system after the Nth term and omit the equations with $n \geq N + 1$, we obtain by inserting the ansatz

$$\tilde{A}_{n+1}(\omega) = R_n^+ \tilde{A}_n(\omega) + C_{n+1} \quad \text{for } n \geq 0, \, \tilde{A}_0(\omega) = C_0 \tag{389}$$

the following recurrence relations:

$$R_{N-1}^+ = -\frac{Q_N^-}{\hat{Q}_N} \quad C_N = -\frac{B_N}{\hat{Q}_N} \tag{390}$$

$$R_n^+ = -\frac{Q_{n+1}^-}{\hat{Q}_{n+1} + Q_{n+1}^+ R_{n+1}^+} \quad \text{for } 0 \leq n \leq N - 2 \tag{391}$$

$$C_n = -\frac{B_n + Q_n^+ C_{n+1}}{\hat{Q}_n + Q_n^+ R_n^+} \quad \text{for } 0 \leq n \leq N - 1 \tag{392}$$

Thus, all $\tilde{A}_n(\omega)$ with $n \geq 1$ can now be obtained by iteration. Finally, $\tilde{A}_0(\omega)$ follows from Eq. (385) with $n = 0$; i.e., $\tilde{A}_0(\omega) = C_0$. This method can be used for numerical calculations.[20] The exact calculation of the linear junction impedance based on a numerical solution of the hierarchy of the differential-difference equations (384) obtained from the Fokker-Planck equation has been given in Ref. 53 (see also the discussion of this results in Ref. 54).

However, there is another representation of the exact solution. One may show on successive elimination of the variables in the inhomogeneous

Eq. (384) that the solution is[22]

$$\tilde{A}_1(\omega) = -2i\bigg[\tilde{S}_1(\omega)\langle r \rangle_0 - \tilde{S}_1(\omega)\tilde{S}_2(\omega)\langle r^2 \rangle_0$$

$$+\tilde{S}_1(\omega)\tilde{S}_2(\omega)\tilde{S}_3(\omega)\langle r^3 \rangle_0 - \cdots - (-1)^{n+1}\langle r^n \rangle_0 \prod_{k=1}^{n}\tilde{S}_k(\omega) + \cdots \bigg]$$

where the $\tilde{S}_n(\omega)$ are given by Eq. (387).

On taking into account Eqs. (372) and (388), the above equation becomes

$$\tilde{A}_1(\omega) = -2i\tilde{S}_1(\omega)\tilde{S}_1\big(0\big(1 - \tilde{S}_2(\omega)\tilde{S}_2(0)\big(1 - \tilde{S}_3(\omega)\tilde{S}_3(0)(1 - \cdots)\big)\big)$$

$$\tag{393}$$

We can show in the same way that the response of $\langle r^{-1} \rangle$ where

$$\langle r^{-1} \rangle = \langle r^{-1} \rangle_0 + \tilde{A}_{-1}(\omega)I_m e^{-i\omega t}/I + \cdots$$

is given by

$$\tilde{A}_{-1}(\omega) = 2i\tilde{S}_1^*(-\omega)\tilde{S}_1^*(0)\big(1 - \tilde{S}_2^*(-\omega)\tilde{S}_2^*(0)\big(1 - \tilde{S}_3^*(-\omega)\tilde{S}_3^*(0)$$

$$\times\big(1 - \tilde{S}_4^*(-\omega)\tilde{S}_4^*(0)(1 - \cdots)\big)\big)$$

$$= A_1^*(-\omega) \tag{394}$$

where the asterisk indicates the complex conjugate. Equations (393) and (394) allow us to calculate $\langle \sin\phi - \langle \sin\phi \rangle_0 \rangle$ and $\langle \cos\phi - \langle \cos\phi \rangle_0 \rangle$ and thus the linear ac response

$$\langle \sin\phi - \langle \sin\phi \rangle_0 \rangle = D_s(\omega)\frac{I_m e^{-i\omega t}}{I} \tag{395}$$

$$\langle \cos\phi - \langle \cos\phi \rangle_0 \rangle = D_c(\omega)\frac{I_m e^{-i\omega t}}{I} \tag{396}$$

where

$$D_{\mathrm{s}}(\omega) = \frac{i}{2}\left[\tilde{A}_1(\omega) - \tilde{A}_{-1}(\omega)\right] \qquad (397)$$

$$D_{\mathrm{c}}(\omega) = \frac{1}{2}\left[\tilde{A}_1(\omega) + \tilde{A}_{-1}(\omega)\right] \qquad (398)$$

Equations (393)–(396) are very convenient for numerical calculations.

In particular one may use Eq. (397) to evaluate the impedance $Z(\omega)$ offered by the junction. To accomplish this we recall that the averaged current balance equation in the presence of the ac is

$$I_{\mathrm{dc}} + I_m\, e^{-i\omega t} - I\langle\sin\phi\rangle - \frac{\langle V\rangle}{R} = 0$$

We have supposed that

$$\langle\sin\phi\rangle = \langle\sin\phi\rangle_0 + \langle\sin\phi\rangle_1$$
$$\langle V\rangle = \langle V\rangle_0 + \langle V\rangle_1$$

where the subscript 0 on the angular braces denotes the average in the absence of the ac, and the subscript 1 denotes the portion of the average that is linear in I_m. Thus, on dividing across the current balance equation by the supercurrent amplitude I so that

$$x + \frac{I_m}{I}\, e^{-i\omega t} - \left[\langle\sin\phi\rangle_0 + \langle\sin\phi\rangle_1\right] - \left(\langle\nu\rangle_0 + \langle\nu\rangle_1\right) = 0$$

and recalling that

$$x - \langle\sin\phi\rangle_0 - \langle\nu\rangle_0 = 0$$

we have

$$\langle\nu\rangle_1 = \frac{I_m}{I}\, e^{-i\omega t} - \langle\sin\phi\rangle_1$$

so that on reverting to the original variables

$$\langle V\rangle_1 = R\left[1 - D_{\mathrm{s}}(\omega)\right]I_m\, e^{-i\omega t} = Z(\omega)I_m\, e^{-i\omega t}$$

where

$$Z(\omega) = R[1 - D_s(\omega)] \qquad (399)$$

is the impedance of the junction.

E. The Effective Eigenvalues for the Josephson Junction

Let us suppose that a strong dc current I_{dc} had been applied to the junction in the infinite past and that at $t = 0$, I_{dc} is incremented by a small current $U(t)\Delta$, where $U(t)$ is the unit step function so that the total current is $I_t = I_{dc} + U(t)\Delta$. Now we are only interested in the response linear in Δ. We therefore assume that

$$\langle r^n \rangle = \langle r^n \rangle_{eq} + \langle r^n \rangle_1 \qquad (400)$$

where the subscript 1 denotes the portion of the statistical average which is linear in Δ and the subscript eq denotes the statistical average in the stationary state computed using the stationary distribution function $W_0(\phi)$ of Eq. (371) where

$$U(\phi) = -\frac{\hbar[\cos\phi + (I_{dc} + \Delta)\phi]}{2e} \qquad (401)$$

As $t \to \infty$ we have

$$\lim_{t \to \infty} \langle r^n \rangle = \langle r^n \rangle_{eq} \qquad \lim_{t \to \infty} \langle r^n \rangle_1 = 0$$

On substituting Eq. (400) into Eq. (349) we obtain

$$\left(n^2 + \frac{in\hbar}{2ekT}(I_{dc} + \Delta) \right) \langle r^n \rangle_{eq} + \frac{n\hbar I}{4ekT}(\langle r^{n+1} \rangle_{eq} - \langle r^{n-1} \rangle_{eq}) = 0 \quad (402)$$

and

$$\frac{d}{dt}\langle r^n \rangle_1 + \frac{1}{\tau_0}\left(n^2 + \frac{in\hbar}{2ekT}I_{dc} \right)\langle r^n \rangle_1 = \frac{n\hbar I}{4\tau_0 ekT}(\langle r^{n-1} \rangle_1 - \langle r^{n+1} \rangle_1)$$
$$-\frac{i\Delta n\hbar}{2e\tau_0 kT}\langle r^n \rangle_{eq} U(t) \qquad (403)$$

Equation (403) represents a three-term recurrence relation driven by a forcing function, namely the $U(t)$ term. To determine the effective eigenvalue we consider the unforced part of Eq. (403) and reduce it to an eigenvalue problem:

$$\frac{\mathrm{d}}{\mathrm{d}t}\langle r \rangle_1 + \lambda_{\mathrm{eff}}^+ \langle r \rangle_1 = 0 \tag{404}$$

In this case

$$\langle r \rangle_1 = \langle \mathrm{e}^{-i\phi} \rangle_1 \tag{405}$$

is a complex variable, so that λ_{eff} is also complex:

$$\lambda_{\mathrm{eff}}^+ = \lambda = \lambda' + i\lambda'' \tag{406}$$

The real part of λ_{eff}^+ when inverted will give the effective relaxation time, while the imaginary part will give the frequency of oscillation.

Equation (404) yields

$$\lambda_{\mathrm{eff}}^+ = -\frac{(\mathrm{d}/\mathrm{d}t)\langle r \rangle_1}{\langle r \rangle_1} \tag{407}$$

The effective eigenvalue method suggests that Eq. (407) may be replaced by its initial value (i.e., its value at $t = 0$). We therefore have

$$\lambda_{\mathrm{eff}}^+ = -\frac{(\mathrm{d}/\mathrm{d}t)\langle r \rangle_1}{\langle r \rangle_1} \quad t = 0 \tag{408}$$

On substituting Eq. (403) into Eq. (408) for $n = 1$, we obtain

$$\lambda_{\mathrm{eff}}^+ = \frac{(1/\tau_0)(1 + i\hbar I_{\mathrm{dc}}/2ekT)\langle r \rangle_1 + (eI/\hbar)R\langle r^2 \rangle_1}{\langle r \rangle_1} \quad t = 0 \tag{409}$$

Equation (409) may be further simplified to

$$\lambda_{\mathrm{eff}}^+ = \frac{1}{\tau_0} + \frac{i2eR}{\hbar}I_{\mathrm{dc}} + \frac{\langle r^2 \rangle_1}{\langle r \rangle_1}\frac{eIR}{\hbar} \quad t = 0 \tag{410}$$

Now from Eq. (400) we have

$$\langle r^n \rangle_1 = \langle r^n \rangle - \langle r^n \rangle_{eq} \qquad (411)$$

with

$$\lim_{t \to 0} \langle r^n \rangle_1 = \langle r^n \rangle_0 - \langle r^n \rangle_{eq} \qquad (412)$$

Equation (410) with the aid of the Eqs. (411) and (412) simplifies further to

$$\lambda_{\text{eff}}^+ = \frac{1}{\tau_0} + \frac{i2eR}{\hbar} I_{dc} + \left[\frac{\langle r^2 \rangle_{eq} - \langle r^2 \rangle_0}{\langle r \rangle_{eq} - \langle r \rangle_0} \right] \frac{eIR}{\hbar} \qquad (413)$$

where the averages $\langle r \rangle_{eq}$ and $\langle r^2 \rangle_{eq}$ are over the stationary distribution $W_0(\phi)$ of Eq. (371) with the perturbed potential of Eq. (401). However, remembering that we are interested only in the linear response of Δ, we can express $\langle r^n \rangle_{eq}$ as

$$\langle r^n \rangle_{eq} = \langle r^n \rangle_0 + \Delta \frac{\partial}{\partial \Delta} \langle r^n \rangle_0 + O(\Delta^2) \qquad (414)$$

Thus, we need to evaluate only $(\partial/\partial\Delta)\langle r \rangle_0$ and $(\partial/\partial\Delta)\langle r^2 \rangle_0$, which may be done as follows.

On using Eq. (414) in Eq. (402), we find that the linear approximation in Δ is given by the following set of equations:

$$\left(\frac{2n}{\gamma} + ix \right) \langle r^n \rangle_0 + \frac{1}{2} (\langle r^{n+1} \rangle_0 - \langle r^{n-1} \rangle_0) = 0 \qquad (415)$$

$$\left(\frac{2n}{\gamma} + ix \right) \frac{\partial}{\partial \Delta} \langle r^n \rangle_0 + \frac{1}{2} \left(\frac{\partial}{\partial \Delta} \langle r^{n+1} \rangle_0 - \frac{\partial}{\partial \Delta} \langle r^{n-1} \rangle_0 \right) = -i22 \langle r^n \rangle_0$$

$$(416)$$

The solution of Eq. (415) may again be given in terms of an infinite continued fraction (see Section IV.C). We have from Eq. (415)

$$\frac{\langle r^n \rangle_0}{\langle r^{n-1} \rangle_0} = \frac{0.5}{\dfrac{2n}{\gamma} + ix + \dfrac{1}{2}\dfrac{\langle r^{n+1} \rangle_0}{\langle r^n \rangle_0}} \tag{417}$$

Thus, noting that

$$\langle r^0 \rangle_0 = 1$$

we obtain Eq. (376), viz.,

$$\langle r \rangle_0 = \cfrac{0.5}{\dfrac{2}{\gamma} + ix + \cfrac{0.25}{\dfrac{4}{\gamma} + ix + \cfrac{0.25}{\dfrac{6}{\gamma} + ix + \cfrac{0.25}{\dfrac{8}{\gamma} + ix + \cdots}}}} \tag{418}$$

The other quantities $\langle r^n \rangle_0$ with $n \geq 2$ can be obtained from the recurrence relation of Eq. (415) by iteration; for example,

$$\langle r^2 \rangle_0 = 1 - 2\left(\frac{2}{\gamma} + ix\right)\langle r \rangle_0 \tag{419}$$

On substituting $\langle r^n \rangle_0$ into Eq. (416) and successively eliminating the variables we obtain

$$\frac{\partial}{\partial \Delta}\langle r \rangle_0 = -\frac{2i}{I}\Bigg[S_1\langle r \rangle_0 - S_1 S_2\langle r^2 \rangle_0 + S_1 S_2 S_3\langle r^3 \rangle_0$$

$$- \cdots + (-1)^{n-1}\prod_{k=1}^{n} S_k\langle r^n \rangle_0 + \cdots \Bigg] \tag{420}$$

$$= -\frac{2i}{I}S_1^2\big(1 - S_2^2\big(1 - S_3^2\big(1 - S_4^2(1 - \cdots)\big)\big)\big)$$

where

$$S_k = \cfrac{0.5}{\cfrac{2k}{\gamma} + ix + \cfrac{0.25}{\cfrac{2(k+1)}{\gamma} + ix + \cfrac{0.25}{\cfrac{2(k+2)}{\gamma} + ix + \cdots}}} \tag{421}$$

Noting that Eq. (417) allows us to express S_n in terms of $\langle r^n \rangle_0$ as

$$\langle r^n \rangle_0 = \langle r^{n-1} \rangle_0 S_n \tag{422}$$

we obtain

$$\frac{\partial}{\partial \Delta} \langle r \rangle_0 = -\frac{2i}{I} \left[\langle r \rangle_0^2 - \langle r^2 \rangle_0^2 + \langle r^3 \rangle_0^2 - \langle r^4 \rangle_0^2 + \cdots \right]$$
$$= \frac{2i}{I} \sum_{n=1}^{\infty} (-1)^n \langle r^n \rangle_0^2 \tag{423}$$

The quantity $(\partial/\partial\Delta)\langle r^2 \rangle_0$ can be obtained from Eq. (416) at $n = 1$, whence

$$\frac{\partial}{\partial \Delta} \langle r^2 \rangle_0 = -\frac{2i}{I} \langle r \rangle_0 - 2\left(\frac{2}{\gamma} + ix \right) \frac{\partial}{\partial \Delta} \langle r \rangle_0$$
$$= -\frac{2i}{I} \langle r \rangle_0 + \frac{4i}{I} \left(\frac{2}{\gamma} + ix \right) \tag{424}$$
$$\times \left[\langle r \rangle_0^2 - \langle r^2 \rangle_0^2 + \langle r^3 \rangle_0^2 - \langle r^4 \rangle_0^2 + \cdots \right]$$

On substituting Eqs. (423) and (424) into Eq. (413) we obtain

$$\lambda_{\text{eff}}^+ = \frac{1}{\tau_0} + \frac{i2RI_{\text{dc}}}{\hbar}$$

$$+ \frac{eIR}{\hbar} \frac{\langle r \rangle_0 - 2(2/\gamma + ix)\left[\langle r \rangle_0^2 - \langle r^2 \rangle_0^2 + \langle r^3 \rangle_0^2 - \cdots \right]}{\langle r \rangle_0^2 - \langle r^2 \rangle_0^2 + \langle r^3 \rangle_0^2 - \cdots} \tag{425}$$

$$= \frac{1}{\tau_0} + \frac{i2RI_{\text{dc}}}{\hbar} - \frac{2eIR}{\hbar} \left(\frac{2}{\gamma} + ix \right) + \frac{eIR}{\hbar} \frac{\langle r \rangle_0}{\sum_{n=1}^{\infty} (-1)^{n+1} \langle r^n \rangle_0^2} \tag{426}$$

where

$$\Omega = \frac{2\omega\tau_0}{\gamma} \qquad (439)$$

The above equations have a simple physical interpretation. If $x < 1$ the junction behaves like an inductance, then

$$L_j = \frac{\hbar/2eI}{\sqrt{1 - x^2}}$$

in parallel with the resistance R, yielding the admittance

$$Y(\omega) = \frac{1}{R} - \frac{1}{i\omega L}$$

which gives the impedance $Z(\omega) = Y^{-1}(\omega)$ from Eq. (437).[55] If $x > 1$ the impedance is entirely real with a singularity at $x_s = \sqrt{1 + \Omega^2}$. This singularity vanishes in the presence of noise, as is evident from Figs. 14–19. Such behavior is even more pronounced in Figs. 20–23. There we

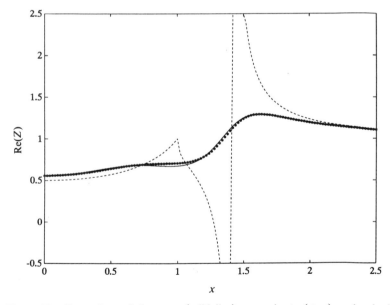

Figure 20. Comparison of the exact (solid line), approximate (stars), and noise-less solutions (dashed line) for the real part of the normalized impedance Re{Z} versus x at $\gamma = 10$.

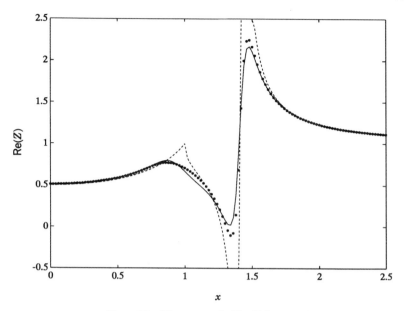

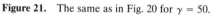

Figure 21. The same as in Fig. 20 for $\gamma = 50$.

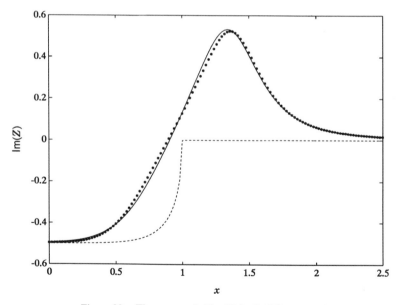

Figure 22. The same as in Fig. 20 for $\mathrm{Im}\{Z\}$ at $\gamma = 10$.

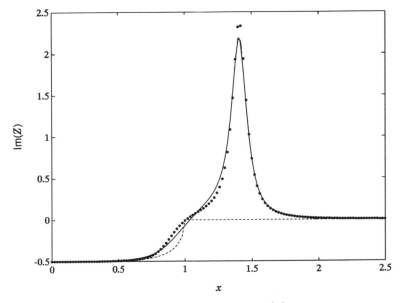

Figure 23. The same as in Fig. 20 for Im{Z} at $\gamma = 50$.

have plotted the normalized impedance $Z(\omega)/R$ as a function of the bias parameter x and have compared it with the noiseless case. It is apparent from Figs. 15 and 17 that for weak noise (large γ) Eqs. (437) and (438) yield a satisfactory description of the impedance excluding the region in the vicinity of the singular point $x_s = \sqrt{1 + \Omega^2}$. However, at $\gamma = 10$, there is a striking difference between the solutions with and without noise. This is particularly apparent in Fig. 20, where close to the singular point x_s the noiseless solution (in contrast to that including noise) possesses a negative real part, which is an indication of amplification or oscillation, which may occur if the junction is inserted in an appropriate microwave circuit.[56]

G. Calculation of the Beat Signal Spectrum of the Ring Laser Gyro

In the previous sections we have considered the Josephson junction as an example of the Brownian motion in a tilted cosine potential. However, the above results can also be applied to an analogous system, namely the ring laser gyro.[19-21, 47] We recall that the dynamical behavior of the ring laser gyro operating in the steady state is described by a similar Langevin equation to that of the Josephson junction[19, 47]:

$$\dot{\phi}(t) - b \sin \phi(t) = a + \Gamma(t) \tag{440}$$

where ϕ is the relative phase between the clockwise and counterclockwise modes of the laser,

$$a = -\frac{8\pi}{\lambda L}(\mathbf{A} \cdot \mathbf{\Omega}) \tag{441}$$

$|\mathbf{A}|$ is the area covered by the optical path of length L along the ring cavity, $\mathbf{\Omega}$ is the rotation rate of the gyro, and λ is the wavelength of the laser; b is the backscattering coefficient and $\Gamma(t)$ is the noise source with properties

$$\overline{\Gamma(t)} = 0 \qquad \overline{\Gamma(0)\Gamma(t)} = 2D\delta(t) \tag{442}$$

The Langevin equation (357) for the Josephson junction has the same mathematical form as Eq. (440) with appropriate change of parameters, viz.,

$$\frac{2eIR}{\hbar} \rightarrow -b \tag{443}$$

$$\frac{2eRI_{dc}}{\hbar} \rightarrow a \tag{444}$$

$$kTR\left(\frac{2e}{\hbar}\right)^2 \rightarrow D \tag{445}$$

Thus, on redefining the parameters x, γ, and τ_0 as

$$x = -\frac{a}{b} \tag{446}$$

$$\gamma = -\frac{2b}{D} \tag{447}$$

$$\tau_0 = \frac{1}{D} \tag{448}$$

we obtain equations for the ring laser gyro in the same form as those for the Josephson junction. However, both x and γ are now negative quantities.

For the ring laser gyro the quantity of interest is the spectrum of the beat signal,[19] defined as

$$\alpha(\omega) = \int_{-\infty}^{\infty} \langle \cos \phi(0) \cos \phi(t) \rangle_0 \, e^{i\omega t} \, dt$$
$$= 2 \, \text{Re} \left\{ \int_0^{\infty} \langle \cos \phi(0) \cos(t) \rangle_0 \, e^{i\omega t} \, dt \right\} \tag{449}$$

where

$$C(t) = \langle \cos \phi(0) \cos \phi(t) \rangle_0 \tag{450}$$

is the beat signal autocorrelation function $\cos \phi(t)$, which is a measure of the total detected intensity.[19, 47] A numerical method of exact calculation of $\alpha(\omega)$ has been given in Ref. 20.

Another representation of the exact solution can be obtained in the following manner. On multiplying the both sides of Eq. (364) by $\cos \phi(0)$ and averaging the equation so obtained over the stationary distribution function $W_0(\phi(0))$, we arrive at the set of differential-difference equations:

$$\frac{d}{dt} \psi_n(t) + \frac{1}{\tau_0} \left(n^2 + \frac{inx\gamma}{2} \right) \psi_n(t) = \frac{n\gamma}{4\tau_0} (\psi_{n-1}(t) - \psi_{n+1}(t)) \tag{451}$$

where

$$\psi_n(t) = \langle \cos \phi(0) r^n(t) \rangle_0 \tag{452}$$

It is obvious that $C(t)$ from Eq. (450) is related to $\psi_1(t)$ by

$$C(t) = \text{Re}\{\psi_1(t)\} \tag{453}$$

Let us introduce instead of $\psi_n(t)$ the true correlation functions $C_n(t)$, defined as

$$C_n(t) = \psi_n(t) - \psi_n(\infty)$$
$$= \langle \cos \phi(0) r^n(t) \rangle_0 - \langle \cos \phi(0) \rangle_0 \langle r^n(0) \rangle_0 \tag{454}$$

Then we obtain from Eq. (451)

$$\frac{d}{dt} C_n(t) + \frac{1}{\tau_0} \left(n^2 + \frac{inx\gamma}{2} \right) C_n(t) = \frac{n\gamma}{4\tau_0} (C_{n-1}(t) - C_{n+1}(t)) \tag{455}$$

It should be noted that now

$$C_0(t) = 0 \qquad (456)$$

in contrast to the calculations previously described. We have taken into account here that $\psi_n(\infty)$ satisfies the recurrence relation

$$\left(n^2 + \frac{inx\gamma}{2}\right)\psi_n(\infty) = \frac{n\gamma}{4}\left(\psi_{n-1}(\infty) - \psi_{n+1}(\infty)\right) \qquad (457)$$

With the help of the one-sided Fourier transform we may now derive from Eq. (455) the usual three-term recurrence relation:

$$\left(-i\omega\tau_0 + n^2 + \frac{inx\gamma}{2}\right)\tilde{C}_n(\omega) = \frac{\gamma n}{4}\left(\tilde{C}_{n-1}(\omega) - \tilde{C}_{n+1}(\omega)\right) + \tau_0 C_n(0) \qquad (458)$$

where

$$\tilde{C}_n(\omega) = \int_0^\infty e^{i\omega t} C_n(t)\, dt \qquad (459)$$

Equation (458) has the same form as Eq. (384) and hence has a solution of similar form:

$$
\begin{aligned}
\tilde{C}_1(\omega) &= \frac{4\tau_0}{\gamma}\left(\tilde{S}_1(\omega)C_1(0) - \frac{1}{2}\tilde{S}_1(\omega)\tilde{S}_2(\omega)C_2(0) \right. \\
&\quad + \frac{1}{3}\tilde{S}_1(\omega)\tilde{S}_2(\omega)\tilde{S}_3(\omega)C_3(0) - \cdots \\
&\quad \left. + \frac{1}{n}(-1)^{n+1}C_n(0)\prod_{i=1}^n \tilde{S}_i(\omega) + \cdots \right) \\
&= \frac{4\tau_0}{\gamma}\tilde{S}_1(\omega)C_1(0)\left[1 - \frac{\tilde{S}_2(\omega)C_2(0)}{2C_1(0)}\left(1 - \frac{2\tilde{S}_3(\omega)C_3(0)}{3C_2(0)}\right.\right. \\
&\quad \left.\left. \times \left(1 - \frac{3\tilde{S}_4(\omega)C_4(0)}{4C_3(0)}(1 - \cdots)\right)\right)\right]
\end{aligned}
\qquad (460)
$$

where as before $\tilde{S}_n(\omega)$ is determined by Eq. (387). The quantity $C_n(0)/C_{n-1}(0)$ appearing in Eq. (460) can be expressed in terms of S_1 and

S_n as follows. We have

$$
\begin{aligned}
C_n(0) &= \tfrac{1}{2}\langle r^{n+1}(0) + r^{n-1}(0)\rangle_0 - \mathrm{Re}(S_1)\langle r^n(0)\rangle_0 \\
&= \tfrac{1}{2} S_1 S_2 \cdots S_{n-1}(1 + S_n S_{n+1}) - \mathrm{Re}(S_1) S_1 S_2 \cdots S_n
\end{aligned}
\tag{461}
$$

Hence, we have for $n \geq 2$

$$
\frac{C_n(0)}{C_{n-1}(0)} = \frac{S_{n-1}(1 + S_n S_{n+1}) - 2S_{n-1}S_n\,\mathrm{Re}(S_1)}{1 + S_{n-1}S_n - 2S_{n-1}\,\mathrm{Re}(S_1)}
\tag{462}
$$

or on taking into account Eq. (373),

$$
\frac{C_n(0)}{C_{n-1}(0)} = \frac{1 - S_n(\mathrm{i}x + 2n/\gamma) - S_n\,\mathrm{Re}(S_1)}{S_n + \mathrm{i}x + 2(n-1)/\gamma - \mathrm{Re}(S_1)} \qquad n \geq 2 \tag{463}
$$

We also note that

$$
C_1(0) = \frac{1}{2}(S_2 S_1 + 1) - S_1\,\mathrm{Re}(S_1) = 1 - S_1\!\left(\mathrm{i}x + \frac{2}{\gamma}\right) - S_1\,\mathrm{Re}(S_1)
\tag{464}
$$

We can show in the same way that

$$
\tilde{C}_{-1}(\omega) = \frac{4\tau_0}{\gamma}\tilde{S}_1^*(-\omega)C_1^*(0)\Bigg(1 - \frac{\tilde{S}_2^*(-\omega)C_2^*(0)}{2C_1^*(0)}\Bigg(1 - \frac{2\tilde{S}_3^*(-\omega)C_3^*(0)}{3C_2^*(0)}
$$
$$
\times\Bigg(1 - \frac{3\tilde{S}_4^*(-\omega)C_4^*(0)}{4C_3^*(0)}(1 - \cdots)\Bigg)\Bigg)\Bigg)
\tag{465}
$$

Thus, on using Eqs. (453), (454), (459), (460), and (465), we obtain the beat signal spectrum as

$$
\alpha(\omega) = 2\pi\,\mathrm{Re}^2(S_1)\delta(\omega) + \mathrm{Re}\{T(\omega) + T^*(-\omega)\}
\tag{466}
$$

where the function $T(\omega)$ is given by

$$
T(\omega) = \frac{4\tau_0}{\gamma}\tilde{S}_1(\omega)C_1(0)\left(1 - \frac{\tilde{S}_2(\omega)C_2(0)}{2C_1(0)}\left(1 - \frac{2\tilde{S}_3(\omega)C_3(0)}{3C_2(0)}\right.\right.
$$
$$
\left.\left.\times\left(1 - \frac{3\tilde{S}_4(\omega)C_4(0)}{4C_3(0)}(1 - \cdots)\right)\right)\right)
\tag{467}
$$

Equations (466) and (467) are very convenient for numerical calculations and constitute a simpler algorithm than Eq. (389), which has been used in Ref. 20 to evaluate $\alpha(\omega)$. Equations (466) and (467) allow the exact calculation of $\alpha(\omega)$.

We shall now show how the effective eigenvalue method can be used to evaluate $\alpha(\omega)$. To implement it we need, just as in the application to ferromagnetic resonance, to keep equations for $C_1(t)$ and $C_2(t)$ of Eq. (455). Thus, the effective eigenvalue method requires that $C_1(t)$ and $C_2(t)$ obey the coupled equations

$$
\frac{d}{dt}C_1(t) + \lambda_1 C_1(t) + \frac{\gamma}{4\tau_0}C_2(t) = 0
\tag{468}
$$

$$
\frac{d}{dt}C_2(t) + \lambda_2 C_2(t) - \frac{\gamma}{2\tau_0}C_1(t) = 0
\tag{469}
$$

where $\lambda_1 = (1 + i x \gamma/2)/\tau_0$ and λ_2 is the effective eigenvalue to be determined.

According to Eq. (455), at $n = 2$, λ_2 is determined as

$$
\lambda_2 = -\frac{\dot{C}_2(0) - (\gamma/2\tau_0)C_1(0)}{C_2(0)}
\tag{470}
$$

where

$$
\dot{C}_2(0) = -\frac{1}{\tau_0}(4 + i x \gamma)C_2(0) - \frac{\gamma}{2\tau_0}(C_3(0) - C_1(0))
\tag{471}
$$

and $C_2(0), C_3(0)$ are given by Eq. (461).

On substituting Eq. (471) into Eq. (470) and using Eq. (462), we obtain

$$
\lambda_2 = \frac{1}{\tau_0}\left[4 + i x \gamma + \frac{\gamma S_2}{2}\left(\frac{S_4 S_3 + 1 - 2S_3 \operatorname{Re}(S_1)}{1 + S_2[(S_3 - 2\operatorname{Re}(S_1)]}\right)\right]
\tag{472}
$$

The solution of Eqs. (468) and (469) is easily found using the one-sided Fourier transform. The result is (where we emphasis that $C_1(0)$, $C_2(0)$, λ_1 and λ_2 are complex)

$$\tilde{C}_1(\omega) = \frac{C_1(0)(\lambda_2 - i\omega) - \gamma C_2(0)/4\tau_0}{(\lambda_1 - i\omega)(\lambda_2 - i\omega) + \gamma^2/8\tau_0^2} \qquad (473)$$

where

$$\tilde{C}_1(\omega) = \int_0^\infty e^{i\omega t} C_1(t)\, dt \qquad (474)$$

Thus, we may now calculate in a simple analytical form from Eqs. (449), (453), and (473) the beat signal spectrum:

$$\alpha(\omega) = 2\pi \, \mathrm{Re}^2(S_1)\delta(\omega) + \mathrm{Re}\{\tilde{C}_1(\omega) + \tilde{C}_1^*(-\omega)\} \qquad (475)$$

The spectrum $\alpha(\omega)$ yielded by Eqs. (466) and (474) contains two parts, a coherent δ-function spectrum and an incoherent broad spectrum.[20] The incoherent part of the laser-gyro beat signal spectrum is presented in Figs. 24–26 for different values of the parameters a/b and γ/b. Our exact

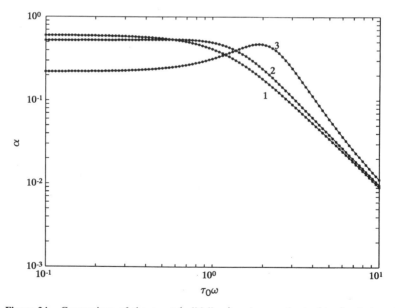

Figure 24. Comparison of the exact (solid lines) and approximate (stars) solutions for "incoherent" part of the beat signal spectrum α of the ring laser gyro versus $\tau_0\omega$. 1, $\gamma = -2$, $x = -0.5$; 2, $x = -0.9$; 3, $x = -2.0$.

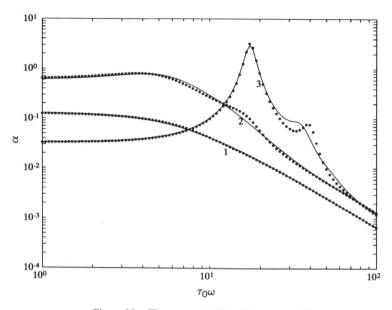

Figure 25.   The same as in Fig. 24 for $\gamma = -20$.

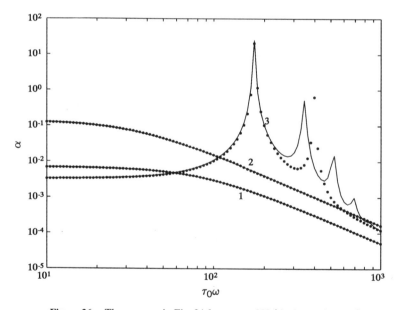

Figure 26.   The same as in Fig. 24 for $\gamma = -200$ (the low noise case).

We now make the transformation

$$\mu_x(t) \rightarrow x(t) \tag{A.5}$$

$$\mu_y(t) \rightarrow y(t) \tag{A.6}$$

$$\mu_x(t)\mu_z^2(t) \rightarrow z(t) \tag{A.7}$$

so that Eqs. (A.2)–(A.4) become

$$\dot{x}(t) = \frac{1}{\zeta}\left(\lambda_y(t)\sqrt{\frac{z(t)}{x(t)}} - \lambda_z(t)y(t)\right) - \frac{z(t)}{\mu\zeta}E_0 \tag{A.8}$$

$$\dot{y}(t) = \frac{1}{\zeta}\left(\lambda_z(t)x(t) - \lambda_x(t)\sqrt{\frac{z(t)}{x(t)}}\right) - \frac{y(t)}{\mu\zeta}\frac{z(t)}{x(t)}E_0 \tag{A.9}$$

$$\dot{z}(t) = \frac{1}{\zeta}\left[2\lambda_x(t)y(t)\sqrt{\frac{z(t)}{x(t)}} + \lambda_y(t)\left(\sqrt{\frac{z^3(t)}{x^3(t)}} - 2\sqrt{x^3(t)z(t)}\right)\right.$$

$$\left. -\lambda_z(t)\frac{y(t)z(t)}{x(t)}\right] + \frac{z(t)}{\zeta\mu}\left(2\mu^2 - 3\frac{z(t)}{x(t)}\right)E_0 \tag{A.10}$$

We now apply formula (39) for the noise-induced drift terms to Eq. (A.10), just as in Section II.B. The relevant components of the tensor g_{ij} in Eq. (36) are

$$g_{11} = 0 \qquad g_{12} = \frac{\sqrt{z/x}}{\zeta} \qquad g_{13} = \frac{-y}{\zeta}$$

$$g_{21} = \frac{-\sqrt{z/x}}{\zeta} \qquad g_{22} = 0 \qquad g_{23} = \frac{x}{\zeta} \tag{A.11}$$

$$g_{31} = \frac{2y\sqrt{z/x}}{\zeta} \qquad g_{32} = \frac{1}{\zeta}\left(\sqrt{\frac{z^3}{x^3}} - 2\sqrt{x^3z}\right) \qquad g_{33} = -\frac{yz}{\zeta x}$$

Hence, after tedious algebra we obtain

$$kT\zeta\left(g_{kj}\frac{\partial}{\partial x_k}g_{3j}\right) = \frac{2kT}{\zeta}(x\mu^2 - 6z) \tag{A.12}$$

This is the noise-induced part of z. On averaging over the distribution $W(\{\mu\}, t)$ at time t the deterministic contribution to z is

$$\left\langle\frac{\mu_x\mu_z^2}{\mu\zeta}(2\mu^2 - 3\mu_z^2)E_0\right\rangle \tag{A.13}$$

Thus, the complete equation averaged over $W(\{\mu\}, t)$ is

$$\frac{d}{dt}\langle z\rangle + \frac{12kT}{\zeta}\langle z\rangle = \frac{2kT}{\zeta}\langle\mu_x\rangle\mu^2 + \frac{E_0}{\mu\zeta}\langle\mu_x\mu_z^2(2\mu^2 - 3\mu_z^2)\rangle \tag{A.14}$$

We note that

$$X_{31} = \frac{3}{2}(5\mu_z^2 - 1)\mu_x = \frac{3}{2}\left(5\frac{z}{\mu^3} - X_{11}\right) \tag{A.15}$$

so that

$$\langle\dot{X}_{31}\rangle = \frac{15}{2\mu^3}\langle\dot{z}\rangle - \frac{3}{2}\langle\dot{X}_{11}\rangle \tag{A.16}$$

where $\langle\dot{X}_{11}\rangle$ is given by Eq. (51). Therefore, Eq. (A.14) becomes

$$\frac{d}{dt}\langle X_{31}\rangle + \frac{6}{\tau_D}\langle X_{31}\rangle\left(1 - \frac{\sigma}{30}\right) = \frac{48\sigma}{35\tau_D}\langle X_{11}\rangle - \frac{4\sigma}{7\tau_D}\langle X_{51}\rangle \tag{A.17}$$

where the spherical harmonic X_{51} is defined as[25]

$$X_{51} = \tfrac{1}{8}\sin\vartheta\cos\phi(315\cos^4\vartheta + 210\cos^2\vartheta + 15)$$

Thus, the calculation shows how the hierarchy of equations may be formed.

APPENDIX B: DERIVATION OF THE EQUATION FOR $\langle H_{ef} M_y \rangle$ FROM THE STOCHASTIC LANDAU-LIFSHITZ-GILBERT EQUATION

To calculate the transverse magnetic susceptibility $\chi_\perp(\omega)$ we first require the equation of motion for $\langle H_{ef} M_y \rangle$. The effective field $H_{ef}(\{M\})$ for the mean field potential

$$V(\{M\}) = -K \cos^2 \vartheta + K$$

is from Eq. (175)

$$\mathbf{H}_{ef} = -\mathbf{k}\frac{\partial V}{\partial M_z} = \mathbf{k}\frac{2K}{M_s^2}M_z \qquad (B.1)$$

Thus, on noting that

$$\frac{d}{dt}\left(H_{ef}(\{\mathbf{M}(t)\})M_y(t)\right) = \frac{2K}{M_s^2}\left(M_z(t)\frac{d}{dt}M_y(t) + M_y(t)\frac{d}{dt}M_z(t)\right) \qquad (B.2)$$

we can derive the Langevin equation for $H_{ef}M_y$ from Eqs. (162) and (163). On introducing new variables

$$X(t) = M_x(t) \qquad (B.3)$$

$$Y(t) = M_s H_{ef}(\{\mathbf{M}(t)\}M_y(t)/2K \qquad (B.4)$$

$$Z(t) = M_z(t) \qquad (B.5)$$

we obtain from Eqs. (161)–(163)

$$
\frac{d}{dt}X(t) = 2g'KY(t) - \frac{2h'K}{M_s^2}X(t)Z^2(t) + F_x(t)H(t)
$$

$$
+ h'\left(M_s^2 - X^2(t)\right)h_x(t)
$$

$$
-\left(g'M_sZ(t) + h'M_s\frac{X(t)Y(t)}{Z(t)}\right)h_y(t) \qquad \text{(B.6)}
$$

$$
+\left(g'M_s^2\frac{Y(t)}{Z(t)} - h'X(t)Z(t)\right)h_z(t)
$$

$$
\frac{d}{dt}Y(t) = -2g'\frac{K}{M_s^2}Z^2(t)X(t) - 4h'\frac{K}{M_s^2}Z^2(t)Y(t) + 2h'KY(t)
$$

$$
+\left[F_y(t)\frac{1}{M_s}Z(t) + \frac{Y(t)}{Z(t)}F_z(t)\right]H(t)
$$

$$
+\left[g'\left(Z^2(t) - M_s^2\frac{Y^2(t)}{Z^2(t)}\right) - 2h'X(t)Y(t)\right]h_x(t) \quad \text{(B.7)}
$$

$$
+\left[h'M_s\left(Z(t) - \frac{2Y^2(t)}{Z(t)}\right) + g'M_s\frac{X(t)Y(t)}{Z(t)}\right]h_y(t)
$$

$$
+\left[\frac{h'Y(t)}{Z(t)}\left(M_s^2 - 2Z^2(t)\right) - g'X(t)Z(t)\right]h_z(t)
$$

$$
\frac{d}{dt}Z(t) = \frac{2h'K}{M_s^2}Z(t)\left(M_s^2 - Z^2(t)\right) + F_z(t)H(t)
$$

$$
+ h'\left(M_s^2 - Z^2(t)\right)h_z(t) - \left(g'M_s^2\frac{Y(t)}{Z(t)} + h'X(t)Z(t)\right)h_x(t)
$$

$$
+\left(g'M_sX(t) - h'M_sY(t)\right)h_y(t) \qquad \text{(B.8)}
$$

We may now apply Eq. (39) to calculate the noise induced drift terms in Eqs. (B.6)–(B.8), just as in Section II.B and Appendix A. The tensor g_{ik} is given by Equation B.9.

$$g_{ik} = \begin{bmatrix} h'(M_s^2 - X^2) & -\left(g'M_s Z + \dfrac{h'M_s XY}{Z}\right) & \left(\dfrac{g'M_s^2 Y}{Z} - h'XZ\right) \\[2ex] \left[g'\left(Z^2 - \dfrac{M_s^2 Y^2}{Z^2}\right) - 2h'XY\right] & \left[h'M_s\left(Z - \dfrac{2Y^2}{Z}\right) + \dfrac{g'M_s XY}{Z}\right] & \left[\dfrac{h'Y}{Z}(M_s^2 - 2Z^2) - g'XZ\right] \\[2ex] -\left(\dfrac{g'M_s^2 Y}{Z} + h'XZ\right) & (g'M_s X - h'M_s Y) & h'(M_s^2 - Z^2) \end{bmatrix} \tag{B.9}$$

The quantity of interest, namely

$$g_{jk}\frac{\partial}{\partial x_j}g_{2k}$$

is

$$g_{jk}\frac{\partial}{\partial x_j}g_{2k} = g_{11}\frac{\partial}{\partial x_1}g_{21} + g_{21}\frac{\partial}{\partial x_2}g_{21} + g_{31}\frac{\partial}{\partial x_3}g_{21}$$

$$+ g_{12}\frac{\partial}{\partial x_1}g_{22} + g_{22}\frac{\partial}{\partial x_2}g_{22} + g_{32}\frac{\partial}{\partial x_3}g_{22} \qquad (B.10)$$

$$+ g_{13}\frac{\partial}{\partial x_1}g_{23} + g_{23}\frac{\partial}{\partial x_2}g_{23} + g_{33}\frac{\partial}{\partial x_3}g_{23}$$

where $x_1 = X$, $x_2 = Y$, and $x_3 = Z$, can be evaluated after considerable algebra. The result is

$$g_{jk}\frac{\partial}{\partial x_j}g_{2k} = -6(g'^2 + h'^2)M_s^2 Y \qquad (B.11)$$

Thus, on returning to the old variables we obtain the equation for the starting value of $H_{ef}M_y$ at time t

$$\frac{M_s}{2K}\frac{d}{dt}(H_{ef}M_y) = -g'H_{ef}M_zM_x - \frac{2h'M_s^2 H_{ef}M_y}{M_s} + h'M_sM_y H_{ef}$$

$$+ \left(F_y\frac{M_z}{M_s} + \frac{M_s}{2K}\frac{H_{ef}M_y}{M_z}\right)H(t) \qquad (B.12)$$

$$- \frac{3\mu}{2K}(h'^2 + g'^2)M_s^3 H_{ef}M_y$$

or

$$\frac{d}{dt}(H_{ef}M_y) + \left(\frac{3}{\tau_N} - \frac{\nu K}{kT}\right)M_y H_{ef}$$

$$= -H_{ef}^2(g'M_sM_x + 2h'M_zM_y) + \left(F_y + \frac{F_zM_y}{M_z}\right)H_{ef}H(t) \qquad (B.13)$$

where

$$\tau_N^{-1} = \mu\left(h'^2 + g'^2\right)M_s^2$$

In the case when the ac probe field $\mathbf{H}(t)$ is applied along the x axis

$$F_y + \frac{F_z M_y}{M_z} = h'\left(M_s^2 - M_y^2 - M_x M_y\right) - \frac{g' M_s M_y^2}{M_z} \qquad (B.14)$$

On introducing the dimensionless variables

$$u_x = \frac{M_x}{M_s} = \sin\vartheta\cos\phi \qquad u_y = \frac{M_y}{M_s} = \sin\vartheta\sin\phi \qquad u_z = \frac{M_z}{M_s} = \cos\vartheta$$

$$(B.15)$$

and

$$\xi(t) = \frac{\nu M_s H(t)}{kT} \qquad \alpha = \frac{h'}{g'} \qquad \sigma = \frac{\nu K}{kT} \qquad (B.16)$$

we obtain from Eqs. (170) and (B.13) and (B.14) (having been averaged) the equations governing ferromagnetic resonance:

$$\frac{d}{dt}\langle u_x\rangle + \frac{1}{\tau_N}\langle u_x\rangle = \frac{\sigma}{\alpha\tau_N}\langle u_z u_y\rangle - \frac{\sigma}{\tau_N}\langle u_x u_z^2\rangle + \frac{\xi(t)}{2\tau_N}\left(1 - \langle u_x^2\rangle\right)$$

$$(B.17)$$

$$\frac{d}{dt}\langle u_z u_y\rangle + \frac{1}{\tau_N}(3 - \sigma)\langle u_z u_y\rangle$$

$$= -\frac{\sigma}{\alpha\tau_N}\langle u_x u_z^2\rangle - \frac{2\sigma}{\tau_N}\langle u_z^3 u_y\rangle$$

$$+ \frac{\xi(t)}{2\tau_N}\left[\left\langle u_z\left(1 - u_x^2 - u_x u_y\right)\right\rangle - \frac{1}{\alpha}\langle u_y^2\rangle\right] \qquad (B.18)$$

Equations (B.17) and (B.18) can also be written in terms of spherical harmonics X_{nm} as

$$\frac{d}{dt}\langle X_{11}\rangle + \frac{1}{\tau_N}\left(1 + \frac{\sigma}{5}\right)\langle X_{11}\rangle = -\frac{2\sigma}{15\tau_N}\langle X_{31}\rangle + \frac{\sigma}{3\alpha\tau_N}\langle X_{2,-1}\rangle$$

$$+\frac{\xi(t)}{6\sigma_N}\left[2 - \frac{1}{2}\langle X_{22}\rangle + \langle P_2\rangle\right] \quad (B.19)$$

$$\frac{d}{dt}\langle X_{2,-1}\rangle + \frac{1}{\tau_N}\left(3 - \frac{\sigma}{7}\right)\langle X_{2,-1}\rangle$$

$$= -\frac{2\sigma}{5\alpha\tau_N}\langle X_{31}\rangle - \frac{3\sigma}{5\alpha\tau_N}\langle X_{11}\rangle - \frac{12\sigma}{35\tau_N}\langle X_{4,-1}\rangle$$

$$+\frac{\xi(t)}{\tau_N}\left[\frac{6}{5}\langle P_1\rangle + \frac{3}{10}\langle P_3\rangle - \frac{1}{20}(\langle X_{32}\rangle + \langle X_{3,-2}\rangle)\right.$$

$$\left. -\frac{1}{2\alpha}\left(1 - \langle P_2\rangle - \frac{1}{2}\langle X_{22}\rangle\right)\right] \quad (B.20)$$

where

$$\langle X_{nm}\rangle = \langle P_n^{|m|}\cos m\phi\rangle \quad \langle X_{n,-m}\rangle = \langle P_n^{|m|}\sin m\phi\rangle \quad m \le n$$

P_n is the Legendre polynomial of order n and $P_n^{|m|}$ is the associated Legendre function. In the absence of the probe field $\mathbf{H}(t)$, Eqs. (B.19) and (B.20) can be derived from the hierarchy of differential-difference equations obtained in Refs. 11 and 51 from the underlying Fokker-Planck equation. The above procedure shows how this hierarchy is formed.

APPENDIX C: A FORTRAN PROGRAM FOR CALCULATING THE IMPEDANCE OF A JOSEPHSON JUNCTION

This simple program has been used to calculate the impedance of a Josephson junction. In the program:

X = bias parameter x.
G = barrier height parameter γ.
KM = level of truncation of the continued fractions.
　　(KM depends on x and γ).
A(AE) = normalized impedance $Z(\omega)/R$ for the exact (approximate) solutions respectively.
w = normalized frequency $\omega\tau_0$.

```
SUBROUTINE MCAW(A,AE,w)
REAL X,G,KMI,G1
COMPLEX A,E,S,AE,Z,A1,E1,S1,Z1,XG,w1
```

```
        COMMON KM,X,G
        XG = X*G*(0.0,1.0)
        w1 = (0.0,- 2.0)*w / KM
        G1 = 0.5*G
        Z = G1 / (2.0*KM + XG)
        E = G1 / (XG + w1 + 2.0*KM)
        E1 = G1 / (w1- XG + 2.0*KM)
        A = 1.0- Z*E
        A1 = 1.0- CONJG(Z)*E1
        S = 1.0- Z*Z
C       CALCULATION OF CONTINUED FRACTIONS
        DO 10 I = 1,KM- 1
        KMI = 2.0*(KM- I)
        w1 = (0.0,- 4.0)*w / KMI
        Z = G1 / (XG + KMI + Z*G1)
        E = G1 / (XG + w1 + KMI + E*G1)
        E1 = G1 / (w1- XG + KMI + E1*G1)
        S = 1.0- Z*Z*S
        A = 1- Z*E*A
        A1 = 1- CONJG(Z)*E1*A1
10ENDDO
        S = 0.5*G1*Z / (1.0- S)
        S1 = CONJG(S)
        Z1 = CONJG(Z)
        A = -1.0 + A1 + A
        AE = 1.0- 0.5*G1*(Z / ((0.0,- 1.0)*w + S) + Z1 /
          ((0.0,- 1.0)*w + S1))
        RETURN
        END
```

Acknowledgments

We thank D. S. F. Crothers, M. San Miguel, and M. I. Shliomis for helpful discussions. We thank the British Council, HEDCO, The Institute of Radio Engineering and Electronics of the Russian Academy of Sciences and Trinity College Dublin for financial support for this work. E. S. Massawe wishes to thank The University of Dar-Es-Salaam for granting him study leave. We would also like to thank John Waldron for the numerical calculations in Section II.J.

References

1. M. A. Leontovich, *Statisticheskaya Fizikia* [*Statistical Physics*] Gostekhizdat, Moscow, 1944.

2. K. Kaminishi, R. Roy, R. Short, and L. Mandel, *Phys. Rev.* **A24**, 370 (1981).

3. H. Risken, *The Fokker-Planck Equation*, Springer, Berlin, 1984.

4. M. San Miguel, L. Pesquara, M. A. Rodriquez, and A. Hernández-Machado, *Phys. Rev.* **A35**, 208 (1987).

5. M. A. Martsenyuk, Yu. L. Raĭkher, and M. I. Shliomis, *Sov. Phys. JETP* **38**, 413 (1974).

6. Yu. L. Raĭkher and M. I. Shliomis, *Sov. Phys. JETP* **40**, 526 (1974).

7. A. Morita, *J. Phys. D* **11**, 1357 (1978).

8. M. I. Shliomis and Yu. L. Raĭkher, *Sov. Phys. JETP* **47**, 918 (1978).

9. G. Meier, Thesis, Freiburg i., Br., 1960.

10. C. J. F. Böttcher and P. Bordewijik, *Theory of Electric Polarisation*, 2d ed., Vol. 2, Elsevier, Amsterdam, 1978.

11. W. T. Coffey, P. J. Cregg, and Yu. P. Kalmykov, *Adv. Chem. Phys.*, Vol. 83, *On the Theory of Debye and Néel Relaxation of Single Domain Ferromagnetic Particles*, Wiley Interscience, New York, 1993, p. 263.

12. W. T. Coffey, *J. Chem. Phys.* **93**, 724 (1990).

13. W. T. Coffey, J. L. Déjardin, Yu. P. Kalmykov, and K. P. Quinn, *Chem. Phys.* **64**, 357 (1992).

14. A. Morita and S. Watanabe, *Adv. Chem. Phys.* **56**, 255 (1984).

15. W. F. Brown, Jr., *Phys. Rev.* **130** 1677 (1963).

16. A. J. Martin, G. Meier, and A. Saupe, *Symp. Faraday Soc.* **5**, 119 (1971).

17. J. R. McConnell, *Rotational Brownian Motion and Dielectric Theory*, Academic, New York, 1980.

18. H. Haken, *Synergetics*, 2d ed., Springer, Berlin, 1978.

19. J. D. Cresser, W. H. Louisell, P. Meystre, W. Schleich, and M. O. Scully, *Phys. Rev.* **25A**, 2214 (1982).

20. J. D. Cresser, D. Hammonds, W. H. Louisell, P. Meystre, and H. Risken, *Phys. Rev.* **25A**, 2226 (1982).

21. J. D. Cresser, *Phys. Rev.* **26A**, 398 (1982).

22. W. T. Coffey, in M. W. Evans (Ed.), *Adv. Chem. Phys.*, Vol. 63, *Development and Application of the Theory of the Brownian Motion in Dynamical Processes in Condensed Matter*, Wiley Interscience, New York, 1985.

23. G. Barone and A. Paterno, *Physics and Application of the Josephson Effect*, Wiley Interscience, New York, 1982.

24. W. T. Coffey, Yu P. Kalmykov and K. P. Quinn, *J. Chem. Phys.* **96**, 5471 (1992).

25. H. Bateman, *Partial Differential Equations*, Cambridge University Press, Cambridge, UK, 1932. Reprinted by Dover, New York.

26. M. Abramowitz and I. A. Stegun (Eds.), *Handbook of Mathematical Functions*, Dover, New York, 1964.

27. M. I. Shliomis and Yu. L. Raĭkher, *IEEE Trans. Magn.* **16**, 237 (1980).

28. W. H. de Jeu, *Physical Properties of Liquid Crystalline Materials*, Gordon & Breach, New York, 1980.

29. Yu. P. Kalmykov, *Liquid Crystals* **10**, 519 (1991).

30. B. A. Storonkin, *Kristallografiya* **30**, 841 (1985).

31. M. Davis, R. Moutran, A. H. Price, M. S. Beevers, and G. Williams, *J. Chem. Soc. Faraday Trans. 2* **72**, 1447 (1976).

32. E. A. Milne, *Vectorial Mechanics*, Methuen, London, 1948.

33. W. F. Brown, Jr., *IEEE Trans. Magn.*, **15**, 1196 (1979).

34. P. Debye, *Polar Molecules*, Chemical Catalog, 1929, Reprinted by Dover, New York.

35. R. Kikuchi, *J. Appl. Phys.* **27**, 1352 (1956).

36. L. D. Landau and E. M. Lifshitz, *Phys. Z. Sowjetunion* **8**, 153 (1935).

37. C. J. Reid, *Mol. Phys.* **49**, 331 (1983).

38. J. I. Lauritzen and R. Zwanzig, *Adv. Mol. Rel. Interact. Proc.* **5**, 339 (1973).

39. F. M. Arscott, *Periodic Differential Equations*, Pergamon, Oxford, 1964.

40. E. L. Ince, *Proc. London Math. Soc*, (2) **23**, 56 (1924).

41. S. Chandrasekhar, *Rev. Mod. Phys.* **15**, 1 (1943).

42. P. S. Hubbard, *Phys. Rev.* **A8**, 1429 (1973).

43. W. T. Coffey, *J. Chem. Phys.* **95**, 2026 (1991).

44. I. N. Sneddon, *Special Functions of Mathematical Physics and Chemistry*, 2d ed., Oliver & Boyd, Edinburgh, 1961.

45. V. Ambegaokar and B. I. Halperin, *Phys. Rev. Lett.* **22**, 1364 (1969).

46. A. J. Viterbi, *Principles of Coherent Communication*, McGraw Hill, New York, 1966.

47. W. W. Chow, J. Gea-Banacloche, L. M. Pedrotti, V. E. Sanders, W. Schleich, and M. O. Scully, *Rev. Mod. Phys.* **57**, 61 (1985).

48. H. Fröhlich, *Theory of Dielectrics*, 2d ed., Oxford University Press, Oxford, 1958.

49. B. K. P. Scaife, *Principles of Dielectrics*, Oxford University Press, Oxford, 1989.

50. T. Van Duzer and C. W. Turner, *Principles of Superconductive Devices and Circuits*, Edward Arnold, London, 1981; simultaneously published by Elsevier, New York.

51. M. I. Stepanov and M. I. Shliomis, *Izv. Akad. Nauk. Ser. Fiz.* **55**, 1042 (1991).

52. F. T. Arecchi, M. Giglio and A. Sona, *Phys. Lett.* **25A**, 341 (1967).

53. K. K. Likharev and V. K. Semenov, *Radiotekh. Elektron.* **18**, 1757 (1973).

54. A. N. Vystavkin, V. N. Gubankov, L. S. Kuzmin, K. K. Likharev, V. V. Migulin, and V. K. Semenov, *Rev. Phys. Appl.* **9**, 79 (1974).

55. F. Auracher and T. Van Duzer, *J. Appl. Phys.* **44**, 848 (1973).

56. C. V. Stancampiano, *IEEE Trans., Electron. Dev.* **27**, 1934 (1980).

57. M. J. Stephen, *Phys. Rev. Lett.* **21**, 1629 (1968).

58. M. J. Stephen, *Phys. Rev.* **182**, 531 (1969).

AUTHOR INDEX

Numbers in parentheses are reference numbers and indicate that the author's work is referred to although his name is not mentioned in the text. Numbers in *italic* show the pages on which the complete references are listed.

Happer, W., 328(93), *357*
Haque, M. A., 401(85), 404(85), 406(85), *413*
Harget, A., 441(302), *483*
Harlow, M. J., 576(31), 595(31), *604*
Harmuth, H. F., 312(6), 317(6,31), 354(6), *354-355*
Haroche, S., 649(49), *666*
Harris, C. B., 593(61), *605*
Harris, S. E., 416(22-23), *475*
Hartmann, S. R., 589(49), *604*
Harvey, A. H., 335(128), *358*
Hasanein, A. A., 465(72-475), *488*
Haselbach, E., 441(301-302), *483*
Hasted, J. B., 335(129), *358*
Hauchecorne, G., 416(11), 424(11), *475*
Haverkort, J. E. H., *489*
Hawkes, J. F. B., 579(34), *604*
Hay, P. J., 437(208), 453-454(208), 457(208, 428), 468(208), *481, 487*
He, Q., 353(182), *359*
He, Q. C., 32(84), 37(101-102), 38-39(105), 42(110,112), 45(116), *49-50*, 62-63(32), 65-66(32), *96*
Head-Gordon, M., 456-457(421), *486*
Healy, E. F., 441(312), *484*
Heaton, J., 16(24,29,31) 17(29), *47*
Hecht, L., 509(29), *541*
Hehre, W. J., 455(412-413), 465(412-413), *486*
Heiman, D., 9(8), *47*
Heinz, T. F., 570(23), 590(23,53), 593(63), *604-605*
Heisenberg, W., 420(45), *476*
Heitler, W., 116(17), *126*, 197(6), 202(6), 230, 232-235(6), 242, 284(22), *309*
Helfrich, W., 384(51), *412*
Heller, D. F., 465-466(477), *488*
Hellwarth, R. W., 9(8), *47*
Hellworth, R., 386(58), *413*
Hendrickson, B. M., 42-43(113), 45(113), *50*
Hernandez-Machado, A., 668(4), *791*
Herriau, J. P., 18(37-38), *48*
Hetherington, W. M., 598(75), *605*
Higgs, P. G., 328(86), *357*
Hild, E., 348(169), *359*
Hill, A. E., 546(1), *603*
Hill, E. L., 281, 283, 286, 288, 292, 297, 301, 306(12), *309*
Hill, N. E., 423(114), *478*
Hillegas, C., 62-63(32), 65-66(32), 96, 353(182), *359*
Hillion, P., 354(183-184), *359*
Hinchliffe, A., 463(456-458,460-465), 464(457,459-465), 465(457,478), 466(478), *487-488*
Hinds, E. A., 649(49), *666*
Hine, J., 421(77), *477*
Hobbs, C. P., 424(119), *478*

Hochstadt, H., 341(154), 347(154), *359*
Hoffmann, R., 439(243), *482*
Hohm, U., 328(99), *357*
Holtz, M., 534(104), 536-537(104), *543*
Homans, S. W., 63(35), 71(35), 76-77(35), 79-82(35), *96*
Hong, J., 34(98), *49*
Hong, K. H., 464(467), *488*
Hopf, F. A., 592(60), *605*
Horowitz, M., 16(27), 28(27), 36(27), *47*
Howell, I., 407-408(91), *414*
Hu, Yu-Kuang, 496(11), *540*
Hubbard, P. S., 731(42), *792*
Hückel, E., 438(221), *481*
Hudis, J. A., 454(383), *485*
Hughes, T. P., 423(109), *478*
Huignard, J. P., 3(1,2), 10(14), 17(14), 18(37-38,43,46,49), 22(1,2), *46-48*
Huiszoom, C., 452(328), *484*
Huppert, D., 312(18), *355*
Hurst, J. B., 455-456(408), *486*
Hurst, R. P., 420(58), 424(58), *476*
Huse, D. A., 124(28), *126*
Hush, N. S., 437(210,212,214-215), 439(251-253), 442(252), 453-454(210, 212,214-215), 468(210), *481-482*
Huttner, W., 364(28), *412*
Huzinaga, S., 454-455(398), *486*

Iacopini, E., 376(45), *412*
Ichihara, A., 452(361), 460-461(361), 468(361), *485*
Ikeda, O., 40(106), *49*
Ince, E. L., 728(40), *792*
Indenbom, V. L., 320(44), *355*
Infeld, L., 219(35), *231*
Ingold, C. K., 421(76), *477*
Innocenti, G., 649(49), *666*
Ishikawa, K., 534(101), *543*
Ishimaru, A., 329(111), 335(111), *357*
Isnard, P., 425(142), *479*
Itoh, R., 452(361), 460-461(361), 468(361), *485*

Jackel, S., 34(96), *49*
Jackson, J. D., 198(8), 200(8), 204(8), 206(8), 211-212(8), 214-218(8), 224(8), 226(8), 228(8), *230*, 232-235(7), *242*, 244-245(10), 252-253(10), *255*, 281, 283, 286, 288, 292, 297, 301, 306(1), *308*, 324(56), 327(56), 329(56), 332-333(56), 345-346(56), *356*, 552(10), *603*
Jacobovitz, G. R., 649(49), *666*
Jacobs, A. A., 533(93), *542*
Jacquier, Ph., 515(58), *541*
Jaffé, H. H., 441-442(276,281-282), 460(281), *482-483*
Jagannath, H., 34(91), 37(104), *49*

Xu, K., 34(98), *49*

Yaghijian, A. D., 323(55), 326(55), *356*
Yakubovich, V. A., 351–352(176), *359*
Yamaguchi, Y., 441–442(307), 454(384–386), *483, 485, 489*
Yang, S.-K., 339(149), *358*
Yariv, A., 3(40), 18(40), 19–20(50–51), 26(40), 31(81), 41(107), *48–49*, 362(4), *411*
Yeager, D. L., 460(443–444), *487–488*
Yeazell, J. A., 68(45), 77(45), *96*
Yeh, P., 3(17), 14(17), 18(36,48), 26–27(17), 29(75,80), 32(86), 34(80), 35(100), 37(102), *47–49*
Yoshimine, M., 453(368), 455(402), *485–486*
Yoshioka, K., 364(15), *411*
You, T., 335(132), *358*
Young, L., 10(11,15), *47*
Yu, J., 499(16), *540*, 581(38), *604*
Yuratich, M. A., 71(40), *96*, 431(187), *480*, 559(13), 564(13), *603*
Yurke, B., 614(12), *665*

Zamani-Khamini, O., 438(230–231,235–236), 439(236), *481*
Zanasi, R., 455(407), 458(407), *486*
Zanassi, R., 468(484), *488*

Zare, R., 52(12), *95*, 121(20), *126*, 162(41), *166*
Zavattini, E., 328(92), *357*, 376(45), *412*
Zawodny, R., 171(31), *172*, 181(3,5–6), 184–185(5–6), *187*, 203(20), 219(40), 224(44), *231*, 259–260(17), *281*, 425(165–166), 426(175), *479–480*, 511(39), 513(43), *541*, 586(46–48), 603(88), *604, 606*
Zeiss, G. D., 461(448), *487*
Zel'dovich, B. Ya, 181(4), *187*
Zeldovich, B. Y., 19(52), *48*
Zerner, M. C., 441(298), 455(298), *483*
Zhang, W., 624(23), 637(31), *665*
Zigler, A., 34(96), *49*
Ziyuan, Li., 534(103), *543*
Zoebisch, E. G., 441(312), *484*
Zoller, P., 617(13), 624(21), 628(13), 630(13), 638(32), 658(54), *665–666*
Zorn, J. C., 416(1), 424(1), *475*
Zou, Y. H., 312(17), 327(17), *355*
Zozulya, A. A., 29(77), 34–35(77,92), *49*
Zürcher, R. F., 422(90–91), *477*
Zwanzig, R., 728(38), *792*
Zwanziger, D., 328(79), *356*
Zyss, J., 313(24), *355*, 437(217), 441(295–296, 298), 453(217), 455(298), *481, 483*, 546(2), 569(2), 572(24), 597(2,74), *603–605*

SUBJECT INDEX

811